GEOMETRIC MODELING

GEOMETRIC MODELING

Michael E. Mortenson

John Wiley & Sons

New York   Chichester   Brisbane   Toronto   Singapore

Library of Congress Cataloging in Publication Data:

Mortenson, Michael E., 1939–
   Geometric modeling.

   Includes index.
   1. Geometry—Data processing.   2. Mathematical
models.   3. CAD/CAM systems.   4. Computer graphics.
I. Title.
QA447.M62   1985   516′.0028′54   84-29940
ISBN 0-471-88279-8

Printed in the United States of America

10 9 8 7 6 5 4 3 2 1

Ianeke
Aloha Nui Loa

PREFACE

Geometric modeling is the technique we use to describe the shape of an object or to simulate dynamic processes. Much of the power of contemporary geometric modeling resides in its techniques for synthesizing, allowing us to easily describe complex shapes as arrangements of simpler ones. Geometric modeling provides a description or model that is analytical, mathematical, and abstract rather than concrete. We create a model because it is a convenient and economical substitute for the real object or process. It is often easier and more practical to analyze a model than to test or measure or experiment with the real object. Beyond the advantages of analysis, the model is also a useful, if not necessary, way of conveying information; for example, a model can be used to transmit design information between engineering and manufacturing functions in industry.

Geometric modeling brings together and applies analytic geometry, vector calculus, topology, set theory, and an arsenal of computation methods. This combination of mathematical tools and the potential complexity of the model require the computational power of a computer. Even though the computer, with its own special requirements and influence, is not always explicit in our studies here, it is ever present, if only in the background, as our "geometry engine." Furthermore, although geometric modeling is often associated with what is called solid modeling, it is this and more because it applies to forms and functions of any dimension.

The importance of geometric modeling is rapidly increasing in many fields. It is a primary ingredient in computer-aided design and computer-aided manufacturing (CAD/CAM) systems, computer graphics, computer art, animation, simulation, computer vision, and robotics. Advances in any of these fields depend on how well we can create effective geometric models.

We must sometimes create realistic geometric models of things that already exist. Sometimes we must create geometric models of things existing in only our imaginations, as in design and art. In either case, the demand is increasing for greater facility and sophistication in our ability to "do" geometric modeling.

This text, *Geometric Modeling*, presents the nature, development, and application of the basic concepts of geometric modeling. While this text is written for upper-division and graduate students in engineering, computer science, or applied mathematics, it is also useful for those already practicing in these and related fields. You should already be familiar with elementary calculus, analytic geometry, and vectors and matrix methods, since we simply review some results in these areas, without the burden of proofs, to emphasize certain important computational features.

The history of the development of geometric modeling is one of unique cooperation between industry and university, and thus material presented here is drawn from both areas. Most of the material has been developed and refined by many contributors; the most important contributions are referenced, and a complete bibliography is appended. Finally, not to be forgotten are the many anonymous workers in this field whose valuable contributions have passed second and third hand into the general body of geometric-modeling technology.

Geometric Modeling consists of three parts: parametric geometry, solid modeling, and applications. The first eight chapters lay the foundation with the parametric geometry of curves, surfaces, and solids and show how particular formulations facilitate calculating geometric properties. The next three chapters explore the fundamentals of solid model construction and analysis from many points of view. The final chapters discuss three important areas of application: computer graphics, CAD and CAM. The last chapter is a broad view of the frontiers of geometric modeling.

Fifty years ago, David Hilbert observed two tendencies in any scientific research:

> On the one hand, the tendency toward abstraction seeks to crystallize the logical relations inherent in the maze of material that is being studied, and to correlate the material in a systematic and orderly manner. On the other hand, the tendency toward intuitive understanding fosters a more immediate grasp of the objects one studies, a live rapport with them, so to speak, which stresses the concrete meaning of their relations.

Geometric Modeling strives to crystallize and unify the diverse analytical aspects of geometric modeling, while at the same time establishing an intuitive rapport with these aspects. Of all the branches of mathematics, geometry appeals most directly to our intuition. I have tried to strengthen this appeal with appropriate examples of applications, exercises, and a generous use of illustrations, all of which form an integral part of the text.

There are two types of exercises. The first tests understanding of concepts and facility with their quantitative aspects. Answers to some of these exercises appear at the end of the text. The second type of exercise requires developing and writing procedures. At the instructor's discretion, these procedures can range in complexity and applicability from a general outline of an algorithm to a ready-to-run subroutine in whatever programming language is convenient. If the latter approach to writing the suggested procedures is taken, then after working through the entire text, a set of routines will be accumulated that will form the core of a geometric-modeling program. Since this text does not discuss data base design, display generation, or interactive procedures, these "details" must necessarily be accommodated "locally." As you might suspect, no answers are provided for the second type of exercise. You are urged to read all of the exercises whether or not they are assigned to be worked. They will give you a taste of the types of problems you will encounter in real-world modeling situations.

You will notice that many of the procedures required in the exercises and described in the text are highly specialized, usually calling for some sort of analysis of a particular type of curve or surface (that is, pc, Bezier, *B*-spline, and so on). A sophisticated general-purpose modeling system would not necessarily exhibit such a proliferation of specialized routines. Instead, it would most likely incorporate a parametric evaluator to compute specific points, tangents, and so forth. The data are then passed on to general-purpose routines for computing local and global properties, intersections, display data, and so on. Thus, a single computational strategy can be developed for each analysis requirement, independent of the special nature of any geometric element.

Special procedures are emphasized in the exercises to enhance their didactic value, giving students a deeper understanding of the nuts-and-bolts concepts of geometric modeling before requiring them to come to grips with the larger, though nonetheless important, considerations of computational efficiency and system architecture of very large-scale modeling systems.

Because you should be aware of what is out there in the real world of geometric modeling, Appendices A, B, and C describe current working modeling systems that are representative examples of different approaches successfully combining hardware and software. Appendix D presents Reeve's seminal work in particle systems, a new approach to modeling such complex physical phenomena as fire, clouds, and grass.

Finally, a warning: The model of an object or the simulation of a process is not the object or the process. When we ask the model questions, when we analyze it to better understand or predict the behavior of what is represented, we must remember that the answers and the results pertain to only the model. If and only if the model closely corresponds to the object or process, can we safely infer the object's or process' behavior from the model's behavior. The accuracy of representation methods is an art and science in itself and a subject sufficient in itself for another textbook.

ACKNOWLEDGMENTS

Geometric Modeling, like most other textbooks, necessarily draws heavily from the works of others. Therefore, I first acknowledge with deep respect all those both well known and anonymous who worked and contributed so diligently over the past 25 years to enrich and infuse new excitement into the art and science of geometry, those who have created the new discipline that we now call geometric modeling. The bibliography is only a small mirror reflecting the field of geometric modeling. I hope that my organization and interpretation of this field has not altered its original brilliance and intention.

Thanks to my friends and professional colleagues who read the complete manuscript and made many valuable suggestions: Art Eshleman of Douglas Aircraft Company, Harry Meriwether of McDonnell Automation Company, Bill Miller of Townsend Microware, Ron Nowek of McDonnell Douglas Astronautics Company, and Hank Timmer of Northrop Corporation. For the most part, their reviews were done over a very hectic Christmas holiday. True friends.

Thanks to Professors Brian Barsky of the University of California, Berkeley, Martin Kaliski of Northeastern University, Lawrence Kaatz of Idaho State University, Kenneth Magel of North Dakota State University, Lawrence Miller of the University of California, Los Angeles, Dr. Bruce Naylor of the Georgia Institute of Technology, and Professor Sam Uselton of the University of Tulsa for their careful reviews and valuable suggestions. A heroic task for a manuscript of this size and complexity. This textbook, and thus ultimately the student who uses it, has greatly benefited because of their efforts.

Thanks to K. H. Auger, B. Fox, and C. R. Lambson, all of Evans and Sutherland, for making available information describing the ROMULUS system; to T. J. Beecher and A. E. Doelling of McDonnell Aircraft Company for making available information on the CADD system; to W. T. Reeves of

Lucasfilm for his kind permission to reprint his work on particle systems; and to J. W. Boyse and J. E. Gilchrist of General Motors for their kind permission to reprint their description of GMSOLID.

Thanks to the following people for their generous support by making available their work or that of their institution or corporation: Prof. N. L. Badler of the University of Pennsylvania, Prof. B. A. Barsky of the University of California, Berkeley, Dr. J. F. Blinn of JPL, California Institute of Technology, Dr. D. P. Greenberg of Cornell University, S. Mardesich of Digital Productions, Dr. D. J. Meagher of Phoenix Data Systems, Inc.), J. Rosebush of Digital Effects, Inc., A. R. Smith of Lucasfilm, and Prof. H. B. Voelcker and A. A. G. Requicha of the University of Rochester's Production Automation Project.

Thanks to the staff at John Wiley & Sons for their very professional guidance and support throughout this project, particularly Carol Beasley, Computer Science Editor, Marilyn Buckingham, who very creatively and competently was responsible for the enormous job of copy editing, Sheila Granda, who designed the text format, Bruce Safford, Senior Editor, Philip McCaffrey, Production Supervisor, Dean Gonzalez, Illustration Supervisor, and Lorraine Mellon, Senior Administrative Assistant.

Thanks to Alvy Ray Smith of Lucasfilm for permission to use his *white sand* as the cover artwork. The flowering plants were "grown" in three dimensions from a single cell using an algorithmic computer model written by the artist and based on mathematics by A. Lindenmayer, P. Hogeweg, and B. Hesper. The grasses were contributed by William Reeves, using his procedural modeling technique called particle systems. The compositing software is by Thomas Porter and the hidden-surface sofrware by Loren Carpenter. The picture was rendered using an Ikonas graphics processor and framebuffers. The resolution is 512 × 488, 24 bits per pixel. The Chinese (also Japanese) in the upper left-hand corner is the artist's name and is part of the piece.

There is a small-town library in a far corner of the Olympic Peninsula in Washington that can get a researcher and writer almost anything, anytime, from anyplace. Thanks to librarian Bev Lamb and her staff at the Port Townsend Library.

Very special thanks to my wife, Janet, who now knows more about geometric modeling than I do. Her editorial talent and mastery of a not-always-willing word processor kept this project moving.

Michael E. Mortenson
Port Townsend, Washington
March 1985

CONTENTS

1

INTRODUCTION

1 GEOMETRIC MODELING: WHAT IT IS AND WHAT IT IS NOT

The term **geometric modeling** first came into use in the early 1970s with the rapidly developing computer graphics and computer-aided design and manufacturing technologies. It refers to a collection of methods used to define the shape and other geometric characteristics of an object. Geometric modeling embraces an area often called **computational geometry** and extends beyond this to the newer field of solid modeling, creating an elegant synthesis of geometry and the computer.

When we construct a model of something, we create a substitute of that thing—a representation. We cast it into a more convenient form, easier for us to use and analyze. If the model is a good one, it will respond to our proddings and questions in the same way that the original would. We try to abstract only the information essential for our objectives while ignoring other information. For many applications, the geometric model of a physical object may require the complete description of surface reflectance properties, texture, and color; or it may include only information on the elastic properties of the object's material. We can determine the detail required in a model by the uses and operations to which we intend to subject it. If the model is rich enough in descriptive detail, we can perform operations on it that yield the same results as operations performed on the object itself.

We use geometric-modeling methods to construct a precise mathematical description of the shape of a real object or to simulate some process. The construction itself is usually a computer-aided operation, with the model stored in, and analyzed by, a computer. Using a computer is, in fact, central to the entire geometric-modeling process. Without computational power, we would be unable to construct and analyze sophisticated models of any practical importance.

Geometric-modeling methods are a synthesis of techniques from many fields: analytic and descriptive geometry, topology, set theory, numerical analysis, vector calculus, and matrix methods, to name a few. Originally, only a loose ad hoc assembly, they have coalesced into an independent discipline with a well-defined language and logic.

Where we were once satisfied with two-dimensional representations of physical objects, we now demand and achieve topologically and analytically complete three-dimensional models. Furthermore, these models allow us to derive quickly and automatically any geometric property or attribute that the object is likely to possess. Obviously, appearance is not the sole concern of geometric modeling, so we carefully distinguish between analytical models and rendering. We also require a mathematical setting that allows us to calculate relationships between models—closest approach, intersections, shadowing, and so forth. In many situations, we are interested in representing arbitrary shape information—shapes that do not have special names and whose characteristics are not well defined. The technologies serviced by geometric-modeling schemes continue to stretch those schemes to the limits of their capabilities.

We identify three distinct aspects of geometric modeling: (1) **Representation**: The physical shape of an object is given and presumed to be fixed; we compute a mathematical approximation once. (2) **Design**: We must create a new shape to satisfy some operational or aesthetic objectives. We manipulate the variables defining the shape until we meet the objectives. (3) **Rendering**: We usually need an image of the model to visually interpret it. These three aspects are, of course, closely related; for example, when we create the geometric model of a new product, at first, the model represents a physically nonexistent object. The model must be suitable for analysis and evaluation. After selecting a specific design, we can use its geometric model to guide the manufacture of the object. When production is complete, the model finally represents the physical shape of an existing object. At any point in this process, the geometric model provides information for rendering visual images of the object, including engineering drawings and computer graphics displays.

Computer graphics, computer-aided design (**CAD**), and computer-aided manufacturing (**CAM**) have been the driving forces behind the rapid development of geometric-modeling schemes. And, naturally, it is in these very areas that we can apply geometric modeling most effectively. Robotics, computer vision, and artificial intelligence are also making increasing demands on geometric-modeling capabilities. Let us review the traditional applications.

Computer graphics systems now routinely produce realistic two-dimensional color-shaded renderings of three-dimensional objects. Solid modeling, polygonal, and sculptured-surface techniques have made this possible: They now produce the appearance of texture and translucence and the effects of multiple light sources. Art, animation, and reality simulation are contributing their own impetus to the state of the art, as in several popular feature films and, of course, in generating scenes for aircraft flight simulators.

Work continues on more sophisticated computer-aided design and drafting systems. The ultimate objective is a virtually paperless engineering department. In design and drafting, the engineer's activities will soon be more like sculpting than drafting, enabling designers to rapidly create highly complex models by using combinations of simple solid shapes.

Engineering analysis is another area undergoing rapid change with the advent of solid modelers. Solid modelers permit rapid construction of finite-element models. They also permit automatic static and dynamic structural analysis of mechanical parts subjected to a variety of loading conditions. We can display the effects of these loads with new graphics capabilities. Analytical results presented graphically can provide clearer and easier interpretation. Computer-aided kinematic analysis also becomes possible, because we can move parts of a solid model independently and easily check clearances between these parts, either visually or analytically. It is also now possible to ask other often-neglected questions, such as, Is there sufficient clearance for a wrench or screwdriver to reach and turn a fastener whose location allows only limited access?

Perhaps one of the most fertile applications is computer-aided manufacturing. Geometric modeling makes possible process planning and machine-tool path-verification systems that are completely automatic. Engineering designers will furnish, via a computer data base, complete and unambiguous geometric models of parts to be manufactured. Computers in the manufacturing department will then interpret these models and generate production and assembly instructions for the computerized and robotized machine shop and assembly line. Automated inspection of finished parts is now becoming possible, resulting in increased quality assurance.

Designers of geometric-modeling systems now offer their products for application to robotics and computer vision, both important areas related to and supporting artificial intelligence systems. Future geometric-modeling systems will encourage a synthesis of these three fields to produce intelligent robots that move in three dimensions and recognize objects from visual, tactile, and other sensory information. Ultimately, robots will create their own geometric model of their environment by using artificial senses and powers of inference.

What does the architecture of a geometric-modeling system look like? What is it physically? The major components are the geometric-modeling software (including the model builder), a computer (from mainframe to micro), a user interface device (usually graphic), a data base for storing the model, a display generator for creating the graphic output, and usually an array of application programs. Appendices A, B, and C describe three contemporary modeling systems.

In the sections that follow in this introductory chapter, we briefly explore the history of modeling; review applicable mathematical techniques; explicit, implicit, and intrinsic equations; parametric equations; and coordinate systems. Finally, we discuss the notation schemes and general conventions used in this text and elsewhere.

EXERCISES

The following questions are intended to make you think "geometrically." They are not quantitative; they require no computations, and you will not find answers to them in any textbook.

1. How many different solids can the wireframe model in Fig. 1.1 represent? (**Note**: A wireframe model represents an object by its edges only. The physical equivalent of a wireframe model is simply an appropriate collection of interconnected wires or slender sticks.) What additional information do you need to ensure a unique, unambiguous model of a solid?

2. Think of optical illusions you may have seen. Could any of them be constructed as a wireframe model? Do you think they could be described by mathematical equations?

3. Study some common object—a coffee cup, for example. What simpler shapes does the object consist of? Try to develop and write down a complete, unambiguous description of the selected object's shape. How do the simpler component shapes fit together (in other words, are they joined with abruptly intersecting surfaces or with smoothly blending surfaces)? Look closely!

4. Identify several objects in each of the following three categories: manufactured (machine-made), handmade, and natural. What are the distinguishing geometric features of each category? In addition to general shape characteristics, also consider symmetry and surface texture.

5. Select one object from each of the three categories in Ex. 4. For each object, first develop a complete verbal definition of its shape. Second, using modeling clay, construct an accurate as possible scale model. Third, develop an orthographic (three-view) mechanical drawing. Compare each form of representation. Consider such things as completeness, ambiguity, quantity and quality of information.

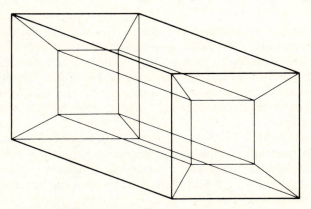

FIGURE 1.1 Wireframe ambiguity.

6. Compare how geometric information is presented in various art forms, such as sculpting, painting, and ceramics. Compare impressionistic styles with realistic and abstract. Can you decide what minimum information is required to communicate a geometric form and the unique identification associated with it?

2 HISTORICAL DEVELOPMENT

The roots of geometric modeling are found in the earliest computer graphic systems developed for computer-aided design and manufacturing. Ivan Sutherland (1963) is among the earliest pioneers in this field with his Sketchpad system. The modeling of that era tended to emphasize appearance and rendition on the design side, with the notable exception of the finite-element models used in structural analysis. On the manufacturing side, history begins with the simple "subtractive" modeling of numerical control processes.

Computers were introduced into manufacturing to compute and control the cutter motions of machine tools. This required a new way of understanding and extracting the shape information of a part design from engineering drawings. Such a task was not possible until special languages were developed to translate the shape information from the drawing into a computer-compatible format. This descriptive information was then transformed into instructions for computer-controlled machine tools. This marked the beginning of the numerically controlled (NC) machinery revolution in industry. Pressman and Williams (1977) present a thorough treatment of numerical control technology and computer-aided manufacturing.

In the mid-1960s, D. T. Ross (1967) of MIT developed an advanced compiler language for graphics programming and generating problem-oriented languages. S. A. Coons (1963, 1965), also at MIT, and J. C. Ferguson (1964) at Boeing, at this time began important work in sculptured surfaces. At about this time, General Motors developed their DAC-1 system. Several other companies, notably Douglas, Lockheed, and McDonnell also made significant developments.

Early limitations of the NC programming languages stimulated work on the mathematics and applications of sculptured surfaces and solid modeling. This was reinforced by a growing need in the automotive and aircraft industries. Detroit was interested in the geometry of styling—body shapes, aesthetics, and highlights. In Seattle, St. Louis, and Southern California, aircraft engineering departments were looking for better structural and aerodynamic shape-modeling tools. At this time, new ground was broken in the mathematics of parametric geometry, including Coons' bicubic patches and Bezier's (1975) special surfaces.

Meanwhile, workers in computer graphics and certain areas of computer-aided design (including architecture) began work in two areas of geometric modeling we now call wireframe and polygonal schemes. These systems were initially two dimensional and intended as drafting tools or methods for reducing and interpreting digitized data. A wireframe model is composed of lines and curves defining the edges of an object, and it is usually constructed interactively, although in much the same way as an engineering drawing. Each line or curve element is separately and independently constructed. The model is stored in the computer as a list of lines and curves.

Current wireframe-modeling systems are capable of constructing three-dimensional representations and have considerably enhanced interactive procedures. However, all pure wireframe systems exhibit well-known, often severe, deficiencies. For example, three-dimensional wireframe models are frequently ambiguous; Fig. 1.1 is a good example of this ambiguity, since there are three equally likely passages through the object. It is also easy to create nonsense objects, since these systems usually have no internal logical tests to prohibit this; Fig. 1.2a shows a classical example. Implied faces interpenetrate in a way that makes the interpretation of a physically realizable solid impossible.

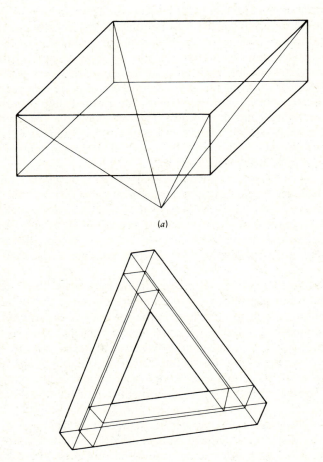

(a)

(b)

FIGURE 1.2 Wireframe nonsense objects.

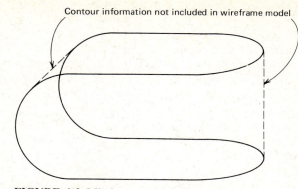

Contour information not included in wireframe model

FIGURE 1.3 Missing contour information.

Notice that the wireframe in Fig. 1.2*b* is a different kind of nonsense object: It cannot be constructed or interpreted in three-dimensional space; the art of M. C. Escher furnishes many more examples. Another deficiency is the lack of contour or profile information for surfaces inferred between the wireframe lines and curves; Fig. 1.3 illustrates this limitation.

Polygon-modeling schemes were initially developed to create pictures or renderings. There is a considerable grey area between polygon and wireframe systems, and the distinction is not always clear. The polygonal data structure is straightforward, consisting of topologically cross-referenced lists of vertices, edges, and faces; however, the picture-generating and picture-manipulating algorithms are often highly sophisticated. This is true in part because polygonal schemes have been extensively developed and used as research tools for graphic display technology, with animation and hidden surface or "visibility" techniques at the leading edge.

Sculptured surfaces were developed to replace the classic lofting techniques of the shipbuilding, automotive, and aircraft industries. The well-known parametric cubic patches of Coons and Ferguson and the techniques of Bezier have been most successful. Their emphasis is on accurate representation of surfaces and on interpolating techniques to meet a variety of design criteria. A synthesis of sculptured surfaces and wireframe methods, as just mentioned, offers considerable promise as a future solid modeler.

True solid modeling is a relative newcomer. It is intended to overcome the limitations of the other schemes when representing and analyzing three-dimensional objects. The goal of solid modeling is to create unambiguous and complete geometric representations of objects. There have been several different approaches: boundary representations, constructive solid geometry (Boolean operations on simple "solids" to build more complex solids), sweep representations, logical collections of half-spaces, and others. We will say much more about these in later chapters. Current geometric modeling systems are discussed throughout the text and in Appendices A, B, and C.

3 APPLICABLE MATHEMATICAL TECHNIQUES

In our study of geometric modeling, we will make extensive use of several important mathematical techniques. The most important are: linear algebra, vectors, matrix methods, determinants, set theory, polynomial interpolation, and numerical approximation. Let us briefly review some of these.

Perhaps the single most important mathematical device we will use is the vector. We can think of vectors as conceptual entities in their own right, because they seem to fit our geometric intuition of displacement. This is easily obscured if we are limited to working only with their components separately.

Vectors offer a distinct advantage over classic analytic geometry by minimizing our dependence on a specific coordinate system. At the very least, vectors allow us to postpone the choice of a coordinate system until the later stages of the problem-solving process. Then we can often choose more appropriate coordinates than might have been evident earlier. In addition, vectors carry inherent geometric meaning, such as length and direction. Vector operations allow us to readily determine perpendicularity or parallelism. These operations support algebraic operations while retaining geometric meaning. Last, but not least, vector equations handle several component equations at once. Section 3.1 presents a review of vectors and vector geometry.

Next we turn to matrix methods. The array of numbers that makes up a matrix can represent simply an orderly way of storing numbers pertinent to some problem or perhaps a set of polynomial equation coefficients. The rules of matrix algebra define allowable operations on these arrays.

Another use of a matrix is as an operator. Here, the matrix performs a geometric transformation on a set of points by operating on the position vectors that define these points. The rules of allowable operations are in turn governed by the rules of matrix algebra. This interpretation of a matrix as a geometric operator is the foundation of most geometric-modeling computations; Section 3.2 reviews the basic procedures.

Since we will encounter determinants in many operations and expressions in this book, it is a good idea to become reacquainted with their special properties (see Section 3.3).

Now let us look at polynomial interpolation. A great deal of practical numerical analysis depends on techniques called numerical interpolation. Here, we discover the powerful but simple theorem that a straight line can be defined by, and passed through, two points, a conic curve through three points, a cubic through four, and so on. We use polynomials for interpolation, because they can be evaluated, differentiated, and integrated easily and in a finite number of steps by using just the basic arithmetic operations of addition, subtraction, and multiplication.

A polynomial of degree n is a function of the form

$$f(x) = a_0 + a_1 x + \cdots + a_n x^n \tag{1.1}$$

Polynomial interpolation is very sensitive to the choice of interpolation points. At appropriately chosen points, it produces an approximation differing very little from the best approximant by polynomials of the same order. However, if the function to be approximated is badly behaved anywhere in the interval of approximation, then the approximation is poor everywhere. We can avoid this global dependence on local properties if we use piecewise polynomials. That is, we construct a composite curve by fitting successive low-degree polynomial curves to successive groups of data points (for example, see Chapter 2, Section 12).

Numerical analysis is necessary because the algorithms we use are not processed on ideal computing machines with unlimited precision and capacity. Computations cannot be executed or carried out exactly. We will discuss numerical analysis, interpolation, and approximation in later chapters in the context of specific applications.

3.1   Vectors

Vectors are quantities that have direction and magnitude and obey laws that follow. We will symbolically denote a vector with boldface, lowercase letters and graphically with an arrow. A vector with zero magnitude is a null vector and denoted by **0** (no direction is defined for it). The elementary laws governing vectors are (also see Fig. 1.4):

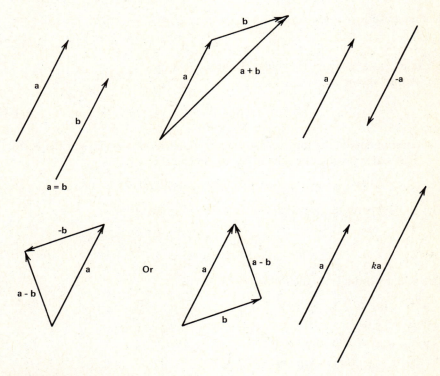

FIGURE 1.4 Vectors.

1. *Equality* Two vectors are equal when they have the same magnitude and direction. We can represent them with parallel lines of equal length when drawn to the same scale. Position is unimportant for equality.

2. *Addition* Given two vectors **a** and **b**, their sum **a** + **b** is graphically defined by joining the tail of **b** to the head of **a**. Then the line from the tail of **a** to the head of **b** is the sum **a** + **b**.

3. *Negation* The vector −**a** has the same magnitude as **a** but opposite direction.

4. *Subtraction* From laws 2 and 3, we define **a** − **b** = **a** + (−**b**).

5. *Scalar multiplication* The vector k**a** has the same direction as **a**, with a magnitude k times that of **a**. We call k a scalar.

From these laws, we determine the following properties. Given vectors **p**, **q**, **r**, and scalars k and l, we find,

1. $\mathbf{p} + \mathbf{q} = \mathbf{q} + \mathbf{p}$
2. $\mathbf{p} + (\mathbf{q} + \mathbf{r}) = (\mathbf{p} + \mathbf{q}) + \mathbf{r}$
3. $k(l\mathbf{p}) = kl\mathbf{p}$
4. $(k + l)\mathbf{p} = k\mathbf{p} + l\mathbf{p}$
5. $k(\mathbf{p} + \mathbf{q}) = k\mathbf{p} + k\mathbf{q}$

The magnitude of a vector is

$$|\mathbf{p}| = \sqrt{p_x^2 + p_y^2 + p_z^2} \tag{1.2}$$

where p_x, p_y, and p_z are the scalar components of **p**.

A unit vector is

$$\mathbf{n} = \frac{\mathbf{p}}{|\mathbf{p}|} \tag{1.3}$$

The components of the unit vector are also the direction cosines of the vector.

The scalar product has the following properties:

$$\mathbf{p} \cdot \mathbf{r} = p_x r_x + p_y r_y + p_z r_z = |\mathbf{p}||\mathbf{r}| \cos \theta \tag{1.4}$$

The angle θ between two vectors **p** and **r** is computed from

$$\cos \theta = \frac{\mathbf{p} \cdot \mathbf{r}}{|\mathbf{p}||\mathbf{r}|} \tag{1.5}$$

The scalar product has the following properties:

1. $\mathbf{p} \cdot \mathbf{p} = |\mathbf{p}|^2$
2. $\mathbf{p} \cdot \mathbf{r} = \mathbf{r} \cdot \mathbf{p}$
3. $\mathbf{p} \cdot (\mathbf{r} + \mathbf{q}) = \mathbf{p} \cdot \mathbf{r} + \mathbf{p} \cdot \mathbf{q}$
4. $(k\mathbf{p}) \cdot \mathbf{r} = \mathbf{p} \cdot (k\mathbf{r}) = k(\mathbf{p} \cdot \mathbf{r})$
5. If **p** is perpendicular to **r**, then $\mathbf{p} \cdot \mathbf{r} = 0$

The vector, or cross product, is

$$\mathbf{p} \times \mathbf{r} = (p_y r_z - p_z r_y)\mathbf{i} + (p_z r_x - p_x r_z)\mathbf{j} + (p_x r_y - p_y r_x)\mathbf{k} \qquad (1.6)$$

where $\mathbf{i}, \mathbf{j},$ and $\mathbf{k}$ are unit vectors in the x, y, and z directions. The vector product has the following properties:

1. $\mathbf{p} \times \mathbf{r} = \mathbf{s}$, $\mathbf{s}$ perpendicular to $\mathbf{p}$ and $\mathbf{r}$

2. $\mathbf{p} \times \mathbf{r} = \begin{vmatrix} \mathbf{i} & \mathbf{j} & \mathbf{k} \\ p_x & p_y & p_z \\ r_x & r_y & r_z \end{vmatrix}$

3. $\mathbf{p} \times \mathbf{r} = |\mathbf{p}||\mathbf{r}|\mathbf{n} \sin \theta$, $\mathbf{n}$ perpendicular to the plane of $\mathbf{p}$ and $\mathbf{r}$

4. $\mathbf{p} \times \mathbf{r} = -\mathbf{r} \times \mathbf{p}$

5. $\mathbf{p} \times (\mathbf{r} + \mathbf{s}) = \mathbf{p} \times \mathbf{r} + \mathbf{p} \times \mathbf{s}$

6. $(k\mathbf{p}) \times \mathbf{r} = \mathbf{p} \times (k\mathbf{r}) = k(\mathbf{p} \times \mathbf{r})$

7. $\mathbf{i} \times \mathbf{j} = \mathbf{k}, \mathbf{j} \times \mathbf{k} = \mathbf{i}, \mathbf{k} \times \mathbf{i} = \mathbf{j}$

8. If $\mathbf{p}$ is parallel to $\mathbf{r}$, then $\mathbf{p} \times \mathbf{r} = \mathbf{0}$.

A point $\mathbf{p}$ is

$$\mathbf{p} = p_x\mathbf{i} + p_y\mathbf{j} + p_z\mathbf{k} = [p_x \quad p_y \quad p_z][\mathbf{i} \quad \mathbf{j} \quad \mathbf{k}]^T \qquad (1.7)$$

It is often convenient to drop the matrix $[\mathbf{i} \ \mathbf{j} \ \mathbf{k}]^T$ and simply state

$$\mathbf{p} = [p_x \quad p_y \quad p_z] \qquad (1.8)$$

A line segment between points $\mathbf{p}_0$ and $\mathbf{p}_1$ results from

$$\mathbf{p}(u) = \mathbf{p}_0 + u(\mathbf{p}_1 - \mathbf{p}_0) \qquad \forall u \in [0, 1] \qquad (1.9)$$

Note that at $u = 0$, we use $\mathbf{p}(0) = \mathbf{p}_0$, and at $u = 1$, we use $\mathbf{p}(1) = \mathbf{p}_1$.

A line through $\mathbf{p}_0$ and in the direction of (parallel to) $\mathbf{p}_1$ is

$$\mathbf{p}(u) = \mathbf{p}_0 + u\mathbf{p}_1 \qquad \forall u \in [-\infty, +\infty] \qquad (1.10)$$

A plane containing the point $\mathbf{p}_0$ and perpendicular to the vector $\mathbf{p}_1 \times \mathbf{p}_2$ is

$$\mathbf{p}(u, w) = \mathbf{p}_0 + u\mathbf{p}_1 + w\mathbf{p}_2 \qquad \forall u, w \in [-\infty, +\infty] \qquad (1.11)$$

An interesting aspect of the vector product is that many geometric-modeling problems require that we find a vector perpendicular to two given vectors. If $\mathbf{p}$ is perpendicular to $\mathbf{a}$ and $\mathbf{b}$, then

$$\mathbf{p} \cdot \mathbf{a} = 0 \quad \text{and} \quad \mathbf{p} \cdot \mathbf{b} = 0 \qquad (1.12)$$

so that

$$p_x a_x + p_y a_y + p_z a_z = 0 \qquad (1.13)$$

and

$$p_x b_x + p_y b_y + p_z b_z = 0 \qquad (1.14)$$

From these, we derive the ratios

$$\frac{p_x}{a_y b_z - a_z b_y} = \frac{p_y}{a_z b_x - a_x b_z} = \frac{p_z}{a_x b_y - a_y b_x} \tag{1.15}$$

Clearly, the vector $[(a_y b_z - a_z b_y)\ (a_z b_x - a_x b_z)\ (a_x b_y - a_y b_x)]$ is proportional to **p** and thus perpendicular to **a** and **b**. It is the vector, or cross product, **a** × **b**.

Any three linearly independent vectors form a **basis**. To be linearly independent requires that the three vectors are not colinear and do not lie in a common plane. Thus, the three vectors, one in each of the three principal coordinate directions, form a basis. Any vector can be expressed in terms of a linear combination of the basis vectors.

3.2   Matrix Methods

Any array of numbers or other mathematical elements arranged in m rows and n columns is called a **matrix**. We denote a matrix with a boldface, uppercase letter, such as **B**, or by an indexed, lowercase letter, b_{ij} or b_i^j. When we use the index notation, the first subscript represents the row, and the second subscript or superscript represents the column.

A matrix with one row is a row vector, and a matrix with one column is a column vector. Matrix elements in this book are real numbers or real-valued functions and vectors; for example, a row matrix composed of vector elements might look like

$$\mathbf{B} = [\mathbf{p}_0 \quad \mathbf{p}_1 \quad \mathbf{p}_0^u \quad \mathbf{p}_1^u] \tag{1.16}$$

(**Note**: Superscript u denotes differentiation and is defined in Section 9 and later sections.)

We can arrange the same set of vector elements in a column matrix

$$\mathbf{B} = \begin{bmatrix} \mathbf{p}_0 \\ \mathbf{p}_1 \\ \mathbf{p}_0^u \\ \mathbf{p}_1^u \end{bmatrix} \tag{1.17}$$

An example of a scalar rectangular array is

$$\mathbf{A} = a_{ij} = \begin{bmatrix} a_{11} & a_{12} & a_{13} \\ a_{21} & a_{22} & a_{23} \\ a_{31} & a_{32} & a_{33} \\ a_{41} & a_{42} & a_{43} \end{bmatrix} \tag{1.18}$$

Two $m \times n$ matrices are equal if all corresponding elements are equal; however, there can be no equality between matrices of different sizes.

The sum of two $m \times n$ matrices $\mathbf{A}$ and $\mathbf{B}$ is the $m \times n$ matrix formed by the sum of corresponding elements of $\mathbf{A}$ and $\mathbf{B}$. Matrices of different sizes cannot be added. If

$$\mathbf{A} = \begin{bmatrix} 3 & 5 & 1 \\ 1 & -2 & 2 \end{bmatrix} \quad \text{and} \quad \mathbf{B} = \begin{bmatrix} 1 & 2 & 4 \\ 9 & 4 & 1 \end{bmatrix} \tag{1.19}$$

then

$$\mathbf{A} + \mathbf{B} = \begin{bmatrix} 4 & 7 & 5 \\ 10 & 2 & 3 \end{bmatrix} \tag{1.20}$$

We can find the difference between two matrices similarly.

When we multiply a matrix by a number or function, we multiply each individual element by that number or function; thus,

$$c\mathbf{A} = ca_{ij} \tag{1.21}$$

This operation is called **scalar multiplication**.

From these laws of matrix addition and scalar multiplication, we obtain the following properties:

1. $\mathbf{A} + \mathbf{B} = \mathbf{B} + \mathbf{A}$
2. $\mathbf{A} + (\mathbf{B} + \mathbf{C}) = (\mathbf{A} + \mathbf{B}) + \mathbf{C}$
3. $b(\mathbf{A} + \mathbf{B}) = b\mathbf{A} + b\mathbf{B}$
4. $(b + d)\mathbf{A} = b\mathbf{A} + d\mathbf{A}$
5. $b(d\mathbf{A}) = (bd)\mathbf{A} = d(b\mathbf{A})$

The product of two matrices a_{ij} and b_{jk} (or a_i^j and b_j^k) is a new matrix c_{ik} (or c_i^k). Note that the number of columns of a_{ij} must agree with the number of rows of b_{jk}. Thus, we sum on the index j to obtain

$$c_{ik} = a_{ij}b_{jk} \tag{1.22}$$

The product is also written as

$$\mathbf{C} = \mathbf{AB} \tag{1.23}$$

In general, the product $\mathbf{AB}$ of an $m \times n$ matrix $\mathbf{A}$ and an $n \times p$ matrix $\mathbf{B}$ is an $m \times p$ matrix $\mathbf{C}$. The properties of matrix multiplication are

1. $(\mathbf{AB})\mathbf{C} = \mathbf{A}(\mathbf{BC})$
2. $\mathbf{A}(\mathbf{B} + \mathbf{C}) = \mathbf{AB} + \mathbf{AC}$
3. $(\mathbf{A} + \mathbf{B})\mathbf{C} = \mathbf{AC} + \mathbf{BC}$
4. $\mathbf{A}(k\mathbf{B}) = k(\mathbf{AB}) = (k\mathbf{A})\mathbf{B}$

We define a unit or identity matrix δ_{ij} as an $m \times m$ square matrix whose elements satisfy the following equations:

$$\delta_{ij} = 1 \quad \text{for } i = j \tag{1.24}$$

and

$$\delta_{ij} = 0 \quad \text{for } i \neq j \tag{1.25}$$

δ_{ij} is also known as the Kronecker delta. An example of an identity matrix is

$$I = \begin{bmatrix} 1 & 0 & 0 \\ 0 & 1 & 0 \\ 0 & 0 & 1 \end{bmatrix} \tag{1.26}$$

The transpose of a matrix A is denoted by A^T. We obtain it by interchanging the rows and columns of A. Transpose matrices obey the following laws:

1. $(A + B)^T = A^T + B^T$
2. $(kA)^T = kA^T$
3. $(AB)^T = B^T A^T$

We observe here that an identity matrix is its own transpose.

The inverse A^{-1} of a square matrix A is a matrix that satisfies the conditions

$$AA^{-1} = A^{-1}A = I \tag{1.27}$$

If the elements of a matrix A are functions of some variable, say, x,

$$A(x) = \begin{bmatrix} a_{11}(x) & a_{12}(x) & \cdots & a_{1n}(x) \\ a_{21}(x) & a_{22}(x) & \cdots & a_{2n}(x) \\ \cdots & \cdots & \cdots & \cdots \\ a_{m1}(x) & a_{m2}(x) & \cdots & a_{mn}(x) \end{bmatrix}$$

Then, the integral of A with respect to x is

$$\int A(x)\, dx = \begin{bmatrix} \int a_{11}\, dx & \int a_{12}\, dx & \cdots & \int a_{1n}\, dx \\ \int a_{21}\, dx & \int a_{22}\, dx & \cdots & \int a_{2n}\, dx \\ \cdots & \cdots & \cdots & \cdots \\ \int a_{m1}\, dx & \int a_{m2}\, dx & \cdots & \int a_{mn}\, dx \end{bmatrix}$$

And the derivative of A with respect to x is

$$\frac{dA(x)}{dx} = \begin{bmatrix} \dfrac{da_{11}}{dx} & \dfrac{da_{12}}{dx} & \cdots & \dfrac{da_{1n}}{dx} \\[2ex] \dfrac{da_{21}}{dx} & \dfrac{da_{22}}{dx} & \cdots & \dfrac{da_{2n}}{dx} \\[2ex] \cdots & \cdots & \cdots & \cdots \\[2ex] \dfrac{da_{m1}}{dx} & \dfrac{da_{m2}}{dx} & \cdots & \dfrac{da_{mn}}{dx} \end{bmatrix}$$

3.3 Determinants

A determinant is a square array of elements that we can reduce to a single value. The 2×2 determinant of the square matrix

$$\begin{bmatrix} a_{11} & a_{12} \\ a_{21} & a_{22} \end{bmatrix}$$

is written

$$\begin{vmatrix} a_{11} & a_{12} \\ a_{21} & a_{22} \end{vmatrix}$$

where

$$\begin{vmatrix} a_{11} & a_{12} \\ a_{21} & a_{22} \end{vmatrix} = a_{11}a_{22} - a_{12}a_{21} \tag{1.28}$$

For a 3×3 determinant, we obtain

$$\begin{vmatrix} a_{11} & a_{12} & a_{13} \\ a_{21} & a_{22} & a_{23} \\ a_{31} & a_{32} & a_{33} \end{vmatrix} = a_{11}\begin{vmatrix} a_{22} & a_{23} \\ a_{32} & a_{33} \end{vmatrix} - a_{12}\begin{vmatrix} a_{21} & a_{23} \\ a_{31} & a_{33} \end{vmatrix} + a_{13}\begin{vmatrix} a_{21} & a_{22} \\ a_{31} & a_{32} \end{vmatrix} \tag{1.29}$$

A matrix must be square to have a determinant. The determinant of matrix $\mathbf{A}$ is denoted by $|\mathbf{A}|$.

We use the following properties of determinants:

1. The determinant of a matrix is equal to the determinant of its transpose; thus,

$$|\mathbf{A}| = |\mathbf{A}^T| \tag{1.30}$$

2. If we interchange any two rows (or columns) of $\mathbf{A}$, we change the sign of the determinant. If we denote the changed matrix as $\mathbf{B}$, then

$$|\mathbf{B}| = -|\mathbf{A}| \tag{1.31}$$

3. If we obtain matrix $\mathbf{B}$ by multiplying one row or column of $\mathbf{A}$ by a constant c, then

$$|\mathbf{B}| = c|\mathbf{A}| \tag{1.32}$$

4. If two rows or columns of $\mathbf{A}$ are identical, then

$$|\mathbf{A}| = 0 \tag{1.33}$$

5. If we derive $\mathbf{B}$ from $\mathbf{A}$ by adding a multiple of one row or column of $\mathbf{A}$ to another, then

$$|\mathbf{B}| = |\mathbf{A}| \tag{1.34}$$

6. If $\mathbf{A}$ and $\mathbf{B}$ are both $n \times n$ matrices, then the determinant of the product is

$$|\mathbf{AB}| = |\mathbf{A}||\mathbf{B}| \tag{1.35}$$

7. For each element of a determinant, we define a minor. We obtain the minor by striking out the row and column where the element is found. The determinant of the array remaining is the minor of the element. There is an algebraic sign associated with each element. For element a_{ij} the sign is $(-1)^{i+j}$. The product of the minor and the sign give the cofactor of an element a_{ij}, denoted by A_{ij}.

8. The elements of the inverse matrix $\mathbf{C}$ of a matrix $\mathbf{A}$ can be shown to be

$$c_{ij} = \frac{A_{ji}}{|\mathbf{A}|} \tag{1.36}$$

where A_{ij} is the ij^{th} cofactor of $\mathbf{A}$. The inverse does not exist if $|\mathbf{A}| = 0$, in which case the matrix is singular.

EXERCISES

1. Take the vector product of two vectors $\mathbf{a}$ and $\mathbf{b}$ in two dimensions. The resulting vector has a magnitude $ab \sin \theta$. What about its direction? Where does it point? Remember, this is a two-dimensional problem.

2. Repeat Ex. 1 for the vector product in four dimensions.

3. A triangle ABC has sides of length a, b, c. Use vector products to prove that

$$\frac{a}{\sin A} = \frac{b}{\sin B} = \frac{c}{\sin C}$$

4. Show that $(\mathbf{p} - \mathbf{q}) \cdot (\mathbf{p} + \mathbf{q}) = |\mathbf{p}|^2 - |\mathbf{q}|^2$ and that $(\mathbf{p} - \mathbf{q}) \times (\mathbf{p} + \mathbf{q}) = 2\mathbf{p} \times \mathbf{q}$. Interpret the results with an appropriate sketch.

5. Given the coordinates of the end points of a straight line (x_0, y_0, z_0) and (x_1, y_1, z_1), find the coordinates of the midpoint, the length of the line, and the direction cosines.

6. Expand the two following expressions:

$$\sum_{i=0}^{3} a_i u^i \text{ and } \sum_{i=0}^{3} \sum_{j=0}^{3} a_{ij} u^i w^j$$

7. Given $r = at^3 + bt^2 + ct + d$, find d^2r/dt^2.

8. Given $f(s) = s^3 + s$, find $f(2)$.

9. Given $r(s, t) = as^3 + bs^3t + cst^2 + dt^3 + es^2 + ft^2 + g$, find $\partial^2r/\partial t^2$.

10. Using the results from Ex. 9, evaluate $\partial^2r/\partial t^2$ at $s = 1, t = 2$.

11. Given two matrices $\mathbf{A}$ and $\mathbf{B}$, find the product $\mathbf{AB}$, where $\mathbf{A} = a_{ij}$; $i = 1$; $j = 1, 2, 3$; and $\mathbf{B} = b_{kl}$; $k = 1, 2, 3$; $l = 1, 2$.

12. Given vectors $\mathbf{a} = 6\mathbf{i} + 10\mathbf{j} + 2\mathbf{k}$ and $\mathbf{b} = \mathbf{i} - 2\mathbf{j} + 6\mathbf{k}$, find $\mathbf{c} = \mathbf{a} + \mathbf{b}$ ($\mathbf{i}, \mathbf{j}, \mathbf{k}$ are unit vectors in the x, y, z directions, respectively).

13. Given two matrices **A** and **B**,

$$\mathbf{A} = \begin{bmatrix} 7 & 3 & -1 \\ 2 & -5 & 6 \end{bmatrix} \qquad \mathbf{B} = \begin{bmatrix} 1 & 5 & 6 \\ -4 & -2 & 3 \end{bmatrix}$$

find **C** = **A** + **B** and **D** = **A** − **B**.

14. Given $\mathbf{A} = \begin{bmatrix} 3 & 5 \\ 4 & 2 \end{bmatrix}$, find 1.5**A**.

15. Given $\mathbf{A} = \begin{bmatrix} 5 & 3 & 8 \\ -1 & 4 & 7 \\ 0 & 1 & 1 \end{bmatrix}$ and $\mathbf{B} = \begin{bmatrix} 6 & 7 \\ 10 & 9 \\ 2 & -3 \end{bmatrix}$ find their product **A** = **AB**.

16. Given $\mathbf{A} = \begin{bmatrix} 2 & 1 \\ 3 & 4 \end{bmatrix}$ and $\mathbf{B} = \begin{bmatrix} 6 \\ 3 \end{bmatrix}$, find **C** = **AB**.

17. If $\mathbf{A} = \begin{bmatrix} 4 & 0 & 7 \\ 5 & 1 & 2 \end{bmatrix}$, find its transpose $\mathbf{A}^T$.

18. Write a procedure to compute the vector product of two vectors. Denote this as **VPROD**(V1, V2, V3), where

V1(3) is the input array of components of $\mathbf{p}_1$;

V2(3) is the input array of components of $\mathbf{p}_2$; and

V3(3) is the output array of components of the vector product $\mathbf{p}_1 \times \mathbf{p}_2$.

19. Write a procedure to transpose a 4 × 4 matrix. Denote this as **MATRNS**(*MI*, *MO*), where

MI(4, 4) is the input matrix **M** and

MO(4, 4) is the output matrix $\mathbf{M}^T$.

20. Write a procedure to invert a 4 × 4 matrix. Denote this as **MATINV**(MI, MO), where

MI(4, 4) is the input matrix **M** and

MO(4, 4) is the output matrix $\mathbf{M}^{-1}$.

21. Write a procedure to evaluate a 3 × 3 determinant. Denote this as **DETVAL**(DI, VAL), where

DI(3, 3) is the input array and

VAL is the output value of the determinant.

22. Write a procedure to compute the scalar product of two vectors. Denote this as **SPROD**(V1, V2, S), where

V1(3) is the input array of components of $\mathbf{p}_1$;

V2(3) is the input array of components of $\mathbf{p}_2$; and

S is the output value of the scalar product $\mathbf{p}_1 \cdot \mathbf{p}_2$.

23. Write a procedure to compute the sum of two vectors. Denote this as **VSUM(V1, V2, V3)**, where

V1(3) is the input array of components of $\mathbf{p}_1$;

V2(3) is the input array of components of $\mathbf{p}_2$; and

V3(3) is the output array of components of the sum $\mathbf{p}_1 + \mathbf{p}_2$.

24. Write a procedure to compute the magnitude of a vector. Denote this as **VMAG(V, MAG)**, where

V(3) is the input array of components of $\mathbf{p}_1$ and

MAG is the output value of $|\mathbf{p}_1|$.

25. Write a procedure to compute the unit vector of a vector. Denote this as **VUNIT(V, U)**, where

V(3) is the input array of components of $\mathbf{p}_1$ and

U(3) is the output array of components of $\mathbf{p}_1/|\mathbf{p}_1|$.

26. Write a procedure to compute the angle between two vectors. Denote this as **VANG(V1, V2, A)**, where

V1(3) is the input array of components of $\mathbf{p}_1$;

V2(3) is the input array of components of $\mathbf{p}_2$; and

A is the output value of the angle between $\mathbf{p}_1$ and $\mathbf{p}_2$.

27. Show algebraically that the equation $\mathbf{p}_0 \cdot \mathbf{n} = d$ represents the equation of a plane passing through $\mathbf{p}_0$, normal to the unit vector $\mathbf{n}$, where d is the perpendicular distance from the origin to the plane. Also interpret this equation graphically with an appropriate sketch.

28. Show that the intersection of three planes is given by

$$\mathbf{p} = \frac{d_1(\mathbf{n}_2 \times \mathbf{n}_3) + d_2(\mathbf{n}_3 \times \mathbf{n}_1) + d_3(\mathbf{n}_1 \times \mathbf{n}_2)}{\mathbf{n}_1 \cdot \mathbf{n}_2 \times \mathbf{n}_3}$$

where $\mathbf{n}_1$, $\mathbf{n}_2$, and $\mathbf{n}_3$ are unit normals to each of the three planes and d_1, d_2, and d_3 are the respective perpendicular distances of each from the origin.

4   NAMEABLE AND UNNAMEABLE GEOMETRIC SHAPES

We classify geometric elements—the simpler shapes into which we can decompose a complex object—in two ways, nameable or unnameable. The nameable elements are those of classical geometry: planes, spheres, circles, ellipses, parabolas, straight lines, and so forth. The shapes used by industry range from objects with planar and cylindrical faces to objects with much more elaborate forms. Not only are the shapes diverse, but reasons for using any particular shape are also varied. An automotive designer blends aerodynamic efficiency with aesthetic appeal. An aeronautical engineer designing ducting blends strength-to-weight characteristics with aerodynamic flow criteria. An industrial designer is concerned with the aesthetics and manufacturability of a telephone handset shell. A design shape may be of some classical analytical form or a more general nature.

Classical mathematics does not provide adequate methods for creating new curves and surfaces that will satsify various design criteria. Today, we must characterize special properties of shapes, such as smoothness, fairness (a subjective measure of the appearance of a curve or surface, usually indicating an absence of kinks, abrupt changes in direction, and the like), and continuity. We must frequently balance the attainment of these properties against one another and with the cost and complexity of the model. For example, a fillet surface (usually a surface whose cross section is a circular arc and acts as a transition between two intersecting flat surfaces) is not fair by most definitions, nor is it continuous in the second degree. Yet, a fillet surface is an important feature of most mechanical parts.

Until recently, curves and surfaces used in engineering were determined by the tools available to the draftsman. These, in turn, were close analogs of the tools and materials of manufacturing. For example, circles are widely used because they are easily drawn by compass and readily made with a lathe or drill press. The draftsman's flexible spline was long used in shipbuilding and then in the early days of aircraft. The spline is an accurate small-scale simulator of the geometric behavior (elastic deformation) of the structural systems and materials used to build ships and planes. In addition, there are many classical techniques available to construct conic curves.

Because drafting is a two-dimensional medium, all these curves are planar, even though many curves in engineering are inherently three dimensional. CAD/CAM systems now routinely generate fully three-dimensional shapes. And, as we shall see, these shapes are increasingly nonclassical and unnameable. So, forewarned is fairly warned: Be prepared for curves and surfaces whose characteristics we can categorize only generally.

5 EXPLICIT AND IMPLICIT EQUATIONS

We can use either parametric or nonparametric equations to represent geometric elements mathematically. Nonparametric equations are expressed in explicit or implicit form. For a plane curve, the explicit nonparametric equation takes the general form

$$y = f(x) \tag{1.37}$$

In this form, there is only one y value for each x value. The explicit form cannot represent closed or multiple-valued curves. We can overcome this limitation by using an implicit equation of the general form

$$f(x, y) = 0 \tag{1.38}$$

Both explicit and implicit nonparametric elements are axis dependent. Thus, the choice of the coordinate system affects the ease of using the elements and calculating their properties. In spite of this limitation, for appropriate applications, nonparametric equations can be used with success.

We write a general second-degree, implicit equation as

$$ax^2 + 2bxy + cy^2 + 2dx + 2ey + f = 0 \qquad (1.39)$$

If this curve passes through the origin of the coordinate system, then $f = 0$. By specifying the coefficients a, b, c, d, e, and f, we can produce several types of plane curves. We can use boundary conditions to establish a particular curve through specific points.

If we let $c = 1$ in the general equation, then, to define a curve segment between two points, we must specify five independent conditions, or degrees of freedom, to determine the values of the remaining five coefficients. One choice is to specify the position of the two end points, the slope of the curve segment at each end point, and an intermediate point through which the curve must pass.

We might take another approach. This time, we define $b = 0$ and $c = 1$. Now the analytical description of the resulting curve is fixed by specifying only four additional conditions, such as two end points and the two end slopes.

We can define an even simpler curve by setting $a = 1$, $b = 0$, and $c = 1$. Then, the form of the curve is

$$x^2 + y^2 + 2dx + 2ey + f = 0 \qquad (1.40)$$

Here, the three conditions required to fix d, e, and f can be the two end points and either the slope at the beginning or the slope at the end of the curve segment. Of course there are other choices of conditions available to us. For example, we might specify the two end points and a third intermediate point through which the curve must pass.

We obtain a straight line by setting $a = b = c = 0$; the equation reduces to

$$dx + ey + f = 0 \qquad (1.41)$$

To create an inflection point in the curve segment, we must use a higher order curve, such as a cubic; for example,

$$a + bx + cx^2 + dx^3 - y = 0 \qquad (1.42)$$

Specifying the two end points and the slopes at these points fixes the values of the four coefficients (see Chapter 2).

We can create a more complex curve by joining various types of curve segments. We define each segment by the values of the coefficients fixed by the specified boundary conditions and the algebraic form of the curve. Sometimes, we must require continuity in position and slope at the common point joining two adjacent curve segments. We compute any point on a specified segment by finding the correct root of the appropriate algebraic equation; however, to determine roots and gradients of implicit curves often requires lengthy calculations. Another difficulty when using axis-dependent, nonparametric curves arises when the end point of a curve has a vertical slope relative to the chosen coordinate system. We cannot use the resulting infinite slope as a numerical boundary condition. We must change the coordinate system or approximate an infinite slope with a large positive or negative value.

We will investigate other aspects of explicit and implicit equations for curves and surfaces in Chapters 2 and 3.

6 INTRINSIC EQUATIONS

An intrinsic property is one that depends on only the figure in question, not the figure's relation to a frame of reference. The fact that a rectangle has four equal angles is intrinsic to the rectangle, but the fact that a particular rectangle has two vertical sides is extrinsic, because an external frame of reference is required to determine which direction is vertical.

Imagine that your are standing in a large open space, free of any obstructions or convenient reference points. You want to do some experimental exercises in intrinsic geometry. First, you decide to walk in a square path; this is easy, because the intrinsic definition of a square is simple. From any initial position, walk ahead, say, ten paces; turn right 90°, and walk ten more paces; turn right 90° again, and walk ten more paces. Repeat this one more time, and you should be back at your starting position. The 90° turns certainly do not require any external reference system, since we can easily imagine various "local" methods to determine the turn, and counting to ten is obviously a purely local operation.

You have just exercised a procedural intrinsic definition of a square figure; try this with a circle. Again, start anywhere. How many different intrinsic procedures can you find for walking in a circle? We will see that intrinsic procedures often prove to be very useful in geometric modeling, particularly in interactive design. We find that intrinsic definitions are more local than traditional Cartesian coordinate definitions. Intrinsic definitions deal with geometry a little piece at a time.

Let us now investigate an example of an intrinsic equation. A curve requires two intrinsic equations, one expressing the curvature $1/\rho$ and one the torsion τ of the curve as functions of its arc length s

$$\frac{1}{\rho} = f(s) \qquad \tau = g(s) \tag{1.43}$$

Note that torsion is a measure of how much a space curve deviates from a plane curve (how much it tries to twist out of plane), and arc length is the length measured along the curve.

The theory of curves proceeds from the intrinsic equations. It is interesting to make a distinction between intrinsic equations, as just defined, and natural equations, defined in the following way: A natural equation of a curve is any equation connecting the curvature $1/\rho$, the torsion τ, and the arc length s of the curve. We have

$$f\left(\frac{1}{\rho}, \tau, s\right) = 0 \tag{1.44}$$

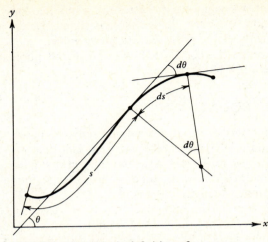

FIGURE 1.5 Intrinsic definition of a curve.

A natural equation of a curve imposes a condition on the curve, so that the curve has certain special properties, but there may be many curves having these properties. For example, $\tau = 0$ is a natural equation characterizing all plane curves, and $1/\rho = 0$ is also a natural equation characterizing all straight lines. An additional independent natural equation $g = 0$ of the curve determines the curve still more. If we solve the two natural equations $f = 0$ and $g = 0$ simultaneously for $1/\rho$ and τ as functions of s, intrinsic equations result. We conclude that two natural equations ordinarily determine a curve uniquely, except for its position in space.

We now investigate a slightly different approach. We will use Fig. 1.5 and limit our discussion to plane curves. Given the initial point of a curve, then the variation of θ with arc length s will completely define the curve. Here, θ is the angle subtended by the tangent to the curve with the x-axis. A relationship between s and θ is called an intrinsic equation of the curve.

We obtain curvature ($\kappa = 1/\rho$) from the intrinsic equation

$$\kappa = \frac{d\theta}{ds} \qquad (1.45)$$

We can also describe a curve parametrically in terms of the arc length, with the equations $x = x(s)$ and $y = y(s)$. The functions $x(s)$ and $y(s)$ must then be related by the equations

$$\frac{dx}{ds} = \cos\theta \quad \text{and} \quad \frac{dy}{ds} = \sin\theta \qquad (1.46)$$

By differentiating these equations with respect to s and substituting κ for $d\theta/ds$, dx/ds for $\cos\theta$, and dy/ds for $\sin\theta$, we obtain the simultaneous differential equations

$$\frac{d^2x}{ds^2} + \frac{\kappa(s)\,dy}{ds} = 0 \tag{1.47}$$

and

$$\frac{d^2y}{ds^2} - \frac{\kappa(s)\,dx}{ds} = 0 \tag{1.48}$$

With appropriate numerical procedures, we can solve these two second-order equations for $x(s)$ and $y(s)$ for any given curvature function $\kappa(s)$.

7 PARAMETRIC EQUATIONS

We cannot express shapes required for geometric modeling with ordinary, single-valued functions, such as $y = f(x)$. There are many reasons for this; first, the shapes of most objects we want to model are intrinsically independent of any coordinate system. If we must fit a curve or surface through a set of points, we quickly see that it is the relationship between the points themselves that determines the resulting shape, not the relationship between these points and some arbitrary coordinate system. In fact, most modeling applications require that the choice of a coordinate system not affect the shape. Second, any "solid," closed object will have vertical tangent lines or planes with respect to any choice of coordinate system. This results in infinite slopes or other ill-defined mathematical properties. Finally, the curves and surfaces of geometric modeling are often nonplanar and bounded in some sense and cannot be represented by an ordinary nonparametric function at all.

For these reasons and many others related to ease of programming and computability, the dominant means of representing shapes in geometric modeling is with parametric equations. For example, a two-dimensional curve is expressed not by a single ordinary function like $y = f(x)$, but by a set of two functions $x = x(u)$, $y = y(u)$ of a parameter u. A point on such a curve is represented by the vector

$$\mathbf{p} = [x(u) \quad y(u)] \tag{1.49}$$

Similarly, a point on a space curve is given by the vector

$$\mathbf{p} = [x(u) \quad y(u) \quad z(u)] \tag{1.50}$$

and a point on a surface is represented by the vector

$$\mathbf{p} = [x(u, w) \quad y(u, w) \quad z(u, w)] \tag{1.51}$$

(Remember, for convenience, we are omitting the multiplying matrix $[\mathbf{i} \quad \mathbf{j} \quad \mathbf{k}]^T$.)

This parametric representation not only avoids the previously mentioned problems associated with nonparametric functions, but it is also the most adequate description of the way curves are drawn by a plotter or on a computer graphics display screen; here, we apply two time functions $x(t)$ and $y(t)$ as a driving function to the servo system of the plotter or the electron beam deflection system of the display screen's cathode ray tube (CRT), causing the pen or electron beam to move on the appropriate curve.

The **point** is the basic element of parametric geometry and geometric modeling in general. It is defined as an ordered set of real numbers, usually the coordinates giving position in space. A vector may be defined in much the the same way. Frequently, the terms point and vector are interchangeable (for example, radius vector, position vector).

The mathematical functions that generate the sets of points defining curves, surfaces, and other geometric elements are parametric equations. The following equations are an example of the parametric representation of a curve. This curve is plotted in Fig. 1.6; of course, it is not possible to plot the curve for all values of u from $-\infty$ to $+\infty$. We must select an interval that has some significance to us. The curve in the figure happens to be plotted for an interval on u of $u \in [-1, 1]$.

$$x = 3u^2$$

$$y = u^3 - u + 1 \qquad\qquad (1.52)$$

$$z = 2u + 3$$

Here, u is the parametric variable and also the independent variable. When we substitute a specific value of u into each of these equations, we produce

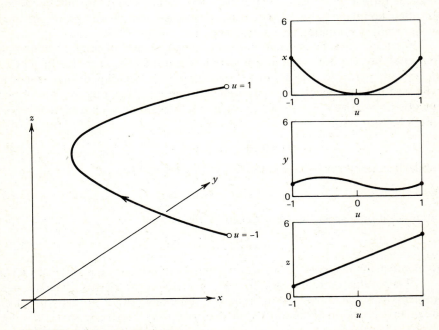

FIGURE 1.6 A parametric curve.

specific values of x, y, and z. Thus, each value of u generates a point on the curve. Three other curves are produced: (x, u), (y, u), and (z, u), also shown in Fig. 1.6.

It is convenient for us to normalize the parametric variable, which means limiting its value to the closed interval between 0 and 1, inclusive. We express this symbolically as $u \in [0, 1]$, This restriction gives rise to curve and surface boundaries, which we will investigate later.

Parametric equations have many advantages over direct or explicit forms. Let us review some of those we have already discussed and look at some new ones.

1. Parametric equations usually offer more degrees of freedom for controlling the shape of curves and surfaces than do nonparametric forms. For example, a two-dimensional explicit curve of the form

$$y = ax^3 + bx^2 + cx + d \tag{1.53}$$

has four coefficients that we may vary to control the curve. A two-dimensional parametric cubic curve of the form

$$x = au^3 + bu^2 + cu + d$$
$$y = eu^3 + fu^2 + gu + h \tag{1.54}$$

has eight coefficients available.

2. Transformations (for example, translation and rotation) can be performed directly on parametric equations.

3. Parametric forms readily handle infinite slopes without breaking down computationally, since

$$\frac{dy}{dx} = \frac{dy/du}{dx/du} \tag{1.55}$$

Using the notation $y^x = dy/dx$ and $x^u = dx/du$, and so on, then $x^u = 0$ indicates $y^x = \infty$.

4. Parametric equations completely separate the roles of the dependent and independent variable, both algebraically and geometrically, and allow any number of variables. For example, we can "extend" a curve in two-dimensional space into three- or four-dimensional spaces or higher without compromising its shape or any of its geometric properties in the lower spaces. This separation allows us to process geometric elements in their **geometric form**. Through the mathematical device of blending functions (introduced in Chapter 2), the geometric form offers us much greater insight into the control and behavior of curves and surfaces than is otherwise available with the classical algebraic formulation.

5. Parametrically defined geometric elements are inherently bounded, as mentioned earlier, because the parametric variables are normalized. We do not have to carry along extra geometric data to define or imply boundaries.

6. Geometric elements are easy to express in the form of vectors and matrices. This allows us to use relatively simple computation techniques, so that we can use computers economically.

7. Other advantages accrue when parametric functions themselves meet certain special conditions, such as a common form for all curves and a common form for all surfaces. There are additional advantages if the forms for curves and surfaces are analogous. Consider the advantages of the cubic form in this regard: A parametric cubic polynomial function is the simplest form capable of describing a curve that can twist in space and have points of inflection. Higher degree functions are also available, for example, a quintic polynomial, but they introduce cumbersome complexities for most applications.

8. A parametric bicubic polynomial function is the simplest form capable of describing a surface patch whose boundaries and interior are defined by a mesh of parametric cubic curves. The use of a single, common mathematical parametric format for representing curves and surfaces drastically reduces the number and complexity of subroutines required by any computer program designed to solve geometry problems.

So now we must ask what specifically is parametric geometry. A very short answer might be that it is a mathematical technique for describing curves and surfaces in a form readily applicable to geometric modeling.

Parametric geometry is easy to express in the form of vectors and matrices, which lets us use relatively simple techniques to solve very complex analytical geometry problems. Again, it is this simplicity of form that permits the economical use of computers and provides a common mathematical format for representing any curve or surface. A common mathematical form drastically reduces the number and complexity of subroutines required by any computer program designed to solve geometric-modeling problems.

There are three important facts to consider when using computers for geometric modeling. First, in engineering design, for example, we ordinarily use many different kinds of nameable curves and surfaces, such as circles, straight lines, planes, cylinders, conics, airfoils, spheres, splines — not to mention an unlimited variety of unnameable shapes. This means that if we have n kinds of curves to deal with and if, at one time or another, we may want the computer to compute, say, the length of one of them, we would need n different formulae for length programmed and ready to apply. If we want the points of intersection between any two curves, we must have a subroutine coded for that specific combination. This means being prepared for all possible combinations of intersections, all $0.5n^2$ of them. If $n = 10$, then 50 subroutines are required to cover all possible intersection combinations; if $n = 20$, then 200 are required. To make matters even worse, the computer must decide at the outset what kinds of curves it is dealing with. This means curve data must include information to identify it.

Second, the curves and surfaces we are most likely to use for geometric modeling are almost always bounded: The surface of a fender is bounded, the shape of a wing or nose cone is bounded. The classical methods used to describe shape geometry are inadequate, since they do not offer inherently bounded elements. The only immediate way around this problem is to add more mathematical data to define the boundaries.

Third, when we sketch curves or surfaces with pencil and paper, we can, using the methods of descriptive geometry, correctly show these curves or surfaces from any point of view. It has proven to be difficult, or at least cumbersome, to program a computer to mathematically rotate and otherwise manipulate the great variety of curves demanded by the classical approach without running into problems. As we have mentioned before, slopes (derivatives) and other properties may take on zero or infinite values in certain orientations, for example, if curves are expressed as $y = f(x)$.

In an environment of curves and surfaces of many special forms, all inherently unbounded and all mathematically formulated by equations difficult to operate on with simple geometric transformations, the complexity and resulting cost of creating and using any computer program for solving geometric-modeling problems become considerable.

Is there a single mathematical form that fits identically, or very nearly so, any curve or surface useful to us in geometric modeling? Yes. Parametric cubic and bicubic equations offer one such form. The following chapters will focus on this and certain related forms of parametric equations, including Bezier and *B*-spline forms.

EXERCISES

1. There are four coefficients, *a*, *b*, *c*, and *d*, in Eq. 1.53. When specific values are assigned to these coefficients, a specific curve is defined. Usually, we know something about the curve we want to define that helps us determine the coefficients. For example, we may want the ends of the curve segment to pass through end points (7, 0) and (1, 8) and through intermediate points (9, 4) and (0, 6). Use these points to determine *a*, *b*, *c*, and *d* in Eq. 1.53. Plot the results. Is it what you expected? If not, why not?

2. Repeat Ex. 1 for a curve with the same end points (7, 0) and (1, 8), but instead of specifying intermediate points, specify the slope at each end. Use $dy/dx = 1$ at (7, 0) and $dy/dx = -1$ at (1, 8). Again, determine *a*, *b*, *c*, and *d*, and plot the results. Is it what your expected? If not, why not?

3. Repeat Ex. 1 for any set of four points. Select specific points and the order in which you intend the curve to pass through them. Compute *a*, *b*, *c*, and *d*. Plot the resulting curve. Did the curve pass through the points in the order you expected? Change the order; how does this affect the curve? Can it affect the curve? What factors are involved?

4. In Eq. 1.54, what are the *x*, *y* coordinates of the point when $u = 0$; when $u = 0.5$; when $u = 1$? Find dy/dx when $u = 0$ and $u = 1$.

5. Using Eq. 1.54, select *x*, *y* coordinates for an ordered sequence of four points. Assign a value of $u = 0$ to the first point, $u = 1/3$ to the second point, $u = 2/3$

to the third, and $u = 1$ to the fourth. Now construct and solve the two sets of four simultaneous linear equations for the coefficients a–h. Plot the resulting parametric curve for the interval $u \in [0, 1]$; increments of $\Delta u = 0.1$ should be sufficient.

6. Repeat Ex. 5 with the order of the second and third points interchanged.

7. Repeat Ex. 5 with the order of all four points reversed. What can you conclude from the results?

8. Plot u, $x(u)$ and u, $y(u)$ over the interval $u \in [0, 1]$ for the parametric equation found in Ex. 5.

9. Plot u, $x(u)$ and u, $y(u)$ over the interval $u \in [0, 1]$ for the parametric equation found in Ex. 6.

10. Plot u, $x(u)$ and u, $y(u)$ over the interval $u \in [0, 1]$ for the parametric equation found in Ex. 7.

11. Repeat Ex. 8 for u, $x(u)$ only, but double the value of the coefficient d. What effect does this have on the x, y plot?

12. Repeat Ex. 8 for u, $x(u)$ only, but double the value of the coefficient c. What effect does this have on the x, y plot?

8   COORDINATE SYSTEMS

There are many coordinate systems relevant to geometric modeling. In addition to Cartesian, cylindrical, and spherical, we will want to construct coordinate systems on curves and surfaces, and there are the special projective or display coordinates. For these and other transformations, we use what are called homogeneous coordinates. When we investigate geometric modeling in certain computer-aided manufacturing applications, we refer to machine-axis coordinates. Every situation requires its own analysis of appropriate coordinate systems. However, most non-Cartesian systems are useful in only special circumstances. For example, let us investigate the disadvantages of polar coordinates.

1. There is no simple way of relating two sets of polar coordinates when each refers to a different origin. Thus, we might expect difficulty in developing efficient transformation schemes and subroutines for rotating, translating, and scaling geometric models.

2. The equations of tangents and normals cannot be simply described in polar form.

3. The polar angle of a point $\mathbf{p}(x, y)$ can only be found by using inverse trigonometric functions.

Thus, we must determine how easily we can transform between different coordinate systems and within a single system. We must study the effects of coordinate system characteristics on the expression of the various differential properties. Finally, we should verify the efficiency of computing points and their parametric inverse solutions. Such an analysis should be applied to any co-ordinate system we consider potentially useful to us in geometric modeling. We will study these and related subjects in Chapter 8.

9 NOTATION SCHEME AND GENERAL CONVENTIONS

Here is a list of the more common symbols and notations used in this text. In most cases, they will also be explained when first used. As for conventions, the right-handed Cartesian coordinate system is assumed unless explicitly noted otherwise.

Scalars, constant coefficients: lower case letters — a, b, c, and so forth.

Vectors: boldface, lower case letters — $\mathbf{a}$, $\mathbf{b}$, $\mathbf{p}$, $\mathbf{r}$, $\mathbf{s}$, and so forth.

Length of a vector: $|\mathbf{a}|$.

Matrices: boldface, upper case letters — $\mathbf{A}$, $\mathbf{B}$, $\mathbf{M}$, and so forth.

Matrix inverse: $\mathbf{A}^{-1}$, $\mathbf{B}^{-1}$, $\mathbf{M}^{-1}$, and so forth.

Matrix transpose: $\mathbf{A}^T$, $\mathbf{B}^T$, $\mathbf{M}^T$, and so forth.

Inverse of a transposed matrix: $\mathbf{A}^{T-1}$, $\mathbf{B}^{T-1}$, $\mathbf{M}^{T-1}$, and so forth.

Transpose of an inverted matrix: $\mathbf{A}^{-T}$, $\mathbf{B}^{-T}$, $\mathbf{M}^{-T}$, and so forth.

Transformed vector or matrix: $\mathbf{a}^*$, $\mathbf{A}^*$, and so forth.

Determinants: $|\mathbf{A}|$ or $|A|$.

Parametric variables: s, t, u, v, w.

Cartesian coordinates: x, y, z; or x_1, x_2, $x_3 \ldots x_i \ldots$ and so forth.

Boolean operators: $\cup$ (union), $\cap$ (intersection), $-$ (difference).

a is an element of b: $a \in b$.

For all: $\forall$.

Closed interval, continuous values: $[a, b]$.

Closed interval, integral or discrete values: $[a: b]$.

Open interval, continuous values: (a, b).

Differentiation: $y^x = dy/dx$;
$y^{xx} = d^2y/dx^2$;
$x^u = dx/du$;
$x^{uw} = \partial^2 x/\partial u\, \partial w$, and so forth.

2

CURVES

A curve segment is a point-bounded collection of points whose coordinates are given by continuous, one-parameter, single-valued mathematical functions of the form

$$x = x(u) \qquad y = y(u) \qquad z = z(u) \tag{2.1}$$

The parametric variable u is constrained to the interval $u \in [0, 1]$, and the positive sense on a curve is the sense in which u increases. The curve is point bounded because it has two definite end points, one at $u = 0$ and the other at $u = 1$.

We treat the coordinates of any point on a parametric curve as the components of a vector $\mathbf{p}(u)$. Fig. 2.1 illustrates this and other important vector elements. Note that ordinarily the tangent vectors will not be drawn to scale. Here, $\mathbf{p}(u)$ is the vector to the point $x(u)$, $y(u)$, $z(u)$, and $\mathbf{p}^u(u)$ is the tangent vector to the curve at the same point. It is found by differentiating $\mathbf{p}(u)$; thus,

$$\mathbf{p}^u(u) = \frac{d\mathbf{p}(u)}{du} \tag{2.2}$$

and the vector components are

$$x^u = \frac{dx(u)}{du} \qquad y^u = \frac{dy(u)}{du} \qquad z^u = \frac{dz(u)}{du} \tag{2.3}$$

These are the parametric derivatives. Note: When u appears as a superscript, it indicates differentiation with respect to u. The relationship between the parametric derivatives and the ordinary derivatives of Cartesian space is

$$\frac{dy}{dx} = \frac{dy/du}{dx/du} \tag{2.4}$$

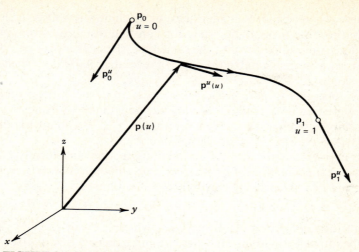

FIGURE 2.1 Vector elements of a parametric curve.

and similarly for dy/dz and dz/dx. We omit the functional notation when such a reference is obvious from the context, so that, for example,

$$\mathbf{p} = \mathbf{p}(u) \qquad \mathbf{p}^u = \mathbf{p}^u(u) \qquad x^u = x^u(u) \tag{2.5}$$

Referring to Fig. 2.1 again, the vectors $\mathbf{p}(0)$, $\mathbf{p}^u(0)$, $\mathbf{p}(1)$, and $\mathbf{p}^u(1)$ are the boundary conditions. The curve is bounded by the point $\mathbf{p}(0)$ at $u = 0$ and by the point $\mathbf{p}(1)$ at $u = 1$.

So far, we have presumed that $\mathbf{p}(u)$ is a three-component vector describing a curve in ordinary three-dimensional Cartesian space; however, it is not restricted to three components. Thus, in general, we have

$$\mathbf{p}(u) = [x_1(u), x_2(u), \ldots, x_i(u), \ldots, x_n(u)] \tag{2.6}$$

for an n-component curve in an n-dimensional space.

Let us investigate some examples of parametric equations of curves. First, here is a set of linear parametric equations

$$x = a + lu \qquad y = b + mu \qquad z = c + nu \tag{2.7}$$

where a, b, c, and l, m, n are constants. This particular curve is a straight line starting at point $\mathbf{p}(0) = [a \quad b \quad c]$ and ending at $\mathbf{p}(1) = [(a + l) \ (b + m) \ (c + n)]$, with direction cosines proportional to l, m, n.

Another example is

$$x = u \qquad y = u^2 \qquad z = u^3 \tag{2.8}$$

This curve is a cubical parabola and also one form of a twisted cubic.

The final example is

$$x = a \cos u \qquad y = a \sin u \qquad z = bu \tag{2.9}$$

This curve is a left-handed circular helix, or machine screw. It is the locus of a point that revolves around the z-axis at a constant distance a from it (in other words, a helix of radius $= a$). At the same time, the helix moves parallel to the z-axis at a rate proportional to the angle of revolution u. If $b < 0$, then the helix is right handed.

The versatility of parametric geometry is apparent in these examples. It is also interesting how it adapts to curve-defining functions expressing a direct relationship between the Cartesian variables x, y, z. The following discussion reveals this not always obvious relationship.

We know there are ways of representing a curve analytically other than with parametric equations. For example, one equation in x, y, z represents a surface, and two independent simultaneous equations in x, y, z, say,

$$F(x, y, z) = 0 \qquad G(x, y, z) = 0 \tag{2.10}$$

represent the intersection of two surfaces, which is locally a curve. They are called the implicit equations of a curve. A curve defined this way is inherently unbounded; however, only a bounded part of it may be of interest.

If we solve the implicit equations for two of the variables in terms of the third, say, for y and z in terms of x, then the results are

$$y = y(x) \qquad z = z(x) \tag{2.11}$$

These equations represent the same curve as Eqs. 2.10, and they, or the equations similarly expressing any two of the coordinates of a variable point on the curve as functions of the third coordinate, are the explicit equations of the curve. Each of Eqs. 2.11 separately represents a cylinder projecting the curve onto one of the principal planes, so these equations are a special form of Eqs. 2.10 for which the two surfaces are projecting cylinders.

If we solve the first of the three parametric Eqs. 2.1 of a curve for u as a function of x, $u(x)$ and substitute the result into the two remaining expressions, we then obtain the explicit Eqs. 2.11 of the curve. From one point of view, these explicit equations, when supplemented by the identity $x = x$, are also parametric equations of the curve

$$x = x \qquad y = y(x) \qquad z = z(x) \tag{2.12}$$

The parameter now is the coordinate x.

The difficulty with this approach is the obviously unacceptable limitation on the range of x, since the parametric variable must be normalized to the interval $x \in [0, 1]$ by our established convention. We easily resolve the problem by introducing a parametric function of the form

$$u = \frac{x - x_0}{x_1 - x_0} \tag{2.13}$$

This allows an explicit range of x from x_0 to x_1, provided $u \in [0, 1]$, through the parametric functions $x(u)$, where

$$x = x(u) = x_0 + (x_1 - x_0)u \tag{2.14}$$

This satisfies the normalization constraint on the parametric variable without compromising the range of x. If we substitute this relationship into

Eqs. 2.12, we obtain

$$x = x(u) \qquad y = y[x(u)] \qquad z = z[x(u)] \tag{2.15}$$

which simplifies to

$$x = x(u) \qquad y = y(u) \qquad z = z(u) \tag{2.16}$$

Clearly, these are the parametric forms introduced in Eqs. 2.1. This means that we can convert a large class of explicit functions into a parametric form.

Bezier (1972) correctly identified the fundamental property of parametric curves: Their shape depends on only the relative position of the points defining their characteristic vectors and is independent of the position of the total set of points with respect to the coordinate system in use. This is an essential characteristic for many applications, such as CAD/CAM modeling. In general, to transform an axis-dependent curve, we must first compute the coordinates of every point required in the original system, then transform each into the new system. For axis-independent curves, it is sufficient to transform the points defining the characteristic vectors from one system to another. Matrix formulation greatly simplifies these operations, as we will see in later sections.

Chapter 2 presents curves fully expressed in three-dimensional space; plane curves are a special case. Many considerations enter into the decision of how to represent a curve: The form of computer input desired, the type of manipulations required, and the display device used for output all introduce considerations and limitations. We investigate next the mathematics of curves in more detail, which will enable us to make the best decision for their use.

EXERCISES

1. Find the coefficients of Eq. 2.7 for the straight-line segment between point $(0, 0, 0)$ at $u = 0$ and point $(4, 4, 3)$ at $u = 1$. Verify the results at $u = 0.5$.

2. Find the coefficients of Eq. 2.7 for the straight-line segment between point $(1, 1, 2)$ at $u = 0$ and point $(1, 1, -4)$ at $u = 1$. Verify the results at $u = 0.5$.

3. Plot (x, u), (y, u), and (z, u) for the parametric components of the line segment in Ex. 1. Plot u from 0 to 1 on the abcissa.

4. Plot (x, u), (y, u), and (z, u) for the parametric components of the line segment in Ex. 2. Compare these plots to those obtained in Ex. 3.

5. Sketch the curve resulting from Eq. 2.8 for the parametric interval $u \in [0, 1]$. Plot the parametric components.

6. Find dy/dx at $u = 0$ and at $u = 1$ for Eq. 2.8.

7. Let $a = b = 1$ in Eq. 2.9. Sketch the resulting curve for the parametric interval $u \in [0, 1]$ where $u = 2\omega/\pi$, with ω in radians. Plot the parametric components.

8. Find dy/dx and dz/dx at $u = 0$ and at $u = 1$ for Eq. 2.9 (again, let $u = 2\omega/\pi$, ω in radians).

1   ALGEBRAIC AND GEOMETRIC FORM

The algebraic form of a parametric cubic (pc) curve segment is given by the following three polynomials:

$$x(u) = a_{3x}u^3 + a_{2x}u^2 + a_{1x}u + a_{0x}$$
$$y(u) = a_{3y}u^3 + a_{2y}u^2 + a_{1y}u + a_{0y} \qquad (2.17)$$
$$z(u) = a_{3z}u^3 + a_{2z}u^2 + a_{1z}u + a_{0z}$$

The parameter u, the independent variable, is restricted by definition to values in the interval 0 to 1, inclusive. This restriction makes the curve segment bounded.

A unique set of 12 constant coefficients, called **algebraic coefficients**, determines a unique pc curve; it determines the size and shape of a curve and its position in space. Two curves of the same shape have different algebraic coefficients if they occupy different positions in space. Note that shape characteristics include size or scale considerations.

We can easily write the three equations in Eqs. 2.17 in the more compact form of a vector equation. Vector equations are not only less cumbersome to read and write, but they are also capable of denoting an arbitrary number of dimensions. The value of the intrinsic geometric nature of vector notation was mentioned earlier. Thus, in vector notation, Eqs. 2.17 become

$$\mathbf{p}(u) = \mathbf{a}_3 u^3 + \mathbf{a}_2 u^2 + \mathbf{a}_1 u + \mathbf{a}_0 \qquad (2.18)$$

where $\mathbf{p}(u)$ is the position vector of any point on the curve, and $\mathbf{a}_0, \mathbf{a}_1, \mathbf{a}_2, \mathbf{a}_3$ are the vector equivalents of the scalar algebraic coefficients. The components of $\mathbf{p}(u)$ correspond to the Cartesian coordinates of the point.

The restriction on the parametric variable u(a scalar) is expressed as

$$u \in [0, 1] \qquad (2.19)$$

The algebraic coefficients are not always the most convenient way of controlling the shape of a curve in typical modeling situations, nor do they contribute much to an intuitive sense of a curve. The geometric form of a pc space curve seems to fulfill these needs. We can readily define a pc curve in terms of conditions at its end points, or boundaries. For a space curve, there are several conditions to choose from: end point coordinates, tangents, curvature, torsion, plus any number of conditions dependent on higher order derivatives. Using the two end points $\mathbf{p}(0)$ and $\mathbf{p}(1)$ and the corresponding tangent vectors $d\mathbf{p}(0)/du = \mathbf{p}^u(0)$ and $d\mathbf{p}(1)/du = \mathbf{p}^u(1)$, we obtain the following four equations from Eq. 2.18:

$$\mathbf{p}(0) = \mathbf{a}_0$$
$$\mathbf{p}(1) = \mathbf{a}_0 + \mathbf{a}_1 + \mathbf{a}_2 + \mathbf{a}_3$$
$$\mathbf{p}^u(0) = \mathbf{a}_1 \qquad (2.20)$$
$$\mathbf{p}^u(1) = \mathbf{a}_1 + 2\mathbf{a}_2 + 3\mathbf{a}_3$$

To obtain $\mathbf{p}(0)$, we simply substitute $u = 0$ into Eq. 2.18. To obtain $\mathbf{p}(1)$, substitute $u = 1$ into the equation. Finally, differentiate $\mathbf{p}(u)$ with respect to u, and again substitute $u = 0$ and $u = 1$ into the resulting equation for $\mathbf{p}^u(u)$ to obtain $\mathbf{p}^u(0)$ and $\mathbf{p}^u(1)$.

By solving this set of four simultaneous equations in four unknowns, we can redefine the algebraic coefficients in terms of the boundary conditions

$$\mathbf{a}_0 = \mathbf{p}(0)$$

$$\mathbf{a}_1 = \mathbf{p}^u(0)$$

$$\mathbf{a}_2 = -3\mathbf{p}(0) + 3\mathbf{p}(1) - 2\mathbf{p}^u(0) - \mathbf{p}^u(1) \tag{2.21}$$

$$\mathbf{a}_3 = 2\mathbf{p}(0) - 2\mathbf{p}(1) + \mathbf{p}^u(0) + \mathbf{p}^u(1)$$

Substitute these expressions for the algebraic coefficients in Eq. 2.18, and rearrange terms to obtain

$$\mathbf{p}(u) = (2u^3 - 3u^2 + 1)\mathbf{p}(0) + (-2u^3 + 3u^2)\mathbf{p}(1)$$
$$+ (u^3 - 2u^2 + u)\mathbf{p}^u(0) + (u^3 - u^2)\mathbf{p}^u(1) \tag{2.22}$$

This equation is simplified by the following substitutions:

$$F_1(u) = 2u^3 - 3u^2 + 1$$

$$F_2(u) = -2u^3 + 3u^2$$

$$F_3(u) = u^3 - 2u^2 + u \tag{2.23}$$

$$F_4(u) = u^3 - u^2$$

Thus, we rewrite Eq. 2.22 as

$$\mathbf{p}(u) = F_1(u)\mathbf{p}(0) + F_2(u)\mathbf{p}(1) + F_3(u)\mathbf{p}^u(0) + F_4(u)\mathbf{p}^u(1) \tag{2.24}$$

We again further simplify by dropping the functional notation and using superscripts and subscripts to represent differentiation with respect to u and the end-point u value, respectively. Thus, Eq. 2.24 simplifies to

$$\mathbf{p} = F_1\mathbf{p}_0 + F_2\mathbf{p}_1 + F_3\mathbf{p}_0^u + F_4\mathbf{p}_1^u \tag{2.25}$$

This is the geometric form, and $\mathbf{p}_0$, $\mathbf{p}_1$, $\mathbf{p}_0^u$, $\mathbf{p}_1^u$ are called **geometric coefficients**. The F terms are blending functions, about which we will say more later. Figure 2.1 presents the basic elements of the vector geometric expression of a pc space curve.

Notice that we could have selected any two points and tangent vectors as long as we specified the corresponding value of the parametric variable u. In fact, we could select four tangent vectors and no points or four points and no tangent vectors or even three tangent vectors and one point. Each point must have a different u value and similarly for each tangent vector. Note that a point and tangent vector can have the same u value. The point is that there are 12 degrees of freedom (four vectors × three components each) to be specified to define fully and unambiguously the equations of the curve. As we will see later in more detail, it is most convenient to specify these variables at the end points of a curve where the end points are at $u = 0$ and $u = 1$. First, it is computationally easy to evaluate curve equations at $u = 0$ and $u = 1$. Second, when we join many curves together end to end to form more complex curves, we will want to control what is going on at these ends.

Note that pc curves are directional as a result of being generated by successive values of the parametric variable u. We will discuss details of this mathematical relationship later, but it is important to understand that an identical curve can be generated with the sequence of u values reversed. The choice of which end of a curve to define as p_0 and which as p_1 is arbitrary unless there is some external (computational, and so on) constraint dictating a preferred direction. There are no mathematical constraints.

Earlier, we asserted that the matrix form is the most compact form for representing a pc curve and thus geometric operations and analyses are simple matrix manipulations. Now, it is time to demonstrate these assertions. Equation 2.18 can be written as the product of two matrices

$$\mathbf{p} = [u^3 \quad u^2 \quad u \quad 1][\mathbf{a}_3 \quad \mathbf{a}_2 \quad \mathbf{a}_1 \quad \mathbf{a}_0]^T \qquad (2.26)$$

Let $\mathbf{U} = [u^3 \quad u^2 \quad u \quad 1]$ and $\mathbf{A} = [\mathbf{a}_3 \quad \mathbf{a}_2 \quad \mathbf{a}_1 \quad \mathbf{a}_0]^T$, then rewrite Eq. 2.26 as

$$\mathbf{p} = \mathbf{UA} \qquad (2.27)$$

Treat the geometric form similarly to obtain

$$\mathbf{p} = [F_1 \quad F_2 \quad F_3 \quad F_4][\mathbf{p}_0 \quad \mathbf{p}_1 \quad \mathbf{p}_0^u \quad \mathbf{p}_1^u]^T \qquad (2.28)$$

Now let $\mathbf{F} = [F_1 \quad F_2 \quad F_3 \quad F_4]$ and $\mathbf{B} = [\mathbf{p}_0 \quad \mathbf{p}_1 \quad \mathbf{p}_0^u \quad \mathbf{p}_1^u]^T$, and rewrite Eq. 2.28 as

$$\mathbf{p} = \mathbf{FB} \qquad (2.29)$$

$\mathbf{A}$ is the matrix of algebraic coefficients, and $\mathbf{B}$ is the matrix of geometric coefficients or boundary conditions. Now, using matrix methods, we can easily develop a relationship between the algebraic and geometric forms. From Eq. 2.23, we obtain

$$\mathbf{F} = [(2u^3 - 3u^2 + 1) \quad (-2u^3 + 3u^2) \quad (u^3 - 2u^2 + u) \quad (u^3 - u^2)] \qquad (2.30)$$

By inspection, we can express the right side of Eq. 2.30 as

$$\mathbf{F} = [u^3 \quad u^2 \quad u \quad 1] \begin{bmatrix} 2 & -2 & 1 & 1 \\ -3 & 3 & -2 & -1 \\ 0 & 0 & 1 & 0 \\ 1 & 0 & 0 & 0 \end{bmatrix} \qquad (2.31)$$

The 4×4 matrix is called the **universal transformation matrix** and denoted by $\mathbf{M}$, thus,

$$\mathbf{F} = \mathbf{UM} \qquad (2.32)$$

Substitute Eq. 2.32 for Eq. 2.29 to obtain

$$\mathbf{p} = \mathbf{UMB} \tag{2.33}$$

Thus,

$$\mathbf{A} = \mathbf{MB} \tag{2.34}$$

and, conversely,

$$\mathbf{B} = \mathbf{M}^{-1}\mathbf{A} \tag{2.35}$$

where

$$\mathbf{M}^{-1} = \begin{bmatrix} 0 & 0 & 0 & 1 \\ 1 & 1 & 1 & 1 \\ 0 & 0 & 1 & 0 \\ 3 & 2 & 1 & 0 \end{bmatrix} \tag{2.36}$$

Therefore, Eqs. 2.34 and 2.35 allow ready conversion between algebraic and geometric forms.

Equation 2.33 is the form we most often use for curves in this text. For completeness, we have

$$\mathbf{p}(u) = \mathbf{UMB} \qquad u \in [0, 1] \tag{2.37}$$

Cubic curves of this form, defined by the coordinates and tangent vectors at their end points, are often called **Hermite curves**; we will occasionally use the term here. It is a convenient way of distinguishing this form from Bezier and *B*-spline curves, which we will discuss later. Hermite, by the way, was a nineteenth-century French mathematician of considerable accomplishment in the area of cubic and quintic polynomial equations.

Finally, notice that matrices **U**, **F**, and **M** are identical for all pc curves. Only the **A** and **B** matrices vary from curve to curve, depending on shape and position. This means that it is often most efficient to denote a specific curve by simply giving its algebraic or geometric matrix, which is done throughout the text.

EXERCISES

1. Find the algebraic coefficients of a degenerate pc curve of zero length located at point $(2, 3.5, -1.2)$.

2. Find the algebraic coefficients of a pc curve whose coordinates at $u = 0$ are $(-1, -2, -1)$ and at $u = 1$ are $(4, 2, 4)$ and whose direction cosines are constant at all points on the curve (also, find these direction cosines).

3. Find the algebraic coefficients of a pc curve lying in the $x = 10$ plane whose geometric characteristics at $u = 0$ are $(10, 2, 2)$ and $dz/dy = 2$, at $u = 0.5$ are $(10, 6, 6)$, and at $u = 1$ are $(10, 10, 2)$ and $dz/dy = -2$.

4. Find the geometric coefficients for the curve in Ex. 3.

5. Using the geometric coefficients found in Ex. 4, derive a new set with the only difference that components of the tangent vectors $\mathbf{p}_0^u$ and $\mathbf{p}_1^u$ are multiplied by 4. Then calculate the corresponding algebraic coefficients, and plot the results in object space (that is, x, y, z Cartesian coordinate space) and the three parametric spaces [that is, (x, u), (y, u), and (z, u)].

6. Using the geometric coefficients found in Ex. 4, derive a new set with the only difference that components of $\mathbf{p}_0^u$ are multiplied by 4. Then calculate the corresponding algebraic coefficients, and plot the results in object space and the three parametric spaces.

7. Using the geometric coefficients found in Ex. 4, derive a new set with the only difference that components of $\mathbf{p}_0^u$ are multiplied by 8. Then calculate the corresponding algebraic coefficients, and plot the results in object space and the three parametric spaces.

8. Again using the geometric coefficients found in Ex. 4, derive a new set with the only difference that components of the tangent vector $\mathbf{p}_1^u$ are made equal to the corresponding components of $\mathbf{p}_0^u$. Then calculate the corresponding algebraic coefficients, and plot the results in object space and the three parametric spaces.

9. Write the algebraic expressions for $x(u)$, $y(u)$, and $z(u)$ for quintic polynomials in u. How many degrees of freedom are there?

10. Derive the geometric form for quintics. Use

$$\mathbf{B} = [\mathbf{p}_0 \quad \mathbf{p}_1 \quad \mathbf{p}_0^u \quad \mathbf{p}_1^u \quad \mathbf{p}_0^{uu} \quad \mathbf{p}_1^{uu}]^T$$

Find the blending functions and the universal transformation matrix $\mathbf{M}$.

11. Write a procedure to convert pc curve coefficients in algebraic form into geometric form. Denote this as $\mathbf{ATG}(CI, CO)$, where

CI(4, 3) is the input array of algebraic coefficients and

CO(4, 3) is the output array of geometric coefficients.

12. Write a procedure to convert the pc curve coefficients in geometric form into algebraic form. Denote this as $\mathbf{GTA}(CI, CO)$, where

CI(4, 3) is the input array of geometric coefficients and

CO(4, 3) is the ouput array of algebraic coefficients.

13. Write a procedure to compute the x, y, z coordinates of a point on a pc curve for a specific value of the parametric variable u. Use the geometric form to evaluate the point. Denote this as $\mathbf{PCRV}(CI, U, P)$, where

CI(4, 3) is the input array of geometric coefficients defining the pc curve;

U is the input value of u; and

P(3) is the ouput array of the x, y, z coordinates.

14. Write a procedure to compute the x, y, z coordinates of a point on a pc curve. Use the algebraic form to evaluate the point. Denote this as **PCRVA(CI, U, P)**, where

> CI(4, 4, 3) is the input array of algebraic coefficients;
>
> U is the input value of u; and
>
> P(3) is the output array of the x, y, z coordinates.

15. Compare the procedures **PCRV** and **PCRVA**.

16. Write a procedure to compute the x, y, z coordinates of points along a pc curve at a specified incremental value of the parametric variable u. Make use of procedures **GTA** and **PCRV**. Denote this as **PPCRV(CI, NU, P)**, where

> CI(4, 3) is the input array of geometric coefficients defining the pc curve;
>
> NU is the input number of increments of the parametric variable; and
>
> P(4, NU + 1) is the output array of u values and the corresponding x, y, z coordinates, starting at $u = 0$ and proceeding to $u = 1$ in NU equal increments.

17. Write a procedure to find the roots of a cubic equation in a given interval. Sort the roots in ascending order. Use an iterative technique to find a solution, where the cubic equation formed is assumed to be $(COF_1)u^3 + (COF_2)u^2 + (COF_3)u + COF_4 = 0$. Denote this as **CUBIC(COF, UMIN, UMAX, NU, U)**, where

> COF(4) is the input array of the four coefficients of the cubic equation;
>
> UMIN is the input minimum acceptable value of a root;
>
> UMAX is the input maximum acceptable value of a root;
>
> NU is the output number of roots found within the acceptable interval; and
>
> U(3) is the output array of roots found in the interval UMIN to UMAX.

2 TANGENT VECTORS

The tangent vectors p_0^u and p_1^u introduced in the development of Eqs. 2.20 need considerable elaboration. The tangent vector at an arbitrary point on a curve is given by $dp(u)/du$. What this means in terms of the more conventional concepts of slopes and direction cosines, as well as curve behavior, we will now learn.

Specifying the end points and end slopes (or direction cosines) of a pc curve accounts for only 10 of the 12 degrees of freedom implied by Eqs. 2.17. Six are supplied by the two end-point coordinates (x_0, y_0, z_0) and (x_1, y_1, z_1), the components of p_0 and p_1. Four are supplied by the slopes or direction cosines, two from each end. Remember that there are only two independent slopes or direction cosines at any point on a curve. The third direction cosine may be derived from any given two. (For example, the sum of the squares of the three direction cosines at a point equals 1.) Thus, two more degrees of freedom are available to control the behavior of a curve.

The slopes, as represented by tangent vectors in the geometric form $\mathbf{p}_0^u$ and $\mathbf{p}_1^u$, are those in parametric space. (Section 3 discusses the parametric space of a curve.) The components are

$$x_0^u = \frac{dx(0)}{du} \qquad x_1^u = \frac{dx(1)}{du} \qquad (2.38)$$

and similarly for y and z. They are directly related to the ordinary, or object-space, slopes y^x and z^x as follows:

$$y^x = \frac{y^u}{x^u} \qquad z^x = \frac{z^u}{x^u} \qquad (2.39)$$

Of course, subscripts are added as needed to indicate an evaluation at, or reference to, a specific u value. **Note:** The term **object space** is used to denote ordinary, nonparametric x, y, z Cartesian coordinate space.

The concept of a **tangent vector** resolves the problem of the extra two degrees of freedom in the following way: A vector is assumed to operate at each end of the pc curve in object space, as in Fig. 2.2. These vectors have the same direction

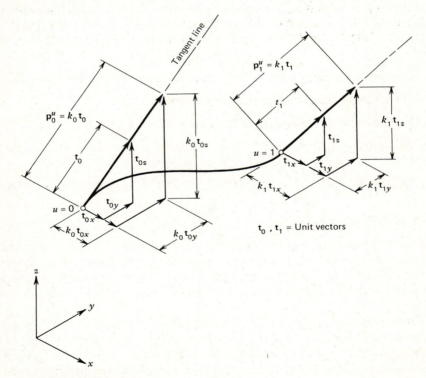

FIGURE 2.2 Tangent vectors.

as the tangents at each end, hence the name tangent vector. Each vector also has a magnitude associated with it. The magnitudes supply the eleventh and twelfth degrees of freedom and contribute to the control of the shape of a curve's interior.

The direction cosines of the tangent to a point on a curve comprise the components of a unit vector $\mathbf{t}$, where $\mathbf{t} = \mathbf{p}^u/|\mathbf{p}^u|$ and the direction cosines are denoted by t_x, t_y, t_z. Then, clearly,

$$|\mathbf{t}| = \sqrt{t_x^2 + t_y^2 + t_z^2} = 1 \qquad (2.40)$$

where $t_x, t_y, t_z \in [0, 1]$. Further, let

$$k = |\mathbf{p}^u| \qquad (2.41)$$

Then,

$$\mathbf{p}^u = k\mathbf{t} \qquad (2.42)$$

must result, where k represents the magnitude of $\mathbf{p}^u$. Again, subscripts may be added to indicate a specific point, as denoted by its u value, giving the tangent vectors for the two end points of a curve as

$$\mathbf{p}_0^u = k_0\mathbf{t}_0 \quad \text{and} \quad \mathbf{p}_1^u = k_1\mathbf{t}_1 \qquad (2.43)$$

Because of the condition Eq. 2.40 places on the three direction cosines of each tangent vector, only two are independent, free variables. The matrix of geometric coefficients can now be expressed as

$$\mathbf{B} = [\mathbf{p}_0 \quad \mathbf{p}_1 \quad \mathbf{p}_0^u \quad \mathbf{p}_1^u]^T \qquad (2.44)$$

or

$$\mathbf{B} = [\mathbf{p}_0 \quad \mathbf{p}_1 \quad k_0\mathbf{t}_0 \quad k_1\mathbf{t}_1]^T \qquad (2.45)$$

This is a very interesting state of affairs, since we see that it is a property of pc curves that many different curves (in fact, an infinite family of them) may be obtained, all of which have the same end points and slopes yet entirely different interior shapes, depending on the magnitudes k_0 and k_1 of the tangent vectors. By varying k_0 and k_1, we can control the curvature at each end, or we can fix the location of some intermediate point. We will explore the full power of this property of tangent vectors in subsequent sections.

Figures 2.3 and 2.4 show the effect of varying only k_0 and k_1. We hold constant all the other geometric coefficients, which have the following values:

$x_0 = 4.0$	$y_0 = 4.0$	$z_0 = 0$
$x_1 = 24.0$	$y_1 = 4.0$	$z_1 = 0$
$t_{0x} = 0.8320$	$t_{0y} = 0.5547$	$t_{0z} = 0$
$t_{1x} = 0.8320$	$t_{1y} = -0.5547$	$t_{1z} = 0$

The curves in Fig. 2.3 are symmetrical around a common vertical axis, and we varied their respective values of k_0 and k_1 in a symmetrical way. In Fig. 2.4, we varied k_0 while holding k_1 constant, which resulted in the skewed

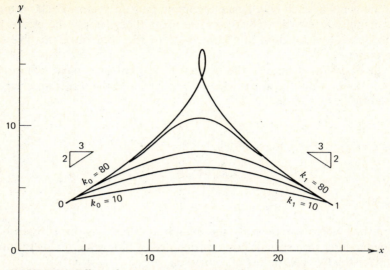

FIGURE 2.3 Effect of tangent vector magnitude on curve shape.

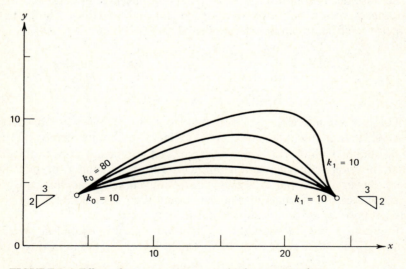

FIGURE 2.4 Effect of tangent vector magnitude on curve shape.

family of curves. Usually, when we model a curve, we must meet some specific criteria. The illustrations here are meant to explain curve behavior and do not represent particular examples of applications to engineering problems.

When the magnitudes of the tangent vectors begin to exceed some multiple of the chord length between the end points of the curve segment, the curve begins to exhibit (sometimes) undesirable characteristics, such as loops and cusps. Conversely, very low magnitudes produce a relatively flat curve. The chord length $|\mathbf{p}_1 - \mathbf{p}_0| = 20$ for every curve in Fig. 2.3, and a loop is present in the curve with $k_0 = k_1 = 80$ (that is, four times the chord length).

EXERCISES†

1. Show that $\mathbf{p}_1 - \mathbf{p}_0, \mathbf{p}_0^u$, and $\mathbf{p}_1^u$ are linearly independent for nonplanar curves.

2. Given unit tangent vectors $\mathbf{t}_0$ and $\mathbf{t}_1$, where $k_0\mathbf{t}_0 = \mathbf{p}_0^u$ and $k_1\mathbf{t}_1 = \mathbf{p}_1^u$, show that any point on a nonplanar pc curve can be expressed as

$$\mathbf{p}(u) = \mathbf{p}_0 + \lambda(u)(\mathbf{p}_1 - \mathbf{p}_0) + \mu(u)\mathbf{t}_0 + v(u)\mathbf{t}_1$$

where $\lambda(u)$, $\mu(u)$, and $v(u)$ are scalars.

3. Show that for a particular point $\mathbf{p}(u)$, the scalars $\lambda(u)$, $\mu(u)$, and $v(u)$ are given by

$$\lambda(u) = \frac{[\mathbf{p}(u) - \mathbf{p}_0] \cdot (\mathbf{t}_0 \times \mathbf{t}_1)}{(\mathbf{p}_1 - \mathbf{p}_0) \cdot (\mathbf{t}_0 \times \mathbf{t}_1)}$$

$$\mu(u) = \frac{[\mathbf{p}(u) - \mathbf{p}_0] \cdot \mathbf{t}_1 \times (\mathbf{p}_1 - \mathbf{p}_0)}{(\mathbf{p}_1 - \mathbf{p}_0) \cdot (\mathbf{t}_0 \times \mathbf{t}_1)}$$

$$v(u) = \frac{[\mathbf{p}(u) - \mathbf{p}_0] \cdot (\mathbf{p}_1 - \mathbf{p}_0) \times \mathbf{t}_0}{(\mathbf{p}_1 - \mathbf{p}_0) \cdot (\mathbf{t}_0 \times \mathbf{t}_1)}$$

4. Show that the denominator is nonzero in each of the expressions given in Ex. 3. (*Hint:* See Ex. 1.)

5. Show that Eq. 2.25 can be rewritten as

$$\mathbf{p}(u) = \mathbf{p}_0 + F_2(u)(\mathbf{p}_1 - \mathbf{p}_0) + k_0 F_3(u)\mathbf{t}_0 + k_1 F_4(u)\mathbf{t}_1$$

(*Hint:* Use the identity $F_1(u) + F_2(u) = 1$.)

6. Compare the equation in Ex. 5 to the one in Ex. 2; show that

$$\lambda(u) = F_2(u)$$

$$\mu(u) = aF_3(u)$$

$$v(u) = bF_4(u)$$

7. In Ex. 6, $\lambda(u) = F_2(u)$ is a cubic equation, with $\lambda(u)$ determined by specifying a particular point and solving the equation for $\lambda(u)$, given by the equation in Ex. 3. This cubic has one or three real roots. Show that additional restrictions on the roots for an acceptable solution are

$$0 < u < 1$$

$$k_0 > 0$$

$$k_1 > 0$$

† Exercises 1–11 were inspired by a very brief but elegant paper by A. A. Ball (1978).

8. Let $f(u) = F_2(u) - \lambda(u)$, and show that there exists at most one root of the equation $f(u) = 0$ in the open interval $u \in (0, 1)$ (that is, if a root exists, it is unique). Show that as a consequence, $f(0)$ and $f(1)$ must have opposite signs and this means $1 > \lambda(u) > 0$.

9. Now show that if $u > 0$, then $F_3(u) > 0$ and that $k_0 > 0$ if and only if $\mu(u) > 0$.

10. Show that if $u < 1$, then $F_4(u) < 0$; and show that it follows that $k_1 > 0$ if and only if $v(u) < 0$.

11. Show that the results in Ex. 1–10 lead us to conclude that $\mathbf{p}_0$, $\mathbf{p}(u)$, $\mathbf{p}_1$, $\mathbf{t}_0$, and $\mathbf{t}_1$ specify a nonplanar pc curve segment if and only if

$$(\mathbf{p}_1 - \mathbf{p}_0) \cdot (\mathbf{t}_0 \times \mathbf{t}_1) = 0$$

$$1 > \lambda(u) > 0$$

$$\mu(u) > 0$$

$$v(u) < 0$$

If these conditions exist, then a unique solution to the cubic equation is computed that lies in the interval $u \in (0, 1)$. Use this value of u in $\mu(u)$ and $v(u)$ to find k_0 and k_1, the tangent vector magnitudes.

12. Refer to Fig. 2.3. Plot the curves of the parametric components for the curve produced by $k_0 = k_1 = 80$. What characteristics of the parametric component curves might indicate the presence of a loop in the curve segment?

13. What conditions must exist for the tangent vector at any point on a curve to be zero? If $k_0 = k_1$, what is their value that produces a cusp?

14. Show that $\mathbf{p}(0.5) = 0.5(\mathbf{p}_0 + \mathbf{p}_1) + 0.125(\mathbf{p}_0^u - \mathbf{p}_1^u)$ for any pc curve.

15. Write a procedure to find the parametric slopes and direction cosines at a point on a pc curve by computing x^u, y^u, and z^u at a specified value of the parametric variable u. Denote this as **TVCRV**(CI, U, PU, DC), where

CI(4, 3) is the input array of geometric coefficients;

U is the input value of the parametric variable;

PU(3) is the output array of the tangent vector components x^u, y^u, z^u; and

DC(3) is the output array of direction cosines.

16. If $\mathbf{t}(u) = \mathbf{r}(u)/|\mathbf{r}(u)|$, derive $\mathbf{t}^u$ and $\mathbf{t}^{uu}$.

3   PARAMETRIC SPACE OF A CURVE

Object space is the three-dimensional space defined by Cartesian coordinates x, y, and z. It is the space in which a geometric model is fully developed. The **parametric space** of a curve is the set of two-dimensional spaces defined by (x, u), (y, u), and (z, u) coordinates. We can decompose any parametric curve

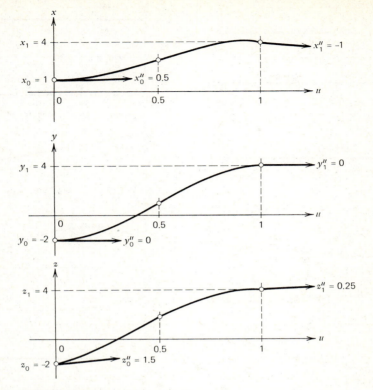

FIGURE 2.5 Parametric space of a curve.

into its three components in parametric space. Plots of pc curves in parametric space are often useful in visualizing their behavior in object space.

Figure 2.5 demonstrates the decomposition of a specific curve whose algebraic representation is

$$x(u) = -6.5u^3 + 9u^2 + 0.5u + 1$$
$$y(u) = -12u^3 + 18u^2 - 2 \qquad (2.46)$$
$$z(u) = -10.25u^3 + 14.75u^2 + 1.5u - 2$$

The corresponding geometric coefficients are

$$\mathbf{B} = \begin{bmatrix} x_0 & y_0 & z_0 \\ x_1 & y_1 & z_1 \\ x_0^u & y_0^u & z_0^u \\ x_1^u & y_1^u & z_1^u \end{bmatrix} = \begin{bmatrix} 1 & -2 & -2 \\ 4 & 4 & 4 \\ 0.5 & 0 & 1.5 \\ -1 & 0 & 0.25 \end{bmatrix}$$

Notice that the parametric components are only plotted in the interval $u \in [0, 1]$. Obviously, we can plot these equations over any range of u values from $+\infty$ to $-\infty$. However, we will restrict our interest here to the range from

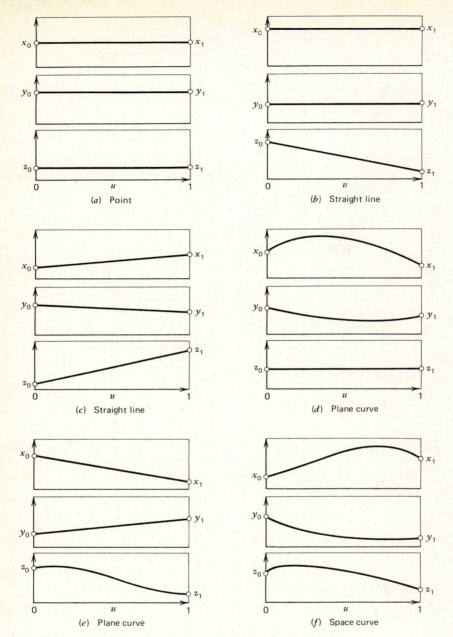

FIGURE 2.6 Special curves shown in parametric space.

$u = 0$ to $u = 1$. These are mathematically convenient delimiting values of the parametric variable, although any interval on the u line can be used without affecting the shape or position of the curve if appropriate adjustments are made to the algebraic or geometric coefficients. Section 5 will discuss how we can change this interval with a procedure called reparametrization. Finally, notice how the different horizontal and vertical scales affect the apparent slope of the tangent vectors. Replot them so that the x, y, z, and u-axis scales are equal.

Remember, all pc curves have identical algebraic and geometric forms and are distinguished from one another by only the value of their algebraic and geometric coefficients. These coefficients control the position of a curve in space and its size and shape. Every value of u (for example, u_i in Fig. 2.5) defines a unique point on the curve in object space. We obtain the entire pc curve by mapping or plotting the successive sets of x, y, z values generated by successive u values. Clearly, each variable x, y, and z is independently controlled, and an entirely different curve results even if only one coefficient is changed.

The graphs in parametric space show the behavior of each Cartesian coordinate from one end of the pc curve to the other as the parametric variable u varies over its domain from 0 to 1. The graphs also show the behavior of the parametric derivatives as represented by the slopes at points on these curves (for example, dx/du at $u = u_i$). Each Cartesian coordinate varies continuously over its range (that is, from $\mathbf{p}_0$ to $\mathbf{p}_1$) as the parametric variable u varies continuously over its domain. We say that x, y, and z are the dependent variables whose values are determined by the independent variable u.

We can infer many characteristics of a pc curve by inspecting the graphs of the component curves. Figure 2.6 shows several special cases that you should verify by sketching the resulting curves in object space. Here, we see the pc curve as a point (a degenerate curve of zero length), a straight line, a plane curve, or a space curve, depending on the values assigned to its coefficients. Notice that the Cartesian coordinates x, y, and z are "stacked" in the figure. Another convenient graphic or display technique is to superimpose them. Thus, we plot the value of the parametric variable along the horizontal axis and plot and superimpose the object-space variables along a common vertical axis.

EXERCISES

1. Refer to Fig. 2.6a. Specify the geometric coefficients for a pc curve whose parametric space component curves are analogous to those in the figure. Verify by plotting them.

2. Same as Ex. 1, but refer to Fig. 2.6b.

3. Same as Ex. 1, but refer to Fig. 2.6c.

4. Same as Ex. 1, but refer to Fig. 2.6d.

5. Same as Ex. 1, but refer to Fig. 2.6e.

6. Same as Ex. 1, but refer to Fig. 2.6f.

4 BLENDING FUNCTIONS

The blending functions first appear in the derivation of the geometric form from the algebraic form. They are the F values in Eqs. 2.23–2.25, namely,

$$\mathbf{p} = F_1\mathbf{p}_0 + F_2\mathbf{p}_1 + F_3\mathbf{p}_0^u + F_4\mathbf{p}_1^u \tag{2.47}$$

where

$$\begin{aligned}
F_1 &= 2u^3 - 3u^2 + 1 \\
F_2 &= -2u^3 + 3u^2 \\
F_3 &= u^3 - 2u^2 + u \\
F_4 &= u^3 - u^2
\end{aligned} \tag{2.48}$$

Note that

$$F_2 = 1 - F_1 \tag{2.49}$$

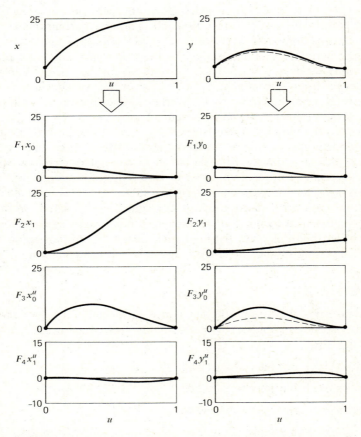

FIGURE 2.7 Curve decomposed into the four orthogonal curves of the blending functions.

Think of these functions as blending the effects or contributions of the end points and tangent vectors (that is, boundary conditions) to produce the intermediate point coordinate values over the domain of u.

First, decompose a pc curve into its parametric components $x(u)$, $y(u)$, and $z(u)$. Next, decompose each component curve into the four orthogonal curves of the blending functions. An explanation of the term **orthogonal** curves or functions follows, and Fig. 2.7 shows an example. Here, the skewed curve of Fig. 2.4 ($k_0 = 80$, $k_1 = 10$) decomposes into its parametric and orthogonal constituents. The dashed curve shows the effect of multiplying y_0^u by 0.5.

The obvious advantages of the blending functions are, first, that these functions hold for all pc curves and are identical for each of the three object-space coordinates, since they are dependent on only u. Second, the constituent boundary condition coefficients are in essence decoupled from each other. Thus, we can selectively modify the boundary conditions to alter or shape the interior of a curve in a way easily adapted to CAM techniques.

A situation analogous to that described for the position vector **p** holds for the tangent vector and second derivative vector $\mathbf{p}^u$ and $\mathbf{p}^{uu}$, respectively; here, we have

$$\mathbf{p}^u = \frac{d\mathbf{p}(u)}{du} \quad \text{and} \quad \mathbf{p}^{uu} = \frac{d^2\mathbf{p}(u)}{du^2} \tag{2.50}$$

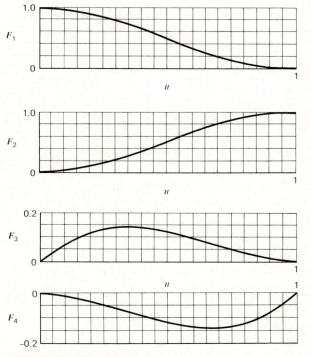

FIGURE 2.8 Blending-function curves.

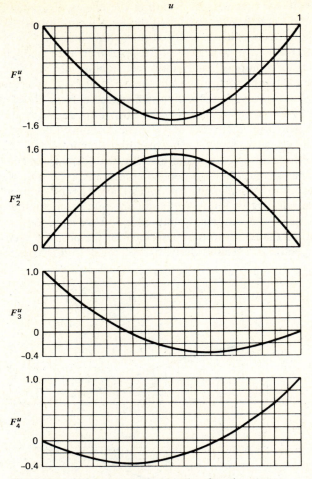

FIGURE 2.9 First-derivative blending-function curves.

We define these vectors, too, at any point on a curve by sets of blending functions operating on the geometric coefficients; they are

$$\mathbf{p}^u = F_1^u \mathbf{p}_0 + F_2^u \mathbf{p}_1 + F_3^u \mathbf{p}_0^u + F_4^u \mathbf{p}_1^u \tag{2.51}$$

and

$$\mathbf{p}^{uu} = F_1^{uu} \mathbf{p}_0 + F_2^{uu} \mathbf{p}_1 + F_3^{uu} \mathbf{p}_0 + F_4^{uu} \mathbf{p}_1^u \tag{2.52}$$

Or, we can write them more compactly as

$$\mathbf{p}^u = \mathbf{F}^u \mathbf{B} \tag{2.53}$$

and

$$\mathbf{p}^{uu} = \mathbf{F}^{uu} \mathbf{B} \tag{2.54}$$

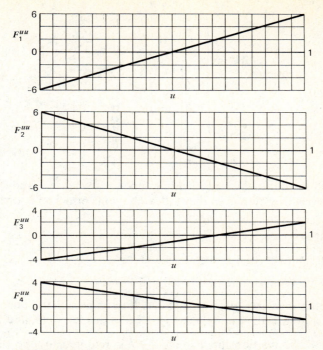

FIGURE 2.10 Second-derivative blending-function curves.

It is very easy to show that

$$\mathbf{F}^u = [(6u^2 - 6u) \quad (-6u^2 + 6u) \quad (3u^2 - 4u + 1) \quad (3u^2 - 2u)] \quad (2.55)$$

and

$$\mathbf{F}^{uu} = [(12u - 6) \quad (-12u + 6) \quad (6u - 4) \quad (6u - 2)] \quad (2.56)$$

simply by performing the indicated differentiation of the **F**-matrix elements.

The complete set of orthogonal blending function components for $\mathbf{F}$, $\mathbf{F}^u$, and $\mathbf{F}^{uu}$ are graphed in Figs. 2.8–2.10. These functions are orthogonal at only $u = 0$ and $u = 1$ and are so called because at these values each of the geometric coefficients in turn is independently the sole contributor to local curve behavior. For example, at $u = 0$ only F_1 determines the value of $\mathbf{p}(u)$, since here $F_1 = 1$ and $F_2 = F_3 = F_4 = 0$. The influence of F_1 gradually diminishes to 0 as the value of u gradually increases to 1. As another example, at $u = 1$, only F_4^u determines the value of $\mathbf{p}^u$, since here $F_4^u = 1$ and $F_1^u, F_2^u, F_3^u = 0$.

Consider what a powerful device these blending functions are. Using them, we can immediately calculate the coordinates and first and second derivatives of any point on any pc curve. Furthermore, Eqs. 2.53 and 2.54 can be made analogous to Eq. 2.33

$$\mathbf{p}^u = \mathbf{U}\mathbf{M}^u\mathbf{B} \qquad u \in [0, 1] \qquad (2.57)$$

$$\mathbf{p}^{uu} = \mathbf{U}\mathbf{M}^{uu}\mathbf{B} \qquad u \in [0, 1] \qquad (2.58)$$

where

$$\mathbf{M}^u = \begin{bmatrix} 0 & 0 & 0 & 0 \\ 6 & -6 & 3 & 3 \\ -6 & 6 & -4 & -2 \\ 0 & 0 & 1 & 0 \end{bmatrix} \tag{2.59}$$

$$\mathbf{M}^{uu} = \begin{bmatrix} 0 & 0 & 0 & 0 \\ 0 & 0 & 0 & 0 \\ 12 & -12 & 6 & 6 \\ -6 & 6 & -4 & -2 \end{bmatrix} \tag{2.60}$$

We will see quite different blending functions when we study Bezier and B-spline curves. However, even they follow the same principal of orthogonality.

EXERCISES

1. Derive $\mathbf{F}^u$ and $\mathbf{F}^{uu}$ (that is, Eqs. 2.55 and 2.56).

2. Derive $\mathbf{F}^{uuu}$. What do the results tell you?

3. Decompose the curve given by Eq. 2.46 into its parametric and orthogonal components. Follow the example in Fig. 2.7. Do this for $\mathbf{F}$, $\mathbf{F}^u$, and $\mathbf{F}^{uu}$.

4. Show that the relationship $F_2 = 1 - F_1$ is required for the condition of orthogonality to hold (see Eq. 2.49). Study Eq. 2.48 and determine what other conditions are required for orthogonality.

5. What conditions are required for the orthogonality of the $\mathbf{F}^u$ and $\mathbf{F}^{uu}$ functions?

6. Create a different set of blending functions, say, $\mathbf{G}$, where

$$\mathbf{p}(u) = G_1(u)\mathbf{p}_0 + G_2(u)\mathbf{p}_1 + G_3(u)\mathbf{p}_0^u + G_4(u)\mathbf{p}_1^u$$

For example, use trigonometric functions for $G_1(u)$, $G_2(u)$, $G_3(u)$, and $G_4(u)$. Select and design these functions to achieve orthogonality.

7. Continue Ex. 6 by computing $\mathbf{G}^u$ and $\mathbf{G}^{uu}$. Are the elements of $\mathbf{G}^u$ orthogonal? How about $\mathbf{G}^{uu}$? Can $\mathbf{G}$ be modified to achieve orthogonality if these elements are not orthogonal?

8. Plot the blending functions for quintics (see Ex. 10 in Section 1). Find and plot the components of $\mathbf{F}^u$ and $\mathbf{F}^{uu}$ for quintics. Compare these to the cubic blending functions.

9. Write a procedure that generates a pc curve that represents the first derivative at any point along a specified pc curve. Denote this as **CDU(CI, CO)**, where

CI(4, 3) is the input array of geometric coefficients of a pc curve, and

CO(4, 3) is the output array of geometric coefficients of a pc curve representing the first derivative of any point on the first pc curve.

Note that in this exercise, you should verify or prove that a pc curve can represent the first derivative at points along a specified pc curve. The position coordinates of a point and the first-derivative components at that point are determined at equivalent values of the parametric variable on the two curves.

10. Write a procedure to calculate the four components of the blending function **F** at any point u. Denote this as **BF(U, F)**, where

U is the input value of the parametric variable u, and

F(4) is the output array of values of the blending function **F** at u (that is, F_1, F_2, F_3, and F_4).

11. Write a procedure to calculate the four components of the blending function $\mathbf{F}^u$ at any point u. Denote this as **BFU(U, FU)**, where

U is the input value of the parametric variable u, and

FU(4) is the output array of values of the blending function $\mathbf{F}^u$ at u (that is, F_1^u, F_2^u, F_3^u, and F_4^u).

12. Write a procedure to calculate the four components of the blending function $\mathbf{F}^{uu}$ at any point u. Denote this as **BFUU(U, FUU)**, where

U is the input value of the parametric variable, and

FUU(4) is the output array of values of the blending function $\mathbf{F}^{uu}$ at u (that is, $F_1^{uu}, F_2^{uu}, F_3^{uu}$, and F_4^{uu}).

5   REPARAMETRIZATION

Reparametrization of a curve produces a change in the parametric interval in a carefully controlled way, so that neither the shape nor the position of the curve is changed. A function $v = f(u)$ describes the precise way this interval is changed. For example, conditions often arise that make it desirable to reverse the direction of parametrization of a pc curve, which is the simplest form of reparametrization. This is quite easy to do, and as required, it does not change the shape or position of a curve; nor does it change any of the curve's analytic properties. In this case, $v = -u$, where v is the new parametric variable.

Figure 2.11 illustrates a parametric curve segment with its end points and tangent vectors for the two possible directions of parametrization. The transformation from the condition illustrated in Fig. 2.11a into that in Fig. 2.11b requires first interchanging elements in rows 1 and 2 of the matrix of geometric coefficients $\mathbf{B}_1$ to form rows 1 and 2 of matrix $\mathbf{B}_2$ and second interchanging elements in rows 3 and 4 and reversing their signs.

Consider two pc curves, each defined by its respective **B** matrix of geometric coefficients

$$\mathbf{B}_1 = [\mathbf{p}_0 \quad \mathbf{p}_1 \quad \mathbf{p}_0^u \quad \mathbf{p}_1^u]^T \tag{2.61}$$

$$\mathbf{B}_2 = [\mathbf{q}_0 \quad \mathbf{q}_1 \quad \mathbf{q}_0^u \quad \mathbf{q}_1^u]^T \tag{2.62}$$

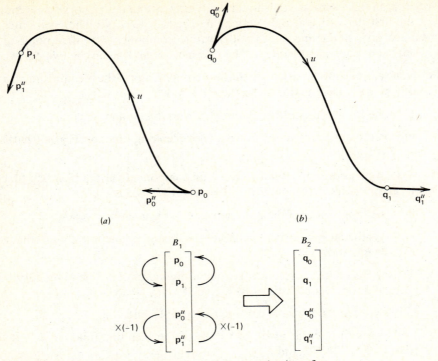

FIGURE 2.11   Two possible directions of parametrization of a curve.

If

$$\mathbf{q}_0 = \mathbf{p}_1$$

$$\mathbf{q}_1 = \mathbf{p}_0$$

$$\mathbf{q}_0^u = -\mathbf{p}_1^u$$ (2.63)

$$\mathbf{q}_1^u = -\mathbf{p}_0^u$$

then the two curves are identical except that they have opposite directions of parametrization. We can rewrite $\mathbf{B}_2$ in terms of the coefficients of $\mathbf{B}_1$

$$\mathbf{B}_2 = [\mathbf{p}_1 \quad \mathbf{p}_0 \quad -\mathbf{p}_1^u \quad -\mathbf{p}_0^u]^T \tag{2.64}$$

Next, consider the more general form of reparametrization, such as that shown in Fig. 2.12. Here, we see two possible states of parametrization for a single curve. Assume that the curve is initially parametrized from u_i to u_j. Suppose we want to change this so that the parametric variable ranges from v_i to v_j. To begin, let the geometric coefficients in the first case be

$$\mathbf{B}_1 = [\mathbf{p}_i \quad \mathbf{p}_j \quad \mathbf{p}_i^u \quad \mathbf{p}_j^u]^T$$

and in the second case $\mathbf{B}_2 = [\mathbf{q}_i \quad \mathbf{q}_j \quad \mathbf{q}_i^v \quad \mathbf{q}_j^v]^T$. There is a simple relationship between the elements of $\mathbf{B}_1$ and $\mathbf{B}_2$.

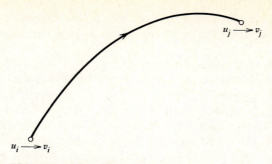

FIGURE 2.12 Reparametrization.

The end-point coordinates are related in a rather straightforward way. They must be invariant or insensitive to any change of parametrization so that $\mathbf{q}_i = \mathbf{p}_i$ and $\mathbf{q}_j = \mathbf{p}_j$, otherwise the reparametrized curve does not satisfy the constant position requirement. The tangent vectors are another matter. Since they are defined by the first derivative of the parametric functions, they are sensitive to the functional relationship between u and v, that is $v = f(u)$. A linear relationship is required to preserve the cubic form of the parametric equations and the direction of the tangent vector; thus,

$$v = au + b \tag{2.65}$$

and so

$$dv = a \, du \tag{2.66}$$

Since we know that $v_i = au_i + b$ and $v_j = au_j + b$, we can easily find a and b and, subsequently, the relationship between the tangent vectors, which is simply

$$\mathbf{q}^v = \frac{u_j - u_i}{(v_j - v_i)} \mathbf{p}^u \tag{2.67}$$

Now we can state the complete relationship between the two sets of geometric coefficients as

$$\mathbf{q}_i = \mathbf{p}_i$$

$$\mathbf{q}_j = \mathbf{p}_j$$

$$\mathbf{q}_i^v = \frac{u_j - u_i}{(v_j - v_i)} \mathbf{p}_i^u \tag{2.68}$$

$$\mathbf{q}_j^v = \frac{u_j - u_i}{(v_j - v_i)} \mathbf{p}_j^u$$

This result tells us that the tangent vector magnitudes must change to accommodate a change in the range of the parametric variable. We see that these magnitudes are simply scaled by the ratio of the ranges of the parametric variables. This preserves the directions of the tangent vectors and the shape of the curve.

Observe that if u_i and u_j are successive pairs of integers, then $u_j - u_i = 1$, and the same holds true for v_i and v_j. This is very useful when dealing with strings or curve segments or composite curves (Section 12), since individual segments originally parametrized to the interval $u_i = 0$ and $u_j = 1$ may be reparametrized to allow relatively simple curve-segment identifying schemes. For example, the nth curve segment can be parametrized from $u_i = n - 1$ to $u_j = n$, with the result that $u_j - u_i = 1$. This means that the **B** matrix for the segment is unchanged by this form of reparametrization.

EXERCISES

1. Show that a linear relationship between u and v is required to preserve the cubic form of the equations when reparametrizing.

2. Show that the relationship $u = 1 - v$ reverses the direction of parametrization of $\mathbf{p}(u)$.

3. Show that the linear relationship $u = av + b$ preserves the directions of the tangent lines.

4. Reverse the direction of parametrization of the curve given by Eq. 2.46.

5. Reparametrize Eq. 2.46 so that the parametric interval $u \in [0, 1]$ is replaced by the interval $v \in [0.5, 1]$.

6. Given the geometric coefficients of a curve

$$\mathbf{B} = [\mathbf{p}_0 \quad \mathbf{p}_1 \quad \mathbf{p}_0^u \quad \mathbf{p}_1^u]^T$$

for $u \in [0, 1]$, reparametrize this curve and find the geometric coefficients for $v \in [1, 3]$.

7. What effect does reparametrization have on the geometric coefficients of a parametric quintic curve?

8. Write a procedure to reverse the direction of parametrization of a pc curve. Denote this as **INVCRV(CI, CO)**, where

CI(4, 3) is the input array of geometric coefficients of the original pc curve, and

CO(4, 3) is the output array of geometric coefficients of the reversed pc curve.

9. Write a procedure to reparametrize a pc curve. Denote this as **RPRCRV(CI, U, CO)**, where

CI(4, 3) is the input array of geometric coefficients;

U(4) is the input array of u values, with the interval $u \in [U(1), U(2)]$ reparametrized to the interval $u \in [U(3), U(4)]$; and

CO(4, 3) is the output array of geometric coefficients of the reparametrized curve.

6 TRUNCATING, EXTENDING, AND SUBDIVIDING

The reparametrization formulas developed in Section 5 allow us to compute a new **B** matrix for a truncated or extended pc curve. Later sections describing composite curves and solid-modeling techniques demonstrate the need for these capabilities as modeling elements are intersected, limited, and joined together to form more complex objects. First, let us investigate the case of a truncated pc curve.

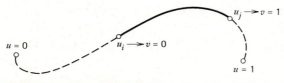

FIGURE 2.13 Reparametrization of a truncated curve.

Figure 2.13 illustrates a curve truncated at u_i and u_j; that is, the segments from u_0 to u_i and from u_j to u_1 are eliminated. We can represent the remaining segment as a full pc curve, parametrized from $v = 0$ to $v = 1$, by proceeding as follows: Compute $\mathbf{p}_i$ and $\mathbf{p}_j$ using $\mathbf{p} = \mathbf{UMB}$, and $\mathbf{p}_i^u$ and $\mathbf{p}_j^u$ using $\mathbf{p}^u = \mathbf{UM}^u\mathbf{B}$. The ratio of parametric interval lengths $(u_j - u_i)/(v_j - v_i)$ given by Eq. 2.68 reduces to $u_j - u_i$, since $v_j - v_i = 1$. If $\mathbf{q}$ represents elements of the truncated curve and $\mathbf{p}$ elements of the original, then

$$\mathbf{q}_0 = \mathbf{p}_i$$
$$\mathbf{q}_1 = \mathbf{p}_j$$
$$\mathbf{q}_0^v = (u_j - u_i)\mathbf{p}_i^u$$
$$\mathbf{q}_1^v = (u_j - u_i)\mathbf{p}_j^u$$

$$(2.69)$$

The same relationships apply when $u_i < 0$ and/or $u_j > 1$. Caution is advised, however, when extending a curve, since we do not usually monitor its behavior in regions outside the interval $u \in [0, 1]$ during original shaping of the curve in this unit interval.

To subdivide a pc curve into n successive segments of arbitrary length and generate n new pc curves, assume that u_i, $\mathbf{p}_i$, and $\mathbf{p}_i^u$ at the segment boundaries are given or readily computed. Then, the geometric coefficients, or **B** matrix elements, of the ith segment are

$$\mathbf{B}_i = [\mathbf{p}_{i-1} \quad \mathbf{p}_i \quad (u_i - u_{i-1})\mathbf{p}_{i-1}^u \quad (u_i - u_{i-1})\mathbf{p}_i^u]^T \qquad (2.70)$$

This is clearly an application of Eq. 2.69, where $v_j - v_{j-1} = 1$.

If a curve is divided into n equal segments (that is, segment boundaries occur at equal intervals of the parametric variable), then the **B** matrix of the ith segment is

$$\mathbf{B}_i = \left[\mathbf{p}_{(i-1)/n} \quad \mathbf{p}_{i/n} \quad \frac{1}{n}\mathbf{p}^u_{(i-1)/n} \quad \frac{1}{n}\mathbf{p}^u_{i/n} \right]^T \tag{2.71}$$

EXERCISES

1. Given a pc curve whose geometric coefficients are

$$\mathbf{B} = [\mathbf{p}_0 \quad \mathbf{p}_1 \quad \mathbf{p}^u_0 \quad \mathbf{p}^u_1]^T$$

truncate the curve at $u = 0.2$ and $u = 0.7$, and reparametrize the remaining segment so that $v \in [0, 1]$. Find the relationship between the geometric coefficients of the truncated curve and those of the original.

2. Given a pc curve whose geometric coefficients are

$$\mathbf{B} = \begin{bmatrix} 1 & 1 & 1 \\ 4 & 2 & 4 \\ 1 & 1 & 0 \\ 1 & 1 & 1 \end{bmatrix}$$

subdivide this curve into three segments with joints at $u = 1/3$ and $u = 2/3$. Reparametrize each segment over a unit interval, and compute the three sets of geometric coefficients. Compare the geometric coefficients of adjacent curves.

3. Write a procedure to truncate a pc curve at two specified values of the parametric variable u. Denote this as **TRNCRV**(CI, U1, U2, CO), where

CI(4, 3) is the input array of geometric coefficients;

U1 is the input u value for truncation of the input curve that will correspond to the $u = 0$ end of the output curve;

U2 is the input u value for truncation of the input curve that will correspond to the $u = 1$ end of the output curve, and

CO(4, 3) is the output array of geometric coefficients of the truncated pc curve.

Compare this procedure to procedure **RPRPCC**, written in Section 5. Can they be combined? (See Ex. 4.)

4. Write a simple procedure to reparametrize a pc curve that includes truncating, extending, and subdividing.

7 SPACE CURVE

We are now ready to examine a method of defining a nonplanar pc curve, a space curve. Begin with the following information:

1. Two end points $\mathbf{p}_0$ and $\mathbf{p}_1$.

2. Unit tangent vectors t_0 and t_1.

3. An intermediate point p_i.

4. An unspecified parametric variable u_i.

The fourth condition saves the problem from being overconstrained. Our problem now is to determine p_0'' and p_1'' required to fill the **B** matrix defining the curve.

Let $p_0'' = k_0 t_0$ and $p_1'' = k_1 t_1$. Find k_0 and k_1 such that the curve passes through p_i. Begin by using Eq. 2.25 to express p_i in terms of the given conditions.

$$p_i = F_1(u_i)p_0 + F_2(u_i)p_1 + k_0 F_3(u_i)t_0 + k_1 F_4(u_i)t_1 \qquad (2.72)$$

Since $F_1 = 1 - F_2$, we obtain

$$p_i - p_0 = F_2(u_i)(p_1 - p_0) + k_0 F_3(u_i)t_0 + k_1 F_4(u_i)t_1 \qquad (2.73)$$

Expand Eq. 2.73 in terms of each of the real-space variables

$$x_i - x_0 = F_2(u_i)(x_1 - x_0) + k_0 F_3(u_i)t_{x0} + k_1 F_4(u_i)t_{x1}$$
$$y_i - y_0 = F_2(u_i)(y_1 - y_0) + k_0 F_3(u_i)t_{y0} + k_1 F_4(u_i)t_{y1} \qquad (2.74)$$
$$z_i - z_0 = F_2(u_i)(z_1 - z_0) + k_0 F_3(u_i)t_{z0} + k_1 F_4(u_i)t_{z1}$$

Solve the first two equations of Eq. 2.74 in terms of $k_0 F_3(u_i)$ and $k_1 F_4(u_i)$, using matrix notation

$$\begin{bmatrix} k_0 F_3(u_i) \\ k_1 F_4(u_i) \end{bmatrix} = \begin{bmatrix} t_{x0} & t_{x1} \\ t_{y0} & t_{y1} \end{bmatrix}^{-1} \left\{ \begin{bmatrix} (x_i - x_0) \\ (y_i - y_0) \end{bmatrix} - \begin{bmatrix} (x_1 - x_0) \\ (y_1 - y_0) \end{bmatrix} F_2(u_i) \right\} \qquad (2.75)$$

From the third equation of Eq. 2.74, we have

$$[z_i - z_0] = [z_1 - z_0]F_2(u_i)$$
$$+ [t_{z0} t_{z1}][k_0 F_3(u_i) k_1 F_4(u_i)]^T \qquad (2.76)$$

Substitute Eq. 2.75 into Eq. 2.76 to obtain

$$[z_i - z_0] = [z_1 - z_0]F_2(u_i) + [t_{z0} t_{z1}]$$
$$\times \begin{bmatrix} t_{x0} & t_{x1} \\ t_{y0} & t_{y1} \end{bmatrix}^{-1} \left\{ \begin{bmatrix} x_i - x_0 \\ y_i - y_0 \end{bmatrix} - \begin{bmatrix} x_1 - x_0 \\ y_1 - y_0 \end{bmatrix} F_2(u_i) \right\} \qquad (2.77)$$

thus leaving only u_i as an unknown.

After working through some rather tedious, though straightforward, matrix algebra, we finally obtain

$$\begin{vmatrix} (x_1 - x_0) & (y_1 - y_0) & (z_1 - z_0) \\ t_{x0} & t_{y0} & t_{z0} \\ t_{x1} & t_{y1} & t_{z1} \end{vmatrix} F_2(u_i) = \begin{vmatrix} x_i - x_0 & y_i - y_0 & z_i - z_0 \\ t_{x0} & t_{y0} & t_{z0} \\ t_{x1} & t_{y1} & t_{z1} \end{vmatrix}$$

$$(2.78)$$

where $F_2(u_i) = -2u_i^3 + 3u_i^2$ (from Eq. 2.48) and the initial constraints are expressed in determinant arrays. This is a key equation in the development.

When we determine the value of u_i by solving the indicated cubic equation, we can compute the values of k_0 and k_1 using two equations from, say, Eq. 2.74. Then we easily determine $\mathbf{p}_0^u$ and $\mathbf{p}_1^u$. Note that an acceptable solution of the cubic equation must satisfy these constraints

$$0 < u_i < 1$$
$$k_0 > 0 \tag{2.79}$$
$$k_1 > 0$$

Review Exs. 1–11 in Section 2 for a similar approach. These are not the only ways of specifying a pc space curve. Section 8 presents another very convenient way of doing it.

EXERCISES

1. Given Eq. 2.74, show all the mathematical steps to arrive at Eq. 2.76.

2. Discuss the problem of constructing a pc curve through three arbitrary points and tangent to one, two, or three specified planes where the points of tangency correspond successively to the three arbitrary points. Denote the three points as $\mathbf{p}_1$, $\mathbf{p}_2$, and $\mathbf{p}_3$, and denote the three planes by their normal vectors: $\mathbf{n}_1$, $\mathbf{n}_2$, and $\mathbf{n}_3$. Case 1 is a pc curve through $\mathbf{p}_1$, $\mathbf{p}_2$, and $\mathbf{p}_3$ and tangent to a plane through $\mathbf{p}_1$ whose unit normal is $\mathbf{n}_1$. Case 2 is a pc curve through $\mathbf{p}_1$, $\mathbf{p}_2$, and $\mathbf{p}_3$ and tangent to a plane at $\mathbf{p}_1$ and another plane at $\mathbf{p}_2$, whose unit normals are $\mathbf{n}_1$ and $\mathbf{n}_2$, respectively. Case 3 is a pc curve through the three points and tangent to a different plane at each of these points. Is the problem under- or overconstrained? How would you resolve the constraint problem?

3. Show that it is possible for the projection of a space curve to form a loop without the curve itself being self-intersecting.

4. Write a procedure to compute the geometric coefficients of a pc curve given its two end points and slopes and an intermediate point. Denote this as **SPCRV**(P0, P1, S0, S1, PINT, CO), where

P0(3) is the input array of coordinates of the point at the $u = 0$ end of the curve;

P1(3) is the input array of coordinates of the point at the $u = 1$ end of the curve;

S0(3) is the input array of components of a unit tangent vector in the direction of the slope at the $u = 0$ end of the curve;

S1(3) is the input array of components of a unit tangent vector in the direction of the slope at the $u = 1$ end of the curve;

PINT(3) is the input array of coordinates of an intermediate point through which the curve is to pass; and

CO(4, 3) is the output array of geometric coefficients of the desired pc curve.

5. Given

$$\mathbf{p}_0 = [0 \quad 0 \quad 0]$$

$$\mathbf{p}_1 = [1 \quad 0 \quad 0]$$

$$\mathbf{t}_0 = [1 \quad 0 \quad 0]$$

$$\mathbf{t}_1 = [0 \quad 1 \quad 0]$$

and

$$\mathbf{p}_i = [0 \quad 1 \quad 0]$$

find u_i, $\mathbf{p}_0^u$, and $\mathbf{p}_1^u$. Accurately graph this pc curve.

6. Repeat Ex. 5 for $\mathbf{t}_1 = [0 \quad 0 \quad 1]$.

8   FOUR-POINT FORM

It is quite easy for us now to define a pc curve that passes through four points in terms of those four points. Naturally, this is called the **four-point form**. It is comparable to the algebraic and geometric forms in its usefulness, and it is a derivative of the latter. Figure 2.14 illustrates this simple situation.

We state the problem as follows: Specify four distinct points in space $[\mathbf{p}_1 \quad \mathbf{p}_2 \quad \mathbf{p}_3 \quad \mathbf{p}_4]$; assign to each of them a successive u value, so that $u_1 < u_2 < u_3 < u_4$. Next, find a 4×4 matrix of constants that, when it premultiplies the matrix of coordinates of the four points, produces the familiar **B** matrix of

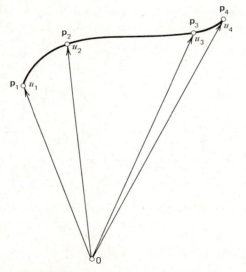

FIGURE 2.14 Four-point form of a curve.

geometric coefficients. Thus,

$$[\mathbf{q}_0 \quad \mathbf{q}_1 \quad \mathbf{q}_0^u \quad \mathbf{q}_1^u]^T = \mathbf{K}[\mathbf{p}_1 \quad \mathbf{p}_2 \quad \mathbf{p}_3 \quad \mathbf{p}_4]^T \tag{2.80}$$

$\mathbf{K}$ is the 4×4 matrix of constants; it depends solely on the u values assigned to the four points. (For the moment, we will treat all four as having arbitrary values.) Begin with the following true expression:

$$[\mathbf{p}_1 \quad \mathbf{p}_2 \quad \mathbf{p}_3 \quad \mathbf{p}_4]^T = [\mathbf{U}_1 \quad \mathbf{U}_2 \quad \mathbf{U}_3 \quad \mathbf{U}_4]^T \mathbf{MB} \tag{2.81}$$

where $\mathbf{U}_1 = [u_1^3 \quad u_1^2 \quad u_1 \quad 1]$ and similarly for $\mathbf{U}_2, \mathbf{U}_3, \mathbf{U}_4$. In fact, for clarity let us write down the completely expanded matrix of u values

$$[\mathbf{U}_1 \quad \mathbf{U}_2 \quad \mathbf{U}_3 \quad \mathbf{U}_4]^T = \begin{bmatrix} u_1^3 & u_1^2 & u_1 & 1 \\ u_2^3 & u_2^2 & u_2 & 1 \\ u_3^3 & u_3^2 & u_3 & 1 \\ u_4^3 & u_4^2 & u_4 & 1 \end{bmatrix} \tag{2.82}$$

Equation 2.81 is solved directly for $\mathbf{B}$ to obtain

$$\mathbf{B} = \mathbf{M}^{-1}[\mathbf{U}_1 \quad \mathbf{U}_2 \quad \mathbf{U}_3 \quad \mathbf{U}_4]^{T-1}[\mathbf{p}_1 \quad \mathbf{p}_2 \quad \mathbf{p}_3 \quad \mathbf{p}_4]^T \tag{2.83}$$

from which we extract the desired $\mathbf{K}$ matrix

$$\mathbf{K} = \mathbf{M}^{-1}[\mathbf{U}_1 \quad \mathbf{U}_2 \quad \mathbf{U}_3 \quad \mathbf{U}_4]^{T-1} \tag{2.84}$$

We now rewrite Eq. 2.83 as

$$\mathbf{B} = \mathbf{K}[\mathbf{p}_1 \quad \mathbf{p}_2 \quad \mathbf{p}_3 \quad \mathbf{p}_4]^T \tag{2.85}$$

Conversely, given the geometric form (that is, the $\mathbf{B}$ matrix), find the equivalent four-point form, which is found directly

$$[\mathbf{p}_1 \quad \mathbf{p}_2 \quad \mathbf{p}_3 \quad \mathbf{p}_4]^T = \mathbf{K}^{-1}\mathbf{B} \tag{2.86}$$

If we take a standard approach and choose equally distributed u values, that is $u_1 = 0$, $u_2 = 1/3$, $u_3 = 2/3$, and $u_4 = 1$, then $\mathbf{K}$ and $\mathbf{K}^{-1}$ are

$$\mathbf{K} = \begin{bmatrix} 1 & 0 & 0 & 0 \\ 0 & 0 & 0 & 1 \\ -11/2 & 9 & -9/2 & 1 \\ -1 & 9/2 & -9 & 11/2 \end{bmatrix} \tag{2.87}$$

$$\mathbf{K}^{-1} = \begin{bmatrix} 1 & 0 & 0 & 0 \\ 20/27 & 7/27 & 4/27 & -2/27 \\ 7/27 & 20/27 & 2/27 & -4/27 \\ 0 & 1 & 0 & 0 \end{bmatrix} \tag{2.88}$$

We can extract a slightly different expression as follows: Let

$$[\mathbf{p}_1 \quad \mathbf{p}_2 \quad \mathbf{p}_3 \quad \mathbf{p}_4]^T = \mathbf{P}$$

Then, since $\mathbf{p} = \mathbf{UMB}$ and $\mathbf{B} = \mathbf{KP}$, by substitution, we obtain

$$\mathbf{p} = \mathbf{UMKP} \tag{2.89a}$$

Let $\mathbf{MK} = \mathbf{N}$, so that Eq. 2.89a becomes

$$\mathbf{p} = \mathbf{UNP} \tag{2.89b}$$

with

$$\mathbf{N} = \begin{bmatrix} -9/2 & 27/2 & -27/2 & 9/2 \\ 9 & -45/2 & 18 & -9/2 \\ -11/2 & 9 & -9/2 & 1 \\ 1 & 0 & 0 & 0 \end{bmatrix} \tag{2.90a}$$

The algebraic coefficients are found from

$$\mathbf{A} = \mathbf{NP} \tag{2.90b}$$

Expand Eq. 2.89a to obtain

$$\begin{aligned} \mathbf{p} = \ & (-4.5u^3 + 9u^2 - 5.5u + 1)\mathbf{p}_1 \\ & + (13.5u^3 - 22.5u^2 + 9u)\mathbf{p}_2 \\ & + (-13.5u^3 + 18u^2 - 4.5u)\mathbf{p}_3 \\ & + (4.5u^3 - 4.5u^2 + u)\mathbf{p}_4 \end{aligned} \tag{2.90c}$$

Notice that we have created new blending functions, say, G_i, so that

$$\mathbf{p} = G_1\mathbf{p}_1 + G_2\mathbf{p}_2 + G_3\mathbf{p}_3 + G_4\mathbf{p}_4 \tag{2.90d}$$

Find the elements of $\mathbf{G}^u$. What about the orthogonality of these blending functions $\mathbf{G}$ and $\mathbf{G}^u$?

EXERCISES

1. Verify the dimensions of the matrices of Eqs. 2.80 and 2.81.

2. Find $\mathbf{K}$ for $u_1 = 0$, $u_2 = 1/4$, $u_3 = 3/4$, and $u_4 = 1$.

3. Does changing $\mathbf{p}_4$ change the tangent vector at $\mathbf{p}_1$? Use blending functions to demonstrate your answer.

4. Does changing $\mathbf{p}_3$ change the tangent vector at $\mathbf{p}_1$? Again, use blending functions to demonstrate your answer.

5. Find $\mathbf{K}$ for a parametric quintic curve given $\mathbf{p}_1$, $\mathbf{p}_2$, $\mathbf{p}_3$, $\mathbf{p}_4$, $\mathbf{p}_5$, and $\mathbf{p}_6$, with $u_1 = 0$, $u_2 = 0.2$, $u_3 = 0.4$, $u_4 = 0.6$, $u_5 = 0.8$, and $u_6 = 1$.

6. Find $\mathbf{N}$ for a parametric quintic curve. Use the same conditions outlined in Ex. 5.

7. Write a procedure to compute the scalar magnitudes of the tangent vectors of a planar pc curve so that the curve will pass through two given intermediate points. Assume that the curve is parallel to one of the three principal coordinate planes and that the end points and slopes of the curve are fixed. Note that the output scalar magnitudes can be negative. Denote this procedure as

PUM(Q, CI, I1, I2, UI, M0, M1), where

> **Q(4)** is an input array of coordinates for the two intermediate points;
>
> $Q(1) =$ coordinate along the I1 axis for the first intermediate point;
>
> $Q(2) =$ coordinate along the I2 axis for the first intermediate point;
>
> $Q(3) =$ coordinate along the I1 axis for the second intermediate point;
>
> $Q(4) =$ coordinate along the I2 axis for the second intermediate point.

> CI(4, 3) is the input array of geometric coefficients, with direction cosines in place of the tangent vector components.

> I1 is an input flag identifying the first component axis of the coordinate plane parallel to the curve, with

> $I1 = 1$ for the x-axis;
> $ = 2$ for the y-axis;
> $ = 3$ for the z-axis.

> I2 is an input flag identifying the second component axis of the coordinate plane parallel to the curve, with

> $I2 = 1$ for the x-axis;
> $ = 2$ for the y-axis;
> $ = 3$ for the z-axis.

> UI(2) is an input array of initial estimates of the value of the parametric variable u for each of the two intermediate points. (Hint: Use ratios of the chord lengths to total chord length.)

> MO is the output scalar magnitude of the tangent vector at $u = 0$.

> MI is the output scalar magnitude of the tangent vector at $u = 1$.

8. Write a procedure to compute the geometric coefficients of a pc curve given four points. Denote this as **FPTOG(P, CO)**, where

> P(4, 3) is the input array of coordinates of four points, and

> CO(4, 3) is the output array of geometric coefficients.

9. Write a procedure to convert the geometric coefficients of a pc curve into the four-point form. Denote this as **GTOFP(CI, P)**, where

> CI(4, 3) is the input array of geometric coefficients, and

> P(4, 3) is the output array of points at $u = 0, 1/3, 2/3$, and 1.

9 GRAPHIC CONSTRUCTION AND INTERPRETATION

A brief digression from the rigors of mathematics is appropriate now. We might still encounter curves in the rather old-fashioned way, on paper as a graph, drawing, or sketch. The ability to interpret them as parametric curves is often useful; the development of the ability itself is even more useful, because it enhances intuitive understanding of the behavior of parametric curves. We are restricted to two-dimensional curves on paper, but the technique described here applies equally well to three-dimensional curves.

We can determine the geometric coefficients of a pc curve by using classical graphic construction tools, that is, a pencil, compass, and straight-edge ruler. Conversely, knowing the geometric coefficients, we can draw an approximation of the corresponding pc curve. The construction technique begins with Eq. 2.47

$$\mathbf{p} = F_1\mathbf{p}_0 + F_2\mathbf{p}_1 + F_3\mathbf{p}_0^u + F_4\mathbf{p}_1^u \tag{2.91}$$

The values of the blending functions at $u = 0.5$ are

$$F_1 = 0.5 \qquad F_2 = 0.5 \qquad F_3 = 0.125 \qquad F_4 = -0.125 \tag{2.92}$$

Substitute these equations into Eq. 2.91 to obtain

$$\mathbf{p}_{0.5} = 0.5\mathbf{p}_0 + 0.5\mathbf{p}_1 + 0.125\mathbf{p}_0^u - 0.125\mathbf{p}_1^u \tag{2.93}$$

Sometimes, it is more convenient to write Eq. 2.93 as

$$\mathbf{p}_{0.5} = 0.5(\mathbf{p}_0 + \mathbf{p}_1) + 0.125(\mathbf{p}_0^u - \mathbf{p}_1^u) \tag{2.94}$$

Remember that this mathematical expression is the vector geometric form of a pc curve. It means that the vector pointing to the real-space point on the curve corresponding to $u = 0.5$ is equal to the sum of the four vectors appearing on the right-hand side of the expression. Inspection and a little thought show that the sum of the first two terms on the right, $0.5(\mathbf{p}_0 + \mathbf{p}_1)$, is the vector to the midpoint of a line joining the end points of the curve. (see Ex. 1).

Figure 2.15 illustrates the vector graphic interpretation of Eq. 2.94 for the case of a curve lying in the x, y plane. Note that since F_4 is negative, $-0.125\mathbf{p}_1^u$ is in the opposite direction from $\mathbf{p}_1^u$.

Now let us proceed to graphically construct a pc curve given the following array of geometric coefficients:

$$\mathbf{B} = \begin{bmatrix} \mathbf{p}_0 \\ \mathbf{p}_1 \\ \mathbf{p}_0^u \\ \mathbf{p}_1^u \end{bmatrix} = \begin{bmatrix} x_0 & y_0 & z_0 \\ x_1 & y_1 & z_1 \\ x_0^u & y_0^u & z_0^u \\ x_1^u & y_1^u & z_1^u \end{bmatrix} = \begin{bmatrix} 4 & 8 & 0 \\ 24 & 16 & 0 \\ 20 & 24 & 0 \\ 12 & -8 & 0 \end{bmatrix} \tag{2.95}$$

We also know that

$$\mathbf{p} = \mathbf{FB} = \begin{bmatrix} F_1 & F_2 & F_3 & F_4 \end{bmatrix} \begin{bmatrix} 4 & 8 & 0 \\ 24 & 16 & 0 \\ 20 & 24 & 0 \\ 12 & -8 & 0 \end{bmatrix} \tag{2.96}$$

which further expands to

$$\mathbf{p} = F_1 \begin{bmatrix} 4 \\ 8 \\ 0 \end{bmatrix} + F_2 \begin{bmatrix} 24 \\ 16 \\ 0 \end{bmatrix} + F_3 \begin{bmatrix} 20 \\ 24 \\ 0 \end{bmatrix} + F_4 \begin{bmatrix} 12 \\ -8 \\ 0 \end{bmatrix} \tag{2.97}$$

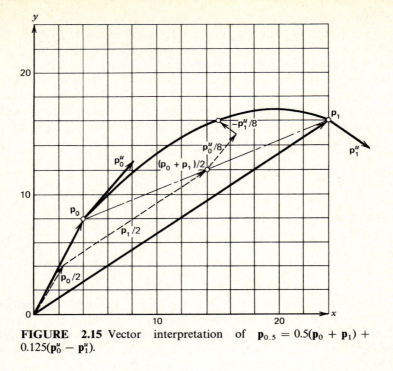

FIGURE   2.15 Vector interpretation of $\mathbf{p}_{0.5} = 0.5(\mathbf{p}_0 + \mathbf{p}_1) + 0.125(\mathbf{p}_0^u - \mathbf{p}_1^u)$.

Figure 2.15 shows the four vectors that define boundary conditions for this curve. It is important to note here that to obtain the proper vector magnitudes, we must draw the vectors to the same scale as the coordinate system where they are located.

To begin the graphic construction, first plot the curve end points given by vectors $\mathbf{p}_0$ and $\mathbf{p}_1$, which are simply vectors to the points $(4, 8, 0)$ and $(24, 16, 0)$. Again, the conventional tools of descriptive geometry are all that are required: compass, straight-edge, ruler, pencil, and so forth. Continue to refer to Figs. 2.15 and 2.16 while following this development.

Next, construct the end slopes using the directions of the tangent vectors $\mathbf{p}_0^u$ and $\mathbf{p}_1^u$, and superimpose or lay out the magnitudes of these vectors on their respective slope lines, beginning the vectors at the appropriate curve end points. This latter operation is merely for clarity and adds emphasis to the end slopes. In this position, we can then use the end slopes to help sketch in the final curve. The coordinates of the far ends of the tangent vectors at these locations become $(\mathbf{p}_0 + \mathbf{p}_0^u)$ and $(\mathbf{p}_1 + \mathbf{p}_1^u)$, respectively.

Because it is simple to do, locating the point at $u = 0.5$ follows. At the midpoint of the line between $\mathbf{p}_0$ and $\mathbf{p}_1$, construct a line parallel to $\mathbf{p}_0^u$ and equal to one-eighth of its length. Label this line $0.125\mathbf{p}_0^u$, and at the end of it, construct a line parallel to $\mathbf{p}_1^u$. In a direction opposite to $\mathbf{p}_1^u$, lay off a length equal to one-eighth of its length; label this second line $-0.125\mathbf{p}_1^u$. The point $\mathbf{p}(0.5)$ is located at the end of this line, and this part of the construction is complete.

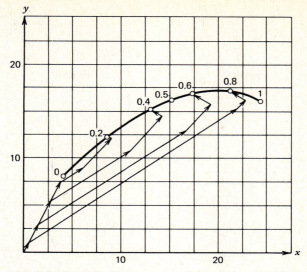

FIGURE 2.16 Graphic construction of intermediate points on the curve.

Knowing the end points and slopes and the intermediate point $\mathbf{p}(0.5)$, we can now sketch a smooth curve conforming to this minimum set of constraints. Or, to obtain a more accurate conformation to the pc curve with the given geometric coefficients, we can determine additional intermediate points from the blending functions at selected values of u and then plot them. This has been done at $u = 0.2, 0.4, 0.6,$ and $0.8,$ and Fig. 2.16 shows the result. Note the graphic confirmation of the effect of u on vector magnitudes on points near the ends of the curve. For low values of u, $\mathbf{p}_0^u$ predominates. For high values of u (that is, $u \to 1$), $\mathbf{p}_1^u$ predominates in determining points on the curve. A further refinement to this method is to compute the tangents at these intermediate points.

The preceding development shows how to construct a pc curve when we know the geometric coefficients. There is also a simple graphic technique for determining approximate values for the geometric coefficients given only a sketch or drawing of a curve. We do this now for the curve shown in Fig. 2.17, which is a plane curve and presumed to be located and scaled by the superimposed coordinate system.

We define one end of the curve to correspond to the $u = 0$ point, making the other end $u = 1$. This choice is entirely arbitrary, as we will demonstrate following the present development. We first draw the end slopes, or tangent vector lines, and the line joining the end points. Find the midpoint $\mathbf{a}$ next. This is simply $0.5(\mathbf{p}_0 + \mathbf{p}_1)$. Read the coordinates of the end points directly from the graph; they are the vector components of $\mathbf{p}_0$ and $\mathbf{p}_1$

$$\mathbf{p}_0 = [7 \quad 5 \quad 0] \qquad \mathbf{p}_1 = [28 \quad 13 \quad 0] \qquad (2.98)$$

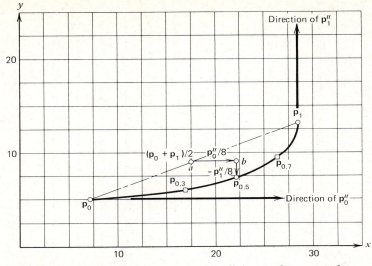

FIGURE 2.17 Determining geometric coefficients of a curve from a drawing of the curve.

Next comes the tricky part, estimating the location of the $u = 0.5$ point, which is generally not the point midway along the length of the curve. Usually, the curve must be symmetrical for this to occur. Both the directions and relative magnitudes of $\mathbf{p}_0^u$ and $\mathbf{p}_1^u$ control the location of this point, with the point usually closer to the end of the curve with the smallest tangent vector magnitude. The sensitivity of estimating the location of this point when matching the shape of the given curve from the tangent vectors graphically derived from it depends on how dramatically shaped the curve is. Generally, the greater the difference between the tangent vector magnitudes and the greater the total change in slope between them, the more dramatic is the curve. For the curve in question, we estimate that

$$\mathbf{p}_{0.5} = \begin{bmatrix} 22 & 7.4 & 0 \end{bmatrix} \qquad (2.99)$$

The coordinates of point **a** are

$$0.5(\mathbf{p}_0 + \mathbf{p}_1) = \begin{bmatrix} 17.5 & 9 & 0 \end{bmatrix} \qquad (2.100)$$

Once point **a** and $\mathbf{p}(0.5)$ are plotted, construct a line through **a** parallel to the direction of $\mathbf{p}_0^u$ and another through $\mathbf{p}(0.5)$ parallel to the direction of $\mathbf{p}_1^u$. Call their intersection point **b**. Clearly, from our earlier discussions, the line segment from **a** to **b** is equal to the vector $0.125\mathbf{p}_0^u$, and the line segment from **b** to $\mathbf{p}(0.5)$ is equal to the vector $-0.125\mathbf{p}_1^u$. Measuring the components of these vectors, we find

$$0.125\mathbf{p}_0^u = \begin{bmatrix} 4.5 & 0 & 0 \end{bmatrix} \qquad (2.101)$$

$$-0.125\mathbf{p}_1^u = \begin{bmatrix} 0 & -1.6 & 0 \end{bmatrix} \qquad (2.102)$$

and therefore

$$\mathbf{p}_0^u = [36 \quad 0 \quad 0] \tag{2.103}$$

$$\mathbf{p}_1^u = [0 \quad 12.8 \quad 0] \tag{2.104}$$

At this point, we have determined all the geometric coefficients, so that we have

$$\mathbf{B} = \begin{bmatrix} 7 & 5 & 0 \\ 28 & 13 & 0 \\ 36 & 0 & 0 \\ 0 & 12.8 & 0 \end{bmatrix} \tag{2.105}$$

We can compute and plot intermediate points to check the accuracy of these coefficients insofar as they generate a curve to match the shape of the given curve. We did this for points at $u = 0.3$ and $u = 0.7$, whose coordinates were

$$\mathbf{p}_{0.3} = [16.8 \quad 5.9 \quad 0] \tag{2.106}$$

and

$$\mathbf{p}_{0.7} = [25.7 \quad 9.4 \quad 0] \tag{2.107}$$

Plotting these points, we see by their closeness to the initial curve that the graphically determined geometric coefficients appear to be reasonably acceptable. Of course, we can use a larger plotting scale and make a more accurate measure of the distance between the check points and the curve. Also, the specific problem situation will usually quantitatively define what error or deviation is acceptable. If the check points fall unacceptably far from the curve, we must choose another point for $\mathbf{p}(0.5)$ and then repeat the procedure.

This example illustrates an important feature and strength of parametric geometry. *Note*: In Fig. 2.17, the slope at $u = 0$ is horizontal (that is, $dy/dx = 0$) and the slope at $u = 1$ is vertical (that is, $dy/dx = \infty$). Our mathematics has no difficulty whatsoever in dealing with these slope values, whereas the classical explicit mathematical expressions usually break down, resulting in singularities, discontinuities, or computer programs that "do not compute"— they blow up, so to speak. Usually, this is the result of an attempt to divide by zero. However, with the parametric formulation, we find that at $u = 0$

$$y^x = \frac{y_0^x}{x_0^u} = \frac{0}{36} = 0 \tag{2.108}$$

and at $u = 1$

$$y^x = \frac{y_1^u}{x_1^u} = \frac{12.8}{0} = \infty \tag{2.109}$$

Figure 2.18 demonstrates graphically how we control the location of $\mathbf{p}_{0.5}$ by varying the magnitudes of $\mathbf{p}_0^u$ and $\mathbf{p}_1^u$. Mathematically, this control is exercised by introducing the scalar multipliers k_0 and k_1 (see Eq. 2.72)

$$\mathbf{p}_{0.5} = F_1(0.5)\mathbf{p}_0 + F_2(0.5)\mathbf{p}_1 + k_0 F_3(0.5)\mathbf{p}_0^u + k_1 F_4(0.5)\mathbf{p}_1^u \tag{2.110}$$

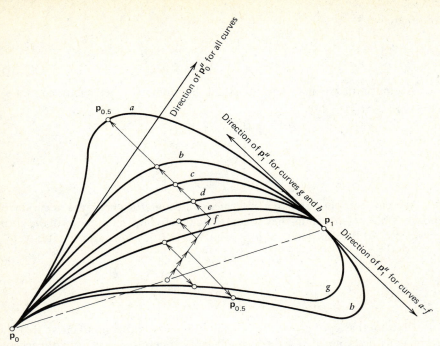

FIGURE 2.18 Effect of tangent vector magnitude on the location of $\mathbf{p}_{0.5}$.

When the scalars k_0 and k_1 change, the tangent vector magnitudes change correspondingly while maintaining constant slopes at the ends of a curve. However, we can force this interior shape of the curve into dramatic variations.

The techniques we have discussed apply to both two- and three-dimensional curves, although the specific examples explored have been planar curves. From the family of curves in Fig. 2.18, it appears that there is no restriction on the location of $\mathbf{p}_{0.5}$. By manipulating the magnitudes (including signs) of $\mathbf{p}_0^u$ and $\mathbf{p}_1^u$ while maintaining their directions constant, we seem able to force $\mathbf{p}_{0.5}$ to any location. This is true for any plane curve whether or not it lies in a plane coincident with, or parallel to, a principal coordinate-system plane.

There is, however, a property that is a restriction readily apparent only for curves that twist in space (nonplanar curves). If the magnitudes of the tangent vectors are varied, $\mathbf{p}_{0.5}$ will always lie in a plane that is parallel to both of the tangent vectors $\mathbf{p}_0^u$ and $\mathbf{p}_1^u$ and passes through the midpoint of the line joining the end points $0.5(\mathbf{p}_0 + \mathbf{p}_1)$. (Remember, the direction lines of the tangent vectors must remain constant for this to be true.)

A casual proof of this follows.

The vectors $0.125\mathbf{p}_0^u$ and $-0.125\mathbf{p}_1^u$ intersect by construction and define a plane (that is, any two intersecting lines in space define a plane). The midpoint of the line joining the points $\mathbf{p}_0$ and $\mathbf{p}_1$ and the point $\mathbf{p}_{0.5}$ lies in this plane, since each point lies on one or the other of the lines defining the plane. By construction, $0.125\mathbf{p}_0^u$ and $-0.125\mathbf{p}_1^u$ are parallel to $\mathbf{p}_0^u$ and $\mathbf{p}_1^u$, respectively. The plane is parallel to $\mathbf{p}_0^u$, because it contains $0.125\mathbf{p}_0^u$, and it is parallel to $\mathbf{p}_1^u$, because it also contains $-0.125\mathbf{p}_1^u$. Therefore, no matter what their magnitudes, the plane is always parallel to both $\mathbf{p}_0^u$ and $\mathbf{p}_1^u$, which fix its orientation. Furthermore, since by construction $0.125\mathbf{p}_0^u$ is always "attached" to the midpoint, the plane will always pass through this point, which fixes its location. Thus, the plane's location and orientation are both determined independently of the magnitudes of $\mathbf{p}_0^u$ and $\mathbf{p}_1^u$. Varying the magnitudes of $\mathbf{p}_0^u$ and $\mathbf{p}_1^u$ merely changes the location of $\mathbf{p}_{0.5}$ within the plane.

This property also applies to planar curves, although it is obscured because the plane in which the curve lies coincides with the plane to which $\mathbf{p}_{0.5}$ is restricted. Inspection shows that this plane passes through point $0.5(\mathbf{p}_0 + \mathbf{p}_1)$ and is parallel to both $\mathbf{p}_0^u$ and $\mathbf{p}_1^u$.

Furthermore, using projective geometry, we can show that the same graphic construction and vector properties that hold for an arbitrary pc curve also hold for any plane projection of a pc curve. The simplest example of this and one that demonstrates other unique and interesting features of these curves is the construction of the three principal plane projections of some arbitrary pc curve. Let us do this given a set of geometric coefficients $\mathbf{B}$. We accomplish this by merely ignoring (that is, substituting zeros for) the coefficients or matrix elements of the dependent variable x, y, or z whose coordinate axis is normal to the plane of projection. Figure 2.19 illustrates this.

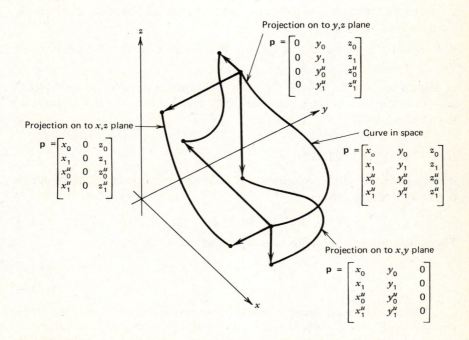

FIGURE 2.19 Projections of a curve onto the principal coordinate planes.

The curve's projection onto the x, y plane is

$$\mathbf{B}_{xy} = \begin{bmatrix} x_0 & y_0 & 0 \\ x_1 & y_1 & 0 \\ x_0^u & y_0^u & 0 \\ x_1^u & y_1^u & 0 \end{bmatrix} \tag{2.111}$$

and similarly for the other two principal planes. We should now be able to deduce the properties that are invariant under plane-projection operations. Although the same properties are invariant when a curve is projected onto some arbitrary skewed plane, the actual process of projection is more complex. We will discuss many of these issues in later sections.

EXERCISES

1. Use the well-known parallelogram technique of vector addition to demonstrate that $(\mathbf{p}_0 + \mathbf{p}_1)/2$ lies on the midpoint of the line joining $\mathbf{p}_0$ to $\mathbf{p}_1$. (*Hint:* Use properties of the intersection of the parallelogram diagonals.)

2. Using the graphic construction techniques discussed in this section, find the geometric coefficients of a pc curve that approximates a 90° circular arc. Assume the arc is centered at the origin, has a unit radius, and is drawn in the positive x, y quadrant.

3. Compute $\mathbf{p}(0.25)$ and $\mathbf{p}(0.5)$ for the pc curve found in Ex. 2. Compare the results to the coordinates of points on the circular arc 22.5° and 45°, respectively, from the x-axis.

4. Repeat Ex. 2 for a 180° circular arc that sweeps from the $+x$-axis through the $+y$-axis to the $-x$-axis.

5. Repeat Ex. 3 for the circular arc in Ex. 4, comparing pc curve points at $\mathbf{p}(0.25)$ and $\mathbf{p}(0.5)$ to points on the circular arc at 45° and 90°, respectively (as measured from the $+x$-axis).

10   STRAIGHT LINES

A straight line results if we impose the following conditions on the geometric coefficients of a pc curve:

$$\mathbf{p} = \mathbf{F}[\mathbf{p}_0 \quad \mathbf{p}_1 \quad (\mathbf{p}_1 - \mathbf{p}_0) \quad (\mathbf{p}_1 - \mathbf{p}_0)]^T \tag{2.112}$$

The tangent vectors are a function of the end-point vectors

$$\mathbf{p}_0^u = (\mathbf{p}_1 - \mathbf{p}_0) \tag{2.113}$$

$$\mathbf{p}_1^u = (\mathbf{p}_1 - \mathbf{p}_0) \tag{2.114}$$

These coefficients force a linear relationship between the parametric variable u and the vector $\mathbf{p}$. We demonstrate this mathematically as follows:

$$\mathbf{p} = F_1\mathbf{p}_0 + F_2\mathbf{p}_1 + F_3(\mathbf{p}_1 - \mathbf{p}_0) + F_4(\mathbf{p}_1 - \mathbf{p}_0) \tag{2.115}$$

Next, rearrange terms to obtain

$$\mathbf{p} = (F_1 - F_3 - F_4)\mathbf{p}_0 + (F_2 + F_3 + F_4)\mathbf{p}_1 \qquad (2.116)$$

Substitute the expressions for the F values given by Eq. 2.23 into Eq. 2.116, and do the necessary arithmetic to find

$$\mathbf{p} = \mathbf{p}_0 + u(\mathbf{p}_1 - \mathbf{p}_0) \qquad Q.E.D. \qquad (2.117)$$

There are several interesting and possibly useful alternative forms of a straight line. We derive them from a simple modification of Eq. 2.112

$$\mathbf{p} = \mathbf{F}[\mathbf{p}_0 \quad \mathbf{p}_1 \quad a(\mathbf{p}_1 - \mathbf{p}_0) \quad b(\mathbf{p}_1 - \mathbf{p}_0)]^T \qquad (2.118)$$

When $a = b = 1$, we obtain the conventional "linear" pc straight line; see Fig. 2.20 for an example. However, there are many other possibilities. First, we must show that Eq. 2.118 does result in a straight line for all real values of a and b. We begin with the expression

$$\mathbf{p} = F_1\mathbf{p}_0 + F_2\mathbf{p}_1 + aF_3(\mathbf{p}_1 - \mathbf{p}_0) + bF_4(\mathbf{p}_1 - \mathbf{p}_0) \qquad (2.119)$$

Once again, we use the relationship $F_1 = 1 - F_2$. Substitute this relationship into Eq. 2.119 and rearrange terms

$$\mathbf{p} = \mathbf{p}_0 + (F_2 + aF_3 + bF_4)(\mathbf{p}_1 - \mathbf{p}_0) \qquad (2.120)$$

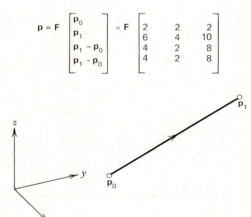

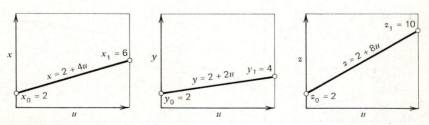

FIGURE 2.20 Straight lines.

Again, use the F values from Eq. 2.23 in Eq. 2.120, and rearrange terms

$$\mathbf{p} = \mathbf{p}_0 + [(a + b - 2)u^3 - (2a + b - 3)u^2 + au](\mathbf{p}_1 - \mathbf{p}_0) \quad (2.121)$$

This expression is analogous to the classical slope-intercept form of a straight line, usually written as $y = mx + b$. The $\mathbf{p}$ corresponds to the dependent variable y, $(\mathbf{p}_1 - \mathbf{p}_0)$ to the slope m, $[f(u)]$ to the independent variable x, and $\mathbf{p}_0$ to the intercept b. Since the slope $(\mathbf{p}_1 - \mathbf{p}_0)$ is a constant, all the points $\mathbf{p}$ generated by values of u lie on a straight line of that slope passing through $\mathbf{p}_0$.

What are the differences between Eqs. 2.117 and 2.121 and the straight lines each generates? There are several. In Eq. 2.117, the relationships between x and u, y and u, and z and u are linear. Here, for example, we find that the real-space distance between a point at $u = 0.1$ and a point at $u = 0.2$ is exactly half the distance between points at $u = 0.5$ and $u = 0.7$. This is true when $a = b = 1$. But for other sets of values of a and b, we see that both the rate and direction at which the set of points are generated in Cartesian space along the line through $\mathbf{p}_0$ and $\mathbf{p}_1$ are neither uniform nor constant. Let us examine some of the combinations in more detail.

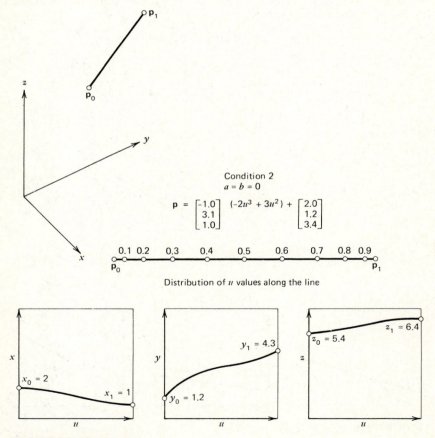

Condition 2
$a = b = 0$

$$\mathbf{p} = \begin{bmatrix} -1.0 \\ 3.1 \\ 1.0 \end{bmatrix} (-2u^3 + 3u^2) + \begin{bmatrix} 2.0 \\ 1.2 \\ 3.4 \end{bmatrix}$$

Distribution of u values along the line

FIGURE 2.21 Straight lines.

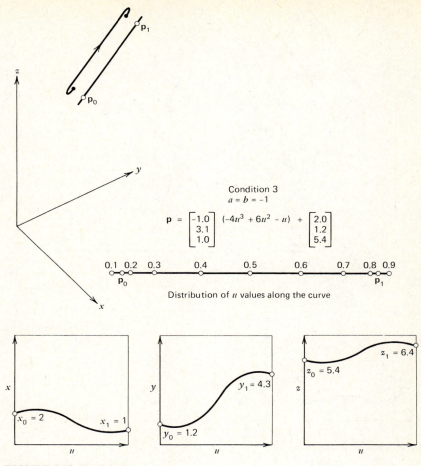

Condition 3
$a = b = -1$

$$\mathbf{p} = \begin{bmatrix} -1.0 \\ 3.1 \\ 1.0 \end{bmatrix} (-4u^3 + 6u^2 - u) + \begin{bmatrix} 2.0 \\ 1.2 \\ 5.4 \end{bmatrix}$$

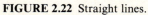

Distribution of u values along the curve

FIGURE 2.22 Straight lines.

First, list the possible conditions on a and b

Condition 1. $a = 1$,   Condition 6. $a > 0$,
             $b = 1$. $b \leq 0$.

Condition 2. $a = 0$,   Condition 7. $a < 0$,
             $b = 0$. $b \geq 0$.

Condition 3. $a = -1$,   Condition 8. $a \leq 0$,
             $b = -1$. $b > 0$.

Condition 4. $a > 0$,   Condition 9. $a \geq 0$,
             $b > 0$. $b < 0$.

Condition 5. $a < 0$,
             $b < 0$.

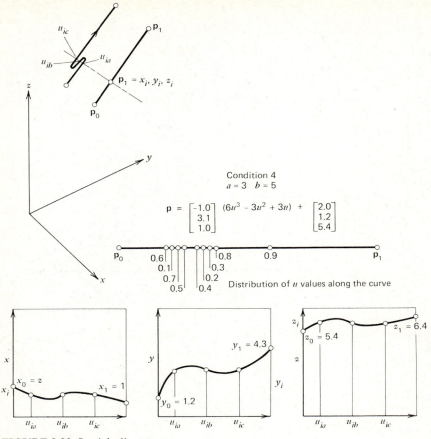

FIGURE 2.23 Straight lines.

We will use the example of a line passing through

$$\mathbf{p}_0 = [2 \quad 1.2 \quad 5.4] \text{ and } \mathbf{p}_1 = [1 \quad 4.3 \quad 6.4]$$

For condition 1, Eq. 2.121 reduces to $\mathbf{p} = \mathbf{p}_0 + u(\mathbf{p}_1 - \mathbf{p}_0)$ or

$$\mathbf{p} = [2 \quad 1.2 \quad 5.4] + u[-1 \quad 3.1 \quad 1] \qquad (2.122)$$

For condition 2, we find that $\mathbf{p} = \mathbf{p}_0 + (-2u^3 + 3u^2)(\mathbf{p}_1 - \mathbf{p}_0)$, and, since $F_2 = (-2u^3 + 3u^2)$, we can write

$$\mathbf{p} = [2 \quad 1.2 \quad 5.4] + F_2[-1 \quad 3.1 \quad 1] \qquad (2.123)$$

Figure 2.21 plots the resulting straight line; here, the parametric component curves are also shown. Most enlightening is the distribution of points at u values incremented equally by 0.1 along the line.

Figure 2.22 reveals the effects of condition 3, where $a = b = -1$. Notice that the straight line is forced beyond points $\mathbf{p}_0$ and $\mathbf{p}_1$. We could predict this

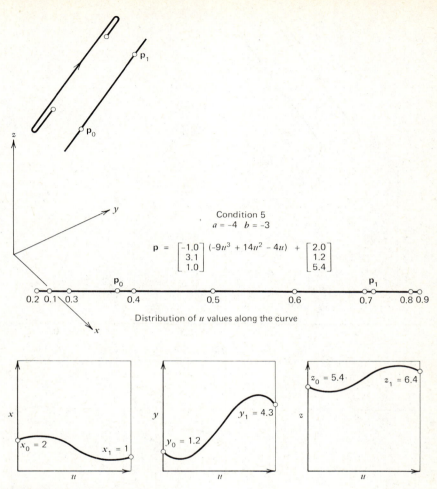

Condition 5
$a = -4 \quad b = -3$

$$p = \begin{bmatrix} -1.0 \\ 3.1 \\ 1.0 \end{bmatrix} (-9u^3 + 14u^2 - 4u) + \begin{bmatrix} 2.0 \\ 1.2 \\ 5.4 \end{bmatrix}$$

Distribution of u values along the curve

FIGURE 2.24 Straight lines.

by investigating the plots in parametric space. For example, in the graph of y versus u, we see that the curve dips below the $y_0 = 1.2$ line and also extends above $y_1 = 4.3$.

Study carefully the conditions shown in Figs. 2.23–2.25. In each figure, to the left of the straight-line representation in object space, a curve with an arrow is added to indicate the direction and sequence of point generation as u increases from 0 to 1. In condition 4, Fig. 2.23, for example, three different u values each generate the same point in object space; p_i is an example of such a point. Inspecting Fig. 2.23, we see that this is true over only a limited portion of the line. We can think of a plotting machine starting to draw a straight line from p_0 to p_1. At some point in its motion toward p_1, the stylus slows, stops, and reverses its direction, retracing part of its path. At some later value of u,

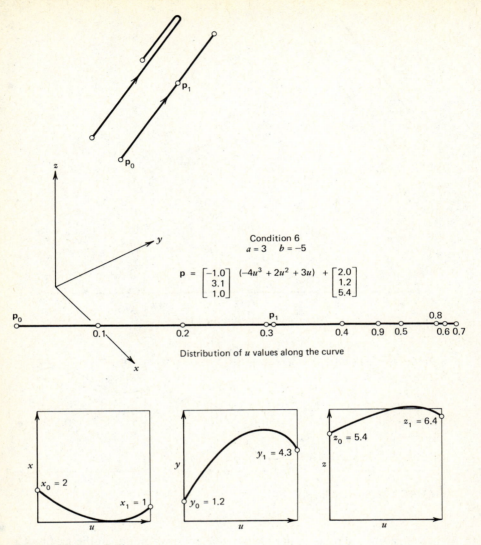

Figure 2.25 Straight lines.

it will again slow its backward motion, stop, and resume its course toward $\mathbf{p}_1$, again retracing part of the line. These "kinks" are avoidable. If the product of the end tangent vector magnitudes is less than or equal to 9 × (length of line)^2, then the line will not have a kink.

The only appeal to our intuition short of manually graphing the functions or obtaining computer graphics displays of them lies in the fact that the endpoint tangent vectors differ by only a scale factor; that is, they are linearly dependent. We can detect this relationship easily by simple mathematical computations on the vector components.

EXERCISES

1. Derive the general equation for $\mathbf{p}^u$ for the straight-line form from Eq. 2.121. Compute the tangent vectors $\mathbf{p}_0^u$ and $\mathbf{p}_1^u$ for condition 1 ($a = b = 1$), condition 2 ($a = b = 0$), and condition 3 ($a = b = -1$). Compare results and discuss any differences.

2. Compute $dy/dx, dy/dz$, and dz/dx at $u = 0, u = 0.5$, and $u = 1$ for the specific curve shown in Fig. 2.23. Compare results and discuss any differences.

3. Find the value of u at the turn-around points for the curve in Fig. 2.24.

4. Replot the component curves shown in Fig. 2.25 so that the axes of the parametric variable and each Cartesian coordinate are to the same scale.

5. Write a procedure to compute the geometric coefficients of a pc straight line between two points. Denote this as **PCLIN(P, CO)**, where

P(6) is the input array of the coordinates of the two points, and

CO(4, 3) is the output array of geometric coefficients.

11   CONICS

The classical manual method of constructing a conic curve is worth studying so that we can more appropriately use parametric curves to represent conics.

We can state the problem as: Given two end points and corresponding tangent lines and an intermediate point, all in a common plane, find a set of intermediate points sufficient to plot a segment of a conic curve.

In Fig. 2.26, A and B are the end points, and C is the intermediate point. The tangent at A is represented by line AD and at B by BD. (D is the point of intersection of the tangent lines.) Find a set of points on the curve as follows:

1. Draw lines AC and BC. Extend both lines past C.

2. Draw any line through D that intersects both lines AC and BC; a and b are such intersection points.

3. Draw a line through points A and b.

4. Draw line aB, and extend it until it intersects Ab. The point of intersection lies on the conic curve.

Repeat these four steps as many times as necessary, each time starting with a different line through D. If we construct a line through D intersecting line AB at its midpoint E, then we can demonstrate that the tangent to the conic at point F is parallel to AB. Furthermore, we can determine the type of conic by computing the ratio of the length of line segment EF to the length of line segment DE. This ratio is called **rho** (ρ). If ρ is less than 0.5, then the curve is a segment of an ellipse. If $\rho = 0.5$, the curve is a parabolic segment. If $0.5 < \rho \le 1$, the curve is a segment of a hyperbola.

Alternatively, construct line DE (see Fig. 2.27) so that it intersects AB at its midpoint. Now locate point C to yield ρ appropriate to the type of conic

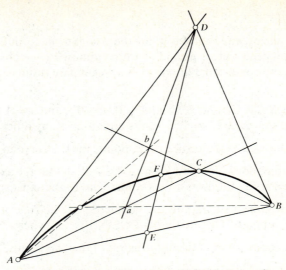

FIGURE 2.26 Classical construction technique of a conic curve.

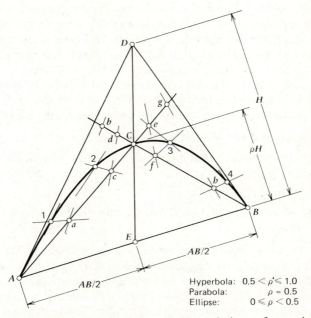

Hyperbola: $0.5 < \rho \leqslant 1.0$
Parabola: $\rho = 0.5$
Ellipse: $0 \leqslant \rho < 0.5$

FIGURE 2.27 Classical construction technique of a conic curve.

required. For example, if the segment of a parabola has end points at A and B and end slopes corresponding to lines AD and BD, then locate C to make $\rho = 0.5$, that is, at the midpoint of DE. Next, proceed with the steps previously outlined in order to generate additional points on the curve. Notice that if the perpendicular distance from point D to line AB is denoted as H, then the perpendicular distance from point C to this line is ρH.

We can define a pc curve whose end points and slopes correspond to A and B, and AD and BD, respectively, and that passes through the intermediate point C. We derive only the case where point C lies on the line passing through D and the midpoint E of line AB; Fig. 2.28 illustrates this. Keep in mind that all the elements are coplanar. (We can show that a polynomial of degree n is always contained in an n-dimensional space.) In Fig. 2.28a, we see a relationship derived earlier, namely,

$$\mathbf{p}_{0.5} = 0.5(\mathbf{p}_0 + \mathbf{p}_1) + 0.125(\mathbf{p}_1^u - \mathbf{p}_0^u) \tag{2.124}$$

Point C is located on DE, a distance ρDE from point E. In Fig. 2.28b, lines EG and GC are parallel to AD and BD, respectively. Triangle BEF is similar to BAD; since $BE = AB/2$, we have $EF = AD/2$ and $BF = BD/2$. Triangle CEG is similar to DEF; since $CE = \rho DE$, we have $CG = \rho DF/2$ and $EG = \rho EF/2$. Furthermore, since $\rho DF = \rho(BD - BF) = \rho BD/2$, we find that

$$0.125|\mathbf{p}_0^u| = 0.5 \, \rho AD \tag{2.125}$$

From Fig. 2.28c, we obtain

$$AD = |\mathbf{p}_2 - \mathbf{p}_0| \tag{2.126}$$

Therefore,

$$|\mathbf{p}_0^u| = 4\rho|\mathbf{p}_2 - \mathbf{p}_0| \tag{2.127}$$

Similarly,

$$0.125|\mathbf{p}_1^u| = 0.5\rho BD \tag{2.128}$$

where

$$BD = |\mathbf{p}_1 - \mathbf{p}_2| \tag{2.129}$$

and

$$|\mathbf{p}_1^u| = 4\rho|\mathbf{p}_1 - \mathbf{p}_2| \tag{2.130}$$

But $\mathbf{p}_0^u$ is in the direction of $\mathbf{p}_2 - \mathbf{p}_0$, and $\mathbf{p}_1^u$ is in the direction of $\mathbf{p}_1 - \mathbf{p}_2$, so that

$$\mathbf{p}_0^u = 4\rho(\mathbf{p}_2 - \mathbf{p}_0) \quad \text{and} \quad \mathbf{p}_1^u = 4\rho(\mathbf{p}_1 - \mathbf{p}_2) \tag{2.131}$$

Therefore, the pc equation for a conic is

$$\mathbf{p} = \mathbf{F}[\mathbf{p}_0 \quad \mathbf{p}_1 \quad 4\rho(\mathbf{p}_2 - \mathbf{p}_0) \quad 4\rho(\mathbf{p}_1 - \mathbf{p}_2)]^T \tag{2.132}$$

Figure 2.29 illustrates this and also shows that $\mathbf{p}^u(0.5)$ is parallel to $\mathbf{p}_1 - \mathbf{p}_0$. Try to verify this to your own satisfaction.

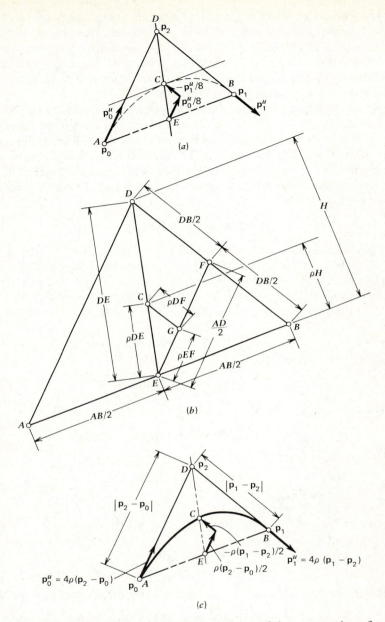

FIGURE 2.28 Parametric cubic approximation of the construction of a conic curve.

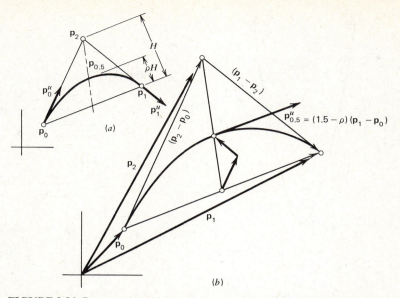

FIGURE 2.29 Parametric cubic approximation of the construction of a conic curve.

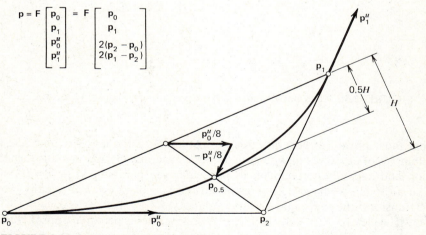

FIGURE 2.30 A pc curve exactly matching a segment of a parabola.

Now let us investigate the accuracy of Eq. 2.132 as it applies to various types of conics. When $\rho = 0.5$, a parabola is generated, and Eq. 2.43 is reduced to

$$\mathbf{p} = \mathbf{F}[\mathbf{p}_0 \quad \mathbf{p}_1 \quad 2(\mathbf{p}_2 - \mathbf{p}_0) \quad 2(\mathbf{p}_1 - \mathbf{p}_2)]^T \tag{2.133}$$

A pc curve defined by these boundary conditions is a conic, and, moreover, it is exactly a segment of a parabola. Figure 2.30 illustrates the general conditions

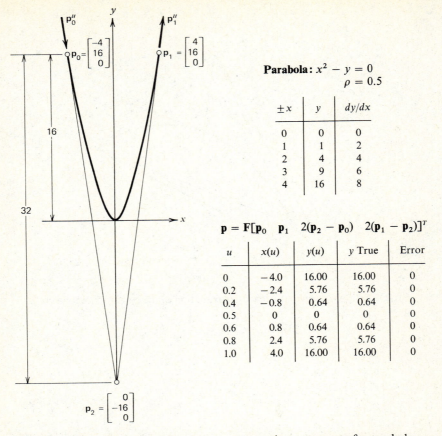

$$\text{Parabola: } x^2 - y = 0$$
$$\rho = 0.5$$

$\pm x$	y	dy/dx
0	0	0
1	1	2
2	4	4
3	9	6
4	16	8

$$\mathbf{p} = F[\mathbf{p}_0 \quad \mathbf{p}_1 \quad 2(\mathbf{p}_2 - \mathbf{p}_0) \quad 2(\mathbf{p}_1 - \mathbf{p}_2)]^T$$

u	$x(u)$	$y(u)$	y True	Error
0	−4.0	16.00	16.00	0
0.2	−2.4	5.76	5.76	0
0.4	−0.8	0.64	0.64	0
0.5	0	0	0	0
0.6	0.8	0.64	0.64	0
0.8	2.4	5.76	5.76	0
1.0	4.0	16.00	16.00	0

FIGURE 2.31 An example of a pc curve representing a segment of a parabola.

that must prevail to generate this type of curve, and Figs. 2.31 and 2.32 are examples of pc curves that demonstrate the exact match between this formulation and a parabolic segment, in particular the parabola $x^2 - y = 0$. Note that although we have been working in the x, y plane, we can readily create or transform the curve into any skew plane.

If we let $0.5 < \rho < 1$, then Eq. 2.132 generates a pc curve approximating a segment of a hyperbola. The approximation improves as the hyperbolic segment becomes smaller, that is, as the difference between the slopes at each end decreases. In another sense, the more pc curves strung end to end that we use to approximate a given hyperbolic segment, the better the match—the smaller the error or deviation of the pc curve from the true hyperbola. Figures 2.33 and 2.34 demonstrate this for different segments of the hyperbola given by the equation $x^2 - y^2 + 1 = 0$.

In Fig. 2.33, a pc curve approximates the segment of the hyperbola from $\mathbf{p}_0 = [-10 \quad 10.05 \quad 0]$ to $\mathbf{p}_1 = [10 \quad 10.05 \quad 0]$. This segment is symmetrical

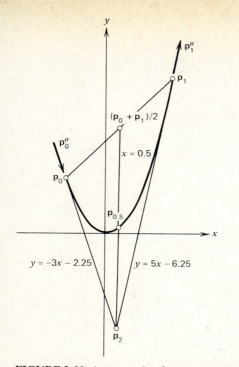

Parabola: $x^2 - y = 0$
$\rho = 0.5$

x	y	$\dfrac{dy}{dx}$
-1.5	2.25	-3
-1	1	-2
0	0	0
1	1	2
2	4	4
2.5	6.25	5

$$\mathbf{p} = \mathbf{F}[\mathbf{p}_0 \quad \mathbf{p}_1 \quad 2(\mathbf{p}_2 - \mathbf{p}_0) \quad 2(\mathbf{p}_1 - \mathbf{p}_2)]^T$$

u	$x(u)$	$y(u)$	y True	Error
0	-1.5	2.25	2.25	0
0.2	-0.7	0.49	0.49	0
0.4	0.1	0.01	0.01	0
0.5	0.5	0.25	0.25	0
0.6	0.9	0.81	0.81	0
0.8	1.7	2.89	2.89	0
1	2.5	6.25	6.25	0

FIGURE 2.32 An example of a pc curve representing a segment of a parabola.

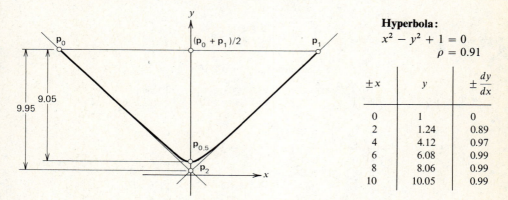

Hyperbola:
$x^2 - y^2 + 1 = 0$
$\rho = 0.91$

$\pm x$	y	$\pm \dfrac{dy}{dx}$
0	1	0
2	1.24	0.89
4	4.12	0.97
6	6.08	0.99
8	8.06	0.99
10	10.05	0.99

$$\mathbf{p} = \mathbf{F}[\mathbf{p}_0 \quad \mathbf{p}_1 \quad 4\rho(\mathbf{p}_2 - \mathbf{p}_0) \quad 4\rho(\mathbf{p}_1 - \mathbf{p}_2)]^T$$

u	$x(u)$	$y(u)$	y True	Error
0	-10	10.05	10.05	0
0.2	-4.43	4.26	4.54	0.28
0.4	-1.21	1.86	1.57	0.21
0.5	0	1	1.00	0
0.6	1.21	1.36	1.57	0.21
0.8	4.43	4.26	4.54	0.28
1	10	10.05	10.05	0

FIGURE 2.33 An example of a pc curve approximating a segment of a hyperbola.

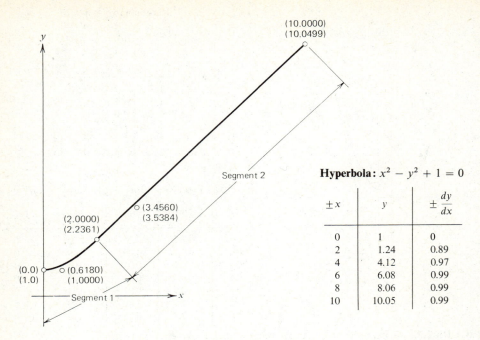

Hyperbola: $x^2 - y^2 + 1 = 0$

$\pm x$	y	$\pm \dfrac{dy}{dx}$
0	1	0
2	1.24	0.89
4	4.12	0.97
6	6.08	0.99
8	8.06	0.99
10	10.05	0.99

Segment 1, $\quad \rho = 0.5601$

$$\mathbf{p}_0 = \begin{pmatrix} 0.0 \\ 1.0 \end{pmatrix} \quad \mathbf{p}_1 = \begin{pmatrix} 2.0000 \\ 2.2361 \end{pmatrix} \quad \mathbf{p}_2 = \begin{pmatrix} 0.6180 \\ 1.0009 \end{pmatrix}$$

Segment 2, $\quad \rho = 0.5679$

$$\mathbf{p}_0 = \begin{pmatrix} 2.0000 \\ 2.2361 \end{pmatrix} \quad \mathbf{p}_1 = \begin{pmatrix} 10.0000 \\ 10.0499 \end{pmatrix} \quad \mathbf{p}_2 = \begin{pmatrix} 3.4560 \\ 3.5384 \end{pmatrix}$$

	u	$x(u)$	$y(u)$	y True	Error
Segment 1	0	0	1	1	0
	0.2	0.2861	1.0400	1.0401	0.0001
	0.4	0.6062	1.1692	1.1694	0.0002
	0.5	0.7861	1.2719	1.2719	0
	0.6	0.9831	1.4022	1.4023	0.0001
	0.8	1.4400	1.7530	1.7532	0.0002
	1	2.0000	2.2361	2.2361	0
Segment 2	0	2.0000	2.2361	2.2361	0
	0.2	2.7796	2.9541	2.9540	0.0001
	0.4	3.8652	3.9926	3.9924	0.0002
	0.5	4.5552	4.6637	4.6637	0
	0.6	5.3609	5.4534	4.4534	0
	0.8	7.3710	7.4386	7.4385	0.0001
	1	10.0000	10.0499	10.0499	0.0001

FIGURE 2.34 An example of a pc curve approximating a segment of a hyperbola.

around the y-axis. We find the point on this curve corresponding to $u = 0.5$ by inspection to be $\mathbf{p}_{0.5} = \begin{bmatrix} 0 & 1 & 0 \end{bmatrix}$. The point $0.5(\mathbf{p}_0 + \mathbf{p}_1) = \begin{bmatrix} 0 & 10 & 0 \end{bmatrix}$. We find $\mathbf{p}_2$ by determining the point of intersection of the slopes or tangents at $\mathbf{p}_0$ and $\mathbf{p}_1$. Since we know these points, we next calculate $\rho = 0.91$. When we substitute these values into Eq. 2.132, we obtain

$$\mathbf{B} = \begin{bmatrix} -10 & 10.05 & 0 \\ 10 & 10.05 & 0 \\ 36.4 & -36.22 & 0 \\ 36.4 & 36.22 & 0 \end{bmatrix} \tag{2.134}$$

We compute and plot the coordinates of points on the pc curve for a set of u values. We substitute the values of $x(u)$ into the classical equation for the hyperbola and compute the corresponding true values of y. Then, we compare these to the $y(u)$ values and find a deviation or approximation error and tabulate it. Note that the tabulated error is simply the difference between y values. This is greater than the perpendicular distance between the two curves, which is a more appropriate measure of the error.

Figure 2.34 shows a similar approach. Here, we represent the right side of the hyperbola by two pc curves. Compute the ρ values and geometric coefficients as before. Notice that the error or deviation of the pc from the true hyperbola is much less than that of the model in Fig. 2.33.

Finally, we investigate the circular arc. As you might suspect, we can use a pc curve to approximate a circular arc. The match is not exact, but the smaller the angle subtended by the circular arc, the better the approximation. Figure 2.37 shows that for angles less than $45°$, $\delta R/R$ is less than 5×10^{-6}, more than adequate for most applications.

There are two different approaches that our investigation can take, both of which are based on the relationship expressed in Eq. 2.132. Figure 2.35 shows the first approach; it puts $\mathbf{p}_{0.5}$ exactly on the true circle. The geometry of this situation is very simple, and we can compute the value of ρ easily

$$\rho = \frac{R(1 - \cos\theta)}{R\tan\theta\sin\theta} = \frac{1 - \cos\theta}{\tan\theta\sin\theta} \tag{2.135}$$

A few simple trigonometric substitutions quickly reduces Eq. 2.135 to

$$\rho = \frac{\cos\theta}{1 + \cos\theta} \tag{2.136}$$

This allows us to write directly the pc equation approximating a circular arc

$$\mathbf{p} = \mathbf{F}\begin{bmatrix} \mathbf{p}_0 & \mathbf{p}_1 & \dfrac{4\cos\theta}{1 + \cos\theta}(\mathbf{p}_2 - \mathbf{p}_0) & \dfrac{4\cos\theta}{1 + \cos\theta}(\mathbf{p}_1 - \mathbf{p}_2) \end{bmatrix}^T \tag{2.137}$$

We can also express this equation in terms of $\mathbf{p}_0$, $\mathbf{p}_1$, and the center point of the arc $\mathbf{p}_c$ (see Ex. 10).

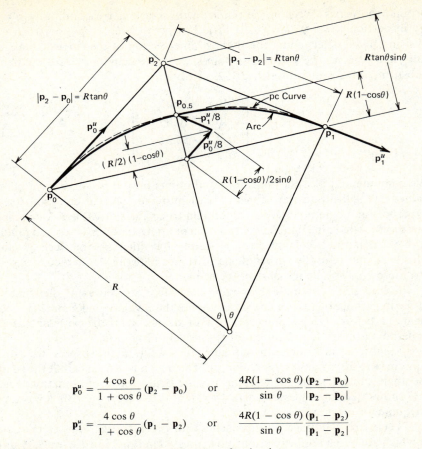

$$\mathbf{p}_0^u = \frac{4 \cos \theta}{1 + \cos \theta} (\mathbf{p}_2 - \mathbf{p}_0) \quad \text{or} \quad \frac{4R(1 - \cos \theta)}{\sin \theta} \frac{(\mathbf{p}_2 - \mathbf{p}_0)}{|\mathbf{p}_2 - \mathbf{p}_0|}$$

$$\mathbf{p}_1^u = \frac{4 \cos \theta}{1 + \cos \theta} (\mathbf{p}_1 - \mathbf{p}_2) \quad \text{or} \quad \frac{4R(1 - \cos \theta)}{\sin \theta} \frac{(\mathbf{p}_1 - \mathbf{p}_2)}{|\mathbf{p}_1 - \mathbf{p}_2|}$$

FIGURE 2.35 A pc curve approximation of a circular arc.

The alternative approach is also shown in Fig. 2.35; it uses the now tried-and-true relationship $\mathbf{p}_{0.5} = 0.5(\mathbf{p}_0 + \mathbf{p}_1) + 0.125(\mathbf{p}_0^u - \mathbf{p}_1^u)$. Again, the geometry of the elements is very simple, and we find

$$0.125\mathbf{p}_0^u = \frac{R(1 - \cos \theta)}{2 \sin \theta} \left(\frac{\mathbf{p}_2 - \mathbf{p}_0}{|\mathbf{p}_2 - \mathbf{p}_0|} \right) \tag{2.138}$$

$$\mathbf{p}_0^u = \frac{4R(1 - \cos \theta)}{\sin \theta} \left(\frac{\mathbf{p}_2 - \mathbf{p}_0}{|\mathbf{p}_2 - \mathbf{p}_0|} \right) \tag{2.139}$$

The same reasoning yields

$$\mathbf{p}_1^u = \frac{4R(1 - \cos \theta)}{\sin \theta} \left(\frac{\mathbf{p}_1 - \mathbf{p}_2}{|\mathbf{p}_1 - \mathbf{p}_2|} \right) \tag{2.140}$$

We then assemble all the elements into the full pc equation

$$\mathbf{p} = \mathbf{F}\left[\mathbf{p}_0 \quad \mathbf{p}_1 \quad \frac{4R(1 - \cos\theta)}{\sin\theta}\left(\frac{\mathbf{p}_2 - \mathbf{p}_0}{|\mathbf{p}_2 - \mathbf{p}_0|}\right) \quad \frac{4R(1 - \cos\theta)}{\sin\theta}\left(\frac{\mathbf{p}_1 - \mathbf{p}_2}{|\mathbf{p}_1 - \mathbf{p}_2|}\right)\right]^T$$

(2.141)

Notice that in Fig. 2.35, the maximum deviation of the pc curve from the true circle occurs at two places. The magnitude of the deviation depends on θ (see Fig. 2.37). Adjusting the magnitudes of the tangent vectors so that $\mathbf{p}_{0.5}$ falls slightly inside the circular arc reduces the deviation somewhat. This is equivalent to reducing the ρ value. Therefore, from Fig. 2.36, we obtain

$$\rho = \frac{(1 - \cos\theta - \delta R/R)\cos\theta}{\sin^2\theta}$$

(2.142)

For any specific angle θ, an increase of $\delta R/R$ at $\mathbf{p}_{0.5}$ causes a decrease in the maximum deviation in the interval $u = 0$ to $u = 0.5$ and in the interval $u = 0.5$ to $u = 1$. We can find a value of $\delta R/R$ that makes the three maximum deviations equal. This value is different for every value of θ. The smaller the

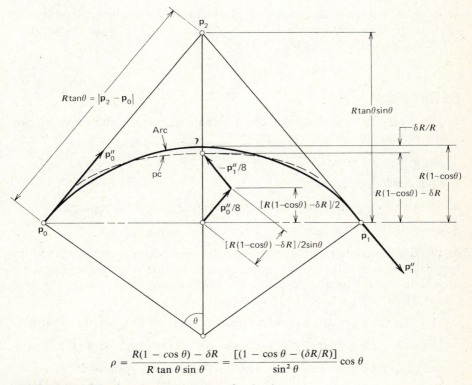

$$\rho = \frac{R(1 - \cos\theta) - \delta R}{R\tan\theta\sin\theta} = \frac{[(1 - \cos\theta - (\delta R/R)]}{\sin^2\theta}\cos\theta$$

FIGURE 2.36 A pc curve approximation of a circular arc.

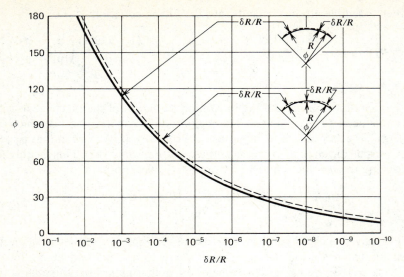

FIGURE 2.37 Deviation of a pc curve from a true circular arc.

angle an arc subtends, the smaller the deviation of the pc curve from the true circular arc. This relationship is also plotted in Fig. 2.37. If $\delta R/R = 0$, then Eq. 2.142 reduces to Eq. 2.136.

EXERCISES

1. Equation 2.132 was developed using graphic techniques. Show how we can obtain the same result analytically.

2. Show how to derive Eq. 2.135.

3. Show how to derive Eq. 2.138.

4. Show that a pc curve approximating a conic always lies in a plane.

5. Verify that for a conic, $\mathbf{p}''(0.5)$ is parallel to $\mathbf{p}_1 - \mathbf{p}_0$.

6. Write a procedure to compute the geometric coefficients of a pc curve that approximates a circular arc of a given radius and center point. Assume that the arc lies in a plane parallel to one of the principal coordinate planes. Denote this as **ARCPC(P0, P1, CP, CO)**, where

P0(3) is the input array of coordinates of the point at $u = 0$ on the arc;

P1(3) is the input array of coordinates of the point at $u = 1$ on the arc;

CP(3) is the input array or coordinates of the center point of the arc, and

CO(4, 3) is the output array of geometric coefficients.

7. How would you modify procedure **ARCPC** in Ex. 6 if the arc does not lie in a plane parallel to one of the principal coordinate planes?

8. Write a procedure to compute the geometric coefficients of a pc curve that approximates a general conic. Denote this as **PCNC(P0, P1, P2, RHO, CO)**, where

P0(3) is the input array of coordinates of the point $\mathbf{p}_0$ at $u = 0$;

P1(3) is the input array of coordinates of the point $\mathbf{p}_1$ at $u = 1$;

P2(3) is the input array of coordinates of the point $\mathbf{p}_2$ as shown in Fig. 2.29;

RHO is the input ρ value; and

CO(4, 3) is the output array of geometric coefficients.

9. Compute the geometric coefficients of a pc curve that approximates 1/4 of a sine wave, say, from $x = 0$ to $x = \pi/2$ for the function $y = \sin x$. Show that the remaining quarter segments of the sine wave are simple transformations of these coefficients. Compare the pc representation with the original function, and find the deviation of the pc curve from the true sine curve at a series of points (say, at intervals of $u = 0.1$). Give the approximate maximum deviation.

10. Express Eq. 2.137 in terms of the end points $\mathbf{p}_0$ and $\mathbf{p}_1$ and the center point of the circular arc $\mathbf{p}_c$.

12   COMPOSITE CURVES

When two or more curve segments are joined together, they form a continuous **composite curve**. Let us start the discussion of composite curves and continuity by investigating the blending of a new curve between two existing curves to form a composite curve consisting of three segments. Refer to Fig. 2.38.

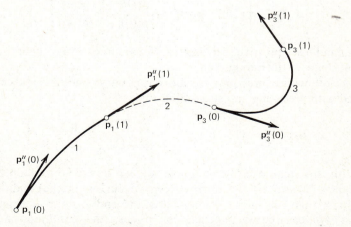

FIGURE 2.38 Blending a curve between two existing curves.

We can state the problem as: Given the geometric coefficients of two disjoint curves $\mathbf{B}_1$ and $\mathbf{B}_3$, find the geometric coefficients $\mathbf{B}_2$ of a curve joining them so that the two curve segments meeting at each of the joints are tangent to the same line. Let

$$\mathbf{B}_1 = [\mathbf{p}_1(0) \quad \mathbf{p}_1(1) \quad \mathbf{p}_1^u(0) \quad \mathbf{p}_1^u(1)]^T \tag{2.143}$$

and

$$\mathbf{B}_3 = [\mathbf{p}_3(0) \quad \mathbf{p}_3(1) \quad \mathbf{p}_3^u(0) \quad \mathbf{p}_3^u(1)]^T \tag{2.144}$$

The notation scheme is modified to permit the use of a subscript to identify a specific curve segment. Later, we will modify this further so that each point is uniquely identified. First, the end points must coincide, so that

$$\mathbf{p}_2(0) = \mathbf{p}_1(1) \quad \text{and} \quad \mathbf{p}_2(1) = \mathbf{p}_3(0) \tag{2.145}$$

Furthermore, the tangent at each end of the new curve must match the tangent of the adjoining curve. The magnitudes of adjoining pc tangent vectors can be different. In fact, we note that an infinite number of pc segments satisfy the tangency conditions. However, the unit tangent vectors must be equal. Thus,

$$\mathbf{p}_2^u(0) = a \frac{\mathbf{p}_1^u(1)}{|\mathbf{p}_1^u(1)|} \tag{2.146}$$

and

$$\mathbf{p}_2^u(1) = b \frac{\mathbf{p}_3^u(0)}{|\mathbf{p}_3^u(0)|}$$

The fully assembled $\mathbf{B}_2$ matrix, in terms of the adjacent curves, is

$$\mathbf{B}_2 = \left[\mathbf{p}_1(0) \quad \mathbf{p}_3(0) \quad a \frac{\mathbf{p}_1^u(1)}{|\mathbf{p}_1^u(0)|} \quad b \frac{\mathbf{p}_3^u(1)}{|\mathbf{p}_3^u(0)|}\right]^T \tag{2.147}$$

where a and b are positive scale factors. We are free to vary them to change the internal shape of the new curve.

The following expression describes any pc curve smoothly and continuously blended with preceding and succeeding curves:

$$\mathbf{B}_i = \left[\mathbf{p}_{i-1}(1) \quad \mathbf{p}_{i+1}(0) \quad a \frac{\mathbf{p}_{i-1}^u(1)}{|\mathbf{p}_{i-1}^u(1)|} \quad b \frac{\mathbf{p}_{i+1}^u(0)}{|\mathbf{p}_{i+1}^u(0)|}\right]^T \tag{2.148}$$

With this small exercise complete, we can now elaborate on the general conditions of continuity at a joint between two curves and the conditions arising when more than two curves are joined end to end forming a string of curves, called a composite curve. In general, a curve is either continuous at all points, or it has one or more points of discontinuity. The simplest kind of continuity a curve can have ensures that there are no gaps or breaks between its beginning and ending points. This is called C^0 continuity. If what is alleged to be a single curve does not have C^0 continuity, then it must really be two or more curves, depending on how many gaps there are.

Two pc curves joined at a common end point have at least C^0 continuity at their joint. They comprise the simplest kind of composite curve. Note that the term *curve* alone will often mean a curve segment or composite curve. The meaning should be clear from the context.

We see in Fig. 2.39 examples of composite curves and C^0 continuity. Fig. 2.39*a* obviously consists of two distinct and separate curves, whereas in Fig. 2.39*b*, they are joined and form a compound curve with C^0 continuity at the joint. Figure 2.39*c* is a compound curve consisting of three segments, again with C^0 continuity at the joints. Figure 2.39*d* is an arbitrary shape we can define with an arbitrary set of segments to form the closed composite curve in Fig. 2.39*e*.

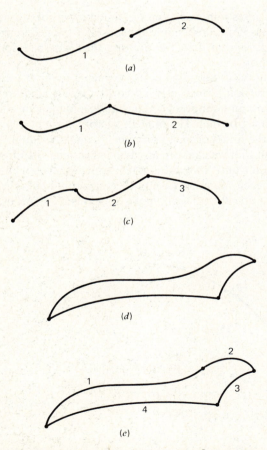

FIGURE 2.39 Composite curves and C^0 continuity.

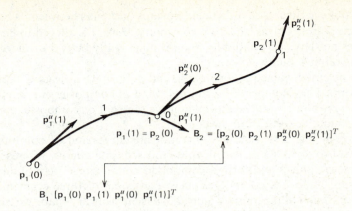

FIGURE 2.40 Conditions required for C^0 continuity.

When two curves join end to end at a common point, as in Fig. 2.40, the coordinates of that point are a common subset of the geometric coefficients of both curves. Observe that the C^0 continuity thus obtained does not affect the shape of either curve.

A C^1 continuity between two curve segments requires a common tangent line at their joining point; obviously, C^1 continuity also requires C^0 continuity. Figure 2.41 presents an example of C^1 continuity and indicates the mathematical conditions that must be met. The tangent vectors at $\mathbf{p}_1(1) = \mathbf{p}_2(0)$ do not have to be equal, but the ratios of their components must be equal.

It is apparent that we can join any two curves and then rotate them about their common point until they are tangent to each other while preserving the internal shape of each. Whereas C^0 and C^1 continuity imply sharing a common point and tangent line at the junction of two curves, C^2 continuity also requires

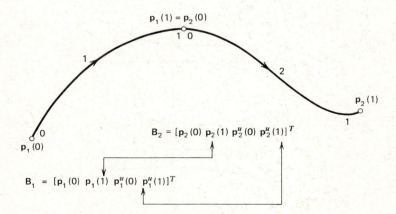

FIGURE 2.41 Conditions required for C^0 and C^1 continuity.

that the two curves possess equal curvature at their joint. To ensure this, the following conditions must exist.

$$\mathbf{p}_i(1) = \mathbf{p}_{i+1}(0) \tag{2.149}$$

$$\mathbf{p}_i''(1) = k_1 \mathbf{p}_{i+1}''(0) \tag{2.150}$$

$$\mathbf{p}_i'''(1) = k_2 \mathbf{p}_{i+1}'''(0) \tag{2.151}$$

The relationship between the parametric second derivatives ensures that $(y^{xx})_i = (y^{xx})_{i+1}$ at their common point.

One more condition must be met for C^2 continuity: The osculating planes of the two curves must coincide or their binormals (Chapter 5, Section 1.4) must be colinear. However, the usefulness of C^2 continuity is limited. Most mechanical parts do not require it, since fillets or rounded edges usually blend directly into plane faces.

Now let us investigate further why composite curves are so useful in geometric modeling and show how they resolve certain problems arising from classical polynomial interpolation methods—methods developed to fit a single curve to a set of data points. Lagrange and Hermite interpolation schemes are useful in illustrating the more important concepts. We limit the discussion to two-dimensional plane curves.

First, we demonstrate empirically how to define a polynomial passing through just three given points (x_0, y_0), (x_1, y_1), and (x_2, y_2), constrained by monotonicity in x, that is, $x_0 < x_1 < x_2$.

Define an expression

$$L_0(x) = \frac{(x - x_1)(x - x_2)}{(x_0 - x_1)(x_0 - x_2)} \tag{2.152}$$

When $x = x_0$, we find that $L_0(x_0) = 1$. However, if $x = x_1$ or $x = x_2$, we find that $L_0(x_1) = L_0(x_2) = 0$. Similar expressions can be defined so that

$$L_1(x) = \frac{(x - x_0)(x - x_2)}{(x_1 - x_0)(x_1 - x_2)} = 1 \qquad \text{when } x = x_1$$
$$= 0 \qquad \text{when } x = x_0$$
$$\qquad \text{or } x = x_2 \tag{2.153}$$

$$L_2(x) = \frac{(x - x_0)(x - x_1)}{(x_2 - x_0)(x_2 - x_1)} = 1 \qquad \text{when } x = x_2$$
$$= 0 \qquad \text{when } x = x_0$$
$$\qquad \text{or } x = x_1 \tag{2.154}$$

Next, we use these three functions to construct a function incorporating the y coordinates

$$f_L(x) = L_0(x)y_0 + L_1(x)y_1 + L_2(x)y_2 \tag{2.155}$$

Observe that the L terms act almost like blending functions, and each term is a quadratic polynomial. The L terms yield $f_L(x_0) = y_0$, $f_L(x_1) = y_1$, and $f_L(x_2) = y_2$. The function $f_L(x)$ interpolates the three points exactly.

To interpolate $(n + 1)$ points (x_i, y_i), we readily generalize this procedure. Again, we must guarantee that $x_0 < x_1 < \cdots < x_n$, and we find we need an nth-degree polynomial

$$f_L(x) = \sum_{i=0}^{n} L_i(x)y_i \tag{2.156}$$

where

$$
\begin{aligned}
L_i(x) &= \frac{(x - x_0)(x - x_1) \cdots (x - x_{i-1})(x - x_{i+1}) \cdots (x - x_n)}{(x_i - x_0)(x_i - x_1) \cdots (x_i - x_{i-1})(x_i - x_{i+1}) \cdots (x_i - x_n)} \\
&= \prod_{\substack{j=0 \\ j \neq i}}^{n} \left(\frac{x - x_j}{x_i - x_j} \right)
\end{aligned}
\tag{2.157}
$$

This is Lagrange's interpolating formula, and the $L_i(x)$ terms are called Lagrange interpolation coefficients. If we use the formula for the simplest case of $n = 1$, we interpolate two points (x_0, y_0) and (x_1, y_1) with

$$
\begin{aligned}
f_L(x) &= \frac{(x - x_1)y_0}{x_0 - x_1} + \frac{(x - x_0)y_1}{x_1 - x_0} \\
&= y_0 + \frac{(y_1 - y_0)(x_0 - x)}{x_0 - x_1}
\end{aligned}
\tag{2.158}
$$

We immediately recognize that this is the equation of a straight line connecting the two points. In any interpolation problem where all $(n + 1)$ points happen to lie on a straight line, the higher degree terms cancel, and what remains is a first-degree polynomial, a linear equation.

The Lagrange polynomials are not the only polynomials capable of solving the interpolation problem. There are infinitely many, but the Lagrange form is the unique form of degree $\leq n$ solving the problem. There is, however, a serious problem with this scheme: As the degree of the Lagrange polynomial increases, so does the tendency for ripples or oscillations to occur. Thus, any attempt to increase the accuracy of the curve fit by increasing the number of interpolation points usually increases the number of oscillations. This problem can be alleviated somewhat by careful attention to the spacing of the points, but this is not a very efficient or practical approach for most geometric-modeling applications.

The oscillations occur at curve maxima or minima, of which there may be as many as $(n - 1)$ if the roots of $f_L(x) = 0$ all happen to be real. We can avoid this problem by constructing a composite curve by fitting low-degree polynomials to successive groups of data points. The resulting piecewise polynomial will have at least C^0 continuity but may have C^1 discontinuity at the joints between the curve segments—usually another unacceptable situation.

Hermite interpolation is another possibility. This scheme requires values for the slope or first derivatives at each interpolation point. This form turns out to be rather sensitive to changes in these derivative values, and oscillations are again likely to occur if a high-degree interpolating polynomial is used. For these reasons, piecewise Hermite interpolation using low-degree polynomials is preferred. We have done just that in the composite pc curves described earlier in this section. (Remember, we will occasionally use the term **Hermite** to distinguish pc curves from other forms, such as Bezier and *B*-spline.) In Section 13, we see how splines remove the requirement for prespecification of slopes at interior points of a composite curve.

EXERCISES

1. A C^i is often used to denote continuity, where i indicates continuity through the ith derivative. State the conditions required for C^3 continuity between two quintic curve segments.

2. Discuss the continuity limitations present at a joint between a cubic and quintic curve segment.

3. Show how two circular arcs with the same radius can be joined with C^1 continuity, but not C^2 continuity.

4. Change the notation of the three points in Fig. 2.41 as follows: Let $\mathbf{p}_0 = \mathbf{p}_1(0)$, $\mathbf{p}_1 = \mathbf{p}_1(1) = \mathbf{p}_2(0)$, and $\mathbf{p}_2 = \mathbf{p}_2(1)$. Add (sketch) a third curve segment joined with C^1 continuity at $\mathbf{p}_2$ and extending to $\mathbf{p}_3$. Use this new point notation to express the elements of the **B** matrix for each curve segment. Use an appropriately subscripted scale factor k to denote the ratio of the two colinear tangent vectors at a joint.

5. Given n pc curve segments joined with C^1 continuity, use the point notation introduced in Ex. 4 to find $\mathbf{B}_i$ (that is, the matrix of geometric coefficients of the ith curve segment). Note that there are $n + 1$ points. Again, use an appropriately subscripted scale factor k to denote the ratio of two colinear tangent vectors at a joint.

6. Repeat Ex. 5 for a closed composite curve and find $\mathbf{B}_1$ and $\mathbf{B}_n$. How many points are there?

7. Use four pc curve segments joined end to end to approximate the curve $y = \sin x$ for $x \in [-\pi, +\pi]$. Joints should be at $x = -\pi/2, 0$ and $\pi/2$. Note the inflection point at $x = 0$ (curvature is zero here). Find the four sets of geometric coefficients; they should produce the inflection point at $x = 0$. Sketch the composite curve. Each pc segment can be parametrized on the unit interval.

8. Compute the deviation of the composite curve found in Ex. 7 from the true sine curve at intervals of $\Delta u = 0.2$ on each of the four pc curve segments.

9. Reparametrize the composite curve found in Ex. 7 so that the entire composite curve is parametrized on the unit interval.

10. Write a procedure for computing the geometric coefficients of a pc curve that spans the gap between two given discontinuous pc curves so that the resulting three-segment curve string has C^1 continuity throughout. The intermediate curve string should begin at the $u = 1$ end of the first curve and terminate at the $u = 0$ end of the other. Denote this as **BLNDC**(CI1, CI2, CO), where

CI1(4, 3) is the input array of geometric coefficients of a pc curve;

CI2(4, 3) is the input array of geometric coefficients of another pc curve; and

CO(4, 3) is the output array of geometric coefficients of the spanning pc curve.

The scalar magnitudes of the tangent vectors of the spanning pc curve offer two additional degrees of freedom. How can they be used, and how does the procedure accommodate them?

13   SPLINE CURVES

The spline curve is perhaps the single most important curve in both the aircraft and shipbuilding industries. A drafting tool called a **spline** is a strip of plastic or other material that is easily flexed to pass through a series of key design points (control points) already established on a drawing. Weights called **ducks** hold the spline in place while the draftsman uses the spline as a guide to draw a smooth curve formed by it through the design points. A spline curve can be drawn through any set of n points that imply a smooth curve. The rate of change of curvature is gradual, and there are no kinks.

A spline behaves structurally exactly like a beam, with bending deflections forming it into a smooth curve. As long as the distribution of control points and the material and stiffness of the spline allow the spline to deform elastically, any spline will form the same curve for the same set of control points. This curve is often called an elastic curve, or minimum-energy curve.

The most commonly used spline curve is a plane curve, and the material of the most common type of spline tool produces a linear relationship between stress and strain in the elastic range. This results in a bending deflection curve that can be defined by a cubic function. Thus, the development that follows addresses plane cubic spline curves and how we represent them exactly by piecewise parametric cubic curves.

13.1   The Second-Derivative Form of a PC Curve

We must first develop the second-derivative form of a pc equation. With this form, we can show mathematically an exact analogy with the formulas for the beam deformations of the classical spline curve. Derive the second-derivative form as follows: Given the geometric form $\mathbf{p} = \mathbf{FB}$ with $\mathbf{B} = [\mathbf{p}_0 \quad \mathbf{p}_1 \quad \mathbf{p}_0^u \quad \mathbf{p}_1^u]^T$, find an expression for $\mathbf{p}$ in terms of $\mathbf{p}_0, \mathbf{p}_1, \mathbf{p}_0^{uu}, \mathbf{p}_1^{uu}$, where $\mathbf{p}^{uu} = d^2\mathbf{p}/du^2$.

We know from Eqs. 2.52 and 2.56 that at $u = 0$ and $u = 1$,

$$\mathbf{p}_0^{uu} = -6\mathbf{p}_0 + 6\mathbf{p}_1 - 4\mathbf{p}_0^u - 2\mathbf{p}_1^u \tag{2.159}$$

and

$$\mathbf{p}_1^{uu} = 6\mathbf{p}_0 - 6\mathbf{p}_1 + 2\mathbf{p}_0^u + 4\mathbf{p}_1^u \tag{2.160}$$

Solve these two equations for $\mathbf{p}_0^u$ and $\mathbf{p}_1^u$ to obtain

$$\mathbf{p}_0^u = -\mathbf{p}_0 + \mathbf{p}_1 - \tfrac{1}{6}(2\mathbf{p}_0^{uu} + \mathbf{p}_1^{uu}) \tag{2.161}$$

and

$$\mathbf{p}_1^u = -\mathbf{p}_0 + \mathbf{p}_1 + \tfrac{1}{6}(\mathbf{p}_0^{uu} + 2\mathbf{p}_1^{uu}) \tag{2.162}$$

Substitute Eqs. 2.161 and 2.162 into Eq. 2.47, collect terms, and expand the blending functions to yield the second-derivative form

$$\mathbf{p} = (1 - u)\mathbf{p}_0 + u\mathbf{p}_1 + \tfrac{1}{6}(-u^3 + 3u^2 - 2u)\mathbf{p}_0^{uu} + \tfrac{1}{6}(u^3 - u)\mathbf{p}_1^{uu} \tag{2.163}$$

This is the result we want, and we see that substituting $\mathbf{p}_0^{uu}$ and $\mathbf{p}_1^{uu}$ for $\mathbf{p}_0^u$ and $\mathbf{p}_1^u$, respectively, gives us a new set of blending functions

$$
\begin{aligned}
F_1 &= (1 - u) \\
F_2 &= u \\
F_3 &= \tfrac{1}{6}(-u^3 + 3u^2 - 2u) \\
F_4 &= \tfrac{1}{6}(u^3 - u)
\end{aligned}
\tag{2.164}
$$

We can now write Eq. 2.163 more compactly as

$$\mathbf{p} = F_1\mathbf{p}_0 + F_2\mathbf{p}_1 + F_3\mathbf{p}_0^{uu} + F_4\mathbf{p}_1^{uu} \tag{2.165}$$

or in even more compact matrix form as

$$\mathbf{p} = F[\mathbf{p}_0 \quad \mathbf{p}_1 \quad \mathbf{p}_0^{uu} \quad \mathbf{p}_1^{uu}]^T \tag{2.166}$$

This completes the derivation of the second-derivative form.

13.2 Equations Describing the Elastic Behavior of a Beam

Next, we demonstrate that the equations of structural mechanics describing the elastic behavior of a simply supported beam are analogous in form to Eq. 2.163. Figure 2.42 shows a beam with a deflected shape caused by applying a bending moment at each end. The elastic beam equations, expressed in terms of a local coordinate system, are shown alongside their pc analogs. It is important to note the sign convention, particularly the slope α and the deflection δ. Positive slope is measured counterclockwise from the x-axis, and positive deflection is measured in the positive y direction. The deflected beam lies in the x, y plane.

We derive the analogous pc curve equations from the second-derivative form by first rewriting Eq. 2.163 in terms of the x and y components

$$x = (1 - u)x_0 + ux_1 + \tfrac{1}{6}(-u^3 + 3u^2 - 2u)x_0^{uu} + \tfrac{1}{6}(u^3 - u)x_1^{uu} \tag{2.167}$$

and

$$y = (1 - u)y_0 + uy_1 + \tfrac{1}{6}(-u^3 + 3u^2 - 2u)y_0^{uu} + \tfrac{1}{6}(u^3 - u)y_1^{uu} \tag{2.168}$$

Beam theory assumes that there is no deflection in the y direction at the supports, no change in length, and that deflections are small enough so that

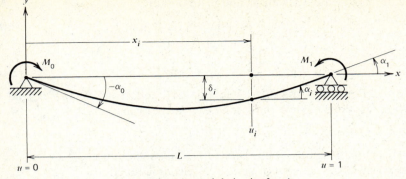

pc equations (second-derivative form)

$$y_i = \frac{1}{6} \{[(1 - u_i)^3 - (1 - u_i)]y_0^{uu} + (u_i^3 - u_i)y_1^{uu}$$

$$\left(\frac{dy}{dx}\right)_i = \frac{1}{6L} \{[-3(1 - u_i)^2 + 1]y_0^{uu} + (3u_i^2 - 1)y_1^{uu}\}$$

$$\left(\frac{d^2y}{dx^2}\right)_i = \frac{1}{L^2} \{(1 - u_i)y_0^{uu} + (u_i)y_1^{uu}\}$$

$$\text{where } u_i = \frac{x_i}{L}$$

Elastic-beam equations

$$\delta_i = \frac{L^2}{6EI} \left\{\left[\left(1 - \frac{x_i}{L}\right)^3 - \left(1 - \frac{x_i}{L}\right)\right]M_0 + \left[\left(\frac{x_i}{L}\right)^3 - \frac{x_i}{L}\right]M_1\right\}$$

$$\alpha_i = \frac{L}{6EI} \left\{\left[-3\left(1 - \frac{x_i}{L}\right)^2 + 1\right]M_0 + \left[3\left(\frac{x_i}{L}\right)^2 - 1\right]M_1\right\}$$

$$\frac{1}{R_i} = \frac{1}{EI} \left[\left(1 - \frac{x_i}{L}\right)M_0 + \frac{x_i}{L}M_1\right]$$

FIGURE 2.42 Deflected beam.

x is a linear function of the beam's span length L. These assumptions are boundary conditions; we express them mathematically as

$$\begin{array}{ll} x_0 = 0 & x_1 = L \\ y_0 = 0 & y_1 = 0 \\ x_0^{uu} = 0 & x_1^{uu} = 0 \end{array}$$

(2.169)

The last two equations, namely, $x_0^{uu} = 0$ and $x_1^{uu} = 0$, follow from the assumption that x is a linear function of the beam span's length; see Eq. 2.170.

Next, we substitute these boundary conditions into Eqs. 2.167 and 2.168 to obtain

$$x = uL \quad \text{or} \quad u = \frac{x}{L}$$

(2.170)

and

$$y = \tfrac{1}{6}\{[(1 - u)^3 - (1 - u)]y_0^{uu} + (u^3 - u)y_1^{uu}\}\tag{2.171}$$

We know that

$$y^x = \frac{y^u}{x^u}\tag{2.172}$$

and from Eqs. 2.170 and 2.171, we can compute

$$x^u = L\tag{2.173}$$

and

$$y^u = \frac{1}{6}\{[-3(1 - u)^2 + 1]y_0^{uu} + (3u^2 - 1)y_1^{uu}\}\tag{2.174}$$

Therefore,

$$y^x = \frac{1}{6L}\{[-3(1 - u)^2 + 1]y_0^{uu} + (3u^2 - 1)y_1^{uu}\}\tag{2.175}$$

Remember, $y^x = dy/dx$ and $y^{xx} = d^2y/dx^2$.

Taking the derivative of a quotient, we know from elementary calculus that

$$y^{xx} = \frac{(y^u/x^u)^u}{x^u} = \frac{x^u y^{uu} - x^{uu} y^u}{(x^u)^3}\tag{2.176}$$

And since

$$x^{uu} = (x^u)^u$$

$$= L^u\tag{2.177}$$

$$= 0$$

and

$$y^{uu} = (y^u)^u = (1 - u)y_0^{uu} + uy_1^{uu}\tag{2.178}$$

then substituting as appropriate into Eq. 2.176, we obtain

$$y^{xx} = \frac{1}{L^2}[(1 - u)y_0^{uu} + uy_1^{uu}]\tag{2.179}$$

Finally, we substitute $u = x/L$ into Eqs. 2.171, 2.175, and 2.179 to obtain

$$y = \frac{1}{6}\left\{\left[\left(1 - \frac{x}{L}\right)^3 - \left(1 - \frac{x}{L}\right)\right]y_0^{uu} + \left[\left(\frac{x}{L}\right)^3 - \frac{x}{L}\right]y_1^{uu}\right\}\tag{2.180}$$

$$y^x = \frac{1}{6L}\left\{\left[-3\left(1 - \frac{x}{L}\right)^2 + 1\right]y_0^{uu} + \left[3\left(\frac{x}{L}\right)^2 - 1\right]y_1^{uu}\right\}\tag{2.181}$$

$$y^{xx} = \frac{1}{L^2}\left[\left(1 - \frac{x}{L}\right)y_0^{uu} + \left(\frac{x}{L}\right)y_1^{uu}\right]\tag{2.182}$$

Compare these three equations to the beam equations; their forms are analogous. They are identical if we stipulate the following relationships:

$$y_0^{uu} = \frac{M_0 L^2}{EI} \tag{2.183}$$

and

$$y_1^{uu} = \frac{M_1 L^2}{EI} \tag{2.184}$$

We find that the y coordinate of a pc curve is identified with the beam deflection δ, the first derivative with the beam slope α, and the second derivative with the beam curvature $(1/R)$. We have thus demonstrated that we can use a pc curve or set of curves as an exact geometric model of a traditional spline.

13.3   Derivation of the Composite PC Spline Curve

We start with a set of points whose coordinates in some arbitrary coordinate system are known. We want to find a set of pc curves, joined end to end to form a composite curve that passes through these points in exactly the same way a spline would. Note that the set of points may be the result of measurements taken of some physical shape, such as a metal, wood, or clay model of a new automobile body, ship's hull, or wing airfoil. Or they may be determined mathematically from theoretical models describing some phenomenon, such as points of equal aerodynamic pressure over a surface or points defining the relationship between temperature and enthalpy. However the points are determined, we assume that we can order them so that a smooth composite spline curve passes sequentially through each of them. Figure 2.43 illustrates a set of n points that lie in a plane and are ordered 1 through n.

Figure 2.44 shows another set of n points, sequentially ordered and lying in a plane. The figure also shows certain geometric relationships we can easily obtain from operations on the point coordinates. We will use these relationships to compute the set of pc curves passing through these points. Therefore, our

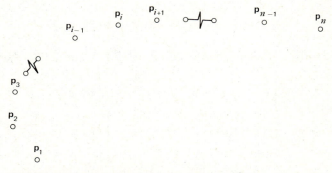

FIGURE 2.43 An ordered set of points in a plane.

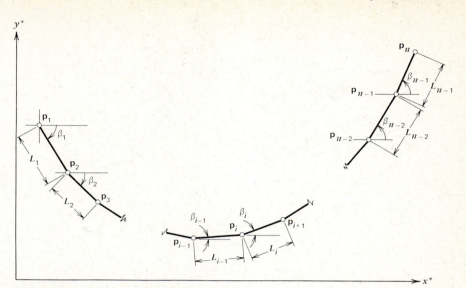

FIGURE 2.44 An ordered set of points in a plane.

first job is to define these relationships in a suitable way. We measure the angle β_i from a line through point $\mathbf{p}_i$, parallel to the x-axis, to the line joining $\mathbf{p}_i$ to $\mathbf{p}_{i+1}$. L_i is the distance between $\mathbf{p}_i$ and $\mathbf{p}_{i+1}$.

Figure 2.45 is a close-up view of an arbitrary point $\mathbf{p}_i$ presumed to lie on a spline curve, and the points immediately adjacent to it, $\mathbf{p}_{i-1}$ and $\mathbf{p}_{i+1}$. Observe that case 1 is a situation where the curve near $\mathbf{p}_i$ is generally concave on the upper side and in case 2, concave downward. These two cases (to which we can add concave right and left) illustrate a sign convention necessary to later development, which we will discuss soon.

The pc curve segment between each sequential pair of points is equivalent to the elastic curve of a deflected spline. We treat each segment as a simply supported beam behaving as though under the influence of bending moments applied to each end. The lines L_i represent the unloaded, undeflected beams. Note that both the elastic beam equations and the pc equations are in terms of the local coordinate system of each beam (that is, curve segment). For each curve segment, L_i is colinear with the local x-axis and positive in the direction from $\mathbf{p}_i$ to $\mathbf{p}_{i+1}$ with the origin at $\mathbf{p}_i$. The local y-axis is normal to L_i at $\mathbf{p}_i$. Beam theory assumes the spline undergoes only small deflections, perhaps max $\delta_i < 0.1L_i$. A set of points will satisfy this restriction if they are sufficiently closely spaced.

Since the curve we will try to model must be smooth and continuous through all the points as a result of the structural behavior of the original spline itself, two other conditions must be met. First, two pc curve segments meeting at a point $\mathbf{p}_i$ must have a common tangent. Second, the curvature must be continuous through this point.

We assign the end of each curve segment a value of u; one is $u = 0$, and the other is $u = 1$. Thus, for the segment between $\mathbf{p}_i$ and $\mathbf{p}_{i+1}$, we assign $\mathbf{p}_i$ a value of

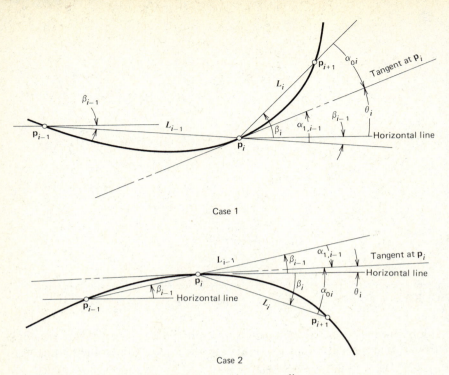

FIGURE 2.45 Coordinate geometry at a point on a spline curve.

$u = 0$ and $\mathbf{p}_{i+1}$ a value of $u = 1$. If this is done sequentially, we see that the $u = 0$ end of a particular segment coincides with the $u = 1$ end of the previous segment and that the $u = 1$ end coincides with the $u = 0$ end of the succeeding segment. The angle $\alpha_{0,i}$ is the slope at the $u = 0$ end of the elastic curve of the beam analogous to the ith segment. We can measure this angle in the local coordinate system of segment i.

   If we know the α's or can determine them, then we can compute the coefficients of the pc curves representing the spline curve. We must define a pc curve for each segment of the spline curve, and the total set of these pc curves represents the full spline curve. If we express mathematically the constraints (continuity and elasticity) at each of the n points, we obtain n simultaneous equations. The solution of these equations yields the geometric coefficients of the pc curves, and we reach our goal. We proceed as follows.

   From Fig. 2.45, θ_i is the angle from a line through point i parallel to the x-axis of the global coordinate system to the tangent line at that point on the particular curve we are attempting to define. In case 1, we write

$$\theta_i = \beta_{i-1} + \alpha_{1,i-1} \tag{2.185}$$

and

$$\theta_i = \beta_i + \alpha_{0,i} \tag{2.186}$$

At this point, we should briefly discuss the sign convention mentioned earlier. The + sign in Eq. 2.186 is justified, since in the local coordinate system for segment i, the angle $\alpha_{0,i}$ is negative for case 1. Refer to the sign relationship between β_i and $\alpha_{0,i}$ in case 2 to clarify this. Here, $\alpha_{0,i}$ is positive in its local system convention, and it is easy to see that $\theta_i = \beta_i + \alpha_{0,i}$. Thus, the + sign in the general relationship in Eq. 2.186 is correct, since it allows the local system to determine the sign of the α's and the global system to determine the sign of the β's, both of which depend on the global orientation, the direction of concavity, and the ordering sequence chosen. Therefore, from Eqs. 2.185 and 2.186, we obtain

$$\beta_{i-1} + \alpha_{1,i-1} = \beta_i + \alpha_{0,i} \tag{2.187}$$

Furthermore, we can state that

$$\tan(\alpha_{1,i-1} - \alpha_{0,i}) = \tan(\beta_i - \beta_{i-1}) \tag{2.188}$$

A common trigonometric identity

$$\tan(\theta \pm v) = \frac{\tan\theta \pm \tan v}{1 \mp \tan\theta \tan v} \tag{2.189a}$$

allows us to write

$$\tan(\beta_i - \beta_{i-1}) = \frac{\tan\alpha_{1,i-1} - \tan\alpha_{0,i}}{1 + \tan\alpha_{1,i-1}\tan\alpha_{0,i}} \tag{2.189b}$$

Elastic beam theory assumes small deflections; therefore, we let $\alpha_{1,i-1}$, $\alpha_{0,i} \ll 1$, which means that $\tan\alpha_{1,i-1}\tan\alpha_{0,i} \ll 1$. Equation 2.189b becomes

$$\tan(\beta_i - \beta_{i-1}) = \tan\alpha_{1,i-1} - \tan\alpha_{0,i} \tag{2.190}$$

Since we know $y^x = \tan\alpha$ and $y^u/x^u = y^x$, then

$$\tan\alpha_{1,i-1} = \frac{y^u_{1,i-1}}{x^u_{1,i-1}}$$

and $\tag{2.191}$

$$\tan\alpha_{0,i} = \frac{y^u_{0,i}}{x^u_{0,i}}$$

Substituting into Eq. 2.190 and rearranging terms,

$$\frac{y^u_{1,i-1}}{x^u_{1,i-1}} - \frac{y^u_{0,i}}{x^u_{0,i}} = \tan(\beta_i - \beta_{i-1}) \tag{2.192}$$

Solving the right side of Eq. 2.181 first with $x/L = 0$ and then with $x/L = 1$, we find, respectively,

$$\frac{y^u_{0,i}}{x^u_{0,i}} = -\frac{1}{6L_i}(2y^{uu}_{0,i} + y^{uu}_{1,i}) \tag{2.193}$$

$$\frac{y^u_{1,i-1}}{x^u_{1,i-1}} = \frac{1}{6L_{i-1}}(y^{uu}_{0,i-1} + 2y^{uu}_{1,i-1}) \tag{2.194}$$

Substitute Eqs. 2.193 and 2.194 into Eq. 2.192 to find that

$$\tan(\beta_i - \beta_{i-1}) = \frac{1}{6L_{i-1}}(y^{uu}_{0,i-1} + 2y^{uu}_{1,i-1})$$

$$+ \frac{1}{6L_i}(2y^{uu}_{0,i} + y^{uu}_{1,i}) \tag{2.195}$$

Because the second derivative of a function is invariant under a rigid-body transformation, we know

$$y^{uu}_{i-1} = y^{uu}_{0,i-1} \tag{2.196}$$

$$y^{uu}_i = y^{uu}_{0,i} = y^{uu}_{1,i-1} \tag{2.197}$$

$$y^{uu}_{i+1} = y^{uu}_{1,i} \tag{2.198}$$

Substitute Eqs. 2.196–2.198 into Eq. 2.195, and rearrange the results into the more convenient form

$$6\tan(\beta_i - \beta_{i-1}) = \left(\frac{1}{L_{i-1}}\right)y^{uu}_{i-1} + \left(\frac{2}{L_{i-1}} + \frac{2}{L_i}\right)y^{uu}_i + \left(\frac{1}{L_i}\right)y^{uu}_{i+1} \tag{2.199}$$

Now, we know the β angles and lengths L at each of the n points, and there are n unknowns, namely the y^{uu}_i values for $i \in [1, n]$. If we write Eq. 2.199 for each point, we obtain n simultaneous equations and n unknowns. These equations are

$$6\tan(\beta_1 - \beta_0) = \left(\frac{1}{L_0}\right)y^{uu}_0 + \left(\frac{2}{L_0} + \frac{2}{L_1}\right)y^{uu}_1 + \left(\frac{1}{L_1}\right)y^{uu}_2$$

$$6\tan(\beta_2 - \beta_1) = \left(\frac{1}{L_1}\right)y^{uu}_1 + \left(\frac{2}{L_1} + \frac{2}{L_2}\right)y^{uu}_2 + \left(\frac{1}{L_2}\right)y^{uu}_3$$

$$6\tan(\beta_3 - \beta_2) = \left(\frac{1}{L_2}\right)y^{uu}_2 + \left(\frac{2}{L_2} + \frac{2}{L_3}\right)y^{uu}_3 + \left(\frac{1}{L_3}\right)y^{uu}_4$$

$$\vdots \tag{2.200}$$

$$6\tan(\beta_{n-1} - \beta_{n-2}) = \left(\frac{1}{L_{n-2}}\right)y^{uu}_{n-2} + \left(\frac{2}{L_{n-2}} + \frac{2}{L_{n-1}}\right)y^{uu}_{n-1} + \left(\frac{1}{L_{n-1}}\right)y^{uu}_n$$

$$6\tan(\beta_n - \beta_{n-1}) = \left(\frac{1}{L_{n-1}}\right)y^{uu}_{n-1} + \left(\frac{2}{L_{n-1}} + \frac{2}{L_n}\right)y^{uu}_n + \left(\frac{1}{L_n}\right)y^{uu}_{n+1}$$

Inspection reveals what appear to be ambiguities in the first and last equations of the set. What do y^{uu}_0 and y^{uu}_{n+1} mean? These terms, as well as L_0 and β_0, refer to nonexistent points. A convenient way of removing these ambiguities is to specify boundary conditions.

We encounter three basic sets of conditions most often, and they are the most useful. In the first condition, the spline is unrestrained at points 1 and n.

That is, the spline curve is free to assume a slope at each end point that is solely a result of the distribution of the interior points, $2-(n-1)$. This means the curvatures $1/R_1$ and $1/R_n$ at points 1 and n are both equal to zero. Thus,

$$y_1^{uu} = 0 \tag{2.201}$$

$$y_n^{uu} = 0 \tag{2.202}$$

We rewrite the set of Eqs. 2.200 using Eqs. 2.201 and 2.202 to obtain

$$y_1^{uu} = 0$$

$$6 \tan (\beta_2 - \beta_1) = \left(\frac{2}{L_1} + \frac{2}{L_2}\right) y_2^{uu} + \left(\frac{1}{L_2}\right) y_3^{uu}$$

$$6 \tan (\beta_3 - \beta_2) = \left(\frac{1}{L_2}\right) y_2^{uu} + \left(\frac{2}{L_2} + \frac{2}{L_3}\right) y_3^{uu} + \left(\frac{1}{L_3}\right) y_4^{uu}$$

$$\vdots \tag{2.203}$$

$$6 \tan (\beta_{n-1} - \beta_{n-2}) = \left(\frac{1}{L_{n-2}}\right) y_{n-2}^{uu} + \left(\frac{2}{L_{n-2}} + \frac{2}{L_{n-1}}\right) y_{n-1}^{uu} + \left(\frac{1}{L_{n-1}}\right) y_n^{uu}$$

$$y_n^{uu} = 0$$

In matrix form, Eq. 2.203 becomes

$$
\begin{bmatrix}
1 & 0 & 0 & \cdot & & \cdot & 0 \\
0 & (2/L_1) + (2/L_2) & (1/L_2) & 0 & & \cdot & \\
0 & (1/L_2) & (2/L_2) + (2/L_3) & (1/L_3) & 0 & & \cdot \\
\vdots & & & \vdots & & & \vdots \\
\cdot & \cdot & \cdot & (1/L_{n-2}) & (2/L_{n-2}) + (2/L_{n-1}) & (1/L_{n-1}) \\
0 & \cdot & & \cdot & & 1 \\
\end{bmatrix}
\begin{bmatrix}
y_1^{uu} \\
y_2^{uu} \\
y_3^{uu} \\
\vdots \\
y_{n-1}^{uu} \\
y_n^{uu} \\
\end{bmatrix}
$$

$$
=
\begin{bmatrix}
0 \\
6 \tan (\beta_2 - \beta_1) \\
6 \tan (\beta_3 - \beta_2) \\
6 \tan (\beta_4 - \beta_3) \\
\vdots \\
6 \tan (\beta_{n-1} - \beta_{n-2}) \\
0 \\
\end{bmatrix}
\tag{2.204}
$$

We can specify another set of boundary conditions by specifying the slopes at $i = 1$ and $i = n$. For example, let

$$\alpha_{0,1} = \theta_1 - \beta_1 \tag{2.205}$$

$$\alpha_{1,n-1} = \theta_n - \beta_{n-1} \tag{2.206}$$

Then, we must rewrite Eq. 2.204

$$
\begin{bmatrix}
(2/L_1) & (1/L_1) & 0 & & \cdot & & 0 \\
(1/L_1) & (2/L_1)+(2/L_2) & (1/L_2) & 0 & & \cdot & \\
0 & (1/L_2) & (2/L_2)+(2/L_3) & (1/L_3) & 0 & & \cdot \\
\vdots & & & \vdots & & & \vdots \\
\cdot & \cdot & \cdot & (1/L_{n-2}) & (2/L_{n-2})+(2/L_{n-1}) & (1/L_{n-1}) & y_{n-1}^{uu} \\
0 & \cdot & \cdot & \cdot & & (1/L_{n-1}) & (2/L_{n-1})
\end{bmatrix}
\begin{bmatrix}
y_1^{uu} \\ y_2^{uu} \\ y_3^{uu} \\ \vdots \\ y_{n-1}^{uu} \\ y_n^{uu}
\end{bmatrix}
$$

$$
=
\begin{bmatrix}
6\tan(\theta_1-\beta_1) \\
6\tan(\beta_2-\beta_1) \\
6\tan(\beta_3-\beta_2) \\
6\tan(\beta_4-\beta_3) \\
\vdots \\
6\tan(\beta_{n-1}-\beta_{n-2}) \\
6\tan(\theta_n-\beta_{n-1})
\end{bmatrix}
\tag{2.207}
$$

The third and final set of boundary conditions available to us asserts a constant curvature throughout the length of each end segment. This means

$$
y_1^{uu} = y_2^{uu} \tag{2.208}
$$

$$
y_{n-1}^{uu} = y_n^{uu} \tag{2.209}
$$

and we again rewrite 2.204

$$
\begin{bmatrix}
1 & -1 & 0 & & \cdot & & 0 \\
(1/L_1) & (2/L_1)+(2/L_2) & (1/L_2) & 0 & & \cdot & \\
0 & (1/L_2) & (2/L_2)+(2/L_3) & (1/L_3) & 0 & & \cdot \\
\vdots & & & \vdots & & & \vdots \\
\cdot & \cdot & \cdot & (1/L_{n-2}) & (1/L_{n-2})+(2/L_{n-1}) & (1/L_{n-1}) & \\
0 & \cdot & \cdot & \cdot & 1 & -1
\end{bmatrix}
\begin{bmatrix}
y_1^{uu} \\ y_2^{uu} \\ y_3^{uu} \\ \vdots \\ y_{n-1}^{uu} \\ y_n^{uu}
\end{bmatrix}
$$

$$
=
\begin{bmatrix}
6\tan(\theta_1-\beta_1) \\
6\tan(\beta_2-\beta_1) \\
6\tan(\beta_3-\beta_2) \\
6\tan(\beta_4-\beta_3) \\
\vdots \\
6\tan(\beta_{n-1}-\beta_{n-2}) \\
6\tan(\theta_n-\beta_{n-1})
\end{bmatrix}
\tag{2.210}
$$

We determine the $y_1^{uu}, y_2^{uu}, \ldots, y_n^{uu}$ by solving the set of simultaneous equations appropriate to the boundary conditions we have chosen. Using the y_i^{uu}, we can compute the geometric coefficients in the global coordinate system (x^*, y^*) for each of the $(n-1)$ pc curve segments. We do this as follows: For any curve segment i, we know

$$\tan \alpha_{0,i} = \frac{y_{0,i}^u}{x_{0,i}^u}$$

$$= -\frac{1}{6L_i}(2y_{0,i}^{uu} + y_{1,i}^{uu}) \qquad (2.211)$$

$$\tan \alpha_{1,i} = \frac{y_{1,i}^u}{x_{1,i}^u}$$

$$= \frac{1}{6L_i}(y_{0,i}^{uu} + 2y_{1,i}^{uu}) \qquad (2.212)$$

and since $x = uL$, $x^u = L$, we know that

$$x_{0,i}^u = L_i \qquad (2.213)$$

$$x_{1,i}^u = L_i \qquad (2.214)$$

$$y_{0,i}^u = -\tfrac{1}{6}(2y_{0,i}^{uu} + y_{1,i}^{uu}) \qquad (2.215)$$

$$y_{1,i}^u = \tfrac{1}{6}(y_{0,i}^{uu} + 2y_{1,i}^{uu}) \qquad (2.216)$$

Figure 2.46 shows how to express vector components in one coordinate system (x, y) in terms of another system (x^*, y^*), given the angle β between them.

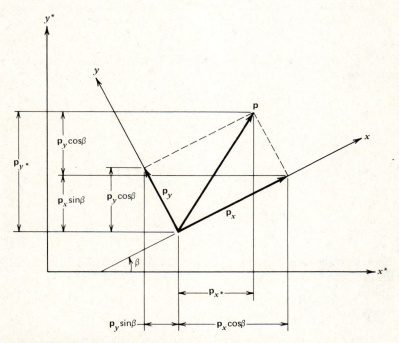

FIGURE 2.46 Coordinate transformation of vector components.

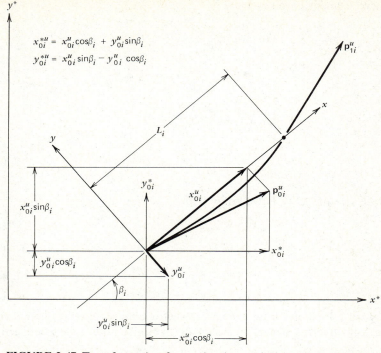

FIGURE 2.47 Transformation from a local to a global coordinate system for a spline curve segment.

Figures 2.47 and 2.48 show relationships for segment i of a spline curve. From them, we see

$$\cos \beta_i = \frac{x^*_{1,i} - x^*_{0,i}}{L_i} \tag{2.217}$$

$$\sin \beta_i = \frac{y^*_{1,i} - y^*_{0,i}}{L_i} \tag{2.218}$$

$$x^{*u}_{0,i} = x^u_{0,i} \cos \beta_i - y^u_{0,i} \sin \beta_i \tag{2.219}$$

$$y^{*u}_{0,i} = x^u_{0,i} \sin \beta_i + y^u_{0,i} \cos \beta_i \tag{2.220}$$

$$x^{*u}_{1,i} = x^u_{1,i} \cos \beta_i - y^u_{1,i} \sin \beta_i \tag{2.221}$$

$$y^{*u}_{1,i} = x^u_{1,i} \sin \beta_i + y^u_{1,i} \cos \beta_i \tag{2.222}$$

Now, substituting into Eqs. 2.221–2.224 from Eqs. 2.219 and 2.220 and Eqs. 2.215–2.218 yields

$$x^{*u}_{0,i} = x^*_{1,i} - x^*_{0,i} + \frac{1}{6L_i} (2y^{uu}_{0,i} + y^{uu}_{1,i})(y^*_{1,i} - y^*_{0,i}) \tag{2.223}$$

$$y^{*u}_{0,i} = y^*_{1,i} - y^*_{0,i} - \frac{1}{6L_i} (2y^{uu}_{0,i} + y^{uu}_{1,i})(x^*_{1,i} - x^*_{0,i}) \tag{2.224}$$

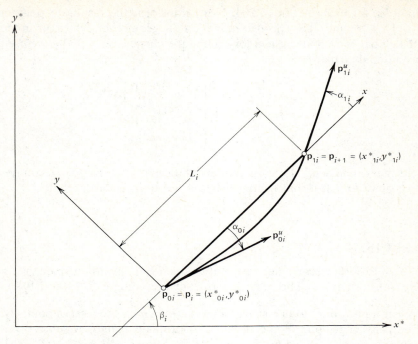

$$x_{0i}^{*u} = x_{1i}^* - x_{0i}^* + \frac{1}{6L_i}(2y_{0i}^{uu} + y_{1i}^{uu})(y_{1i}^* - y_{0i}^*)$$

$$y_{0i}^{*u} = y_{1i}^* - y_{0i}^* - \frac{1}{6L_i}(2y_{0i}^{uu} + y_{1i}^{uu})(x_{1i}^* - x_{0i}^*)$$

$$x_{1i}^{*u} = x_{1i}^* - x_{0i}^* - \frac{1}{6L_i}(y_{0i}^{uu} + 2y_{1i}^{uu})(y_{1i}^* - y_{0i}^*)$$

$$y_{1i}^{*u} = y_{1i}^* - y_{0i}^* + \frac{1}{6L_i}(y_{0i}^{uu} + 2y_{1i}^{uu})(x_{1i}^* - x_{0i}^*)$$

FIGURE 2.48 Transformation from a local to a global coordinate system for a spline curve segment.

$$x_{1,i}^{*u} = x_{1,i}^* - x_{0,i}^* - \frac{1}{6L_i}(y_{0,i}^{uu} + 2y_{1,i}^{uu})(y_{1,i}^* - y_{0,i}^*) \qquad (2.225)$$

$$y_{1,i}^{*u} = y_{1,i}^* - y_{0,i}^* + \frac{1}{6L_i}(y_{0,i}^{uu} + 2y_{1,i}^{uu})(x_{1,i}^* - x_{0,i}^*) \qquad (2.226)$$

These are transformation equations from a local to a global coordinate system.

Let us briefly summarize the techniques just described for fitting a sequential set of pc curves to a set of points so that the pc curves form a single continuous spline curve. First, solve the appropriate set of simultaneous equations, for example, Eq. 2.204. Next, transform the results to determine $x_{0,i}^{*u}$, $y_{0,i}^{*u}$, $x_{1,i}^{*u}$, and $y_{1,i}^{*u}$. This determines all the geometric coefficients in terms of the global coordinate system for each of the n pc curves. Finally, we use the conventional pc matrix

$$\mathbf{p}_i^* = \mathbf{FB}_i^* = \mathbf{F} \begin{bmatrix} x_0^* & y_0^* & 0 \\ x_1^* & y_1^* & 0 \\ x^{*u} & y_0^{*u} & 0 \\ x_1^{*u} & y_1^{*u} & 0 \end{bmatrix} \tag{2.227}$$

A further refinement is the reparametrization of the complete composite spline curve so that it runs from $u = 0$ to $u = 1$ over its entire length.

EXERCISES

1. Show that the second derivative of a function is invariant under a rigid-body transformation. (This assertion is made preceding Eqs. 2.196–2.198.)

2. Write a procedure to fit a string of pc curve segments to a set of points lying in, or parallel to, one of the principal coordinate planes. Assume that the maximum number of points is ten. Denote this as **SPLNPC(N, P, IBFLG, IPFLG, SLP, CO)**, where

N is the input number of points;

P(3, N) is the input array of point coordinates;

IBFLG is the input flag specifying boundary conditions;

IBFLG $= 0$ for unconstrained slopes at the spline end points;
IBFLG $= 1$ for specified slopes at the spline end points;

IPFLG is the input flag specifying the principal plane that the spline is parallel to;

IPFLG $= 1$ for the x, y plane,
IPFLG $= 2$ for the y, z plane, and
IPFLG $= 3$ for the z, x plane;

SLP(3, 2) is the input array of direction cosines of the slopes at the spline end points when IBFLG $= 1$; and

CO(4, 3, N $- 1$) is the output array of geometric coefficients of the (N $- 1$) pc curves.

3. Show that changing one point on a spline requires recomputing the entire curve. This is the disadvantage of using this type of curve for highly interactive design situations.

14 BEZIER CURVES

Some curve-defining techniques **interpolate** a given set of points, which means that the curve produced passes exactly through the points. An alternative approach defines a curve that only **approximates** or approaches the given points. In the techniques we have just discussed, curves are constrained to pass through all the specified points and are, therefore, interpolation techniques.

Interpolation techniques have certain disadvantages when incorporated into an interactive CAD program. Specifically, we do not get a strong intuitive feel for how to change or control the shape of a curve. For example, if we try to change the shape of a spline-interpolated curve by moving one or more of the interpolating points, we may produce unexpected, if not undesirable, perturbations and inflections, both locally and remotely. It is much easier if we can control curve shape in a predictable way by changing only a few simple parameters. **Bezier's curve** partially satisfies this need.

P. Bezier, of the French automobile company of Renault, was familiar with the work of Ferguson and Coons and their parametric cubic curve and bicubic surface interpolating techniques. Bezier set out in the early 1960s to find what he hoped would be a mathematical format more amenable to the designer and the design process. The result of Bezier's work was the UNISURF system, used by Renault since 1972 to design the sculptured surfaces of many of their automobile bodies. At the heart of the UNISURF system are the curves and surfaces that bear his name.

Bezier started with the principle that any point on a curve segment must be given by a parametric function of the following form:

$$\mathbf{p}(u) = \sum_{i=0}^{n} \mathbf{p}_i f_i(u) \qquad u \in [0, 1] \tag{2.228}$$

where the vectors $\mathbf{p}_i$ represent the $n + 1$ vertices of a **characteristic polygon**; see Fig. 2.49. These vertices are also called **control points**. He then set forth certain properties that the $f_i(u)$ blending functions must have and then looked for specific functions to meet these requirements. Let us briefly review these properties and the reasons for them. (Note that Bezier's presentation is more elegant and more rigorous. The interested student should read his work.)

1. The functions must interpolate the first and last vertex points; that is, the curve segment must start on $\mathbf{p}_0$ and end on $\mathbf{p}_n$. It is up to us to control the starting and ending points of a Bezier curve.

2. The tangent at $\mathbf{p}_0$ must be given by $\mathbf{p}_1 - \mathbf{p}_0$, and the tangent at $\mathbf{p}_n$ by $\mathbf{p}_n - \mathbf{p}_{n-1}$. This, of course, gives us direct control of the tangent to the curve at each end.

3. This requirement is generalized for higher derivatives at the curve's end points. Thus, the second derivative at $\mathbf{p}_0$ must be determined by $\mathbf{p}_0$, $\mathbf{p}_1$, and $\mathbf{p}_2$. In general, the rth derivative at an end point must be determined by its r neighboring vertices. This allows us virtually unlimited control of the continuity at the joints between curve segments of a composite Bezier curve.

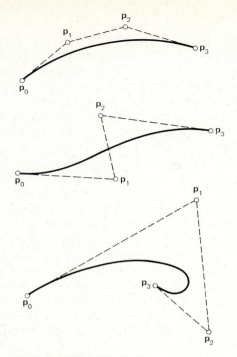

FIGURE 2.49 Bezier curves.

4. The functions $f_i(u)$ must be symmetric with respect to u and $(1 - u)$. This means that we can reverse the sequence of the vertex points defining the curve without changing the shape of the curve. In effect, this reverses the direction of parametrization.

Bezier chose a family of functions called Bernstein polynomials to satisfy these conditions simply and directly. He originally chose a form of vector representation that used the sides of the characteristic polygon. However, we will use the notation introduced by Forrest (1971), which uses the vectors defining the polygon vertices; it is a more compact scheme, with greater intuitive appeal. It turns out that the functions Bezier selected depend on the number of vertices used to specify a particular curve. To indicate this, Eq. 2.228 becomes

$$\mathbf{p}(u) = \sum_{i=0}^{n} \mathbf{p}_i B_{i,n}(u) \qquad u \in [0, 1] \tag{2.229}$$

where

$$B_{i,n}(u) = C(n, i)u^i(1 - u)^{n-i} \tag{2.230}$$

and where $C(n, i)$ is the familiar binomial coefficient

$$C(n, i) = \frac{n!}{i!(n - i)!} \tag{2.231}$$

Note: When i and $u = 0$, $u^i = 1$; also $0! = 1$. Observe that for $(n + 1)$ vertices, $B_{i,n}(u)$ yields an nth-degree polynomial.

Let us expand Eq. 2.229 for curves defined by three, four, five, and six points, so that we become familiar with the polynomial forms produced.

For three points, $n = 2$, and

$$\mathbf{p}(u) = (1 - u)^2\mathbf{p}_0 + 2u(1 - u)\mathbf{p}_1 + u^2\mathbf{p}_2 \qquad (2.232)$$

For four points, $n = 3$, and

$$\mathbf{p}(u) = (1 - u)^3\mathbf{p}_0 + 3u(1 - u)^2\mathbf{p}_1 + 3u^2(1 - u)\mathbf{p}_2 + u^3\mathbf{p}_3 \qquad (2.233)$$

For five points, $n = 4$, and

$$\mathbf{p}(u) = (1 - u)^4\mathbf{p}_0 + 4u(1 - u)^3\mathbf{p}_1 + 6u^2(1 - u)^2\mathbf{p}_2 + 4u^3(1 - u)\mathbf{p}_3 + u^4\mathbf{p}_4$$

$$(2.234)$$

For six points, $n = 5$, and

$$\mathbf{p}(u) = (1 - u)^5\mathbf{p}_0 + 5u(1 - u)^4\mathbf{p}_1 + 10u^2(1 - u)^3\mathbf{p}_2$$

$$+ 10u^3(1 - u)^2\mathbf{p}_3 + 5u^4(1 - u)\mathbf{p}_4 + u^5\mathbf{p}_5 \qquad (2.235)$$

Figure 2.50 shows two examples of cubic Bezier curves; in Fig. 2.50a, the disposition of points generates a smooth, uninflected curve, while in Fig. 2.50b, the disposition of points generates an inflected curve. Notice that in both cases, the curves are tangent to the lines defined by $\mathbf{p}_1 - \mathbf{p}_0$ and $\mathbf{p}_3 - \mathbf{p}_2$.

As we learned for pc curves, the blending functions are the key to the behavior of Bezier curves. Figure 2.51 shows blending-function curves $B_{i,n}$ for $n = 2, 3, 4$, and 5. For our examples in Fig. 2.50, the first control point $\mathbf{p}_0$, whose contribution to the curve's shape is propagated by $B_{0,3}(u)$, is the most influential when $u = 0$. The other control points do not contribute to $\mathbf{p}(u)$ for $u = 0$, since their associated blending functions are each zero. A symmetrical situation occurs for $\mathbf{p}_3$ when $u = 1$. Control points $\mathbf{p}_1$ and $\mathbf{p}_2$ are most influential when $u = 1/3$ and $2/3$, respectively.

For any Bezier curve, each $\mathbf{p}_i$ is weighted by its associated blending function; as we have seen, when $u = 0$, $\mathbf{p}_0$ is given a weight of 1.0 and $\mathbf{p}_1$ through $\mathbf{p}_n$ a weight of zero. Less weight is given to $\mathbf{p}_0$ and more to each succeeding $\mathbf{p}_i$ as u increases, reaching a maximum weight for each $\mathbf{p}_i$ when u becomes i/n. Then all other weights decay gradually to 0 as the weight of $\mathbf{p}_n$ reaches 1 when $u = 1$. So, we observe a shift in the influence of each point (each polygon vertex) as the parametric variable u moves through its range from 0 to 1.

The facility to modify a curve that the Bezier formulation provides is demonstrated in Fig. 2.52. In Fig. 2.52a, moving point $\mathbf{p}_1$ to $\mathbf{p}_1^*$ "pulls" the curve toward that vertex. The effect can be weaker or stronger depending on the distance (or direction) the point is moved.

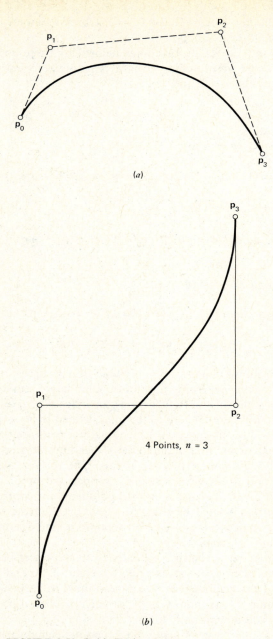

(a)

$\mathbf{p}_3$

$\mathbf{p}_1$

$\mathbf{p}_2$

4 Points, $n = 3$

$\mathbf{p}_0$

(b)

FIGURE 2.50 Cubic Bezier curves.

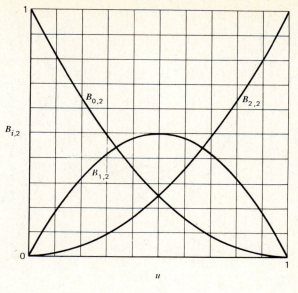

(a)

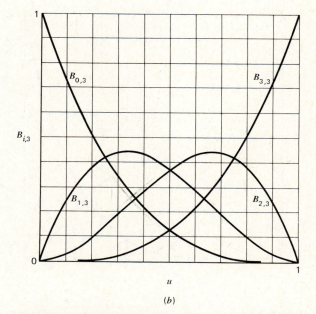

(b)

FIGURE 2.51 Bezier blending functions. (a) Three points, $n = 2$; (b) Four points, $n = 3$.

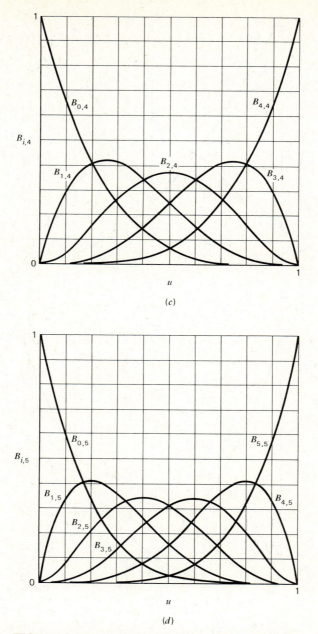

FIGURE 2.51 (*Continued*) Bezier blending functions. (*c*) Five points $n = 4$; (*d*) Six points, $n = 5$.

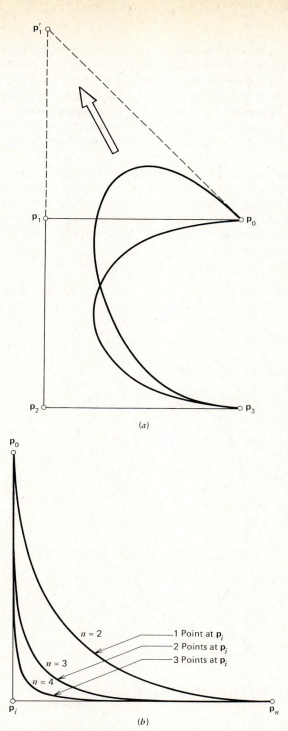

(a)

(b)

FIGURE 2.52 Cubic Bezier curves and their modification.

In Fig. 2.52*b*, we observe another interesting degree of freedom. By specifying multiple coincident points at a vertex, we pull the curve in closer and closer to that vertex. To do this, a curve-generating program might simply require us to specify a positive integer for each vertex. The integer indicates the number of coincident points at a vertex. By adding points coincident with existing vertex points, we increase the degree of the polynomial representing the curve without changing the number of sides and shape of the characteristic polygon.

If the first and last points of the characteristic polygon coincide, we produce a closed curve. The closed curve in Fig. 2.53*a* has C^1 continuity (that is, first derivative or slope continuity) where the ends of the curve join together at $\mathbf{p}_0, \mathbf{p}_5$. This continuity is produced by making points $\mathbf{p}_1, \mathbf{p}_0, \mathbf{p}_5$, and $\mathbf{p}_4$ colinear. A similar condition exists in Fig. 2.53*c*. However, the curve in Fig. 2.53*b* exhibits only C^0 continuity.

We can write Eq. 2.233 for $n = 3$ in matrix form as follows:

$$\mathbf{p}(u) = [(1 - u)^3 \quad 3u(1 - u)^2 \quad 3u^2(1 - u) \quad u^3][\mathbf{p}_0 \quad \mathbf{p}_1 \quad \mathbf{p}_2 \quad \mathbf{p}_3]^T \quad (2.236)$$

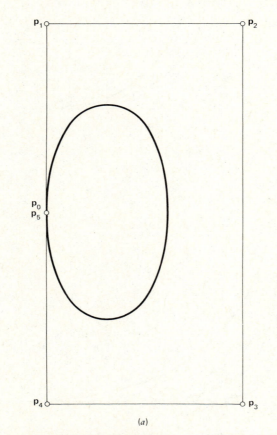

(*a*)

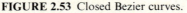

FIGURE 2.53 Closed Bezier curves.

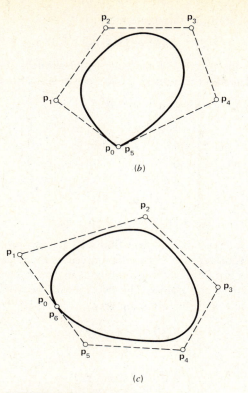

(b)

(c)

FIGURE 2.53 (*Continued*)

We can express these more generally as

$$\mathbf{p}(u) = \mathbf{F}_n \mathbf{B}_n \qquad (2.237)$$

Here, $\mathbf{F}_n$ is the blending-function matrix, and $\mathbf{B}_n$ is a matrix of geometric coefficients (for Bezier curves, the polygon vertex points). The subscript n is necessary to distinguish the degree of the Bezier curve intended. Notice the distinctions (semantic and notational) between $B_{i,n}$ and $\mathbf{B}_n$. The former denotes a set of functions, and the latter is a matrix of points. This is similar to the pc formulation. We can easily enhance the similarity by developing Eq. 2.236 in even more familiar terms. Expanding the matrix of u functions yields

$$\mathbf{p}(u) = [(1 - 3u + 3u^2 - u^3) \quad (3u - 6u^2 + 3u^3)$$
$$(3u^2 - 3u^3) \quad u^3][\mathbf{p}_0 \quad \mathbf{p}_1 \quad \mathbf{p}_2 \quad \mathbf{p}_3]^T \qquad (2.238)$$

Rewrite this as

$$\mathbf{p}(u) = [u^3 \quad u^2 \quad u \quad 1] \begin{bmatrix} -1 & 3 & -3 & 1 \\ 3 & -6 & 3 & 0 \\ -3 & 3 & 0 & 0 \\ 1 & 0 & 0 & 0 \end{bmatrix} \begin{bmatrix} \mathbf{p}_0 \\ \mathbf{p}_1 \\ \mathbf{p}_2 \\ \mathbf{p}_3 \end{bmatrix} \qquad (2.239)$$

Now generalize and contract this expression to

$$p(u) = UM_nB_n \qquad (2.240)$$

This is remarkably like Eq. 2.33 for pc curves. Obviously, the composition of these matrices varies with the number of vertices $(n + 1)$. So, for $n = 4$,

$$U = [u^4 \quad u^3 \quad u^2 \quad u \quad 1] \qquad (2.241)$$

$$M_4 = \begin{bmatrix} 1 & -4 & 6 & -4 & 1 \\ -4 & -12 & -12 & 4 & 0 \\ 6 & -12 & 6 & 0 & 0 \\ -4 & 4 & 0 & 0 & 0 \\ 1 & 0 & 0 & 0 & 0 \end{bmatrix} \qquad (2.242)$$

$$B_4 = [p_0 \quad p_1 \quad p_2 \quad p_3 \quad p_4]^T \qquad (2.243)$$

At this point, it may occur to you that we can also represent a pc curve with a sequence of four vertices. Figure 2.54 presents the rather simple graphic details. The vertices are denoted sequentially as p_0, p_1, p_2, and p_3; obviously, p_0 and p_3 are the curve end points. The two interior points contribute to the required tangent vectors in the following way:

$$p^u(0) = k(p_1 - p_0) \qquad (2.244)$$

and

$$p^u(1) = k(p_3 - p_2) \qquad (2.245)$$

where k is an arbitrary scale factor introduced to control the scale of the polygon. The matrix of geometric coefficients is simply

$$B = [p_0 \quad p_3 \quad k(p_1 - p_0) \quad k(p_3 - p_2)]^T \qquad (2.246)$$

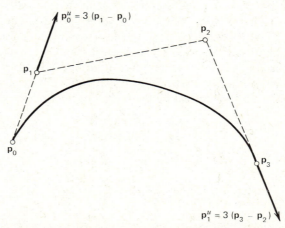

FIGURE 2.54 pc equivalent of a cubic Bezier curve.

For $k = 3$, the pc curve is identical to the cubic Bezier curve defined by $\mathbf{p}_0$, $\mathbf{p}_1$, $\mathbf{p}_2$, and $\mathbf{p}_3$, as we will soon see.

Vary the position of $\mathbf{p}_1$ or $\mathbf{p}_2$ and the curve's shape varies. The only real difference between this and the Bezier formulation is that the pc is limited to four vertices. Otherwise, this technique offers many of the same advantages when we construct continuous composite curves.

Let us explore more deeply the continuity criteria discussed earlier. First, differentiate Eq. 2.236 and rearrange terms to obtain

$$\mathbf{p}''(u) = 3[(1 - u)^2 \quad 2u(1 - u) \quad u^2][(\mathbf{p}_1 - \mathbf{p}_0) \quad (\mathbf{p}_2 - \mathbf{p}_1) \quad (\mathbf{p}_3 - \mathbf{p}_2)]^T$$

$$(2.247)$$

Next, evaluate this function at $u = 0$ and $u = 1$, obtaining

$$\mathbf{p}''(0) = 3(\mathbf{p}_1 - \mathbf{p}_0) \tag{2.248}$$

and

$$\mathbf{p}''(1) = 3(\mathbf{p}_3 - \mathbf{p}_2) \tag{2.49}$$

This demonstrates that the tangent at $u = 0$ is determined by the line between $\mathbf{p}_0$ and $\mathbf{p}_1$, and at $u = 1$ by the line between $\mathbf{p}_2$ and $\mathbf{p}_3$. This reveals that certain continuity conditions between adjacent Bezier segments of a composite curve are relatively simple to specify. Define one segment by vertices $\mathbf{p}_i$ and the adjacent segment by vertices $\mathbf{q}_i$. C^1 continuity is established by making $\mathbf{p}_{n-1}$, $\mathbf{p}_n = \mathbf{q}_0$, and $\mathbf{q}_1$ colinear; Fig. 2.55 illustrates this.

A condition for C^2 continuity at a joint between curves is that the five vertices $\mathbf{p}_{n-2}$, $\mathbf{p}_{n-1}$, $\mathbf{p}_n = \mathbf{q}_0$, $\mathbf{q}_1$, and $\mathbf{q}_2$ must be coplanar. To achieve this might require us to increase the number of vertices in one or more segments.

The equations for the curvature at each end of a cubic Bezier curve, $\kappa(0)$ and $\kappa(1)$, follow without derivation. We will discuss curvature in more detail in Chapter 5, Section 1.8.

$$\kappa(0) = \frac{2|(\mathbf{p}_1 - \mathbf{p}_0) \times (\mathbf{p}_2 - \mathbf{p}_1)|}{3|\mathbf{p}_1 - \mathbf{p}_0|^3} \tag{2.250}$$

and

$$\kappa(1) = \frac{2|(\mathbf{p}_2 - \mathbf{p}_1) \times (\mathbf{p}_3 - \mathbf{p}_2)|}{3|\mathbf{p}_3 - \mathbf{p}_2|^3} \tag{2.251}$$

We can easily generalize these equations for Bezier curves with $n > 3$.

As Faux and Pratt (1980) point out, the advantage of higher order Bezier curves is that we can achieve correspondingly higher orders of continuity between segments of compound curves. For example, a fifth-order, or quintic, Bezier polynomial permits us to specify end points, end tangents, and curvature at each end. But how do higher degree polynomial functions affect the computation of geometric properties and relationships? If we have quintic curves and the related biquintic surfaces, for example, how much more complex are

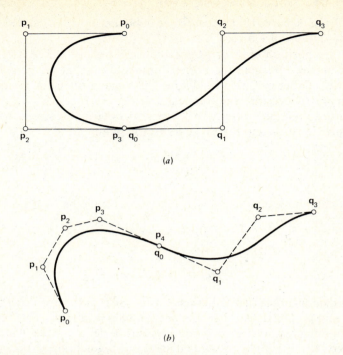

FIGURE 2.55 Composite Bezier curves

intersection computations for these than for, say, cubic curves and bicubic surfaces? We will address these questions in later chapters.

When we use the Bezier curve as a modeling technique in a CAD system, our objective is usually not to try to approximate the Bezier polygon. The polygon serves two functions: It establishes the initial shape of the curve and then furnishes a framework for altering the curve.

Let us review the properties of a Bezier curve that make it an unusually effective interactive design tool. First, the curve has end points in common with the polygon (the other vertices are usually not on the curve). Second, the slope of the tangent vectors at the end points equals the slope of the first and last segments of the polygon. Third, the curve lies entirely within the convex figure defined by the extreme points of the polygon (often called the convex hull) and generally mimics the gross features of the polygon. Fourth, Bezier curves are variation-diminishing. This means that they never oscillate wildly away from their defining control points. Fifth, when we compare Bezier curves to conventional polynomials or splines, we find that Bezier curves do not require us to input slopes, just data points. Finally, the parametric formulation allows a curve to represent multiple-valued shapes. In fact, as we have seen, if the first and last points concide (that is, if $\mathbf{p}_0 = \mathbf{p}_n$), then the curve is closed.

Bezier curves are special cases of the more general B-spline curves that we consider next.

EXERCISES

1. Show mathematically that a straight line results whenever the vertices of a characteristic polygon are colinear. (*Hint:* $\sum_{i=0}^{n} B_{i,n} = 1$.)

2. Show that for a Bezier curve with $n = 3$, a parabolic arc with end points $\mathbf{p}_0$ and $\mathbf{p}_3$ results, if the end tangents intersect at $\mathbf{p}^*$ such that $\mathbf{p}_1 = (\mathbf{p}_0 + 2\mathbf{p}^*)/3$ and $\mathbf{p}_2 = (\mathbf{p}_3 + 2\mathbf{p}^*)/3$.

3. Evaluate the first, second, and third derivatives of a quintic Bezier curve at $u = 0$ and $u = 1$. What vertex points control or contribute to these derivatives? Is the result in keeping with Bezier's initial requirements for his curve functions?

4. Generalize the curvature Eqs. 2.250 and 2.251 for Bezier curves with $n > 3$.

5. Write a procedure to compute the coordinates of a point on a Bezier curve. Denote this as **BZCRV**(N, CI, IFLG, U, P), where

  N is the input number of control points, where $N \in [3:6]$;

  CI(N 3) is the input array of coordinates of the control points;

  IFLG is an input flag specifying that the curve is open (IFLG = 0) or closed (IFLG = 1);

  U is the input value of the parametric variable at the point; and

  P(3) is the output array of coordinates of the point on the curve.

The IFLG variable is not strictly necessary. Describe conditions whereby it can be omitted. Note that here $N = n + 1$. See Eq. 2.229.

6. Modify **BZCRV** to eliminate the restriction on the number of control points as specified in Ex. 5.

7. Develop a formula for a cubic Bezier curve that best approximates a circular arc, given the end points of the arc and its center point. Assume the Bezier curve must pass through the end points and its tangent vectors at these points must be perpendicular to their respective radius vectors. Show how the approximation varies as the angle subtended by the arc increases.

15 *B*-SPLINE CURVES

Most curve-defining techniques do not provide for local control of shape. Consequently, local changes (for example, a small change in the position of a point on a spline curve or of a vertex of a characteristic polygon of a Bezier curve) tend to be strongly propagated throughout the entire curve. This is sometimes described as a **global** propagation of change. The *B*-spline curve avoids this problem by using a special set of blending functions that has only local influence and depends on only a few neighboring control points.

B-spline curves are similar to Bezier curves in that a set of blending functions combines the effects of $n + 1$ control points $\mathbf{p}_i$ given by

$$\mathbf{p}(u) = \sum_{i=0}^{n} \mathbf{p}_i N_{i,k}(u) \tag{2.252}$$

Compare Eq. 2.252 to Eq. 2.229 for Bezier curves. The most important difference is the way the blending functions $N_{i,k}(u)$ are formulated. For Bezier curves, the number of control points determines the degree of the blending function polynomials. For B-spline curves, the degree of these polynomials is specially controlled by a parameter k and usually independent of the number of control points, except as limited by Eq. 2.257. The B-spline blending functions are defined recursively by the following expressions:

$$N_{i,1}(u) = 1 \quad \text{if } t_i \leq u < t_{i+1}$$
$$= 0 \quad \text{otherwise} \tag{2.253}$$

and

$$N_{i,k}(u) = \frac{(u - t_i)N_{i,k-1}(u)}{t_{i+k-1} - t_i} + \frac{(t_{i+k} - u)N_{i+1,k-1}(u)}{t_{i+k} - t_{i+1}} \tag{2.254}$$

where k controls the degree $(k - 1)$ of the resulting polynomial in u and thus also controls the continuity of the curve. The t_i are called **knot values**. They relate the parametric variable u to the $\mathbf{p}_i$ control points. For an open curve, the t_i are

$$t_i = 0 \qquad \text{if } i < k$$
$$t_i = i - k + 1 \quad \text{if } k \leq i \leq n \tag{2.255}$$
$$t_i = n - k + 2 \quad \text{if } i > n$$

with

$$0 \leq i \leq n + k \tag{2.256}$$

The range of the parametric variable u is

$$0 \leq u \leq n - k + 2 \tag{2.257}$$

Since the denominators in Eq. 2.254 can be zero, we define $0/0 = 1$.

Let us see how these equations compute the blending functions $N_{i,1}(u)$, $N_{i,2}(u)$, and $N_{i,3}(u)$ for $k = 1, k = 2$, and $k = 3$, given six control points $(n = 5)$. For the $N_{i,1}(u)$ blending functions with $n = 5$ and $k = 1$, we find that

$$0 \leq u \leq 6 \tag{2.258}$$

and

$$
\begin{aligned}
t_0 &= 0 & t_4 &= 4 \\
t_1 &= 1 & t_5 &= 5 \\
t_2 &= 2 & t_6 &= 6 \\
t_3 &= 3
\end{aligned}
\tag{2.259}
$$

From Eq. 2.253 we obtain directly

$$
\begin{aligned}
N_{0,1}(u) &= 1 && \text{if } 0 \le u < 1 \\
&= 0 && \text{otherwise} \\[4pt]
N_{1,1}(u) &= 1 && \text{if } 1 \le u < 2 \\
&= 0 && \text{otherwise} \\[4pt]
N_{2,1}(u) &= 1 && \text{if } 2 \le u < 3 \\
&= 0 && \text{otherwise} \\[4pt]
N_{3,1}(u) &= 1 && \text{if } 3 \le u < 4 \\
&= 0 && \text{otherwise} \\[4pt]
N_{4,1}(u) &= 1 && \text{if } 4 \le u < 5 \\
&= 0 && \text{otherwise} \\[4pt]
N_{5,1}(u) &= 1 && \text{if } 5 \le u \le 6 \\
&= 0 && \text{otherwise}
\end{aligned}
\tag{2.260}
$$

These six blending functions are plotted in Fig. 2.56. Notice that the blending functions are obviously identical in form and each is limited to a different, successive unit interval of the parametric variable u. If we apply these blending functions to any set of six control points using Eq. 2.252, we discover that $\mathbf{p}(u)$ must be given by a different equation for each unit interval in u, so that

$$
\begin{aligned}
\mathbf{p}(u) &= \mathbf{p}_0 && \text{for } 0 \le u < 1 \\[4pt]
\mathbf{p}(u) &= \mathbf{p}_1 && \text{for } 1 \le u < 2 \\[4pt]
\mathbf{p}(u) &= \mathbf{p}_2 && \text{for } 2 \le u < 3 \\[4pt]
\mathbf{p}(u) &= \mathbf{p}_3 && \text{for } 3 \le u < 4 \\[4pt]
\mathbf{p}(u) &= \mathbf{p}_4 && \text{for } 4 \le u < 5 \\[4pt]
\mathbf{p}(u) &= \mathbf{p}_5 && \text{for } 5 \le u < 6
\end{aligned}
\tag{2.261}
$$

The resulting "curve" is clearly not a curve, but merely the initial set of control points. (We could also interpret Eq. 2.261 as six disjoint curve segments of zero length, each segment concentrated at a control point.)

Next, for the $N_{i,2}(u)$ blending functions with $n = 5$ and $k = 2$, we find that

$$
0 \le u \le 5 \tag{2.262}
$$

and

$$
\begin{aligned}
t_0 &= 0 & t_4 &= 3 \\
t_1 &= 0 & t_5 &= 4 \\
t_2 &= 1 & t_6 &= 5 \\
t_3 &= 2 & t_7 &= 5
\end{aligned}
\tag{2.263}
$$

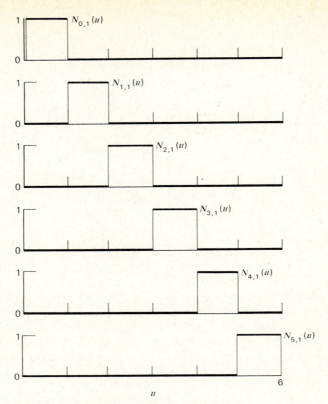

FIGURE 2.56 Blending functions for $n = 5, k = 1$.

Equation 2.254 indicates that the $N_{i,2}(u)$ depend on the $N_{i,1}(u)$, so before we can compute the $N_{i,2}(u)$, we must first compute another set of $N_{i,1}(u)$ corresponding to the knot values in Eq. 2.263. This computation produces

$$
\begin{aligned}
N_{0,1}(u) &= 1 && \text{if } u = 0 \\
&= 0 && \text{otherwise} \\
N_{1,1}(u) &= 1 && 0 \le u < 1 \\
&= 0 && \text{otherwise} \\
N_{2,1}(u) &= 1 && 1 \le u < 2 \\
&= 0 && \text{otherwise} \\
N_{3,1}(u) &= 1 && 2 \le u < 3 \\
&= 0 && \text{otherwise} \\
N_{4,1}(u) &= 1 && 3 \le u < 4 \\
&= 0 && \text{otherwise} \\
N_{5,1}(u) &= 1 && 4 \le u < 5 \\
&= 0 && \text{otherwise}
\end{aligned}
\tag{2.264}
$$

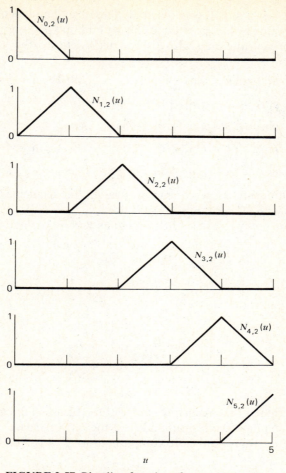

FIGURE 2.57 Blending functions for $n = 5, k = 2$.

Substitute this set as required into Eq. 2.254 to obtain

$$N_{0,2}(u) = (1 - u)N_{1,1}(u)$$
$$N_{1,2}(u) = uN_{1,1}(u) + (2 - u)N_{2,1}(u)$$
$$N_{2,2}(u) = (u - 1)N_{2,1}(u) + (3 - u)N_{3,1}(u)$$
$$N_{3,2}(u) = (u - 2)N_{3,1}(u) + (4 - u)N_{4,1}(u) \tag{2.265}$$
$$N_{4,2}(u) = (u - 3)N_{4,1}(u) + (5 - u)N_{5,1}(u)$$
$$N_{5,2}(u) = (u - 4)N_{5,1}(u)$$

Notice that the $N_{i,1}(u)$ functions act like switches, taking on a value of 1 or 0, depending on the value of u. These blending functions are plotted in Fig. 2.57; they are identical in size and shape, except at the end segments spanning

$0 \leq u < 1$ and $4 \leq u \leq 5$. If we now apply these blending functions to any set of six control points by means of Eq. 2.252, we find that $\mathbf{p}(u)$ is given by a different equation for each unit interval in u, through the action of the $N_{i,1}(u)$ switches, so that

$$
\begin{aligned}
\mathbf{p}(u) &= (1 - u)\mathbf{p}_0 + u\mathbf{p}_1 && \text{for } 0 \leq u < 1 \\
\mathbf{p}(u) &= (2 - u)\mathbf{p}_1 + (u - 1)\mathbf{p}_2 && \text{for } 1 \leq u < 2 \\
\mathbf{p}(u) &= (3 - u)\mathbf{p}_2 + (u - 2)\mathbf{p}_3 && \text{for } 2 \leq u < 3 \\
\mathbf{p}(u) &= (4 - u)\mathbf{p}_3 + (u - 3)\mathbf{p}_4 && \text{for } 3 \leq u < 4 \\
\mathbf{p}(u) &= (5 - u)\mathbf{p}_4 + (u - 4)\mathbf{p}_5 && \text{for } 4 \leq u \leq 5
\end{aligned}
\tag{2.266}
$$

Here, the resulting curve is a sequence of straight line segments connecting the control points with C^0 continuity at the joints.

Finally, for the $N_{i,3}(u)$ blending functions with $n = 5$ and $k = 3$, we find that

$$
0 \leq u \leq 4 \tag{2.267}
$$

and

$$
\begin{aligned}
t_0 &= 0 & t_5 &= 3 \\
t_1 &= 0 & t_6 &= 4 \\
t_2 &= 0 & t_7 &= 4 \\
t_3 &= 1 & t_8 &= 4 \\
t_4 &= 2
\end{aligned}
\tag{2.268}
$$

Again, because of the recursive nature of the blending function equations, we must first compute the $N_{i,1}(u)$ and then the $N_{i,2}(u)$, strictly in that order, and the new knot values t_i, before computing the $N_{i,3}(u)$. For the $N_{i,1}(u)$, we obtain

$$
\begin{aligned}
N_{0,1}(u) &= 1 && \text{for } u = 0 \\
&= 0 && \text{otherwise} \\
N_{1,1}(u) &= 1 && \text{for } u = 0 \\
&= 0 && \text{otherwise} \\
N_{2,1}(u) &= 1 && \text{for } 0 \leq u < 1 \\
&= 0 && \text{otherwise} \\
N_{3,1}(u) &= 1 && \text{for } 1 \leq u < 2 \\
&= 0 && \text{otherwise} \\
N_{4,1}(u) &= 1 && \text{for } 2 \leq u < 3 \\
&= 0 && \text{otherwise} \\
N_{5,1}(u) &= 1 && \text{for } 3 \leq u \leq 4 \\
&= 0 && \text{otherwise}
\end{aligned}
\tag{2.269}
$$

Substitute these equations as required into Eq. 2.254 to obtain

$$N_{0,2}(u) = 0$$
$$N_{1,2}(u) = (1 - u)N_{2,1}(u)$$
$$N_{2,2}(u) = uN_{2,1}(u) + (2 - u)N_{3,1}(u)$$
$$N_{3,2}(u) = (u - 1)N_{3,1}(u) + (3 - u)N_{4,1}(u)$$
$$N_{4,2}(u) = (u - 2)N_{4,1}(u) + (4 - u)N_{5,1}(u)$$
$$N_{5,2}(u) = (u - 3)N_{5,1}(u)$$

(2.270)

Notice how the new knot values cause Eq. 2.70 to differ from the $N_{i,2}$ in Eq. 2.265. Now substitute from Eqs. 2.270 into Eq. 2.254 to obtain the $N_{i,3}(u)$

$$N_{0,3}(u) = (1 - u)^2 N_{2,1}(u)$$
$$N_{1,3}(u) = \tfrac{1}{2}u(4 - 3u)N_{2,1}(u) + \tfrac{1}{2}(2 - u)^2 N_{3,1}(u)$$
$$N_{2,3}(u) = \tfrac{1}{2}u^2 N_{2,1}(u) + \tfrac{1}{2}(-2u^2 + 6u - 3)N_{3,1}(u)$$
$$\qquad + \tfrac{1}{2}(3 - u)^2 N_{4,1}(u)$$
$$N_{3,3}(u) = \tfrac{1}{2}(u - 1)^2 N_{3,1}(u) + \tfrac{1}{2}(-2u^2 + 10u - 11)N_{4,1}(u)$$
$$\qquad + \tfrac{1}{2}(4 - u)^2 N_{5,1}(u)$$
$$N_{4,3}(u) = \tfrac{1}{2}(u - 2)^2 N_{4,1}(u) + \tfrac{1}{2}(-3u^2 + 20u - 32)N_{5,1}(u)$$
$$N_{5,3}(u) = (u - 3)^2 N_{5,1}(u)$$

(2.271)

Again, the $N_{i,1}(u)$ functions act like switches, turning on and off the terms that they control. These blending functions are plotted in Fig. 2.58. Notice the symmetries and congruences of the shapes of these functions; we will explore this phenomenon in more detail a little later. If we now apply these blending functions to any set of six control points by means of Eq. 2.252, we find that $\mathbf{p}(u)$ is again given by a different equation for each unit interval in u through the action of the $N_{i,1}(u)$ switches, so that

$$\mathbf{p}_1(u) = (1 - u)^2 \mathbf{p}_0 + \tfrac{1}{2}u(4 - 3u)\mathbf{p}_1 + \tfrac{1}{2}u^2 \mathbf{p}_2 \qquad \text{for } 0 \leq u < 1$$
$$\mathbf{p}_2(u) = \tfrac{1}{2}(2 - u)^2 \mathbf{p}_1 + \tfrac{1}{2}(-2u^2 + 6u - 3)\mathbf{p}_2 + \tfrac{1}{2}(u - 1)^2 \mathbf{p}_3 \qquad \text{for } 1 \leq u < 2$$
$$\mathbf{p}_3(u) = \tfrac{1}{2}(3 - u)^2 \mathbf{p}_2 + \tfrac{1}{2}(-2u^2 + 10u - 11)\mathbf{p}_3 + \tfrac{1}{2}(u - 2)^2 \mathbf{p}_4 \qquad \text{for } 2 \leq u < 3$$
$$\mathbf{p}_4(u) = \tfrac{1}{2}(4 - u)^2 \mathbf{p}_3 + \tfrac{1}{2}(-3u^2 + 20u - 32)\mathbf{p}_4 + (u - 3)^2 \mathbf{p}_5 \qquad \text{for } 3 \leq u < 4$$

(2.272)

The resulting curve is a composite sequence of four curve segments connected with C^1 continuity; an example is shown in Fig. 2.59. Notice that the curve passes through only the first and last points, $\mathbf{p}_0$ and $\mathbf{p}_5$, and it is tangent to $\mathbf{p}_1 - \mathbf{p}_0$ and $\mathbf{p}_5 - \mathbf{p}_4$ at these same points. Also, notice that this curve is tangent to each successive side of the characteristic polygon (only for $k = 3$ curves). This tangency occurs at the joints between curve segments (that is, at integral values of u). These joints are indicated by tick marks. In the figure, point $\mathbf{p}_1$ is moved to $\mathbf{p}_1^*$, and the effect on the curve is plotted. This local change affects only two segments of the curve; it is also evident in Eqs. 2.272, because $\mathbf{p}_1$ occurs in only the first and second equations.

From these same equations, we infer that only three control points influence each curve segment. Conversely, a control point can influence the shape of at most only three curve segments. Similar observations apply to $k = 1$ and $k = 2$ curves. In fact, we can generalize this observation: Each segment of a B-spline curve is influenced by only k control points, and conversely each control point influences only k curve segments.

What happens if we increase the number of control points? To answer this, we must review Figs. 2.56–2.58. In Fig. 2.56, we see no evidence whatsoever

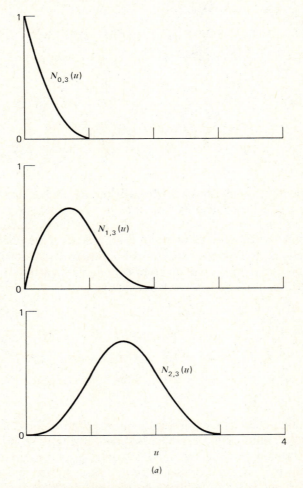

(a)

FIGURE 2.58 Blending functions for $n = 5, k = 3$.

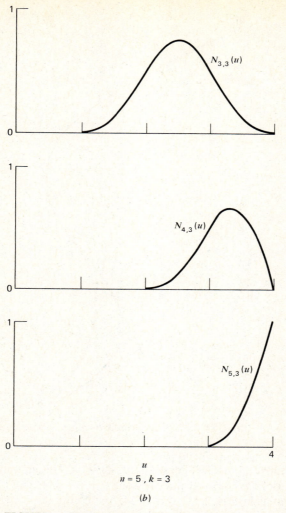

$n = 5, k = 3$

(b)

FIGURE 2.58 (*Continued*) Blending functions for $n = 5, k = 3$.

to indicate an influence on the shape of the blending-function curves that we can attribute to the number of control points. In both Figs. 2.57 and 2.58, we also correctly infer that, except for the blending functions influenced by the end points or those very near them (depending on k), the "interior" blending functions are independent of n. For example, if $n = 7, k = 3$, we would find that $N_{2,3}(u) = N_{3,3}(u) = N_{4,3}(u) = N_{5,3}(u)$.

Let us pursue this idea of blending-function independence of n and at the same time try to develop a more convenient and more familiar matrix notation. We will do this for $k = 3$. First, compute $N_{i,3}(u)$, $N_{i+1,3}(u)$, and $N_{i+2,3}(u)$. The reason for selecting these three will become clear as our investigation

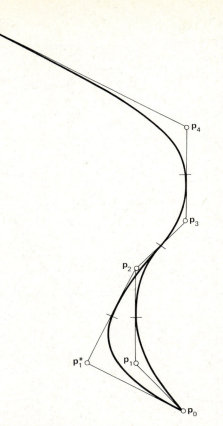

FIGURE 2.59 Non-periodic B-spline curves: $n = 5, k = 3$.

progresses. Choose an interval in i so that $k \leq i \leq n$. This assumption simplifies the calculation of the t_i knot values so that $t_i = i - k + 1 = i - 2$. Using the recursive formulas in Eqs. 2.553 and 2.554, first compute

$$
\begin{aligned}
N_{i,1}(u) &= 1 && \text{for } i - 2 \leq u < i - 1 \\
&= 0 && \text{otherwise} \\[4pt]
N_{i+1,1}(u) &= 1 && \text{for } i - 1 \leq u < i \\
&= 0 && \text{otherwise} \\[4pt]
N_{i+2,1}(u) &= 1 && \text{for } i \leq u < i + 1 \\
&= 0 && \text{otherwise} \\[4pt]
N_{i+3,1}(u) &= 1 && \text{for } i + 1 \leq u < i + 2 \\
&= 0 && \text{otherwise} \\[4pt]
N_{i+4,1}(u) &= 1 && \text{for } i + 2 \leq u < i + 3 \\
&= 0 && \text{otherwise}
\end{aligned}
\tag{2.273}
$$

Next, compute

$$N_{i,2}(u) = (u - i + 2)N_{i,1}(u) + (i - u)N_{i+1,1}(u)$$

$$N_{i+1,2}(u) = (u - i + 1)N_{i+1,1}(u) + (i + 1 - u)N_{i+2,1}(u)$$

$$N_{i+2,2}(u) = (u - i)N_{i+2,1}(u) + (i + 2 - u)N_{i+3,1}(u)$$

$$N_{i+3,2}(u) = (u - i - 1)N_{i+3,1}(u) + (i + 3 - u)N_{i+4,1}(u)$$

(2.274)

And finally compute

$$N_{i,3}(u) = \tfrac{1}{2}[(u - i + 2)^2 N_{i,1}(u) + (u - i + 2)(i - u)N_{i+1,1}(u)$$
$$+ (i + 1 - u)(u - i + 1)N_{i+1,1}(u) + (i + 1 - u)^2 N_{i+2,1}(u)]$$

$$N_{i+1,3}(u) = \tfrac{1}{2}[(u - i + 1)^2 N_{i+1,1}(u)$$
$$+ (u - i + 1)(i + 1 - u)N_{i+2,1}(u)$$
$$+ (i + 2 - u)(u - i)N_{i+2,1}(u) + (i + 2 - u)^2 N_{i+3,1}(u)]$$

$$N_{i+2,3}(u) = \tfrac{1}{2}[(u - i)^2 N_{i+2,1}(u) + (u - i)(i + 2 - u)N_{i+3,1}(u)$$
$$+ (i + 3 - u)(u - i - 1)N_{i+3,1}(u)$$
$$+ (i + 3 - u)^2 N_{i+4,1}(u)]$$

(2.275)

Now, we have enough information to find an expression for $\mathbf{p}(u)$ over an arbitrary segment of the curve, say, for the interval $i \le u < i + 1$. We again use the $N_{i,1}(u)$ as switches. Only $N_{i+2,1}(u) = 1$ in this interval. If we collect those terms from Eqs. 2.275 that $N_{i+2,1}(u)$ switches on, then we obtain from Eq. 2.252

$$\mathbf{p}(u) = \tfrac{1}{2}(i + 1 - u)^2 \mathbf{p}_i + \tfrac{1}{2}[(u - i + 1)(i + 1 - u)$$
$$+ (i + 2 - u)(u - i)]\mathbf{p}_{i+1} + \tfrac{1}{2}(u - i)^2 \mathbf{p}_{i+2}$$

(2.276)

There are computational advantages to reparametrizing the interval so that $0 \le u < 1$ and then identifying the interval in some way, for example, by subscripting $\mathbf{p}(u)$ as $\mathbf{p}_i(u)$ for the ith interval. To reparametrize Eq. 2.276, replace u by $u + i$, so that

$$\mathbf{p}_i(u) = \tfrac{1}{2}[(1 - u)^2 \mathbf{p}_i + (-2u^2 + 2u + 1)\mathbf{p}_{i+1} + u^2 \mathbf{p}_{i+2}]$$

(2.277)

where $0 \le u < 1$. Note that as long as we maintain the full functional notation, we will not confuse the function $\mathbf{p}_i(u)$ with the control point $\mathbf{p}_i$.

We can now easily rewrite Eq. 2.277 in matrix notation, replacing i by $i - 1$ (i now denotes the curve segment number). So for a *B*-spline with $k = 3$, we obtain

$$\mathbf{p}_i(u) = \tfrac{1}{2}[u^2 \quad u \quad 1]\begin{bmatrix} 1 & -2 & 1 \\ -2 & 2 & 0 \\ 1 & 1 & 0 \end{bmatrix}\begin{bmatrix} \mathbf{p}_{i-1} \\ \mathbf{p}_i \\ \mathbf{p}_{i+1} \end{bmatrix} \qquad \text{for } i \in [1{:}n - 1] \quad (2.278)$$

The analogous form for cubic B-splines ($k = 4$) is

$$\mathbf{p}_i(u) = \tfrac{1}{6}[u^3 \quad u^2 \quad u \quad 1]\begin{bmatrix} -1 & 3 & -3 & 1 \\ 3 & -6 & 3 & 0 \\ -3 & 0 & 3 & 0 \\ 1 & 4 & 1 & 0 \end{bmatrix}\begin{bmatrix} \mathbf{p}_{i-1} \\ \mathbf{p}_i \\ \mathbf{p}_{i+1} \\ \mathbf{p}_{i+2} \end{bmatrix} \quad \text{for } i \in [1:n-2]$$

(2.279)

Use the following substitutions to simplify these two equations,

$$\mathbf{U}_3 = [u^2 \quad u \quad 1] \tag{2.280}$$

$$\mathbf{M}_3 = \frac{1}{2}\begin{bmatrix} 1 & -2 & 1 \\ -2 & 2 & 0 \\ 1 & 1 & 0 \end{bmatrix} \tag{2.281}$$

$$\mathbf{U}_4 = [u^3 \quad u^2 \quad u \quad 1] \tag{2.282}$$

and

$$\mathbf{M}_4 = \frac{1}{6}\begin{bmatrix} -1 & 3 & -3 & 1 \\ 3 & 6 & 3 & 0 \\ -3 & 0 & 3 & 0 \\ 1 & 4 & 1 & 0 \end{bmatrix} \tag{2.283}$$

so that for $k = 3$,

$$\mathbf{p}_i(u) = \mathbf{U}_3 \mathbf{M}_3 \begin{bmatrix} \mathbf{p}_{i-1} \\ \mathbf{p}_i \\ \mathbf{p}_{i+1} \end{bmatrix} \quad i \in [1:n-1] \text{ for open curves} \tag{2.284}$$

and for $k = 4$,

$$\mathbf{p}_i(u) = \mathbf{U}_4 \mathbf{M}_4 \begin{bmatrix} \mathbf{p}_{i-1} \\ \mathbf{p}_i \\ \mathbf{p}_{i+1} \\ \mathbf{p}_{i+2} \end{bmatrix} \quad i \in [1:n-2] \text{ for open curves} \tag{2.285}$$

Caution: The number of segments i is determined here for open curves. We will investigate closed B-spline curves later in this section.

Equations 2.284 and 2.285 are for specific k values. The general formulation is

$$\mathbf{p}_i(u) = \mathbf{U}_k \mathbf{M}_k \mathbf{P}_k \quad i \in [1:n+2-k] \tag{2.286}$$

where

$$\mathbf{U}_k = [u^{k-1} \quad u^{k-2} \quad \cdots \quad u \quad 1] \tag{2.287}$$

and

$$\mathbf{P}_k = [\mathbf{p}_j] \quad j \in [i-1:i+k-2] \text{ for open curves} \tag{2.288}$$

The matrix $\mathbf{M}_k$ is determined in the same way as $\mathbf{M}_3$ was earlier in this section.

In many modeling situations, we may not require the curve to pass through the end points; in those cases, we can simply apply a formulation as that in Eq. 2.278 or 2.279 to the set of control points. Let us do this for the set of control points in Fig. 2.59, first for $k = 3$ and then for $k = 4$.

For $n = 5$ and $k = 3$, $i \in [1:4]$, we find

$$\mathbf{p}_1(u) = \mathbf{U}_3 \mathbf{M}_3 \begin{bmatrix} \mathbf{p}_0 \\ \mathbf{p}_1 \\ \mathbf{p}_2 \end{bmatrix}$$

$$\mathbf{p}_2(u) = \mathbf{U}_3 \mathbf{M}_3 \begin{bmatrix} \mathbf{p}_1 \\ \mathbf{p}_2 \\ \mathbf{p}_3 \end{bmatrix}$$

$$\mathbf{p}_3(u) = \mathbf{U}_3 \mathbf{M}_3 \begin{bmatrix} \mathbf{p}_2 \\ \mathbf{p}_3 \\ \mathbf{p}_4 \end{bmatrix}$$

$$\mathbf{p}_4(u) = \mathbf{U}_3 \mathbf{M}_3 \begin{bmatrix} \mathbf{p}_3 \\ \mathbf{p}_4 \\ \mathbf{p}_5 \end{bmatrix}$$

(2.289)

For $n = 5$ and $k = 4$, $i \in [1:3]$, we find

$$\mathbf{p}_1(u) = \mathbf{U}_4 \mathbf{M}_4 \begin{bmatrix} \mathbf{p}_0 \\ \mathbf{p}_1 \\ \mathbf{p}_2 \\ \mathbf{p}_3 \end{bmatrix}$$

$$\mathbf{p}_2(u) = \mathbf{U}_4 \mathbf{M}_4 \begin{bmatrix} \mathbf{p}_1 \\ \mathbf{p}_2 \\ \mathbf{p}_3 \\ \mathbf{p}_4 \end{bmatrix}$$

(2.290)

$$\mathbf{p}_3(u) = \mathbf{U}_4 \mathbf{M}_4 \begin{bmatrix} \mathbf{p}_2 \\ \mathbf{p}_3 \\ \mathbf{p}_4 \\ \mathbf{p}_5 \end{bmatrix}$$

The *B*-spline curves in Fig. 2.60 result when we complete the remaining computations. Notice that neither the $k = 3$ curve nor the $k = 4$ curve passes through any of the control points. This method produces a **periodic** *B*-spline curve. It is periodic because the blending function repeats itself identically over successive intervals of the parametric variable. On the other hand, the curves in Fig. 2.59 are **nonperiodic** *B*-spline curves.

Study the arrays of control points in Eqs. 2.289 and 2.290. These equations show which control points influence a particular curve segment, and they show the influence of a single control point.

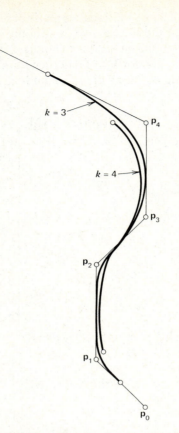

FIGURE 2.60 Periodic *B*-spline curves:
$n = 5, k = 3; n = 5, k = 4$.

Figure 2.61 presents the blending functions for $k = 2$, $k = 3$, and $k = 4$ periodic *B*-spline curves. Note that these blending functions span two, three, and four unit intervals on the parametric line, respectively. Compare these functions to those in Figs. 2.57 and 2.58.

To compare the periodic *B*-spline blending functions to the pc blending functions, we must first do some more mathematics. Define

$$\mathbf{F}_3 = \mathbf{U}_3 \mathbf{M}_3 \tag{2.291}$$

and

$$\mathbf{F}_4 = \mathbf{U}_4 \mathbf{M}_4 \tag{2.292}$$

where $\mathbf{F}_3$ and $\mathbf{F}_4$ are the blending function matrices for $k = 3$ and $k = 4$, respectively. Perform the matrix multiplications for $k = 3$

$$\mathbf{F}_3 = [F_{1,3}(u) \quad F_{2,3}(u) \quad F_{3,3}(u)] \tag{2.293}$$

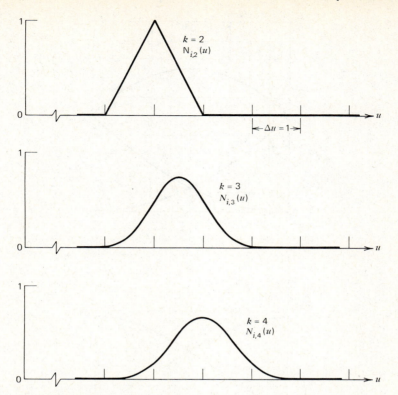

FIGURE 2.61 Periodic blending functions $N_{i,2}(u)$, $N_{i,3}(u)$, and $N_{i,4}(u)$.

where

$$F_{1,3}(u) = \tfrac{1}{2}(u^2 - 2u + 1)$$
$$F_{2,3}(u) = \tfrac{1}{2}(-2u^2 + 2u + 1) \qquad (2.294)$$
$$F_{3,3}(u) = \tfrac{1}{2}u^2$$

and for $k = 4$,

$$\mathbf{F}_4 = [F_{1,4}(u) \quad F_{2,4}(u) \quad F_{3,4}(u) \quad F_{4,4}(u)] \qquad (2.295)$$

where

$$F_{1,4}(u) = \tfrac{1}{6}(-u^3 + 3u^2 - 3u + 1)$$
$$F_{2,4}(u) = \tfrac{1}{6}(3u^3 - 6u^2 + 4)$$
$$F_{3,4}(u) = \tfrac{1}{6}(-3u^3 + 3u^2 + 3u + 1) \qquad (2.296)$$
$$F_{4,4}(u) = \tfrac{1}{6}u^3$$

Figure 2.62 plots these blending functions.

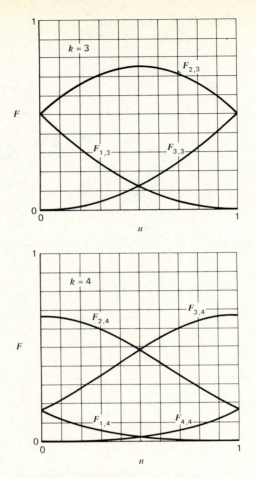

FIGURE 2.62 Periodic blending functions on the unit interval for $k = 3$ and $k = 4$.

Now rewrite Eqs. 2.284 and 2.285 as follows for $k = 3$

$$\mathbf{p}_i(u) = F_{1,3}(u)\mathbf{p}_{i-1} + F_{2,3}(u)\mathbf{p}_i + F_{3,3}(u)\mathbf{p}_{i+1} \quad i \in [1:n-1] \text{ for open curves}$$
$$(2.297)$$

and for $k = 4$

$$\mathbf{p}_i(u) = F_{1,4}(u)\mathbf{p}_{i-1} + F_{2,4}(u)\mathbf{p}_i + F_{3,4}(u)\mathbf{p}_{i+1} + F_{4,4}(u)\mathbf{p}_{i+2}$$
$$i \in [1:n-2] \text{ for open curves} \quad (2.298)$$

The periodic B-spline curves are particularly well suited to produce closed curves. Equations 2.284 and 2.285 are easily adapted by simple modifications of the segment number range and the subscripts on the control points.

For closed curves, rewrite Eqs. 2.284 and 2.285 as follows for $k = 3$

$$\mathbf{p}_i(u) = \mathbf{U}_3 \mathbf{M}_3 \begin{bmatrix} \mathbf{P}_{(i-1)\bmod(n+1)} \\ \mathbf{P}_{i\bmod(n+1)} \\ \mathbf{P}_{(i+1)\bmod(n+1)} \end{bmatrix} \qquad i \in [1{:}n+1] \text{ for closed curves} \quad (2.299)$$

and for $k = 4$

$$\mathbf{p}_i(u) = \mathbf{U}_4 \mathbf{M}_4 \begin{bmatrix} \mathbf{P}_{(i-1)\bmod(n+1)} \\ \mathbf{P}_{i\bmod(n+1)} \\ \mathbf{P}_{(i+1)\bmod(n+1)} \\ \mathbf{P}_{(i+2)\bmod(n+1)} \end{bmatrix} \qquad i \in [1{:}n+1] \text{ for closed curves} \quad (2.300)$$

where mod $(n + 1)$ is the remaindering operator (that is, 5 mod 4 = 1, 8 mod 3 = 2, 4 mod 4 = 0, and so on). The curve in Fig. 2.63 is a good example to work through. Here, $n = 5$ and $k = 4$; using these values in Eq. 2.300, we obtain

$$\mathbf{p}_i(u) = \mathbf{U}_4 \mathbf{M}_4 \begin{bmatrix} \mathbf{P}_{(i-1)\bmod 6} \\ \mathbf{P}_{i\bmod 6} \\ \mathbf{P}_{(i+1)\bmod 6} \\ \mathbf{P}_{(i+2)\bmod 6} \end{bmatrix} \qquad i \in [1{:}6] \quad (2.301)$$

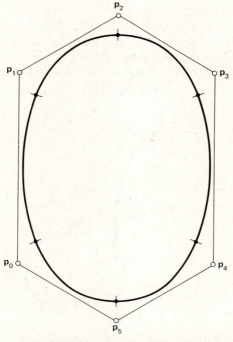

FIGURE 2.63 Closed, periodic B-spline curve with $n = 5$, $k = 4$.

Expand this equation to obtain

$$\mathbf{p}_1(u) = \mathbf{U}_4\mathbf{M}_4[\mathbf{p}_0 \quad \mathbf{p}_1 \quad \mathbf{p}_2 \quad \mathbf{p}_3]^T$$
$$\mathbf{p}_2(u) = \mathbf{U}_4\mathbf{M}_4[\mathbf{p}_1 \quad \mathbf{p}_2 \quad \mathbf{p}_3 \quad \mathbf{p}_4]^T$$
$$\mathbf{p}_3(u) = \mathbf{U}_4\mathbf{M}_4[\mathbf{p}_2 \quad \mathbf{p}_3 \quad \mathbf{p}_4 \quad \mathbf{p}_5]^T$$
$$\mathbf{p}_4(u) = \mathbf{U}_4\mathbf{M}_4[\mathbf{p}_3 \quad \mathbf{p}_4 \quad \mathbf{p}_5 \quad \mathbf{p}_0]^T \qquad (2.302)$$
$$\mathbf{p}_5(u) = \mathbf{U}_4\mathbf{M}_4[\mathbf{p}_4 \quad \mathbf{p}_5 \quad \mathbf{p}_0 \quad \mathbf{p}_1]^T$$
$$\mathbf{p}_6(u) = \mathbf{U}_4\mathbf{M}_4[\mathbf{p}_5 \quad \mathbf{p}_0 \quad \mathbf{p}_1 \quad \mathbf{p}_2]^T$$

We would then evaluate each of these curve segments for some sequence of u values and, of course, specific control-point coordinates to produce the plot of the resulting B-spline curve. Notice the subscript sequences. Figure 2.64 is another example of a closed curve where the same conditions hold. It is clearly possible to generate a self-intersecting curve.

In Fig. 2.65, moving $\mathbf{p}_2$ to $\mathbf{p}_2^*$ produces a more elongated closed curve. Since only four control points define this curve, the perturbation of $\mathbf{p}_2$ affects the entire curve, although the most drastic change occurs near $\mathbf{p}_2^*$, while the original curve is only slightly disturbed near $\mathbf{p}_0$.

We can define B-spline curves with multiply-coincident control points. The effect obtained is similar to that for Bezier curves: The curve's shape is pulled in closer to the multiple control point. The example in Fig. 2.66 shows the effect of one, two, and three control points at $\mathbf{p}_4$ (that is, at location $\mathbf{p}_4$, we have $\mathbf{p}_4$ alone; $\mathbf{p}_4$ and $\mathbf{p}_5$; or $\mathbf{p}_4$, $\mathbf{p}_5$, and $\mathbf{p}_6$, where $\mathbf{p}_4$, $\mathbf{p}_5$, and $\mathbf{p}_6$ have identical coordinate values). This makes it possible to define B-spline curves with sharp corners, an advantage in many geometric-modeling situations.

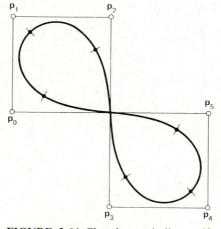

FIGURE 2.64 Closed, periodic, self-intersecting B-spline curve with $n = 5$, $k = 4$.

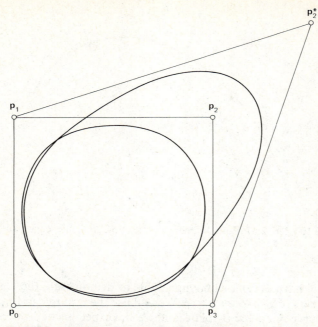

FIGURE 2.65 Closed, periodic *B*-spline curves with $n = 3$, $k = 4$.

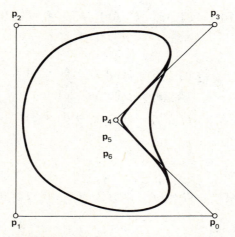

FIGURE 2.66 Closed, periodic *B*-spline curves with multiply-coincident control points and $k = 4$.

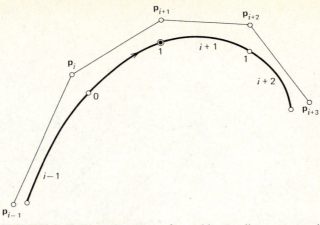

FIGURE 2.67 Four segments of a cubic *B*-spline curve and $k = 4$.

We now have developed enough tools to demonstrate the continuity between segments of a *B*-spline curve. We will use the example in Fig. 2.67. Since $k = 4$, the curve is a cubic *B*-spline, and we can expect second-order continuity throughout the curve. Let us examine the first- and second-derivative continuity of this curve at point $u = 1$ of segment i, corresponding to point $u = 0$ of segment $(i + 1)$. First, compute $\mathbf{F}_4^u$ and $\mathbf{F}_4^{uu}$:

$$\mathbf{F}_4^u = [F_{1,4}^u(u) \quad F_{2,4}^u(u) \quad F_{3,4}^u(u) \quad F_{4,4}^u(u)] \tag{2.303}$$

where

$$
\begin{aligned}
F_{1,4}^u(u) &= \tfrac{1}{6}(-3u^2 + 6u - 3) \\
F_{2,4}^u(u) &= \tfrac{1}{6}(9u^2 - 12u) \\
F_{3,4}^u(u) &= \tfrac{1}{6}(-9u^2 + 6u + 3) \\
F_{4,4}^u(u) &= \tfrac{1}{2}u^2
\end{aligned}
\tag{2.304}
$$

and

$$\mathbf{F}_4^{uu} = [F_{1,4}^{uu}(u) \quad F_{2,4}^{uu}(u) \quad F_{3,4}^{uu}(u) \quad F_{4,4}^{uu}(u)] \tag{2.305}$$

where

$$
\begin{aligned}
F_{1,4}^{uu}(u) &= -u + 1 \\
F_{2,4}^{uu}(u) &= 3u - 2 \\
F_{3,4}^{uu}(u) &= -3u + 1 \\
F_{4,4}^{uu}(u) &= u
\end{aligned}
\tag{2.306}
$$

Now, we can write

$$\mathbf{p}_i^u(u) = F_{1,4}^u(u)\mathbf{p}_{i-1} + F_{2,4}^u(u)\mathbf{p}_i$$
$$+ F_{3,4}^u(u)\mathbf{p}_{i+1} + F_{4,4}^u(u)\mathbf{p}_{i+2} \tag{2.307}$$

and

$$\mathbf{p}_i^{uu}(u) = F_{1,4}^{uu}(u)\mathbf{p}_{i-1} + F_{2,4}^{uu}(u)\mathbf{p}_i$$
$$+ F_{3,4}^{uu}(u)\mathbf{p}_{i+1} + F_{4,4}^{uu}(u)\mathbf{p}_{i+2} \tag{2.308}$$

Next, evaluate Eq. 2.298 for $\mathbf{p}_i(1)$ and $\mathbf{p}_{i+1}(0)$ to obtain

$$\mathbf{p}_i(1) = \tfrac{1}{6}(\mathbf{p}_i + 4\mathbf{p}_{i+1} + \mathbf{p}_{i+2}) \tag{2.309}$$

and

$$\mathbf{p}_{i+1}(0) = \tfrac{1}{6}(\mathbf{p}_i + 4\mathbf{p}_{i+1} + \mathbf{p}_{i+2}) \tag{2.310}$$

Equation 2.310 shows that $\mathbf{p}_i(1) = \mathbf{p}_{i+1}(0)$, as we would expect. Now, examine first-derivative continuity by evaluating Eq. 2.307 for $\mathbf{p}_i^u(1)$ and $\mathbf{p}_{i+1}^u(0)$. This yields

$$\mathbf{p}_i^u(1) = \tfrac{1}{2}(-\mathbf{p}_i + \mathbf{p}_{i+2}) \tag{2.311}$$

and

$$\mathbf{p}_{i+1}^u(0) = \tfrac{1}{2}(-\mathbf{p}_i + \mathbf{p}_{i+2}) \tag{2.312}$$

We see that $\mathbf{p}_i^u(0) = \mathbf{p}_{i+1}^u(1)$, demonstrating first-derivative continuity at the joint. Finally, examine second-derivative continuity by evaluating Eq. 2.308 for $\mathbf{p}_i^{uu}(1)$ and $\mathbf{p}_{i+1}^u(0)$. This yields

$$\mathbf{p}_i^{uu}(1) = \mathbf{p}_i - 2\mathbf{p}_{i+1} + \mathbf{p}_{i+2} \tag{2.313}$$

and

$$\mathbf{p}_{i+1}^{uu}(0) = \mathbf{p}_i - 2\mathbf{p}_{i+1} + \mathbf{p}_{i+2} \tag{2.314}$$

We see that $\mathbf{p}_i^{uu}(1) = \mathbf{p}_{i+1}^{uu}(0)$, demonstrating second-derivative continuity at the joint. Of course, we can easily extend this analysis for higher degree *B*-spline curves.

B-spline curves and Bezier curves have many advantages in common: Control points influence curve shape in a predictable, natural way, making them good candidates for use in an interactive environment. Both types of curve are variation diminishing, axis independent, and multivalued, and both exhibit the convex hull property. However, it is the local control of curve shape possible with *B*-splines that gives the technique an advantage over the Bezier technique, as does the ability to add control points without increasing the degree of the curve.

There is a great deal of accessible literature on the use of *B*-splines and their development. The works of some of the contributors to this literature are listed in the bibliography: Ahuja (1968), Catmull and Rom (1974), Clark (1975, 1976), Coons (1974), Cox (1971), deBoor (1972), Forrest (1972), Gordon and Riesenfeld (1974), and Riesenfeld (1972).

Barsky (1981) introduced a generalization of the uniform cubic B-spline called the Beta-spline. He uses parametric discontinuities in a way that preserves continuity of the unit tangent and curvature vectors at joints, and provides two additional parameters called bias and tension, that are independent of the control vertices, and by which the shape of a curve or surface can be manipulated. See the listings for Barsky *et al.* in the bibliography for some excellent presentations on this more advanced subject.

EXERCISES

1. For a given k value, what happens as the number of control points is decreased for open curves; for closed curves?

2. What does the symmetry of the blending function curves in Figs. 2.56–2.58 and 2.61 imply about the direction of parametrization? Support your observations and conclusions analytically.

3. Find the polynomial expressions for $N_{0,4}(u)$, $N_{1,4}(u)$, and $N_{2,4}(u)$ in a form analogous to those for $N_{i,3}(u)$, as in Eq. 2.271. Next, graph each blending function. (A suitable increment of the parametric variable u for this problem is $\Delta u = 0.1$.)

4. Show that each of the three Eqs. 2.275 produces an identical blending function curve.

5. What effects are likely when nonperiodic B-splines are used to create closed curves?

6. Derive the matrix forms for the nonperiodic B-spline curves with $k = 3$ and $k = 4$. Do this for $\mathbf{p}_1(u)$ for $k = 3$, and for $\mathbf{p}_1(u)$ for $k = 4$. Does the number of control points $n + 1$ affect the results?

7. Does the approximation of a B-spline curve to the characteristic polygon defined by its control points grow stronger or weaker as k increases? Analyze both open and closed curves (both periodic and nonperiodic forms). Demonstrate your conclusions both analytically and graphically.

8. Describe the closed periodic B-spline curve produced by $n = 2$, $k = 4$ and by $n = 1$, $k = 4$.

9. In light of the results from Ex. 8, search for degenerate or pathological conditions that might be produced when no constraints are placed on n or k or the multiply-coincident control points.

10. Derive the general matrix expression for closed periodic B-spline curves that is analogous to Eq. 2.286 for open curves.

11. Write a procedure to compute the coordinates of a point on a B-spline curve. Denote this as **BSPLNC(N, CI, K, I1, I2, U, P)**, where

N is the input number of control points. (*Note:* N $= n + 1$.)

CI(N, 3) is the input array of x, y, z coordinates of the control points;

K is the input degree of the curve. (*Note:* $K = k - 1$.)

I1 is the input flag specifying if the curve is periodic (I1 = 0) or nonperiodic (I1 = 1);

I2 is the input flag specifying if the curve is open (I2 = 0) or closed (I2 = 1);

U is the input value of the parametric variable; and

P(3) is the output array of coordinates of the point.

12. Generalize **BSPLNC** so that a series of points can be efficiently computed along the curve at specified equal increments of the parametric variable.

13. Compare the Hermite (pc), Bezier, and *B*-spline cubic functions. Can you find a function for each formulation that will produce an equivalent curve?

16 RATIONAL POLYNOMIALS

This section only briefly introduces curves defined by **rational polynomials**, that is, by the algebraic ratio of two polynomial functions. Although the concept is certainly not new [for example, see Rowin (1964), Roberts (1965), Coons (1967), and Forrest (1968)], its development has been somewhat sporadic, and as yet it has not gained so wide an acceptance and use as a geometric-modeling technique as perhaps it should and ultimately will. A complete formal derivation of the principles involved requires the use of theorems from projective geometry and homogeneous functions that are beyond the scope of this text. A simple empirical introduction is offered here, and, even so, it anticipates the use of homogeneous coordinates presented in Chapter 5, Section 5.

The functions discussed in earlier sections can only approximate conic curves and the more general quadric functions. The rational polynomials can exactly represent these curves; in fact, the ordinary polynomials are a subset of the rational polynomials. The latter are thus more general and offer more degrees of freedom for shaping a curve or surface. There is a wide range of applications, since one form can represent both conic and cubic curves.

We begin with rational parametric cubic polynomials by reinterpreting and rewriting Eq. 2.27 as follows:

$$\mathbf{p}_h = \mathbf{UA}_h \tag{2.315}$$

where the subscript h indicates the use of homogeneous coordinates (see Chapter 5, Section 5). Here, a fourth coordinate is introduced, so that

$$\mathbf{p}_h = [hx \quad hy \quad hz \quad h] \tag{2.316}$$

or

$$\mathbf{p}_h = h\mathbf{p} \tag{2.317}$$

so that

$$\mathbf{p} = [x \quad y \quad z \quad 1] \tag{2.318}$$

The normal 4×3 **A** matrix is now expanded to the 4×4 $\mathbf{A}_h$ matrix

$$\mathbf{A}_h = \begin{bmatrix} a_x & a_y & a_z & a_h \\ b_x & b_y & b_z & b_h \\ c_x & c_y & c_z & c_h \\ d_x & d_y & d_z & d_h \end{bmatrix} \tag{2.319}$$

With appropriate substitution, expand Eq. 2.315 to obtain

$$[hx \quad hy \quad hz \quad h] = [u^3 \quad u^2 \quad u \quad 1] \begin{bmatrix} a_x & a_y & a_z & a_h \\ b_x & b_y & b_z & b_h \\ c_x & c_y & c_z & c_h \\ d_x & d_y & d_z & d_h \end{bmatrix} \tag{2.320}$$

where $u \in [0, 1]$, as usual. Thus hx, hy, hz, and h are each a cubic polynomial in u. For hx, we have

$$hx = a_x u^3 + b_x u^2 + c_x u + d_x \tag{2.321}$$

with, of course, similar equations for hy, hz, and h. Next, we find that since $x = hx/h$, we obtain

$$x(u) = \frac{a_x u^3 + b_x u^2 + c_x u + d_x}{a_h u^3 + b_h u^2 + c_h u + d_h} \tag{2.322}$$

Again, there are similar functions for $y(u)$ and $z(u)$. Notice that when $a_h = b_h = c_h = 0$ and $d_h = 1$, we obtain the ordinary parametric cubic equations.

The geometric form is simply

$$\mathbf{p}_h = \mathbf{UMB}_h \tag{2.323}$$

where

$$\mathbf{B}_h = \begin{bmatrix} h_0 \mathbf{p}_0 \\ h_1 \mathbf{p}_1 \\ h_0^u \mathbf{p}_0 + h_0 \mathbf{p}_0^u \\ h_1^u \mathbf{p}_1 + h_1 \mathbf{p}_1^u \end{bmatrix} \tag{2.324}$$

The tangent vectors are obtained by differentiating $\mathbf{p}_h = h\mathbf{p}$

$$\frac{d(h\mathbf{p})}{du} = \frac{\mathbf{p}\,dh}{du} + \frac{h\,d\mathbf{p}}{du} \tag{2.325}$$

Remember, h is a function of u, so $h = h(u)$. With appropriate account taken of the extra homogeneous coordinate, the rational parametric cubic formulation can be developed in the same way as the ordinary pc form was in earlier sections of this chapter.

Tiller (1983) presents an excellent introduction to rational *B*-splines. Express Eq. 2.252 in four-dimensional homogeneous space as

$$\mathbf{p}_h(u) = \sum_{i=0}^{n} \mathbf{p}_{hi} N_{i,k}(u) \tag{2.326}$$

where $\mathbf{p}_{hi}$ are the control points in homogeneous space. The curve in three-dimensional space is, again, obtained by dividing each of the first three coordinates of a point given by Eq. 2.326 by that point's homogeneous coordinate. We express this as

$$\mathbf{p}(u) = \frac{\displaystyle\sum_{i=0}^{n} h_i \mathbf{p}_i N_{i,k}(u)}{\displaystyle\sum_{i=0}^{n} h_i N_{i,k}(u)} \tag{2.327}$$

The curve expressed by Eq. 2.327 is called a rational *B*-spline curve.

EXERCISES

1. Derive expressions for the blending functions of the rational parametric cubic curve.

2. Show that a rational parametric cubic curve can produce an exact segment of a circular arc. (A complete circle cannot be represented by a single rational curve. For this exercise, choose an arc centered at the origin and in the first quadrant of the *x*, *y* plane.)

3. Derive the vector-formulation equivalent of the graphic construction of a point on a conic curve as presented in Section 11. Show that this results in a rational quadric expression. Hint: Chapter 7, Section 1 presents useful vector expressions of two straight lines.

4. Show that by setting $h_i = 1$, Eq. 2.327 is reduced to Eq. 2.252. (*Hint:* Show that for all u $\sum_{i=0}^{u} N_{i,k}(u) = 1$.)

17 DEGENERATE CURVES AND PATHOLOGICAL CONDITIONS

Be aware that in a complex and powerful interactive geometric-modeling system, it is likely that we will create degenerate or pathological conditions. They may arise by design, coincidence, or malfunction. The modeling system must recognize these conditions and deal with them. We will take only a brief look at some of the more obvious cases of degenerate curves. We will focus on the geometric coefficients of a single pc curve.

The simplest case is when $\mathbf{B} = [0 \quad 0 \quad 0 \quad 0]^T$; which is simply a point at the origin; that is, for any value of *u* substituted into $\mathbf{p}(u) = \mathbf{UMB}$, $\mathbf{p}(u) = 0$. When $\mathbf{p}_1 = \mathbf{p}_0$, and $\mathbf{p}_0^u = \mathbf{p}_1^u = 0$, $\mathbf{B} = [\mathbf{p}_0 \quad \mathbf{p}_0 \quad 0 \quad 0]^T$. This curve degenerates to a point at $\mathbf{p}_0$.

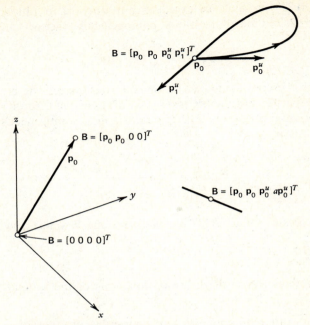

FIGURE 2.68 Degenerate pc curves.

The curve becomes something of a straight line if $\mathbf{B} = [\mathbf{p}_0 \quad \mathbf{p}_0 \quad \mathbf{p}_0^u \quad a\mathbf{p}_0^u]^T$. The scalar a can be positive or negative, or we obtain a closed curve if

$$\mathbf{B} = [\mathbf{p}_0 \quad \mathbf{p}_0 \quad \mathbf{p}_0^u \quad \mathbf{p}_1^u]^T$$

Figure 2.68 illustrates these cases.

As you have no doubt surmised, special conditions exist when one or more of the geometric coefficients are zero or linearly related to another coefficient. Such conditions are not necessarily pathological, but a modeling system must be able to detect them and handle them.

3

SURFACES

The simplest mathematical element we use to model a surface is a **patch**. A patch is a curve-bounded collection of points whose coordinates are given by continuous, two-parameter, single-valued mathematical functions of the form

$$x = x(u, w) \qquad y = y(u, w) \qquad z = z(u, w) \qquad (3.1)$$

The parametric variables u and w are constrained to the intervals $u, w \in [0, 1]$.

Fixing the value of one of the parametric variables results in a curve on the patch in terms of the other variable, which remains free. By continuing this process first for one variable and then the other for any number of arbitrary values in the allowed interval, we form a parametric net of two one-parameter families of curves on the patch so that just one curve of each family passes through each point $\mathbf{p}(u, w)$. Again, the positive sense on any curve is the sense in which the free parameter increases.

Associated with every patch is a set of boundary conditions; see Fig. 3.1. The most obvious of these are the four corner points and the four curves defining its edges. Others of importance are the tangent vectors and twist vectors, which we will discuss later. For an ordinary patch, there are always four and only four corner points and edge curves. This follows from the possible combinations of the two limits of the two parametric variables. We find the corner points by substituting these four combinations of 0 and 1 into $\mathbf{p}(u, w)$ to obtain $\mathbf{p}(0, 0)$ $\mathbf{p}(0, 1)$, $\mathbf{p}(1, 0)$, and $\mathbf{p}(1, 1)$. On the other hand, the edge or boundary curves are functions of one of the two parametric variables. We obtain these by allowing one of the variables to remain free while fixing the other to its limiting values. This procedure results in four and only four possible combinations yielding the functions of the four parametric boundary curves $\mathbf{p}(u, 0)$, $\mathbf{p}(u, 1)$, $\mathbf{p}(0, w)$, and $\mathbf{p}(1, w)$.

Some examples will help you to better understand general parametric surface patches and their formulating equations. The simplest example is a

151

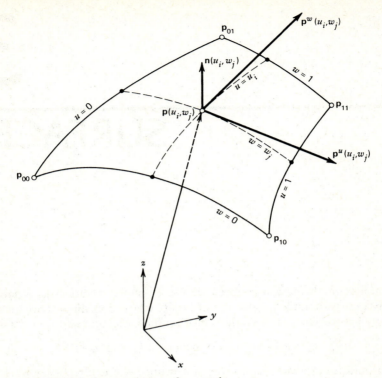

FIGURE 3.1 A parametric surface patch.

plane. The following parametric equations represent a rectangular segment of the x, y plane.

$$x = (c - a)u + a \qquad y = (d - b)w + b \qquad z = 0 \qquad (3.2)$$

Figure 3.2 illustrates this patch, showing both the parametric coordinates and the x, y coordinates of each corner point. The curves of constant w are straight lines running parallel to the x-axis. These curves are functions of u. An analogous situation exists for the lines of constant u, which are straight lines parallel to the y-axis.

The next simplest surface is the sphere, the locus of a point moving at a constant distance from a fixed point. The parametric equations of a sphere of radius r, centered at point (x_0, y_0, z_0) are

$$x = x_0 + r \cos u \cos w \qquad u \in \left[-\frac{\pi}{2}, \frac{\pi}{2}\right]$$

$$y = y_0 + r \cos u \sin w \qquad w \in [0, 2\pi] \qquad (3.3)$$

$$z = z_0 + r \sin u$$

where u is analogous to latitude and w to longitude, with both angles in radians. Obviously, curves of constant u are parallels of latitude, so that a circle is generated in a plane parallel to the x, y plane as w varies. Curves of constant w are

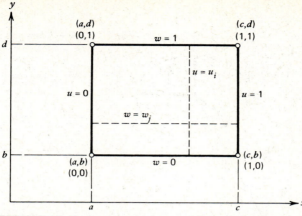

FIGURE 3.2 Parametric and x, y coordinates of a plane.

meridians of longitude, generating semicircles as u varies. Figure 3.3 shows these features.

An ellipsoid centered at (x_0, y_0, z_0) has the following parametric equations:

$$x = x_0 + a \cos u \cos w \qquad u \in \left[-\frac{\pi}{2}, \frac{\pi}{2} \right]$$

$$y = y_0 + b \cos u \sin w \qquad w \in [0, 2\pi] \qquad (3.4)$$

$$z = z_0 + c \sin u$$

Equations 3.4 are illustrated in Fig. 3.4.

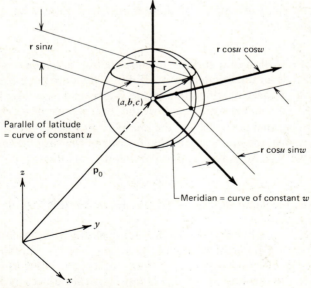

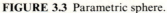

FIGURE 3.3 Parametric sphere.

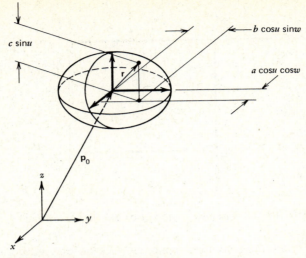

FIGURE 3.4 Parametric ellipsoid.

Finally, Fig. 3.5 shows a parametric surface of revolution; for clarity, we see only part of the surface. Notice that it is the curve defined by the function $x(u)$, $z(u)$ that is revolved around the z-axis. The equations are

$$
\begin{aligned}
x &= x(u) \cos w \\
y &= x(u) \sin w \qquad w \in [0, 2\pi] \\
z &= z(u)
\end{aligned}
\tag{3.5}
$$

There are many examples of nonparametric surfaces; for example, an equation of the form

$$
F(x, y, z) = 0
\tag{3.6}
$$

is the implicit equation of a surface. If this equation is linear in all variables, then the surface is an unbounded plane. If it is a second-degree equation, then the surface is a quadric, of which the sphere is a special case. Finally, if one of the variables is missing from the equation, the surface must be a cylinder whose generators are parallel to the axis of the missing variable.

When we solve the implicit equation for one of the variables as a function of the other two, say, for z as a function of x and y, we obtain

$$
z = f(x, y)
\tag{3.7}
$$

This represents the same surface as Eq. 3.6. However, this form is the explicit equation of the surface.

Although the implicit and explicit forms are useful, in representing surfaces they suffer from an inherent weakness: their inability to represent an easily transformable and bounded surface. However, we can either directly adapt or closely approximate most, if not all, of these classical surfaces by a parametric formulation.

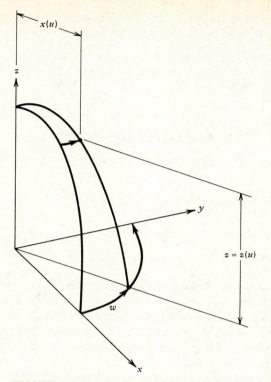

FIGURE 3.5 Parametric surface of revolution.

Another advantage of the parametric representation of a surface is the complete control we have of the domain of a surface-modeling operation simply by an appropriate choice of the parametrization scheme. By carefully specifying subsets of a particular domain $[u_{min}, u_{max}] \times [w_{min}, w_{max}]$, we can readily define certain sections of a surface. This feature is useful whenever a surface is composed of several patches. We use this freedom of parametrization to choose $[0, 1] \times [0, 1]$ as the domain of a surface-interpolating operation. Any other domain can be normalized as necessary to this unit square in parametric space. Also, note that we will often use the terms patch and surface interchangeably. Although strictly speaking, a patch is a limited region of a larger surface, in the context of composite surfaces or patches, the term *patch* has more significance and helps us distinguish these limited regions.

EXERCISES

1. Write the vector form of Eq. 3.1.

2. Using Eq. 3.2, write the equation of a plane patch whose corner point coordinates are $\mathbf{p}(0, 0) = [2, 1, 0]$, $\mathbf{p}(1, 0) = [6, 1, 0]$, $\mathbf{p}(0, 1) = [2, 5, 0]$, and $\mathbf{p}(1, 1) = [6, 5, 0]$.

3. Using Eq. 3.3, write the equation of a sphere whose radius is 2 and has a center at $(-1, 1, 4)$. Find the equation of the equator.

4. Find the equations for the surface produced when a Bezier curve is revolved about the x-axis. The curve is defined by $\mathbf{p}_0 = [2, 0, 1]$, $\mathbf{p}_1 = [3, 0, 2]$, $\mathbf{p}_2 = [5, 0, 1]$, and $\mathbf{p}_3 = [6, 0, 2]$.

1 ALGEBRAIC AND GEOMETRIC FORM

The algebraic form of a bicubic patch is given by

$$\mathbf{p}(u, w) = \sum_{i=0}^{3} \sum_{j=0}^{3} \mathbf{a}_{ij} u^i w^j \tag{3.8}$$

The restriction on the parametric variables is

$$u, w \in [0, 1] \tag{3.9}$$

The $\mathbf{a}_{ij}$ vectors are called the **algebraic coefficients** of the surface. The reason for the term *bicubic* is obvious, since both parametric variables can be cubic terms.

Notice that as with curves, the parametric variables u and w are restricted by definition to values in the interval 0 to 1, inclusive. This restriction makes the surface bounded in a regular way; we will discuss this restriction and the use of irregular boundaries later.

Next, expand Eq. 3.8, and arrange the terms in descending order; the result is similar to Eq. 2.18 for curves

$$
\begin{aligned}
\mathbf{p}(u, w) = {} & \mathbf{a}_{33} u^3 w^3 + \mathbf{a}_{32} u^3 w^2 + \mathbf{a}_{31} u^3 w + \mathbf{a}_{30} u^3 \\
& + \mathbf{a}_{23} u^2 w^3 + \mathbf{a}_{22} u^2 w^2 + \mathbf{a}_{21} u^2 w + \mathbf{a}_{20} u^2 \\
& + \mathbf{a}_{13} u w^3 + \mathbf{a}_{12} u w^2 + \mathbf{a}_{11} u w + \mathbf{a}_{10} u \\
& + \mathbf{a}_{03} w^3 + \mathbf{a}_{02} w^2 + \mathbf{a}_{01} w + \mathbf{a}_{00}
\end{aligned}
\tag{3.10}
$$

This 16-term polynomial in u and w defines the set of all points lying on the surface. It is the algebraic form of a bicubic patch. Since each of the vector coefficients $\mathbf{a}$ has three independent components, there is a total of 48 algebraic coefficients, or 48 degrees of freedom. Thus, each vector component is simply

$$x(u, w) = a_{33x} u^3 w^3 + a_{32x} u^3 w^2 + a_{31x} u^3 w + \cdots + a_{00x} \tag{3.11}$$

There are similar expressions for $y(u, w)$ and $z(u, w)$.

The algebraic form in matrix notation is

$$\mathbf{p} = \mathbf{U} \mathbf{A} \mathbf{W}^T \tag{3.12}$$

where

$$\mathbf{U} = [u^3 \quad u^2 \quad u \quad 1] \tag{3.13}$$

and

$$\mathbf{W} = [w^3 \quad w^2 \quad w \quad 1] \tag{3.14}$$

$$\mathbf{A} = \begin{bmatrix} \mathbf{a}_{33} & \mathbf{a}_{32} & \mathbf{a}_{31} & \mathbf{a}_{30} \\ \mathbf{a}_{23} & \mathbf{a}_{22} & \mathbf{a}_{21} & \mathbf{a}_{20} \\ \mathbf{a}_{13} & \mathbf{a}_{12} & \mathbf{a}_{11} & \mathbf{a}_{10} \\ \mathbf{a}_{03} & \mathbf{a}_{02} & \mathbf{a}_{01} & \mathbf{a}_{00} \end{bmatrix} \tag{3.15}$$

Note that the subscripts of the vector elements in the $\mathbf{A}$ matrix correspond to those in Eq. 3.10. They have no direct relationship to the normal indexing convention for matrices. Since the $\mathbf{a}$ elements are three-component vectors, the $\mathbf{A}$ matrix is a $4 \times 4 \times 3$ array.

As we found with curves, the algebraic coefficients of a patch determine its shape and position in space. However, patches of the same size and shape have a different set of coefficients if they occupy different positions in space. Change any one of the 48 coefficients, and a completely different patch results. A point on the patch is generated each time we insert a specific pair of u, w values into Eq. 3.12. And, although the u, w values are restricted by Eq. 3.9, the range of the object-space variables x, y, and z is unrestricted because the range of the algebraic coefficients is unrestricted.

A patch consists of an infinite number of points given by their x, y, z coordinates. There is also an infinite number of pairs of u, w values in the corresponding parametric space. Clearly, there is a unique pair of u, w values associated with each point in object space. Figure 3.6 shows a bicubic patch mapped into object space from its components in parametric space.

Observe that a bicubic patch is bounded by four curves and each boundary curve is obviously a pc curve. Each of these curves is named as follows: u_0, u_1, w_0, w_1 (for $u = 0$, $u = 1$, and so on), because they arise at the constant limiting values of the parametric variables. Another way of noting the boundary curves is by an appropriately subscripted vector $\mathbf{p}$. Thus, we use $\mathbf{p}_{0w}$, $\mathbf{p}_{1w}$, $\mathbf{p}_{u0}$, and $\mathbf{p}_{u1}$; their interpretation should be obvious. There are also four unique corner points $\mathbf{p}_{00}$, $\mathbf{p}_{10}$, $\mathbf{p}_{01}$, and $\mathbf{p}_{11}$.

As with the pc curve, reversing the sequence of u, w parametrization does not change the shape of a surface. And, aside from problem-specific computation constraints (for example, the direction of surface normals), we have almost complete freedom to assign u, $w = 0$ or 1 to the boundary curves. The exception is that $\mathbf{p}_{0w}$ must be opposite $\mathbf{p}_{1w}$, and $\mathbf{p}_{u0}$ opposite $\mathbf{p}_{u1}$. Figure 3.7 presents a useful nomenclature for a bicubic patch.

It is a simple exercise to show that a boundary is a pc curve. Let us investigate the curve $\mathbf{p}_{u0}$. First, the subscript indicates $w = 0$, so all terms containing w vanish, and Eq. 3.10 becomes

$$\mathbf{p}_{u0} = \mathbf{a}_{30} u^3 + \mathbf{a}_{20} u^2 + \mathbf{a}_{10} u + \mathbf{a}_{00} \tag{3.16}$$

This is the expression for a pc curve. *Q.E.D.*

As we saw for curves, algebraic coefficients are not the most convenient way of defining and controlling the shape of a patch, and they do not contribute

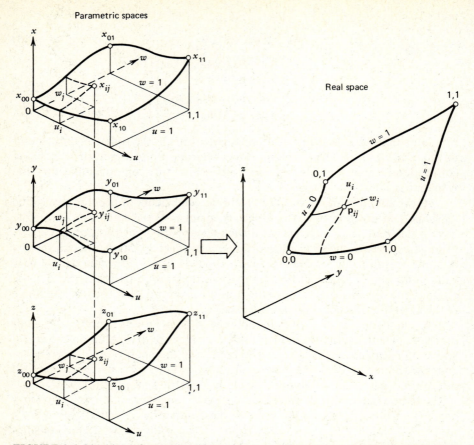

FIGURE 3.6 Bicubic surface mapped into object space from its components in parametric space. Four surfaces are shown.

to our understanding of surface behavior. Therefore, we must turn to the geometric form.

We derive the geometric form of representing a bicubic patch from a relationship between certain boundary conditions and the algebraic form. The four patch corner points $\mathbf{p}_{00}$, $\mathbf{p}_{10}$, $\mathbf{p}_{01}$, $\mathbf{p}_{11}$ and the eight tangent vectors $\mathbf{p}_{00}^u$, $\mathbf{p}_{00}^w$, $\mathbf{p}_{10}^u$, $\mathbf{p}_{10}^w$, $\mathbf{p}_{01}^u$, $\mathbf{p}_{01}^w$, $\mathbf{p}_{11}^u$, $\mathbf{p}_{11}^w$ are some of the conditions we will use. They, of course, define the boundary curves

$$\mathbf{p}_{u0} = \mathbf{F}[\mathbf{p}_{00} \quad \mathbf{p}_{10} \quad \mathbf{p}_{00}^u \quad \mathbf{p}_{10}^u]^T \qquad (3.17)$$

$$\mathbf{p}_{u1} = \mathbf{F}[\mathbf{p}_{01} \quad \mathbf{p}_{11} \quad \mathbf{p}_{01}^u \quad \mathbf{p}_{11}^u]^T \qquad (3.18)$$

$$\mathbf{p}_{0w} = \mathbf{F}[\mathbf{p}_{00} \quad \mathbf{p}_{01} \quad \mathbf{p}_{00}^w \quad \mathbf{p}_{01}^w]^T \qquad (3.19)$$

$$\mathbf{p}_{1w} = \mathbf{F}[\mathbf{p}_{10} \quad \mathbf{p}_{11} \quad \mathbf{p}_{10}^w \quad \mathbf{p}_{11}^w]^T \qquad (3.20)$$

These curves provide 12 of the 16 vectors we need to specify the 48 geometric coefficients required to define a bicubic patch. We need four additional vectors.

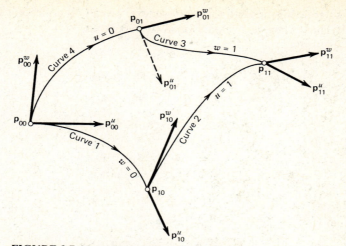

FIGURE 3.7 Nomenclature for a bicubic surface.

We will use the so-called twist vectors, one at each of the four corner points. Mathematically, we express the twist vectors as follow:

$$\mathbf{p}_{00}^{uw} = \frac{\partial^2 \mathbf{p}(u, w)}{\partial u\, \partial w} \qquad \text{at } u = 0, w = 0 \tag{3.21}$$

$$\mathbf{p}_{10}^{uw} = \frac{\partial^2 \mathbf{p}(u, w)}{\partial u\, \partial w} \qquad \text{at } u = 1, w = 0 \tag{3.22}$$

$$\mathbf{p}_{01}^{uw} = \frac{\partial^2 \mathbf{p}(u, w)}{\partial u\, \partial w} \qquad \text{at } u = 0, w = 1 \tag{3.23}$$

$$\mathbf{p}_{11}^{uw} = \frac{\partial^2 \mathbf{p}(u, w)}{\partial u\, \partial w} \qquad \text{at } u = 1, w = 1 \tag{3.24}$$

We calculate the mixed partial derivative of the function $\mathbf{p}(u, w)$ and then evaluate the result at the indicated u, w values. Thus, from Eq. 3.10, we obtain

$$\frac{\partial^2 \mathbf{p}(u, w)}{\partial u\, \partial w} = 9\mathbf{a}_{33} u^2 w^2 + 6\mathbf{a}_{32} u^2 w + 3\mathbf{a}_{31} u^2$$
$$+ 6\mathbf{a}_{23} u w^2 + 4\mathbf{a}_{22} u w + 2\mathbf{a}_{21} u$$
$$+ 3\mathbf{a}_{13} w^2 + 2\mathbf{a}_{12} w + \mathbf{a}_{11} \tag{3.25}$$

When we evaluate this equation at the corner points, we obtain

$$\mathbf{p}_{00}^{uw} = \mathbf{a}_{11} \tag{3.26}$$

$$\mathbf{p}_{10}^{uw} = 3\mathbf{a}_{31} + 2\mathbf{a}_{21} + \mathbf{a}_{11} \tag{3.27}$$

$$\mathbf{p}_{01}^{uw} = 3\mathbf{a}_{13} + 2\mathbf{a}_{12} + \mathbf{a}_{11} \tag{3.28}$$

$$\mathbf{p}_{11}^{uw} = 9\mathbf{a}_{33} + 6\mathbf{a}_{32} + 3\mathbf{a}_{31}$$
$$+ 6\mathbf{a}_{23} + 4\mathbf{a}_{22} + 2\mathbf{a}_{21}$$
$$+ 3\mathbf{a}_{13} + 2\mathbf{a}_{12} + \mathbf{a}_{11} \tag{3.29}$$

Having gone this far with the twist vectors, we can do no less for the other 12 vectors. Thus, when we evaluate Eq. 3.10, we obtain

$$\mathbf{p}_{00} = \mathbf{a}_{00} \tag{3.30}$$

$$\mathbf{p}_{10} = \mathbf{a}_{30} + \mathbf{a}_{20} + \mathbf{a}_{10} + \mathbf{a}_{00} \tag{3.31}$$

$$\mathbf{p}_{01} = \mathbf{a}_{03} + \mathbf{a}_{02} + \mathbf{a}_{01} + \mathbf{a}_{00} \tag{3.32}$$

$$\begin{aligned} \mathbf{p}_{11} = \mathbf{a}_{33} + \mathbf{a}_{32} + \mathbf{a}_{31} &+ \mathbf{a}_{30} \\ + \mathbf{a}_{23} + \mathbf{a}_{22} + \mathbf{a}_{21} &+ \mathbf{a}_{20} \\ + \mathbf{a}_{13} + \mathbf{a}_{12} + \mathbf{a}_{11} &+ \mathbf{a}_{10} \\ + \mathbf{a}_{03} + \mathbf{a}_{02} + \mathbf{a}_{01} &+ \mathbf{a}_{00} \end{aligned} \tag{3.33}$$

$$\mathbf{p}_{00}^{u} = \mathbf{a}_{10} \tag{3.34}$$

$$\mathbf{p}_{00}^{w} = \mathbf{a}_{01} \tag{3.35}$$

$$\mathbf{p}_{10}^{u} = 3\mathbf{a}_{30} + 2\mathbf{a}_{20} + \mathbf{a}_{10} \tag{3.36}$$

$$\mathbf{p}_{10}^{w} = \mathbf{a}_{31} + \mathbf{a}_{21} + \mathbf{a}_{11} + \mathbf{a}_{01} \tag{3.37}$$

$$\mathbf{p}_{01}^{u} = \mathbf{a}_{13} + \mathbf{a}_{12} + \mathbf{a}_{11} + \mathbf{a}_{10} \tag{3.38}$$

$$\mathbf{p}_{01}^{w} = 3\mathbf{a}_{03} + 2\mathbf{a}_{02} + \mathbf{a}_{01} \tag{3.39}$$

$$\begin{aligned} \mathbf{p}_{11}^{u} = 3\mathbf{a}_{33} + 3\mathbf{a}_{32} + 3\mathbf{a}_{31} &+ 3\mathbf{a}_{30} \\ + 2\mathbf{a}_{23} + 2\mathbf{a}_{22} + 2\mathbf{a}_{21} &+ 2\mathbf{a}_{20} \\ + \mathbf{a}_{13} + \mathbf{a}_{12} + \mathbf{a}_{11} &+ \mathbf{a}_{10} \end{aligned} \tag{3.40}$$

$$\begin{aligned} \mathbf{p}_{11}^{w} = 3\mathbf{a}_{33} + 2\mathbf{a}_{32} &+ \mathbf{a}_{31} \\ + 3\mathbf{a}_{23} + 2\mathbf{a}_{22} &+ \mathbf{a}_{21} \\ + 3\mathbf{a}_{13} + 2\mathbf{a}_{12} &+ \mathbf{a}_{11} \\ + 3\mathbf{a}_{03} + 2\mathbf{a}_{02} &+ \mathbf{a}_{01} \end{aligned} \tag{3.41}$$

Now that this rather exhaustive process is complete, let us return to the twist vectors. What is their geometric interpretation? Study Fig. 3.8; it illustrates our investigation of the twist vectors as we continue to explore the geometric form.

A bicubic patch consists of two mutually orthogonal sets of curves—orthogonal in parametric space. They are the pc curves defined by constant values of the parametric variables u_i and w_i. One set contains the u_0 and u_1 and an infinite number of intermediate u_i curves, sweeping across a surface from one boundary to the other. A similar set of w_i pc curves overlays the u_i's, forming an orthogonal net. Equations 3.17–3.20 are sufficient to define the boundary curves, but how do we define intermediate curves, such as w_i, that is, $\mathbf{p}_{ui}$?

First, find the end points of $\mathbf{p}_{ui}$ by computing $\mathbf{p}_{0i}$ and $\mathbf{p}_{1i}$ at $w = i$ from Eqs. 3.19 and 3.20. Next, we must find the tangent vectors $\mathbf{p}_{0i}^{u}$ and $\mathbf{p}_{1i}^{u}$. To do this, we observe, for example, that for the general case, $\mathbf{p}_{00}^{u}$ and $\mathbf{p}_{01}^{u}$ are not equal. Thus, $\mathbf{p}_{0w}^{u}$ changes as we change w from w_0 through w_i to w_1. The same thing happens to the tangent vectors $\mathbf{p}_{1w}^{u}$ along curve $\mathbf{p}_{1w}$. Mathematically, we ex-

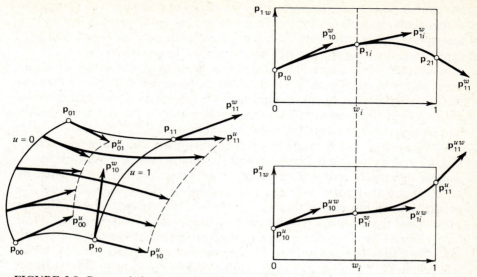

FIGURE 3.8 Geometric interpretation of twist vectors.

press this change in the tangent vectors as $(\partial/\partial w)(\mathbf{p}_{0w}^u)$ and $(\partial/\partial w)(\mathbf{p}_{1w}^u)$, but

$$(\partial/\partial w)(\mathbf{p}_{0w}^u) = (\partial^2 \mathbf{p}/\partial u\, \partial w)_{0w}$$
$$= \mathbf{p}_{0w}^{uw} \tag{3.42}$$

and

$$(\partial/\partial w)(\mathbf{p}_{1w}^u) = (\partial^2 \mathbf{p}/\partial u\, \partial w)_{1w}$$
$$= \mathbf{p}_{1w}^{uw} \tag{3.43}$$

At the patch corners, these terms are $\mathbf{p}_{00}^{uw}$, $\mathbf{p}_{01}^{uw}$, $\mathbf{p}_{10}^{uw}$, $\mathbf{p}_{11}^{uw}$, the twist vectors of the set of boundary conditions we saw earlier.

There is an analogy between the way we handle a simple pc curve and the way we use the twist vectors to find the intermediate tangent vectors. Just as we specify points on a pc curve by two end points and tangent vectors, we specify intermediate tangent vectors along a boundary curve by the two end tangent vector values and corresponding twist vectors. We find points along a boundary curve $\mathbf{p}_{0w}$, for example, by using Eq. 3.19

$$\mathbf{p}_{0w} = \mathbf{F}[\mathbf{p}_{00} \quad \mathbf{p}_{01} \quad \mathbf{p}_{00}^w \quad \mathbf{p}_{01}^w]^T \tag{3.44}$$

We find we can use an identical form to determine an intermediate tangent vector along, but across, a curve boundary

$$\mathbf{p}_{0w}^u = \mathbf{F}[\mathbf{p}_{00}^u \quad \mathbf{p}_{01}^u \quad \mathbf{p}_{00}^{uw} \quad \mathbf{p}_{01}^{uw}]^T \tag{3.45}$$

Equation 3.45 describes a "curve" formed by the "arrowhead" end points of the intermediate tangent vectors as w varies along the unit interval. Meriwether, who first proposed this interpretation, calls this an **auxiliary curve**. Note that there are four auxiliary curves: $\mathbf{p}_{0w}^u$, $\mathbf{p}_{1w}^u$, $\mathbf{p}_{0u}^w$, $\mathbf{p}_{1u}^w$. These curves are an abstract property of the general mathematical form of the bicubic patch.

We cannot point to them on any physical model of a surface. The parametric components of the boundary curve $u = 1$ and its associated auxiliary curve are shown in Fig. 3.8. Note that the ordinates are labeled as $\mathbf{p}_{1w}$ and $\mathbf{p}_{1w}^u$, expediently denoting the x, y, or z components of these vectors.

Another interesting mathematical property of twist vectors is that at any point on the bicubic patch, $\mathbf{p}^{uw} = \mathbf{p}^{wu}$. That is, the order in which we take the mixed partial derivative is not important. This property means that u and w are completely interchangeable and no experiment on a physical manifestation of a bicubic patch can distinguish the u, w order.

Since the boundary conditions permit a complete and unambiguous definition of a bicubic patch as two orthogonal and infinite sets of pc curves, so can we investigate any interior point or geometric property of a patch by selecting and analyzing the proper set of pc curves. Now, we begin to see the geometric form in terms of simple pc curves in their geometric form. They, in turn, define the necessary patch boundary conditions. Bicubic patches of this form, defined by the coordinates and tangent vectors at their corner points, are often called **Hermite patches**. This name also applies to pc curves, which will help us distinguish this form from other parametric surfaces, such as Bezier and B-spline surfaces.

We can now summarize these patch-defining curves by assembling them in a 4×4 matrix. Begin by filling the first two rows with the geometric coefficients of curves $\mathbf{p}_{0w}$ and $\mathbf{p}_{1w}$

$$\text{Row 1:} \quad \mathbf{p}_{0w} \rightarrow \mathbf{p}_{00} \quad \mathbf{p}_{01} \quad \mathbf{p}_{00}^w \quad \mathbf{p}_{01}^w$$

$$\text{Row 2:} \quad \mathbf{p}_{1w} \rightarrow \mathbf{p}_{10} \quad \mathbf{p}_{11} \quad \mathbf{p}_{10}^w \quad \mathbf{p}_{11}^w$$

Next, fill the first two columns with the geometric coefficients of curves $\mathbf{p}_{u0}$ and $\mathbf{p}_{u1}$. Observe that some of the coefficients are already entered.

$$\text{Column 1:} \quad \mathbf{p}_{u0}$$

$$\text{Column 2:} \quad \mathbf{p}_{u1}$$

$$
\begin{matrix}
\mathbf{p}_{00} & \mathbf{p}_{01} & \mathbf{p}_{00}^w & \mathbf{p}_{01}^w \\
\mathbf{p}_{10} & \mathbf{p}_{11} & \mathbf{p}_{10}^w & \mathbf{p}_{11}^w \\
\mathbf{p}_{00}^u & \mathbf{p}_{01}^u & & \\
\mathbf{p}_{10}^u & \mathbf{p}_{11}^u & &
\end{matrix}
$$

We have 12 of the 16 vectors entered into the matrix. Now, we must use the auxiliary curves to finish filling the matrix.

$$
\begin{matrix}
 & \mathbf{p}_{u0} & \mathbf{p}_{u1} & \mathbf{p}_{u0}^w & \mathbf{p}_{u1}^w \\
 & \downarrow & \downarrow & \downarrow & \downarrow \\
\mathbf{p}_{0w} \rightarrow & \mathbf{p}_{00} & \mathbf{p}_{01} & \mathbf{p}_{00}^w & \mathbf{p}_{01}^w \\
\mathbf{p}_{1w} \rightarrow & \mathbf{p}_{10} & \mathbf{p}_{11} & \mathbf{p}_{10}^w & \mathbf{p}_{11}^w \\
 & & & & \\
\mathbf{p}_{0w}^u \rightarrow & \mathbf{p}_{00}^u & \mathbf{p}_{01}^u & \mathbf{p}_{00}^{uw} & \mathbf{p}_{01}^{uw} \\
\mathbf{p}_{1w}^u \rightarrow & \mathbf{p}_{10}^u & \mathbf{p}_{11}^u & \mathbf{p}_{10}^{uw} & \mathbf{p}_{11}^{uw}
\end{matrix}
$$

From this completed matrix, we can easily extract the geometric coefficients of the four boundary curves and the four auxiliary curves. When we divide the

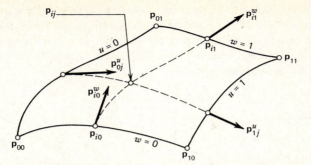

FIGURE 3.9 Determining the coordinates of a point on a bicubic surface.

matrix into quadrants, we observe that the four vectors in the upper left define the four corner points. The upper-right quadrant contains the tangent vectors with respect to w at the corner points, while those with respect to u are in the lower left. Finally, the lower-right quadrant contains the twist vector or cross-derivatives at the corner points. Note the pattern of repetition of these vectors' subscripts.

This is the matrix of geometric coefficients for a bicubic patch. Given these coefficients, let us see how we evaluate a point on it at a specific pair of u, w values, say, u_i and w_j. Figure 3.9 illustrates the basic ingredients of this situation. Here, we see that the point of interest obviously lies at the intersection of the curves $\mathbf{p}_{iw}$ and $\mathbf{p}_{uj}$. Notice that this problem is easily reduced to finding a point on a pc curve at a given value of a parametric variable. We can work with either curve, so we arbitrarily choose to begin with $\mathbf{p}_{iw}$. First, we determine the geometric coefficients of this curve, $\mathbf{p}_{i0}$, $\mathbf{p}_{i1}$, $\mathbf{p}_{i0}^w$, $\mathbf{p}_{i1}^w$. With these coefficients, we can then determine the coordinates of the point $\mathbf{p}_{ij}$.

We compute $\mathbf{p}_{i0}$ using curve $\mathbf{p}_{u0}$

$$\mathbf{p}_{i0} = F_1(u_i)\mathbf{p}_{00} + F_2(u_i)\mathbf{p}_{10} + F_3(u_i)\mathbf{p}_{00}^u + F_4(u_i)\mathbf{p}_{10}^u \qquad (3.46)$$

Compute $\mathbf{p}_{i1}$ using $\mathbf{p}_{u1}$

$$\mathbf{p}_{i1} = F_1(u_i)\mathbf{p}_{01} + F_2(u_i)\mathbf{p}_{11} + F_3(u_i)\mathbf{p}_{01}^u + F_4(u_i)\mathbf{p}_{11}^u \qquad (3.47)$$

Compute $\mathbf{p}_{i0}^w$ using $\mathbf{p}_{u0}^w$, an auxiliary curve

$$\mathbf{p}_{i0}^w = F_1(u_i)\mathbf{p}_{00}^w + F_2(u_i)\mathbf{p}_{10}^w + F_3(u_i)\mathbf{p}_{00}^{uw} + F_4(u_i)\mathbf{p}_{10}^{uw} \qquad (3.48)$$

Finally, compute $\mathbf{p}_{i1}^w$ using $\mathbf{p}_{u1}^w$, another auxiliary curve

$$\mathbf{p}_{i1}^w = F_1(u_i)\mathbf{p}_{01}^w + F_2(u_i)\mathbf{p}_{11}^w + F_3(u_i)\mathbf{p}_{01}^{uw} + F_4(u_i)\mathbf{p}_{11}^{uw} \qquad (3.49)$$

Now that we have the geometric coefficients of $\mathbf{p}_{iw}$, we evaluate it at w_j. We do this by substituting from Eqs. 3.46–3.49 as follows:

$$\begin{aligned}
\mathbf{p}_{ij} = {} & F_1(w_j)[F_1(u_i)\mathbf{p}_{00} + F_2(u_i)\mathbf{p}_{10} + F_3(u_i)\mathbf{p}_{00}^u + F_4(u_i)\mathbf{p}_{10}^u] \\
& + F_2(w_j)[F_1(u_i)\mathbf{p}_{01} + F_2(u_i)\mathbf{p}_{11} + F_3(u_i)\mathbf{p}_{01}^u + F_4(u_i)\mathbf{p}_{11}^u] \\
& + F_3(w_j)[F_1(u_i)\mathbf{p}_{00}^w + F_2(u_i)\mathbf{p}_{10}^w + F_3(u_i)\mathbf{p}_{00}^{uw} + F_4(u_i)\mathbf{p}_{10}^{uw}] \\
& + F_4(w_j)[F_1(u_i)\mathbf{p}_{01}^w + F_2(u_i)\mathbf{p}_{11}^w + F_3(u_i)\mathbf{p}_{01}^{uw} + F_4(u_i)\mathbf{p}_{11}^{uw}] \quad (3.50)
\end{aligned}$$

We would obtain the same result by using curve $\mathbf{p}_{uj}$ as the initial working curve. Equation 3.50 can certainly be reduced to matrix form. We will do this while at the same time generalizing the parametric variables by dropping the subscripts. We cannot yet drop the functional notation; do you see why? Equation 3.50 can be abbreviated to

$$\mathbf{p}(u, w) = [F_1(u) \quad F_2(u) \quad F_3(u) \quad F_4(u)]$$

$$\times \begin{bmatrix} \mathbf{p}_{00} & \mathbf{p}_{01} & \mathbf{p}_{00}^w & \mathbf{p}_{01}^w \\ \mathbf{p}_{10} & \mathbf{p}_{11} & \mathbf{p}_{10}^w & \mathbf{p}_{11}^w \\ \mathbf{p}_{00}^u & \mathbf{p}_{01}^u & \mathbf{p}_{00}^{uw} & \mathbf{p}_{01}^{uw} \\ \mathbf{p}_{10}^u & \mathbf{p}_{11}^u & \mathbf{p}_{10}^{uw} & \mathbf{p}_{11}^{uw} \end{bmatrix} [F_1(w) \quad F_2(w) \quad F_3(w) \quad F_4(w)]^T \quad (3.51)$$

Denote the 4×4 matrix of geometric coefficients with $\mathbf{B}$, and simplify the $\mathbf{F}$ arrays to $\mathbf{F}(u)$ and $\mathbf{F}(w)^T$ to obtain

$$\mathbf{p}(u, w) = \mathbf{F}(u)\mathbf{B}\mathbf{F}(w)^T \quad (3.52)$$

Now, we will eliminate the functional notation and use the universal transformation matrix introduced in Chapter 2, Section 1. From Eq. 2.32, we know that $\mathbf{F}(u) = \mathbf{UM}$. Since u and w are symmetrical, interchangeable parametric variables, we also know that $\mathbf{F}(w) = \mathbf{WM}$. From matrix algebra, $\mathbf{F}(w)^T = \mathbf{M}^T\mathbf{W}^T$. With appropriate substitution into Eq. 3.52, we derive the conventional geometric form of a bicubic patch

$$\mathbf{p} = \mathbf{UMBM}^T\mathbf{W}^T \quad (3.53)$$

Compare Eq. 3.53 to Eq. 3.12; the relationship between the algebraic and geometric form is apparent. We convert from one form into the other by using the following formulas:

$$\mathbf{A} = \mathbf{MBM}^T \quad (3.54)$$

and

$$\mathbf{B} = \mathbf{M}^{-1}\mathbf{AM}^{T-1} \quad (3.55)$$

EXERCISES

1. Use the summation notation to define a parametric biquintic surface patch. How many algebraic coefficients does a biquintic patch have?

2. Show that if $\mathbf{p}(u, w)$ is continuous in u and w, then $\mathbf{p}_{ij}^{uw} = \mathbf{p}_{ij}^{wu}$. Give a function discontinuous in u and w, and show the contrary, that is, $\mathbf{p}_{ij}^{uw} \neq \mathbf{p}_{ij}^{wu}$.

3. Write the three other equations for cross-boundary tangent vectors corresponding to Eq. 3.45.

4. Develop the algebraic, geometric, and matrix forms for parametric biquintic patches. What special limitations are they subject to? Is there a vector comparable to the twist vector?

5. What is the significance of higher order derivatives, such as p^{uuw}?

6. Verify the dimensions of the matrix expressions in Eqs. 3.53–3.55.

7. What can you say about the bicubic patch whose geometric coefficients are

$$\mathbf{B}_x = \begin{bmatrix} 0 & 10 & 16 & 0 \\ 0 & 18 & 24 & 0 \\ 0 & 8 & 8 & 0 \\ 0 & 8 & 8 & 0 \end{bmatrix}$$

$$\mathbf{B}_y = \begin{bmatrix} 0 & 0 & 0 & 0 \\ 10 & 10 & 0 & 0 \\ 10 & 10 & 0 & 0 \\ 10 & 10 & 0 & 0 \end{bmatrix}$$

$$\mathbf{B}_z = \begin{bmatrix} 10 & 0 & 0 & -16 \\ 8 & 0 & 0 & -14 \\ -2 & 0 & 0 & 2 \\ -2 & 0 & 0 & 2 \end{bmatrix}$$

8. Compute the coordinates of the points at $u = 0.5$, $w = 0$, and at $u = 0.5$, $w = 1$, for the bicubic patch defined in Ex. 7.

9. Write a procedure to convert bicubic patch coefficients in algebraic form to geometric form. Denote this as **ATGS(CI, CO)**, where

CI(4, 4, 3) is the input array of algebraic coefficients, and

CO(4, 4, 3) is the output array of geometric coefficients.

10. Write a procedure to convert bicubic patch coefficients in geometric form to algebraic form. Denote this as **GTAS(CI, CO)**, where

CI(4, 4, 3) is the input array of geometric coefficients, and

CO(4, 4, 3) is the output array of algebraic coefficients.

11. Write a procedure to compute the x, y, z coordinates of a point on a bicubic patch for specific values of the parametric variables u and w. Use the geometric form to evaluate the point. Denote this as **PSRF(CI, U, W, P)**, where

CI(4, 4, 3) is the input array of geometric coefficients defining the bicubic patch;

U is the input value of the parametric variable u;

W is the input value of the parametric variable w; and

P(3) is the output array of x, y, z coordinates of the point.

12. Write a procedure to compute the x, y, z coordinates of a point on a bicubic patch. Use the algebraic form to evaluate the point. Denote this as **PSRFA**(CI, U, W, P), where

CI(4, 4, 3) is the input array of algebraic coefficients defining the bicubic patch;

U is the input value of the parametric variable u;

W is the input value of the parametric variable w; and

P(3) is the output array of x, y, z coordinates of the point.

13. Compare procedures **PSRF** and **PSRFA**. Are there any obvious advantages or disadvantages in one or the other?

14. Write a procedure to compute the x, y, z coordinates of a rectangular array of points on a bicubic patch at a specified incremental value of the parametric variables u and w. Denote this as **PPSRF**(CI, NUW, P), where

CI(4, 4, 3) is the input array of geometric coefficients defining the bicubic patch?

NUW is the input number of increments of the parametric variables u and w. (*Note*: Each variable has the same number of increments, namely, NUW.)

P(5, NUW + 1, NUW + 1) is the output array of u, w values and the corresponding x, y, z coordinates, starting at $u = 0$, $w = 0$ and proceeding to $u = 1$, $w = 1$ in NUW equal increments.

15. Modify procedure **PPSRF** defined in Ex. 14 so that u and w can be incremented separately.

2   TANGENT AND TWIST VECTORS

Computing and understanding the significance of the parametric derivatives of the bicubic patch function is similar to these same processes in the pc curve. The difference is that the bicubic function is expressed in two independent variables instead of just the one for curves. This means that we must deal with partial derivatives. In Section 1, we developed expressions for the tangent and twist vectors and extended the notation system introduced for curves. Let us briefly review these procedures.

The tangent vectors are

$$\mathbf{p}_{uw}^u = \frac{\partial \mathbf{p}(u, w)}{\partial u} \tag{3.56}$$

and

$$\mathbf{p}_{uw}^w = \frac{\partial \mathbf{p}(u, w)}{\partial w} \tag{3.57}$$

The twist vector is

$$\mathbf{p}_{uw}^{uw} = \frac{\partial^2 \mathbf{p}(u, w)}{\partial u \, \partial w} \tag{3.58}$$

We can use various characteristics of these vectors to change the shape of a patch in a controlled, predictable way. We take the same approach here that we did with curves. Since the patch is defined by a curve net, inducing changes in these curves changes the patch. To investigate the effect of manipulating the tangent and twist vectors, recompose the **B** matrix by expressing these vectors in terms of unit vectors and scalar magnitudes

$$\mathbf{B} = \begin{bmatrix} \mathbf{p}_{00} & \mathbf{p}_{01} & e\mathbf{t}_{00}^{w} & f\mathbf{t}_{01}^{w} \\ \mathbf{p}_{10} & \mathbf{p}_{11} & g\mathbf{t}_{10}^{w} & h\mathbf{t}_{11}^{w} \\ a\mathbf{t}_{00}^{u} & c\mathbf{t}_{01}^{u} & i\mathbf{t}_{00}^{uw} & j\mathbf{t}_{01}^{uw} \\ b\mathbf{t}_{10}^{u} & d\mathbf{t}_{11}^{u} & k\mathbf{t}_{10}^{uw} & l\mathbf{t}_{11}^{uw} \end{bmatrix} \tag{3.59}$$

Notice that we must now add a superscript u, w, or uw to the unit tangent vector **t**, since there are two independent variables and two distinct directions on the patch.

We change one or more of the boundary curves by changing the appropriate set of scalars a–h. But notice that a change to any of the scalar multipliers of the twist vectors does not change any of the four boundary curves. This is interesting. We can fix the four corner points and boundary curves and still affect the interior shape of a patch through operations on the twist vector magnitudes i–l. Figure 3.10 shows some of these features. This figure indicates

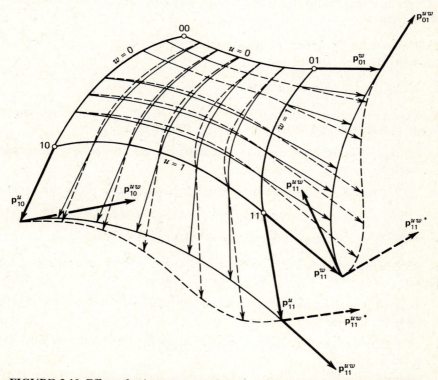

FIGURE 3.10 Effect of twist vectors on the interior of a bicubic surface. The dashed curves on the surface show how the surface changes when boundary conditions change.

the result of a change in the twist vector at $\mathbf{p}_{11}$: $\mathbf{p}_{11}^{uw} \rightarrow \mathbf{p}_{11}^{uw}*$. The dashed curve net on the patch highlights the effect of this change on the interior; note that the boundary curves are unaffected.

We can define a patch with all twist vectors equal to zero. This type of patch is called an **F-patch**, after Ferguson, its developer. Only first-order, C^1, continuity is possible across the boundaries of adjacent F-patches because these patches are constrained to have zero cross-derivatives at their corners. This can lead to surfaces that are not smooth enough for some applications. But it is easy to construct and modify F-patches, and they are adequate for many geometric-modeling situations. The matrix of geometric coefficients then becomes

$$\mathbf{B} = \begin{bmatrix} \mathbf{p}_{00} & \mathbf{p}_{01} & \mathbf{p}_{00}^{w} & \mathbf{p}_{01}^{w} \\ \mathbf{p}_{10} & \mathbf{p}_{11} & \mathbf{p}_{10}^{w} & \mathbf{p}_{11}^{w} \\ \mathbf{p}_{00}^{u} & \mathbf{p}_{01}^{u} & 0 & 0 \\ \mathbf{p}_{10}^{u} & \mathbf{p}_{11}^{u} & 0 & 0 \end{bmatrix} \tag{3.60}$$

EXERCISES

1. Show mathematically that only C^1 continuity is possible across the boundary of an F-patch.

2. Demonstrate a number of conditions that can cause a patch to be self-intersecting.

3. Derive the equivalent algebraic equations for Eqs. 3.56–3.58.

4. Write a procedure to compute the two tangent vectors at a point on a bicubic patch. Denote this as **TVSRF(CI, UW, PU, PW)**, where

CI(4, 4, 3) is the input array of geometric coefficients defining the bicubic patch;

UW(2) are the input values of the parametric variables u and w at the point;

PU(3) is the output array of tangent vector components $\mathbf{p}_{uw}^{u}$; and

PW(3) is the output array of tangent vector components $\mathbf{p}_{uw}^{w}$.

5. Write a procedure to compute the twist vector at a point on a bicubic patch. Denote this as **TWSTV(CI, UW, PUW)**, where

CI(4, 4,3) is the input array of geometric coefficients defining the bicubic patch;

UW(2) are the input values of the parametric variables u and w at the point; and

PUW(3) is the output array of twist vector components $\mathbf{p}_{uw}^{uw}$.

3   NORMALS

At any point $\mathbf{p}(u, w)$ on a bicubic patch, we can construct a vector normal (that is, perpendicular) to the patch. We easily find a unit normal vector $\mathbf{n}(u, w)$ by computing the vector product of the tangent vectors $\mathbf{p}^u$ and $\mathbf{p}^w$ at the point.

$$\mathbf{n}(u, w) = \frac{\mathbf{p}^u \times \mathbf{p}^w}{|\mathbf{p}^u \times \mathbf{p}^w|} \tag{3.61a}$$

It proves convenient to express this as $k_n \mathbf{n}(u, w) = \mathbf{p}^u \times \mathbf{p}^w$ where $k_n = |\mathbf{p}^u \times \mathbf{p}^w|$.

The order in which we take the vector product determines the direction of $\mathbf{n}(u, w)$. Notice that we can interpret $\mathbf{n}(u, w)$ itself as a patch, the **normals patch**. We will see later that this is a very useful interpretation.

Figure 3.11 shows a convention for assigning identifying numbers to the corner points and boundary curves. Here, we see that if the fingers of the right hand curl around the patch in the direction of ascending curve or corner-point numbering, then the thumb points in the direction of the positive or outward surface normal as defined by Eq. 3.61a. This convention gives a consistent algebraic sign when summing the area of several surface patches.

The unit normal is indispensable in almost all phases of geometric modeling, and in most applications, a consistent normal direction is required. For example, we usually want the normal to point outward from the surface of a solid model. Computing silhouette curves, hidden surfaces, and shading effects are just a few more examples of model analysis that require information about surface normals. Equation 3.61a can be expressed in a more convenient form, one that will take advantage of existing algorithms for the bicubic form. We rewrite this equation as

$$k_n \mathbf{n} = [x^u \quad y^u \quad z^u] \times [x^w \quad y^w \quad z^w]$$
$$= [(y^u z^w - y^w z^u) \quad (z^u x^w - x^u z^w) \quad (x^u y^w - y^u x^w)] \tag{3.61b}$$

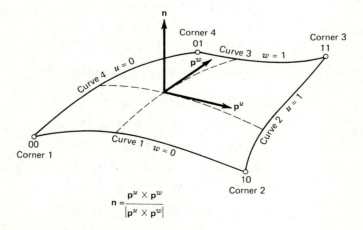

FIGURE 3.11 Normal vector to a surface.

The x component of $\mathbf{n}$ is

$$n_x = \frac{y^u z^w - y^w z^u}{k_n} \qquad (3.61c)$$

We can rewrite each of the terms y^u, z^w, y^w, and z^u in their matrix form. To do this, we must move ahead a bit and make use of Eq. 3.65, $\mathbf{p}^u = \mathbf{UM}^u\mathbf{BM}^T\mathbf{W}^T$, and Eq. 3.66, $\mathbf{p}^w = \mathbf{UMBM}^{wT}\mathbf{W}^T$, in Section 5. Using these we find

$$n_x = \frac{1}{k_n}[(\mathbf{UM}^u\mathbf{B}_y\mathbf{M}^T\mathbf{W}^T)(\mathbf{UMB}_z\mathbf{M}^{wT}\mathbf{W}^T) - (\mathbf{UMB}_y\mathbf{M}^{wT}\mathbf{W}^T)(\mathbf{UM}^u\mathbf{B}_z\mathbf{M}^T\mathbf{W}^T)] \qquad (3.61d)$$

and similarly for n_y and n_z. Equation 3.61d is a biquintic in u and w. However, if the surface is well behaved (not self-intersecting, without undulations, and so on), we can then approximate the normal at any point on the surface with a bicubic expression, so that $\mathbf{n}$ is

$$\mathbf{n} = \mathbf{UMB}_n\mathbf{M}^T\mathbf{W}^T \qquad (3.61e)$$

where

$$\mathbf{B}_{nx} = \begin{bmatrix} n_x & \vdots & n_x^w \\ \hline n_x^u & \vdots & n_x^{uw} \end{bmatrix}, \qquad \mathbf{B}_{ny} = \begin{bmatrix} n_y & \vdots & n_y^w \\ \hline n_y^u & \vdots & n_y^{uw} \end{bmatrix}, \qquad \mathbf{B}_{nz} = \begin{bmatrix} n_z & \vdots & n_z^w \\ \hline n_z^u & \vdots & n_z^{uw} \end{bmatrix} \qquad (3.61f)$$

We have already derived n_x, n_y, and n_z. We find the three remaining components of $\mathbf{B}_{nx}$ by appropriate differentiation of n_x

$$n_x^u = \frac{1}{k_n}[(\mathbf{UM}^{uu}\mathbf{B}_y\mathbf{M}^T\mathbf{W}^T)(\mathbf{UMB}_z\mathbf{M}^{wT}\mathbf{W}^T)$$

$$+ (\mathbf{UM}^u\mathbf{B}_y\mathbf{M}^T\mathbf{W}^T)(\mathbf{UM}^u\mathbf{B}_z\mathbf{M}^{wT}\mathbf{W}^T)$$

$$- (\mathbf{UM}^u\mathbf{B}_y\mathbf{M}^{wT}\mathbf{W}^T)(\mathbf{UM}^u\mathbf{B}_z\mathbf{M}^T\mathbf{W}^T)$$

$$- (\mathbf{UMB}_y\mathbf{M}^{wT}\mathbf{W}^T)(\mathbf{UM}^{uu}\mathbf{B}_z\mathbf{M}^T\mathbf{W}^T)]$$

$$n_x^w = \frac{1}{k_n}[(\mathbf{UM}^u\mathbf{B}_y\mathbf{M}^{wT}\mathbf{W}^T)(\mathbf{UMB}_z\mathbf{M}^{wT}\mathbf{W}^T)$$

$$+ (\mathbf{UM}^u\mathbf{B}_y\mathbf{M}^T\mathbf{W}^T)(\mathbf{UMB}_z\mathbf{M}^{wwT}\mathbf{W}^T)$$

$$- (\mathbf{UMB}_y\mathbf{M}^{wwT}\mathbf{W}^T)(\mathbf{UM}^u\mathbf{B}_z\mathbf{M}^T\mathbf{W}^T)$$

$$- (\mathbf{UMB}_y\mathbf{M}^{wT}\mathbf{W}^T)(\mathbf{UM}^u\mathbf{B}_z\mathbf{M}^{wT}\mathbf{W}^T)] \qquad (3.61g)$$

$$n_x^{uw} = \frac{1}{k_n}[(\mathbf{UM}^{uu}\mathbf{B}_y\mathbf{M}^{wT}\mathbf{W}^T)(\mathbf{UMB}_z\mathbf{M}^{wT}\mathbf{W}^T)$$

$$+ (\mathbf{UM}^{uu}\mathbf{B}_y\mathbf{M}^T\mathbf{W}^T)(\mathbf{UMB}_z\mathbf{M}^{wwT}\mathbf{W}^T)$$

$$+ (\mathbf{UM}^u\mathbf{B}_y\mathbf{M}^{wT}\mathbf{W}^T)(\mathbf{UM}^u\mathbf{B}_z\mathbf{M}^{wT}\mathbf{W}^T)$$

$$+ (\mathbf{UM}^u\mathbf{B}_y\mathbf{M}^T\mathbf{W}^T)(\mathbf{UM}^u\mathbf{B}_z\mathbf{M}^{wwT}\mathbf{W}^T)$$

$$- (\mathbf{UM}^u\mathbf{B}_y\mathbf{M}^{wwT}\mathbf{W}^T)(\mathbf{UM}^u\mathbf{B}_z\mathbf{M}^T\mathbf{W}^T)$$

$$- (\mathbf{UM}^u\mathbf{B}_y\mathbf{M}^{wT}\mathbf{W}^T)(\mathbf{UM}^u\mathbf{B}_z\mathbf{M}^{wT}\mathbf{W}^T)$$

$$- (\mathbf{UMB}_y\mathbf{M}^{wwT}\mathbf{W}^T)(\mathbf{UM}^{uu}\mathbf{B}_z\mathbf{M}^T\mathbf{W}^T)$$

$$- (\mathbf{UMB}_y\mathbf{M}^{wT}\mathbf{W}^T)(\mathbf{UM}^{uu}\mathbf{B}_z\mathbf{M}^T\mathbf{W}^T)]$$

Evaluating each of these expressions at $u, w \in [0:1]$ [that is, $(0, 0)$, $(1, 0)$, $(0, 1)$, $(1, 1)$] determines the 16 elements of $\mathbf{B}_{nx}$. We can derive similar matrices for $\mathbf{B}_{ny}$ and $\mathbf{B}_{nz}$, thus determining all 48 elements of $\mathbf{B}_n$.

EXERCISES

1. Verify the differentiation of n_x to obtain n_x^u, n_x^w, and n_x^{uw}.

2. Verify the dimensions of the matrix expressions in Eq. 3.61g.

3. Can you think of an alternative way of expressing the normals patch? (*Hint:* See Section 8.)

4. Write a procedure to compute the unit normal vector at a point on a bicubic patch. Denote this as **UNV(CI, UW, N)**, where

CI(4, 4, 3) is the input array of geometric coefficients defining the bicubic patch;

UW(2) are the input values of u and w at the point; and

N(3) is the output array of components of the unit normal vector at the point.

5. Write a procedure to compute the geometric coefficients $\mathbf{B}_n$ of a bicubic patch representing the unit normal vectors at any point on a given bicubic patch. Denote this as **BN(BI, BO)**, where

BI(4, 4, 3) is the input array of geometric coefficients defining a bicubic patch, and

BO(4, 4, 3) is the output array of geometric coefficients $\mathbf{B}_n$.

4 PARAMETRIC SPACE OF A SURFACE

The parametric space of a surface patch differs from that of a curve, since we must account for an additional parametric variable. Thus, we have a set of three-dimensional parametric spaces defined by (u, w, x), (u, w, y), and (u, w, z) coordinates. As with pc curves, decomposing a bicubic patch into its parametric space components helps us understand its behavior in object space.

The surface in Fig. 3.12 has the following properties: $\mathbf{p}_{00}$ and $\mathbf{p}_{01}$ lie in the y, z plane; $\mathbf{p}_{10}$ and $\mathbf{p}_{11}$ lie in the x, y plane; each boundary curve is planar. The graphs on the left show the patch's components in parametric space. Shown at the top left in the u, w, x coordinate system is a component patch given by the equation $x = x(u, w)$; the u, w, y and u, w, z patches are given by similar equations. Again, notice that the parametric variables are constrained to the unit square $u, w \in [0, 1]$.

Holding one of the parametric variables constant defines a specific parametric curve on the patch; Figure 3.12 shows $\mathbf{p}_{iw}$. Every point on this

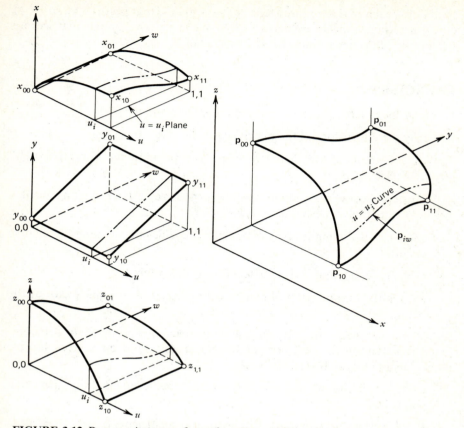

FIGURE 3.12 Parametric space of a surface. Four surfaces are shown.

curve has the same u value, and in parametric space, we see this curve decomposed into its components. These parametric curves are revealed by passing the plane $u = u_i$ through the component patches. Obviously, these curves are pc curves, and we treat them as such.

Again, all bicubic patches have the identical algebraic form, and what distinguishes one patch from another are the coefficients. These coefficients control the position of a patch in space and its size and shape. Thus, each set of values of u, w defines a unique point in x, y, z space on the bicubic patch. We obtain the entire bicubic patch in object space by plotting successive u, w values. Remember that each dependent variable x, y, z is independently controlled and that an entirely different patch results if we change only one algebraic coefficient of one of these variables.

Now for a bit of philosophy: Although we can generate as many points on the patch as we want, depending only on the size of the u, w increment we choose, these points are not the patch. These points are certainly a very convenient way of describing a patch. In the abstract, a patch (Eq. 3.12, for example) is a real,

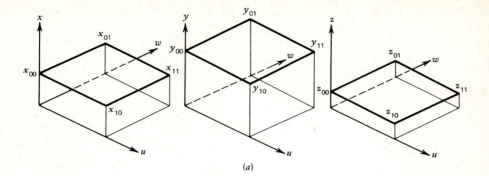

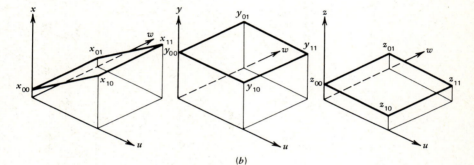

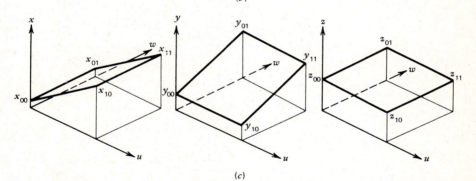

FIGURE 3.13 Parametric component surfaces of three special surfaces. Nine plane faces are shown in three groups of three.

analytic, continuous function. We operate on this function to generate points, slopes, and curves on the patch to determine a variety of analytic properties. It is practical to speak of points because to plot a patch, to make it visible, we must use and understand its point model.

As with curves, we deduce many characteristics of a bicubic patch through a careful inspection of the parametric component patches. Figure 3.13 shows three special cases; we call Figs. 3.13a and 3.13b degenerate patches. In Fig. 3.13a, a point is the result in object space; in Fig. 3.13b, a straight line; and in Fig. 3.13c, a plane. Figure 3.14 shows all three results in object space.

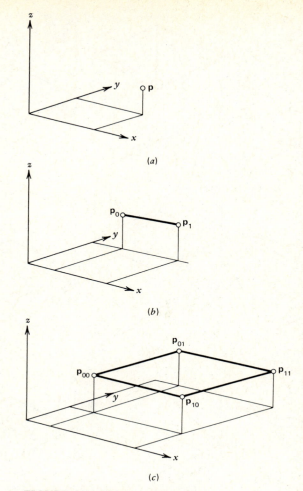

FIGURE 3.14 Three special surfaces.

5 BLENDING FUNCTIONS

The blending functions for the bicubic patch have the same form and serve the same purpose as curve blending functions. Since the patch has two independent parametric variables, we use the blending functions twice. We saw this in Eq. 3.52 and, of course, in the development preceding it.

We can apply the same tangent vector blending functions in Eq. 2.55 to patches. We obtain the two principal tangent vectors and the twist vector directly

$$\mathbf{p}^u(u, w) = \mathbf{F}^u(u)\mathbf{BF}(w)^T \tag{3.62}$$

$$\mathbf{p}^w(u, w) = \mathbf{F}(u)\mathbf{BF}^w(w)^T \tag{3.63}$$

$$\mathbf{p}^{uw}(u, w) = \mathbf{F}^u(u)\mathbf{BF}^w(w)^T \tag{3.64}$$

Or, in a form similar to Eq. 3.51, we have

$$\mathbf{p}^u = \mathbf{UM}^u\mathbf{BM}^T\mathbf{W}^T \tag{3.65}$$

$$\mathbf{p}^w = \mathbf{UMBM}^{wT}\mathbf{W}^T \tag{3.66}$$

$$\mathbf{p}^{uw} = \mathbf{UM}^u\mathbf{BM}^{wT}\mathbf{W}^T \tag{3.67}$$

Remember that $\mathbf{M}^u$ (and $\mathbf{M}^w$) are the universal transformation matrix defined in Eq. 2.59.

The Bezier and *B*-spline expressions for curves and surfaces also have blending functions. In fact, as you may have gathered by now, blending functions are the distinguishing feature of the various parametric forms. We are essentially free to construct whatever functions might apply to a modeling situation, with attention, of course, to the choice of appropriate boundary conditions and orthogonality.

6 REPARAMETRIZATION OF A SURFACE PATCH

The reparametrization of a surface patch proceeds in much the same way as for a pc curve. The simplest form of reparametrization is a reversal of the direction of one or both of the parametric variables u and w. Again, as with curves, we can do this easily, and it does not change the shape of the patch. Figure 3.15 shows the initial parametrization of a patch and the three possible

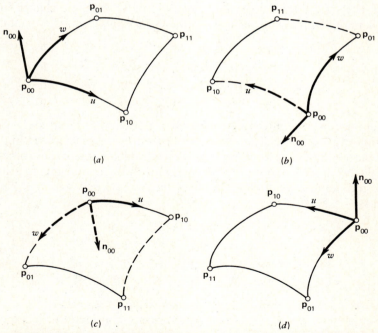

(a) (b)

(c) (d)

FIGURE 3.15 Reparametrization of a surface patch: three possible types of reversal. Four surfaces are shown.

cases of reversal. We will confine our attention to the **B** matrix, since this is the only element that changes in our formulation.

The matrix of geometric coefficients for the initial cases will be $\mathbf{B}_1$, where

$$\mathbf{B}_1 = \begin{bmatrix} \mathbf{p}_{00} & \mathbf{p}_{01} & \mathbf{p}_{00}^w & \mathbf{p}_{01}^w \\ \mathbf{p}_{10} & \mathbf{p}_{11} & \mathbf{p}_{10}^w & \mathbf{p}_{11}^w \\ \mathbf{p}_{00}^u & \mathbf{p}_{01}^u & \mathbf{p}_{00}^{uw} & \mathbf{p}_{01}^{uw} \\ \mathbf{p}_{10}^u & \mathbf{p}_{11}^u & \mathbf{p}_{10}^{uw} & \mathbf{p}_{11}^{uw} \end{bmatrix} , \tag{3.68}$$

First, let us reverse the parametric variable u, as in Fig. 3.15b. This means that all curves of constant w will change their direction of parametrization. To do this, first simply interchange rows 1 and 2 of the $\mathbf{B}_1$ matrix. Next, interchange rows 3 and 4, and multiply each coefficient in these rows by -1. Review the equations in Chapter 2, Section 5, and verify the relationship.

$$\mathbf{B}_2 = \begin{bmatrix} \mathbf{q}_{00} = \mathbf{p}_{10} & \mathbf{q}_{01} = \mathbf{p}_{11} & \mathbf{q}_{00}^w = \mathbf{p}_{10}^w & \mathbf{q}_{01}^w = \mathbf{p}_{11}^w \\ \mathbf{q}_{10} = \mathbf{p}_{00} & \mathbf{q}_{11} = \mathbf{p}_{01} & \mathbf{q}_{10}^w = \mathbf{p}_{00}^w & \mathbf{q}_{11}^w = \mathbf{p}_{01}^w \\ \mathbf{q}_{00}^u = -\mathbf{p}_{10}^u & \mathbf{q}_{01}^u = -\mathbf{p}_{11}^u & \mathbf{q}_{00}^{uw} = -\mathbf{p}_{10}^{uw} & \mathbf{q}_{01}^{uw} = -\mathbf{p}_{11}^{uw} \\ \mathbf{q}_{10}^u = -\mathbf{p}_{00}^u & \mathbf{q}_{11}^u = -\mathbf{p}_{01}^u & \mathbf{q}_{10}^{uw} = -\mathbf{p}_{00}^{uw} & \mathbf{q}_{11}^{uw} = -\mathbf{p}_{01}^{uw} \end{bmatrix} \tag{3.69}$$

Notice two things: First, the coefficients of the auxiliary curves are interchanged and multiplied by -1; second, the directions of the patch normals are reversed.

Next, starting with $\mathbf{B}_1$, reverse the parametric variable w, as in Fig. 3.15c. This operation will change the direction of parametrization of curves of constant u. This time, interchange columns 1 and 2 of $\mathbf{B}_1$. Also interchange columns 3 and 4, multiplying each coefficient in these rows by -1; thus, we obtain $\mathbf{B}_3$

$$\mathbf{B}_3 = \begin{bmatrix} \mathbf{r}_{00} = \mathbf{p}_{01} & \mathbf{r}_{01} = \mathbf{p}_{00} & \mathbf{r}_{00}^w = -\mathbf{p}_{01}^w & \mathbf{r}_{01}^w = -\mathbf{p}_{00}^w \\ \mathbf{r}_{10} = \mathbf{p}_{11} & \mathbf{r}_{11} = \mathbf{p}_{10} & \mathbf{r}_{10}^w = -\mathbf{p}_{11}^w & \mathbf{r}_{11}^w = -\mathbf{p}_{10}^w \\ \mathbf{r}_{00}^u = \mathbf{p}_{01}^u & \mathbf{r}_{01}^u = \mathbf{p}_{00}^u & \mathbf{r}_{00}^{uw} = -\mathbf{p}_{01}^{uw} & \mathbf{r}_{01}^{uw} = -\mathbf{p}_{00}^{uw} \\ \mathbf{r}_{10}^u = \mathbf{p}_{11}^u & \mathbf{r}_{11}^u = \mathbf{p}_{10}^u & \mathbf{r}_{10}^{uw} = -\mathbf{p}_{11}^{uw} & \mathbf{r}_{11}^{uw} = -\mathbf{p}_{10}^{uw} \end{bmatrix} \tag{3.70}$$

Again, notice that the directions of the patch normals are reversed.

Finally, reverse the direction of both parametric variables; see Fig. 3.15d. Again, start with $\mathbf{B}_1$; we can reverse either u or w first. Let us choose u, resulting in Eq. 3.69 or $\mathbf{B}_2$. Next, reverse w by operating on $\mathbf{B}_2$, interchanging columns 1 and 2, and 3 and 4, multiplying each coefficient in 3 and 4 by -1. Thus, we obtain $\mathbf{B}_4$

$$\mathbf{B}_4 = \begin{bmatrix} \mathbf{s}_{00} = \mathbf{p}_{11} & \mathbf{s}_{01} = \mathbf{p}_{10} & \mathbf{s}_{00}^w = -\mathbf{p}_{11}^w & \mathbf{s}_{01}^w = -\mathbf{p}_{10}^w \\ \mathbf{s}_{10} = \mathbf{p}_{01} & \mathbf{s}_{11} = \mathbf{p}_{00} & \mathbf{s}_{10}^w = -\mathbf{p}_{0w}^w & \mathbf{s}_{11}^w = -\mathbf{p}_{00}^w \\ \mathbf{s}_{00}^u = -\mathbf{p}_{11}^u & \mathbf{s}_{01}^u = -\mathbf{p}_{10}^u & \mathbf{s}_{00}^{uw} = \mathbf{p}_{11}^{uw} & \mathbf{s}_{01}^{uw} = \mathbf{p}_{10}^{uw} \\ \mathbf{s}_{10}^u = -\mathbf{p}_{01}^u & \mathbf{s}_{11}^u = -\mathbf{p}_{00}^u & \mathbf{s}_{10}^{uw} = \mathbf{p}_{01}^{uw} & \mathbf{s}_{11}^{uw} = \mathbf{p}_{00}^{uw} \end{bmatrix} \tag{3.71}$$

In this case, the directions of the patch normals are not reversed.

For a more generalized reparametrization, consider the bicubic patch in Fig. 3.16a. Here, the patch is parametrized from u_i to u_j and from w_k to w_l. Figure 3.16b shows the same patch except that the parametric variables range

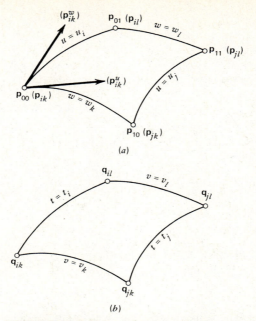

FIGURE 3.16 General reparametrization of a
surface.

from t_i to t_j and from v_k to v_l. Let $\mathbf{B}_1$ denote the matrix of geometric coefficients in the first case and $\mathbf{B}_2$ the matrix for the second. Fully expressed, they are

$$\mathbf{B}_1 = \begin{bmatrix} \mathbf{p}_{ik} & \mathbf{p}_{il} & \mathbf{p}_{ik}^w & \mathbf{p}_{il}^w \\ \mathbf{p}_{jk} & \mathbf{p}_{jl} & \mathbf{p}_{jk}^w & \mathbf{p}_{jl}^w \\ \mathbf{p}_{ik}^u & \mathbf{p}_{il}^u & \mathbf{p}_{ik}^{uw} & \mathbf{p}_{il}^{uw} \\ \mathbf{p}_{jk}^u & \mathbf{p}_{jl}^u & \mathbf{p}_{jk}^{uw} & \mathbf{p}_{jl}^{uw} \end{bmatrix} \tag{3.72}$$

$$\mathbf{B}_2 = \begin{bmatrix} \mathbf{q}_{ik} & \mathbf{q}_{il} & \mathbf{q}_{ik}^v & \mathbf{q}_{il}^v \\ \mathbf{q}_{jk} & \mathbf{q}_{jl} & \mathbf{q}_{jk}^v & \mathbf{q}_{jl}^v \\ \mathbf{q}_{ik}^t & \mathbf{q}_{il}^t & \mathbf{q}_{ik}^{tv} & \mathbf{q}_{il}^{tv} \\ \mathbf{q}_{jk}^t & \mathbf{q}_{jL}^t & \mathbf{q}_{jk}^{tv} & \mathbf{q}_{jl}^{tv} \end{bmatrix} \tag{3.73}$$

The relationships between the elements of $\mathbf{B}_1$ and $\mathbf{B}_2$ are the same as those developed for curves, and the same reasoning applies. For patches, as for curves, the corner points are related directly.

$$\begin{aligned} \mathbf{q}_{ik} &= \mathbf{p}_{ik} \\ \mathbf{q}_{jk} &= \mathbf{p}_{jk} \\ \mathbf{q}_{il} &= \mathbf{p}_{il} \\ \mathbf{q}_{jl} &= \mathbf{p}_{jl} \end{aligned} \tag{3.74}$$

Again, the tangent vectors are a different matter. To preserve the bicubic form of the patch equations, we assume a linear relationship between u and t

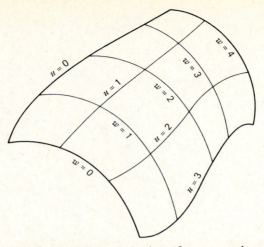

FIGURE 3.17 Parametrization of a rectangular array of surfaces.

and between w and v. When we do this, we obtain results similar to those expressed by the tangent vectors in Eqs. 2.70. In particular, we find that

$$\mathbf{q}^t = \frac{u_j - u_i}{t_j - t_i} \mathbf{p}^u \tag{3.75}$$

$$\mathbf{q}^v = \frac{w_l - w_k}{v_l - v_k} \mathbf{p}^w \tag{3.76}$$

For the cross-derivatives, we obtain

$$\mathbf{q}^{tv} = \frac{(u_j - u_i)(w_l - w_k)}{(t_j - t_i)(v_l - v_k)} \mathbf{p}^{uw} \tag{3.77}$$

If u_j and u_i are successive pairs of integers, then $u_j - u_i = 1$; similarly for $t_j - t_i$, $w_l - w_k$, and $v_l - v_k$. As we noted for curves, this relationship is useful when dealing with arrays of contiguous patches. In Fig. 3.17, individual patches in the array are readily identified by the array indexing scheme of successive integers.

EXERCISES

1. Reparametrize a patch from $u, w \in [0, 1]$ to $u^*, w^* \in [0, 1/3]$. How does this affect the geometric coefficients? How does this affect the algebraic coefficients in Eq. 3.10?

2. How does reparametrization affect the normal vector $\mathbf{n}(u, w)$?

3. Write a procedure to reverse the direction of parametrization of a bicubic patch. Denote this as **INVSRF(CI, IFLG, CO)**, where

CI(4, 4, 3) is the input array of geometric coefficients defining the bicubic patch; and

IFLG is an input flag whose value specifies which parametric variable is to be reversed.

IFLG = 1 specifies u is to be reversed,

IFLG = 2 specifies w is to be reversed, and

IFLG = 3 specifies both u and w are to be reversed.

CO(4, 4, 3) is the output array of geometric coefficients of the reparametrized bicubic patch.

4. Write a procedure to reparametrize a bicubic patch. Denote this as **RPSRF(CI, UMIN, UMAX, WMIN, WMAX, CO)**, where

CI(4, 4, 3) is the input array of geometric coefficients defining the bicubic patch;

UMIN is the input minimum value of u (originally $u = 0$);

UMAX is the input maximum value of u (originally $u = 1$);

WMIN is the input minimum value of w (originally $w = 0$);

WMAX is the input maximum value of w (originally $w = 1$); and

CO(4, 4, 3) is the output array of geometric coefficients of the reparametrized bicubic patch.

7 SUBDIVIDING

We can now apply techniques from Section 6 to create a new patch from an existing one. We state the problem as follows: Given a patch whose geometric coefficients are denoted $\mathbf{B}_1$, find the matrix $\mathbf{B}_2$ of a new patch that is a subpatch of the given patch and bounded by curves u_i, u_j, w_k, and w_l. From Fig. 3.18, we easily find the corner points of the new patch

$$\mathbf{q}_{00} = \mathbf{p}_{ik}$$
$$\mathbf{q}_{10} = \mathbf{p}_{jk}$$
$$\mathbf{q}_{01} = \mathbf{p}_{il}$$
$$\mathbf{q}_{11} = \mathbf{q}_{jl}$$

where the $\mathbf{p}$ vectors and $\mathbf{q}$ vectors are elements of $\mathbf{B}_1$ and $\mathbf{B}_2$, respectively.

Using Eqs. 3.65 and 3.76, the tangent vectors are

$$
\begin{aligned}
\mathbf{q}_{00}^t &= (u_j - u_i)\mathbf{p}_{ik}^u & \mathbf{q}_{00}^v &= (w_l - w_k)\mathbf{p}_{ik}^w \\
\mathbf{q}_{10}^t &= (u_j - u_i)\mathbf{p}_{jk}^u & \mathbf{q}_{10}^v &= (w_l - w_k)\mathbf{p}_{jk}^w \\
\mathbf{q}_{01}^t &= (u_j - u_i)\mathbf{p}_{il}^u & \mathbf{q}_{01}^v &= (w_l - w_k)\mathbf{p}_{il}^w \\
\mathbf{q}_{11}^t &= (u_j - u_i)\mathbf{p}_{jl}^u & \mathbf{q}_{11}^v &= (w_l - w_k)\mathbf{p}_{jl}^w
\end{aligned}
\qquad (3.79)
$$

Remember that $t_1 - t_0 = 1$ and $v_1 - v_0 = 1$.

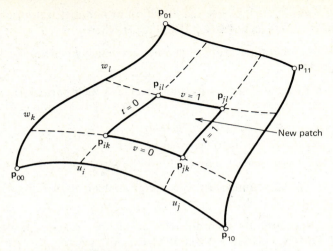

FIGURE 3.18 Subdividing a surface.

We obtain the cross-derivatives or twist vectors by evaluating $\mathbf{p}_{ik}^{uw}$, $\mathbf{p}_{jk}^{uw}$, $\mathbf{p}_{il}^{uw}$, and $\mathbf{p}_{jl}^{uw}$, and using Eq. 3.77

$$
\begin{aligned}
\mathbf{q}_{00}^{tv} &= (u_j - u_i)(w_l - w_k)\mathbf{p}_{ik}^{uw} \\
\mathbf{q}_{10}^{tv} &= (u_j - u_i)(w_l - w_k)\mathbf{p}_{jk}^{uw} \\
\mathbf{q}_{01}^{tv} &= (u_j - u_i)(w_l - w_k)\mathbf{p}_{il}^{uw} \\
\mathbf{q}_{11}^{tv} &= (u_j - u_i)(w_l - w_k)\mathbf{p}_{jl}^{uw}
\end{aligned}
\tag{3.80}
$$

Clearly, in this example the normals do not reverse direction.

EXERCISES

1. Subdivide a bicubic patch into nine subpatches along the constant parametric curves $u = 1/3$, $u = 2/3$, $w = 1/3$, and $w = 2/3$. Reparametrize the resulting 3×3 array of patches so that $u^*, w^* \in [0, 3]$, where u^* and w^* denote the new parametric variables. Express the elements of each of the nine new **B** matrices in terms of the original unsubdivided **B** matrix elements.

2. Reverse the direction of parametrization of w^* in Ex. 1, and repeat that exercise.

3. A bicubic patch is parametrized so that $u, w \in [0, 1]$. Subdivide and reparametrize the part of the patch bounded by $u = 0.1$, $u = 0.4$, $w = 0.6$, and $w = 0.8$ so that $u^*, w^* \in [0, 1]$.

4. Repeat Ex. 3 for $u^* \in [1/3, 2/3]$ and $w^* \in [1/4, 1/2]$.

5. Write a procedure to subdivide a bicubic patch. Assume that the original patch and the subdivided portion of it are parametrized so that $u, w \in [0, 1]$.

Denote this as **SUBSRF(CI, UMIN, UMAX, WMIN, WMAX, CO)**, where

> CI(4, 4, 3) is the input array of geometric coefficients defining the bicubic patch;

> UMIN is the input minimum value of u along which the patch is to be truncated;

> UMAX is the input maximum value of u along which the patch is to be truncated;

> WMIN is the input minimum value of w along which the patch is to be truncated;

> WMAX is the input maximum value of w along which the patch is to be truncated; and

> CO(4, 4, 3) is the output array of geometric coefficients defining the truncated, reparametrized bicubic patch.

8 SIXTEEN-POINT FORM

It is not always easy or practical, if possible at all, to provide the tangent and twist vector values required to define a bicubic patch. Therefore, we will now develop another way of representing a patch. Remember, there are 48 degrees of freedom or algebraic coefficients that we must specify. Consider the 4×4 grid of 16 points shown in Fig. 3.19; these points supply the 48 conditions. However, we must also choose values for u and w at these points. We can do this in several ways: We can estimate the u, w values for points other than the corner points by using a ratio of line segment lengths to get values between 0 and 1, or we can assume u, w values at the one-third points.

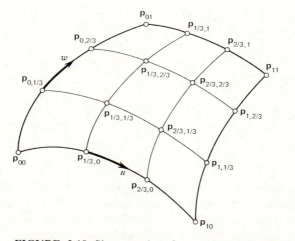

FIGURE 3.19 Sixteen-point form of a bicubic surface.

Now expand the algebraic form of the patch $\mathbf{p} = \mathbf{UAW}^T$ to obtain

$$\mathbf{a}_{33}u^3w^3 + \mathbf{a}_{32}u^3w^2 + \cdots + \mathbf{a}_{00} = \mathbf{p}(u, w) \qquad (3.81)$$

This is, of course, Eq. 3.10. We find we can generate 16 of these equations, one for each of the 16 points. Let us use the u, w values in Fig. 3.19. Thus,

$$\mathbf{a}_{33}(0)^3(0)^3 + \mathbf{a}_{32}(0)^3(0)^2 + \cdots + \mathbf{a}_{00} = \mathbf{p}(0, 0)$$

$$\mathbf{a}_{33}(1/3)^3(0)^3 + \mathbf{a}_{32}(1/3)^3(0)^2 + \cdots + \mathbf{a}_{00} = \mathbf{p}(1/3, 0)$$

$$\vdots \qquad\qquad\qquad\qquad\qquad\qquad \vdots$$

$$\mathbf{a}_{33}(2/3)^3(1/3)^3 + \mathbf{a}_{32}(2/3)^3(1/3)^2 + \cdots + \mathbf{a}_{00} = \mathbf{p}(2/3, 1/3) \qquad (3.82)$$

$$\vdots \qquad\qquad\qquad\qquad\qquad\qquad \vdots$$

$$\mathbf{a}_{33}(1)^3(1)^3 + \mathbf{a}_{32}(1)^3(1)^2 + \cdots + \mathbf{a}_{00} = \mathbf{p}(1, 1)$$

In matrix form, this set of equations becomes

$$\mathbf{Ea} = \mathbf{p} \qquad (3.83a)$$

or

$$\mathbf{a} = \mathbf{E}^{-1}\mathbf{p} \qquad (3.83b)$$

The result is a system of simultaneous linear equations whose unknowns are the elements of $\mathbf{a}$. $\mathbf{E}$ is a 16×16 matrix of uw products; $\mathbf{a}$ is a 16×1 vector of the unknown algebraic coefficients of the $\mathbf{A}$ matrix; and $\mathbf{p}$ is a 16×1 vector of the given data points. Notice that the solution to Eq. 3.83b involves three $\mathbf{A}$ matrices—one for each of the coordinates x, y, and z.

Next, we investigate the 16-point solution for the geometric form. We rewrite $\mathbf{p} = \mathbf{UMBM}^T\mathbf{W}^T$

$$\mathbf{p} = \mathbf{UNPN}^T\mathbf{W}^T \qquad (3.84a)$$

The $\mathbf{B}$ matrix is replaced by a matrix $\mathbf{P}$

$$\mathbf{P} = \begin{bmatrix} \mathbf{p}(0, 0) & \mathbf{p}(0, 1/3) & \mathbf{p}(0, 2/3) & \mathbf{p}(0, 1) \\ \mathbf{p}(1/3, 0) & \mathbf{p}(1/3, 1/3) & \mathbf{p}(1/3, 2/3) & \mathbf{p}(1/3, 1) \\ \mathbf{p}(2/3, 0) & \mathbf{p}(2/3, 1/3) & \mathbf{p}(2/3, 2/3) & \mathbf{p}(2/3, 1) \\ \mathbf{p}(1, 0) & \mathbf{p}(1, 1/3) & \mathbf{p}(1, 2/3) & \mathbf{p}(1, 1) \end{bmatrix} \qquad (3.84b)$$

If we do the indicated algebra, we find $\mathbf{N}$ to be

$$\mathbf{N} = \begin{bmatrix} -9/2 & 27/2 & -27/2 & 9/2 \\ 9 & -45/2 & 18 & -9/2 \\ -11/2 & 9 & -9/2 & 1 \\ 1 & 0 & 0 & 0 \end{bmatrix} \qquad (3.84c)$$

This is, of course, the same matrix we encountered in Chapter 2, Section 8 while investigating the four-point form for curves. With the 16-point formulation, the input data $\mathbf{P}$ are uniformly distributed over the surface of the patch. The $\mathbf{B}$ and $\mathbf{P}$ matrices are related as follows:

$$\mathbf{B} = \mathbf{LPL}^T \qquad (3.84d)$$

where

$$L = M^{-1}N = \begin{bmatrix} 1 & 0 & 0 & 0 \\ 0 & 0 & 0 & 1 \\ -11/2 & 9 & -9/2 & 1 \\ -1 & 9/2 & -9 & 11/2 \end{bmatrix}$$

(3.84e)

which we also encountered in Chapter 2, Section 8.

This approach, requiring only point data for input, is an improvement over the practical application for the surface-fitting problem, but certain characteristics remain that tend to make its use somewhat tedious. For example, when applying this approach to the representation of analytically known surfaces (spherical, parabolic sheets, and so on), a significant deviation arises through nonuniform spacing of the input points. Also, deciding on a patch distribution over a composite surface (Section 16) requires considerable care. Finally, since each patch is fit independently, continuity across the boundaries is neither guaranteed nor likely. For many modeling applications, these flaws are unimportant; still, they motivate us to make further improvements.

EXERCISES

1. Use the 16-point form to represent a unit normals patch. What can you say about the surface that is formed?

2. What effect do you expect a nonuniform distribution of 16 points to have on the shape of a patch they are used to define?

3. Derive a different 16-point form with points distributed as follows: along $w = 0$ and $w = 1$; at $u = 0$, 0.25, 0.5, 0.75, and 1; along $u = 0$ and $u = 1$; at $w = 0.25, 0.5,$ and 0.75. Express $\mathbf{p}$ as in Eq. 3.8b. Find $\mathbf{N}$ and $\mathbf{L}$. Sketch the point distribution on the patch (analogous to Fig. 3.19). Describe any difficulties you encounter and how you resolve them. What effect does this point distribution have on the twist vectors?

4. Given a bicubic patch defined by 16 points, show that if any or all of the four interior points are changed, then only the twist vectors are changed in the equivalent geometric form of the patch and that these changes are given by

$$\Delta\mathbf{p}_{00}^{uw} = (81/4)[4\Delta\mathbf{p}(1/3, 1/3) - 2\Delta\mathbf{p}(1/3, 2/3) - 2\Delta\mathbf{p}(2/3, 1/3) + \Delta\mathbf{p}(2/3, 2/3)]$$
$$\Delta\mathbf{p}_{10}^{uw} = (81/4)[4\Delta\mathbf{p}(2/3, 1/3) - 2\Delta\mathbf{p}(1/3, 1/3) - 2\Delta\mathbf{p}(2/3, 2/3) + \Delta\mathbf{p}(1/3, 2/3)]$$
$$\Delta\mathbf{p}_{01}^{uw} = (81/4)[4\Delta\mathbf{p}(1/3, 2/3) - 2\Delta\mathbf{p}(1/3, 1/3) - 2\Delta\mathbf{p}(2/3, 2/3) + \Delta\mathbf{p}(2/3, 1/3)]$$
$$\Delta\mathbf{p}_{11}^{uw} = (81/4)[4\Delta\mathbf{p}(2/3, 2/3) - 2\Delta\mathbf{p}(2/3, 1/3) - 2\Delta\mathbf{p}(2/3, 1/3) + \Delta\mathbf{p}(1/3, 1/3)]$$

5. Write a procedure to convert the 16-point form of a bicubic patch to the geometric form. Denote this as **P16G(P, G)**, where

P(4, 4, 3) is the input array of x, y, z coordinates of the 16 points (note that the array order should conform to that shown in Eq. 3.84b); and G(4, 4, 3) is the output array of geometric coefficients.

6. Write a procedure to convert the geometric form of a bicubic patch to the 16-point form. Denote this as **G16P(G, P)**, where

G(4, 4, 3) is the input array of geometric coefficients, and

P(4, 4, 3) is the output array of coordinates of 16 points.

7. Write a procedure to compute the coordinates of a point on a bicubic patch using the 16-point form. Compare this to evaluating a point on a bicubic patch using the geometric and algebraic forms. Denote this as **PSRF16(CI, U, W, P)**, where

CI(4, 4, 3) is the input array of x, y, z coordinates of the 16 points;

U is the input value of the parametric variable u;

W is the input value of the parametric variable w; and

P(3) is the output array of x, y, z coordinates of the point.

9   FOUR-CURVE FORM

We may choose to represent a bicubic patch with four pc curves (see Fig. 3.20). If these curves are in geometric form, the following procedure will generate the patch representation.

Step 1. Given four curves in geometric form, assign a u (or w) value to each, corresponding to a curve of constant u (or w) on the bicubic patch. For example, for the 16-point form, choose $u = 0$, $u = 1/3$, $u = 2/3$, and $u = 1$ curves. *Caution*: Do not mix u and w curves.

Step 2. Convert to four pc curves in four-point form.

Step 3. Convert these curves to a bicubic patch in 16-point form.

Step 4. Convert the 16-point form to geometric form (if necessary).

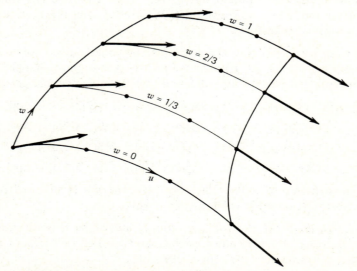

FIGURE 3.20 Four-curve form of a bicubic surface.

EXERCISES

1. Use four curves as boundary curves to define a bicubic patch. How is this an exception to the caution in Step 1 in Section 9? What are the consequences?

2. Write a procedure to compute the geometric coefficients of a bicubic patch, given four pc curves. Denote this as **C4SRF(CRV, IFLG, VAL, CO)**, where

CRV(4, 4, 3) is the input array of geometric coefficients of four pc curves, and

IFLG is an input number whose value indicates how the four curves are to be interpreted.

IFLG = 1 is they are boundary curves of the patch;

IFLG = 2 if they are constant u-value curves on the patch; and

IFLG = 3 if they are constant w-value curves on the patch.

VAL(4) is an input array of parametric values corresponding to constant-valued parametric curves on the patch, and

CO(4, 4, 3) is the output array of geometric coefficients of the patch.

10 PLANE SURFACE

There are several ways of constructing a plane patch; we will investigate three of them here. Figure 3.21 shows the simplest construction, which we interpret as follows:

$$\mathbf{p}(u, w) = \mathbf{p}_{00} + u\mathbf{r} + w\mathbf{s} \qquad u, w \in [0, 1] \tag{3.85}$$

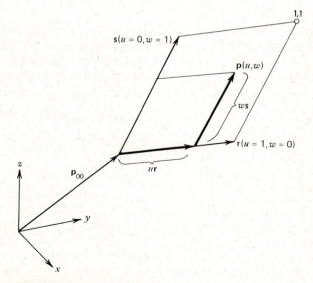

FIGURE 3.21 Vector equation of a plane.

This equation defines a plane patch through the point $\mathbf{p}_{00}$ and parallel to the vectors $\mathbf{r}$ and $\mathbf{s}$; it is a special case of the bicubic patch that becomes quite obvious if we review Eq. 3.10. We set all the algebraic coefficients equal to zero except $\mathbf{a}_{00}$, $\mathbf{a}_{10}$, and $\mathbf{a}_{01}$, resulting in

$$\mathbf{p}(u, w) = \mathbf{a}_{00} + u\mathbf{a}_{10} + w\mathbf{a}_{01} \tag{3.86}$$

Equation 3.86 is identical to Eq. 3.85, and we find $\mathbf{a}_{00} = \mathbf{p}_{00}$, $\mathbf{a}_{10} = \mathbf{r}$, and $\mathbf{a}_{01} = \mathbf{s}$. We can determine the geometric coefficients for this construction as well. From Eqs. 3.30–3.41, we find

$$
\begin{aligned}
\mathbf{p}_{00} &= \mathbf{p}_{00} & \mathbf{p}_{01} &= \mathbf{p}_{00} + \mathbf{s} \\
\mathbf{p}_{10} &= \mathbf{p}_{00} + \mathbf{r} & \mathbf{p}_{11} &= \mathbf{p}_{00} + \mathbf{r} + \mathbf{s} \\
\mathbf{p}_{00}^u &= \mathbf{r} & \mathbf{p}_{01}^u &= \mathbf{r} \\
\mathbf{p}_{10}^u &= \mathbf{r} & \mathbf{p}_{11}^u &= \mathbf{r} \\
\mathbf{p}_{00}^w &= \mathbf{s} & \mathbf{p}_{01}^w &= \mathbf{s} \\
\mathbf{p}_{10}^w &= \mathbf{s} & \mathbf{p}_{11}^w &= \mathbf{s}
\end{aligned}
\tag{3.87}
$$

We obtain the twist vectors from Eqs. 3.26–3.29

$$
\begin{aligned}
\mathbf{p}_{00}^{uw} &= 0 & \mathbf{p}_{01}^{uw} &= 0 \\
\mathbf{p}_{10}^{uw} &= 0 & \mathbf{p}_{11}^{uw} &= 0
\end{aligned}
\tag{3.88}
$$

We assemble the $\mathbf{B}$ matrix directly from these equations

$$
\mathbf{B} = \begin{bmatrix}
\mathbf{p}_{00} & \mathbf{p}_{00} + \mathbf{s} & \mathbf{s} & \mathbf{s} \\
\mathbf{p}_{00} + \mathbf{r} & \mathbf{p}_{00} + \mathbf{r} + \mathbf{s} & \mathbf{s} & \mathbf{s} \\
\mathbf{r} & \mathbf{r} & 0 & 0 \\
\mathbf{r} & \mathbf{r} & 0 & 0
\end{bmatrix}
\tag{3.89}
$$

Figure 3.22 illustrates a second method of construction. Here, we start with three points, $\mathbf{p}_{00}$, $\mathbf{p}_{10}$ and $\mathbf{p}_{01}$. From elementary geometry, we know that three points define a plane. We construct all the other geometric coefficients from these points and in such a way as to guarantee that the patch lies in this plane. Using the elements from the figures, we easily assemble the $\mathbf{B}$ matrix

$$
\mathbf{B} = \begin{bmatrix}
\mathbf{p}_{00} & \mathbf{p}_{01} & \mathbf{p}_{01} - \mathbf{p}_{00} & \mathbf{p}_{01} - \mathbf{p}_{00} \\
\mathbf{p}_{10} & \mathbf{p}_{10} + \mathbf{p}_{01} - \mathbf{p}_{00} & \mathbf{p}_{01} - \mathbf{p}_{00} & \mathbf{p}_{01} - \mathbf{p}_{00} \\
\mathbf{p}_{10} - \mathbf{p}_{00} & \mathbf{p}_{10} - \mathbf{p}_{00} & 0 & 0 \\
\mathbf{p}_{10} - \mathbf{p}_{00} & \mathbf{p}_{10} - \mathbf{p}_{00} & 0 & 0
\end{bmatrix}
\tag{3.90}
$$

The final plane patch construction is considerably more generalized than the first two. It permits us to construct plane figures with curved boundaries. As we see in Fig. 3.23, we do not restrict the boundary curves to being straight lines. Start with the point $\mathbf{p}_{00}$ and the two tangent vectors $\mathbf{p}_{00}^u$ and $\mathbf{p}_{00}^w$. We use these tangent vectors as a basis in the plane of the patch and define the other geometric coefficients as linear combinations of them. Once again, we can

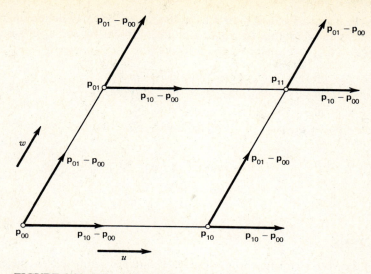

FIGURE 3.22 Special bicubic form of a plane.

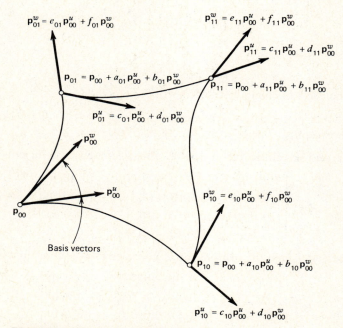

FIGURE 3.23 General bicubic form of a plane.

assemble the elements of a **B** matrix from the figure. We may define the cross-derivatives or twist vectors to be zero or define them by some relationship, such as $\mathbf{p}_{10}^{uw} = (c_{11} - c_{10})\mathbf{p}_{00}^u + (d_{11} - d_{10})\mathbf{p}_{00}^w$.

$$
\mathbf{B} = \begin{bmatrix}
\mathbf{p}_{00} & \mathbf{p}_{00} + a_{01}\mathbf{p}_{00}^u + b_{01}\mathbf{p}_{00}^w \\
\mathbf{p}_{00} + a_{10}\mathbf{p}_{00}^u + b_{10}\mathbf{p}_{00}^w & \mathbf{p}_{00} + a_{11}\mathbf{p}_{00}^u + b_{11}\mathbf{p}_{00}^w \\
\mathbf{p}_{00}^u & c_{01}\mathbf{p}_{00}^u + d_{01}\mathbf{p}_{00}^w \\
c_{10}\mathbf{p}_{00}^u + d_{10}\mathbf{p}_{00}^w & c_{11}\mathbf{p}_{00}^u + d_{11}\mathbf{p}_{00}^w
\end{bmatrix}
$$

$$
\begin{bmatrix}
\mathbf{p}_{00}^w & e_{01}\mathbf{p}_{00}^u + f_{01}\mathbf{p}_{00}^w \\
e_{10}\mathbf{p}_{00}^u + f_{10}\mathbf{p}_{00}^w & e_{11}\mathbf{p}_{00}^u + f_{11}\mathbf{p}_{00}^w \\
0 & 0 \\
0 & 0
\end{bmatrix}
\tag{3.91}
$$

EXERCISES

1. Using the vector construction in Fig. 3.21, find the geometric coefficients of a plane patch with

$$\mathbf{a}_{00} = \begin{bmatrix} 1 & 1 & 0 \end{bmatrix}$$

$$\mathbf{a}_{10} = \begin{bmatrix} 4 & 0 & 0 \end{bmatrix}$$

and

$$\mathbf{a}_{01} = \begin{bmatrix} 0 & 1 & 4 \end{bmatrix}$$

2. Find the geometric coefficients of a plane patch given three points,

$$\mathbf{p}_{00} = \begin{bmatrix} 0 & 0 & 3 \end{bmatrix}$$

$$\mathbf{p}_{10} = \begin{bmatrix} 3 & 0 & 0 \end{bmatrix}$$

and

$$\mathbf{p}_{01} = \begin{bmatrix} 0 & 3 & 0 \end{bmatrix}$$

Describe the effect of changing the order of these points.

3. Show that a plane patch is preserved when $\mathbf{p}^{uw} = 0$ if all the geometric coefficients except $\mathbf{p}_{00}$ are given by functions of the basis vectors $\mathbf{p}_{00}^u$ and $\mathbf{p}_{00}^w$.

4. Describe and sketch the bicubic patch whose geometric coefficients are

$$
\mathbf{B} = \begin{bmatrix}
(0,0,0) & (0,10,0) & (0,10,0) & (0,10,0) \\
(10,0,0) & (10,10,0) & (0,10,0) & (0,10,0) \\
(10,0,0) & (10,0,0) & (0,0,0) & (0,0,0) \\
(10,0,0) & (10,0,0) & (0,0,0) & (0,0,0)
\end{bmatrix}
$$

5. Describe and sketch the bicubic patch whose geometric coordinates are

$$
\mathbf{B} = \begin{bmatrix}
(0,0,0) & (0,9,0) & (0,9,0) & (0,9,0) \\
(10,2,0) & (8,10,0) & (0,8,0) & (-4,8,0) \\
(10,0,0) & (8,2,0) & (0,0,0) & (0,0,0) \\
(10,4,0) & (8,0,0) & (0,0,0) & (0,0,0)
\end{bmatrix}
$$

6. Write a procedure to compute the geometric coefficients of a planar bicubic patch defined by the three vectors expressed in Eq. 3.85, $\mathbf{p}_{00}$, $\mathbf{r}$, and $\mathbf{s}$. Denote this as **PLN(P, R, S, CO)**, where

P(3) is the input array of components of $\mathbf{p}_{00}$;

R(3) is the input array of components of $\mathbf{r}$;

S(3) is the input array of components of $\mathbf{s}$; and

CO(4, 4, 3) is the output array of geometric coefficients of the planar bicubic patch.

11   CYLINDRICAL SURFACE

A cylindrical surface is a special case of a ruled surface. A cylinder is a surface generated by a straight line as it moves parallel to itself along a curve. Here, we will construct a bicubic cylindrical surface patch. In Fig. 3.24, we see a curve whose geometric coefficients are $[\mathbf{p}_0 \quad \mathbf{p}_1 \quad \mathbf{p}_0^u \quad \mathbf{p}_1^u]^T$ and a straight line defined from $\mathbf{p}_0$ to $\mathbf{p}_2$. From Eq. 2.118, the geometric coefficients of the straight line are $[\mathbf{p}_0 \quad \mathbf{p}_2 \quad (\mathbf{p}_2 - \mathbf{p}_0) \quad (\mathbf{p}_2 - \mathbf{p}_0)]^T$. Using these elements, we readily find the **B** matrix for a patch defined by the given curve and line

$$\mathbf{B} = \begin{bmatrix} \mathbf{p}_0 & \mathbf{p}_2 & \mathbf{p}_2 - \mathbf{p}_0 & \mathbf{p}_2 - \mathbf{p}_0 \\ \mathbf{p}_1 & \mathbf{p}_1 + \mathbf{p}_2 - \mathbf{p}_0 & \mathbf{p}_2 - \mathbf{p}_0 & \mathbf{p}_2 - \mathbf{p}_0 \\ \mathbf{p}_0^u & \mathbf{p}_0^u & 0 & 0 \\ \mathbf{p}_1^u & \mathbf{p}_1^u & 0 & 0 \end{bmatrix} \quad (3.92a)$$

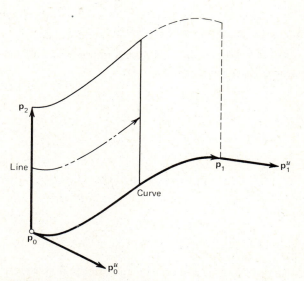

FIGURE 3.24 Cylindrical surface.

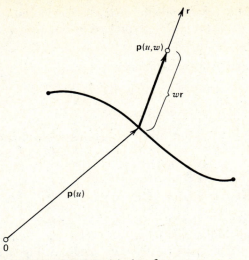

FIGURE 3.25 Cylindrical surface.

A more general expression is

$$\mathbf{p}(u, w) = \mathbf{p}(u) + w\mathbf{r} \qquad (3.92b)$$

where $\mathbf{p}(u)$ is any space curve and $\mathbf{r}$ is the direction vector of the straight-line rulings (see Fig. 3.25). We do not restrict the function $\mathbf{p}(u)$ to the cubic polynomial form.

EXERCISES

1. Why are the twist vectors zero in Eq. 3.92a?

2. Find the expression for a cylindrical patch defined by a Bezier curve whose four-point characteristic polygon is given by

$$\mathbf{p}_0 = [1 \quad 4 \quad 0]$$
$$\mathbf{p}_1 = [4 \quad 8 \quad 0]$$
$$\mathbf{p}_2 = [12 \quad 4 \quad 0]$$

and

$$\mathbf{p}_3 = [16 \quad 8 \quad 0]$$

and where

$$\mathbf{r} = [0 \quad 0 \quad 4]$$

Sketch the expected shape of the patch.

3. Follow instructions in Ex. 2 but consider a periodic *B*-spline curve with $k = 4$ and the additional points $\mathbf{p}_4 = [20 \quad 16 \quad 4]$ and $\mathbf{p}_5 = [24 \quad 18 \quad 6]$.

4. Write a procedure to compute the geometric coefficients of a cylindrical bicubic patch. Denote this as **CYLSRF(CRV, LN, CO)**, where

CRV(4, 3) is the input array of geometric coefficients of a pc curve;

LN(3) is the input array of components of a vector defining the magnitude and direction of the straight-line elements of the cylindrical surface; and

CO(4, 4, 3) is the output array of geometric coefficients.

Note: You must specify if the straight-line elements are lines of constant u or w, which can be done by using an input flag or assuming one or the other.

5. Devise a way to vary in a controlled manner the length of the straight-line elements of the patch defined by **CYLSRF**. Try to do this for the ends of the straight-line elements along only one edge of the patch first, then try it for both edges.

12 RULED SURFACE

We now know that the locus of a moving point with one degree of freedom is a curve. We also know that the locus of a straight line moving with one degree of freedom is a surface. It is a special type of surface, called a **ruled surface**. We can define a ruled surface as a surface such that through each point of it passes at least one straight line lying entirely in the surface.

The simplest of all ruled surfaces are the plane, the cone, and the cylinder, all special cases. The hyperbolic paraboloid and the hyperboloid of one sheet are also special ruled surfaces. Each of these surfaces exhibits two mathematically distinct families of straight lines (also called **rulings**). We will investigate one of these later in this section.

Every developable surface is ruled, but every ruled surface is not developable. A developable surface is one that can be "unrolled" onto a plane without stretching or distorting it; cones and cylinders are good examples. However, the hyperbolic paraboloid is not developable. In a later chapter, we will examine the special testable characteristics of developable surfaces. Figure 3.26 shows the key ingredients of a bicubic ruled surface patch. Given two pc curves $\mathbf{p}(u)$ and $\mathbf{q}(u)$, we construct a ruled surface by joining with a straight line each point on $\mathbf{p}(u)$ to a point on $\mathbf{q}(u)$ having an equivalent u value. We fill in the **B**-matrix elements accordingly

$$\mathbf{B} = \begin{bmatrix} \mathbf{p}_0 & \mathbf{q}_0 & \mathbf{q}_0 - \mathbf{p}_0 & \mathbf{q}_0 - \mathbf{p}_0 \\ \mathbf{p}_1 & \mathbf{q}_1 & \mathbf{q}_1 - \mathbf{p}_1 & \mathbf{q}_1 - \mathbf{p}_1 \\ \mathbf{p}_0^u & \mathbf{q}_0^u & 0 & 0 \\ \mathbf{p}_1^u & \mathbf{q}_1^u & 0 & 0 \end{bmatrix} \tag{3.93}$$

Each parametric curve of constant u value is a straight line.

The hyperbolic paraboloid is a special case, since both families of parametric curves $\mathbf{p}_{ui}$ and $\mathbf{p}_{jw}$ are straight lines. Given the four corner points, we define the

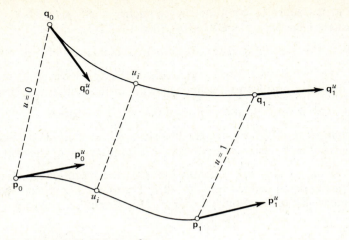

FIGURE 3.26 Ruled surface.

eight tangent vectors as straight lines and the four twist vectors as equal to zero, see Fig. 3.27. From this, the **B** matrix follows directly

$$
\mathbf{B} = \begin{bmatrix}
\mathbf{p}_1 & \mathbf{p}_3 & \mathbf{p}_3 - \mathbf{p}_1 & \mathbf{p}_3 - \mathbf{p}_1 \\
\mathbf{p}_2 & \mathbf{p}_4 & \mathbf{p}_4 - \mathbf{p}_2 & \mathbf{p}_4 - \mathbf{p}_2 \\
\mathbf{p}_2 - \mathbf{p}_1 & \mathbf{p}_4 - \mathbf{p}_3 & 0 & 0 \\
\mathbf{p}_2 - \mathbf{p}_1 & \mathbf{p}_4 - \mathbf{p}_3 & 0 & 0
\end{bmatrix}
\tag{3.94}
$$

Can the representation be improved if we use nonzero twist vectors? How and why? (*Hint*: Try $\mathbf{p}_{00}^{uw} = (\mathbf{p}_4 - \mathbf{p}_3) - (\mathbf{p}_2 - \mathbf{p}_1)$, and so on.)

Let us consider two more ways of expressing a ruled surface by a parametric equation. The first is

$$
\mathbf{p}(u, w) = \mathbf{g}(u) + w\mathbf{d}(u)
\tag{3.95}
$$

where $\mathbf{g}(u)$ is a curve and $\mathbf{d}(u)$ is the direction vector of a straight line at point u on the curve. The parameter w gives the distance ratio of the point $\mathbf{p}(u, w)$ from $\mathbf{g}(u)$; see Fig. 3.28. If $u, w \in [0, 1]$, then $|\mathbf{d}(u)|$ equals the length of the straight line when $w = 1$. The vector functions $\mathbf{g}(u)$ and $\mathbf{d}(u)$ are not restricted to the cubic polynomial form.

An alternative expression based on the straight-line rulings joining corresponding points on two space curves $\mathbf{d}(u)$ and $\mathbf{h}(u)$ is

$$
\begin{aligned}
\mathbf{p}(u, w) &= (1 - w)\mathbf{g}(u) + w\mathbf{h}(u) \\
&= \mathbf{g}(u) + w[\mathbf{h}(u) - \mathbf{g}(u)]
\end{aligned}
\tag{3.96}
$$

where $u, w \in [0, 1]$. Again, the vector functions $\mathbf{g}(u)$ and $\mathbf{h}(u)$ are not restricted to cubic polynomials. Figure 3.29 illustrates the vector construction of a ruled surface via Eq. 3.96.

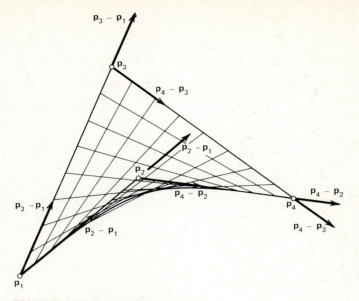

FIGURE 3.27 Hyperbolic paraboloid.

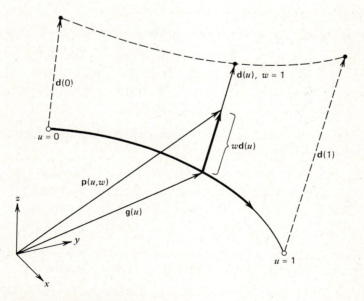

FIGURE 3.28 Ruled surface.

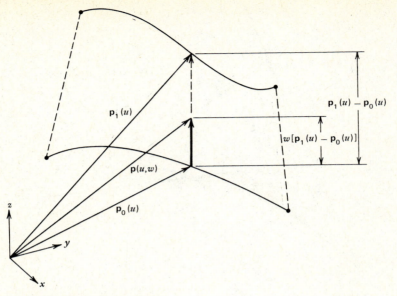

FIGURE 3.29 Ruled surface.

EXERCISES

1. Is it possible to construct a ruled surface between a Bezier and a *B*-spline curve? If so, describe the important steps in doing this.

2. Derive the general parametric equations for a cone.

3. Describe and sketch the bicubic patch whose geometric coefficients are

$$\mathbf{B} = \begin{bmatrix} (0,0,0) & (0,10,4) & (0,10,4) & (0,10,4) \\ (10,0,0) & (10,10,4) & (0,10,-1) & (0,10,-1) \\ (10,0,0) & (10,0,0) & (0,0,0) & (0,0,0) \\ (10,0,0) & (10,0,0) & (0,0,0) & (0,0,0) \end{bmatrix}$$

4. Show that the bicubic patch whose geometric coefficients are given by the following matrix is a parabolic cone:

$$\mathbf{B} = \begin{bmatrix} (0,0,10) & (0,10,5) & (0,10,-5) & (0,10,-5) \\ (10,0,0) & (5,10,0) & (-5,10,0) & (-5,10,0) \\ (10,0,0) & (5,0,0) & (-5,0,0) & (-5,0,0) \\ (10,0,-20) & (5,0,-10) & (-5,0,10) & (-5,0,10) \end{bmatrix}$$

5. Write a procedure to compute the geometric coefficients of a ruled bicubic patch. Denote this as **RULSRF(CRV1, CRV2, IUW, CO)**, where

CRV1(4, 3) is the input array of geometric coefficients of one of the bounding curves of the ruled surface;

CRV2(4, 3) is the input array of geometric coefficients of the other bounding curve; and

IUW is an input number specifying that the rulings are lines of constant *u* or constant *w*.

CO(4, 4 3) is the output array of geometric coefficients of the ruled bicubic patch.

Note: As with **CYLSRF**, you must specify if the rulings are lines of constant *u* or *w*, unless some default condition is assumed.

13 SURFACE OF REVOLUTION

We can generate a surface of revolution by revolving a plane curve around an axis line in its plane (see Fig. 3.30). The plane curve is called the **profile** curve, and in its various positions around the axis, it creates **meridians**. The circles created by each point on this curve are called **parallels**. For the simplest case, let the *z*-axis be the axis of rotation, and let the curve $\mathbf{p}(u) = \mathbf{x}(u) + \mathbf{z}(u)$ be defined in the *x, z* plane. Then, the surface of revolution has the equation

$$\mathbf{p}(u, \theta) = x(u) \cos \theta + x(u) \sin \theta + z(u) \tag{3.97}$$

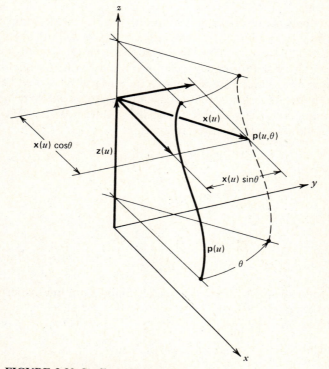

FIGURE 3.30 Surface of revolution.

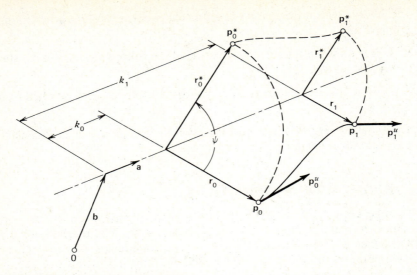

FIGURE 3.31 Surface of revolution.

Of course, the profile curve might be a general curve, elliptic segment, circular arc, and so on.

For a more general surface of revolution, consider the situation depicted in Fig. 3.31. The given or initial elements are a pc curve defined by its geometric coefficients, a unit vector **a** defining the direction of the axis of revolution, a vector **b** locating a point through which the axis passes, and the angle ψ through which the curve is rotated to sweep across the surface. We would like to find a bicubic patch that describes such a surface in terms of these elements. First, we must determine the scalars k_0 and k_1 and the vectors $\mathbf{r}_0$ and $\mathbf{r}_1$. We do this by solving the following sets of vector equations:

$$\mathbf{b} + k_0\mathbf{a} + \mathbf{r}_0 = \mathbf{p}_0 \tag{3.98}$$

$$\mathbf{a} \cdot \mathbf{r}_0 = 0 \tag{3.99}$$

and

$$\mathbf{b} + k_1\mathbf{a} + \mathbf{r}_1 = \mathbf{p}_1 \tag{3.100}$$

$$\mathbf{a} \cdot \mathbf{r}_1 = 0 \tag{3.101}$$

The condition asserted by $\mathbf{a} \cdot \mathbf{r}_0 = 0$ makes **a** and $\mathbf{r}_0$ mutually perpendicular; similarly for **a** and $\mathbf{r}_1$.

From Eqs. 3.98 and 3.99, we obtain

$$k_0 = \mathbf{a} \cdot (\mathbf{p}_0 - \mathbf{b}) \tag{3.102}$$

$$\mathbf{r}_0 = \mathbf{p}_0 - k_0\mathbf{a} - \mathbf{b} \tag{3.103}$$

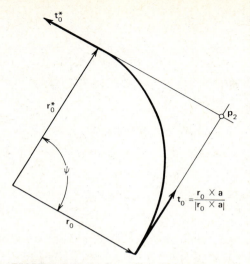

FIGURE 3.32 Surface of revolution: circumferential tangent vectors.

And from Eqs. 3.100 and 3.101, we obtain

$$k_1 = \mathbf{a} \cdot (\mathbf{p}_1 - \mathbf{b}) \tag{3.104}$$

$$\mathbf{r}_1 = \mathbf{p}_1 - k_1\mathbf{a} - \mathbf{b} \tag{3.105}$$

Now, we can proceed to find the other two corner points $\mathbf{p}_0^*$ and $\mathbf{p}_1^*$. Clearly, $\mathbf{p}_0^* = \mathbf{b} + k_0\mathbf{a} + \mathbf{r}_0^*$ and $\mathbf{p}_1^* = \mathbf{b} + k_1\mathbf{a} + \mathbf{r}_1^*$. Thus, we must determine $\mathbf{r}_0^*$ and $\mathbf{r}_1^*$; see Fig. 3.32.

The unit tangents to the circular arc at $\mathbf{p}_0$ and $\mathbf{p}_0^*$ are denoted by $\mathbf{t}_0$ and $\mathbf{t}_0^*$, respectively. We will have to determine these vectors to find the final tangent vectors at these points. The unit vector in the direction of the vector product of $\mathbf{r}_0$ and $\mathbf{a}$ produces $\mathbf{t}_0$

$$\mathbf{t}_0 = \frac{\mathbf{r}_0 \times \mathbf{a}}{|\mathbf{r}_0 \times \mathbf{a}|} \tag{3.106}$$

We can define the vector $\mathbf{r}_0^*$ as the sum of two components, one in the direction of $\mathbf{r}_0$ and one in the direction of $\mathbf{t}_0$. We begin with the observation that the magnitude of $\mathbf{r}_0^*$ equals the magnitude of $\mathbf{r}_0$. Applying elementary trigonometry and vector arithmetic, we obtain

$$\mathbf{r}_0^* = |\mathbf{r}_0| \frac{\mathbf{r}_0}{|\mathbf{r}_0|} \cos \psi + |\mathbf{r}_0| \frac{\mathbf{r}_0 \times \mathbf{a}}{|\mathbf{r}_0 \times \mathbf{a}|} \sin \psi \tag{3.107}$$

or somewhat simplified

$$\mathbf{r}_0^* = \mathbf{r}_0 \cos \psi + |\mathbf{r}_0| \frac{\mathbf{r}_0 \times \mathbf{a}}{|\mathbf{r}_0 \times \mathbf{a}|} \sin \psi \tag{3.108}$$

We find $\mathbf{r}_1^*$ in a similar way

$$\mathbf{r}_1^* = \mathbf{r}_1 \cos \psi + |\mathbf{r}_1| \frac{\mathbf{r}_1 \times \mathbf{a}}{|\mathbf{r}_1 \times \mathbf{a}|} \sin \psi \tag{3.109}$$

We can also define the unit tangent vector $\mathbf{t}_0^*$ as the sum of two components in the directions of $\mathbf{t}_0$ and $\mathbf{r}_0$

$$\mathbf{t}_0^* = |\mathbf{t}_0^*|\mathbf{t}_0 \cos \psi - |\mathbf{t}_0^*| \frac{\mathbf{r}_0}{|\mathbf{r}_0|} \sin \psi \tag{3.110}$$

or

$$\mathbf{t}_0^* = \frac{\mathbf{r}_0 \times \mathbf{a}}{|\mathbf{r}_0 \times \mathbf{a}|} \cos \psi - \frac{\mathbf{r}_0}{|\mathbf{r}_0|} \sin \psi \tag{3.111}$$

We know that $\mathbf{t}_1 = \mathbf{t}_0$ and $\mathbf{t}_1^* = \mathbf{t}_0^*$.

We must now find the magnitudes of these tangent vectors. Recall how we developed the tangent vectors for circular arcs in Chapter 2, Section 11. There, we used a point $\mathbf{p}_2$ lying on the intersection of the tangent lines at the two end points of the arc. We will do the same now. We find $\mathbf{p}_2$ by computing the intersection of the two straight lines $\mathbf{p}_0 + m\mathbf{t}_0$ and $\mathbf{p}_0^* + n\mathbf{t}_0^*$. By symmetry, we see that $n = -m$. Furthermore,

$$\mathbf{p}_0 + m\mathbf{t}_0 = \mathbf{p}_0^* - m\mathbf{t}_0^* \tag{3.112}$$

and

$$m = \frac{|\mathbf{p}_0^* - \mathbf{p}_0|}{|\mathbf{t}_0 + \mathbf{t}_0^*|} \tag{3.113}$$

From these, we obtain

$$\mathbf{p}_2 = \mathbf{p}_0 + \frac{|\mathbf{p}_0^* - \mathbf{p}_0|}{|\mathbf{t}_0 + \mathbf{t}_0^*|} \mathbf{t}_0 \tag{3.114}$$

Using Eq. 2.139 with appropriate substitutions, we find

$$\mathbf{p}_{00}^w = 4|\mathbf{r}_0| \left[\frac{1 - \cos (\psi/2)}{\sin (\psi/2)} \right] \left[\frac{\mathbf{p}_2 - \mathbf{p}_0}{|\mathbf{p}_2 - \mathbf{p}_0|} \right] \tag{3.115}$$

and

$$\mathbf{p}_{01}^w = 4|\mathbf{r}_0| \left[\frac{1 - \cos (\psi/2)}{\sin (\psi/2)} \right] \left[\frac{\mathbf{p}_0^* - \mathbf{p}_2}{|\mathbf{p}_0^* - \mathbf{p}_2|} \right] \tag{3.116}$$

We use this same procedure to find $\mathbf{p}_{10}^w$ and $\mathbf{p}_{11}^w$ at the other end of the patch. Notice that we have changed the nomenclature: We now call point $\mathbf{p}_0^*$, $\mathbf{p}_{01}$, and, of course, $\mathbf{p}_0$ of the initial curve is $\mathbf{p}_{00}$ on the patch. It should help to keep in mind that we are creating a surface patch by sweeping a curve through an arc around an axis.

Next, we compute the tangent vector $\mathbf{p}_{01}^u$, which is, of course, the result of rotating $\mathbf{p}_0^u$ through the angle ψ around $\mathbf{a}$. First, we see that $|\mathbf{p}_{01}^u| = |\mathbf{p}_0^u|$ (see

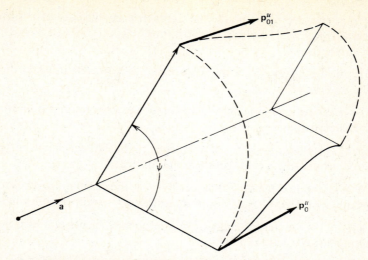

FIGURE 3.33 Surface of revolution: axial tangent vectors.

Fig. 3.33). Also, both vectors have identical projections on the unit vector **a**, so $\mathbf{p}_{01}^u \cdot \mathbf{a} = \mathbf{p}_0^u \cdot \mathbf{a}$.

We use the symmetry of the construction and realize that the projection of $\mathbf{p}_{01}^u$ onto the unit vector in the direction of $\mathbf{r}_0^*$ has the same magnitude as the projection of $\mathbf{p}_0^u$ on to the unit vector in the direction of $\mathbf{r}_0$. Now, we are ready to define $\mathbf{p}_{01}^u$ as the sum of its components along **a** and $\mathbf{r}_0^*$.

$$\mathbf{p}_{01}^u = (\mathbf{p}_0^u \cdot \mathbf{a})\mathbf{a} + \left(\mathbf{p}_0^u \cdot \frac{\mathbf{r}_0}{|\mathbf{r}_0|}\right)\left(\frac{\mathbf{r}_0^*}{|\mathbf{r}_0^*|}\right) \tag{3.117}$$

The tangent vector at the other end $\mathbf{p}_{11}^u$ is found the same way. All the required elements of the patch **B** matrix are determined except the twist vectors. This is left as an exercise at the end of Section 13.

To conclude this discussion on surfaces of revolution, we observe that we could also easily use the 16-point form of a bicubic patch. In this case, we convert the initial curve into the four-point form and then rotate each of these four points to their final positions at $\psi/3$, $2\psi/3$, and ψ.

EXERCISES

1. Develop and discuss the contribution of twist vectors, if any, to the bicubic representation of a surface of revolution.

2. Compare the 16-point form to the Hermite form of bicubic patch representation of a surface of revolution.

3. Show how Bezier and *B*-spline curves can be used to represent surfaces of revolution.

4. Write a procedure to compute the geometric coefficients of a surface of revolution defined by a bicubic patch. Denote this as **SRFREV(A, B, CRV, IUW, CO)**, where

A(3) is the input array of components of the unit vector defining the direction of the axis of revolution;

B(3) is the input array of components of the vector locating a point through which the axis passes;

CRV(4, 3) is the input array of geometric coefficients of the pc curve to be rotated around the axis to produce the surface;

IUW is an input value such that if IUW = 1, the circular arcs are curves of constant u or if IUW = 2, the circular arcs are curves of constant w; and

CO(4, 4, 3) is the output array of geometric coefficients of a bicubic patch defining a surface of revolution.

Note: You must specify if the circular arcs are curves of constant u or w.

14 SPHERICAL SURFACE

We now have several techniques at our disposal for creating a patch of a spherical surface. First, though, let us quickly review the classical parametric equation of a complete sphere, which is illustrated in Fig. 3.34. Here, we are restricted to a

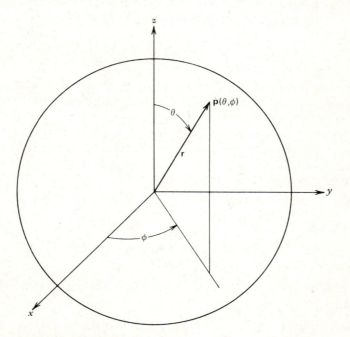

FIGURE 3.34 Spherical surface.

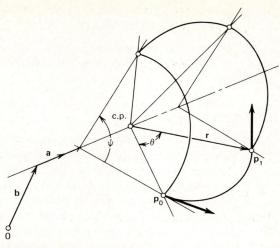

FIGURE 3.35 Spherical surface.

sphere centered at the origin. Let r be the radius of the sphere and θ the angle that the radius vector to the point $\mathbf{p}(\theta, \phi)$ makes with the z-axis. The angle ϕ is taken around the z-axis and measured from the x-axis. The result is

$$\mathbf{p}(\theta, \phi) = (r \sin \theta \cos \phi)\mathbf{i} + (r \sin \theta \sin \phi)\mathbf{j} + (r \cos \theta)\mathbf{k} \qquad (3.118)$$

where $\mathbf{i}$, $\mathbf{j}$, and $\mathbf{k}$ are unit vectors in the x, y, z directions, respectively.

A much more interesting approach is to rotate a circular arc around an axis; see Fig. 3.35. We can approximate the arc by a pc curve. We must meet certain conditions: First, the axis of revolution must be in the plane of the arc and parallel to its chord $\mathbf{p}_1 - \mathbf{p}_0$. Second, the axis must pass through the center point of the arc. The information supplied by these conditions makes it relatively easy for us to determine the geometric coefficients of a bicubic patch that closely approximates a segment of a spherical surface.

Finally, we can also approximate a region of a spherical surface with a bicubic patch by specifying the center point of the sphere and four points on it. Each edge of the bicubic patch is a pc curve approximation of a segment of a great circle on the sphere. Each circular arc is defined by two of the four points and the center point. (See Chapter 2, Section 11, Ex. 10.)

EXERCISES

1. Find the geometric coefficients for a spherical patch of radius R; center point $\mathbf{p}_c = \begin{bmatrix} 0 & 0 & 0 \end{bmatrix}$; and four corner points located at $(\theta = \pi/2, \phi = 0)$, $(\theta = \pi/2, \phi = \phi_1)$, $(\theta = \theta_1, \phi = 0)$, and $(\theta = \theta_1, \phi = \phi_1)$. Interpret θ and ϕ as shown in Fig. 3.34.

2. What is the maximum deviation of the patch defined in Ex. 1 from a true spherical surface?

3. Can a set of patches such as those defined in Ex. 1 be arranged to represent a complete sphere? If so, how? Sketch your answer.

4. Write a procedure to compute the geometric coefficients of a spherical bicubic patch. Denote this as **SPHSRF**, and develop a suitable set of arguments. Use **SRFREV** if possible.

15   CONIC SURFACE

A bicubic patch approximates a conic surface when constructed as shown in Fig. 3.36. We establish the four corner points and four intermediate points $p_1, p_2,$ $p_3,$ and p_4 that define the four boundary curves as conic curves. Figure 3.36 illustrates a very general case, and the resulting surface is something akin to a hyperbolic paraboloid. Figure 3.37 shows other possibilities. We use Chapter 2, Section 11 and Eq. 2.132 in particular to construct the **B** matrix of a conic surface patch. Notice that we assume the cross-derivative vectors to be zero. The geometric coefficients of the boundary curves supply the elements of the **B** matrix, and we obtain

$$
\mathbf{B} = \begin{bmatrix}
\mathbf{p}_{00} & \mathbf{p}_{01} & 4\rho_3(\mathbf{p}_3 - \mathbf{p}_{00}) & 4\rho_3(\mathbf{p}_{01} - \mathbf{p}_3) \\
\mathbf{p}_{10} & \mathbf{p}_{11} & 4\rho_4(\mathbf{p}_4 - \mathbf{p}_{10}) & 4\rho_4(\mathbf{p}_{11} - \mathbf{p}_4) \\
4\rho_1(\mathbf{p}_1 - \mathbf{p}_{00}) & 4\rho_2(\mathbf{p}_2 - \mathbf{p}_{01}) & 0 & 0 \\
4\rho_1(\mathbf{p}_{10} - \mathbf{p}_1) & 4\rho_2(\mathbf{p}_{11} - \mathbf{p}_2) & 0 & 0
\end{bmatrix}
$$

(3.119)

Notice that there are potentially four different ρ values. This means that we could change a parabolic shape at the $u = 0$ edge, say, to a hyperbolic shape at $u = 1$.

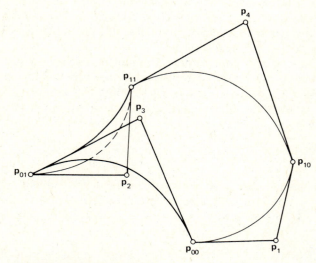

FIGURE 3.36 Conic surface.

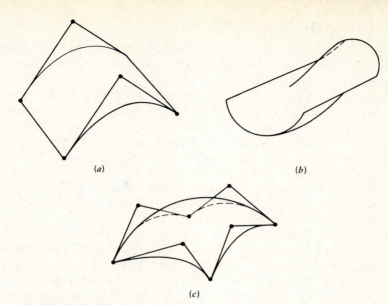

(b)

(c)

FIGURE 3.37 More conic surfaces.

EXERCISES

1. Find the geometric coefficients of a conic surface patch similar to the one in Fig. 3.37b whose $u = $ constant curves approximate parabolas. The end points are

$$\mathbf{p}_{00} = [0 \quad 0 \quad 0]$$
$$\mathbf{p}_{10} = [4 \quad 0 \quad 0]$$
$$\mathbf{p}_{01} = [0 \quad 4 \quad 0]$$

and

$$\mathbf{p}_{11} = [4 \quad 4 \quad 0]$$

The coordinates of the two control points are

$$\mathbf{p}_{02}(0, 0.5) = [0 \quad 2 \quad -2]$$

and

$$\mathbf{p}_{12}(1, 0.5) = [4 \quad 2 \quad 2]$$

2. Can twist vectors be used to control the shape of the intermediate $u = $ constant curves in Ex. 1? If so, demonstrate how this can be done.

3. Write a procedure to compute the geometric coefficients of a bicubic patch that approximates a conic surface. Denote this as **CONSRF**(P, RHO, CO), where

P(8, 3) is the input array of the eight defining points;

RHO(4) is the input array of the four rho values; and

CO(4, 4, 3) is the output array of the geometric coefficients.

16 COMPOSITE SURFACES

A composite surface is a collection of individual surface patches joined to form a continuous, more complex surface. Figure 3.38 shows a composite surface defined by a 4×3 array of 12 patches.

There are two basic phenomena to consider. First, what factors control the shape and, particularly, the continuity of the composite surface? Second, how do we construct such surfaces? We will investigate both in some depth, but we will reserve many aspects of the construction process for later sections and chapters. The bicubic patch is the basis for the investigation that follows.

Let us investigate more closely what effect the twist vector has on the continuity along a shared patch boundary. In particular, we will investigate the continuity of two patches $\mathbf{p}_{uw}$ and $\mathbf{q}_{uw}$ having a common boundary such that the curves $\mathbf{p}_{1w}$ and $\mathbf{q}_{0w}$ are identical; see Fig. 3.39.

To ensure C^1 continuity across this boundary, the coefficients of the respective auxiliary curves must be scalar multiples of each other. Given the coefficients of the auxiliary curve of patch $\mathbf{p}(u, w)$ along $\mathbf{p}_{1w}$, $\mathbf{p}_{10}^u$, $\mathbf{p}_{10}^{uw}$, $\mathbf{p}_{11}^u$, $\mathbf{p}_{11}^{uw}$, then the corresponding coefficients of $\mathbf{q}(u, w)$ are related as follows: $\mathbf{q}_{00}^u = a\mathbf{p}_{10}^u$, $\mathbf{q}_{00}^{uw} = a\mathbf{p}_{10}^{uw}$, $\mathbf{q}_{01}^u = a\mathbf{p}_{11}^u$, $\mathbf{q}_{01}^{uw} = a\mathbf{p}_{11}^{uw}$. When these conditions prevail, we find that the slope $\mathbf{q}_{0w}^u = a\mathbf{p}_{1w}^u$ at any point along $\mathbf{p}_{1w}$ (that is, along $\mathbf{q}_{0w}$).

Figure 3.40 presents the geometric coefficients required to produce C^1 continuity across a common boundary between two patches. This format was introduced by Peters (1974). We conclude that adjacent patches have C^0 and C^1 continuity only if common position-determining rows (or columns) of their $\mathbf{B}$ matrices are identical and if common auxiliary curve-determining rows (or columns) are multiples of each other. Thus, two patches joined to exhibit C^0 and C^1 continuity have a total of 73 degrees of freedom, whereas two disjoint patches have a total of 96.

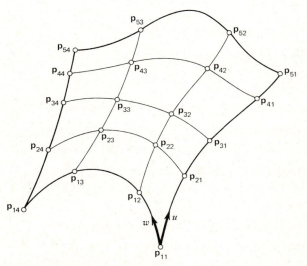

FIGURE 3.38 Composite surface.

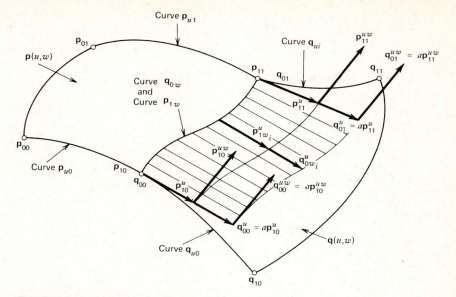

FIGURE 3.39 The effect of the twist vector on continuity.

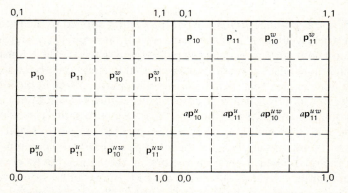

FIGURE 3.40 Geometric coefficients affected by C^1-continuity requirement across a common boundary between two patches.

At a corner of any patch, there are four vectors contributing to its shape through the operation of the blending functions on them. Since the blending functions are invariant, all the shape-determining power vests in these four vectors $\mathbf{p}_{uw}$, $\mathbf{p}_{uw}^u$, $\mathbf{p}_{uw}^w$, and $\mathbf{p}_{uw}^{uw}$. Four patches surround and share an interior corner point of a composite surface. In many modeling situations, it will help us to consider these mesh points or corner points as the centers or foci of the shaping parameters.

Figure 3.41 presents conditions at an arbitrary internal point of a composite surface where four patches meet. The point is denoted as $\mathbf{p}_{ij}$. Each quadrant

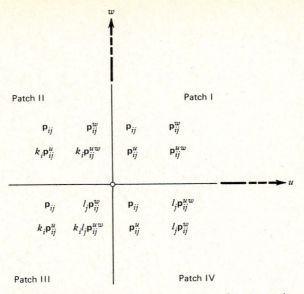

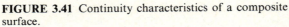

FIGURE 3.41 Continuity characteristics of a composite surface.

surrounding this point contains a list of the vectors applied to the patch in that quadrant. Clearly, all four patches share the common point $\mathbf{p}_{ij}$. Consider next the common boundary curve between patch I and patch II. The tangent vectors in the w direction must be identical to ensure that the respective boundary curves do match. The tangent vectors in the u direction must only be in the same direction (differing only by a scalar multiple k) to ensure C^1 continuity along the curve through the point. And since the cross-derivatives affect the slope along this boundary, as we have seen, they, too, must differ only by a scalar multiple.

By continuing this analysis for the other patch boundary combinations, we generate the relationships between vectors, as the figure shows. We conclude that at an arbitrary point $\mathbf{p}_{ij}$, there are only 14 degrees of freedom available to control the local shape of the composite surface (that is, four vectors and two scalars) if we are to preserve C^0 and C^1 continuity.

Notice that the k and l scale factors are not strictly unique to each point. They are propagated along a curve of constant i or j, as we see in Fig. 3.42. The driving function of this propagation is the requirement of preserving a constant scalar-multiple relationship between adjacent auxiliary curves along a curve of constant u_i or w_j. Also notice that when u and w take on integral numerical values at unit intervals corresponding to the i, j indexing scheme, reparametrization is not necessary, since, for example, $u_{i+1} - u_i = 1$. (In Fig. 3.42, only the k and l scale factors are shown in three of the four quadrants at each i, j node. They multiply the corresponding vectors in the first quadrant of the node.)

Figure 3.43 shows a clearer picture of the distribution of scale factors. Although the arrows at each end of the band of constant scale factor are bidirec-

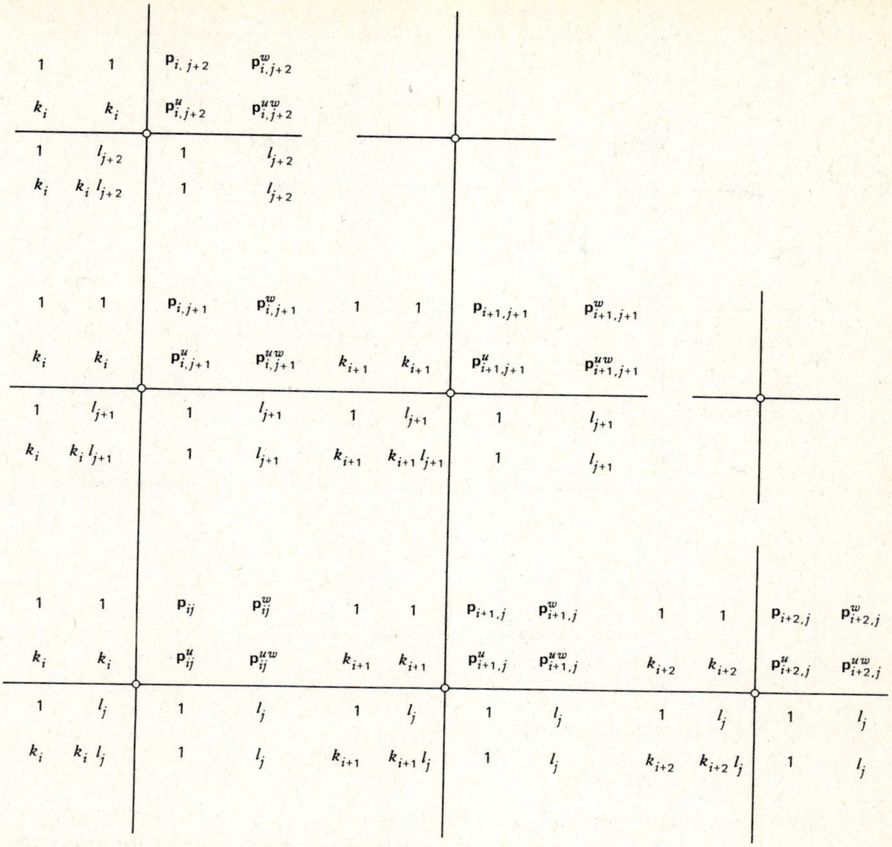

1	1	$\mathbf{p}_{i,j+2}$	$\mathbf{p}^w_{i,j+2}$								
k_i	k_i	$\mathbf{p}^u_{i,j+2}$	$\mathbf{p}^{uw}_{i,j+2}$								
1	l_{j+2}	1	l_{j+2}								
k_i	$k_i l_{j+2}$	1	l_{j+2}								
1	1	$\mathbf{p}_{i,j+1}$	$\mathbf{p}^w_{i,j+1}$	1	1	$\mathbf{p}_{i+1,j+1}$	$\mathbf{p}^w_{i+1,j+1}$				
k_i	k_i	$\mathbf{p}^u_{i,j+1}$	$\mathbf{p}^{uw}_{i,j+1}$	k_{i+1}	k_{i+1}	$\mathbf{p}^u_{i+1,j+1}$	$\mathbf{p}^{uw}_{i+1,j+1}$				
1	l_{j+1}	1	l_{j+1}	1	l_{j+1}	1	l_{j+1}				
k_i	$k_i l_{j+1}$	1	l_{j+1}	k_{i+1}	$k_{i+1} l_{j+1}$	1	l_{j+1}				
1	1	$\mathbf{p}_{ij}$	$\mathbf{p}^w_{ij}$	1	1	$\mathbf{p}_{i+1,j}$	$\mathbf{p}^w_{i+1,j}$	1	1	$\mathbf{p}_{i+2,j}$	$\mathbf{p}^w_{i+2,j}$
k_i	k_i	$\mathbf{p}^u_{ij}$	$\mathbf{p}^{uw}_{ij}$	k_{i+1}	k_{i+1}	$\mathbf{p}^u_{i+1,j}$	$\mathbf{p}^{uw}_{i+1,j}$	k_{i+2}	k_{i+2}	$\mathbf{p}^u_{i+2,j}$	$\mathbf{p}^{uw}_{i+2,j}$
1	l_j	1	l_j	1	l_j	1	l_j	1	l_j	1	l_j
k_i	$k_i l_j$	1	l_j	k_{i+1}	$k_{i+1} l_j$	1	l_j	1	l_j	k_{i+2}	$k_{i+2} l_j$

FIGURE 3.42 Continuity-driven propagation of scale factors.

tional, we must select a single consistent direction for each band for which the scalar relationship applies. Figure 3.43 shows a 4 × 5 array of patches, with seven distinct scale factors.

Once a convention is selected for associating the data at the points with the adjacent or surrounding patches, then we can easily develop a scheme to fill in the **B** matrix for any specific patch. If we identify a patch by the lowest pair of indices of the four sets defining the patch corners, then we readily fill the **B** matrix of (patch)$_{ij}$ as follows:

$$[\mathbf{B}]_{ij} = \begin{bmatrix} \mathbf{p}_{ij} & \mathbf{p}_{i,j+1} & \mathbf{p}^w_{ij} & l_{j+1}\mathbf{p}^w_{i,j+1} \\ \mathbf{p}_{i+1,j} & \mathbf{p}_{i+1,j+1} & \mathbf{p}^w_{i+1,j} & l_{j+1}\mathbf{p}^w_{i+1,j+1} \\ \mathbf{p}^u_{i,j} & \mathbf{p}^u_{i,j+1} & \mathbf{p}^{uw}_{ij} & l_{j+1}\mathbf{p}^w_{i,j+1} \\ k_{i+1}\mathbf{p}^u_{i+1,j} & k_{i+1}\mathbf{p}^u_{i+1,j+1} & k_{i+1}\mathbf{p}^{uw}_{i+1,j} & k_{i+1}l_{j+1}\mathbf{p}^{uw}_{i+1,j+1} \end{bmatrix} \quad (3.120)$$

Here, the indices outside the brackets on **B** identify the patch itself and not an element of the matrix.

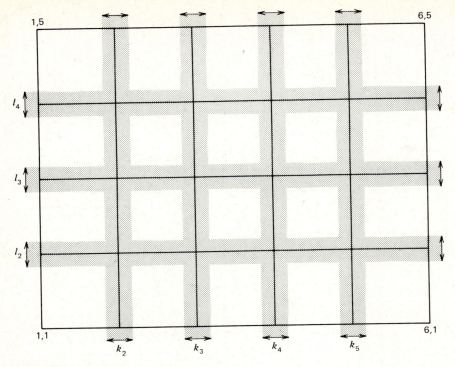

FIGURE 3.43 Distribution of scale factors.

Now let us briefly consider how to construct a composite surface. Applying parametric cubic forms to practical modeling problems grew slowly at first, requiring the user to supply tangent and twist vectors as well as point coordinates at the patch corner points. Soon, there were schemes for defining patches without having to specify tangent vectors. Some schemes replaced the required input tangent vectors with extra point coordinates other than the corner points. This approach was useful for creating individual patches, but most surfaces are sufficiently complex to require a number of patches joined together. Composite surfaces so created usually lacked C^1 continuity, not a satisfactory situation for most applications.

Parametric spline interpolation came into vogue to generate a rectangular network of patches from a corresponding network of points, using a numerical differentiation of the spline interpolation to calculate the derivatives required for the complete set of patch coefficients. Cardinal splines, a more elegant approach, were introduced by Gordon (1969).

To conclude this section, we review Timmer's (1976) application of Gordon's approach to defining a rectangular network of bicubic patches.

Let us consider two families of intersecting curves $\mathbf{q}_i$ and $\mathbf{r}_j$, with $i \in [1:m]$ and $j \in [1:n]$, which combine to form the wireframe surface shown in Fig. 3.44. Note that there are $m \times n$ intersection points.

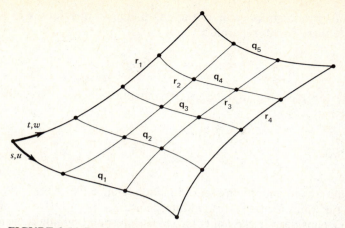

FIGURE 3.44 Rectangular network of intersecting curves.

Express the $\mathbf{q}_i$ family of m curves in terms of the parameter s, and the $\mathbf{r}_i$ family of n curves in terms of the parameter t. Then, for a typical curve,

$$\mathbf{p}(s) = \mathbf{q}_i(s) \qquad s \in [0, S_i] \tag{3.121}$$

and

$$\mathbf{p}(t) = \mathbf{r}_j(t) \qquad t \in [0, T_j] \tag{3.122}$$

Here, $\mathbf{p}$ is the position vector to a point on the indicated curve.

Observe that the range of the parameter is not necessarily identical for all of the $\mathbf{q}_i$ or $\mathbf{r}_j$; therefore, we define a second set of parameters (u, w)

$$u_j = \frac{1}{m} \sum_{i=1}^{m} \frac{s_{ij}}{s_{in}} \tag{3.123}$$

$$w_i = \frac{1}{n} \sum_{j=1}^{n} \frac{t_{ij}}{t_{mj}} \tag{3.124}$$

Note that in Eqs 3.123 and 3.124, the double subscript on s and t denotes their value at the intersection node indicated. The parameters u and w are normalized to map the entire surface into the unit square in u, w space. We find that each s curve corresponds to a single value of w and each t curve corresponds to a single value of u. This procedure has the effect of defining n functions $t = T_j(w)$, $j \in [1:n]$ and m functions $s = S_i(u)$, $i \in [1:m]$. In practice, we express these functions as parametric spline-interpolating functions. As a result, the two families of curves are reparametrized to map them into the unit square.

$$\mathbf{p}(u) = \mathbf{q}_i[S_i(u)] \qquad i \in [1:m], \quad u \in [0, 1] \tag{3.125}$$

$$\mathbf{p}(w) = \mathbf{r}_j[T_j(w)] \qquad j \in [1:n], \quad w \in [0, 1] \tag{3.126}$$

Next, we consider the actual interpolation problem. It is useful to treat this network interpolation as a combination of two problems, interpolating the $\mathbf{q}_i$ and $\mathbf{r}_j$ separately. For example, consider the m curves of $\mathbf{q}_i(u)$. Each of these curves corresponds to a distinct value of the parameter w; therefore, it is natural to interpolate these curves separately.

$$\mathbf{p}(u, w) = \sum_{i=1}^{m} F_i(w)\mathbf{q}_i(u) \qquad (3.127)$$

The $F_i(w)$ are blending or interpolating functions. These have the obvious properties:

$$F_i(w_j) = \delta_{ij}$$
$$= 0 \qquad \text{if } i \neq j$$
$$= 1 \qquad \text{if } i = j \qquad (3.128)$$

We are free to choose any set of functions that satisfies Eq. 3.128 to act as blending functions for the indicated interpolation.

Let us consider the class of cubic spline-interpolating functions for the F_i, often referred to as cardinal splines. Figure 3.45 presents several of the F functions. Notice that the maxima damp quite rapidly as the curve moves away from the defining node. This is significant, since it is often possible to truncate Eq. 3.127 using the F_i in only the immediate neighborhood of w_j.

Specifying values for the dependent variable at each of the interpolating mesh points results in two unspecified conditions. Usually, these conditions are first- or second-derivative constraints applied at each end of the mesh.

For the splines defining the $\mathbf{q}_i(s)$, the constraints reflect the obvious slope and curvature considerations for a space curve. Selecting end-point constraints for the other two spline fits is not so obvious. In the absence of any other constraints, we assume that a constant second derivative in each of the two end segments will be imposed on both $S_i(u)$ and $F_i(w)$. Clearly, we could have performed the interpolation

$$\mathbf{p}(u, w) = \sum_{j=1}^{n} \mathscr{F}_j(u)\mathbf{r}_j(w) \qquad (3.129)$$

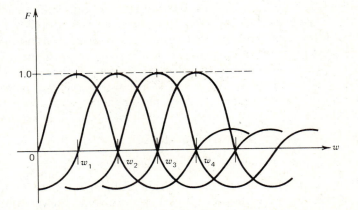

FIGURE 3.45 Cardinal spline interpolating function.

The functions $\mathcal{F}_j(u)$ assume the same role as the $F_i(w)$ in Eq. 4.127. For our example, where we specify both $\mathbf{q}_i(u)$ and $\mathbf{r}_j(w)$, we achieve the optimum interpolation by using

$$\mathbf{p}(u, w) = \sum_{j=1}^{n} \mathcal{F}_j(u)\mathbf{r}_j(w) + \sum_{i=1}^{m} F_i(w)\mathbf{q}_i(u)$$
$$- \sum_{i=1}^{m} \sum_{j=1}^{n} \mathcal{F}_i(w)F_j(u)\mathbf{p}_{ij} \tag{3.130}$$

where $\mathbf{p}_{ij}$ are the values of $\mathbf{p}$ at the i, j mesh points.

It turns out that Eq. 3.130 is an elegant but computationally inefficient way to interpolate two families of curves forming a rectangular mesh. However, Eq. 3.130 may be used to determine the geometric coefficients for a network of bicubic patches. Once we select the patch corner points, we can use the equation to determine values of $\mathbf{p}_{uw}^{u}$, $\mathbf{p}_{uw}^{w}$, and $\mathbf{p}_{uw}^{uw}$.

Finally, according to Timmer, if $\mathbf{q}_i$ and $\mathbf{r}_j$ have C^0, C^1, and C^2 continuity, then Eq. 3.130 guarantees this same continuity throughout the unit square, $u, w \in [0, 1]$. If the unit square is subdivided into a rectangular array of bicubic patches, continuity is preserved, provided the patches are not arbitrarily reparametrized. For example, if each patch is reparametrized to the unit square, a discontinuous relationship between the original parameters and the new parameters is created, and we lose the C^2 continuity of Eq. 3.130.

What are the practical implications of this surface-generating interpolation scheme? First, it is unlikely that the two families of curves will be completely defined at the outset of the modeling problem. It is likely that the initial data will be in the form of discrete points distributed in some semiregular way over the prospective surface, perhaps including slope and curvature data. If this is the case, then the parametric representations of $\mathbf{q}_i(u)$ and $\mathbf{r}_j(w)$ remain to be defined. There are a number of different approaches, and the specific nature of the input data is a controlling factor in selecting a strategy.

Let us consider the case where the point data are in the form of an $m \times n$ array uniformly distributed over the surface, without slope or curvature preconditions. We may now decide to represent $\mathbf{q}_i(u)$ and $\mathbf{r}_j(w)$ as follows:

$$\mathbf{q}_i(u) = \sum_{j=1}^{n} \mathcal{F}_j(u)\mathbf{p}_{ij} \tag{3.131}$$

$$\mathbf{r}_j(w) = \sum_{i=1}^{m} F_i(w)\mathbf{p}_{ij} \tag{3.132}$$

Substitute these equations into Eq. 3.130 to obtain a simpler expression

$$\mathbf{p}(u, w) = \sum_{i=1}^{m} \sum_{j=1}^{n} F_i(w)\mathcal{F}_j(u)\mathbf{p}_{ij} \tag{3.133}$$

If we include slope or curvature constraints, this expression becomes more complex.

Equation 3.130 or 3.133 can be interpolated over the entire surface and mapped into the unit square in u, w space, a process similar to that used for the bicubic patch. The obvious difference is that, for the immediate example, a single equation does not cover the entire unit square. However, each patch element formed by the two families of curves is the equivalent of a bicubic patch in the sense that, within any quadrilateral patch element, the interpolation may be expressed as the familiar

$$\mathbf{p}(u, w) = \sum_{i=0}^{3} \sum_{j=0}^{3} \mathbf{a}_{ij} u^i w^j \qquad (3.134)$$

Clearly, any subelement can be parametrized to the unit square without danger of altering the composite surface. This means that interpolation by Eq. 3.130 or 3.133 is the equivalent of, and may be replaced by, a network of $(m - 1) \times (n - 1)$ conventional bicubic patches without affecting the guaranteed C^0, C^1, and C^2 continuity. A disadvantage of this approach is that it leads to a potentially large number of patches. Thus, a practical alternative would allow us to model the same surface using fewer patches and without changing the original input point data. Figure 3.46 illustrates a scheme to selectively reduce the number of patches used in a composite surface model. Here, what was potentially an 11×15 array of patches is replaced by a 3×3 array.

This reduction in the number of patches, however, could cause a loss of C^2 continuity. This loss becomes apparent when we consider a single curve, for example, the upper boundary curve in the figure. In the second segment, a single pc curve is used to approximate seven pc curves. The single segment will match

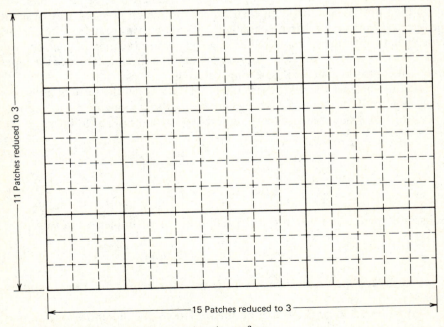

FIGURE 3.46 Patch array on a composite surface.

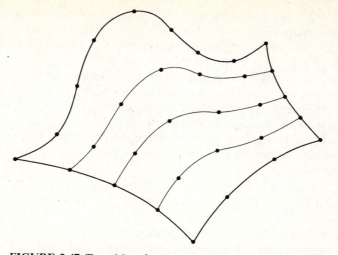

FIGURE 3.47 Transition from a complex to a simple cross section.

the original end points and slopes; however, the curvature may deviate some-what. Freeing the curvature at the end points leads to the discontinuity. *Note*: Since the original composite curve had continuous curvature, the new single-curve approximation should be close and the resulting discontinuities hardly discernible. Note also that, in the course of this reduction in the number of patches, the final network is still rectangular.

A final simplification is available to us. In the preceding example, input data points form a rectangular mesh—each row has the same number of points in it. Now consider the problem of a surface with transitions from a complex cross section to a simple one (Fig. 3.47). The number of points we need to define the simple section is less than for the complex one. To preserve a rectangular mesh of initial data points, we would have to specify more data points for the simple sections than these shapes would otherwise warrant. We conclude that the rectangularity constraint requires us to supply too much data, and thus it is inefficient. To avoid this, Timmer suggests we use the following strategy.

Assume that in most cases we prefer to supply data for a series of cross-sectional curves rather than for a rectangular grid on a surface. Let n_0 denote the number of data points required to define the most complex cross-section curve and m the number of cross sections. A rectangular mesh requires $m \times n_0$ mesh points; however, instead of a rectangular mesh format, we now allow the model input to consist of the exact number of points necessary for each cross section individually. Thus, a linear cross-section curve could be described using only two points. Using this approach, the total number of points required to describe the surface is

$$N = \sum_{i=1}^{m} n_i \qquad (3.135)$$

Obviously, $N \le mn_0$. Using this reduced input data, we interpolate each cross-section curve, using a conventional spline or other form of representation. Next, we compute a set of n_0 points for each curve, and we continue as outlined for a rectangular mesh.

Insofar as composite surfaces are concerned, we have barely scratched the surface (pardon the pun). The literature abounds with a great variety of representational mathematics and construction techniques. A good selection of material is available for most modeling situations.

In the next two sections, we will explore the Bezier and B-spline surfaces, both eminently suitable for modeling complex composite surfaces.

EXERCISES

1. For a biquadric parametric representation, describe conditions at an arbitrary internal point of a composite surface where four patches meet. Assume maximum continuity.

2. Assume the same conditions in Ex. 1 for a biquintic parametric representation.

3. How many degrees of freedom are there at an arbitrary internal point of a composite surface where four patches meet for biquadric, bicubic, and biquintic parametric representations? Again assume maximum continuity.

4. Fill the **B** matrix of an arbitrary (patch)$_{ij}$ in a composite surface for biquadric and biquintic representations similarly to what was done for bicubics in Eq. 3.120.

5. Given nine bicubic patches joined in a 3×3 array with C^0 and C^1 continuity, show that the center patch can be replaced with another 3×3 array of bicubic patches in such a way that C^0 and C^1 continuity are preserved throughout the new 17-patch array. This procedure allows local modification of composite surfaces. Sketch the before-and-after conditions describing this process by showing the patch boundaries, and also indicate these conditions schematically showing the geometric coefficient matrices arranged likewise. Use arrows from elements of one matrix to adjacent matrices to indicate relationships.

6. Generalize the process described in Ex. 5, and show how additional patches can be inserted into a composite array of patches. C^0 and C^1 continuity are preserved.

7. Given nine bicubic patches joined in a 3×3 array with C^0 and C^1 continuity, show that the center patch can be replaced with a 2×3 six-patch array so that C^0 and C^1 continuity are maintained. This demonstrates that the original composite surface can be locally modified internally to blend with any pc space curve, where the space curve defines the boundary between the two middle patches in the 2×3 patch array.

17 BEZIER SURFACE

Just as the Bezier curve has a characteristic polygon, the Bezier surface has a **characteristic polyhedron**. Points on a Bezier surface are given by a simple extension of the general equation for points on a Bezier curve (Eq. 2.229)

$$\mathbf{p}(u, w) = \sum_{i=0}^{m} \sum_{j=0}^{n} \mathbf{p}_{ij} B_{i,m}(u) B_{j,n}(w) \qquad u, w \in [0, 1] \qquad (3.136a)$$

where the $\mathbf{p}_{ij}$ are vertices of the characteristic polyhedron that form an $(m + 1) \times (n + 1)$ rectangular array of points, and $B_{i,m}(u)$ and $B_{j,n}(w)$ are defined as for curves; see Eqs. 2.230 and 2.331. Clearly, the Bezier surface need not be represented by a square array of control points, although we will begin by exploring a 4×4 array of points.

Faux and Pratt (1980) observe that there is a close relationship between the bicubic Ferguson surface patch and the bicubic Bezier surface patch. The latter is a reformulation of the former that avoids the need of specifying tangent vectors.

We express a bicubic Bezier patch in a form similar to the Hermite bicubic patch. Using the binomial representation of the cubic Bezier curve in Chapter 2, Section 14, the matrix equation for a patch defined by a 4×4 array of points is

$$\mathbf{p}(u, w) = [(1 - u)^3 \quad 3u(1 - u)^2 \quad 3u^2(1 - u) \quad u^3] \mathbf{P} \begin{bmatrix} (1 - w)^3 \\ 3w(1 - w)^2 \\ 3w^2(1 - w) \\ w^3 \end{bmatrix}$$

$$(3.136b)$$

where

$$\mathbf{P} = \begin{bmatrix} \mathbf{p}_{11} & \mathbf{p}_{12} & \mathbf{p}_{13} & \mathbf{p}_{14} \\ \mathbf{p}_{21} & \mathbf{p}_{22} & \mathbf{p}_{23} & \mathbf{p}_{24} \\ \mathbf{p}_{31} & \mathbf{p}_{32} & \mathbf{p}_{33} & \mathbf{p}_{34} \\ \mathbf{p}_{41} & \mathbf{p}_{42} & \mathbf{p}_{43} & \mathbf{p}_{44} \end{bmatrix} \qquad (3.137)$$

The matrix $\mathbf{P}$ contains the position vectors for points that define the characteristic polyhedron and, thereby, the Bezier surface patch. Figure 3.48 illustrates these points, the polyhedron, and the resulting patch.

In the Bezier formulation, only the four corner points $\mathbf{p}_{11}, \mathbf{p}_{41}, \mathbf{p}_{14}$, and $\mathbf{p}_{44}$ actually lie on the patch. The points $\mathbf{p}_{21}, \mathbf{p}_{31}, \mathbf{p}_{12}, \mathbf{p}_{13}, \mathbf{p}_{42}, \mathbf{p}_{43}, \mathbf{p}_{24}$, and $\mathbf{p}_{34}$ control the slope of the boundary curves. The remaining four points $\mathbf{p}_{22}, \mathbf{p}_{32}, \mathbf{p}_{23}$, and $\mathbf{p}_{33}$ control the cross slopes along the boundary curves in the same way as the twist vectors of the bicubic patch.

As this figure shows, the Bezier surface is completely defined by a net of design points describing two families of Bezier curves on the surface. Each curve is defined by a polygon of four points or vertices. (A greater number of points could be used, resulting in a higher degree polynomial.)

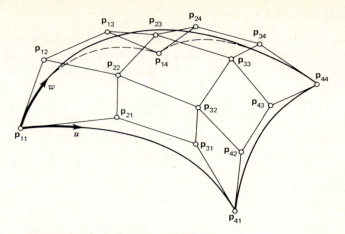

FIGURE 3.48 Cubic Bezier surface.

Consider the generation of a parametric curve on the bicubic Bezier patch for $w = w_i$, a constant. Then the following matrix product produces four points defining the Bezier curve along the patch at w_i:

$$\begin{bmatrix} \mathbf{p}_1 \\ \mathbf{p}_2 \\ \mathbf{p}_3 \\ \mathbf{p}_4 \end{bmatrix} = \mathbf{P} \begin{bmatrix} (1 - w_i)^3 \\ 3w_i(1 - w_i)^2 \\ 3w_i^2(1 - w_i) \\ w_i^3 \end{bmatrix} \tag{3.138}$$

where the single subscripts on $\mathbf{p}_1$, $\mathbf{p}_2$, and so on, indicate that these points are curve defining. The point $\mathbf{p}_1$ lies on the patch boundary curve $\mathbf{p}_{0w}$, and the point $\mathbf{p}_4$ lies on curve $\mathbf{p}_{1w}$. Points $\mathbf{p}_2$ and $\mathbf{p}_3$ control the slopes at either end of the curve and do not lie on it. Now we can write an expression for a point on the patch along the curve $\mathbf{p}(u, w_i)$.

$$\mathbf{p}(u, w_i) = [(1 - u)^3 \quad 3u(1 - u)^2 \quad 3u^2(1 - u) \quad u^3] \begin{bmatrix} \mathbf{p}_1 \\ \mathbf{p}_2 \\ \mathbf{p}_3 \\ \mathbf{p}_4 \end{bmatrix} \tag{3.139}$$

Here, each value of u yields a point on the patch. We use a similar procedure to define the boundary curves. For example, the curve $u = 0$ is

$$\mathbf{p}(0, w) = [\mathbf{p}_{11} \quad \mathbf{p}_{12} \quad \mathbf{p}_{13} \quad \mathbf{p}_{14}] \begin{bmatrix} (1 - w)^3 \\ 3w(1 - w)^2 \\ 3w^2(1 - w) \\ w^3 \end{bmatrix} \tag{3.140}$$

or

$$\mathbf{p}(0, w) = [(1 - w)^3 \quad 3w(1 - w)^2 \quad 3w^2(1 - w) \quad w^3] \begin{bmatrix} \mathbf{p}_{11} \\ \mathbf{p}_{12} \\ \mathbf{p}_{13} \\ \mathbf{p}_{14} \end{bmatrix} \quad (3.141)$$

The boundary curve for $w = 0$ is

$$\mathbf{p}(u, 0) = [(1 - u)^3 \quad 3u(1 - u)^2 \quad 3u^2(1 - u) \quad u^3] \begin{bmatrix} \mathbf{p}_{14} \\ \mathbf{p}_{24} \\ \mathbf{p}_{34} \\ \mathbf{p}_{44} \end{bmatrix} \quad (3.142)$$

The other two boundary curves are defined similarly.

Occasionally, we want to create a patch that is described by a **P** matrix that is not square. This allows a relatively complex surface to be bounded by simple curves or vice versa. For example, the patch shown in Fig. 3.49 has a three-point Bezier curve along boundaries $u = 0$ and $u = 1$ and a five-point curve along boundaries $w = 0$ and $w = 1$. The equation for the boundary along $u = 0$ is

$$\mathbf{p}(0, w) = [(1 - w)^2 \quad 2w(1 - w) \quad w^2] \begin{bmatrix} \mathbf{p}_{11} \\ \mathbf{p}_{12} \\ \mathbf{p}_{13} \end{bmatrix} \quad (3.143)$$

Along $w = 0$, we obtain

$$\mathbf{p}(u, 0) = [(1 - u)^4 \quad 4u(1 - u)^3 \quad 6u^2(1 - u)^2 \quad 4u^3(1 - u) \quad u^4] \begin{bmatrix} \mathbf{p}_{11} \\ \mathbf{p}_{21} \\ \mathbf{p}_{31} \\ \mathbf{p}_{41} \\ \mathbf{p}_{51} \end{bmatrix} \quad (3.144)$$

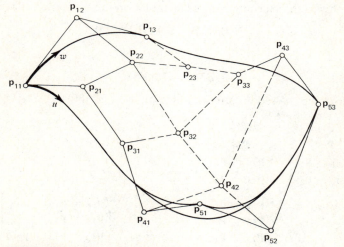

FIGURE 3.49 5 × 3 Bezier surface.

For the patch itself,

$$\mathbf{p}(u, w) = [(1 - u)^4 \quad 4u(1 - u)^3 \quad 6u^2(1 - u)^2 \quad 4u^3(1 - u) \quad u^4]$$

$$\times \begin{bmatrix} \mathbf{p}_{11} & \mathbf{p}_{12} & \mathbf{p}_{13} \\ \mathbf{p}_{21} & \mathbf{p}_{22} & \mathbf{p}_{23} \\ \mathbf{p}_{31} & \mathbf{p}_{32} & \mathbf{p}_{33} \\ \mathbf{p}_{41} & \mathbf{p}_{42} & \mathbf{p}_{43} \\ \mathbf{p}_{51} & \mathbf{p}_{52} & \mathbf{p}_{53} \end{bmatrix} \begin{bmatrix} (1 - w)^2 \\ 2w(1 - w) \\ w^2 \end{bmatrix} \tag{3.145}$$

The advantage of a five-point boundary curve over a four-point curve is that a change in the third or middle point of the curve does not affect the slope at either end. Thus, patch shape can be changed without changing the end slopes and, hence, continuity with adjacent patches.

Let us engage in a comparative study of the bicubic Bezier patch and the Hermite bicubic patch. First we will simplify Eq. 3.136 and express it in terms comparable to those of the Hermite bicubic.

$$[(1 - u)^3 \quad 3u(1 - u^2) \quad 3u^2(1 - u) \quad u^3]$$

$$= [1 \quad u \quad u^2 \quad u^3] \begin{bmatrix} 1 & 0 & 0 & 0 \\ -3 & 3 & 0 & 0 \\ 3 & -6 & 3 & 0 \\ -1 & 3 & -3 & 1 \end{bmatrix} = \mathbf{UM}_z \tag{3.146}$$

The subscript z denotes a Bezier matrix and distinguishes it from the ordinary Hermite bicubic. A similar substitution applies to the matrix of w terms, so that we finally rewrite Eq. 3.136 as

$$\mathbf{p}(u, w) = \mathbf{UM}_z \mathbf{PM}_z^T \mathbf{W}^T \tag{3.147}$$

Compare this to the Hermite bicubic formulation for the same surface

$$\mathbf{UMBM}^T \mathbf{W}^T = \mathbf{UM}_z \mathbf{PM}_z^T \mathbf{W}^T \tag{3.148}$$

or

$$\mathbf{MBM}^T = \mathbf{M}_z \mathbf{PM}_z^T \tag{3.149}$$

After performing the required matrix algebra, we find that

$$\mathbf{B} = \begin{bmatrix} \mathbf{p}_{11} & \mathbf{p}_{14} \\ \mathbf{p}_{41} & \mathbf{p}_{44} \\ 3(\mathbf{p}_{21} - \mathbf{p}_{11}) & 3(\mathbf{p}_{24} - \mathbf{p}_{14}) \\ 3(\mathbf{p}_{41} - \mathbf{p}_{31}) & 3(\mathbf{p}_{44} - \mathbf{p}_{34}) \end{bmatrix}$$

$$\left. \begin{matrix} 3(\mathbf{p}_{12} - \mathbf{p}_{11}) & 3(\mathbf{p}_{14} - \mathbf{p}_{13}) \\ 3(\mathbf{p}_{42} - \mathbf{p}_{41}) & 3(\mathbf{p}_{44} - \mathbf{p}_{43}) \\ 9(\mathbf{p}_{11} - \mathbf{p}_{21} - \mathbf{p}_{12} + \mathbf{p}_{22}) & 9(\mathbf{p}_{13} - \mathbf{p}_{23} - \mathbf{p}_{14} + \mathbf{p}_{24}) \\ 9(\mathbf{p}_{31} - \mathbf{p}_{41} - \mathbf{p}_{32} + \mathbf{p}_{42}) & 9(\mathbf{p}_{33} - \mathbf{p}_{43} - \mathbf{p}_{34} + \mathbf{p}_{44}) \end{matrix} \right] \tag{3.150}$$

Here, we see that the tangent vectors and twist vectors are equivalently expressed in terms of points defining the vertices of the characteristic polyhedron. The tangent vectors are given solely in terms of points on the boundary of the polyhedron, while the interior points contribute to the twist vectors.

EXERCISES

1. Develop and discuss the conditions required for C^0 and C^1 continuity between two bicubic Bezier patches along a common boundary.

2. Consider a composite surface constructed by a 3×3 array of cubic Bezier patches. Define the conditions necessary for C^0 and C^1 continuity throughout.

3. How many degrees of freedom are available in a 3×3 composite array of bicubic Bezier patches? Assume C^0 and C^1 continuity.

4. Assume the same conditions in Ex. 3 but for a 3×3 composite array of biquintic Bezier patches. Assume C^0 and C^1 continuity.

5. Find $\mathbf{p}_{00}^{uw}$, $\mathbf{p}_{10}^{uw}$, $\mathbf{p}_{01}^{uw}$, and $\mathbf{p}_{11}^{uw}$ for a bicubic Bezier patch.

6. Write a procedure to compute the coordinates of a point on a Bezier surface patch. Denote this as **BZSRF(M, N, CI, IFLG, U, W, P)**, where

M is the input number of control points along curves of constant u, where $M \in [3:6]$;

N is the input number of control points along curves of constant w, where $n \in [3:6]$;

CI(M, N, 3) is the input array of coordinates of control points defining the characteristic polygon;

IFLG is an input number whose value specifies that the surface is open or closed;

IFLG = 0 if the surface is open,

IFLG = 1 if the curves of constant u are closed, and

IFLG = 2 if the curves of constant w are closed;

U, W are the input values of the parametric variables u and w; and

P(3) is the output array of x, y, z coordinates of the point.

Note: Here M = $m + 1$ and N = $n + 1$; see Eq. 3.136a.

7. Generalize **BZSRF** so that M and N are not restricted and so that a rectangular set of points can be efficiently computed at specified increments of the parametric variables.

18   *B*-SPLINE SURFACE

The formulation of a *B*-spline surface follows directly from our formulation of *B*-spline curves. This relationship is analogous to that between Bezier curves and surfaces. Furthermore, the *B*-spline surface, like the Bezier surface, is defined in terms of a characteristic polyhedron. The shape of the surface approximates the polyhedron. The approximation is weaker the higher the value of k and l. Thus,

$$\mathbf{p}(u, w) = \sum_{i=0}^{m} \sum_{j=0}^{n} \mathbf{p}_{ij} N_{i,k}(u) N_{j,l}(w) \tag{3.151}$$

The $\mathbf{p}_{ij}$ are the vertices of the defining polyhedron, and the $N_{i,k}(u)$ and $N_{j,l}(w)$ are the blending functions of the same form as those for *B*-spline curves; see Eq. 2.252. The degree of each of the blending-function polynomials $N_{i,k}(u)$ and $N_{j,l}(w)$ is controlled by k and l, respectively.

For a nonperiodic *B*-spline, the values of m, n, k, l, and the t_i and t_j knot values are selected and computed just as for curves. $N_{i,k}(u)$ and $N_{j,l}(w)$ are computed recursively using Eqs. 2.253 and 2.254. Notice that two sets of knot values are required and that the control points form an $(m + 1) \times (n + 1)$ array.

There is a matrix form for the *B*-spline surface; it is similar to the matrix form we developed for the *B*-spline curve. Recall that the *B*-spline curve is computed in segments of a unit interval on the parametric variable u; see Eq. 2.284, 2.285, and 2.297. A unit square on the parametric variables u and w is used to compute patches on the *B*-spline surface. The general matrix form of an open, periodic *B*-spline surface that approximates an $(m + 1) \times (n + 1)$ rectangular array of points is

$$\mathbf{p}_{st}(u, w) = \mathbf{U}_k \mathbf{M}_k \mathbf{P}_{kl} \mathbf{M}_l^T \mathbf{W}_l^T \qquad \begin{aligned} & s \in [1:m + 2 - k] \\ & t \in [1:n + 2 - l] \\ & u, w \in [0, 1] \end{aligned} \tag{3.152}$$

where k and l denote the parameters that control the continuity of the surface and the degree of the blending-function polynomials; s and t identify a particular patch in the surface. The range on s and t is a function of the parameters k and l and the dimensions of the rectangular array of control points. The matrix $\mathbf{U}_k$ is

$$\mathbf{U}_k = [u^{k-1} \quad u^{k-2} \quad \dots \quad u \quad 1] \tag{3.153}$$

and $\mathbf{W}_l$ is

$$\mathbf{W}_l = [w^{l-1} \quad w^{l-2} \quad \dots \quad w \quad 1] \tag{3.154}$$

Elements of the $k \times l$ matrix of control points $\mathbf{p}_{kl}$ depend on the particular patch to be evaluated. Let $\mathbf{p}_{ij}$ denote these matrix elements; then

$$\mathbf{P}_{kl} = \mathbf{p}_{ij} \qquad \begin{aligned} & i \in [s - 1: s + k - 2] \\ & j \in [t - 1: t + l - 2] \end{aligned} \tag{3.155}$$

The matrices $\mathbf{M}_k$ and $\mathbf{M}_l$ are identical to the $\mathbf{M}$ transformation matrix for *B*-spline curves.

If the *B*-spline surface is partially closed (that is, rolled into an open-ended tube), then the ranges on s, t and i, j must reflect this closure. For example, if the $w = 0$ constant curves are closed, then

$$s \in [1:m + 1]$$
$$t \in [1:n + 2 - l]$$
$$i \in [(s - 1) \bmod (m + 1): (s + k - 2) \bmod (m + 1)] \qquad (3.156)$$
$$j \in [t - 1:t + l - 2]$$

Similar expressions apply when the $u = $ constant curves are closed. Later in this section, we will look at completely closed surfaces.

We idealize an array of control points and the patches they produce with a two-dimensional diagram, arranging the points on a rectangular grid. Solid

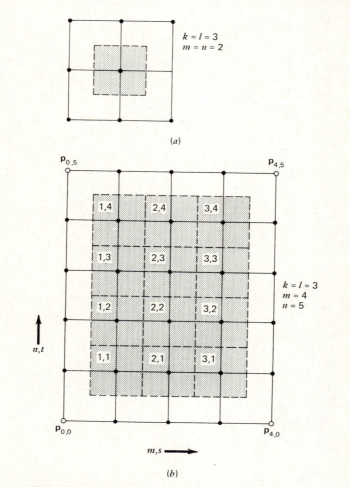

FIGURE 3.50 Open quadric *B*-spline surface idealizations.

lines represent the edges of the faces of the characteristic polyhedron. Dashed lines represent the patch boundaries. For some idealizations, these lines overlap; when this happens, the dashed lines are shown. The patches are lightly shaded to highlight them; this technique is useful for exploring the effects of parameters on the relationship between the control points and patches. Let us look at some examples. All of the following examples are periodic *B*-spline surfaces generated by specific instances of Eq. 3.152.

A quadric *B*-spline surface is produced when $k = l = 3$. This surface exhibits first-derivative continuity throughout. A minimum 3×3 array of

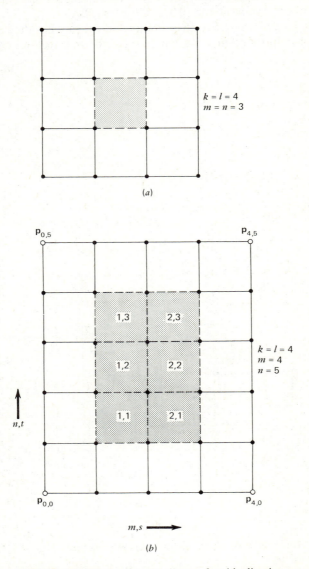

FIGURE 3.51 Open cubic *B*-spline surface idealizations.

control points is required (Fig. 3.50*a*). If $m = 4$ and $n = 5$, a 3×4 array of patches is produced (Fig. 3.50*b*).

A cubic *B*-spline surface is produced when $k = l = 4$. Second-derivative continuity exists at all points on this surface. Here, a minimum 4×4 array of control points is required (Fig. 3.51*a*). Compare this to the bicubic Hermite and Bezier surfaces. If $m = 4$ and $n = 5$, a 2×3 array of patches is produced (Fig. 3.51*b*). Compare this to the result obtained for the quadric surface in Fig. 3.50*b*.

Idealized quartic surfaces are shown in Fig. 3.52 and quintic surfaces in Fig. 3.53. Notice that as the degree of the blending-function polynomials increases, more control points define each patch. This is particularly evident along the patch and control-point-array boundary regions.

It is often more efficient to define a *B*-spline surface with $k = l$. For example, a ruled surface may require second-derivative continuity in one parametric direction, while first-derivative continuity suffices for the other. A cubic-quadric

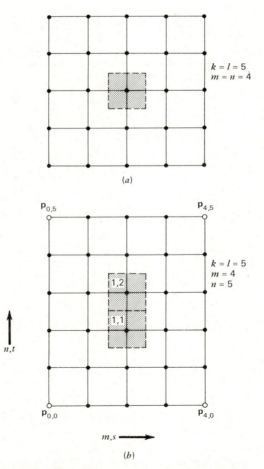

FIGURE 3.52 Open quartic *B*-spline surface idealizations

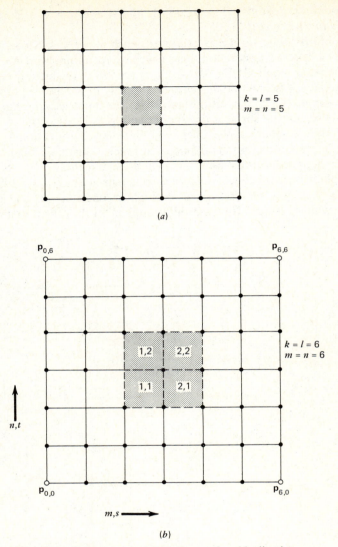

FIGURE 3.53 Open quintic *B*-spline surface idealizations.

B-spline surface is shown in Fig. 3.54. With $m = n = 4$, a 2×3 patch array is produced.

A partially closed *B*-spline surface results when we impose the conditions in Eq. 3.156; Fig. 3.55 is an example. Here, $k = 4$ and $l = 3$, with $m = 3$ and $n = 4$, to produce a 4×3 patch array. The $w = $ constant curves are closed; the $u = $ constant curves are open. A two-dimensional idealization is shown in Fig. 3.55*a* and a three-dimensional one in Fig. 3.55*b*. Only 20 control points produce the complex surface in Fig. 3.55*c*. [Notice the characteristic polyhedron whose edges and vertices (control points) are superimposed.]

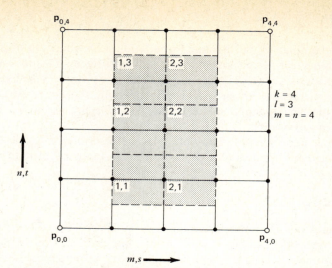

FIGURE 3.54 Open cubic-quadric B-spline surface idealizations.

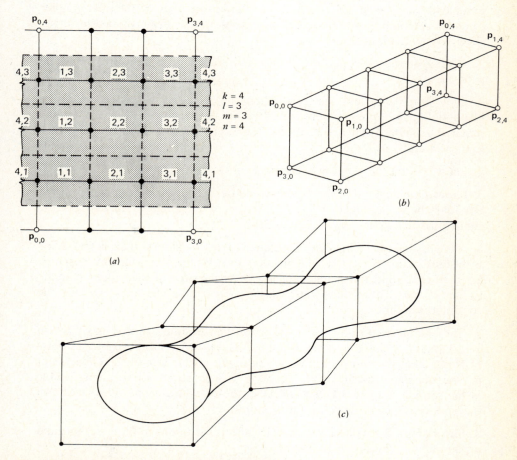

FIGURE 3.55 Partially closed B-spline surface.

One advantage of the *B*-spline formulation is its ability to preserve arbitrarily high degrees of continuity over complex surfaces. This follows directly from the properties of the *B*-spline curves. Also, any change in the local shape of a *B*-spline surface is not propagated throughout the entire surface. These characteristics make the *B*-spline surface suitable for use in an interactive modeling environment.

There is a considerable body of literature describing *B*-spline curves and surfaces and their applications. (Refer to the closing paragraph in Chapter 2, Section 15.)

EXERCISES

1. How many control points are required to produce a 2×2 array of patches for an open periodic bicubic ($k = l = 4$) *B*-spline surface? For a 3×3 array of patches? For a 4×4 array, and so on? Derive an equation that gives the number of points n_p required to produce an $n_a \times n_a$ array of patches for a bicubic *B*-spline. Graph this equation for values of n_a from 1–10. Show graphically and analytically the limiting relationship between the total number of points required and the number of patches in a square array as this array grows arbitrarily larger.

2. Assume the same conditions in Ex. 1 but for square arrays of biquintic patches.

3. Show graphically and analytically how the degree of the periodic *B*-spline blending-function polynomials is related to the number of control points required to define specified sizes of square arrays of patches (as in Ex. 1 and 2 for $k = l = 4$ and $k = l = 6$).

4. Show how partially closing the surfaces affects the results in Ex. 1–3.

5. Show analytically that first- and second-derivative continuity exists on the boundary between patches 1,2 and 2,2 and between patches 1,2 and 1,3 of the cubic-quadric *B*-spline surface in Fig. 3.54.

6. Is it possible to produce the completely closed *B*-spline surface in Fig. 3.56? Is the idealization correct? Can it be improved? What coincident control points are required? Find the elements of $\mathbf{p}_{k,l}$ for each patch of this closed *B*-spline surface. What relationships must exist between these elements? Discuss the continuity conditions.

7. The characteristic polyhedron of a toroidal *B*-spline surface is shown in Fig. 3.57. If $k = 4$ and $l = 3$ and if $m = 3$ and $n = 3$ (as implied by this polyhedron), label the control points in the figure and then find the elements of $\mathbf{p}_{k,l}$ for each patch. Note: Let the curves of $w =$ constant loop through the hole in the toroidal "donut." Draw and label the two-dimensional idealization of this surface. Discuss the continuity conditions.

8. Show how a *B*-spline toroidal surface affects the results in Ex. 1–3. Note and explain all assumptions you make.

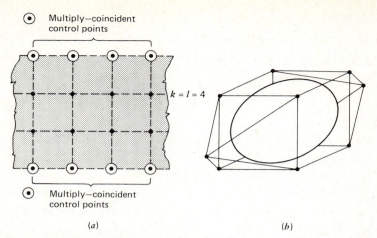

Multiply—coincident control points

$k = l = 4$

Multiply—coincident control points

(a)

(b)

FIGURE 3.56 Closed *B*-spline surface.

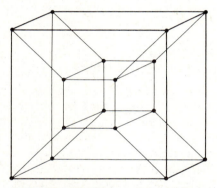

FIGURE 3.57 Characteristic polyhedron of a toroidal *B*-spline surface.

9. Write Eq. 3.152 in expanded form for the specific case of $k = 3$, $l = 4$, $m = 3$, and $n = 4$. Find the ranges of s and t, i and j, and write out matrices $\mathbf{U}_k$ and $\mathbf{W}_l$. Find the elements of $\mathbf{p}_{i,j}$ for patch 2, 1, that is, $\mathbf{p}_{2,1}(u, w)$.

10. Derive the expressions giving the ranges on s and t for open, partially closed, and closed (toroidal) *B*-spline surfaces. Start the derivations with observations of the elementary relationships implied by the relevant idealizations.

11. Write a procedure to compute the coordinates of a point on a *B*-spline surface. Denote this as **BSPLNS**(M, N, Cl, K, L, I1, I2, U, W, P), where

M is the input number of control points on curves of constant u value;

Note: M = m + 1;

N is the input number of control points on curves of constant w value;

Note: $N = n + 1$;

CI(M, N, 3) is the input array of x, y, z coordinates of the control points;

K, L are the input values of the degree of the u and w curves, respectively.

Note: $K = k - 1$ and $L = l - 1$;

I1 is the input flag specifying if the surface is periodic ($I1 = 0$) or non-periodic ($I1 = 1$);

I2 is the input flag specifying if the surface is open or closed:

$I2 = 0$ if open,
$I2 = 1$ if curves of constant u are closed, and
$I2 = 2$ if curves of constant w are closed.

U, W are the input values of the parametric variables u and w; and

P(3) is the output array of coordinates of the point.

12. Generalize **BSPLNS** so that a rectangular set of points can be efficiently computed at specified increments of the parametric variables.

19 DEGENERATE SURFACES AND PATHOLOGICAL CONDITIONS

Degenerate surfaces are at least as likely to occur as are degenerate curves; a review of Chapter 2, Section 17, now, is the best introduction to this section.

The best source of degenerate surfaces is our imagination. Any bizarre form we can conjure up is certain to appear unbidden and at an awkward moment in any sophisticated modeling system.

The conditions we are about to explore can be created with any of the forms we have discussed: bicubic patches, composite surfaces, Bezier or B-spline surfaces. Some of these forms have idiosyncrasies or pathological conditions— superimposed vertices or, better yet, jumbled vertices of characteristic polyhedra.

There is a variety of ways of categorizing degenerate surfaces; for example, those exhibiting degenerate boundary curves, those exhibiting superimposed corners, and those with strictly internal degeneracies. We can easily demonstrate them all using a single bicubic patch.

The simplest degenerate patch is a point where $p_{00} = p_{10} = p_{01} = p_{11}$ and all other coefficients are identically equal to zero. A patch degenerates to a straight line when $p_{00}^u = p_{10}^u = p_{10} - p_{00}$ and all other coefficients are zero.

The degenerate patches in Fig. 3.58 have easily identifiable characteristics in their **B** matrices. For the three-sided patch in Fig. 3.58a, if we assume that $p_{00} = p_{10}$, then the **B** matrix is

$$\mathbf{B}_{(a)} = \begin{bmatrix} p_{00} & p_{01} & p_{00}^w & p_{01}^w \\ p_{00} & p_{11} & p_{10}^w & p_{11}^w \\ 0 & p_{01}^u & p_{00}^{uw} & p_{01}^{uw} \\ 0 & p_{11}^u & p_{10}^{uw} & p_{11}^{uw} \end{bmatrix} \tag{3.157}$$

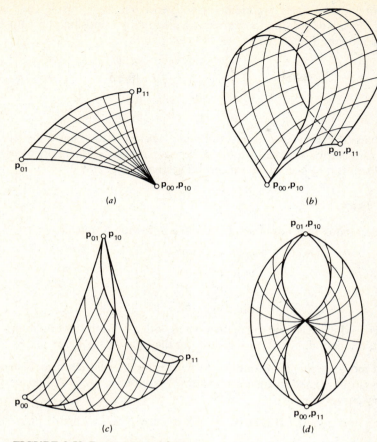

FIGURE 3.58 Degenerate surfaces.

A different kind of three-sided patch is shown in Fig. 3.58*b*. Instead of a zero-length fourth side, one of the sides does double duty. Its **B** matrix might be as follows:

$$\mathbf{B}_{(b)} = \begin{bmatrix} \mathbf{p}_{00} & \mathbf{p}_{01} & \mathbf{p}_{00}^{w} & \mathbf{p}_{01}^{w} \\ \mathbf{p}_{00} & \mathbf{p}_{01} & \mathbf{p}_{00}^{w} & \mathbf{p}_{01}^{w} \\ \mathbf{p}_{00}^{u} & \mathbf{p}_{01}^{u} & \mathbf{p}_{00}^{uw} & \mathbf{p}_{01}^{uw} \\ \mathbf{p}_{10}^{u} & \mathbf{p}_{11}^{u} & \mathbf{p}_{10}^{uw} & \mathbf{p}_{11}^{uw} \end{bmatrix} \qquad (3.158)$$

When points diagonally opposite each other are coincident, the result may be as shown in Fig. 3.58*c*. The **B** matrix would be as follows:

$$\mathbf{B}_{(c)} = \begin{bmatrix} \mathbf{p}_{00} & \mathbf{p}_{10} & \mathbf{p}_{00}^{w} & \mathbf{p}_{01}^{w} \\ \mathbf{p}_{10} & \mathbf{p}_{11} & \mathbf{p}_{10}^{w} & \mathbf{p}_{11}^{w} \\ \mathbf{p}_{00}^{u} & \mathbf{p}_{01}^{u} & \mathbf{p}_{00}^{uw} & \mathbf{p}_{01}^{uw} \\ \mathbf{p}_{10}^{u} & \mathbf{p}_{11}^{u} & \mathbf{p}_{10}^{uw} & \mathbf{p}_{11}^{uw} \end{bmatrix} \qquad (3.159)$$

Finally, in Fig. 3.58d, we create an interesting shape by using the results from Fig. 3.58c and joining the remaining diagonally opposing pair of points, with the following **B** matrix:

$$\mathbf{B}_{(d)} = \begin{bmatrix} \mathbf{p}_{00} & \mathbf{p}_{10} & \mathbf{p}_{00}^{w} & \mathbf{p}_{01}^{w} \\ \mathbf{p}_{10} & \mathbf{p}_{00} & \mathbf{p}_{10}^{w} & \mathbf{p}_{11}^{w} \\ \mathbf{p}_{00}^{u} & \mathbf{p}_{01}^{u} & \mathbf{p}_{00}^{uw} & \mathbf{p}_{01}^{uw} \\ \mathbf{p}_{10}^{u} & \mathbf{p}_{11}^{u} & \mathbf{p}_{10}^{uw} & \mathbf{p}_{11}^{uw} \end{bmatrix} \qquad (3.160)$$

Perhaps the terms *degenerate* and *pathological* are too pejorative, at least for the examples we have just explored, because although some may have little practical value, they clearly have a certain aesthetic appeal.

EXERCISES

1. Sketch a patch that might result from the following **B** matrix:

$$\begin{bmatrix} \mathbf{p}_{00} & \mathbf{p}_{00} & \mathbf{p}_{00}^{w} & \mathbf{p}_{01}^{w} \\ \mathbf{p}_{00} & \mathbf{p}_{00} & \mathbf{p}_{10}^{w} & \mathbf{p}_{11}^{w} \\ \mathbf{p}_{00}^{u} & \mathbf{p}_{01}^{u} & \mathbf{p}_{00}^{uw} & \mathbf{p}_{01}^{uw} \\ \mathbf{p}_{10}^{u} & \mathbf{p}_{11}^{u} & \mathbf{p}_{10}^{uw} & \mathbf{p}_{11}^{uw} \end{bmatrix}$$

2. Sketch and discuss how to create a patch with a self-intersecting interior. In addition, the boundary curves should form the sides of a square.

3. What more specific conditions must be met by the geometric coefficients of the shape in Fig. 3.58d to ensure symmetry? What effect does this have on the boundary curves?

4. Sketch and give the **B** matrix for a two-sided patch with two pairs of co-incident points.

5. Sketch and give the **B** matrix for a two-sided patch with three coincident points.

20   CURVES ON SURFACES

A surface provides a two-dimensional space suitable for supporting the vector analytic representation of curves. The two parametric variables defining the surface also serve to supply the coordinate grid on which embedded curves may be defined. This inherent characteristic of the parametric representation is an invaluable asset to many aspects of geometric modeling, including surfaces with boundaries and the intersection and union of primitive shapes.

Two subjects are of immediate interest—the general formulation of curves embedded in a surface and regular curve networks on surfaces. The curve nets we will investigate are the parametric, orthogonal, and conjugate nets. Our primary goal is to become familiar with certain systems of curves on surfaces that later arise when applying surfaces to modeling problems. We will find that

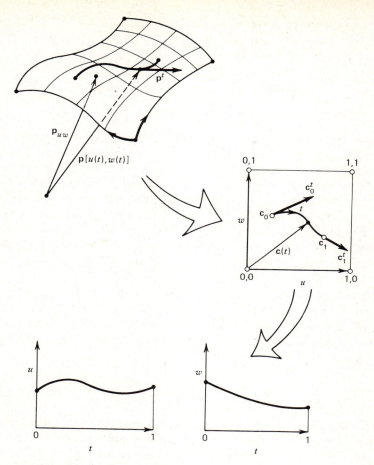

FIGURE 3.59 Curves on surfaces.

the curve nets have special properties. There are many other types of curves on surfaces; among these, lines of constant or equal curvature and geodesics are discussed later in Chapter 5.

Now, consider the curve on the surface in Fig. 3.59. This curve is defined in the u, w parametric space of the patch. Here, we introduce a new parametric variable to distinguish the curve formulation, and we denote the curve as $\mathbf{c}(t)$. Note that in the u, w parametric plane, the curve is necessarily a plane curve. Also note that points on the curve have vector components in the u and w directions. Thus,

$$\mathbf{c}(t) = \mathbf{u}(t) + \mathbf{w}(t) \qquad (3.161)$$

We can define this curve in many ways in the u, w plane: explicitly, implicitly, with Bezier or B-spline forms, with the parametric cubic form, and so on. Let us continue to use the pc curve. Then, we define $\mathbf{c}(t)$ as

$$\mathbf{c}(t) = \mathbf{TMB}_c \qquad (3.162)$$

where

$$\mathbf{T} = [t^3 \quad t^2 \quad t \quad 1]$$

$$\mathbf{B}_c = [\mathbf{c}_0 \quad \mathbf{c}_1 \quad \mathbf{c}_0^t \quad \mathbf{c}_1^t]^T$$

and $\mathbf{M}$ is the standard pc universal transformation matrix. The parametric component curves are sketched in their respective u, t and w, t parametric spaces.

To determine the coordinates of points on this curve in object space, first compute pairs of u, w values for successive values of t. Substitute these u, w values into the surface equation, say, $\mathbf{p}(u, w) = \mathbf{UMB}_s\mathbf{M}^T\mathbf{W}^T$, to obtain a sequence of x, y, z coordinates of points in object space. These points are necessarily on the patch. This process selects only those points on the patch that satisfy the curve equation. We can state this symbolically as

$$\mathbf{p}[u(t), w(t)] = \text{points on the curve, on the patch} \qquad (3.163)$$

We can find an object-space tangent vector $\mathbf{p}_t^t$ to the curve at any point t on it by first computing $\mathbf{c}_t^t$, the tangent vector to the curve at $u(t)$, $w(t)$ in the u, w plane. Determine the direction cosines of this vector (that is, with respect to the u, w coordinate system), and denote them k_u and k_w. Now, compute $\mathbf{p}_{uw}^u$ and $\mathbf{p}_{uw}^w$ at the u, w point determined by $\mathbf{c}(t)$ at t. We find that

$$\mathbf{p}_t^t = k_u \frac{\mathbf{p}_{uw}^u}{|\mathbf{p}_{uw}^u|} + k_w \frac{\mathbf{p}_{uw}^w}{|\mathbf{p}_{uw}^w|} \qquad (3.164)$$

It is the direction of $\mathbf{p}_t^t$ that interests us; the magnitude has no particular meaning.

Of course, the simplest and most obvious curves on a patch are the parametric curves. Earlier, we found them to be those curves on a patch along each of which one of the parameters varies while the other is constant. If the parameter w is held fixed while u varies, the locus of the variable point $\mathbf{p}(u, w)$ is another u curve on the patch. By continuing this process, we can define a one-parameter family of u curves that completely covers the patch. Next, we interchange the roles of u and w and define a one-parameter family of w curves. Thus, the family of u curves and the family of w curves together constitute the parameteric curves.

A curve net on a patch $\mathbf{p}(u, w)$ is two one-parameter families of curves on the patch, constructed so that through each point there passes just one of each family. Furthermore, the two tangents of the curves at the point must be distinct. A net can be covariant to its sustaining patch; that is, it can be defined in a way that is invariant under rigid motion in space.

A parametric net is not necessarily covariant to its surface. However, with a suitably chosen transformation of parameters, the parametric net can be made to coincide with any specified net on the surface. Although we will not develop the proofs here, any covariant net can be transformed to a parametric net. We conclude that the geometry of a surface is independent of the analytical representation used and is, therefore, independent of the choice of parametric nets. If the parametric net is a covariant net, then we find that properties of the net are also properties of the surface itself.

We assert without proof that the only transformations of parameters that leave the parametric net invariant are

$$r = r(u) \qquad s = s(w) \tag{3.165}$$

and

$$r = r(w) \qquad s = s(u) \tag{3.166}$$

from (u, w) to (r, s) or from (u, w) to (s, r). Finally, we see that the normal vector at a point on a surface before and after transformation of parameters is unchanged, except perhaps for its sense.

Next, let us consider a more specialized net, the orthogonal net of curves on a surface. An orthogonal net is a net such that at every point of the surface, the two curves of the net intersect at right angles. We immediately conclude that a necessary and sufficient condition for the parametric net on a patch $\mathbf{p}(u, w)$ to be orthogonal is

$$\mathbf{p}^u \cdot \mathbf{p}^w = 0 \tag{3.167}$$

An orthogonal net is useful in many analytical engineering applications of geometric modeling; for example, in structural analysis, fluid flow, and thermodynamics. Here, the orthogonality of the parametric field of an object's or phenomenon's geometric characteristics is compatible with the orthogonality of the driving and response functions.

The last type of curve net that we will consider is the conjugate net. A net of curves on a surface is a conjugate net when the tangents of the curves of one family of the net at points along each fixed curve of the other family form a developable surface; see Fig. 3.60. The two families of the net do not play symmetrical roles in this definition; however, the two families are interchangeable. A necessary condition for the parametric net on a patch to be a conjugate net is

$$\mathbf{p}^{uw} \cdot \mathbf{n} = 0 \tag{3.168}$$

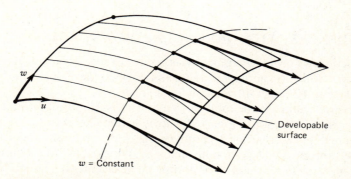

FIGURE 3.60 Conjugate net.

EXERCISES

1. Try to formulate an example of a Bezier curve on a Bezier surface; a *B*-spline curve on a *B*-spline surface. What are the advantages and disadvantages?

2. What happens if in Ex. 1 we mix the type of curve and surface (that is, a Bezier curve on a *B*-spline surface)?

3. Develop expressions for composite pc curves on composite bicubic patches.

4. Given a bicubic patch $\mathbf{p}(u, w)$ and a pc curve $\mathbf{c}(t)$ on it, is there necessarily a pc curve $\mathbf{p}(u)$ in object space (that is, x, y, z coordinates) that exactly represents $\mathbf{c}(t)$?

5. Discuss several alternative methods of transforming the expression of a curve in patch coordinates $\mathbf{p}[u(t), w(t)]$ into an expression in object space coordinates.

6. Give a specific example of an orthogonal parametric net and a conjugate parametric net. Use the Hermite bicubic form, and give only the **B** matrix for each.

7. Write a procedure to compute the coordinates of a point on a curve on a surface. Denote this as **POCOS**(SI, CI, T, PO), where

SI(4, 4, 3) is the input array of geometric coefficients of a bicubic patch;

CI(4, 2) is the input array of geometric coefficients of a pc curve defined on the patch in terms of the u, w parametric variables;

T is the input value of the parametric variable t of a point on the curve; and

PO(5) is the output array of x, y, z, u, w coordinates.

8. Generalize **POCOS** so that the coordinates of a series of points along the curve can be efficiently computed at equal increments of the parametric variable t.

9. Further generalize **POCOS** so that Hermite, Bezier, and *B*-spline curves and surfaces can be processed.

21   SURFACES WITH IRREGULAR BOUNDARIES

In this section, we begin to explore ideas central to geometric modeling of solid objects. We will also anticipate the point of view developed for boundary representations (Chapter 10, Section 7).

We can decompose the total surface of a physical object into a collection of similar surface regions; Fig. 3.61 shows a familiar example. A region is characterized by well-behaved contours, reaching practical limits when we encounter abrupt or discontinuous changes. We establish the boundaries of a region at these discontinuities or abrupt shape changes by intersections with adjacent regions. In general, the intersection curves do not coincide with the

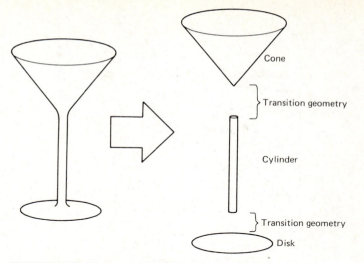

FIGURE 3.61 Decomposition of a complex shape.

isoparametric curves on the surface, so we call them **irregular boundaries of a surface**.

We segment the total surface of an object model in this way because each region is itself then mathematically tractable, whereas the total surface often is not. We map each region and its boundary curves onto a unit square or $m \times n$ rectangle in parametric space. An analytical set of curves for each region is prerequisite to this mapping function.

Earlier, we defined a surface in parametric form by specifying analytical relationships between the object space coordinates and two parameters u and w; thus,

$$\mathbf{p} = [x(u, w) \quad y(u, w) \quad z(u, w)]$$

Think of these parametric variables as surface coordinates. Every point on a surface has not only the familiar triplet of Cartesian coordinates x, y, z, but also a unique pair of parametric coordinates u, w. We have interpreted this as a mapping of the surface onto a unit square in the u, w coordinate plane, where we derived the parametric equations of the surface from the boundary conditions.

Boundary curves of a surface are frequently irregular. For real objects, they may not correspond to the limits of the parametric surface established by the domain of the parametric variables. By construction, the parametric surface always equals or exceeds in extent the object's surface region that we are modeling. Thus, boundary curves delimit a subset of points on the full point set of the parametric surface. The points in this subset are in a one-to-one correspondence with points on the real object's surface.

In Fig. 3.62, an object surface region R is mapped onto the parametric surface S. The boundaries of R are, in part, established by intersections with parametric surfaces S_1 and S_2. They may or may not themselves contain object

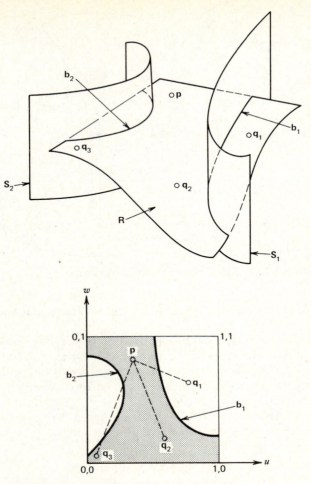

FIGURE 3.62 Surface with irregular boundary.

surface regions. These boundary curves are denoted as $\mathbf{b}_1$ and $\mathbf{b}_2$, and they are mapped onto the same unit square, where they are expressed as parametric equations of the form

$$\mathbf{b}_i = [u_i(t) \quad w_i(t)] \qquad t \in [0, 1] \tag{3.169}$$

After we compute the intersections as sets of u, w points, we apply curve-fitting techniques to define these curves or a set of composite curves.

To obtain the boundary curve in object space, we compute points on the u, w coordinate plane from $u_i(t)$ and $w_i(t)$. We then substitute the values of u and w into the appropriate surface equation

$$\mathbf{p} = [x(u, w) \quad y(u, w) \quad z(u, w)]$$

to obtain (x, y, z) points on the curve.

Many geometric operations require deciding if a given point on the unit square lies within a bounded region. One way to do this is to specify some point **q** that is known to lie in this bounded region. Then, given any other point **p**, we can easily test whether or not it lies in the region as follows: Construct the straight line *pq* in the *u*, *w* plane. Compute the number of intersections this line makes with each of the boundary curves $\mathbf{b}_i$. If it does not intersect any of these curves, then **p** is in an active region. If *pq* intersects any boundary an odd number of times, then *pq* lies outside the region. All even intersections bring the **p** back into the region; Fig. 3.62 shows several examples of this.

Symbolically, one way a region *R* is expressed is as a function of the parametric surface *S* onto which it is mapped, the boundary curves $\mathbf{b}_i$, and a known interior point **q**.

$$\mathbf{R} = \phi(S, \mathbf{b}_i, \mathbf{q}) \tag{3.170}$$

Here, *S* is an $m \times n$ rectangular array of parametric surface patches normalized on a unit square in the parametric plane *u*, *w*. $\mathbf{b}_i$ is a set of parametric curves mapped onto the same plane and expressed as functions of a parameter *t*, that is, $u = u(t)$, $w = w(t)$. **q** is a point on the parametric plane inside the active region.

An alternative, and perhaps more elegant, way of indicating the active region is by specifying the intersection of an appropriate set of two-dimensional half-spaces. Using this approach, we merely test each candidate point against the functions defining the half-spaces; see Fig. 3.63.

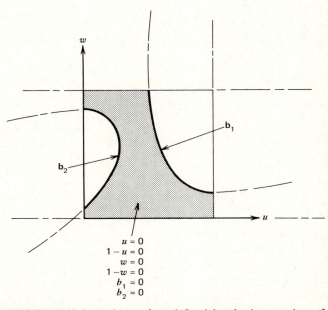

$$
\begin{aligned}
u &= 0 \\
1 - u &= 0 \\
w &= 0 \\
1 - w &= 0 \\
b_1 &= 0 \\
b_2 &= 0
\end{aligned}
$$

FIGURE 3.63 Irregular surface defined by the intersection of half-spaces.

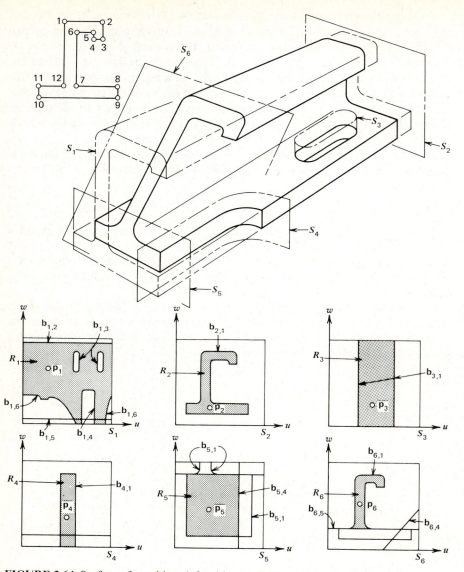

FIGURE 3.64 Surface of an object defined by a collection of smaller regions.

We stated earlier that the surface of any physical object can be decomposed into a collection of smaller surface regions; Fig. 3.64 is an example of this. Here, we see an extrusion that has been slotted and truncated in a variety of ways, yet it is easily decomposed into six regions, each mapped onto a parametric surface.

There are six parametric surfaces. S_1 wraps around the entire cross section and defines the "longitudinal" surface (before "cuts" and "slots" are made). The 12 numbered points shown on the cross section in the upper left of the figure are for reference. We can think of them as the end points of longitudinal lines where S_1 changes direction. This closed curve is an edge of S_1.

S_2 is a plane surface normal to S_1 that completely defines one end plane of the extrusion. S_3 is a closed cylindrical surface intersecting S_1 to form a slot in the extrusion. S_4 is an open cylindrical surface that takes a nick out of the bottom flange. S_5 is a plane normal to S_1 that establishes one facet of the near-end plane. S_6 is also a plane, but it takes an angular cut through S_1.

Each surface is a unique $m \times n$ array of parametric surface patches. We establish each region R on a surface by intersecting the surface with adjacent surfaces. The interiors are defined by six points $\mathbf{p}_1$–$\mathbf{p}_6$. (Or we could use an expression defining the intersecting two-dimensional half-spaces.) Figure 3.64 shows these points and region boundaries mapped onto their respective parametric surfaces in the unit square on the u, w plane.

To complete the definition of the object's total surface, we define a point $\mathbf{p}_v$, which we know to be inside the enclosed volume of the object. This point allows us to test any other point $\mathbf{q}_v$ for the condition of being outside, on, or inside the volume. We do this by connecting $\mathbf{p}_v$ and $\mathbf{q}_v$ with a straight line. By counting intersections that occur between this line and each of the regions, we can determine the test results. If there are no intersections, then $\mathbf{q}_v$ is obviously inside the object. If the line intersects any region an odd number of times, then it lies outside the object. All even intersections bring the point back into the interior of the object. If a point lies exactly on the surface of the object, this is easily determined by testing the distance of $\mathbf{q}_v$ from each region. We will discover other approaches to determining inside/outside status in later chapters.

4

SOLIDS

Many techniques for modeling solids are flawed because they do not represent the interior of a solid. They also do not represent internal properties or offer ways of representing internal behavior. These models assume total internal homogeneity. For many purposes such techniques are adequate; however, in this chapter, we investigate the mathematics and geometry supporting a more complete form of solid representation.

Perhaps the simplest and most direct mathematical element available to model a solid is a **hyperpatch**. A hyperpatch is a patch-bounded collection of points whose coordinates are given by continuous, three-parameter, single-valued mathematical functions of the form

$$x = x(u, v, w) \qquad y = y(u, v, w) \qquad z = z(u, v, w) \tag{4.1}$$

The parametric variables u, v, and w are constrained to the interval $u, v, w \in [0, 1]$. The terms *hyperpatch* and *parametric solid* are used interchangeably, even though historically the term *hyperpatch* had a narrower meaning.

Let us follow a course of development of solids that parallels the one we used for curves and surfaces. In doing this, we find that fixing the value of one of the parametric variables results in a surface within or on the boundary of the hyperpatch or solid in terms of the other two variables, which remain free. We continue this process for any number of arbitrary values in the allowed interval, first for one variable and then for each of the others in turn, forming a parametric network of cells of three two-parameter families of surfaces throughout the solid. Through each point $\mathbf{p}(u, v, w)$, there passes just one surface of each family. As with curves and surfaces, the positive sense on any surface is the sense in which the free parameters increase. *Note:* These surfaces are sometimes called **isoparametric surfaces**, indicating a surface within the solid on which one of the three parametric variables is constant.

Associated with every hyperpatch is a set of boundary elements; see Fig. 4.1. The most obvious of them are the eight corner points, the 12 curves defining the

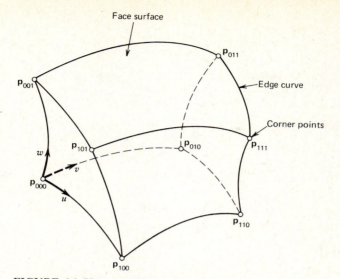

FIGURE 4.1 Hyperpatch boundary elements.

edges, and the six surfaces (patches) defining the faces. Others include the tangent vectors and twist vectors of the edge curves and face patches.

For an ordinary nondegenerate hyperpatch or parametric solid, there are always eight and only eight corner points. This follows from the possible combinations of the two limits of the three parametric variables. Thus, the corner points are found by substituting the eight combinations of 0 and 1 into $\mathbf{p}(u, v, w)$ to obtain $\mathbf{p}(0, 0, 0)$, $\mathbf{p}(1, 0, 0)$, $\mathbf{p}(0, 1, 0)$, $\mathbf{p}(0, 0, 1)$, $\mathbf{p}(1, 1, 0)$, $\mathbf{p}(1, 0, 1)$, $\mathbf{p}(0, 1, 1)$, and $\mathbf{p}(1, 1, 1)$.

The edge curves are functions of one of the three parametric variables. We obtain them by allowing one of the variables to remain free while successively fixing combinations of the other two at their limiting values. This results in the 12 possible combinations yielding the defining functions of the 12 parametric edge curves $\mathbf{p}(u, 0, 0)$, $\mathbf{p}(u, 1, 0)$, $\mathbf{p}(u, 0, 1)$, $\mathbf{p}(u, 1, 1)$, $\mathbf{p}(0, v, 0)$, $\mathbf{p}(1, v, 0)$, $\mathbf{p}(0, v, 1)$, $\mathbf{p}(1, v, 1)$, $\mathbf{p}(0, 0, w)$, $\mathbf{p}(1, 0, w)$, $\mathbf{p}(0, 1, w)$, and $\mathbf{p}(1, 1, w)$.

We conclude that the face patches are functions of two of the three parametric variables. They are obtained by allowing two of the variables to remain free while successively fixing the remaining one to each of its two limiting values. This results in the six possible permutations yielding the defining functions of the six patches $\mathbf{p}(u, v, 0)$, $\mathbf{p}(u, v, 1)$, $\mathbf{p}(u, 0, w)$, $\mathbf{p}(u, 1, w)$, $\mathbf{p}(0, v, w)$, and $\mathbf{p}(1, v, w)$.

The simplest example is a rectangular solid; consider Fig. 4.2. The following parametric equations represent a rectangular solid in an x, y, z coordinate system:

$$x = (b - a)u + a$$
$$y = (d - c)v + c \qquad u, v, w \in [0, 1] \qquad (4.2)$$
$$z = (f - e)w + e$$

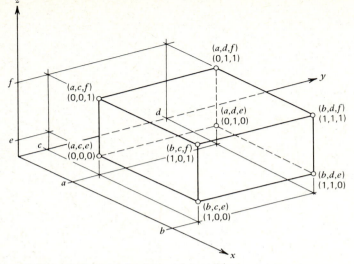

FIGURE 4.2 Rectangular parametric solid.

Although this is an extremely simple example, it illustrates an important feature. These parametric equations define not only the points comprising the bounding elements of the solid, but also all the points interior to it. Also, it is quite easy for us to verify the boundary-defining combinations just described.

Chapter 4 introduces the tricubic parametric solid. The choice of the tricubic form is one of convenience and familiarity as a natural continuation of the cubic and bicubic forms for curves and surfaces we saw earlier.

Although literature on tricubic models is sparse, especially when compared to research on other forms of multivariate interpolation functions, it is a fertile area for development where a greater analytic potential is required of a solid model. This has already been amply demonstrated by Stanton (1974) and Timmer (1970) in the areas of structural mechanics and ablation thermo- and aerodynamics. The earlier natural restraints on research and development of solid modeling techniques of this type were due to the then seemingly massive data requirements. However, current computing resources have nearly eliminated such restraints.

1 ALGEBRAIC AND GEOMETRIC FORM

The algebraic form of a tricubic solid is given by the following polynomial equation:

$$\mathbf{p}(u, v, w) = \sum_{i=0}^{3} \sum_{j=0}^{3} \sum_{k=0}^{3} \mathbf{a}_{ijk} u^i v^j w^k \qquad u, v, w \in [0, 1] \qquad (4.3)$$

The $\mathbf{a}_{ijk}$ vectors are called the **algebraic coefficients** of the solid. Notice the obvious source of the term *tricubic*: Each of the three parametric variables can appear as a cubic term.

The parametric variables u, v, and w are restricted by definition to values in the interval 0 to 1, inclusive. This restriction makes the solid bounded in a regular way by cubic patches. We will explore this situation later in this section.

When we perform the indicated summation and expansion of the cubic polynomials in u, v, and w, for $x(u, v, w)$ we obtain

$$x(u, v, w) = a_{333x}u^3v^3w^3 + a_{332x}u^3v^3w^2 + \cdots + a_{000x} \tag{4.4}$$

There are similar expressions for $y(u, v, w)$ and $z(u, v, w)$. Each expression consists of 64 terms, which means that the total number of algebraic coefficients available is three times this, or 192 coefficients.

Next, we write these coefficients in a more compact form using vector notation. The result is similar to Eq. 2.18 for curves and Eq. 3.10 for surfaces.

$$\begin{aligned}\mathbf{p}(u, v, w) = {} & \mathbf{a}_{333}u^3v^3w^3 + \mathbf{a}_{332}u^3v^3w^2 + \mathbf{a}_{331}u^3v^3w + \mathbf{a}_{330}u^3v^3 \\ & + \mathbf{a}_{323}u^3v^2w^3 + \cdots \\ & \qquad \vdots \\ & + \mathbf{a}_{000}\end{aligned} \tag{4.5}$$

This is an improvement, but perhaps we can do even better if we use matrix notation. We are now confronted with a modest complication, since we have an extra parametric dimension. With three-component vectors and three independent parametric variables, the matrices of algebraic and geometric coefficients are more cumbersome. In any event, let us proceed. We quickly reduce the product of the first and second cubic polynomials in u and v to matrix form as $\mathbf{UA}_{uv}\mathbf{V}^T$. We reduce the third polynomial to $\mathbf{A}_w\mathbf{W}^T$, so that

$$\mathbf{p}(u, v, w) = \mathbf{UA}_{uv}\mathbf{V}^T\mathbf{A}_w\mathbf{W}^T \qquad u, v, w \in [0, 1] \tag{4.6}$$

where

$$\begin{aligned}\mathbf{U} &= [u^3 \quad u^2 \quad u \quad 1] \\ \mathbf{V}^T &= [v^3 \quad v^2 \quad v \quad 1]^T \\ \mathbf{W}^T &= [w^3 \quad w^2 \quad w \quad 1]^T\end{aligned} \tag{4.7}$$

It is evident that Eq. 4.6 is considerably more complex than, say, its equivalent for surfaces, Eq. 3.12.

Now let us look at another choice of indexing, namely,

$$\mathbf{p}(u, v, w) = \sum_{i=1}^{4}\sum_{j=1}^{4}\sum_{k=1}^{4}\mathbf{a}_{ijk}u^{4-i}v^{4-j}w^{4-k} \tag{4.8}$$

Although Eq. 4.8 is not so clean as Eq. 4.3, it is compatible with the equation we will next introduce for the geometric form, and it is a more acceptable form of tensor notation. We can further simplify Eq. 4.8 by dropping the summation signs and adopting the convention that for cubics, the range of the indices is from 1 to 4, so that we rewrite Eq. 4.8 as

$$\mathbf{p} = \mathbf{a}_{ijk}u^{4-i}v^{4-j}w^{4-k} \tag{4.9}$$

We can take the same approach with the geometric form, with the result

$$\mathbf{p} = F_i(u)F_j(v)F_k(w)\mathbf{b}_{ijk} \tag{4.10}$$

The F terms are the now familiar blending functions F_1, F_2, F_3, and F_4, as determined by the subscripted index and in terms of the specified parametric valuable. $\mathbf{b}_{ijk}$ is the array of boundary conditions or geometric vectors. Notice that $\mathbf{a}_{ijk}$ and $\mathbf{b}_{ijk}$ are three-component vectors. Again, there are $4 \times 4 \times 4$ or 64 algebraic vectors and 64 geometric vectors, so that there are 192 algebraic coefficients and 192 geometric coefficients.

Consider the $\mathbf{b}_{ijk}$ array terms. How do we interpret them? We will investigate more formal procedures for doing this in the last section of this chapter, but for now, let us apply some empirical analysis to their interpretation. Since the indices on the $\mathbf{b}$ array are coordinated with those indices on the F terms, we can draw some conclusions from our knowledge of the blending functions as they apply to curves and surfaces.

First, we must have a frame of reference. Let us look at the unit cube in parametric space and establish some nomenclature. Study Fig. 4.3.

At each of the eight corners, we find the following boundary conditions: the corner point itself, three tangent vectors, three twist vectors, and a vector defined by the third-order mixed partial derivative of the function. We find there are eight vectors defining boundary conditions at each of the eight corners, resulting in a total of 64 vectors. An example is in order. We will arbitrarily choose the corner defined by $\mathbf{p}(1, 0, 1)$ and list the eight boundary condition vectors

$$\mathbf{p}_{101} \qquad \text{corner point}$$

$$\left.\begin{array}{l}\mathbf{p}^u_{101} \\ \mathbf{p}^v_{101} \\ \mathbf{p}^w_{101}\end{array}\right\} \quad \text{tangent vectors } \frac{\partial}{\partial u}, \frac{\partial}{\partial v}, \frac{\partial}{\partial w}$$

$$\left.\begin{array}{l}\mathbf{p}^{uv}_{101} \\ \mathbf{p}^{uw}_{101} \\ \mathbf{p}^{vw}_{101}\end{array}\right\} \quad \text{twist vectors } \frac{\partial^2}{\partial u\, \partial v}, \frac{\partial^2}{\partial u\, \partial w}, \frac{\partial^2}{\partial v\, \partial w}$$

$$\mathbf{p}^{uvw}_{101} \qquad \text{triple mixed partial } \frac{\partial^3}{\partial u\, \partial v\, \partial w}$$

Now, let us return to the problem of interpreting the elements of the $\mathbf{b}$ array. Similar representation and expansion for curves and surfaces offer more clues. Thus, for curves we have

$$\mathbf{p} = \sum F_i \mathbf{b}_i \tag{4.11}$$

which we expand to

$$\mathbf{p} = F_1(u)\mathbf{b}_1 + F_2(u)\mathbf{b}_2 + F_3(u)\mathbf{b}_3 + F_4(u)\mathbf{b}_4 \tag{4.12}$$

and interpret as

$$\mathbf{p} = F_1(u)\mathbf{p}_0 + F_2(u)\mathbf{p}_1 - F_3(u)\mathbf{p}^u_0 + F_4(u)\mathbf{p}^u_1 \tag{4.13}$$

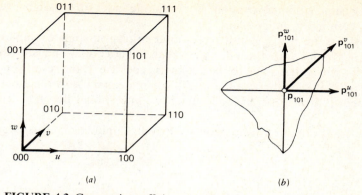

FIGURE 4.3 Geometric coefficients on the unit cube.

Assigning curve boundary-condition vectors to elements of the $\mathbf{b}_i$ array is obvious. Now, look at this development for surfaces

$$\mathbf{p} = F_i(u)F_j(w)\mathbf{b}_{ij} \tag{4.14}$$

which we expand to

$$
\begin{aligned}
\mathbf{p} = \ & F_1(u)F_1(w)\mathbf{b}_{11} + F_1(u)F_2(w)\mathbf{b}_{12} + F_1(u)F_3(w)\mathbf{b}_{13} + F_1(u)F_4(w)\mathbf{b}_{14} \\
& + F_2(u)F_1(w)\mathbf{b}_{21} + F_2(u)F_2(w)\mathbf{b}_{22} + F_2(u)F_3(w)\mathbf{b}_{23} + F_2(u)F_4(w)\mathbf{b}_{24} \\
& + F_3(u)F_1(w)\mathbf{b}_{31} + F_3(u)F_2(w)\mathbf{b}_{32} + F_3(u)F_3(w)\mathbf{b}_{33} + F_3(u)F_4(w)\mathbf{b}_{34} \\
& + F_4(u)F_1(w)\mathbf{b}_{41} + F_4(u)F_2(w)\mathbf{b}_{42} + F_4(u)F_3(w)\mathbf{b}_{43} + F_4(u)F_4(w)\mathbf{b}_{44}
\end{aligned}
$$

$$\tag{4.15}$$

and interpret as

$$
\begin{aligned}
\mathbf{p} = \ & F_1(u)F_1(w)\mathbf{p}_{00} + F_1(u)F_2(w)\mathbf{p}_{01} + F_1(u)F_3(w)\mathbf{p}_{00}^w + F_1(u)F_4(w)\mathbf{p}_{01}^w \\
& + F_2(u)F_1(w)\mathbf{p}_{10} + F_2(u)F_2(w)\mathbf{p}_{11} + F_2(u)F_3(w)\mathbf{p}_{10}^w + F_2(u)F_4(w)\mathbf{p}_{11}^w \\
& + F_3(u)F_1(w)\mathbf{p}_{00}^u + F_3(u)F_2(w)\mathbf{p}_{01}^u + F_3(u)F_3(w)\mathbf{p}_{00}^{uw} + F_3(u)F_4(w)\mathbf{p}_{01}^{uw} \\
& + F_4(u)F_1(w)\mathbf{p}_{10}^u + F_4(u)F_2(w)\mathbf{p}_{11}^u + F_4(u)F_3(w)\mathbf{p}_{10}^{uw} + F_4(u)F_4(w)\mathbf{p}_{11}^{uw}
\end{aligned}
$$

$$\tag{4.16}$$

Again, the assignment of boundary condition vectors to the elements of the $\mathbf{b}_{ij}$ array is obvious by simply comparing Eqs. 4.15 and 4.16 term for term.

Now, we can make some generalizations based on these equations. First, we observe that when the indices of the blending functions of a term are all less than 3, then the associated $\mathbf{b}$ vector denotes a point. Second, when one, and only one, of the indices of a term is 3 or 4, then the associated $\mathbf{b}$ vector denotes a tangent vector. Third, when two, and only two, of the indices are 3 or 4, then the associated $\mathbf{b}$ vector denotes a twist vector (second-order, mixed partial derivative). Fourth, when, in the case of a tricubic solid, three indices of a term are 3 or 4, then the associated $\mathbf{b}$ vector denotes that vector defined by the third-order, mixed partial derivative of the function $\mathbf{p}(u, v, w)$.

Next, we must find some rule to tell us where the $\mathbf{b}$ vector is applied. Our current method of constructing the arrays of geometric coefficients suggests the following empirical interpretation: For each odd index on $\mathbf{b}$, assign a zero, and for each even index, assign a one.

Now, we are ready to test these observations on some examples. For a curve, let us try $\mathbf{b}_2$. Since the index is less than 3, it is a point. And since the index is even, we will assign a one to the geometric coefficient; the result is $\mathbf{p}_1$. For a surface, try $\mathbf{b}_{41}$. Did you obtain $\mathbf{p}_{10}^u$? Do you understand why the superscript u was assigned? For a tricubic solid, try $\mathbf{b}_{131}$; you should find that it denotes $\mathbf{p}_{000}^v$.

The transformation from geometric to algebraic coefficients is given by

$$\mathbf{a}_{ijk} = \sum_{l=1}^{4} \sum_{m=1}^{4} \sum_{n=1}^{4} \mathbf{M}_{il}\mathbf{M}_{jm}\mathbf{M}_{kn}\mathbf{b}_{lmn} \tag{4.17}$$

or more compactly as

$$\mathbf{a}_{ijk} = \mathbf{M}_{il}\mathbf{M}_{jm}\mathbf{M}_{kn}\mathbf{b}_{lmn} \tag{4.18}$$

Here, $\mathbf{M}$ is the universal transformation matrix of Eq. 2.31, and the various subscripted indices denote specific elements of the matrix. The derivation of this transformation equation is left as an exercise.

Stanton (1974) has developed a slightly different approach in which he displays the 64 hyperpatch parameters as four sets of 16 parameters. This allows the blending properties and the relationship between patches and hyperpatches to be more easily understood. He begins by defining three distinct but equivalent forms: the familiar algebraic and geometric forms, and a point form where $\mathbf{p}(u, v, w)$ is given at the one-third points of the parametric variables. The point form thus requires 64 points to be complete. These forms are given as the following triple summations:

$$\mathbf{p}(u, v, w) = F_i^a(u)F_j^a(v)F_k^a(w)\mathbf{a}_{ijk} \tag{4.19}$$

$$\mathbf{p}(u, v, w) = F_i^b(u)F_j^b(v)F_k^b(w)\mathbf{b}_{ijk} \tag{4.20}$$

$$\mathbf{p}(u, v, w) = F_i^P(u)F_j^P(v)F_k^P(w)\mathbf{p}_{ijk} \tag{4.21}$$

$F_i^a(u)$, $F_j^a(v)$, and $F_k^a(w)$ are blending functions applied to the algebraic coefficients, and we define them by

$$\begin{aligned}
F_1^a(u) &= u^3 \\
F_2^a(u) &= u^2 \\
F_3^a(u) &= u \\
F_4^a(u) &= 1
\end{aligned} \tag{4.22}$$

and similarly for v and w.

$F_i^b(u)$ $F_j^b(v)$, and $F_k^b(w)$ are blending functions applied to the geometric coefficients, and we define them by

$$\begin{aligned}
F_1^b(u) &= 2u^3 - 3u^2 + 1 \\
F_2^b(u) &= -2u^3 + 3u^2 \\
F_3^b(u) &= u^3 - 2u^2 + u \\
F_4^b(u) &= u^3 - u^2
\end{aligned} \tag{4.23}$$

and similarly for v and w. Equations 4.23 are simply from $\mathbf{F} = \mathbf{UM}$ (see Eqs. 2.31, 2.32, and 2.48).

Finally, $F_i^P(u)$, $F_j^P(v)$, and $F_k^P(w)$ are blending functions applied to the point coefficients, and we define them by

$$
\begin{aligned}
F_1^P(u) &= (-9/2)u^3 + 9u^2 - (11/2)u + 1 \\
F_2^P(u) &= (27/2)u^3 - (45/2)u^2 + 9u \\
F_3^P(u) &= -(27/2)u^3 + 18u^2 - (9/2)u \\
F_4^P(u) &= (9/2)u^3 - (9/2)u^2 + u
\end{aligned}
\tag{4.24}
$$

and similarly for v and w. Equations 4.24 are from $\mathbf{F} = \mathbf{UN}$ (see Eqs. 2.89 and 2.90a).

We are free to contract any of the three parametric coordinates. If we do this for w, then the geometric coefficients are written in a sequence of four 16-element arrays. We denote these arrays as the matrices

$$
\begin{aligned}
\mathbf{B}_1 &= [\mathbf{b}_{ij1}] \\
\mathbf{B}_2 &= [\mathbf{b}_{ij2}] \\
\mathbf{B}_3 &= [\mathbf{b}_{ij3}] \\
\mathbf{B}_4 &= [\mathbf{b}_{ij4}]
\end{aligned}
\tag{4.25}
$$

We expand them as follows:

$$
\mathbf{B}_1 = \begin{bmatrix}
\mathbf{p}_{000} & \mathbf{p}_{010} & \mathbf{p}_{000}^{v} & \mathbf{p}_{010}^{v} \\
\mathbf{p}_{100} & \mathbf{p}_{110} & \mathbf{p}_{100}^{v} & \mathbf{p}_{110}^{v} \\
\mathbf{p}_{000}^{u} & \mathbf{p}_{010}^{u} & \mathbf{p}_{000}^{uv} & \mathbf{p}_{010}^{uv} \\
\mathbf{p}_{100}^{u} & \mathbf{p}_{110}^{u} & \mathbf{p}_{100}^{uv} & \mathbf{b}_{110}^{uv}
\end{bmatrix}
$$

$$
\mathbf{B}_2 = \begin{bmatrix}
\mathbf{p}_{001} & \mathbf{p}_{011} & \mathbf{p}_{001}^{v} & \mathbf{p}_{011}^{v} \\
\mathbf{p}_{101} & \mathbf{p}_{111} & \mathbf{p}_{101}^{v} & \mathbf{p}_{111}^{v} \\
\mathbf{p}_{001}^{u} & \mathbf{p}_{011}^{u} & \mathbf{p}_{001}^{uv} & \mathbf{p}_{011}^{uv} \\
\mathbf{p}_{101}^{u} & \mathbf{p}_{111}^{u} & \mathbf{p}_{101}^{uv} & \mathbf{p}_{111}^{uv}
\end{bmatrix}
$$

$$
\tag{4.26}
$$

$$
\mathbf{B}_3 = \begin{bmatrix}
\mathbf{p}_{000}^{w} & \mathbf{p}_{010}^{w} & \mathbf{p}_{000}^{vw} & \mathbf{p}_{010}^{vw} \\
\mathbf{p}_{100}^{w} & \mathbf{p}_{110}^{w} & \mathbf{p}_{100}^{vw} & \mathbf{p}_{110}^{vw} \\
\mathbf{p}_{000}^{uw} & \mathbf{p}_{010}^{uw} & \mathbf{p}_{000}^{uvw} & \mathbf{p}_{010}^{uvw} \\
\mathbf{p}_{100}^{uw} & \mathbf{p}_{110}^{uw} & \mathbf{p}_{100}^{uvw} & \mathbf{p}_{110}^{uvw}
\end{bmatrix}
$$

$$
\mathbf{B}_4 = \begin{bmatrix}
\mathbf{p}_{001}^{w} & \mathbf{p}_{011}^{w} & \mathbf{p}_{001}^{vw} & \mathbf{p}_{011}^{vw} \\
\mathbf{p}_{101}^{w} & \mathbf{p}_{111}^{w} & \mathbf{p}_{101}^{vw} & \mathbf{p}_{111}^{vw} \\
\mathbf{p}_{001}^{uw} & \mathbf{p}_{011}^{uw} & \mathbf{p}_{001}^{uvw} & \mathbf{p}_{011}^{uvw} \\
\mathbf{p}_{101}^{uw} & \mathbf{p}_{111}^{uw} & \mathbf{p}_{101}^{uvw} & \mathbf{p}_{111}^{uvw}
\end{bmatrix}
$$

Think of these equations as being organized in a three-dimensional array of vectors, as in Fig. 4.4. Keep in mind, however, that because the elements of each array are vectors, the array is more correctly $4 \times 4 \times 4 \times 3$, a four-dimensional array.

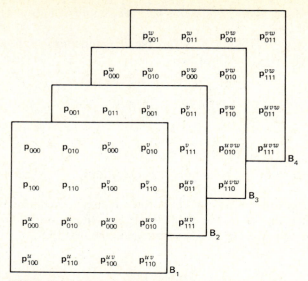

FIGURE 4.4 Three-dimensional array of geometric co-efficients of a hyperpatch.

We take a similar approach with the point form, contracting the third parametric variable w and expressing the point-form coefficients as a sequence of four 16-element arrays. We denote these arrays as the matrices

$$\mathbf{p}_1 = [\mathbf{p}_{ij1}] \qquad \mathbf{p}_2 = [\mathbf{p}_{ij2}] \qquad \mathbf{p}_3 = [\mathbf{p}_{ij3}] \qquad \mathbf{p}_4 = [\mathbf{p}_{ij4}] \qquad (4.27)$$

and expand them as follows:

$$\mathbf{p}_1 = \begin{bmatrix} \mathbf{p}(0, 0, 0) & \mathbf{p}(0, 1/3, 0) & \mathbf{p}(0, 2/3, 0) & \mathbf{p}(0, 1, 0) \\ \mathbf{p}(1/3, 0, 0) & \mathbf{p}(1/3, 1/3, 0) & \mathbf{p}(1/3, 2/3, 0) & \mathbf{p}(1/3, 1, 0) \\ \mathbf{p}(2/3, 0, 0) & \mathbf{p}(2/3, 1/3, 0) & \mathbf{p}(2/3, 2/3, 0) & \mathbf{p}(2/3, 1, 0) \\ \mathbf{p}(1, 0, 0) & \mathbf{p}(1, 1/3, 0) & \mathbf{p}(1, 2/3, 0) & \mathbf{p}(1, 1, 0) \end{bmatrix}$$

$$\mathbf{p}_2 = \begin{bmatrix} \mathbf{p}(0, 0, 1/3) & \mathbf{p}(0, 1/3, 1/3) & \mathbf{p}(0, 2/3, 1/3) & \mathbf{p}(0, 1, 1/3) \\ \mathbf{p}(1/3, 0, 1/3) & \mathbf{p}(1/3, 1/3, 1/3) & \mathbf{p}(1/3, 2/3, 1/3) & \mathbf{p}(1/3, 1, 1/3) \\ \mathbf{p}(2/3, 0, 1/3) & \mathbf{p}(2/3, 1/3, 1/3) & \mathbf{p}(2/3, 2/3, 1/3) & \mathbf{p}(2/3, 1, 1/3) \\ \mathbf{p}(1, 0, 1/3) & \mathbf{p}(1, 1/3, 1/3) & \mathbf{p}(1, 2/3, 1/3) & \mathbf{p}(1, 1, 1/3) \end{bmatrix}$$

$$\mathbf{p}_3 = \begin{bmatrix} \mathbf{p}(0, 0, 2/3) & \mathbf{p}(0, 1/3, 2/3) & \mathbf{p}(0, 2/3, 2/3) & \mathbf{p}(0, 1, 2/3) \\ \mathbf{p}(1/3, 0, 2/3) & \mathbf{p}(1/3, 1/3, 2/3) & \mathbf{p}(1/3, 2/3, 2/3) & \mathbf{p}(1/3, 1, 2/3) \\ \mathbf{p}(2/3, 0, 2/3) & \mathbf{p}(2/3, 1/3, 2/3) & \mathbf{p}(2/3, 2/3, 2/3) & \mathbf{p}(2/3, 1, 2/3) \\ \mathbf{p}(1, 0, 2/3) & \mathbf{p}(1, 1/3, 2/3) & \mathbf{p}(1, 2/3, 2/3) & \mathbf{p}(1, 1, 2/3) \end{bmatrix} \qquad (4.28)$$

$$\mathbf{p}_4 = \begin{bmatrix} \mathbf{p}(0, 0, 1) & \mathbf{p}(0, 1/3, 1) & \mathbf{p}(0, 2/3, 1) & \mathbf{p}(0, 1, 1) \\ \mathbf{p}(1/3, 0, 1) & \mathbf{p}(1/3, 1/3, 1) & \mathbf{p}(1/3, 2/3, 1) & \mathbf{p}(1/3, 1, 1) \\ \mathbf{p}(2/3, 0, 1) & \mathbf{p}(2/3, 1/3, 1) & \mathbf{p}(2/3, 2/3, 1) & \mathbf{p}(2/3, 1, 1) \\ \mathbf{p}(1, 0, 1) & \mathbf{p}(1, 1/3, 1) & \mathbf{p}(1, 2/3, 1) & \mathbf{p}(1, 1, 1) \end{bmatrix}$$

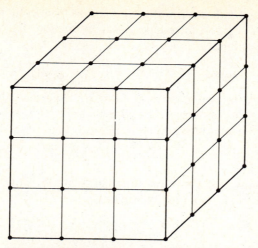

FIGURE 4.5 Sixty-four-point form of a hyper-patch.

Figure 4.5 illustrates the point distribution within the unit cube in parametric space.

Since the algebraic coefficients $\mathbf{a}_{ijk}$ are too cumbersome to present this way, we express them in an unexpanded matrix format. We determine them by operating on the geometric coefficients with the universal transformation matrix **M**. Thus, we obtain

$$\mathbf{a}_{ij1} = \mathbf{M}_{11}\mathbf{MB}_1\mathbf{M}^T + \mathbf{M}_{12}\mathbf{MB}_2\mathbf{M}^T + \mathbf{M}_{13}\mathbf{MB}_3\mathbf{M}^T + \mathbf{M}_{14}\mathbf{MB}_4\mathbf{M}^T$$

$$\mathbf{a}_{ij2} = \mathbf{M}_{21}\mathbf{MB}_1\mathbf{M}^T + \mathbf{M}_{22}\mathbf{MB}_2\mathbf{M}^T + \mathbf{M}_{23}\mathbf{MB}_3\mathbf{M}^T + \mathbf{M}_{24}\mathbf{MB}_4\mathbf{M}^T$$

$$\mathbf{a}_{ij3} = \mathbf{M}_{31}\mathbf{MB}_1\mathbf{M}^T + \mathbf{M}_{32}\mathbf{MB}_2\mathbf{M}^T + \mathbf{M}_{33}\mathbf{MB}_3\mathbf{M}^T + \mathbf{M}_{34}\mathbf{MB}_4\mathbf{M}^T$$

$$\mathbf{a}_{ij4} = \mathbf{M}_{41}\mathbf{MB}_1\mathbf{M}^T + \mathbf{M}_{42}\mathbf{MB}_2\mathbf{M}^T + \mathbf{M}_{43}\mathbf{MB}_3\mathbf{M}^T + \mathbf{M}_{44}\mathbf{MB}_4\mathbf{M}^T$$

(4.29)

where $\mathbf{M}_{lm}$ is a single specific element of the universal transformation matrix and $\mathbf{B}_1$, $\mathbf{B}_2$, $\mathbf{B}_3$, and $\mathbf{B}_4$ are given by Eqs. 4.26. If we let

$$\mathbf{A}_1 = [\mathbf{a}_{ij1}]$$
$$\mathbf{A}_2 = [\mathbf{a}_{ij2}]$$
$$\mathbf{A}_3 = [\mathbf{a}_{ij3}]$$
$$\mathbf{A}_4 = [\mathbf{a}_{ij4}]$$

(4.30)

or, in general,

$$\mathbf{A}_k = [\mathbf{a}_{ijk}]$$

(4.31)

then we rewrite Eq. 4.29 more compactly as

$$\mathbf{A}_k = \mathbf{M}_{ki}\mathbf{MB}_i\mathbf{M}^T$$

(4.32)

where we take the summation over the repeated subscript i.

We must also be able to transform equations from point form to geometric form. To do this, Stanton introduces a patch function that we will denote as $\mathbf{G}(w)$ and define as follows:

$$\mathbf{G}(w) = \mathbf{F}_i^b(w)\mathbf{B}_i \qquad (4.33)$$

This defines a bicubic patch corresponding to a specific value of w given the hyperpatch array $[\mathbf{b}_{ijk}]$ from which the $\mathbf{B}_i$ arrays are extracted. Next, we evaluate this function at the one-third points to obtain

$$\mathbf{G}_j = \mathbf{M}^{-1}\mathbf{N}\mathbf{P}_j\mathbf{N}^T[\mathbf{M}^{-1}]^T \qquad (4.34)$$

where $\mathbf{M}$ is the universal transformation matrix and $\mathbf{N}$ is the array of blending-function coefficients $\mathbf{F}_i^P$ from Eq. 4.24. Thus,

$$\mathbf{N} = \begin{bmatrix} -9/2 & 27/2 & -27/2 & 9/2 \\ 9 & -45/2 & 18 & -9/2 \\ -11/2 & 9 & -9/2 & 1 \\ 1 & 0 & 0 & 0 \end{bmatrix} \qquad (4.35)$$

We now substitute the results from Eq. 4.34 into Eq. 4.33 and solve for $\mathbf{B}_i$ to obtain the final transformation equation

$$\mathbf{B}_i = [\mathbf{F}_{ij}^b]^{-1}\mathbf{M}^{-1}\mathbf{N}\mathbf{P}_j\mathbf{N}^T[\mathbf{M}^{-1}]^T \qquad (4.36)$$

where $[\mathbf{F}_{ij}^b]$ is a matrix whose elements are the blending functions in Eq. 4.23 evaluated at the one-third points. The expanded form of this matrix is

$$[\mathbf{F}_{ij}^b] = \begin{bmatrix} F_1^b(0) & F_2^b(0) & F_3^b(0) & F_4^b(0) \\ F_1^b(1/3) & F_2^b(1/3) & F_3^b(1/3) & F_4^b(1/3) \\ F_1^b(2/3) & F_2^b(2/3) & F_3^b(2/3) & F_4^b(2/3) \\ F_1^b(1) & F_2^b(1) & F_3^b(1) & F_4^b(1) \end{bmatrix} \qquad (4.37)$$

When this matrix is evaluated, it becomes

$$[\mathbf{F}_{ij}^b] = \begin{bmatrix} 1 & 0 & 0 & 0 \\ 20/27 & 7/27 & 4/27 & -2/27 \\ 7/27 & 20/27 & 2/27 & -4/27 \\ 0 & 1 & 0 & 0 \end{bmatrix} \qquad (4.38)$$

and its inverse is

$$[\mathbf{F}_{ij}^b]^{-1} = \begin{bmatrix} 1 & 0 & 0 & 0 \\ 0 & 0 & 0 & 1 \\ -11/2 & 9 & -9/2 & 1 \\ -1 & 9/2 & -9 & 11/2 \end{bmatrix} \qquad (4.39)$$

You will recognize these as the arrays from Eqs. 2.87 and 2.88, respectively.

EXERCISES

1. Find expressions equivalent to Eqs. 4.9 and 4.10 for triquadric, triquartic, and triquintic solids.

2. How many points are required to define a triquadric, triquartic, and triquintic solid?

3. Develop expressions for Bezier and *B*-spline solids.

4. Verify the dimensions of the matrices and their product in Eq. 4.36.

5. Write a procedure to compute the coordinates of a point on or in a tricubic solid. Use the algebraic form. Denote this as **PSOLID**(CI, UVW, P), where

CI(4, 4, 4, 3) is the input array of algebraic coefficients;

UVW(3) are the input values of the parametric variables u, v, and w; and

P(3) is the output array of x, y, z coordinates of the point.

6. Write a procedure to convert tricubic solid coefficients in algebraic form to geometric form. Denote this as **ATGSOL**(CI, CO), where

CI(4, 4, 4, 3) is the input aray of algebraic coefficients, and

CO(4, 4, 4, 3) is the output array of geometric coefficients.

7. Write a procedure to convert tricubic solid coefficients in geometric form to algebraic form. Denote this as **GTASOL**(CI, CO), where

CI(4, 4, 4, 3) is the input array of geometric coefficients, and

CO(4, 4, 4, 3) is the output array of algebraic coefficients.

8. Write a procedure to convert tricubic solid coefficients in point form to geometric form. Denote this as **PTGSOL**(CI, CO), where

CI(4, 4, 4, 3) is the input array of point coordinates, and

CO(4, 4, 4, 3) is the output array of geometric coefficients.

2   TANGENT VECTORS AND TWIST VECTORS

Computing and understanding the significance of the parametric derivatives of the tricubic solid involves processes similar to those applied to the bicubic patch. In fact, as you might now suspect, the parallels continue with higher dimension elements. We express the tricubic function as three independent variables, of course, and we must compute partial derivatives of this function.

We encountered these vectors in Section 1, since they are important constituents of the geometric coefficients. Let us summarize them briefly here.

We now have three species of tangent vectors

$$\mathbf{p}_{uvw}^{u} = \frac{\partial}{\partial u}\, \mathbf{p}(u, v, w)$$

$$\mathbf{p}_{uvw}^{v} = \frac{\partial}{\partial v}\, \mathbf{p}(u, v, w) \qquad (4.40)$$

$$\mathbf{p}_{uvw}^{w} = \frac{\partial}{\partial w}\, \mathbf{p}(u, v, w)$$

three of twist vectors

$$\mathbf{p}_{uvw}^{uv} = \frac{\partial^2}{\partial u\, \partial v}\, \mathbf{p}(u, v, w)$$

$$\mathbf{p}_{uvw}^{uw} = \frac{\partial^2}{\partial u\, \partial w}\, \mathbf{p}(u, v, w) \qquad (4.41)$$

$$\mathbf{p}_{uvw}^{vw} = \frac{\partial^2}{\partial v\, \partial w}\, \mathbf{p}(u, v, w)$$

and the singular triple-mixed partial

$$\mathbf{p}_{uvw}^{uvw} = \frac{\partial^3}{\partial u\, \partial v\, \partial w}\, \mathbf{p}(u, v, w) \qquad (4.42)$$

Just as these vectors change and control the shape of curves and surfaces, so do they also control the shape of a tricubic solid. These vectors control not only the shape of the edges and faces, but also the interior distribution of the parametric variables. At our discretion, this control may work interactively or by

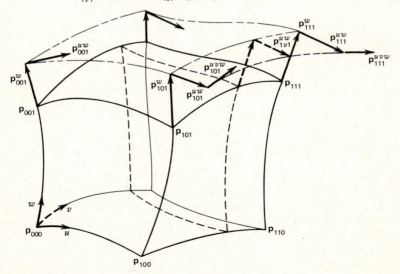

FIGURE 4.6 Hyperpatch tangent vectors.

some performance-analysis program, which we will consider in later chapters. For now, study Fig. 4.6; it illustrates some of the tangent vectors in place on a solid. Notice that the triple-mixed partial vectors $\mathbf{p}_{uvw}^{uvw}$ control the change of the twist vectors through the solid.

Thus, there are various levels of shape control available to us. By fixing the corner points of a solid, we can create concave and convex faces at will by manipulating the tangent and twist vectors. For simple shapes and constrained demands on interior behavior, we can set equal to zero some or all of the mixed vectors defined by mixed partial derivatives, in the manner of the F patch for surfaces.

3 PARAMETRIC SPACE OF A SOLID

The parametric space of a solid differs from that of a curve or surface, since we must account for an additional parametric variable. We now have a set of four-dimensional parametric spaces defined by (u, v, w, x), (u, v, w, y), and (u, v, w, z) coordinates. Contrary to curves and surfaces, decomposing a tricubic solid into its parametric-space components is much more difficult to visualize and, therefore, less likely to contribute to greater understanding of its behavior in object space.

We can, however, apply this technique to the faces in exactly the same way as for bicubic patches. In fact, we can take parametric slices generating surfaces on which one of the parametric variables is constant and then proceed with the familiar bicubic decomposition.

Figure 4.7 shows an interesting exercise. Start with the unit cube. At some convenient distance d from each of the six faces, construct a unit square in

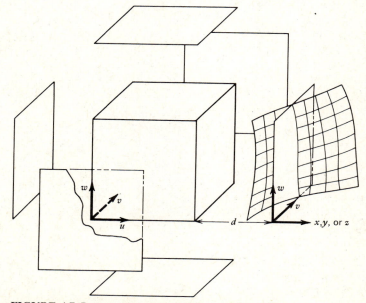

FIGURE 4.7 Parametric space of a hyperpatch.

a plane parallel to the face. This plane is designated as the $x = 0$ plane (or $y = 0$ or $z = 0$). Now, we proceed with the bicubic decomposition for each of the six faces. We might produce some interesting effects if not insights from a dynamic computer-graphics display of the resulting phenomena.

4   CONTINUITY AND COMPOSITE SOLIDS

Consider the possible continuity requirements at the common boundary surface of two joined solids; Fig. 4.8 gives the details of the conditions across the interior of this face. We will have satisfied all required continuity conditions across the adjoining external bounding surfaces of the two solids $\mathbf{p}(u, v, w)$ and $\mathbf{q}(u, v, w)$. Also, we will ignore C^2 (curvature) continuity and focus on only C^0 and C^1 continuity across this surface.

To ensure C^0 continuity, we must meet the following condition on the common surface:

$$\mathbf{p}_{1vw} = \mathbf{q}_{0vw} \tag{4.43}$$

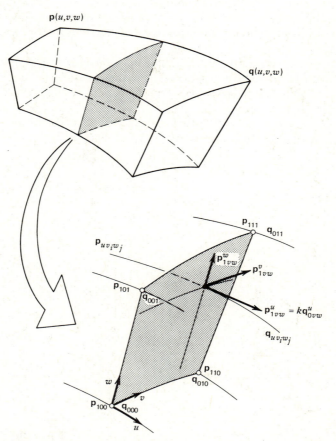

FIGURE 4.8 Continuity conditions.

For C^1 continuity, we must investigate the curves of constant v and w passing through each of these points. The tangent vectors of the curves at these points must be colinear; thus,

$$\mathbf{p}^u_{1vw} = k\mathbf{q}^u_{0vw} \tag{4.44}$$

What does the condition in Eq. 4.45 imply?

$$\mathbf{p}^v_{1vw} \times \mathbf{p}^w_{1vw} = a(\mathbf{q}^v_{0vw} \times \mathbf{q}^w_{0vw}) \tag{4.45}$$

What if we modify this condition so that we now have

$$\mathbf{p}^v_{1vw} \times \mathbf{p}^w_{1vw} = b\mathbf{p}^u_{1vw} \tag{4.46}$$

EXERCISES

1. Discuss the implications of Eq. 4.45.

2. Discuss the implications of Eq. 4.46.

5 SURFACES AND CURVES IN A SOLID

In Chapter 3, Section 20, we found that a parametric surface supports a two-dimensional curvilinear coordinate system on which we can define curves. In a directly analogous way, a parametric solid supports a three-dimensional curvilinear coordinate system on which we can define both curves and surfaces.

The simplest of these are the isoparametric curves and surfaces. An isoparametric curve, for example, results when two of the three tricubic parametric variables are held fixed at some constant values. The curve $\mathbf{p}_{abw}$ in Fig. 4.9 results when we assign $u = a$ and $v = b$, where a and b are constants. In the same figure, we find two isoparametric surfaces $\mathbf{p}_{avw}$ and $\mathbf{p}_{ubw}$.

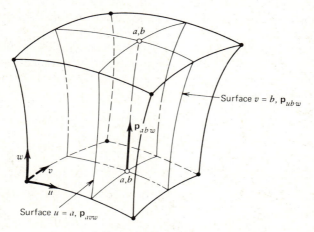

A distorted cubical solid.

FIGURE 4.9 Isoparametric surfaces and curves in a solid.

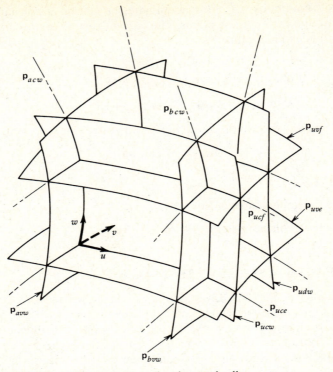

FIGURE 4.10 Parametric and orthogonal cells.

A parametric cell is a subelement of the solid bounded by six isoparametric surfaces. For example, in Fig. 4.10, the six boundary surfaces are $\mathbf{p}_{avw}$, $\mathbf{p}_{bvw}$, $\mathbf{p}_{ucw}$, $\mathbf{p}_{udw}$, $\mathbf{p}_{uve}$, and $\mathbf{p}_{uvf}$. Such a cell is orthogonal if the curve nets of the three families of isoparametric surfaces are defined on orthogonal parametric curve nets.

We define a nonisoparametric curve by introducing an additional parametric variable t. The curve $\mathbf{c}(t)$ is shown mapped in a tricubic solid in both Cartesian space and parametric space in Fig. 4.11. The curvilinear vector components of points on the curve are

$$\mathbf{c}(t) = \mathbf{u}(t) + \mathbf{v}(t) + \mathbf{w}(t) \tag{4.47}$$

In object space, the curve is expressed as

$$\mathbf{c}(x, y, z) = \mathbf{p}[u(t), v(t), w(t)] \tag{4.48}$$

As we see, the procedures are identical to those we use for curves on surfaces in Chapter 3, Section 20.

Because a parametric solid offers a three-dimensional curvilinear coordinate system, we have the additional option of defining nonisoparametric surfaces within the solid. Consider Fig. 4.12. Here, we see a surface patch

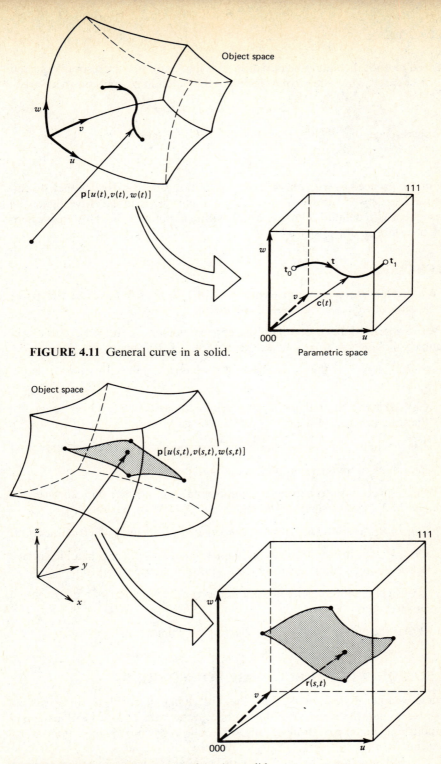

FIGURE 4.11 General curve in a solid.

Parametric space

FIGURE 4.12 Nonisoparametric surfaces in a solid.

Parametric space

mapped in both the unit cube of parametric space and in a corresponding solid in object space. The vector components of the patch in parametric space are

$$\mathbf{r}(s, t) = \mathbf{u}(s, t) + \mathbf{v}(s, t) + \mathbf{w}(s, t) \tag{4.49}$$

In object space, the patch is expressed as

$$\mathbf{r}(x, y, z) = \mathbf{p}[u(s, t), v(s, t), w(s, t)] \tag{4.50}$$

Again, we proceed as with curves in solids and on surfaces. Notice that we can choose a variety of surfaces to model in the parametric solid. Investigate the consequences of defining a Bezier surface within a tricubic solid. How do we map such a surface in object space?

EXERCISES

1. What effect does reparametrization of a hyperpatch have on the parametrization of curves or surfaces embedded in it?

2. Write a procedure to compute the coordinates of a point on a curve in a tricubic solid. Denote this as **PCRSOL(SOL, CI, T, P)**, where

SOL(4, 4, 4, 3) is the input array of geometric coefficients of the tricubic solid;

CI(4, 3) is the input array of geometric coefficients of a pc curve defined in the solid in terms of the u, v, w parametric variables;

T is the input value of the parametric variable t of a point on the curve; and

P(6) is the output array of x, y, z, u, v, w coordinates.

3. Write a procedure to compute the coordinates of a point on a surface in a tricubic solid. Denote this as **PSRSOL(SOL, CI, UW, P)**, where

SOL(4, 4, 4, 3) is the input array of geometric coefficients of a tricubic solid;

CI(4, 4, 3) is the input array of geometric coefficients of a bicubic surface defined in the solid in terms of the u, v, w parametric variables;

UW(2) is the input array of parametric variables s and t of a point on the surface, and

P(6) is the output array of x, y, z, u, v, w coordinates.

6   SOLIDS WITH IRREGULAR BOUNDARIES

Review Chapter 3, Section 21, since the underlying approach to surfaces with irregular boundaries is the natural precursor to the study of solids with irregular boundaries. Our exploration of this subject is brief and merely hints at its potential.

An endess variety of complex solids is possible if we permit them to have

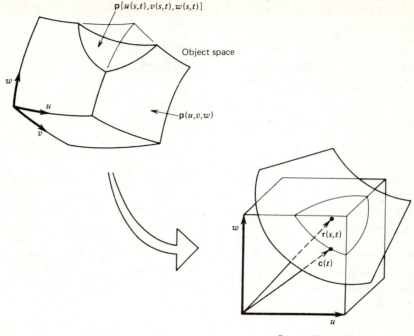

FIGURE 4.13 A solid with an irregular boundary.

irregular or nonisoparametric boundaries. A specific example is shown in Fig. 4.13. The solid has one of its corners clipped and defined by a noniso-parametric surface. The surface is conveniently defined initially in either parametric space or object space, depending on the specific modeling situation. We can determine the validity of points in the solid in a number of ways analogous to those discussed in Chapter 3, Section 21.

7 GENERALIZED NOTATION SCHEME AND HIGHER-DIMENSION ELEMENTS

To introduce each subject, the preceding sections use a classical algebraic notation scheme. Subsequently, incorporating vector and matrix notation significantly reduces the complexity and awkwardness of these formulations. Now, the next logical step is to introduce a more generalized summation scheme that extends to the representation of not only curves and surfaces, but also to solids and higher order geometric forms. The notation convention we use here (largely attributable to Timmer) has these characteristics and others. It can extend to spaces of any dimension; it applies to polycubic interpolation involving an arbitrary number of parametric independent variables and suggests a system that can directly extend to any odd-ordered polynomial form. We will apply the following general symbol definitions:

a = vector elements of the algebraic coefficient array $\mathbf{A}$

b = vector elements of the geometric coefficient array $\mathbf{B}$

$\mathbf{p}$ = point (dependent variable)

u = parametric variable (independent variable)

$\mathbf{M}$ = transformation matrix

The algebraic form is

$$p_i(u_1, u_2, \ldots, u_n) = \sum_{j_1=1}^{4} \sum_{j_2=1}^{4} \cdots \sum_{j_n=1}^{4} (a_{ij_1j_2\cdots j_n} u_1^{4-j_1} u_2^{4-j_2} \cdots u_n^{4-j_n}) \quad (4.51)$$

where $i \in [1:m]$ and p_i is the ith component of a dependent-variable vector of m components, each defined by this n-cubic polynomial parametric function. We can omit the summation symbols. The geometric element described is of nth order, and the equation of each component is of $3n$th degree and has $4n$ terms. There are m dependent variables and n independent or parametric variables. Using the bicubic patch in three-space as an example, we have $m = 3$ and $n = 2$, and we rewrite Eq. 4.51 as

$$p_i(u_1, u_2) = \sum_{j_1=1}^{4} \sum_{j_2=1}^{4} (a_{ij_1j_2} u_1^{4-j_1} u_2^{4-j_2}) \quad i \in [1:3] \quad (4.52)$$

The geometric form is

$$p_i(u_1, u_2, \ldots, u_n)$$

$$= \sum_{j_1=1}^{4} \sum_{j_2=1}^{4} \cdots \sum_{j_n=1}^{4} [F_{j_1}(u_1) F_{j_2}(u_2) \ldots F_{j_n}(u_n) b_{ij_1j_2\cdots j_n}] \quad i \in [1:m]$$

$$(4.53)$$

The $\mathbf{F}$ terms are matrices of the blending functions. For the rth $\mathbf{F}$-term matrix F_{j_r},

$$\begin{aligned} j_r = 1, & \quad F_1(u_r) = 2u_r^3 - 3u_r^2 + 1 \\ j_r = 2, & \quad F_2(u_r) = -2u_r^3 + 3u_r^2 \\ j_r = 3, & \quad F_3(u_r) = u_r^3 - 2u_r^2 + u_r \\ j_r = 4, & \quad F_4(u_r) = u_r^3 - u_r^2 \end{aligned} \quad (4.54)$$

The multisubscripted b array requires more care to interpret. If the following rules are correctly employed, then we can construct a correct and unambiguous b array. The subscript i identifies the vector component. The set of j subscripts conveys two kinds of information. First, it identifies the specific boundary on which the preceding product of blending functions is to operate. Second, it identifies a sequence of n operators applied to p_i, evaluated at the previously determined boundary. Table 4.1 is a summary of the information associated with the four values of the j,th subscript.

TABLE 4.1 Subscript Interpretation

j_r	Boundary	Operator
$j_r = 1$	$u_r = 0$	1
$j_r = 2$	$u_r = 1$	1
$j_r = 3$	$u_r = 0$	$\dfrac{\partial}{\partial u_r}$
$j_r = 4$	$u_r = 1$	$\dfrac{\partial}{\partial u_r}$

The following examples demonstrate how to use this table.

Example 1. $n = 1$, pc Curves

$$b_{i2} = \{1\}p_i(1) = p_i(1)$$

$$b_{i3} = \left\{\frac{\partial}{\partial u_1}\right\}p_i(0) = \frac{\partial p_i(0)}{\partial u_1}$$

Example 2. $n = 2$, Bicubic Patches

$$b_{i14} = \{1\}\left\{\frac{\partial}{\partial u_2}\right\}p_i(0, 1) = \frac{\partial p_i(0, 1)}{\partial u_2}$$

$$b_{i34} = \left\{\frac{\partial}{\partial u_1}\right\}\left\{\frac{\partial}{\partial u_2}\right\}p_i(0, 1) = \frac{\partial^2 p_i(0, 1)}{\partial u_1\, \partial u_2}$$

Example 3. $n = 3$, Tricubic Solids

$$b_{i124} = \{1\}\{1\}\frac{\partial}{\partial u_3}\, p_i(0, 1, 1) = \frac{\partial p_i(0, 1, 1)}{\partial u_3}$$

$$b_{i444} = \left\{\frac{\partial}{\partial u_1}\right\}\left\{\frac{\partial}{\partial u_2}\right\}\left\{\frac{\partial}{\partial u_3}\right\}p_i(1, 1, 1) = \frac{\partial^3 p_i(1, 1, 1)}{\partial u_1\, \partial u_2\, \partial u_3}$$

$$b_{i413} = \left\{\frac{\partial}{\partial u_1}\right\}\{1\}\left\{\frac{\partial}{\partial u_3}\right\}p_i(1, 0, 0) = \frac{\partial^2 p_i(1, 0, 0)}{\partial u_1\, \partial u_3}$$

The braces { } indicate and separate the operators. The operations do not imply multiplication except in the case of {1}.

Finally, the equation for the transformation from geometric to algebraic coefficients is

$$a_{ik_1k_2\cdots k_n} = \sum_{j_1=1}^{4} \sum_{j_2=1}^{4} \cdots \sum_{j_n=1}^{4} (\mathbf{M}_{k_1j_1}\mathbf{M}_{k_2j_2}\cdots \mathbf{M}_{k_nj_n}\mathbf{B}_{ij_1j_2\cdots j_n}) \qquad i \in [1:m]$$

$$(4.55)$$

EXERCISES

1. Modify Eq. 4.51 so that it applies to polyquintic elements.

2. Modify Eq. 4.54 so that it applies to polyquintic elements.

3. Does Eq. 4.53 have to be modified to apply to polycubic Bezier elements? If so, how? Revise Eqs. 4.54 accordingly.

4. Assume the same conditions as Ex. 3 for polycubic, *B*-spline elements.

8   ELEMENT CONSTRUCTION

We have touched lightly on the subject of element construction. We investigated using four points to define a pc curve, 16 points for a bicubic patch, and the special techniques offered by the Bezier and *B*-spline forms. In later chapters, particularly Chapters 10 and 13, we will explore other methods of constructing complex solid geometric models. For the present, though, let us briefly review the elements in this chapter and Chapters 2 and 3 and also see how these elements are hierarchically interrelated. We will see how to move from points to curves to surfaces and finally to solids. The cubic format has been a useful demonstration device, and we will not abandon it now.

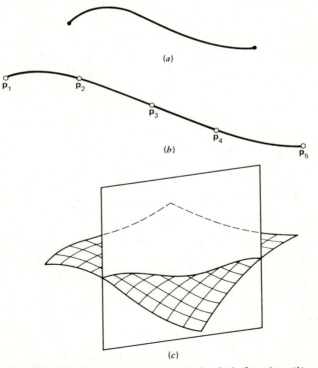

FIGURE 4.14 Curve construction. (*a*) Analytic function. (*b*) Interpolation function. (*c*) Surface intersection.

First, let us consider some new methods and review some familiar ones of constructing pc curves. Note that the number of piecewise cubic segments in a composite curve is not important to our understanding of these methods, so we will ignore it. Figure 4.14 shows three typical constructions. In the first case, we describe the curve by a specific analytic function and a beginning and end point on the function. We might use the function to compute the end tangents and/or curvatures and from the data construct the algebraic- or geometric-coefficient matrix. The second method uses a set of points and an interpolation rule that fits a curve through or near these points. We have not discussed the third method yet, and it requires finding the intersection curve of two surfaces, for example, the intersection of a bicubic patch by a plane. Notice that in all cases, we can increase the number of piecewise cubic curve segments without limit by segmentation.

The cubic polynomial form common to pc curves facilitates the construction of bicubic patches. We recognize that four points will define a curve and four curves or 16 points will define a patch. Thus, patch construction becomes an extension of the curve-construction problem. A common bicubic patch construction consists of nothing more than specifying four pc curves. See Fig. 4.15 for an example of this as well as construction by interpolation through a

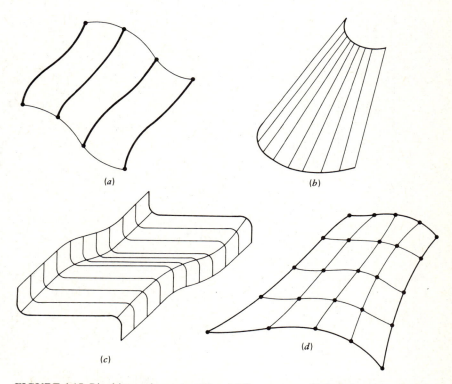

(a)

(b)

(c)

(d)

FIGURE 4.15 Bicubic patch construction. (*a*) Four curves. (*b*) Ruled surface. (*c*) Outline surface. (*d*) Interpolation.

mesh of points. We construct the ruled surface by connecting equal parametric points on the two base curves with straight lines. The last example, and one we will consider more thoroughly later, is called an outline surface. Lossing and Eshleman (1974) have developed a technique to define a sweep surface, based on moving an outline curve along a position-direction curve.

Finally, let us consider the tricubic solid. It should not surprise us to find that there is a close relationship between solids and surfaces. Any tricubic solid can be uniquely defined by four bicubic patches, 16 pc curves, or by 64 points. This means that every solid construction problem can be resolved into a patch construction problem and finally into a curve construction problem. Stanton (1976) has noted natural generalizations by using Hermite bicubic patches to construct a tricubic solid; Fig. 4.16 illustrates some of them. We create the body of revolution by rotating 16 points, four curves, or one patch.

Notice, however, that we cannot model by curve rotation all shapes modeled by point rotation, since the curve does not change shape during rotation. A similar situation holds for curves and patches. Patch rotation requires the least data for construction and is, therefore, more efficient. We should use it when we can. Stanton (1976) also advises that, in general, when equivalent constructions are possible, we use the higher level to reduce data

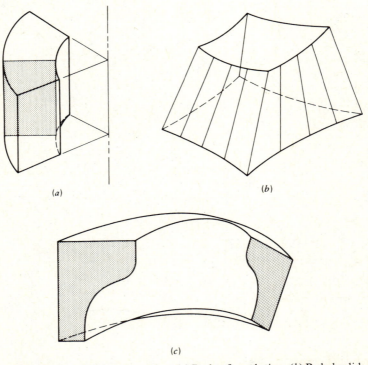

(a)

(b)

(c)

FIGURE 4.16 Solid construction. (a) Body of revolution. (b) Ruled solid. (c) Outline solid.

requirements. We generate the ruled volume in Fig. 4.16 by connecting equal parametric points on the two base patches with "straight line" isoparametric curves. Another approach is to simply specify 16 straight lines. Finally, we generate the outline solid by moving an outline patch along a position-direction curve.

EXERCISES

1. Show how a position-direction curve and surface can be used to sweep out a tricubic solid.

2. List and discuss some degenerate and pathological conditions that might arise when using tricubic solids.

5

ANALYTIC PROPERTIES

Chapter 5 presents methods for determining some of the more useful analytic properties of parametric curves and surfaces. We will explore the ease with which the vector parametric formulation of curves and surfaces lends itself to computational processing. Analytic properties generally fall into one of two categories—intrinsic or extrinsic.

The intrinsic properties of a curve include the principal vectors (tangent, normal, and binormal); the principal planes (normal, osculating, and rectifying); curvature; and torsion. The intrinsic properties of a surface include the normal vector, the tangent plane, and curvature. Intrinsic properties are local properties and they vary from point to point on a curve or surface. Thus, they are only computed at specific points.

We must also consider extrinsic, or global, properties. For a curve, this includes its arc length and whether or not it is a plane curve or a straight line. For a surface, global properties are such things as its area and whether or not it is ruled, developable, or planar. A closed surface also defines a volume. *Note*: A correspondingly closed curve does not define an area unless it lies on a surface. There are tests we can apply to determine some of these global properties. Other properties include the presence or absence of inflection points on curves and flat spots on surfaces and the location of extreme or min/max points.

As a warm-up, try to verify the following theorems for curves:

Theorem 1. $\mathbf{p}(u)$ degenerates to a point $\mathbf{p}(0)$ when $\mathbf{p}''(u) = 0$.

Theorem 2. $\mathbf{p}(u)$ degenerates to a straight line when $\mathbf{p}''(u) \neq 0$ and $\mathbf{p}''(u) \times \mathbf{p}'''(u) = 0$.

Theorem 3. $\mathbf{p}(u)$ is a plane curve, not a straight line, when $\mathbf{p}''(u) \times \mathbf{p}'''(u) \neq 0$ and $|\mathbf{p}''(u) \quad \mathbf{p}'''(u) \quad \mathbf{p}''''(u)| = 0$.

Theorem 4. $\mathbf{p}(u)$ is a space curve, nonplanar, when $\mathbf{p}''(u) \times \mathbf{p}'''(u) \neq 0$ and $|\mathbf{p}''(u) \quad \mathbf{p}'''(u) \quad \mathbf{p}''''(u)| \neq 0$.

Consider Theorem 1 first: We know that for a pc curve,

$$\mathbf{p}^u = F_1^u \mathbf{p}_0 + F_2^u \mathbf{p}_1 + F_3^u \mathbf{p}_0^u + F_4^u \mathbf{p}_1^u$$

or

$$\mathbf{p}^u = (6u^2 - 6u)\mathbf{p}_0 + (-6u^2 + 6u)\mathbf{p}_1 + (3u^2 - 4u + 1)\mathbf{p}_0^u + (3u^2 - 2u)\mathbf{p}_1^u$$

If $\mathbf{p}^u = 0$, then $\mathbf{p}_0^u = \mathbf{p}_1^u = 0$, and we find that

$$(6u^2 - 6u)\mathbf{p}_0 + (-6u^2 + 6u)\mathbf{p}_1 = 0$$

or

$$\mathbf{p}_0 = \mathbf{p}_1 \qquad Q.E.D.$$

Now take a similar approach to the other theorems; you are on your own.

EXERCISES

1. Verify Theorems 2, 3, and 4.

1 INTRINSIC PROPERTIES OF A CURVE

Let us now consider in more detail some of the intrinsic properties of a curve. As usual, we formulate these properties in terms of vectors, matrices, and determinants. We are concerned here with properties at a point $\mathbf{p}_i$ on a curve.

At any point on a curve, we can construct a set of three orthogonal vectors; together, they are called the moving trihedron. This construction is both fundamental to, and arises from, the differential geometry of curves. This construction is also an important ingredient in certain types of sweep representation of solids in geometric modeling.

1.1 Tangent Vector and Line

The tangent vector at point $\mathbf{p}_i$ on a curve is denoted by $\mathbf{p}_i^u$. It is convenient in many situations to work with a unit tangent vector at $\mathbf{p}_i$. We use

$$\mathbf{t}_i = \frac{\mathbf{p}_i^u}{|\mathbf{p}_i^u|} \tag{5.1}$$

The equation of a straight line through $\mathbf{p}_i$ and parallel to $\mathbf{t}_i$ is

$$\mathbf{q} = \mathbf{p}_i + a\mathbf{t}_i \tag{5.2}$$

Here, a is a scalar determining the distance along $\mathbf{t}_i$ from $\mathbf{p}_i$; Fig. 5.1 illustrates this.

1.2 Normal Plane

The normal plane at point $\mathbf{p}_i$ on a curve is a plane through $\mathbf{p}_i$ perpendicular to the unit tangent vector $\mathbf{t}_i$. The equation of the normal plane is given by the scalar product

$$(\mathbf{q} - \mathbf{p}_i) \cdot \mathbf{t}_i = 0 \tag{5.3}$$

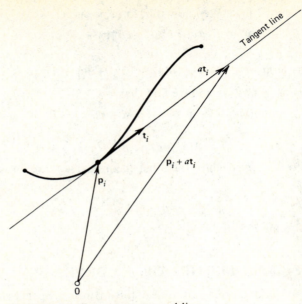

FIGURE 5.1 Tangent vector and line.

where $\mathbf{q}$ is any point on the plane. Since the vector $\mathbf{q} - \mathbf{p}_i$ lies in the plane and $\mathbf{q}$ is any point on the plane, we correctly deduce that, if $\mathbf{q} - \mathbf{p}_i$ is perpendicular to $\mathbf{t}_i$, then the plane is also perpendicular to $\mathbf{t}_i$. Figure 5.2 presents these elements.

Clearly, we can also write Eq. 5.3 as

$$(\mathbf{q} - \mathbf{p}_i) \cdot \mathbf{p}_i^u = 0 \tag{5.4}$$

since $\mathbf{t}_i$ and $\mathbf{p}_i^u$ are parallel. Now, denote the coordinates of the point $\mathbf{q}$ as x, y, z, then rewrite Eq. 5.4 as

$$\begin{bmatrix} x - x_i \\ y - y_i \\ z - z_i \end{bmatrix}^T \begin{bmatrix} x_i^u \\ y_i^u \\ z_i^u \end{bmatrix} = 0 \tag{5.5}$$

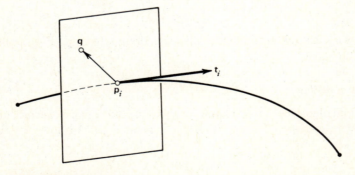

FIGURE 5.2 Normal plane.

or

$$(x - x_i)x_i^u + (y - y_i)y_i^u + (z - z_i)z_i^u = 0 \qquad (5.6)$$

We can rearrange the terms in this equation into a form analogous to the familiar equation for a plane, $ax + by + cz + d = 0$,

$$x_i^u x + y_i^u y + z_i^u z - (x_i x_i^u + y_i y_i^u + z_i z_i^u) = 0 \qquad (5.7)$$

1.3 Principal Normal Vector and Line

The principal normal vector at a point $\mathbf{p}_i$ on a curve is normal to the curve and so must lie in the normal plane; however, it is a special normal vector among the many possible in that it points in the direction the curve is turning and toward the center of curvature for the point. The problem is to find this vector given the parametric expression for a curve.

Study Fig. 5.3; the construction in Fig. 5.3*a* tells us that as we move an infinitesimal distance along the curve at $\mathbf{p}_i$, the tangent vector $\mathbf{p}_i^u$ swings in a

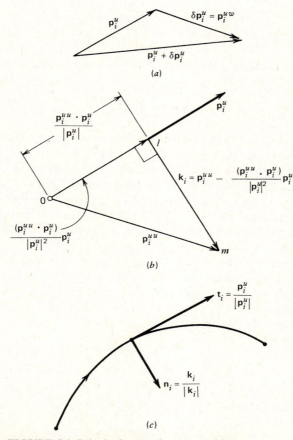

FIGURE 5.3 Principal normal vector and line.

direction given by $\mathbf{p}_i^{uuu}$. This action sketches the merest hint of a plane. (It is called the osculating plane; we will investigate it soon.) It turns out that the principal normal vector lies on the intersection of this plane and the normal plane. Thus, if we can find a vector normal to $\mathbf{p}^u$ and in the plane of $\mathbf{p}_i^u$ and $\mathbf{p}_i^{uu}$, then we will have 90% of the solution.

We proceed as follows: We know from elementary vector geometry that the projection of $\mathbf{p}_i^{uu}$ onto $\mathbf{p}_i^u$ is given by the scalar product

$$\text{length of the projection of } \mathbf{p}_i^{uu} \text{ onto } \mathbf{p}_i^u = \mathbf{p}_i^{uu} \cdot \left(\frac{\mathbf{p}_i^u}{|\mathbf{p}_i^u|} \right) \tag{5.8}$$

and also that the line of projection lm (see Fig. 5.3b) is normal to $\mathbf{p}_i^u$. Using this projected length, a scalar, construct the vector along $0l$ by multiplying the unit vector $\mathbf{p}_i^u/|\mathbf{p}_i^u|$ by the length found in Eq. 5.8

$$\vec{0l} = \frac{\mathbf{p}_i^{uu} \cdot \mathbf{p}_i^u}{|\mathbf{p}_i^u|^2} \, \mathbf{p}_i^u \tag{5.9}$$

Next, let $\mathbf{k}_i$ denote the vector from l to m. Applying simple vector arithmetic, we easily find that

$$\mathbf{k}_i = \mathbf{p}_i^{uu} - \frac{\mathbf{p}_i^{uu} \cdot \mathbf{p}_i^u}{|\mathbf{p}_i^u|^2} \, \mathbf{p}_i^u \tag{5.10}$$

This is 90% of the solution; the remaining 10% is shown in Fig. 5.3c, where the principal unit normal vector is

$$\mathbf{n}_i = \frac{\mathbf{k}_i}{|\mathbf{k}_i|} \tag{5.11}$$

Observe that we might have applied an important and well-known theorem from vector calculus that states that if $\mathbf{t}$ is a unit vector function, then $d\mathbf{t}/du$ is orthogonal to $\mathbf{t}$. From this direct computation, we would then obtain $\mathbf{n}$. The problem with this approach is in establishing a readily differentiable function $\mathbf{t}(u)$, since, in this case, it is defined by $\mathbf{p}^u/|\mathbf{p}^u|$, which must be differentiated before evaluating the expression at u_i. Establishing $\mathbf{t}(u)$ is usually extremely difficult and often impossible, depending on the type of parametric expression used to define the curve (cubic, Bezier, B-spline, and so on).

As a bit of mental exercise on this subject, consider a plane curve. Obviously we can apply Eqs. 5.10 and 5.11 to find the principal normal at a point. There is another approach, although it has limitations; it is analogous to the method for finding the normal plane. For simplicity, assume the curve lies in the x, y plane; then the following must be true:

$$\mathbf{n}_i \cdot \mathbf{t}_i = 0 \tag{5.12}$$

or, of course,

$$\mathbf{n}_i \cdot \mathbf{p}_i^u = 0 \tag{5.13}$$

Writing Eq. 5.12 in expanded component form, we see that

$$n_x t_x + n_y t_y = 0 \tag{5.14}$$

Let $\mathbf{n}_i$ be a unit vector. We know that

$$n_x^2 + n_y^2 = 1 \qquad (5.15)$$

Since we assume to know t_x and t_y, we readily solve for n_x and n_y. However, in doing this, we find that an ambiguity arises in the signs of n_x and n_y. Hence, we see the limitation: We have found the line of action of $\mathbf{n}$, but we cannot determine in which direction along this line the vector is pointing.

Finally, we compute the principal normal line at a point $\mathbf{p}_i$ on a curve

$$\mathbf{q} = \mathbf{p}_i + a\mathbf{n}_i \qquad (5.16)$$

where $\mathbf{q}$ is any point on the line and a is a scalar determining the distance and direction along $\mathbf{n}_i$ from $\mathbf{p}_i$.

1.4   Binormal Vector

We investigate yet another vector normal to the curve at a point $\mathbf{p}_i$ and lying in the normal plane. This time, we make direct use of the principal normal and tangent vectors to define the binormal vector $\mathbf{b}_i$ by simply computing their vector product

$$\mathbf{b}_i = \mathbf{t}_i \times \mathbf{n}_i \qquad (5.17)$$

This formulation ensures that $\mathbf{b}_i$ is normal to $\mathbf{t}_i$ and, thus, lies in the normal plane. It is also normal to $\mathbf{n}_i$, so the three vectors form an orthogonal frame with considerable significance for geometric modeling.

1.5   Osculating Plane

The osculating plane at a point $\mathbf{p}_i$ on a curve is the limiting position of the plane defined by $\mathbf{p}_i$ and two neighboring points on the curve $\mathbf{p}_h$ and $\mathbf{p}_j$ as these neighbor points independently approach $\mathbf{p}_i$; see Fig. 5.4. Notice that the three points cannot be colinear and that the tangent vector $\mathbf{p}_i''$ lies in this plane. The equation of the osculating plane is given by the following determinant:

$$|(\mathbf{q} - \mathbf{p}_i) \quad \mathbf{p}_i^u \quad \mathbf{p}_i^{uu}| = 0 \qquad (5.18)$$

where $\mathbf{q}$, again, is any generic point on the osculating plane. Writing the vectors in terms of their components, we obtain

$$\begin{vmatrix} x - x_i & x_i^u & x_i^{uu} \\ y - y_i & y_i^u & y_i^{uu} \\ z - z_i & z_i^u & z_i^{uu} \end{vmatrix} = 0 \qquad (5.19)$$

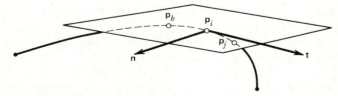

FIGURE 5.4 Osculating plane.

Expand this determinant to obtain

$$(x - x_i)(y_i^u z_i^{uu} - y_i^{uu} z_i^u) - (y - y_i)(x_i^u z_i^{uu} - x_i^{uu} z_i^u) + (z - z_i)(x_i^u y_i^{uu} - x_i^{uu} y_i^u) = 0$$

$$(5.20)$$

The curvature at a point is measured in the osculating plane at that point. Thus, two joined curves have full C^2 continuity only when they have the same radius of curvature and osculating plane at their common point. Eq. 5.18 tells us that if the three vectors $\mathbf{p}_i, \mathbf{p}_i^u, \mathbf{p}_i^{uu}$ associated with curve $\mathbf{p}(u)$ at point i are equal to, respectively, $\mathbf{r}_i, \mathbf{r}_i^u, \mathbf{r}_i^{uu}$ of the curve $\mathbf{r}(u)$ at that point, then the two curves are joined with C^2 continuity. What can we conclude about the osculating plane of a plane curve? Of a straight line?

There is another way of formulating the osculating plane by using the vectors $\mathbf{t}$ and $\mathbf{n}$, which we know from other considerations must lie in the plane. Furthermore, we formulate a parametric expression

$$\mathbf{q} = \mathbf{p}_i + u\mathbf{t}_i + w\mathbf{n}_i \tag{5.21}$$

where $\mathbf{q}$ is any point on the osculating plane and u and w are parametric variables of unspecified bounds.

Incidentally, the word *osculating* means kissing and was introduced by Tinseau around 1780—a Frenchman, of course. The term is well chosen, because it is the plane of closest contact to the curve in the neighborhood of $\mathbf{p}_i$. This is a subject related to concepts of differential geometry, which we will not pursue.

1.6　Rectifying Plane

The rectifying plane at a point $\mathbf{p}_i$ on a curve is the plane through $\mathbf{p}_i$ and perpendicular to the principal normal $\mathbf{n}_i$.

$$(\mathbf{r} - \mathbf{p}_i) \cdot \mathbf{n}_i = 0 \tag{5.22}$$

where $\mathbf{r}$ is a generic point on the rectifying plane. There is another way of formulating this plane, and it is analogous to the method we just developed for the osculating plane. It is

$$\mathbf{r} = \mathbf{p}_i + u\mathbf{t}_i + w\mathbf{b}_i \tag{5.23}$$

where $\mathbf{r}$ is, again, any point on the rectifying plane and u and w are parametric variables of unspecified bounds; see Fig. 5.5.

1.7　Moving Trihedron

We have just developed expressions for three characteristic vectors and planes associated with each point on a curve. They are intrinsic properties, since they vary from point to point on a curve. Figure 5.5 assembles these vectors and planes and makes clear that these elements form a local, three-dimensional orthogonal coordinate system consisting of three axis vectors and three coordinate planes. This coordinate system is called the **moving trihedron** of the curve.

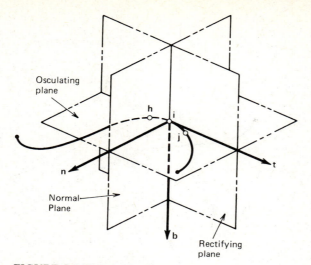

Osculating plane

Normal Plane

Rectifying plane

FIGURE 5.5 The moving trihedron.

Some gross features are apparent. The curve pierces the normal plane and is tangent to the osculating and rectifying planes. In space, a coordinate system defines eight regions called octants, so that the moving trihedron distinguishes eight types of points on the curve. Only one type is considered regular. At a regular point, the curve intersects the normal plane and remains on one side of the rectifying and osculating planes.

Later, we will explore the use of the moving trihedron as a framework for defining the variable cross-section geometry of solids. In so-called sweep representations of solids, a closed generator curve moves along a control curve according to some rules and sweeps out a solid shape.

1.8 Curvature and Torsion

For a space curve of parametric representation, the curvature $1/\rho_i$ at a point $\mathbf{p}_i$ on the curve is

$$\frac{1}{\rho_i} = \frac{|\mathbf{p}_i^u \times \mathbf{p}_i^{uu}|}{|\mathbf{p}_i^u|^3} \tag{5.24}$$

ρ_i is called the radius of curvature. Note that we will also use κ to denote curvature, where $\kappa = 1/\rho$. Curvature is measured in the osculating plane along the principal normal vector $\mathbf{n}_i$. We readily construct the curvature vector $\mathbf{k} = \rho_i \mathbf{n}_i$, which is the vector from $\mathbf{p}_i$ to the center of curvature; see Fig. 5.6.

Let us look at the curvature of a plane curve, say, in the x, y plane. From classic analytic geometry, the curvature is

$$\frac{1}{\rho} = \frac{d^2y/dx^2}{[1 + (dy/dx)^2]^{3/2}} \tag{5.25}$$

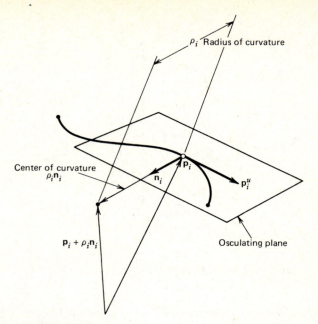

FIGURE 5.6 Curvature.

Now, write Eq. 5.25 in parametric form. The following steps should be self-evident.

$$\frac{dy}{dx} = \frac{y^u}{x^u} \tag{5.26}$$

$$\frac{d^2 y}{dx^2} = \frac{d}{du}\left(\frac{y^u}{x^u}\right)\frac{du}{dx} = \frac{x^u y^{uu} - y^u x^{uu}}{(x^u)^3} \tag{5.27}$$

$$\left[1 + \left(\frac{dy}{dx}\right)^2\right]^{3/2} = \left[1 + \left(\frac{y^u}{x^u}\right)^2\right]^{3/2} = \frac{1}{(x^u)^3}[(x^u)^2 + (y^u)^2]^{3/2} \tag{5.28}$$

Substitute the preceding results into Eq. 5.25 to yield

$$\frac{1}{\rho} = \frac{x^u y^{uu} - x^{uu} y^u}{[(x^u)^2 + (y^u)^2]^{3/2}} \tag{5.29}$$

which is the two-dimensional scalar equivalent of Eq. 5.24.

The torsion at a point $\mathbf{p}_i$ on a curve is the limit of the ratio of the angle between the binormal at $\mathbf{p}_i$ and the binormal at a neighboring point $\mathbf{p}_h$ to the arc $\widehat{hi}$, as $\mathbf{p}_h$ approaches $\mathbf{p}_i$ along the curve (see Fig. 5.7). We can see this clearly by observing the rectifying planes. Note that torsion amounts to a rotation or twist about the tangent vector. It is given by the formula

$$\tau_i = \frac{|\mathbf{p}_i^u \ \mathbf{p}_i^{uu} \ \mathbf{p}_i^{uuu}|}{|\mathbf{p}_i^u \times \mathbf{p}_i^{uu}|^2} \tag{5.30}$$

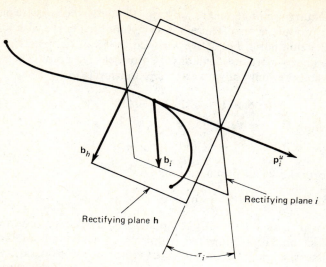

FIGURE 5.7 Torsion.

It is not surprising to find that curvature and torsion are related. From Eq. 5.24,

$$|\mathbf{p}_i^u \times \mathbf{p}_i^{uu}| = \frac{|\mathbf{p}_i^u|^3}{\rho_i} \tag{5.31}$$

so that

$$|\mathbf{p}_i^u \times \mathbf{p}_i^{uu}|^2 = \frac{|\mathbf{p}_i|^6}{\rho_i^2} \tag{5.32}$$

Substitute this equation into Eq. 5.30 to obtain

$$\tau_i = \frac{|\mathbf{p}_i^u \, \mathbf{p}_i^{uu} \, \mathbf{p}_i^{uuu}|}{(1/\rho_i)^2 \, |\mathbf{p}_i^u|^6} \tag{5.33}$$

We can clean up Eq. 5.33 a bit as follows:

$$\tau_i = \frac{|\mathbf{p}_i^u \, \mathbf{p}_i^{uu} \, \mathbf{p}_i^{uuu}|}{|\mathbf{p}_i^u|^6} \, \rho_i^2 \tag{5.34}$$

Note: This relationship holds only if $(1/\rho) \neq 0$. Furthermore, if the curve is a plane curve, then $\tau = 0$.

1.9   Inflection Points

Points on a curve where the curvature equals zero are called inflection points. Curvature is zero when the numerator in Eq. 5.24 is zero. Thus, we find the inflection points of a curve, if any, by solving

$$|\mathbf{p}_i^u \times \mathbf{p}_i^{uu}| = 0 \tag{5.35}$$

From elementary vector geometry, we know that if two vectors are parallel, then their vector product is zero. Therefore, any point on a curve with $\mathbf{p}_i^u$ parallel to $\mathbf{p}_i^{uu}$ is an inflection point. Let us see what Eq. 5.35 can show us. First, drop the subscript i, since we are not dealing with a preselected point. Then rewrite the equation using the well-known vector relationship $|\mathbf{a}| = \sqrt{\mathbf{a} \cdot \mathbf{a}}$.

$$|\mathbf{p}^u \times \mathbf{p}^{uu}| = \sqrt{(\mathbf{p}^u \times \mathbf{p}^{uu}) \cdot (\mathbf{p}^u \times \mathbf{p}^{uu})} = 0 \qquad (5.36)$$

or, more simply,

$$(\mathbf{p}^u \times \mathbf{p}^{uu}) \cdot (\mathbf{p}^u \times \mathbf{p}^{uu}) = 0 \qquad (5.37)$$

Next, perform the scalar product to obtain

$$(\mathbf{p}^u \times \mathbf{p}^{uu})_x^2 + (\mathbf{p}^u \times \mathbf{p}^{uu})_y^2 + (\mathbf{p}^u \times \mathbf{p}^{uu})_z^2 = 0 \qquad (5.38)$$

where $(\mathbf{p}^u \times \mathbf{p}^{uu})_x$ denotes the x component of the vector resulting from the cross-product operation indicated within parentheses and similarly for $(\ )_y$ and $(\ )_z$. Next, we perform the indicated vector-product operation and obtain

$$(y^u z^{uu} - z^u y^{uu})^2 + (z^u x^{uu} - x^u z^{uu})^2 + (x^u y^{uu} - y^u x^{uu})^2 = 0 \qquad (5.39)$$

And you thought this was going to be fun! If you think this equation looks familiar, you are right. Look at the numerator in Eq. 5.27. What Eq. 5.39 tells us is that the sum of the numerators of parametric expressions for $d^2 y/dx^2, d^2 x/dz^2$, and $d^2 z/dy^2$ must equal zero. Wait; there is an even deeper meaning here. Could Eq. 5.39 be saying that $d^2 y/dx^2 = 0$, $d^2 x/dz^2 = 0$ and $dz^2/dy^2 = 0$ and each second-derivative expression is identically equal to zero? Yes. To satisfy this condition, only the numerators (in Eq. 5.27 and the two other similar expressions) must be zero. And, in fact, each of the three terms in Eq. 5.39 must be zero. Thus,

$$y^u z^{uu} - z^u y^{uu} = 0 \qquad (5.40)$$

$$z^u x^{uu} - x^u z^{uu} = 0 \qquad (5.41)$$

$$x^u y^{uu} - y^u x^{uu} = 0 \qquad (5.42)$$

For parametric cubic curves, the expressions reduce to quadratic polynomials. Expanded in pc terms, Eq. 5.42, for example, becomes

$$3(b_x a_y - a_x b_y)u^2 + 3(c_x a_y - a_x c_y)u + (c_x b_y - b_x c_y) = 0 \qquad (5.43)$$

Thus, we must find the roots of three quadratic equations in u in the interval $u \in [0, 1]$. If any roots exist in this interval, they must be compared. Only roots common to all three determine an inflection point. Two roots are possible, and, therefore, two distinct inflection points are also possible.

The example of a pc curve has more to offer. Let us determine what conditions are necessary to have zero curvature at $\mathbf{p}_0$ and $\mathbf{p}_1$. The analysis is simplified if we define the curve tangent vectors in terms of $\mathbf{p}_0$, $\mathbf{p}_1$, and a third point $\mathbf{p}_2$, so that $\mathbf{p}_0^u = a(\mathbf{p}_2 - \mathbf{p}_0)$ and $\mathbf{p}_1^u = b(\mathbf{p}_1 - \mathbf{p}_2)$. Remember that $\mathbf{p}^u$ and $\mathbf{p}^{uu}$ must be

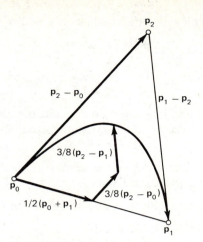

FIGURE 5.8 Inflection points.

parallel at a point of inflection (zero curvature). Starting with $\mathbf{p}_0$ from $\mathbf{p}^u = \mathbf{F}^u \mathbf{B}$ and $\mathbf{p}^{uu} = \mathbf{F}^{uu} \mathbf{B}$, we find

$$\mathbf{p}_0^u = [0 \quad 0 \quad 1 \quad 0][\mathbf{p}_0 \quad \mathbf{p}_1 \quad a(\mathbf{p}_2 - \mathbf{p}_0) \quad b(\mathbf{p}_1 - \mathbf{p}_2)]^T$$
$$= a(\mathbf{p}_2 - \mathbf{p}_0) \tag{5.44}$$

and

$$\mathbf{p}_0^{uu} = [-6 \quad 6 \quad -4 \quad -2][\mathbf{p}_0 \quad \mathbf{p}_1 \quad a(\mathbf{p}_2 - \mathbf{p}_0) \quad b(\mathbf{p}_1 - \mathbf{p}_2)]^T$$
$$= 6(\mathbf{p}_1 - \mathbf{p}_0) - 4a(\mathbf{p}_2 - \mathbf{p}_0) - 2b(\mathbf{p}_1 - \mathbf{p}_2) \tag{5.45}$$

The two vectors are parallel at $u = 0$ if $b = 3$. A similar analysis at $u = 1$ reveals that at that end, the vectors are parallel when $a = 3$. Clearly, we can produce an inflection point simultaneously at $\mathbf{p}_0$ and $\mathbf{p}_1$ if $a = 3$ and $b = 3$. The equation for such a curve is

$$\mathbf{p} = \mathbf{UM}[\mathbf{p}_0 \quad \mathbf{p}_1 \quad 3(\mathbf{p}_2 - \mathbf{p}_0) \quad 3(\mathbf{p}_1 - \mathbf{p}_2)]^T \tag{5.46}$$

The construction in Fig. 5.8 locating the $u = 0.5$ point should be familiar.

EXERCISES

1. What do you conclude about the osculating plane of a plane curve? Of a straight line?

2. Using Eq. 5.24, derive the equations for the curvature at the ends of a cubic Bezier curve (see Eqs. 2.252 and 2.253). Show the steps in the derivation.

3. The derivation of conditions necessary to have zero curvature at $\mathbf{p}_0$ and $\mathbf{p}_1$ of a pc curve and the resulting Eq. 5.46 are for a plane curve, since $\mathbf{p}_0$, $\mathbf{p}_1$, and $\mathbf{p}_2$ define a plane. Can you find a similar derivation for a nonplanar pc curve? Explain your answer.

4. Given a pc curve with geometric coefficients

$$\mathbf{p}_0 = [2 \quad 1 \quad 0]$$
$$\mathbf{p}_1 = [6 \quad 3 \quad 2]$$
$$\mathbf{p}_0^u = \mathbf{p}_1^u = [8 \quad 8 \quad 8]$$

find the tangent vector and line, the normal plane, the principal normal vector and line, the binormal vector, the osculating plane, and rectifying plane at $u = 0.5$.

5. Find the curvature, curvature vector, center of curvature, and torsion at $u = 0.5$ for the pc curve defined in Ex. 4.

6. Find any inflection points on the pc curve defined in Ex. 4.

7. Given a Bezier curve with control points

$$\mathbf{p}_0 = [-2 \quad -2 \quad 4]$$
$$\mathbf{p}_1 = [2 \quad -4 \quad 1]$$
$$\mathbf{p}_2 = [6 \quad -3 \quad 0]$$
$$\mathbf{p}_3 = [10 \quad 0 \quad 0]$$

and

$$\mathbf{p}_4 = [10 \quad 4 \quad 2]$$

find the tangent vector and line, the normal plane, the principal normal vector and line, the binormal vector, the osculating plane, and rectifying plane at $u = 0.5$.

8. Find the curvature, curvature vector, center of curvature, and torsion at $u = 0.5$ for the Bezier curve defined in Ex. 7.

9. Find any inflection points on the Bezier curve defined in Ex. 7.

10. Given a periodic B-spline curve with $n = 5$, $k = 4$, and control points

$$\mathbf{p}_0 = [0 \quad 2 \quad 8]$$
$$\mathbf{p}_1 = [2 \quad 6 \quad 7]$$
$$\mathbf{p}_2 = [6 \quad 10 \quad 4]$$
$$\mathbf{p}_3 = [8 \quad 8 \quad 0]$$
$$\mathbf{p}_4 = [10 \quad 4 \quad -2]$$

and

$$\mathbf{p}_5 = [9 \quad -2 \quad -5)$$

find the tangent vector and line, the normal plane, the principal normal vector and line, the binormal vector, the osculating plane, and rectifying plane at $u = 0.5$ on the second segment.

11. Find the curvature, curvature vector, center of curvature, and torsion for the *B*-spline curve defined in Ex. 10 at $u = 0.5$ on the second segment.

12. Find any inflection points on th *B*-spline curve defined in Ex. 10.

13. Write a procedure to compute the coefficients of the normal, osculating, and rectifying planes at a point on a pc curve. Denote this as **NORPLN**(CI, U, N, O, R), where

> CI(4, 3) is the input array of geometric coefficients defining the pc curve;
>
> U is the input value of the parametric variable *u* at the point;
>
> N(4) is the output array of coefficients of the normal plane;
>
> O(4) is the output array of coefficients of the osculating plane; and
>
> R(4) is the output array of coefficients of the rectifying plane.

14. Generalize **NORPLN** so that it also applies to Bezier and *B*-spline curves.

15. Write a procedure to compute the unit tangent, normal, and binormal vectors at a point on a pc curve. Denote this as **TNBCRV**(CI, U, T, N, B), where

> CI(4, 3) is the input array of geometric coefficients defining the pc curve;
>
> U is the input value of the parametric variable at the point;
>
> T(3) is the output array of components of the unit tangent vector;
>
> N(3) is the output array of components of the unit principal normal vector; and
>
> B(3) is the output array of components of the unit binormal vector.

16. Generalize **TNBCRV** so that it also applies to Bezier and *B*-spline curves.

17. Write a procedure to compute the curvature, radius of curvature, and torsion at a point on a pc curve. Denote this as **CRTCRV**(CI, U, C, R, T), where

> CI(4, 3) is the input array of geometric coefficients defining the pc curve;
>
> U is the input value of the parametric variable *u* at the point; and
>
> C, R, T are the output values of the curvature, radius of curvature, and torsion, respectively, at the point.

18. Generalize **CRTCRV** so that it also applies to Bezier and *B*-spline curves.

19. Write a procedure to compute points of inflection, if any, on a pc curve. Denote this as **POICRV**(CI, N, U), where

> CI(4, 3) is the input array of geometric coefficients defining the pc curve;
>
> N is the output number of points of inflection found in the interval $u \in$ [0, 1]; and
>
> U(2) are the output values of the parametric variable at points of inflection.

20. Generalize **POICRV** so that it also applies to Bezier and *B*-spline curves. Give special attention to the number of inflection points possible and to the range of the parametric variable.

2   INTRINSIC PROPERTIES OF A SURFACE

Intrinsic properties of a surface to consider here include the normal, tangent plane, and principal curvatures at a point. They are the most useful intrinsic properties in geometric modeling. Before discussing them in detail, let us digress a bit.

Earlier, we determined a unique curve by two intrinsic quantities, curvature and torsion, as functions of arc length. We can also determine a unique surface by intrinsic quantities called the **first** and **second fundamental forms**, denoted Form I and Form II, respectively. They are presented here without derivation or proof; see Lane (1940) or O'Neill (1966), among many others, for a discussion. Let $\mathbf{p} = \mathbf{p}(u, w)$ be a parametric surface patch; then the first fundamental form is

$$\text{Form I} = d\mathbf{p} \cdot d\mathbf{p} = E \, du^2 + 2F \, du \, dw + G \, dw^2 \tag{5.47}$$

where

$$E = \mathbf{p}^u \cdot \mathbf{p}^u \tag{5.48}$$

$$F = \mathbf{p}^u \cdot \mathbf{p}^w \tag{5.49}$$

$$G = \mathbf{p}^w \cdot \mathbf{p}^w \tag{5.50}$$

E, F, and G are known as the coefficients of the first fundamental form. In the metric theory of surfaces, the first fundamental form arises in calculating the arc length of a curve on a surface.

Next, denote the unit normal to a surface at a point $\mathbf{p}(u, w)$ as $\mathbf{n}(u, w)$ or, more simply, as $\mathbf{n}$ (see Chapter 3, Section 3). Then, the second fundamental form is

$$\text{Form II} = -d\mathbf{p} \cdot d\mathbf{n} = L \, du^2 + 2M \, du \, dw + N \, dw^2 \tag{5.51}$$

where

$$L = -\mathbf{p}^u \cdot \mathbf{n}^u \tag{5.52}$$

$$M = -1/2(\mathbf{p}^u \cdot \mathbf{n}^w + \mathbf{p}^w \cdot \mathbf{n}^u) \tag{5.53}$$

$$N = -\mathbf{p}^w \cdot \mathbf{n}^w \tag{5.54}$$

L, M, and N are known as the coefficients of the second fundamental form. *Note:* $\mathbf{n}^u$ and $\mathbf{n}^w$ are perpendicular to $\mathbf{n}$. Then, since $\mathbf{p}^u$ and $\mathbf{p}^w$ are perpendicular to $\mathbf{n}$ for all u, w, we can derive alternative expressions for L, M, and N

$$L = \mathbf{p}^{uu} \cdot \mathbf{n} \tag{5.55}$$

$$M = \mathbf{p}^{uw} \cdot \mathbf{n} \tag{5.56}$$

$$N = \mathbf{p}^{ww} \cdot \mathbf{n} \tag{5.57}$$

Note: Coefficients of these forms are not invariant under a parameter transformation. Later sections demonstrate the use of coefficients of the first fundamental form to calculate the arc length of curves on surfaces, angles, and surface area.

2.1   Normal to a Surface

The unit normal at any point on a bicubic patch is given in Chapter 3, Section 3, Eq. 3.61. Clearly, this equation applies equally to any parametric surface of the form $\mathbf{p} = \mathbf{p}(u, w)$. Here is Eq. 3.61a with subscripts omitted, permissible when it is clear that $\mathbf{p} = \mathbf{p}(u, w)$ and $\mathbf{n} = \mathbf{n}(u, w)$

$$\mathbf{n} = \frac{\mathbf{p}^u \times \mathbf{p}^w}{|\mathbf{p}^u \times \mathbf{p}^w|} \tag{5.58}$$

If $\mathbf{p}^u$ and $\mathbf{p}^w$ are perpendicular to each other, $\mathbf{p}^u \cdot \mathbf{p}^w = 0$, then $\mathbf{p}^u$, $\mathbf{p}^w$, and $\mathbf{n}$ form a local orthogonal basis or trihedron at the point. If not, we can easily construct one by any number of conventions. For example, $\mathbf{p}^u$, $\mathbf{q}$, and $\mathbf{n}$ form a basis if we define $\mathbf{q} = \mathbf{n} \times \mathbf{p}^u$.

2.2   Tangent Plane

The equation of a plane tangent to a point on a surface is

$$\begin{vmatrix} X - x & x^u & x^w \\ Y - y & y^u & y^w \\ Z - z & z^u & z^w \end{vmatrix} = 0 \tag{5.59}$$

where (X, Y, Z) are the coordinates of any point $\mathbf{q}$ on the tangent plane; (x, y, z) are the coordinates of the point $\mathbf{p}$ on the surface to which the plane is tangent; and (x^u, y^u, z^u), (x^w, y^w, z^w) are components of the parametric tangent vectors $\mathbf{p}^u$ and $\mathbf{p}^w$ at $\mathbf{p}$. However, in what should now be familiar vector terms, this determinant is

$$(\mathbf{q} - \mathbf{p}) \cdot (\mathbf{p}^u \times \mathbf{p}^w) = 0 \tag{5.60}$$

Figure 5.9 illustrates this situation. Since $\mathbf{p}^u$ and $\mathbf{p}^w$ lie in a plane, their vector product is perpendicular to the plane, defining the direction of the unit normal $\mathbf{n}$. The tangent plane must be perpendicular to the normal, so that any point $\mathbf{q}$ in it defines the vector $\mathbf{q} - \mathbf{p}$; hence, Eq. 5.60 must hold.

The normal at a point of a surface is perpendicular to both parametric tangents at the point. In fact, a remarkable property of the geometry of a surface is that the tangent lines at an ordinary point on the surface of all the curves that lie on it and pass through the point lie in a common plane. This is indeed the tangent plane. There can be extraordinary points on some surfaces that, although they do possess a unique normal and tangent plane, also intersect the surface. We will investigate these special points in Section 2.5.

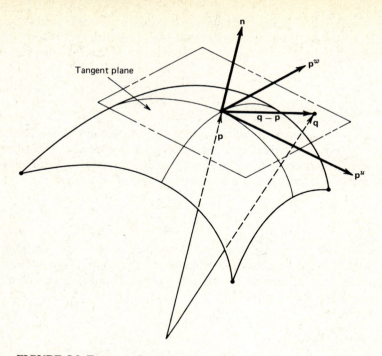

FIGURE 5.9 Tangent plane.

2.3 Principal curvature

Let us now investigate the curvature of all curves C that lie on a surface S and pass through a point $\mathbf{p}$ with the same tangent line t at $\mathbf{p}$. Again, the surface must be parametric and of the form $\mathbf{p} = \mathbf{p}(u, w)$ and a point on the curve of the form $\mathbf{p} = \mathbf{p}[u(t), w(t)]$.

At point $\mathbf{p}$ on the surface $\mathbf{S}$, construct the tangent plane $\mathbf{T}$; see Fig. 5.10. Next, construct any line t through $\mathbf{p}$ that also lies in the tangent plane. Notice that we can assign a direction to t in terms of the parametric variables u, w, namely, dw/du. All the planes $\mathbf{P}$ intersecting the tangent plane $\mathbf{T}$ and containing the line t cut the surface in a one-parameter family of plane curves. Clearly, all these curves pass through $\mathbf{p}$ and are tangent to the line t at $\mathbf{p}$. Furthermore, the osculating plane at $\mathbf{p}$ of any specific curve in this family is, of course, the plane that contains the curve.

There is an infinite number of nonplanar curves on $\mathbf{S}$, through $\mathbf{p}$, tangent to t at $\mathbf{p}$ and having $\mathbf{P}$ as their osculating plane. The remarkable fact is that all these curves also have the same center of curvature, the same radius of curvature, the same curvature vector $\mathbf{k}$, and the same principal normal. The plane curve $\mathbf{c}$ is a typical and readily visualized curve of the family of curves having the plane of $\mathbf{c}$ as their osculating plane.

We can say now that the normal curvature vector to the curve on the surface at $\mathbf{p}$ and denoted by $\mathbf{k}_n$ is the vector projection of the curvature vector $\mathbf{k}$ (see

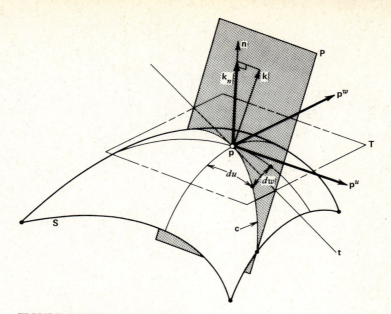

FIGURE 5.10 Normal curvature.

Section 1.8) of the curve at **p** onto the normal **n** at **p**. Thus,

$$\mathbf{k}_n = (\mathbf{k} \cdot \mathbf{n})\mathbf{n} \tag{5.61}$$

where the component of **k** in the direction of **n** is called the normal curvature of **c** at **p** and is denoted by κ_n. In other words,

$$\kappa_n = \mathbf{k} \cdot \mathbf{n} \tag{5.62}$$

We can show that since

$$\kappa_n = \frac{L(du/dt)^2 + 2M(du/dt)(dw/dt) + N(dw/dt)^2}{E(du/dt)^2 + 2F(du/dt)(dw/dt) + G(dw/dt)^2} \tag{5.63}$$

then κ_n depends on only the direction of the tangent line at **p**; that is, the ratio $(du/dt)/(dw/dt)$. Otherwise κ_n is a function of the coefficients of the first and second fundamental forms. We conclude that all curves through a point **p** on a surface tangent to the same line through **p** have the same normal curvature at **p**.

We assert without proof that all curves on a surface through **p** having the same osculating plane through **p** also have the same curvature κ at **p**. (*Note*: This holds provided the osculating plane is not tangent to the surface.) Thus, we conclude that we can obtain all possible values for the curvature of a curve through **p** by considering only the intersection curves of planes through **p**.

Among all these plane intersection curves with the surface through a tangent line at **p**, one is of special interest. This is the intersection curve cut by a plane through the tangent line *t* that also contains the normal **n** to the surface at **p**; see Fig. 5.11*a*. When we rotate the plane around the normal, the curvature κ_n

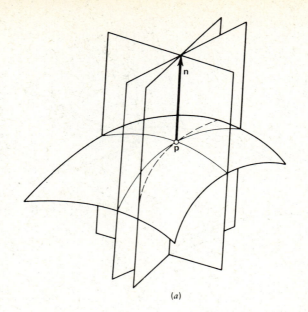

(a)

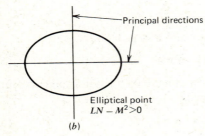

Principal directions

Elliptical point
$LN - M^2 > 0$

(b)

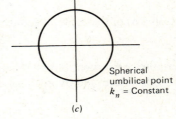

Spherical
umbilical point
k_n = Constant

(c)

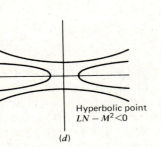

Hyperbolic point
$LN - M^2 < 0$

(d)

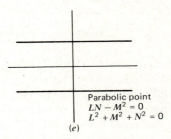

Parabolic point
$LN - M^2 = 0$
$L^2 + M^2 + N^2 = 0$

(e)

FIGURE 5.11 Principal curvature.

varies and has a maximum and a minimum value in two perpendicular directions. We call the extreme values the **principal normal curvatures** and denote them as κ_1 and κ_2. A point on a surface where $\kappa_n =$ constant is called an *umbilical point*, and here every direction is a principal direction.

The principal curvatures are the roots of the following equation:

$$(EG - F^2)\kappa^2 - (EN + GL - 2FM)\kappa + (LN - M^2) = 0 \qquad (5.64)$$

Figure 5.11*b*–5.11*e* shows plots of curvature as the normal plane rotates around the normal line and through the principal directions. We make certain simplifying assumptions concerning the local surface behavior that permit classification of the point in question (see Section 2.6). These assumptions can be found in any text on the differential geometry of surfaces (again, for example, see Lane (1940) or O'Neill (1966)).

Find the principal directions as follows: Let $h = dw/du$, then

$$(FN - GM)h^2 + (EN - GL)h + EM - FL = 0 \qquad (5.65)$$

One of the roots of this equation makes the normal curvature a maximum, and the other makes it a minimum. At an ordinary point on a surface (not a plane or sphere), there is just one direction for which the normal curvature is a maximum and just one for which it is a minimum. These two directions are the directions of the lines of curvature at the point. Remember, at an umbilical point, all directions are principal directions.

Although we will not pursue it here, it can be shown that the normal curvature at a point on a surface in the direction of a tangent line t is

$$\kappa_n = \kappa_1 \cos^2 \alpha + \kappa_2 \sin^2 \alpha \qquad (5.66)$$

where κ_1 and κ_2 are the principal curvatures at the point and α is the angle between t and a tangent line in the principal direction associated with κ_1.

There are two other related measures of curvature for a surface. The first is called the Gaussian curvature K and is

$$K = \kappa_1 \kappa_2 = \frac{LN - M^2}{EG - F^2} \qquad (5.67)$$

which is an invariant property of the surface. The second is called the mean curvature H and is

$$H = \frac{1}{2}(\kappa_1 + \kappa_2) = \frac{EN + GL - 2FM}{2(EG - F^2)} \qquad (5.68)$$

2.4 Geodesics and Geodesic Curvature

Given a curve c_s on a surface S (see Fig. 5.12), at a point p on the curve and surface, we construct a tangent plane T. The curve c_T is the orthogonal projection of c_s onto T. The curvature vector of c_T at p is called the **geodesic curvature vector** of c_s at p and is denoted by k_g.

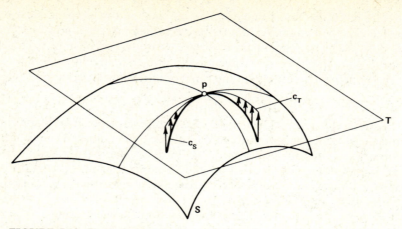

FIGURE 5.12 Geodesic curvature.

Without a rigorous mathematical proof, we correctly surmise that $\mathbf{k}_g$ is in fact the orthogonal or vector projection of the curvature vector $\mathbf{k}$ onto $\mathbf{T}$. This leads us to conclude that

$$\mathbf{k} = \mathbf{k}_g + \mathbf{k}_n \tag{5.69}$$

This relationship allows us to compute $\mathbf{k}_g$ directly as

$$\mathbf{k}_g = \mathbf{k} - \mathbf{k}_n \tag{5.70}$$

Observe that $\mathbf{k}_g$ is independent of the orientation of the surface and the curve on it, since $\mathbf{k}$ and $\mathbf{k}_n$ are independent. Also, the geodesic curvature along a curve on a surface is an intrinsic property of the surface.

Any curve on a surface along which $\mathbf{k}_g = 0$ is a geodesic curve. One of the best known properties of geodesic curves is their minimizing property. If we measure the distance between not-too-distant points on a surface along curves joining them and lying on the surface, the shortest distance between the points is along the geodesic joining them. Other properties are: There is a unique geodesic through a given point in a given direction on a surface, and there is a unique geodesic through two given points not too far apart on a surface.

2.5 Properties of Curves on Surfaces

We will discuss here only a few of the many properties of curves on surfaces; results are given without proof. First, let θ denote the angle between the u- and w-parameter curves at $\mathbf{p}$, in other words, the angle between $\mathbf{p}^u$ and $\mathbf{p}^w$. Then,

$$\cos \theta = \frac{\mathbf{p}^u \cdot \mathbf{p}^w}{|\mathbf{p}^u||\mathbf{p}^w|} = \frac{F}{\sqrt{EG}} \tag{5.71}$$

From this, we conclude that $\mathbf{p}^u$ and $\mathbf{p}^w$ at a point are perpendicular if and only if $F = 0$, which follows from the fact that $F = \mathbf{p}^u \cdot \mathbf{p}^w$.

Finally, at any point on a surface, we can find a coordinate patch such that the directions of the u- and w-parameter curves are principal directions. This is true if and only if $F = M = 0$ at the point. Also, with these conditions the principal curvatures are

$$\kappa_1 = \frac{L}{E} \quad \text{and} \quad \kappa_2 = \frac{N}{G} \tag{5.72}$$

2.6 Point Classification

In many applications of geometric modeling, it is to our advantage to know something about a surface in the neighborhood of a point on it. Is it doubly or singly curved? Does it have positive or negative curvature? In this section, we look at a method for determining some of these properties. It consists primarily of evaluating various expressions composed of coefficients of the second fundamental form.

Given a point $\mathbf{p}$ on a surface and its normal, establish a neighboring point $\mathbf{q}$. These points must be parametric functions such that $\mathbf{p} = \mathbf{p}(u, w)$ and $\mathbf{q} = \mathbf{q}(u, w)$. Let $d = (\mathbf{q} - \mathbf{p}) \cdot \mathbf{n}$ denote the projection of $\mathbf{q} - \mathbf{p}$ onto $\mathbf{n}$ at $\mathbf{p}$; see Fig. 5.13. Let d be positive or negative depending on which side of the tangent plane $\mathbf{q}$ is located. Notice that $|d|$ is the perpendicular distance from $\mathbf{q}$ to the tangent plane. If we let $\mathbf{q}$ approach $\mathbf{p}$, we find that d is increasingly governed by a function of the second fundamental form, which looks like

$$d = f\left[\tfrac{1}{2}(L\, du^2 + 2M\, du\, dw + N\, dw^2)\right] \tag{5.73}$$

This function is called the osculating paraboloid at $\mathbf{p}$; its characteristics determine the nature of the surface in the neighborhood of the point. There are four important classes, depending on the value of $LN - M^2$.

If $LN - M^2 > 0$, the point is an elliptic point. In its neighborhood, the surface lies entirely on one side of the tangent plane, and the surface is locally convex with respect to the plane; see Fig. 5.14a.

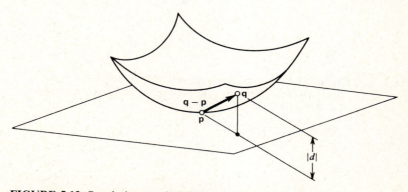

FIGURE 5.13 Osculating paraboloid at a point on a surface.

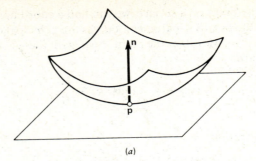

(a)

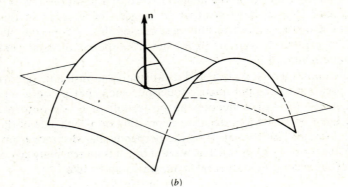

(b)

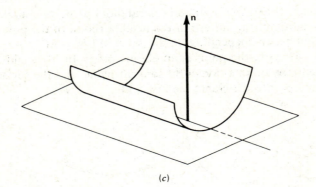

(c)

FIGURE 5.14 Point classification. (a) Elliptic point, $LN - M^2$ > 0. (b) Hyperbolic point, $LN - M^2 < 0$. (c) Parabolic point, $LN - M^2 = 0, L^2 + M^2 + N^2 \neq 0$.

If $LN - M^2 < 0$, the point is a hyperbolic point. Observe that there are two distinct lines through $\mathbf{p}$ in the tangent plane dividing the plane into four regions; see Fig. 5.14*b*. The surface lies on both sides of the tangent plane in the neighborhood of a hyperbolic point, and d is alternately positive and negative.

If $LN - M^2 = 0$ and $L^2 + M^2 + N^2 \neq 0$, the point is a parabolic point. Here, we find a single line in the tangent plane along which $d = 0$; see Fig. 5.14*c*.

Finally, if $L = 0$, $M = 0$, and $N = 0$, then the point is a planar point. Although it is beyond the scope of this text to prove, it is an invariant property of a surface that its classification is independent of its form of representation.

EXERCISES

1. Given a cubic Bezier surface, find another cubic Bezier surface that represents the unit normals on the first.

2. Given a B-spline surface with $m = n = 6$ and $k = l = 4$, find another similar B-spline surface that represents the unit normals on the first.

3. Given a bicubic patch in the 16-point form with

$$\mathbf{P} = \begin{bmatrix} (0,0,0) & (0,4,2) & (0,8,2) & (0,12,0) \\ (4,0,2) & (4,4,4) & (4,8,4) & (4,12,2) \\ (8,0,2) & (8,4,4) & (8,8,4) & (8,12,2) \\ (12,0,0) & (12,4,2) & (12,8,2) & (12,12,0) \end{bmatrix}$$

find the unit normal to the patch, the tangent plane, the principal curvatures, and the principal directions at the point $\mathbf{p}(0.5, 0.5)$.

4. Repeat Ex. 3 for $\mathbf{p}(0.25, 0.5)$ and $\mathbf{p}(0, 0.5)$.

5. Classify each of the points $\mathbf{p}(0, 0.5)$, $\mathbf{p}(0.25, 0.5)$, and $\mathbf{p}(0.5, 0.5)$ on the patch defined in Ex. 3.

6. Given the bicubic patch defined in Ex. 3 and a pc curve defined on it by the four points $\mathbf{p}_1 = \begin{bmatrix} 0.1 & 0.2 \end{bmatrix}$, $\mathbf{p}_2 = \begin{bmatrix} 0.4 & 0.3 \end{bmatrix}$, $\mathbf{p}_3 = \begin{bmatrix} 0.7 & 0.2 \end{bmatrix}$, and $\mathbf{p}_4 = \begin{bmatrix} 0.9 & 0.6 \end{bmatrix}$, find the orthogonal projection of the curve onto the tangent plane, the geodesic curvature vector, and the angle between the u- and w- parameter curves at $\mathbf{p}(0.5, 0.5)$.

7. Repeat Exs. 3–5 for the bicubic patch in the 16-point form with

$$\mathbf{P} = \begin{bmatrix} (0,0,0) & (0,4,2) & (0,8,2) & (0,12,0) \\ (4,0,-2) & (4,4,0) & (4,8,0) & (4,12,-2) \\ (8,0,-2) & (8,4,0) & (8,8,0) & (8,12,-2) \\ (12,0,0) & (12,4,2) & (12,8,2) & (12,12,0) \end{bmatrix}$$

8. Repeat Exs. 3–5 for the bicubic patch in the 16-point form with

$$\mathbf{P} = \begin{bmatrix} (0,0,0) & (0,3,0) & (0,6,0) & (0,9,0) \\ (3,0,0) & (3,3,1) & (3,6,2) & (3,9,3) \\ (6,0,0) & (6,3,2) & (6,6,4) & (6,9,6) \\ (9,0,0) & (9,3,3) & (9,6,6) & (9,9,9) \end{bmatrix}$$

9. Repeat Exs. 3–5 for a Bezier surface defined by the 16-points given in Ex. 3.

10. Write a procedure to compute the coefficients of the first and second fundamental forms at a point on a bicubic patch. Denote this as **EFGLMN(CI, UW, CO)**, where

> CI(4, 4, 3) is the input array of geometric coefficients defining the bicubic patch;
>
> UW(2) is the input array of values of the parametric variables u and w at the point; and
>
> CO(6) is the output array of values of the coefficients E, F, G, L, M, and N.

11. Generalize **EFGLMN** so that it also applies to Bezier and B-spline curves.

12. Write a procedure to compute the coefficients of the tangent plane to a point on a bicubic patch. Denote this as **TANPLN(CI, UW, CO)**, where

> CI(4, 4, 3) is the input array of geometric coefficients defining the bicubic patch;
>
> UW(2) is the input array of values of the parametric variables u and w; and
>
> CO(4) is the output array of plane coefficients.

13. Generalize **TANPLN** so that it also applies to Bezier and B-spline curves.

14. Write a procedure to compute the principal normal curvatures at a point on a bicubic patch. Denote this as **PNC(CI, UW, K)**, where

> CI(4, 4, 3) is the input array of geometric coefficients defining the bicubic surface;
>
> UW(2) is the input array of values of the parametric variables u and w; and
>
> K(2) are the output values of the principal normal curvatures.

15. Generalize **PNC** so that it also applies to Bezier and B-spline curves.

16. Write a procedure to classify a point on a bicubic patch. Denote this as **PCLASS(CI, UW, IC)**, where

> CI(4, 4, 3) is the input array of geometric coefficients defining the bicubic patch;
>
> UW(2) is the input array of values of the parametric variables u and w; and
>
> IC is the output number indicating the point classification;
> IC = 1 if elliptic,
> IC = 2 if hyperbolic,
> IC = 3 if parabolic, and
> IC = 4 if planar.

17. Generalize **PCLASS** so that it also applies to Bezier and B-spline curves.

3 CHARACTERISTIC TESTS

There are several simple tests we can apply to curves and surfaces to determine if they possess special characteristics. Let us begin with the test of whether a curve is a plane curve. We use the intuitive sense that a plane curve has zero torsion or twist, $\tau = 0$. This means that the numerator in Eq. 5.30 must equal zero

$$|\mathbf{p}^u \quad \mathbf{p}^{uu} \quad \mathbf{p}^{uuu}| = 0 \tag{5.74a}$$

which, of course, expands to

$$\begin{vmatrix} x^u & x^{uu} & x^{uuu} \\ y^u & y^{uu} & y^{uuu} \\ z^u & z^{uu} & z^{uuu} \end{vmatrix} = 0 \tag{5.74b}$$

Having determined that a curve is planar, we next wish to know if it is a straight line. In order for this to be true, the curvature must be zero at all points on the curve. This means that the numerator in Eq. 5.24 must equal zero

$$|\mathbf{p}^u \times \mathbf{p}^{uu}| = 0 \tag{5.75}$$

A surface is planar if $\kappa_n = 0$ at all points or spherical if $\kappa_n = $ constant. And, most interestingly, a surface is developable if its Gaussian curvature K is zero everywhere

$$K = \kappa_1 \kappa_2 = \frac{LN - M^2}{EG - F^2} = 0 \tag{5.76}$$

or

$$LN - M^2 = 0 \tag{5.77}$$

EXERCISES

1. Given a pc curve whose geometric coefficients are

$$\mathbf{p}_0 = [1 \quad 2 \quad 0]$$
$$\mathbf{p}_1 = [4 \quad 6 \quad 2]$$
$$\mathbf{p}_0^u = [5 \quad 2 \quad -2]$$

and

$$\mathbf{p}_1^u = [3 \quad 4 \quad 2]$$

determine if it is a plane curve and if it is a straight line.

2. Given a Bezier curve whose control points are

$$\mathbf{p}_0 = [18 \quad -3 \quad -4]$$
$$\mathbf{p}_1 = [7 \quad -2 \quad -1]$$
$$\mathbf{p}_2 = [0 \quad 0 \quad 0]$$
$$\mathbf{p}_3 = [-10 \quad 2 \quad 2]$$

and

$$\mathbf{p}_4 = [-17 \quad 4 \quad 3]$$

determine if it is a plane curve and if it is a straight line.

3. Given a periodic B-spline curve with $k = 4$ and control points

$$\mathbf{p}_0 = [0 \quad 0 \quad 4]$$
$$\mathbf{p}_1 = [0 \quad 0 \quad 0]$$
$$\mathbf{p}_2 = [3 \quad 0 \quad 0]$$
$$\mathbf{p}_3 = [0 \quad 6 \quad 0]$$
$$\mathbf{p}_4 = [-1 \quad 10 \quad 2]$$

and

$$\mathbf{p}_5 = [0 \quad 20 \quad 0]$$

determine if it is a plane curve and a straight line.

4. Given a bicubic patch whose geometric coefficients are

$$\mathbf{B} = \begin{bmatrix} (1,2,0) & (1,4,4) & (0,2,4) & (0,2,4) \\ (4,6,2) & (4,8,6) & (0,2,4) & (0,2,4) \\ (5,2,-2) & (5,2,-2) & 0 & 0 \\ (3,4,2) & (3,4,2) & 0 & 0 \end{bmatrix}$$

determine if it is a plane surface and developable.

5. Write a procedure to determine if a given curve — pc, Bezier, or B-spline — is a planar curve.

6. Write a procedure to determine if a given surface — bicubic, Bezier, or B-spline — is planar, spherical, developable.

4 DIRECT AND INVERSE POINT SOLUTION

To plot a curve or surface, we must first calculate the coordinates of a series of points for different values of the parametric variables. For a curve, we might design a procedure to compute $\mathbf{p}(u)$ at some predetermined increments of u. This means we must evaluate a polynomial in u, which we call the **direct point solution**. Other modeling tasks require us to determine a u value given the point $\mathbf{p}(u)$ in terms of its Cartesian coordinates x, y, and z in object space. This is the **inverse point solution**. Let us investigate methods for the direct point solution first.

As we have seen in previous sections, many curve- and surface-modeling methods, including pc, Bezier, and *B*-spline methods, express *x*, *y*, and *z* as parametric polynomial functions. For these methods, the problem of calculating a point on a curve or surface is reduced to calculating a polynomial.

One simple, straightforward method for calculating polynomials is **Horner's rule**. First, let us take $p(u) = au^3 + bu^2 + cu + d$—our old friend, the cubic polynomial. Although the development that follows is in terms of scalar polynomials, it applies equally to vector polynomials. Rewrite the equation so that only three multiplications and three additions are required to compute the solution for a given value of *u*

$$p(u) = [(au + b)u + c]u + d \tag{5.78}$$

Consider evaluating the general polynomial of degree *n*

$$p(u) = a_n u^n + a_{n-1} u^{n-1} + a_{n-2} u^{n-2} + \cdots + a_1 u + a_0 \tag{5.79}$$

If we are to develop a procedure to compute $p(u)$ efficiently for various values of *u*, we define $a_0, a_1, \ldots a_n$ as the input coefficients along with a value for the independent variable *u*. The output variable is $p(u)$. Return again to Horner's rule, which we easily generalize as follows:

$$p(u) =$$

For $n = 1$	$a_1 u + a_0$
For $n = 2$	$(a_2 u + a_1)u + a_0$
For $n = 3$	$[(a_3 u + a_2)u + a_1]u + a_0$
For $n = 4$	$\{[(a_4 u + a_3)u + a_2]u + a_1\}u + a_0$

and so on.

Horner's rule for polynomials of arbitrary degree *n* should now be evident. For any *n*, we can develop a straight-line program of 2*n* steps to evaluate a general *n*th-degree polynomial. Clearly, *n* multiplications and *n* additions are necessary. Combining Horner's rule with a straight-line program yields for $n = 1$, $n = 2$, and $n = 3$

$$
\begin{array}{ll}
\text{for } n = 1 & t \leftarrow a_1 u \\
 & p \leftarrow t + a_0 \\[4pt]
\text{for } n = 2 & t \leftarrow a_2 u \\
 & t \leftarrow t + a_1 \\
 & t \leftarrow tu \\
 & p \leftarrow t + a_0 \\[4pt]
\text{for } n = 3 & t \leftarrow a_3 u \\
 & t \leftarrow t + a_2 \\
 & t \leftarrow tu \\[4pt]
 & t \leftarrow t + a_1 \\
 & t \leftarrow tu \\
 & p \leftarrow t + a_0
\end{array}
$$

As an alternative, we might use incremental methods to compute coordinates of points on a curve for successive values of the parametric variable. The polynomial format of the parametric functions defining curves and surfaces is conducive to the application of incremental methods. Let us investigate one particular method, the forward difference method, to evaluate a polynomial at equal intervals of the parametric variable.

We begin with the simple case of a linear equation $p(u) = cu + d$ and evaluate $p(u)$ for $(n + 1)$ equally spaced values of u. We intend to find a set of values p_i, where $p_i = p(i/n)$, $i \in [0:1]$ and $u \in [0, 1]$. *Note*: The difference between two successive values of $p(u)$ is constant: $p_{i+1} = p_i + c/n$. Thus, we find successive values of the polynomial $p(u)$ by adding the constant c/n to the previous value. This method is effectively used in the incremental plotting and display of curves.

To apply this method to polynomials of any degree, consider the cubic polynomial $p(u) = au^3 + bu^2 + cu + d$. Here, the forward difference $p_{i+1} - p_i = d_{1,i}$ is not a constant, but a quadratic polynomial in i. If we can evaluate $d_{1,i}$ easily, then we can also evaluate p_i easily. Apply the same process (that is, forward difference) to evaluate $d_{1,i}$. Now, we find $d_{1,i+1} - d_{1,i} = d_{2,i}$, a linear equation in i. We know that this equation has a constant forward difference and is easily evaluated.

We can find an algorithm for evaluating a cubic polynomial by carrying out the algebra to combine the three levels of forward differences

$$d_{1,i} = (3i^2 + 3i + 1)\frac{a}{n^3} + (2i + 1)\frac{b}{n^2} + \frac{c}{n}$$

$$d_{2,i} = 6(i + 1)\frac{a}{n^3} + \frac{2b}{n^2} \tag{5.81}$$

$$d_{3,i} = \frac{6a}{n^3}$$

We initialize these variables and p at $u = 0$, where $i = 0$ also

$$p = d$$

$$d_1 = \frac{a}{n^3} + \frac{b}{n^2} + \frac{c}{n}$$

$$d_2 = \frac{6a}{n^3} + \frac{2b}{n^2} \tag{5.82}$$

$$d_3 = \frac{6a}{n^3}$$

Thus, $p_0 = d$. To calculate p at the next and each successive increment, three additions are required

$$d_2 \leftarrow d_2 + d_3$$
$$d_1 \leftarrow d_1 + d_2 \tag{5.83}$$
$$p \leftarrow p + d_1$$

We must repeat this three-step algorithm n times, generating values for p_1-p_n. Other methods for evaluating cubic polynomials require several multiplications for each evaluation. However, after the computations to initialize the four variables p, d_1, d_2, and d_3, only three additions are required for each point. We can apply this technique to polynomials of any degree, resulting in a loop that uses n additions for polynomials of degree n.

A difficulty with the forward-difference method is that the number of equal increments must be carefully selected to ensure that enough points are evaluated to produce a smooth curve for display purposes or to provide sufficient data for subsequent analysis. It is a good idea to subdivide the parameter range by halves and develop simple expressions for the location of the curve at its midpoint [as we did for pc curves: $\mathbf{p}_{0.5} = 0.5(\mathbf{p}_0 + \mathbf{p}_1) + 0.125(\mathbf{p}_0^u - \mathbf{p}_1^u)$].

Now let us turn to the inverse point solution. Many modeling algorithms require finding the value of the parametric variable corresponding to a given point. In other words, given x, y, z coordinates of a point on (or near) a curve or surface, find the value of u or u, w. This implies finding at least the roots of three cubic equations. For a curve, this looks like

$$a_x u^3 + b_x u^2 + c_x u + d = x \qquad (5.84)$$

with similar equations for y and z. A difficulty arises, though, since it is unlikely that the u values (roots) from each of the three cubic equations $x(u)$, $y(u)$, and $z(u)$ will be exactly equal. Computation methods contribute to the potential disparity, and the source and significance of the point coordinates themselves add to it. In most cases, it is good practice to assume, for example, that a given point $\mathbf{q}$ is not on a curve. Instead, take the following approach: Find the point $\mathbf{p}$ on the curve closest to the given point $\mathbf{q}$. This anticipates matters we will explore in Chapter 6, Section 2, but no harm will come of it.

We confine our attention to a single point $\mathbf{q}$ and a single pc curve; see Fig. 5.15. We determine the point $\mathbf{p}$ on a pc curve closest to a point $\mathbf{q}$ in space by finding the vector $(\mathbf{p} - \mathbf{q})$ from the point $\mathbf{q}$ to the curve that is perpendicular to the tangent of the curve at point $\mathbf{p}$. The problem is to find $\mathbf{p}$ such that this condition is satisfied.

For $(\mathbf{p} - \mathbf{q})$ to be perpendicular to the tangent $\mathbf{p}^u$ at $\mathbf{p}$, their scalar product must be zero

$$(\mathbf{p} - \mathbf{q}) \cdot \mathbf{p}^u = 0 \qquad (5.85)$$

This results in a quintic polynomial in u. Various numerical methods are available to solve such an equation. Only roots in the interval $u \in [0, 1]$ are of interest. Each root value in this interval determines a point $\mathbf{p}$ from which $|\mathbf{p} - \mathbf{q}|$ is computed. The point $\mathbf{p}$ yielding the minimum value for $|\mathbf{p} - \mathbf{q}|$ is the point on the curve closest to $\mathbf{q}$, and $|\mathbf{p} - \mathbf{q}|$ is the distance between them.

Since the inverse point solution is one of the most important analytic operations, let us explore a somewhat more general strategy. If we talk about N-dimensional space, then vectors constructed in it will have N components. To define a surface (hypersurface) in these spaces requires $(N - 1)$ parametric variables.

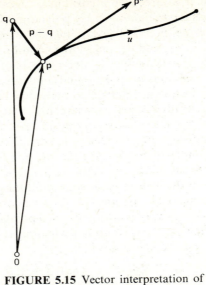

FIGURE 5.15 Vector interpretation of the inverse point solution.

Recall that in the direct-solution mode, the parametric equations yield the N coordinate components of $\mathbf{p}$, as well as their various derivatives with respect to the $(N - 1)$ parametric variables u_j, all at some point on the surface located by specifying a value for each of the parametric variables. However, in many problems of practical interest, a point is located by specifying values for J of the N coordinates of $\mathbf{p}$. For most problems $J \neq N - 1$, so that the inverse may not be unique. In fact, when $J < N - 1$, the inverse degenerates to a surface of $(N - J - 1)$ dimensions. When $J = N$, the problem is somewhat overconstrained, so we simply ignore one of the coordinates to obtain a determinate system.

Let $\mathbf{p}^*$ denote a specific point on the surface (hypersurface) where we are required to compute the first derivatives $\partial \mathbf{p}/\partial u_j$. We know the N coordinates or components of this point, but we do not know the values of the parametric variables u_j at this point; therefore, we cannot evaluate $\partial \mathbf{p}/\partial u_j$ directly. A related type of inversion problem develops if we are given only $(N - 1)$ components of $\mathbf{p}^*$ and must find the Nth component. The resulting matrix algebra required to invert and solve the system of equations defining these types of problems defies solution, and we are forced to use an iterative procedure. To illustrate the problem of finding the tangent vectors at a point $\mathbf{p}^*$ on a surface $\mathbf{p}$, consider the vector $\mathbf{r} = \mathbf{p}^* - \mathbf{p}$. If $|\mathbf{r}|$ is small, then $\mathbf{r}$ approximates a vector tangent to the surface. A good guess is to start the iteration at $\mathbf{p}(1/2, 1/2, 1/2, \ldots, 1/2)$, the parametric center of the hypersurface patch. Now, define a unit vector $\mathbf{t} = \mathbf{r}/|\mathbf{r}|$ so that

$$\mathbf{t} = \sum_{i=1}^{N-1} \frac{\partial \mathbf{p}/\partial u_i}{du_i/ds} \qquad (5.86)$$

or in matrix form

$$[\mathbf{t}] = \left[\frac{\partial \mathbf{p}}{\partial u_i}\right]\left[\frac{du_i}{ds}\right] = \left[\frac{\mathbf{r}}{|\mathbf{r}|}\right] \tag{5.87}$$

or

$$\left[\frac{du_i}{ds}\right] = \left[\frac{\partial \mathbf{p}}{\partial u_i}\right]^{-1}\left[\frac{\mathbf{r}}{|\mathbf{r}|}\right] \tag{5.88}$$

If $\Delta s \cong |\mathbf{r}|$, then the incremental changes Δu_i in the parametric variables are

$$[\Delta u_i] = \left[\frac{\partial \mathbf{p}}{\partial u_i}\right]^{-1}[\mathbf{r}] \tag{5.89}$$

Again, compute the elements of $[\partial \mathbf{p}/\partial u_i]$ at the initial guess

$$\mathbf{p}(1/2, 1/2, 1/2, \ldots, 1/2)$$

and $\mathbf{r}$ from $\mathbf{r} = \mathbf{p}^* - \mathbf{p}$. Next, appropriately increment each of the parametric variables by Δu_i to determine a new $\mathbf{p}$, and continue the procedure until some convergence criterion, $|\mathbf{r}| \leq \epsilon$, is satisfied.

EXERCISES

1. Write a general expression for Horner's rule for a polynomial of degree n.

2. Extend Horner's rule and a straight-line program to obtain a direct solution for points on a bicubic patch.

3. Use the forward-difference method to compute points from $u = 0$ to $u = 1$ at increments of 0.1 on a parametric quintic curve that passes through points

$$\mathbf{p}_1 = [0 \quad 0 \quad 4]$$
$$\mathbf{p}_2 = [0 \quad 0 \quad 0]$$
$$\mathbf{p}_3 = [3 \quad 0 \quad 0]$$
$$\mathbf{p}_4 = [0 \quad 6 \quad 0]$$
$$\mathbf{p}_5 = [-1 \quad 10 \quad 2]$$

and

$$\mathbf{p}_6 = [0 \quad 20 \quad 0]$$

4. Can Horner's rule and the forward-difference method be applied to Bezier and *B*-spline curves? If so, show how by setting up the procedures in general terms.

5. Given the pc curve whose geometric coefficients are

$$\mathbf{p}_0 = [0 \quad 0 \quad 0]$$
$$\mathbf{p}_1 = [10 \quad 0 \quad 0]$$
$$\mathbf{p}_0^u = [16 \quad 16 \quad 0]$$

and

$$\mathbf{p}_1^u = [16 \quad -16 \quad 0]$$

find the value of the parametric variable u corresponding to

$$\mathbf{p}(u) = [5 \quad 4 \quad 0]$$

6. Given the pc curve whose geometric coefficients are

$$\mathbf{p}_0 = [4 \quad 8 \quad 0]$$

$$\mathbf{p}_1 = [24 \quad 16 \quad 0]$$

$$\mathbf{p}_0^u = [20 \quad 24 \quad 0]$$

and

$$\mathbf{p}_1^u = [12 \quad -8 \quad 0]$$

find the values of the parametric variable u corresponding to

$$\mathbf{p}(u) = [8.25 \quad 12.16]$$

and

$$\mathbf{p}(u) = [17.15 \quad 16.64]$$

7. Given the pc curve whose geometric coefficients are

$$\mathbf{p}_0 = [-1.5 \quad 2.25 \quad 0]$$

$$\mathbf{p}_1 = [2.50 \quad 6.25 \quad 0]$$

$$\mathbf{p}_0^u = [4 \quad -12 \quad 0]$$

and

$$\mathbf{p}_1^u = [4 \quad 20 \quad 0]$$

find the values of the parametric variable u corresponding to

$$\mathbf{p}(u) = [-0.7 \quad 0.49 \quad 0]$$

$$\mathbf{p}(u) = [0 \quad 0 \quad 0]$$

and

$$\mathbf{p}(u) = [0.9 \quad 0.81 \quad 0]$$

8. Review procedures you may have written to this point that execute a direct point solution (on curves or surfaces). If you have not applied Horner's rule or forward-difference methods, analyze their possible contributions to procedure efficiency. Incorporate these methods if feasible.

9. Write a procedure to compute the value of the parametric variable at a point on a pc curve. Denote this as **CRVINV(CI, P, U)**, where

CI(4, 3) is the input array of geometric coefficients defining the pc curve;

P(3) is the input array of x, y, z coordinates of the point; and

U is the output value of the parametric variable.

10. Generalize **CRVINV** so that it also applies to Bezier and *B*-spline curves.

11. Write a procedure to compute the value of the parametric variables at a point on a bicubic patch. Denote this as **SRFINV(CI, P, UW)**, where

CI(4, 4, 3) is the input array of geometric coefficients defining the bicubic patch; and

P(3) is the input array of *x*, *y*, *z* coordinates of the point; and

UW(2) is the output array of values of the parametric variables *u* and *w* at the point.

12. Generalize **SRFINV** so that it also applies to Bezier and *B*-spline surfaces.

5 GLOBAL PROPERTIES

Global properties are those that depend on the over-all characteristics of a geometric element. We will focus on the three principal global properties: arc length, surface area, and volume, but we will consider only simple elements — the parametric curve and surface. Later, in Chapter 11, we will explore the more difficult problem of determining the global properties of complex solid models.

5.1 Arc Length

Let us compute the length of a pc curve between u_1 and u_2; see Fig. 5.16. The simplest approach is to first divide this segment into *n* equal parametric intervals. Compute the *x*, *y*, *z* coordinates at the end point of each interval, and then simply compute the sum of the successive straight-line distances between these points. We express this as

$$L = \sum_{i=1}^{n} l_i \qquad (5.90)$$

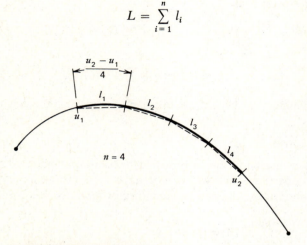

FIGURE 5.16 Arc length.

where

$$l_i = \sqrt{(\mathbf{p}_i - \mathbf{p}_{i-1}) \cdot (\mathbf{p}_i - \mathbf{p}_{i-1})} \quad \text{and} \quad \mathbf{p}_i = \sum_{j=0}^{3} \mathbf{a}_j u_i^j$$

This approach is not usually the most computationally efficient, nor is it particularly accurate unless n is large, since this procedure always returns a somewhat shorter-than-true length. Therefore, let us consider another approach.

The arc length between $\mathbf{p}(u_1)$ and $\mathbf{p}(u_2)$ is

$$L = \int_{u_1}^{u_2} \sqrt{\mathbf{p}^u \cdot \mathbf{p}^u}\, du \quad \text{where } u_2 > u_1 \tag{5.91}$$

When the vector operation is completed, this equation becomes

$$L = \int_{u_1}^{u_2} \sqrt{a_4 u^4 + a_3 u^3 + a_2 u^2 + a_1 u + a_0}\, du \tag{5.92}$$

Define the a_i constants in terms of the algebraic coefficients. Using a functional notation for Eq. 5.92, we obtain

$$L = \int_{u_1}^{u_2} f(u)\, du \tag{5.93}$$

Use Gaussian quadrature to reduce this integral

$$\int_{u_1}^{u_2} f(u)\, du = \sum_{i=1}^{n} w_i f(u_i) \tag{5.94}$$

where n is the number of points used, w_i are the weight values, and u_i are the Gaussian abcissas. The abcissas can be normalized to a more convenient interval $u_i \in [0, 1]$ by using the following transformation:

$$t = \frac{u - u_i}{u_2 - u_1} \tag{5.95}$$

Now we can express L as

$$L = (u_2 - u_1) \int_0^1 f[u_1 + (u_2 - u_1)t]\, dt \tag{5.96}$$

or

$$L = (u_2 - u_1) \sum_{i=1}^{n} w_i g(t_i) \tag{5.97}$$

where the weights and abcissas are taken with respect to the new interval.

5.2 Surface Area

The bicubic surface and other parametric forms are well adapted to computing geometric properties, since we do not have to compute explicit points. To compute surface area, we use an elementary property of vectors shown in Fig. 5.17

$$dA = |\mathbf{n}|\, du\, dw$$
$$= f_1(u, w)\, du\, dw \tag{5.98}$$

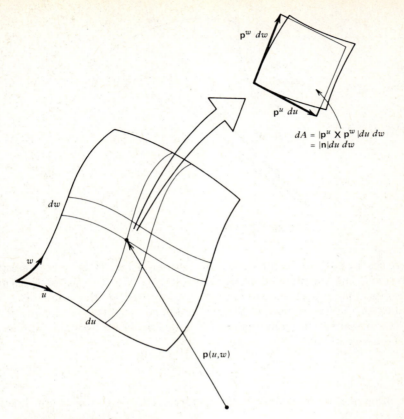

$$\mathbf{p}^w \, dw$$

$$\mathbf{p}^u \, du$$

$$dA = |\mathbf{p}^u \times \mathbf{p}^w| du \, dw$$
$$= |\mathbf{n}| du \, dw$$

$$dw$$

$$w$$

$$u$$

$$du$$

$$\mathbf{p}(u,w)$$

FIGURE 5.17 Surface area.

where $\mathbf{n}(u, w)$ is a vector function defining the patch normals and dA is the scalar element of area. Therefore,

$$A = \int_0^1 \int_0^1 f_1(u, w) \, du \, dw \tag{5.99}$$

Finally, using Gaussian quadrature, we obtain

$$A = \sum_{i=0}^{n} \sum_{j=0}^{n} g_i h_j f_1(u_j, w_i) \tag{5.100}$$

where g_i and h_j are the weight values associated with a specific n-point formula.

5.3 Volume

For the volume of a closed region, we again note from vector calculus that

$$dV = \tfrac{1}{3}[\mathbf{p}(u, w) \cdot \mathbf{n}(u, w)] \, du \, dw$$

$$= f_2(u, w) \, du \, dw \tag{5.101}$$

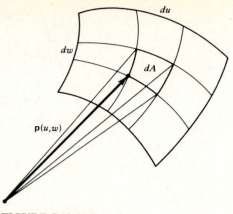

FIGURE 5.18 Volume.

where **p** is the position vector and **n** is the normal vector; dV is the scalar element of volume. The key ingredients are shown in Fig. 5.18. The factor of 1/3 arises from the solid geometry formula for a pyramid (that is, the volume is 1/3 the product of the base area and the perpendicular height). The dot product of the vectors satisfies the requirement of perpendicularity between the base and height. So we obtain

$$V = \int_0^1 \int_0^1 f_2(u, w)\, du\, dw \tag{5.102}$$

and, of course, we may again apply quadrature as we did to obtain Eq. 5.100 from Eq. 5.99.

EXERCISES

1. Find the arc length of the Bezier curve whose control points are

$$\mathbf{p}_0 = [0 \quad -10 \quad 0]$$
$$\mathbf{p}_1 = [0 \quad 0 \quad 0]$$
$$\mathbf{p}_2 = [10 \quad 0 \quad 0]$$

and

$$\mathbf{p}_3 = [10 \quad 0 \quad 10]$$

2. Find the arc length of the Bezier curve whose control points are

$$\mathbf{p}_0 = [0 \quad -10 \quad 0]$$
$$\mathbf{p}_1 = [0 \quad 0 \quad 0]$$
$$\mathbf{p}_2 = [0 \quad 0 \quad 0]$$
$$\mathbf{p}_3 = [10 \quad 0 \quad 0]$$
$$\mathbf{p}_4 = [10 \quad 0 \quad 0]$$

and

$$\mathbf{p}_5 = [10 \quad 0 \quad 10]$$

Notice the multiple points at $\mathbf{p}_1$, $\mathbf{p}_2$ and $\mathbf{p}_3$, $\mathbf{p}_4$. Compare this length with that found for the curve defined in Ex. 1.

3. Find the arc length of the Bezier curve whose control points are

$$\mathbf{p}_0 = [10 \quad 0 \quad 0]$$
$$\mathbf{p}_1 = [0 \quad 10 \quad 0]$$
$$\mathbf{p}_2 = [0 \quad -10 \quad 0]$$

and

$$\mathbf{p}_3 = [10 \quad -10 \quad 0]$$

4. Find the arc length of the closed Bezier curve whose control points are

$$\mathbf{p}_0 = [0 \quad 0 \quad 0]$$
$$\mathbf{p}_1 = [10 \quad 0 \quad 0]$$
$$\mathbf{p}_2 = [0 \quad 10 \quad 0]$$
$$\mathbf{p}_3 = [-10 \quad 0 \quad 0]$$

and

$$\mathbf{p}_4 = [0 \quad 0 \quad 0]$$

5. Find the arc length of the curve defined in Ex. 3 between $u = 0.4$ and $u = 0.7$.

6. Find the arc length of the curve defined in Ex. 4 between $u = 0.8$ and $u = 0.1$.

7. Find the arc length of the periodic B-spline with $k = 3$, whose control points are

$$\mathbf{p}_0 = [0 \quad -10 \quad 0]$$
$$\mathbf{p}_1 = [0 \quad 0 \quad 0]$$
$$\mathbf{p}_2 = [0 \quad 0 \quad 10]$$
$$\mathbf{p}_3 = [10 \quad 0 \quad 10]$$
$$\mathbf{p}_4 = [10 \quad 10 \quad 10]$$
$$\mathbf{p}_5 = [10 \quad 10 \quad 20]$$

and

$$\mathbf{p}_6 = [20 \quad 10 \quad 20]$$

8. Repeat Ex. 7 for $k = 2$.

9. Repeat Ex. 7 for a nonperiodic B-spline. Use only the first six control points. Thus, $k = 3$ and $n = 5$.

10. Find the arc length of the curve defined in Ex. 7 between $u = 1.5$ and $u = 3.5$ (that is, from $u = 0.5$ of segment 2 to $u = 0.5$ of segment 4).

11. Find the arc length of the closed periodic B-spline with $k = 4$, whose control points are

$$\mathbf{p}_0 = [0 \quad 0 \quad 0]$$
$$\mathbf{p}_1 = [10 \quad 0 \quad 0]$$
$$\mathbf{p}_2 = [10 \quad 10 \quad 0]$$

and

$$\mathbf{p}_3 = [0 \quad 10 \quad 0]$$

12. Write a procedure to compute the length of a pc curve. Denote this as **ARCL(CI, L)**, where

CI(4, 3) is the input array of geometric coefficients defining the pc curve, and

L is the output length of the curve.

13. Write a procedure to compute the surface area of a bicubic patch. Denote this as **AREA(CI, A)**, where

CI(4, 4, 3) is the input array of geometric coefficients defining the bicubic patch, and

A is the output area of the surface.

14. Modify **ARCL** and **AREA** so that they also include Bezier and B-spline curves and surfaces.

6

RELATIONAL
PROPERTIES

There are many kinds of relationships between geometric objects. For example, given a group of objects, we might ask which member of the group is nearest or which is farthest from a given point? Or, with respect to an observer, which objects are partially or completely obscured by which other objects? Which is the smallest or largest object in a group? Which objects will fit inside which other objects? And so on.

These and other relational properties are important in geometric modeling, and they must frequently be resolved in computer graphics, CAD/CAM, and other application-problem situations. Underlying these relational properties are two additional relational phenomena of a more fundamental nature. They are the minimum-distance problem and the intersection problem. We will explore the minimum-distance problem in this chapter and the intersection problem in Chapter 7.

Applications, of course, demand numerical solutions. We will see that the minimum-distance problem relies on a variety of numerical techniques and offers subtleties and idiosyncracies sufficient to enthrall the most ardent numerical analyst. Great volumes have been written describing techniques for finding roots of nth-degree polynomials and solving systems of nonlinear equations. We will not discuss these methods here. Our focus is, instead, on the geometric content of each minimum-distance problem.

1 MINIMUM DISTANCE BETWEEN TWO POINTS

The computation of the minimum distance between two points is at once trivial and subtly complex. Consider the four distance problems posed in Fig. 6.1. First, the distance between two arbitrary points in space $\mathbf{p}_1$ and $\mathbf{p}_2$ is of course simply

$$d_{\min} = |\mathbf{p}_2 - \mathbf{p}_1| = \sqrt{(\mathbf{p}_2 - \mathbf{p}_1) \cdot (\mathbf{p}_2 - \mathbf{p}_1)} \tag{6.1}$$

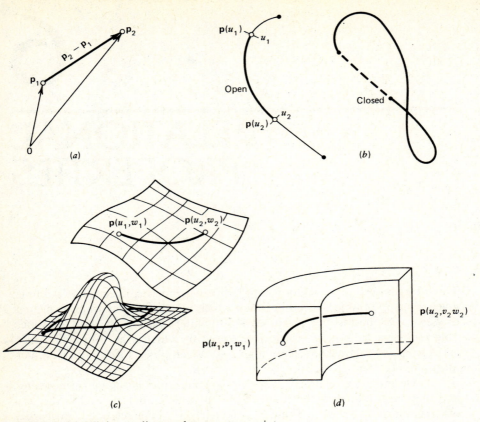

FIGURE 6.1 Minimum distance between two points.

The next step up in complexity requires that we find the minimum distance between two points $p(u_1)$ and $p(u_2)$ on a curve. At once, we realize that the constraints are such that there is only one path between these points along the curve. By default, it is the shortest path or minimum distance, unless (!) the curve is closed, in which case there are two possible candidates. Thus, we are left with finding the arc length, as in Chapter 5, Section 5.1.

To find the minimum distance between two points on a surface or in a solid requires finding a common geodesic curve or curves passing through both points. There is no direct solution, and we must compute by iterative methods and trial and error, as there may be more than one geodesic.

EXERCISES

1. Outline a procedure for finding an approximate geodesic between two points on a surface, given by $p(u, w)$.

2. Find the minimum distance between points **p**(0.1, 0.2) and **p**(0.7, 0.9) on the Bezier surface defined by the following control points:

$$\mathbf{B} = \begin{bmatrix} (2, 2, -2) & (2, 6, 4) & (2, 10, -2) \\ (4, 2, -1) & (4, 6, 3) & (4, 10, -1) \\ (6, 2, 0) & (6, 6, 0) & (5, 10, 0) \\ (8, 2, 1) & (8, 6, -3) & (8, 10, 1) \\ (10, 2, 2) & (10, 6, -4) & (10, 10, 2) \end{bmatrix}$$

2 MINIMUM DISTANCE BETWEEN A POINT AND A CURVE

The minimum distance from a given point **q** to a curve **p**(u) is determined by finding a vector (**p** − **q**) from the point to the curve that is perpendicular to the tangent vector **p**^{u} at **p**. Figure 6.2 shows the vector geometry. Mathematically, we express the required conditions as

$$d_{\min} = |\mathbf{p} - \mathbf{q}| \tag{6.2}$$

when

$$(\mathbf{p} - \mathbf{q}) \cdot \mathbf{p}^u = 0 \tag{6.3}$$

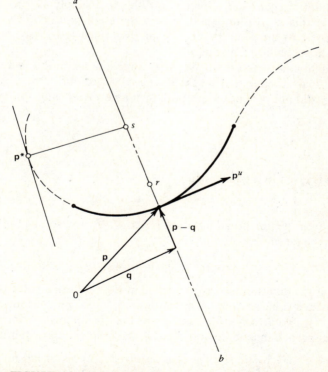

FIGURE 6.2 Minimum distance between a point and a curve.

One solution to this equation was given earlier in Chapter 5, Section 4; let us now approach it from a slightly different angle, suggested by Meriwether. First, we can simplify Eq. 6.3 if we translate the origin of the coordinate system to **q**. This does not affect either the shape of the curve or the minimum distance. Eq. 6.3 becomes

$$\mathbf{p} \cdot \mathbf{p}^u = 0 \tag{6.4}$$

This affects the algebraic coefficients of a pc curve as follows:

$$\mathbf{a}_3 u^3 + \mathbf{a}_2 u^2 + \mathbf{a}_1 u + \mathbf{a}_0 \Rightarrow \mathbf{a}_3 u^3 + \mathbf{a}_2 u^2 + \mathbf{a}_1 u + \mathbf{a}_0 - \mathbf{q} \tag{6.5}$$

The geometric coefficients are transformed similarly

$$[\mathbf{p}_0 \quad \mathbf{p}_1 \quad \mathbf{p}_0^u \quad \mathbf{p}_1^u]^T \Rightarrow [\mathbf{p}_0 - \mathbf{q} \quad \mathbf{p}_1 - \mathbf{q} \quad \mathbf{p}_0^u \quad \mathbf{p}_1^u]^T \tag{6.6}$$

Notice that the algebraic coefficients of $\mathbf{p}^u$ are unchanged by this transformation. Thus,

$$\mathbf{p}^u = 3\mathbf{a}_3 u^2 + 2\mathbf{a}_2 u + \mathbf{a}_1 \tag{6.7}$$

If we now expand Eq. 6.4 in terms of its components, we obtain

$$x(u)x^u(u) + y(u)y^u(u) + z(u)z^u(u) = 0 \tag{6.8}$$

When we insert the appropriate functions, we find that for a pc curve, this is indeed a quintic polynomial, which we will denote simply as $f(u)$. The values of u that set this expression equal to zero represent points on the curve that may be closest to the given point in space. We must test these solutions and eliminate those that are maximum.

We might choose to use the Newton–Raphson method to find the roots of the polynomial. Assume that we are near a root and our initial iterative value of u, say, u_i, did not produce a solution, then a new trial value for u is obtained from

$$u_{i+1} = u_i - \frac{f(u_i)}{f^u(u_i)} \tag{6.9}$$

Since $f(u)$ is given by Eq. 6.8, then

$$f^u(u) = xx^{uu} + x^u x^u + yy^{uu} + y^u y^u + zz^{uu} + z^u z^u \tag{6.10}$$

where we omit the functional notation $[x(u), x^{uu}(u),$ and so on] for brevity and where, for the pc curve,

$$\mathbf{p}^{uu} = 6\mathbf{a}_3 u + 2\mathbf{a}_2 \tag{6.11}$$

Repeat the iteration process indicated in Eq. 6.9 until a value of u is found that produces a value as close to zero as desired. This is not, of course, the whole solution; it just represents the case most often of interest. A more complete practical approach must take into account the following possibilities:

1. There may be more than one occurrence of a normal in the interval $u \in [0, 1]$; if that is the case, compute all normals and select the shortest vector.

Note again that some point–curve relationships can produce solutions to Eq. 6.8 that are maximum distances in the interval, and we must test them to determine if the curve is locally convex or concave with respect to the point in space. If all solutions are maximums, then test the segment end points; one one or both will be the closest distance.

2. There may be no normals in the interval $u \in [0, 1]$; therefore, we must determine if the closest end is an acceptable answer.

3. The point $\mathbf{q}$ may lie on the curve. If so, the normal vector has zero length, and a modeling program should recognize this type of solution.

4. Finally, the curve may be degenerate (that is, a point), in which case, we simply compute the distance between two points, as in Eq. 6.1.

Carefully study Fig. 6.2 again. Notice that points on the normal plane (represented by the line *ab* in the figure) to the curve at $\mathbf{p}$ can have several different relationships to the curve. The curve is locally convex to $\mathbf{q}$ and concave to $\mathbf{r}$ and $\mathbf{s}$. The minimum distance normal from $\mathbf{s}$ to the curve is at $\mathbf{p}^*$ and not $\mathbf{p}$ ($\mathbf{p}^*$ is on the curve but outside the interval $u \in [0, 1]$). Therefore, $\mathbf{p}_0$ or $\mathbf{p}_1$ is the closest point on the curve to $\mathbf{s}$. Clearly, Eqs. 6.2 and 6.3 provide only part of the solution.

EXERCISES

1. If the curve is a straight-line segment, show how Eqs. 6.3, 6.4, and so on, and their solutions are affected.

2. Given a periodic *B*-spline curve with $k = 3$ whose control points are

$$\mathbf{p}_0 = [0 \quad 0 \quad 0]$$
$$\mathbf{p}_1 = [10 \quad 0 \quad 0]$$
$$\mathbf{p}_2 = [10 \quad 10 \quad -5]$$
$$\mathbf{p}_3 = [0 \quad 10 \quad -5]$$
$$\mathbf{p}_4 = [0 \quad 0 \quad -10]$$
$$\mathbf{p}_5 = [10 \quad 0 \quad -10]$$

and

$$\mathbf{p}_6 = [10 \quad 10 \quad -15]$$

find the minimum distance between it and each of the following points:

$$\mathbf{q}_1 = [5 \quad 5 \quad 2]$$
$$\mathbf{q}_2 = [5 \quad 5 \quad -2]$$
$$\mathbf{q}_3 = [5 \quad -5 \quad -2]$$

and

$$\mathbf{q}_4 = [10 \quad 20 \quad -20]$$

3. Write a procedure to compute the minimum distance between a point and a pc curve. Accommodate the four possible relationships discussed at the end of this section. Denote this as **MINDPC(P, CI, NS, U, D)**, where

P(3) is the input array of x, y, z coordinates of the point;

CI(4, 3) is the input array of geometric coefficients defining the pc curve;

NS is an output number specifying the number of solutions;

U(NS) is the output array of values of the parametric variable u at points of minimum distance to P; and

D is the output value of the minimum distance.

4. Generalize **MINDPC** so that it also applies to Bezier and B-spline curves.

3   MINIMUM DISTANCE BETWEEN A POINT AND A SURFACE

An important special case of finding the minimum distance between a point and a surface is finding the minimum distance between a point and a plane. Let the point be denoted by **q** and the plane by $k\mathbf{n}$, where **n** is the unit normal vector to the plane from the origin; see Fig. 6.3. In addition, let **p** denote the point on the plane closest to **q**. Then, we find that

$$d_{min} = |\mathbf{p} - \mathbf{q}| \tag{6.12}$$

where **p** must satisfy

$$(\mathbf{p} - \mathbf{q}) \times \mathbf{n} = 0 \tag{6.13}$$

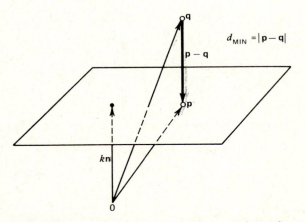

FIGURE 6.3 Minimum distance between a point and a plane.

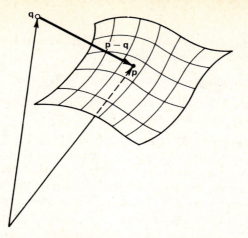

FIGURE 6.4 Minimum distance between a point and a surface.

Equation 6.13 merely asserts that the vector $(\mathbf{p} - \mathbf{q})$ must be parallel to the normal to the plane, $\mathbf{n}$. The point $\mathbf{p}$ must also satisfy the equation of the plane $Ax + By + Cz + D = 0$ or some other equivalent constraint, such as

$$(\mathbf{p} - \mathbf{q}) \cdot (\mathbf{p} - k\mathbf{n}) = 0$$

A convenient alternative is to use $\mathbf{p} = \mathbf{a} + u\mathbf{b} + w\mathbf{c}$ in Eq. 6.13. How is the solution simplified if the point–plane system is translated so that the point coincides with the origin?

Figure 6.4 shows the more general problem. Here, we seek to find the minimum distance between a point and a surface. Again, let the point be denoted by $\mathbf{q}$ and the point on the surface closest to $\mathbf{q}$ by $\mathbf{p}(u, w)$. Then, clearly, $d_{min} = |\mathbf{p} - \mathbf{q}|$. The vector $(\mathbf{p} - \mathbf{q})$ must be in the direction of the surface normal at $\mathbf{p}$, which means that $\mathbf{p}$ must satisfy

$$(\mathbf{p} - \mathbf{q}) \times (\mathbf{p}^u \times \mathbf{p}^w) = 0 \qquad (6.14)$$

or

$$(\mathbf{p} - \mathbf{q}) = a(\mathbf{p}^u \times \mathbf{p}^w) \qquad (6.15)$$

As with points and curves, a numerical analysis method must be employed to determine $\mathbf{p}$. Notice that unless the surface is closed, there may not be a normal in the interval $u, w \in [0, 1]$ or if the surface is composite, within its parametric domain. Again, you may have to check for a concave relationship producing a maximum that also satisfies Eq. 6.14 or 6.15. Or, there may be more than one normal and more than one closest point. Remember to check boundary curves and patch corner points. A final possibility is a locus of equidistant closest points describing a curve on the surface. A modeling system must be capable of detecting and resolving these possibilities.

EXERCISES

1. An unbounded plane intersects the coordinate axes at $x = 7, y = 10, z = 5$. Find the normal to it from the origin.

2. Find the minimum distance from the plane in Ex. 1 to a point at $(3, -1, 4)$.

3. Find the minimum distance from the Bezier surface defined in Ex. 2, Section 1 to each of the following points: $(7, 3, 4)$, $(3, 7, 8)$, and $(9, 5, -6)$.

4. Show how a parametrically defined surface can be represented by a set of triangular plane faces and how these faces can be used to find an approximate solution to the minimum-distance problem. *Hint*: First compute a rectangular array of points over the surface at some equal interval on the parametric variables u and w, then triangulate this array. For example, using an interval of $u = w = 0.5$ over a surface defined on $u, w \in [0, 1]$, the three points $\mathbf{p}(0, 0)$, $\mathbf{p}(0.5, 0)$, and $\mathbf{p}(0, 0.5)$ can be used to define a plane. In fact, for this interval, there are eight triangular plane faces. Notice that as the interval size decreases, the number of triangular plane faces increases and, thus, of course, the approximation becomes more accurate.

5. Write a procedure to compute the minimum distance between a point and a bicubic patch. Denote this as **MINDPS**(P, CI, NS, UW, D), where

> P(3) is the input array of x, y, z coordinates of the point;

> CI(4, 4, 3) is the input array of geometric coefficients defining the bicubic patch;

> NS is an output number specifying the number of solutions;

> UW(NS, 2) is an output array of values of the parametric variables u, w at the points of minimum distance to P; and

> D is the output value of the minimum distance.

6. Generalize **MINDPS** so that it also applies to Bezier and *B*-spline surfaces.

4   MINIMUM DISTANCE BETWEEN TWO CURVES

Given two parametric curves $\mathbf{p}(s)$ and $\mathbf{q}(t)$, we want to find the minimum distance between them; see Fig. 6.5. Let $\mathbf{p}$ and $\mathbf{q}$ denote corresponding points of closest approach on the respective curves; then $d_{min} = |\mathbf{p} - \mathbf{q}|$, and these points satisfy the following conditions:

$$(\mathbf{p} - \mathbf{q}) \cdot \mathbf{p}^s = 0 \tag{6.16}$$

and

$$(\mathbf{p} - \mathbf{q}) \cdot \mathbf{q}^t = 0 \tag{6.17}$$

Two nonlinear simultaneous equations must be solved. Only those solution pairs (s_i, t_i) in the interval $s, t \in [0, 1]$ interest us. Furthermore, we must also test the end points as well as for possible intersections.

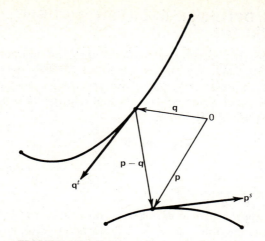

FIGURE 6.5 Minimum distance between two curves.

Notice that there can be more than one pair of points that satisfy Eqs. 6.16 and 6.17. Each solution must be compared to the others to find the minimum; watch for local maximums.

EXERCISES

1. Are Eqs. 6.16 and 6.17 simplified for the case of two coplanar curves? What about two straight-line segments?

2. Write a procedure to compute the minimum distance between two pc curves. Denote this as **MINDCC**(CI, C2, NS, U, DMIN, U1MIN, U2MIN), where

CI(4, 3), C2(4, 3) are input arrays of the geometric coefficients defining the two pc curves;

NS is the output number of solution points on each curve;

U(NS, 2, D) is the output array of values of corresponding pairs of parametric variables on each curve that are solutions to Eqs. 6.16 and 6.17 and the distance between the points;

DMIN is the output value of the minimum distance; and

U1MIN, U2MIN are the output values of the parametric variables of the points on the two curves that are the minimum distance apart.

3. Generalize **MINDCC** so that it also applies to Bezier and *B*-spline curves and any combination of them.

4. Compare the facility of solutions for pc, Bezier, and *B*-spline curves and combinations of them.

5 MINIMUM DISTANCE BETWEEN A CURVE AND A SURFACE

Let us consider first the special problem of finding the minimum distance between a curve and a plane; see Fig. 6.6. Denote a point on the curve $\mathbf{p}(u)$ as $\mathbf{p}$; the plane as $k_n\mathbf{n}$, the normal to the plane from the origin; and $\mathbf{q}$, the point on the plane closest to the curve. Then, as usual, the minimum distance is $d_{min} = |\mathbf{p} - \mathbf{q}|$, and these points satisfy the following conditions:

$$(\mathbf{p} - \mathbf{q}) \cdot \mathbf{p}^u = 0 \tag{6.18}$$

and

$$(\mathbf{p} - \mathbf{q}) \times \mathbf{n} = 0 \tag{6.19}$$

One simple approach is to express Eq. 6.18 as $k_q\mathbf{n} \cdot \mathbf{p}^u = 0$. Since k_q is a scalar, we immediately obtain $\mathbf{n} \cdot \mathbf{p}^u = 0$. Performing vector arithmetic results in a quadratic equation $au^2 + bu + c = 0$, where the constants a, b, c consist of algebraic coefficients of $\mathbf{p}$ and the components of $\mathbf{n}$ (that is, n_x, n_y, n_z). It is possible that both roots of the quadratic are in the interval $u \in [0, 1]$. If so, proceed with both resulting candidate values of $\mathbf{p}$ to the next step, computing $\mathbf{q}$ from Eq. 6.19. Then select the smallest value of $|\mathbf{p} - \mathbf{q}|$ as the correct solution. Be aware that the following conditions are possible: There are no roots in the interval, and one of the end points is closest; the curve and the plane intersect; or the curve and plane are parallel.

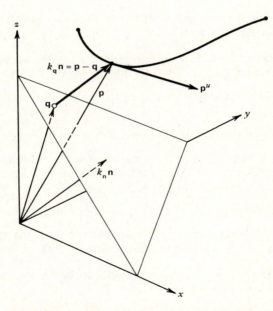

FIGURE 6.6 Minimum distance between a curve and a plane.

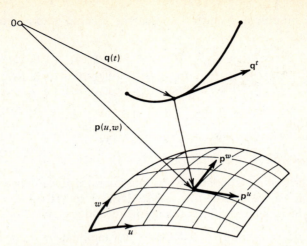

FIGURE 6.7 Minimum distance between a curve and a surface.

Figure 6.7 illustrates the more general problem of finding the minimum distance between a pc curve and a bicubic patch. The vector $(\mathbf{p} - \mathbf{q})$ must satisfy the following conditions if it is a minimum-distance candidate:

$$(\mathbf{p} - \mathbf{q}) \times (\mathbf{p}^u \times \mathbf{p}^w) = 0 \tag{6.20}$$

and

$$(\mathbf{p} - \mathbf{q}) \cdot \mathbf{q}^t = 0 \tag{6.21}$$

Equations 6.20 and 6.21 generate a system of simultaneous, nonlinear equations for whose solution, happily, numerical methods are available. Check for local maximums, patch corner points, and boundary curves and curve end points.

EXERCISES

1. Develop and discuss a solution to the minimum-distance problem between an unbounded plane and a straight-line segment.

2. If the curve in Fig. 6.7 is a straight-line segment, how are Eqs. 6.20 and 6.21 simplified?

3. Find the minimum distance between the plane defined by $k_n\mathbf{n} = [4 \quad 2 \quad 2]$ and the nonperiodic B-spline curve with $k = 4$ and control points

$$\mathbf{p}_0 = [5 \quad 8 \quad 10]$$
$$\mathbf{p}_1 = [5 \quad 5 \quad 5]$$
$$\mathbf{p}_2 = [8 \quad 5 \quad 3]$$
$$\mathbf{p}_3 = [10 \quad 7 \quad 0]$$

and

$$\mathbf{p}_4 = [16 \quad 4 \quad -21]$$

4. Find the minimum distance between the bicubic patch whose geometric coefficients are

$$\mathbf{B}_s = \begin{bmatrix} (10, 0, 0) & (10, 0, 10) & (0, 0, 10) & (0, 0, 10) \\ (0, 10, 0) & (0, 10, 10) & (0, 0, 10) & (0, 0, 10) \\ (0, 8, 0) & (0, 8, 0) & 0 & 0 \\ (-8, 0, 0) & (-8, 0, 0) & 0 & 0 \end{bmatrix}$$

and the pc curve whose geometric coefficients are

$$\mathbf{B}_c = [(0, 20, 10) \quad (20, 0, 0) \quad (7, -12, -7) \quad (7, -12, -7)]^T$$

5. Write a procedure to compute the minimum distance between a pc curve and a bicubic patch. Denote this as **MINDCS**. Use **MINDCC** as a guide for developing the input and output requirements and argument list.

6. How would you modify **MINDCS** to accommodate composite curves and surfaces and accommodate Bezier and *B*-spline curves and surfaces?

6   MINIMUM DISTANCE BETWEEN TWO SURFACES

The minimum distance between two patches $\mathbf{p}(u, w)$ and $\mathbf{q}(s, t)$ is $|\mathbf{p} - \mathbf{q}|$ when these vectors satisfy the following two conditions:

$$(\mathbf{p} - \mathbf{q}) \times (\mathbf{p}^u \times \mathbf{p}^w) = 0 \tag{6.22}$$

and

$$(\mathbf{p} - \mathbf{q}) \times (\mathbf{q}^s \times \mathbf{q}^t) = 0 \tag{6.23}$$

Figure 6.8 illustrates the vector geometry describing this problem.

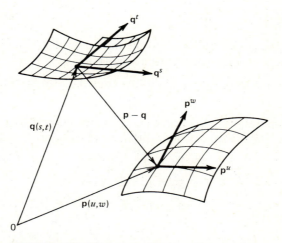

FIGURE 6.8 Minimum distance between two surfaces.

EXERCISES

1 What is the nature of the polynomials implied by Eqs. 6.22 and 6.23 if both surfaces are bicubic?

2. List possible special conditions (that is, multiple solutions, no solutions, intersections, and so on) and their effects on the minimum-distance solution for two surfaces.

3. How are the solutions of Eqs. 6.22 and 6.23 affected if one of the surfaces is a plane?

4. Assume the same conditions as Ex. 3 but both are ruled surfaces.

5. What effect do closed composite surfaces have on implementing a solution?

6. Develop at least two different search strategies for isolating roots of the minimum-distance constraint equations.

7. Outline a procedure to compute the minimum distance between two surfaces. Try to keep the outline general enough so that it applies to all the surfaces we have discussed. How would you adapt it to handle composite surfaces? Show how you would incorporate tests to detect and handle special conditions.

7   NEAREST NEIGHBOR SPATIAL SEARCH

Given a set of points $\{\mathbf{p}_i\}$ in a plane and a point $\mathbf{q}$ not in this set, find the point in $\{\mathbf{p}_i\}$ closest to $\mathbf{q}$. This is a classic problem in geometric sorting and searching. One approach is to compute all the $d_i = |\mathbf{q} - \mathbf{p}_i|$ and then find the minimum. This may be the most effective approach if we are required to solve the problem only once. It is frequently the case, however, that we must repeat the solution for a fixed set $\{\mathbf{p}_i\}$ and many different points $\mathbf{q}$. Now, it is more efficient to compute some sort of structure for the points $\{\mathbf{p}_i\}$ that will facilitate our search.

The structure we choose is the **Voronoi tesselation** (also known as a Dirichlet tesselation); see Fig. 6.9. Its characteristics are such that each point $\mathbf{p}_i$ is surrounded by a convex polygon. All points within this polygon are closer to $\mathbf{p}_i$ than any other points in the set $\{\mathbf{p}_i\}$. This means that, given the Voronoi tesselation of a set of points $\{\mathbf{p}_i\}$ and a point $\mathbf{q}$, we determine the nearest neighboring point to it by finding the polygon in which $\mathbf{q}$ lies.

A Voronoi polygon is defined as

$$V_i = \bigcap_{j \neq i} H(\mathbf{p}_i, \mathbf{p}_j) \tag{6.24}$$

where $H(\mathbf{p}_i, \mathbf{p}_j)$ is the half-plane containing $\mathbf{p}_i$ and bounded by the line that is the perpendicular bisector of the line joining $\mathbf{p}_i$ and $\mathbf{p}_j$. This suggests a direct but somewhat cumbersome algorithm to compute the Voronoi structure: For each of the n points $\mathbf{p}_i$, find the intersection of the $(n-1)$ half-planes. Such an algorithm can be expected to run in a time period proportional to $(n^2 \log n)$ time. Other, more efficient algorithms are available and discussed in the literature.

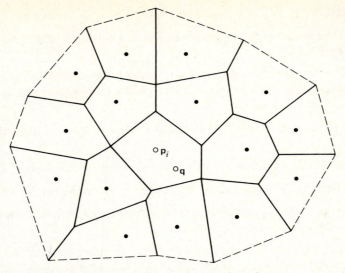

FIGURE 6.9 Nearest neighbor geometry of a point set.

We must now find an algorithm to determine which polygon surrounds **q**. One way is sorting the polygon vertices in ascending order of their y coordinates, thereby partitioning the plane into bands or zones. We easily determine the zone in which **q** lies by a simple search procedure. This dramatically reduces the number of candidate polygons, which we, in turn, can sort and search on an ordering of x coordinates. The result of these two searches produces either a unique solution or reduces the number of candidate polygons to two, with one obvious final test to be made.

EXERCISES

1. How many vertices and edges does a Voronoi tesselation of n points have?

2. Discuss the characteristics of nearest neighbor tesselation procedures for three-dimensional arrays of points (that is, nonplanar).

7

INTERSECTIONS

One of the most important computational tasks of modeling is determining whether or not two geometric objects intersect. If the objects do intersect, we must be able to mathematically describe the intersection. As the number and complexity of objects increases, the need for computational subtlety and complexity increases.

Chapter 7 begins with the simple problem of finding the points of intersection between a straight line and other geometric objects and concludes with the more subtle and complex problem of finding the curve of intersection between two surfaces.

1 INTERSECTIONS WITH STRAIGHT LINES

Two straight-line segments in three-dimensional space may intersect in two different ways if they intersect at all. They may intersect at a single point, as at $\mathbf{r}$ in Fig. 7.1, or they may be colinear (that is, overlap either partially or completely).

Let one line be given by $\mathbf{p}(u) = \mathbf{a} + u\mathbf{b}$ and the other by $\mathbf{q}(w) = \mathbf{c} + w\mathbf{d}$, where $u, w \in [0, 1]$. At their point of intersection $\mathbf{p}(u) = \mathbf{q}(w) = \mathbf{r}$, or

$$\mathbf{a} + u\mathbf{b} = \mathbf{c} + w\mathbf{d} \tag{7.1}$$

Thus, we have three linear equations, one for each component or coordinate direction, and two unknowns, u and w. The system is clearly overconstrained. Solve for u and w using two of the equations, say, $a_x + ub_x = c_x + wd_x$ and $a_y + ub_y = c_y + wd_y$. Then, use the third equation to verify the results. Finally, use either member of the intersection-solution pairs (u_i, w_i) to find $\mathbf{r}$; for example, $\mathbf{r} = \mathbf{a} + u_i\mathbf{b}$.

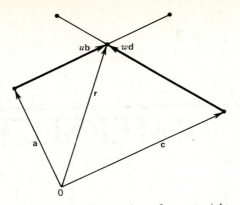

FIGURE 7.1 Intersection of two straight lines.

If we use some fundamental properties of vectors, we can compute u and w directly as follows: To solve Eq. 7.1 for u, take the indicated vector and scalar products

$$(\mathbf{c} \times \mathbf{d}) \cdot (\mathbf{a} + u\mathbf{b}) = (\mathbf{c} \times \mathbf{d}) \cdot (\mathbf{c} + w\mathbf{d}) \tag{7.2}$$

Since $(\mathbf{c} \times \mathbf{d})$ is perpendicular to both $\mathbf{c}$ and $\mathbf{d}$, the right side of Eq. 7.2 becomes zero, and we obtain for u

$$u = -\frac{(\mathbf{c} \times \mathbf{d}) \cdot \mathbf{a}}{(\mathbf{c} \times \mathbf{d}) \cdot \mathbf{b}} \tag{7.3}$$

A similar approach yields w

$$w = -\frac{(\mathbf{a} \times \mathbf{b}) \cdot \mathbf{c}}{(\mathbf{a} \times \mathbf{b}) \cdot \mathbf{d}} \tag{7.4}$$

Now let us consider the intersection of a straight line with an unbounded plane; Fig. 7.2 illustrates this problem. Denote points on the plane by $\mathbf{p}(u, w) = \mathbf{a} + u\mathbf{b} + w\mathbf{c}$, points on the line by $\mathbf{q}(t) = \mathbf{d} + t\mathbf{e}$, and the point of intersection by $\mathbf{r}$. A line segment may intersect a plane at one point, lie in the plane, or not intersect the plane. If they intersect, then $\mathbf{p}(u, w) = \mathbf{q}(t) = \mathbf{r}$, or

$$\mathbf{a} + u\mathbf{b} + w\mathbf{c} = \mathbf{d} + t\mathbf{e} \tag{7.5a}$$

Here, we have three linear equations and three unknowns. Again, we use some fundamental vector properties to isolate, in turn, u, w, and t. For t, we apply $\mathbf{b} \times \mathbf{c}$ as follows:

$$(\mathbf{b} \times \mathbf{c}) \cdot (\mathbf{a} + u\mathbf{b} + w\mathbf{c}) = (\mathbf{b} \times \mathbf{c}) \cdot (\mathbf{d} + t\mathbf{e}) \tag{7.5b}$$

Since $(\mathbf{b} \times \mathbf{c})$ is perpendicular to both $\mathbf{b}$ and $\mathbf{c}$, we obtain

$$(\mathbf{b} \times \mathbf{c}) \cdot \mathbf{a} = (\mathbf{b} \times \mathbf{c}) \cdot \mathbf{d} + (t\mathbf{b} \times \mathbf{c}) \cdot \mathbf{e} \tag{7.6}$$

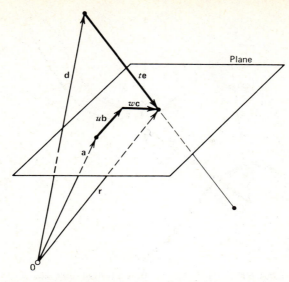

FIGURE 7.2 Intersection of a straight line and a plane.

which we readily solve for t

$$t = \frac{(\mathbf{b} \times \mathbf{c}) \cdot \mathbf{a} - (\mathbf{b} \times \mathbf{c}) \cdot \mathbf{d}}{(\mathbf{b} \times \mathbf{c}) \cdot \mathbf{e}} \qquad (7.7)$$

We find similar expressions for u and w

$$u = \frac{(\mathbf{c} \times \mathbf{e}) \cdot \mathbf{d} - (\mathbf{c} \times \mathbf{e}) \cdot \mathbf{a}}{(\mathbf{c} \times \mathbf{e}) \cdot \mathbf{b}} \qquad (7.8)$$

$$w = \frac{(\mathbf{b} \times \mathbf{e}) \cdot \mathbf{d} - (\mathbf{b} \times \mathbf{e}) \cdot \mathbf{a}}{(\mathbf{b} \times \mathbf{e}) \cdot \mathbf{c}} \qquad (7.9)$$

A straight line may intersect a curve, composite curve, or closed curve (see Fig. 7.3) at more than one point. Denote the straight-line segment by $\mathbf{q}(t) = \mathbf{a} + t\mathbf{b}$ and the curve by $\mathbf{p}(u)$, which may be a cubic function, a Bezier or *B*-spline curve, and so on. Then, an intersection occurs anywhere $\mathbf{p}(u) = \mathbf{q}(t) = \mathbf{r}$, or

$$\mathbf{p}(u) - \mathbf{q}(t) = 0 \qquad (7.10)$$

If $\mathbf{p}(u)$ is a composite curve comprised of piecewise parametric cubic segments, we obtain three nonlinear equations in two unknowns for each segment, an overconstrained system. For any segment, Eq. 7.10 takes the general form

$$\mathbf{c}u^3 + \mathbf{d}u^2 + \mathbf{e}u + \mathbf{f} = \mathbf{a} + \mathbf{b}t \qquad (7.11)$$

Again, some simple vector arithmetic lets us isolate u

$$(\mathbf{a} \times \mathbf{b}) \cdot (\mathbf{c}u^3 + \mathbf{d}u^2 + \mathbf{e}u + \mathbf{f}) = 0 \qquad (7.12)$$

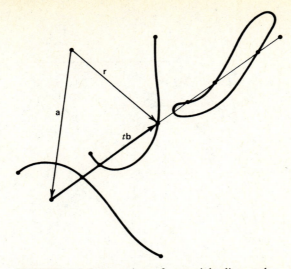

FIGURE 7.3 Intersection of a straight line and a curve.

The roots of Eq. 7.12 are found by numerical analysis methods. Notice that we cannot isolate t, since that would cause a loss of information because u is a cubic function with three roots.

Finally, consider the intersection of a straight-line segment with a surface, as Fig. 7.4 illustrates. Let points on the surface be $\mathbf{p}(u, w)$ and on the straight-line segment, $\mathbf{q}(t) = \mathbf{a} + \mathbf{b}t$. Points of intersection occur when simultaneous sets of u, w, t satisfy

$$\mathbf{p}(u, w) - \mathbf{q}(t) = 0 \tag{7.13}$$

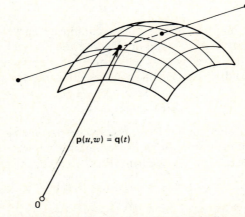

FIGURE 7.4 Intersection of a straight line and a surface.

which represents three simultaneous, nonlinear equations in three unknowns. Once again, the solution requires numerical analysis methods. Of course, we must also anticipate multiple intersections or pathological conditions. How is the solution simplified by rotating and translating the line and surface together so that the line is colinear with the z-axis?

EXERCISES

1. Find the intersections, if any, between the following pairs of straight lines, where $\mathbf{p}(u) = \mathbf{a} + u\mathbf{b}$, and $\mathbf{q}(w) = \mathbf{c} + w\mathbf{d}$, $u, w \in [0, 1]$:

(a) $\mathbf{a} = [2 \quad 2 \quad 4], \mathbf{b} = [2 \quad 1 \quad 2]$
$\mathbf{c} = [-1 \quad 3 \quad 2], \mathbf{d} = [2 \quad 1 \quad 2]$

(b) $\mathbf{a} = [2 \quad 2 \quad 4], \mathbf{b} = [2 \quad 1 \quad 2]$
$\mathbf{c} = [3 \quad 2.5 \quad 5], \mathbf{d} = [2 \quad 1 \quad 2]$

(c) $\mathbf{a} = [2 \quad 2 \quad 4], \mathbf{b} = [2 \quad 1 \quad 2]$
$\mathbf{c} = [6 \quad 1 \quad 0], \mathbf{d} = [-3 \quad 7 \quad 2]$

(d) $\mathbf{a} = [6 \quad -4 \quad -4], \mathbf{b} = [0 \quad 5 \quad 5]$
$\mathbf{c} = [5 \quad 0 \quad 3], \mathbf{d} = [4 \quad 0 \quad -12]$

2. Find the intersections, if any, between each of the eight straight-line segments defined in Ex. 1 and the unbounded plane that passes through the points $(7, 0, 0)$, $(0, 4, 0)$, and $(0, 0, 7)$.

3. Find the intersection between the straight line from point $(6, 5, 8)$ to point $(0, 0, -8)$ and the Bezier surface whose control points are

$$\mathbf{B} = \begin{bmatrix} (-4, 8, 0) & (0, 8, 0) & (2, 8, 0) & (6, 8, 0) \\ (-4, 4, 0) & (0, 4, 4) & (2, 4, 4) & (6, 4, 0) \\ (-4, 2, 0) & (0, 2, 4) & (2, 2, 4) & (6, 2, 0) \\ (-4, 0, 0) & (0, 0, 4) & (2, 0, 4) & (6, 0, 0) \\ (-4, -2, 0) & (0, -2, 4) & (2, -2, 4) & (6, -2, 0) \\ (-4, -6, 0) & (0, -6, 0) & (2, -6, 0) & (6, -6, 0) \end{bmatrix}$$

4. Given the vector equation of a straight-line segment $\mathbf{p}(u) = \mathbf{a} + u\mathbf{b}$, with $u \in [0, 1]$, find $\mathbf{a}$ and $\mathbf{b}$ in terms of the end points $\mathbf{p}(0)$ and $\mathbf{p}(1)$.

5. Write a procedure to compute the intersection between two straight lines. Denote this as **LNLN(LN1, LN2, NINT, P)**, where

LN1(2, 3), LN2(2, 3) are input arrays of the coordinates of the end points of the two lines;

NINT is the output number of intersections computed. Note that for straight lines, NINT = 0 or 1; and

P(3) is the output array of x, y, z coordinates of the point of intersection.

6. Write a procedure to compute the intersection between a straight line and a pc curve. Denote this as **LNCRV(LN, CRV, NINT, P)**, where

> LN(2, 3) is the input array of coordinates of the line;
>
> CRV(4, 3) is the input array of geometric coefficients defining the pc curve;
>
> NINT is the output number of intersections computed; and
>
> P(NINT, 4) is the output array of x, y, z, u coordinates of the points of intersection.

7. Write a procedure to compute the intersection between a straight line and a bicubic patch. Denote this as **LNSRF(LN, SRF, NINT, P)**, where

> LN(2, 3) is the input array of coordinates of the end points of the line;
>
> SRF(4, 4, 3) is the input array of geometric coefficients defining the bicubic patch;
>
> NINT is the output number of intersections computed; and
>
> P(NINT, 5) is the output array of x, y, z, u, w coordinates of the points of intersection.

8. Describe the approach you would take to generalize **LNCRV** and **LNSRF** so that they also applied to Bezier and B-spline curves and surfaces.

2 PLANE INTERSECTIONS

Computing plane intersections with geometric objects is a key capability of CAD/CAM systems and many other geometric-modeling systems. We can obtain most conventional cross sections of parts by computing the appropriate plane intersections.

To begin our study of plane intersections, consider the intersection of two planes $\mathbf{p}(u, w)$ and $\mathbf{q}(s, t)$; see Fig. 7.5. Assume for the moment that the planes

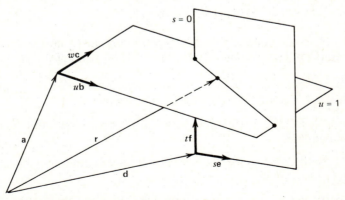

FIGURE 7.5 Intersection of two planes.

are bounded by $u, w, s, t \in [0, 1]$. Furthermore, note that the expected intersection is a straight line. This is revealed mathematically by observing that there are three equations from the intersection equation $\mathbf{p}(u, w) - \mathbf{q}(s, t) = 0$ and four unknowns or variables. The extra degree of freedom implies that the intersection is a curve (a straight line in the case at hand), and so we must introduce an additional constraint (see below). The expanded intersection equation looks like

$$\mathbf{a} + \mathbf{b}u + \mathbf{c}w = \mathbf{d} + \mathbf{e}s + \mathbf{f}t \qquad (7.14)$$

We may now arbitrarily fix as constant one of the variables, either $u, w, s,$ or t. This furnishes an additional equation or constraint (for example, $u = u_i$, where u_i is a constant). Then, we employ techniques from Section 1 to obtain one point on the intersection. By incrementing u_i by some Δu, we find a series of points sufficient to define the intersection.

When the two intersecting planes are bounded, as just described, we can expedite the solution by recognizing first that the line of intersection is bounded and second that the end points of this line each lie on a boundary line of one of the planes (for example, $u = 0$, $u = 1$, and so on). This gives us a clue about how to proceed with a solution.

If the planes are not parallel, coincident, or nonintersecting, they will intersect at a single point or along a bounded line segment. Since the two end points of the line are sufficient to define it, we proceed to solve Eq. 7.14 for each of the additional constraint equations given by the plane boundaries, $u = 0$, $u = 1$, $w = 0$, $w = 1$, $s = 0$, $s = 1$, $t = 0$, $t = 1$. Once two points are found, terminate the algorithm, because the complete solution will have been found.

If an unbounded plane intersects a parametric curve at a point on the curve $\mathbf{p}(u)$, then the scalar product between the normal vector $k_n\mathbf{n}$ to the plane from the origin and the vector $(\mathbf{p} - k_n\mathbf{n})$ lying in the plane must equal zero; see Fig. 7.6. We express this as

$$(\mathbf{p} - k_n\mathbf{n}) \cdot k_n\mathbf{n} = 0 \qquad (7.15)$$

Vector arithmetic yields a single nonlinear equation. If the curve is defined in the interval $u \in [0, 1]$, then only roots in this interval represent valid intersections. Thus, we may find one, two, or three points of intersection for a cubic curve, or we may find none.

Finally, consider the intersection between an unbounded plane and a surface $\mathbf{p}(u, w)$ as Fig. 7.7 shows. Since we already know that a plane is also a two-parameter geometric element, we expect to derive three equations having four degrees of freedom. The extra degree of freedom manifests itself as a curve of intersection; however, since we do not require the values of the parametric variables defining the plane, we define the plane as in Eq. 7.5; that is, by the normal vector $k_n\mathbf{n}$ to the plane from the origin, where $\mathbf{n}$ is the unit normal vector. Then, beginning with the surface boundary curves, we search for intersection points by intersecting these curves and other carefully selected isoparametric curves with the plane, as in Eq. 7.15.

If the plane and object to be intersected are rotated and translated so that the plane is coincident with the $z = 0$ plane, how is the solution simplified?

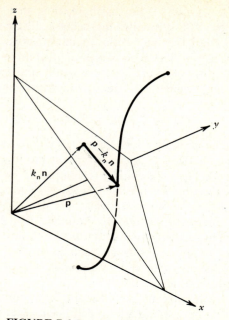

FIGURE 7.6 Intersection of a plane and a curve.

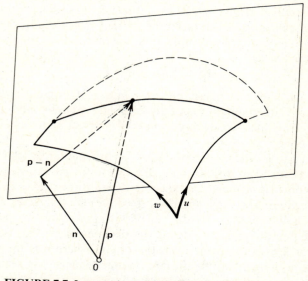

FIGURE 7.7 Intersection of a plane and a surface.

EXERCISES

1. Use Eq. 7.14 to find all the intersections between the three planes whose vector equations are of the form $\mathbf{p}_i(u, w) = \mathbf{a}_i + \mathbf{b}_i u + \mathbf{c}_i w$, with $u, w \in [0, 1]$ and

$$\mathbf{a}_1 = [0 \quad 0 \quad 0] \qquad \mathbf{b}_1 = [10 \quad 8 \quad 0] \qquad \mathbf{c}_1 = [0 \quad 0 \quad 10]$$

$$\mathbf{a}_2 = [-4 \quad 4 \quad 6] \qquad \mathbf{b}_2 = [12 \quad 0 \quad 0] \qquad \mathbf{c}_2 = [4 \quad -4 \quad -6]$$

and

$$\mathbf{a}_3 = [8 \quad -2 \quad 4] \qquad \mathbf{b}_3 = [0 \quad 10 \quad 0] \qquad \mathbf{c}_3 = [-10 \quad 0 \quad 0]$$

2. Find the point of intersection between the plane $\mathbf{p}_1(u, w)$ defined in Ex. 1 and the Bezier curve whose control points are

$$\mathbf{p}_0 = [0 \quad 10 \quad 12]$$

$$\mathbf{p}_1 = [1 \quad 8 \quad 8]$$

$$\mathbf{p}_2 = [3 \quad 4 \quad 8]$$

$$\mathbf{p}_3 = [10 \quad 0 \quad 2]$$

and

$$\mathbf{p}_4 = [6 \quad -4 \quad 2]$$

3. Find the point of intersection between the plane $\mathbf{p}_2(u, w)$ defined in Ex. 1 and the Bezier surface defined in Ex. 3 of Section 1.

4. Write a procedure to compute the intersection between a plane and a straight line. Denote this as **PLNLN(PLN, LN, NINT, P)**, where

PLN(4) is the input array of plane coefficients A, B, C, D;

LN(2, 3) is the input array of coordinates of the end points of the straight line;

NINT is the output number of points of intersection. If NS = 0, there is no intersection; if NS = -1, the line lies in the plane.

P(3) is the output array of x, y, z coordinates of the point of intersection if it exists.

5. Write a procedure to compute the line of intersection between two planes. Denote this as **PLNPLN(PLN1, PLN2, NINT, PA, PB)**, where

PLN1(4), PLN2(4) are the input arrays of plane coefficients;

NINT is the output number of intersections. If the planes are parallel, set NINT = 0; if coincident, set NINT = -1.

PA(3) is the output array of x, y, z coordinates of a point through which the line of intersection passes. Select some appropriate criteria for determining this point. For example, it can be the point of intersection of the line with one of the principal coordinate planes (*Caution*: The line can be parallel to any two of them.); and

PB(3) is the output array of components of the unit tangent vector in the direction of the line of intersection.

6. Write a procedure to compute the intersection between a plane and a pc curve. Denote this as **PLNPC(PLN, PC, NINT, P)**, where

PLN(4) is the input array of plane coefficients A, B, C, D;

PC(4, 3) is the input array of algebraic coefficients defining the pc curve;

NINT is the output number of intersection points computed, or, if the curve lies in the plane, set NINT $= -1$; and

P(NINT, 4) is the output array of x, y, z, u coordinates of the points of intersection.

7. Write a procedure to compute the intersection between a plane and a Bezier curve. Denote this as **PLNBZC(PLN, N, BZ, NINT, P)**, where

PLN(4) is the input array of plane coefficients A, B, C, D;

N is the input number of control points;

BZ(N, 3) is the input array of x, y, z coordinates of the control points;

NINT is the output number of intersection points computed; and

P(NINT, 4) is the output array of x, y, z, u coordinates of the points of intersection.

8. Write a procedure to compute the intersection between a plane and a non-periodic B-spline curve. Denote this as **PLNBSC(PLN, N, K, BS, NINT, P)**, where

PLN(4) is the input array of plane coefficients A, B, C, D;

N is the input number of control points;

K is the input degree of the curve;

BS(N, 3) is the input array of x, y, z coordinates of the control points;

NINT is the output number of intersection points computed; and

P(NINT, 3) is the output array of x, y, z, u coordinates of the points of intersection.

9. How would you change **PLNBSC** to accommodate periodic B-spline curves?

10. Write a procedure to compute the intersection between a plane and a bicubic patch. Denote this as **PLNBCS(PLN, SRF, NINT, CRV)**, where

PLN(4) is the input array of plane coefficients A, B, C, D;

SRF(4, 4, 3) is the input array of geometric coefficients defining the bicubic patch;

NINT is the output number of pc curve segments used to model the curve(s) of intersection; and

CRV(NINT, 4, 3) is the output array of geometric coefficients of the NINT pc curves defining the curve(s) of intersection.

11. Explain the method you used in **PLNBCS** to determine the number of disjoint curves of intersection that might be found.

12. Write a procedure to compute the intersection between a plane and a Bezier surface. Denote this as **PLNBZS**(PLN, M, N, CI, NSEG, NP, P), where

PLN(4) is the input array of plane coefficients A, B, C, D;

M, N are the input dimensions of the array of control points;

CI(M, N, 3) is the input array of coordinates of the control points defining vertices of the characteristic polyhedron;

NSEG is the output number of disjoint segments of the curve of intersection;

NP(NSEG) is the output array specifying the number of points computed on each segment of the curve of intersection; and

$P(\sum NP, 5)$ is the output array of x, y, z, u, w coordinates of points on the curve of intersection.

13. Explain the method you used in **PLNBZS** to determine the spacing of points computed on the curve of intersection.

14. Write a procedure to compute the intersection between a plane and a periodic *B*-spline surface. Denote this as **PLNBSS**(PLN, M, N, K, L, CI, NSEG, NP, P), where

PLN(4) is the input array of plane coefficients A, B, C, D;

M, N are the input dimensions of the array of control points;

K, L are the input values of the degree of the u and w curves, respectively;

CI(M, N, 3) is the input array of coordinates of the control points;

NSEG is the output number of disjoint segments of the curve of intersection;

NP(NSEG) is the output array specifying the number of points computed on each segment of the curve of intersection; and

$P(\sum NP, 5)$ is the output array of x, y, z, u, w coordinates of points on the curve of intersection.

15. Suggest various ways of fitting curves to the point sets generated by **PLNBZS** and **PLNBSS**.

16. Can **PLNBCS**, **PLNBZS**, and **PLNBSS** be easily generalized to a single procedure? Explain your answer.

3 CURVE INTERSECTIONS

Computing the intersection of two curves in three-dimensional space (see Fig. 7.8) requires solving for roots of the equation $\mathbf{p}(u) - \mathbf{q}(t) = 0$, which usually offers three nonlinear equations in two unknowns. If we suppress one of the component equations, say, for z, we then solve the remaining two simultaneous equations for u and t

$$p_x(u) - q_x(t) = 0$$
$$p_y(u) - q_y(t) = 0$$

(7.16)

The solution is really the intersection of the projections of the two curves onto the x, y plane. Verify it by checking $p_z(u) - q_z(t) = 0$. In many cases, this intersection problem is a strictly two-dimensional one, and so we express it as two equations in two unknowns.

Now, consider the problem of finding the intersection between a curve and a surface. We will investigate a solution technique that is the basis for the hunting phase of Timmer's algorithm for computing surface–surface intersections. Figure 7.9 presents the essential ingredients.

We express the point of intersection as a fully specified system of three equations in three unknowns. Denote the curve by $\mathbf{q}(t)$ and the surface by $\mathbf{p}(u, w)$. Then, express the system of equations as

$$\mathbf{r} = \mathbf{q}(t) - \mathbf{p}(u, w) = 0$$

(7.17)

where $\mathbf{r}$ is the minimum distance between the surface and successive points on the curve.

To begin the numerical solution, compute the first variation of this equation

$$d\mathbf{r} = \mathbf{q}^t\, dt - \mathbf{p}^u\, du - \mathbf{p}^w\, dw$$

(7.18)

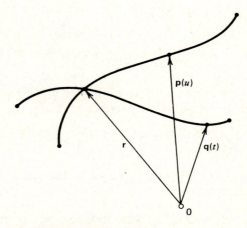

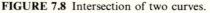

FIGURE 7.8 Intersection of two curves.

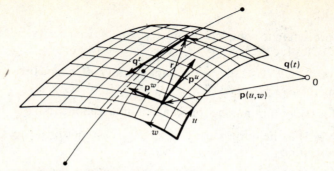

FIGURE 7.9 Intersection of a curve and a surface.

Isolate the variations in the individual parametric variables by imposing appropriate vector and scalar products on Eq. 7.18; for example,

$$\mathbf{p}^u \times d\mathbf{r} = (\mathbf{p}^u \times \mathbf{q}^t)\, dt - (\mathbf{p}^u \times \mathbf{p}^u)\, du - (\mathbf{p}^u \times \mathbf{p}^w)\, dw \qquad (7.19)$$

Since $(\mathbf{p}^u \times \mathbf{p}^u) = 0$, we obtain

$$\mathbf{p}^u \times d\mathbf{r} = (\mathbf{p}^u \times \mathbf{q}^t)\, dt - (\mathbf{p}^u \times \mathbf{p}^w)\, dw \qquad (7.20)$$

Furthermore,

$$\mathbf{q}^t \cdot (\mathbf{p}^u \times d\mathbf{r}) = \mathbf{q}^t \cdot (\mathbf{p}^u \times \mathbf{q}^t)\, dt - \mathbf{q}^t \cdot (\mathbf{p}^u \times \mathbf{p}^w)\, dw \qquad (7.21)$$

Since $\mathbf{q}^t$ is perpendicular to $(\mathbf{p}^u \times \mathbf{q}^t)$, $\mathbf{q}^t \cdot (\mathbf{p}^u \times \mathbf{q}^t) = 0$, and therefore

$$\mathbf{q}^t \cdot (\mathbf{p}^u \times d\mathbf{r}) = -\mathbf{q}^t \cdot (\mathbf{p}^u \times \mathbf{p}^w)\, dw \qquad (7.22)$$

Repeat this method to isolate the variation in the other two parameters, and then construct the following equations, which support the iteration process:

$$t_{i+1} = t_i - \frac{\mathbf{p}_i^u \cdot (\mathbf{p}_i^w \times \mathbf{r}_i)}{D}$$

$$u_{i+1} = u_i + \frac{\mathbf{p}_i^w \cdot (\mathbf{q}_i^t \times \mathbf{r}_i)}{D} \qquad (7.23)$$

$$w_{i+1} = w_i + \frac{\mathbf{q}_i^t \cdot (\mathbf{p}_i^u \times \mathbf{r}_i)}{D}$$

where

$$D = \mathbf{q}_i^t \cdot (\mathbf{p}_i^u \times \mathbf{p}_i^w) = \mathbf{q}^t \cdot \mathbf{n}_i \qquad (7.24)$$

and $\mathbf{n}_i$ is the normal to the surface at $\mathbf{p}_i$.

EXERCISES

1. Write a procedure to complete the intersection of two pc curves. Denote this as **CRV2PC**(CRV1, CRV2, NINT, P), where

CRV1(4, 3), CRV2(4, 3) are the input arrays of geometric coefficients defining the two pc curves;

NINT is the output number of points of intersection; and

P(NINT, 5) is the output array of x, y, z, u_1, u_2 coordinates of the points of intersection.

2. Write a procedure to compute the intersection of two Bezier curves. Denote this as **CRV2BZ**(NI, BZ1, N2, BZ2, NINT, P), where

N1, N2 are the input number of control points of each curve;

BZ1(N1, 3), BZ2(N2, 3) are the input coordinates of the control points for each curve;

NINT is the output number of points of intersection; and

P(NINT, 5) is the output array of x, y, z, u_1, u_2 coordinates of the points of intersection.

3. Write a procedure to compute the intersection of two B-spline curves. Denote this as **CRV2BS**(N1, K1, BS1, N2, K2, BS2, NINT, P), where

N1, N2 are the input number of control points of each curve;

K1, K2 are the input degrees of each curve;

BS1(N1, 3), BS2(N2, 3) are the input coordinates of the control points for each curve;

NINT is the output number of points of intersection; and

P(NINT, 5) is the output array of x, y, z, u_1, u_2 coordinates of the points of intersection.

4. Discuss the problems you would expect to encounter designing a procedure to compute the intersection of any combination of curve types—pc, Bezier, and B-spline.

5. Write a procedure to compute the intersection of a pc curve and a bicubic patch. Denote this as **CRVSRF**(CRV, SRF, NINT, P), where

CRV(4, 3) is the input array of geometric coefficients defining the pc curve;

SRF(4, 4, 3) is the input array of geometric coefficients defining the bicubic patch;

NINT is the output number of points of intersection; and

P(NINT, 6) is the output array of x, y, z, u_c, u_s, w_s coordinates of the points of intersection.

6. Discuss the problems you would expect to encounter designing a procedure to compute the intersection of any combination of curve and surface types — pc or bicubic, Bezier, and *B*-spline.

4 SURFACE INTERSECTIONS

In this section, we will describe a method developed by Timmer (1977) for computing the intersection between two surfaces. There are many other approaches, but this one serves as well as any to illustrate the problem and the sophistication required of an algorithm to solve surface-intersection problems. Because the problem is so difficult and the required analysis somewhat complicated, we begin with a definition of terms and an overview of the strategy we will use.

A surface $\mathbf{p}(u, w)$ is a rectangular mesh of bicubic patches that represent some surface in object space, that is, a physical surface. Figure 7.10 represents a surface defined by a 4 × 5 mesh of patches. A patch-numbering convention is indicated, an important consideration because patches must be stored and retrieved in some logical order. Furthermore, the numbering convention establishes topological connectivity between the individual patches. Another convention, not shown in this figure, is that the vector product $\mathbf{p}^u \times \mathbf{p}^w$ produces an outward surface normal.

If we consider finding the intersection between two surfaces $\mathbf{p}_1(u, w)$ and $\mathbf{p}_2(s, t)$, Timmer suggests that we decompose the procedure into three distinct phases: the hunting phase, the tracing phase, and the sorting phase. He observes

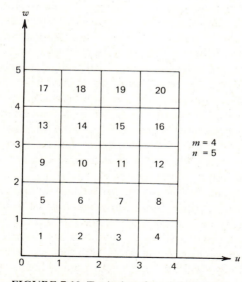

FIGURE 7.10 Typical surface defined by an $m \times n$ mesh of patches.

that this over-all strategy is necessitated by the requirement that the computations be capable of handling relatively complex surfaces. The hunting phase locates discrete starting points required for the curve-tracing operation. The tracing phase creates strings of points lying on the intersection. Finally, the sorting phase orders the point strings and separates them into disjoint segments or loops.

Most of the mathematics for the solution of the surface-intersection problem involves solving three simultaneous equations, one for each of the Cartesian coordinates or vector components. The distinguishing characteristic of this problem is the number of unknown variables. The general surface–surface-intersection problem exhibits four unknown variables, the parametric coordinates of each surface (u, w) and (s, t). We expect the general solution to be a space curve, which implies that we determine these four unknowns as a function of a single new parameter. Compare this problem to that of determining the point on a space curve closest to another given point in space (see Chapter 6, Section 2). The latter problem involves only one unknown, the parametric coordinate of the point on the curve.

4.1 The Hunting Phase

The objective of the hunting phase is to locate a number of points on the intersection to serve as starting points for the tracing phase. We must find at least one point on each isolated loop of the total intersection during the hunting phase, because if a loop is undetected, we will not get a complete solution. Thus, we need a flexible procedure that can adjust to the most complex of problems. We can obtain this by imposing a hunting grid on what we arbitrarily designate as the primary surface, say, $\mathbf{p}_1(u, w)$. Figure 7.11 illustrates an example of a primary surface with a hunting grid superimposed; a single integer n_G is sufficient to define the grid. The intersection shown in the figure consists of three loops. Notice that a smaller value of n_G might have been used without missing any of the loops. However, if a sufficiently small value of n_G is chosen (for example, $n_G = 3$), the loop in the upper left corner escapes detection. The rectangular regions defined by the hunting grid are numbered in the same way as the surface patches (see Fig. 7.10) but, of course, do not necessarily coincide.

In the hunting phase, we compute the possible intersections of each curve defined by the grid with $\mathbf{p}_2(s, t)$. By first taking curves of constant u and then curves of constant w, the intersection points are computed sequentially, as indicated by their numbers in the figure.

At this point, we again address the problem of computing the intersection between a curve and a surface. Specifically, we must compute the intersection between a curve from the hunting grid on $\mathbf{p}_1(u, w)$ and the surface $\mathbf{p}_2(s, t)$. Our major concern is ensuring that we do not miss any intersection points. Any scheme we devise must isolate the roots and then determine them. Timmer suggests a method for isolating intersection points that is a refinement of the hunting-grid philosophy. Thus, the curve is segmented by selecting an integer m_G whose role is analogous to that of n_G in the hunting grid. m_G must be suffi-

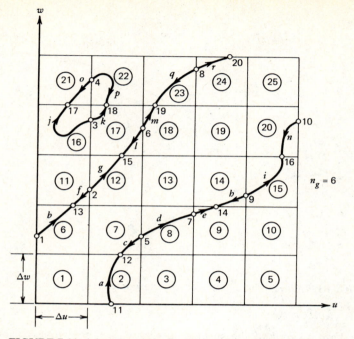

FIGURE 7.11 A hunting grid superimposed on a primary surface with $n_g = 6$.

ciently large (that is, the segments small enough) so that two intersection points do not occur within the same segment. We must, therefore, choose a function that identifies the presence of an intersection point in a simple, easy-to-compute way. Equation 7.25 is a good candidate:

$$\mathbf{r} = \mathbf{p}_c - \mathbf{p}_s \tag{7.25}$$

where $\mathbf{p}_c$ is a point on the curve, $\mathbf{p}_s$ is a point on the surface closest to $\mathbf{p}_c$ (see Chapter 6, Section 3), and $\mathbf{r}$ is the directed distance between these two points (see Fig. 7.12). Let (X, Y, Z) be the components of $\mathbf{r}$. If an intersection with the surface occurs between nodes i and $(i + 1)$ on the curve, then all of the following conditions must hold:

$$X_i X_{i+1} < 0$$
$$Y_i Y_{i+1} < 0 \tag{7.26}$$
$$Z_i Z_{i+1} < 0$$

This is indeed a very simple concept. Since the effectiveness of this algorithm does not depend on the precise determination of s and t so that $|\mathbf{r}|$ is a minimum, we can use a somewhat cruder approximation. Specifying a value of the curve parameter u or w determine a point $\mathbf{p}$. A small number of iterations of the surface-inversion algorithm (Section 4.5) will satisfy the requirement for the closest

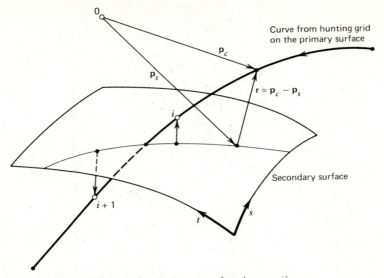

FIGURE 7.12 Hunting phase: curve–surface intersection.

distance from a point to the surface, $\mathbf{p}_c - \mathbf{p}_s = 0$. Once we have isolated an intersection point, that is, when we know the intersection is between node i and $(i + 1)$ on the curve, say,

$$(u \text{ or } w)_i \leq (u \text{ or } w) \leq (u \text{ or } w)_{i+1} \tag{7.27}$$

then the complete three-variable problem $\mathbf{r} = 0$ is solved, subject to the constraint on the curve parameter indicated in Eq. 7.27.

The hunting phase produces a table of intersection points lying on the grid boundaries of the primary surface $\mathbf{p}(u, w)$. The order of the point-value entries in the table depends on the order in which the grid curves are taken, as well as the direction of march along these curves. To demonstrate this, in Fig. 7.11, the grid-intersection points are numbered in the order in which they would appear in the table.

4.2 The Tracing Phase

The hunting phase produces a number of discrete intersection points lying on the boundaries of the hunting grid. We see in Fig. 7.11 that the order of these points must be changed so that they can be properly interpolated to form the complete intersection. Timmer suggests the following procedure for tracing the intersection curves across the interior regions of the hunting grid.

Analyze the regions one at a time. Notice that the region numbering or ordering is logically similar to the patch-numbering scheme. First, compare the region boundaries to the tabulated points. If a match is found, trace the portion of the particular loop containing that boundary point and lying within

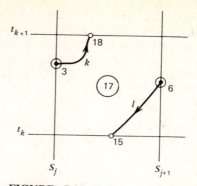

FIGURE 7.13 Curves within a hunting-grid region.

that region; Fig. 7.13 illustrates this. Here, point 3 is found on a boundary of region 17 and is traced to its terminus at point 18 on another boundary of the region. The direction of the trace across a region is established by the point occurring first in the table. The ending point is excluded from the rest of the search for the region. This prevents retracing the same segment and ensures that regions containing more than one segment are properly processed.

At the starting point of a loop segment through a region, obtain the tangent vector to the segment (that is, intersection curve) by taking the vector product between the normals to the surfaces (see Fig. 7.14). Establish the order of the vector product so that the tangent vector points toward the interior of the region through which the intersection curve is to be traced. This tangent vector provides

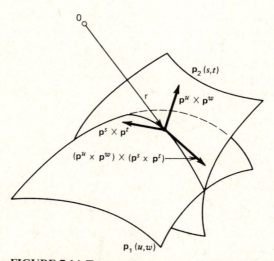

FIGURE 7.14 Tangent vector to the intersection curve between two surfaces.

the derivatives of the four parameters u, w, s, and t with respect to arc length σ along the intersection curve, as will be explained in Section 4.4. These parametric variables are incremented using a step size based on curvature and its relation to slope change along the intersection curve; for example,

$$u_{i+1} = u_i + \left(\frac{du}{d\sigma}\right)_i \Delta\sigma_i \tag{7.28}$$

This step places the next point in the neighborhood of the intersection curve, and we require a secondary iterative convergence to achieve a designated tolerance. Because there are four unknowns and only three equations, hold one of the unknowns constant, making the system fully determinate. A solution is then obtained using Newton's method. Timmer continues this two-level marching until reaching a region boundary. Tracing continues within a region for any other tabulated points on its boundaries, and so on for successive regions until this process is exhausted on the last region, ending the tracing phase. Notice that the arrows in Fig. 7.11 indicate the direction of tracing and the order of point tabulation.

4.3   The Ordering Phase

The tracing phase generates a table of seemingly disjointed point strings. These segments are labeled in Fig. 7.11 as they might have been computed, in an order reflecting the order of the hunting-grid-region numbering scheme. This ordering in no way reflects the connectivity of the strings forming the loops. We must now devise a scheme to reorder the strings to establish their proper connectivity. Timmer observes that while this scheme might be regarded as mere bookkeeping, it is significant enough to justify further discussion.

The result of the first two phases is a collection of n_G point strings ($n_G = 18$ in Fig. 7.11). For any sorting operation, only the first and last points in each string are important. Beginning with the first string, compare the beginning and end points to the remaining strings. When a match is found, save that string number in an ordering array as a signed integer. A minus sign indicates a reversal in the order of the string points when they are joined with others to form a loop. For our example in Fig. 7.11, the ordering arrays are

Loop 1 $a, -c, d, e, -h, i, -n$

Loop 2 $b, -f, g, -l, m, -q, r$

Loop 3 $-k, j, -o, p$

Loop 3 requires further explanation: When Loop 2 is completed, only strings j, k, o, and p remain. The scan to order Loop 3 begins with j, which has boundary point 3 as its beginning and point 17 as its end. Next, the scan proceeds to string k, and a match is found with the starting point of Loop 3 (at this point, consisting only of string j). Since the direction of Loop 3 has already been established by j, k must be reversed and made to precede j in the ordering table. Now, the loop has point 18 as its beginning and 17 as its end. Next, o matches

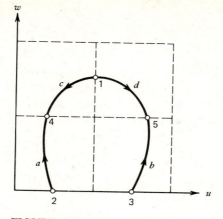

FIGURE 7.15 Ordering phase: an example.

j at point 17, but its direction must be reversed. Finally, p closes the loop, and no strings remain, so the ordering phase is complete.

Figure 7.15 offers a final example of the ordering phase. The loop ordering starts at string a and runs from point 2 to point 4. When we compare string b to the loop (so far containing just a), we find that it is disjoint and discard it. Thus, completing the first sorting pass through the table of point strings, the ordering array for the loop is a, $-c$, d. This failure is a result of the search order and is easily remedied by performing a second sorting pass, which would correctly append $-b$ to the ordering array.

4.4 Computation of Parametric Derivatives

Recall that in Section 4.2, we used the vector product of the two surface normals at a point on the intersection curve to produce a vector tangent to that curve. We then used this tangent vector to obtain the derivatives of the parametric coordinates with respect to arc length along the intersection curve. Let this tangent vector $\mathbf{T}$ be given by

$$\mathbf{T} = (\mathbf{p}^u \times \mathbf{p}^w)_1 \times (\mathbf{p}^s \times \mathbf{p}^t)_2 \tag{7.29}$$

and

$$\tau = \frac{\mathbf{T}}{|\mathbf{T}|} \tag{7.30}$$

The unit vector τ satisfies the equation

$$\tau = \frac{d\mathbf{p}}{d\sigma} \quad \text{or} \quad \mathbf{p}^{\sigma} \tag{7.31}$$

For $\mathbf{p}_1(u, w)$, this becomes

$$\tau = \mathbf{p}_1^u \frac{du}{d\sigma} + \mathbf{p}_1^w \frac{dw}{d\sigma} \tag{7.32}$$

We obtain a similar expression for $\mathbf{p}_2(s, t)$. Each of these vector equations represents a set of three nonlinear equations in two unknowns. Therefore, we expect one degree of freedom in their solution, meaning that any set of two equations is sufficient. We isolate $du/d\sigma$ and $dw/d\sigma$ by taking the appropriate vector products on both sides of Eq. 7.32, which yields

$$\frac{du}{d\sigma} = \frac{\mathbf{p}^w \times \tau}{\mathbf{p}^u \times \mathbf{p}^w}$$

$$\frac{dw}{d\sigma} = \frac{\mathbf{p}^u \times \tau}{\mathbf{p}^u \times \mathbf{p}^w} \tag{7.33}$$

How are we to interpret the quotient of two vectors as a scalar? We evaluate the equations by selecting as constant any one of the three components, for example,

$$\frac{du}{d\sigma} = -\frac{(\partial y/\partial w)(\partial z/\partial \sigma) - (\partial z/\partial w)(\partial y/\partial \sigma)}{(\partial y/\partial u)(\partial z/\partial w) - (\partial y/\partial w)(\partial z/\partial u)} \tag{7.34}$$

Here, the x component was selected.

4.5   Surface Inversion

In Chapter 5, Section 4, we investigated the inverse point solution for curves and the more general n-dimensional case. For the surface–surface-intersection problem, we require the inverse point solution for a surface. Geometrically, this is the equivalent of finding a point on a surface closest to a given point in space, with the distance equal to zero (see Eq. 7.25). Like the problem of computing the parametric derivatives just discussed, this problem is overspecified, since there are three equations and only two unknowns. The surface-inversion problem is clearly expressed by the following vector equation:

$$\mathbf{r} = \mathbf{p}_1(u, w) - \mathbf{p}_c \tag{7.35}$$

which we solve for $\mathbf{r} = 0$; thus,

$$\mathbf{p}_1(u, w) - \mathbf{p}_c = 0 \tag{7.36}$$

Because we cannot directly invert these equations analytically, we need an iteration procedure. Compute the first variation of Eq. 7.35 as

$$d\mathbf{r} = \mathbf{p}_1^u \, du + \mathbf{p}_1^w \, dw \tag{7.37}$$

Assume an initial guess u_i, w_i so that $\mathbf{r}(u_i, w_i) = \mathbf{r}_i$. Then, interpret the variations in Eq. 7.37 as

$$\mathbf{r}_{i+1} - \mathbf{r}_i = (\mathbf{p}_1^u)_i(u_{i+1} - u_i) + (\mathbf{p}_1^w)_i(w_{i+1} - w_i) \tag{7.38}$$

If we set $r_{i+1} = 0$ and take the appropriate vector products, we obtain

$$u_{i+1} = u_i - \frac{(\mathbf{p}_1^w \times \mathbf{r})_i}{(\mathbf{n}_1)_i}$$

$$\text{(7.39)}$$

$$w_{i+1} = w_i - \frac{(\mathbf{p}_1^u \times \mathbf{r})_i}{(\mathbf{n}_1)_i}$$

where $\mathbf{n}_i = (\mathbf{p}_1^u \times \mathbf{p}_1^w)_i$ and the quotient of the two vectors is interpreted as in Eq. 7.33. If $\mathbf{p}_c$ lies on the surface, then convergence can be attained to within any given tolerance. If $\mathbf{p}_c$ does not lie on the surface, however, the solution tends to converge to a point on the surface closest to $\mathbf{p}_c$. In this event, the iterative process cut-off criterion can be as simple as executing no more than a specified maximum number of iterations, or perhaps satisfying

$$|\mathbf{r}_{i+1} - \mathbf{r}_i| \le \varepsilon \qquad \text{(7.40)}$$

4.6   Surface–Surface Intersection

In Section 4.2, it was implied that final convergence in the tracing phase involved an underspecified system, namely, three equations in four unknowns

$$\mathbf{r} = \mathbf{p}_1(u, w) - \mathbf{p}_2(s, t) \qquad \text{(7.41)}$$

with $\mathbf{r} = 0$, so that

$$\mathbf{p}_1(u, w) - \mathbf{p}_2(s, t) = 0 \qquad \text{(7.42)}$$

We found we could resolve this problem by holding one of the unknowns constant. This reduces the problem to the curve–surface-intersection problem discussed in Section 3. The possible criteria available to determine which variable to fix is of some interest. Perhaps the most obvious criterion is fixing the parametric variable having the lowest value of the derivative with respect to arc length at the last converged point.

Another approach evolves from the similarity between the overspecified and the underspecified problems. In both cases, we must make a decision before proceeding with the computation. With an overspecified system of equations, the decision is based on maximizing the magnitude of the denominator (see Eq. 7.23). Any decision in the underspecified problem reduces it to a fully specified system of equations whose solution also involves a denominator, and the same criterion is applicable. Thus, that variable is fixed that produces the maximum denominator in the resulting fully specified condition. Clearly, this criterion is essentially an intuitive one. The only obvious advantage is that we avoid potential numerical problems with a very small value in the denominator.

The resulting four sets of equations for $u = $ constant are

$$u_{i+1} = u_i$$

$$w_{i+1} = w_i = \frac{[\mathbf{p}_2^s \cdot (\mathbf{p}_2^t \times \mathbf{r})]_i}{D_u}$$

$$s_{i+1} = s_i + \frac{[\mathbf{p}_2^t \cdot (\mathbf{p}_1^w \times \mathbf{r})]_i}{D_u}$$

$$t_{i+1} = t_i + \frac{[\mathbf{p}_1^w \cdot (\mathbf{p}_2^s \times \mathbf{r})]_i}{D_u} \qquad (7.43)$$

where

$$D_u = [\mathbf{p}_1^w \cdot (\mathbf{p}_2^s \times \mathbf{p}_2^t)]_i$$

For $w = $ constant,

$$u_{i+1} = u_i - \frac{[\mathbf{p}_2^s \cdot (\mathbf{p}_2^t \times \mathbf{r})]_i}{D_w}$$

$$w_{i+1} = w_i$$

$$s_{i+1} = s_i + \frac{[\mathbf{p}_2^t \cdot (\mathbf{p}_1^u \times \mathbf{r})]_i}{D_w}$$

$$t_{i+1} = t_i + \frac{[\mathbf{p}_1^u \cdot (\mathbf{p}_2^s \times \mathbf{r})]_i}{D_w} \qquad (7.44)$$

where

$$D_w = [\mathbf{p}_1^u \cdot (\mathbf{p}_2^s \times \mathbf{p}_2^t)]_i$$

For $s = $ constant,

$$u_{i+1} = u_i + \frac{[\mathbf{p}_1^w \cdot (\mathbf{p}_2^t \times \mathbf{r})]_i}{D_s}$$

$$w_{i+1} = w_i + \frac{[\mathbf{p}_2^t \cdot (\mathbf{p}_1^u \times \mathbf{r})]_i}{D_s}$$

$$s_{i+1} = s_i$$

$$t_{i+1} = t_i - \frac{[\mathbf{p}_1^u \cdot (\mathbf{p}_1^w \times \mathbf{r})]_i}{D_s} \qquad (7.45)$$

where

$$D_s = -[\mathbf{p}_2^t \cdot (\mathbf{p}_1^u \times \mathbf{p}_1^w)]_i$$

For $t = $ constant,

$$u_{i+1} = u_i + \frac{[\mathbf{p}_1^w \cdot (\mathbf{p}_2^s \times \mathbf{r})]_i}{D_t}$$

$$w_{i+1} = w_i + \frac{[\mathbf{p}_2^s \cdot (\mathbf{p}_1^u \times \mathbf{r})]_i}{D_t}$$

$$s_{i+1} = s_i - \frac{[\mathbf{p}_1^u \cdot (\mathbf{p}_1^w \times \mathbf{r})]_i}{D_t}$$

$$t_{i+1} = t_i \tag{7.46}$$

where

$$D_t = -[\mathbf{p}_2^s \cdot (\mathbf{p}_1^u \times \mathbf{p}_1^w)]_i$$

4.7   Step-Size Selection for the Tracing Phase

Finally, we consider Timmer's step-size-selection procedure for the tracing phase, which allows the step size to be computed based entirely on local surface properties. The geometry for this calculation appears in Fig. 7.16. At a point on the intersection curve, the curvature is expressed in terms of the angular rate of change

$$\kappa = \frac{-d\theta}{d\sigma} \tag{7.47}$$

If we select a nominal value of the angular change, we then compute the step size $\Delta\sigma$

$$\Delta\sigma = \frac{\Delta\theta}{|\kappa|} \tag{7.48}$$

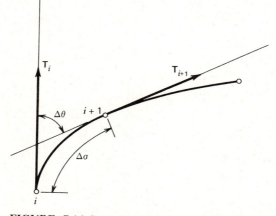

FIGURE 7.16 Step-size selection for the tracing phase.

Denote the tangent vector to the intersection curve by

$$\mathbf{T} = v(\mathbf{p}^u \times \mathbf{p}^w) \times (\mathbf{p}^s \times \mathbf{p}^t) \qquad \text{where } v = \pm 1 \tag{7.49}$$

v is a scalar multiplier to correct the directional sense of the tangent vector.

We can define the curvature $|\kappa|$ as $|d\tau/d\sigma|$, where $\tau = \mathbf{T}/|\mathbf{T}|$. When we expand this expression, we obtain

$$\frac{d\tau}{d\sigma} = \left(\frac{1}{|\mathbf{T}|}\right)\left(\frac{d\mathbf{T}}{d\sigma}\right) - \left(\frac{\mathbf{T} \cdot d\mathbf{T}}{d\sigma}\right)\tau \tag{7.50}$$

Since τ is a function of all four of the independent variables u, w, s, and t, we find that

$$\left(\frac{1}{v}\right)\left(\frac{d\mathbf{T}}{d\sigma}\right) = [(\mathbf{p}^{uu} \times \mathbf{p}^w + \mathbf{p}^u \times \mathbf{p}^{uw}) \times (\mathbf{p}^s \times \mathbf{p}^t)]\left(\frac{du}{d\sigma}\right)$$

$$+ [(\mathbf{p}^{uw} \times \mathbf{p}^w + \mathbf{p}^u \times \mathbf{p}^{ww}) \times (\mathbf{p}^s \times \mathbf{p}^t)]\left(\frac{dw}{d\sigma}\right)$$

$$+ [(\mathbf{p}^u \times \mathbf{p}^w) \times (\mathbf{p}^{ss} \times \mathbf{p}^t + \mathbf{p}^s \times \mathbf{p}^{st})]\left(\frac{ds}{d\sigma}\right)$$

$$+ [(\mathbf{p}^u \times \mathbf{p}^w) \times (\mathbf{p}^{st} \times \mathbf{p}^t + \mathbf{p}^s \times \mathbf{p}^{tt})]\left(\frac{dt}{d\sigma}\right) \tag{7.51}$$

Q.E.D.

The step size may be determined entirely from the local surface properties.

EXERCISES

1. Outline a method for computing the intersection of two surfaces based on approximating each surface using planar triangulation.

2. Does the application of Timmer's method depend on the types of surfaces for which intersections are sought? Compare and critique the method as it applies to bicubic patches, Bezier surfaces, and B-spline surfaces.

3. Find the intersection between the Bezier surface whose control points are

$$\mathbf{B}_1 = \begin{bmatrix} (-4, 8, 0) & (0, 8, 0) & (2, 8, 0) & (6, 8, 0) \\ (-4, 4, 0) & (0, 4, 4) & (2, 4, 4) & (6, 4, 0) \\ (-4, -2, 0) & (0, -2, 4) & (2, -2, 4) & (6, -2, 0) \\ (-4, -6, 0) & (0, -6, 0) & (2, -6, 0) & (6, -6, 0) \end{bmatrix}$$

and the Hermite bicubic patch whose geometric coefficients are

$$\mathbf{B}_2 = \begin{bmatrix} (5, 0, 0) & (5, 0, 5) & (0, 0, 5) & (0, 0, 5) \\ (0, 5, 0) & (0, 5, 5) & (0, 0, 5) & (0, 0, 5) \\ (0, 4, 0) & (0, 4, 0) & 0 & 0 \\ (-4, 0, 0) & (-4, 0, 0) & 0 & 0 \end{bmatrix}$$

8

TRANSFORMATIONS

To be of any practical value, the geometric model of an object must be easy to manipulate or transform. Transforming an object implies changing it in either position, orientation, or shape. Since a geometric model must be stored in a computer as numerical data in terms of some coordinate system, we must devise ways of transforming the data to represent changing an object's position and orientation.

The simplest changes are the so-called rigid-body transformations, such as translation and rotation. We can operate directly on the parametric representation of geometric objects, such as points, vectors, curves, and surfaces, to effect these changes. We will see how this is done by the simple expediency of matrix multiplication and addition.

The full set of transformations that we consider here includes translation, rotation, scaling, symmetry, and reflection. We will consider these transformations first in their classical form and then as homogeneous transformations. We will conclude with a discussion of multiple or concatenated transformations, coordinate system transformations, and general nonlinear transformations. Projective transformations are discussed in Chapter 12, Section 1, where the central topic is the projective geometry of computer graphic displays.

1 TRANSLATION

The rigid-body translation of a geometric object implies that every point on the object is moved equally a given distance in a given direction. We may specify translation by a vector, say, $\mathbf{t}$; Fig. 8.1 shows a curve translated by $\mathbf{t}$. Every point $\mathbf{p}$ on the curve is translated by an amount $\mathbf{t}$, so that a point $\mathbf{p}^*$ on the fully translated curve is given by

$$\mathbf{p}^* = \mathbf{p} + \mathbf{t} \tag{8.1}$$

Notice that an asterisk denotes a transformed vector or matrix quantity.

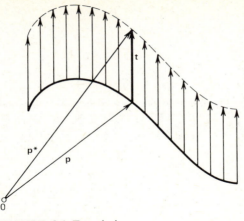

FIGURE 8.1 Translation

Equation 8.1 represents the direct pointwise transformation of **p** and requires that we retain a definition of the curve through either the **B** matrix of geometric coefficients or the **A** matrix of algebraic coefficients. In most applications, it is preferable to transform the **A** or **B** matrices and then discard the old definition.

The **B** matrix for a pc curve is transformed as follows:

$$\mathbf{B^*} = \mathbf{B} + \mathbf{T} \tag{8.2}$$

where

$$\mathbf{T} = [\mathbf{t} \quad \mathbf{t} \quad 0 \quad 0]^T \tag{8.3a}$$

For a bicubic patch,

$$\mathbf{T} = \begin{bmatrix} \mathbf{t} & \mathbf{t} & 0 & 0 \\ \mathbf{t} & \mathbf{t} & 0 & 0 \\ 0 & 0 & 0 & 0 \\ 0 & 0 & 0 & 0 \end{bmatrix} \tag{8.3b}$$

EXERCISES

1. Given a translation vector **t**, find **A*** for a pc curve.

2. Given a translation vector **t**, find **T** for the four-point form of a pc curve.

3. Write a translation transformation equation for a Bezier curve; for a *B*-spline surface.

4. Write a translation transformation equation for a general *n*-cubic hypersurface. What conclusions can you reach?

5. Given a pc curve whose geometric coefficients are **B** = [(4, 3, 7) (8, 10, 12) (4, 3, 0) (8, 10, 0)]^T; find **B*** if **t** = [−4   −3   −7].

6. Given the geometric coefficients of a pc curve, find **T** such that $\mathbf{p}_0$ is translated to the origin.

7. Write a procedure to translate a pc curve. Denote this as **TRN PC(CI, T, CO)**, where

>   CI(4, 3) is the input array of geometric coefficients of a pc curve;
>
>   T(3) is the input array of components of a vector defining the translation; and
>
>   CO(4, 3) is the output array of geometric coefficients of the translated pc curve.

8. Write a procedure to translate a Bezier curve. Denote this as **TRN BZC(N, CI, T, CO)**, where

>   N is the input number of control points;
>
>   CI(N, 3) is the input array of coordinates of the control points;
>
>   T(3) is the input array of components of a vector defining the translation; and
>
>   CO(N, 3) is the output array of coordinates of the control points of the translated Bezier curve.

Will this procedure work equally well for *B*-spline curves?

9. Write procedures to translate bicubic, Bezier, and *B*-spline surfaces. Compare them. They are easily combined into a single procedure; show how this can be done.

2   ROTATION

There are several ways of rotating an object. We will investigate rotation around the origin (that is, the principal axes) and an arbitrary line in space. The rigid-body rotation **R** of a geometric object must meet the following conditions:

1. Relative distances and angles between points and slopes on an object are maintained.

2. A right-hand convention applies to signs of rotation angles.

3. The embedding coordinate system is a right-hand system.

4. For the case of rotation around the origin, a rotation has three possible components: γ, β, θ; where γ is the angle of rotation around the x-axis, β the angle around the y-axis, and θ the angle around the z-axis. In keeping with the right-hand convention, θ is positive in a counterclockwise sense when viewed from a point on the $+z$-axis and toward the origin; β is positive in a counterclockwise sense when viewed from a point on the $+y$-axis; and γ is positive in a counterclockwise sense when viewed from a point on the $+x$-axis.

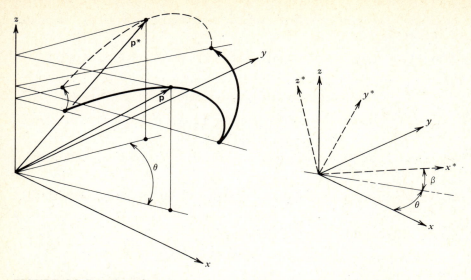

FIGURE 8.2 Rotation of a curve.

5. When the rotation of an object is specified by all three components, the order is important. In the absence of other constraints, the following convention is useful.

(a) First, rotate around the z-axis if $\theta \neq 0$.

(b) Next, rotate around the y-axis if $\beta \neq 0$.

(c) Finally, rotate around the x-axis if $\gamma \neq 0$.

Figure 8.2a shows a curve rotated through the angle θ in the positive direction around the z-axis. Notice that every point on the curve under-goes a rotation θ around the z-axis and that its trajectory lies in planes perpendicular to the z-axis and parallel to the x, y plane. For a parametric curve, the rotation changes (transforms) the functions $x(u)$ and $y(u)$, but not $z(u)$. θ is positive when in the same direction as that assumed for the vector product of unit vectors that produces $\mathbf{i} \times \mathbf{j} = \mathbf{k}$; that is, rotating the positive x-axis into the y-axis when rotation is around the z-axis.

Conversely, Fig. 8.2b shows a rotation sequence applied to the coordinate system. Here, θ and β are nonzero and $\gamma = 0$. Rotation through the angle θ is executed first and then β. As shown, both rotational components are negative, in accordance with the preceding conventions, with respect to objects that we assume to remain fixed relative to the initial, unrotated system.

We will now develop rigid-body rotations of points and vectors within a fixed coordinate system and show how they are related. Points can be rotated and translated with respect to the origin, and vectors can be rotated. A curve or other geometric object defined in terms of points, or points and tangent vectors, is clearly subject to general rigid-body transformations consisting of both translation and rotation, as expressed by

$$\mathbf{p}^* = \mathbf{p}\mathbf{R} + \mathbf{t} \tag{8.4}$$

where

$\mathbf{p}^*$ = the transformed point.

$\mathbf{p}$ = the initial point.

$\mathbf{R}$ = a 3×3 rotation matrix.

$\mathbf{t}$ = a translation vector.

Notice that Eq. 8.4 implies that the point is first rotated—in other words, we first computed $\mathbf{p}\mathbf{R}$—and then translated. It is possible to perform the transformation in the reverse order; that is, translate first and then rotate.

$$\mathbf{p}^* = [\mathbf{p} + \mathbf{t}]\mathbf{R} \tag{8.5}$$

However, in general, $\mathbf{p}\mathbf{R} + \mathbf{t} \neq [\mathbf{p} + \mathbf{t}]\mathbf{R}$.

Now, let us derive the elements of the $\mathbf{R}$ matrix. Figure 8.3 shows the geometry describing the rotation of point $\mathbf{p}$ through an angle θ around the z-axis, so that given $\mathbf{p}$ and θ, we can find $\mathbf{p}^*$, the transformed point. The relationship between the components of $\mathbf{p}$ and $\mathbf{p}^*$ is

$$x^* = |\mathbf{p}| \cos (\alpha + \theta) \quad \text{and} \quad y^* = |\mathbf{p}| \sin (\alpha + \theta) \tag{8.6}$$

where $|\mathbf{p}| = |\mathbf{p}^*| = \sqrt{x^2 + y^2}$. From elementary trigonometry,

$$\begin{aligned} \cos (\alpha + \theta) &= \cos \alpha \cos \theta - \sin \alpha \sin \theta \\ \sin (\alpha + \theta) &= \sin \alpha \cos \theta + \cos \alpha \sin \theta \end{aligned} \tag{8.7}$$

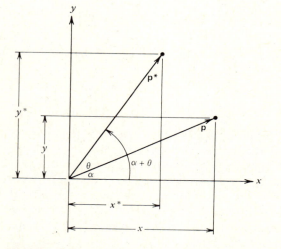

FIGURE 8.3 Rotating a point or vector about the z axis.

Furthermore,

$$\cos \alpha = \frac{x}{|\mathbf{p}|} \quad \text{and} \quad \sin \alpha = \frac{y}{|\mathbf{p}|} \tag{8.8}$$

Substituting appropriately, we obtain

$$\cos (\alpha + \theta) = \left(\frac{z}{|\mathbf{p}|}\right) \cos \theta - \left(\frac{y}{|\mathbf{p}|}\right) \sin \theta \tag{8.9}$$

$$\sin (\alpha + \theta) = \left(\frac{x}{|\mathbf{p}|}\right) \sin \theta + \left(\frac{y}{|\mathbf{p}|}\right) \cos \theta$$

Substitute Eqs. 8.9 into Eq. 8.6 to yield

$$x^* = x \cos \theta - y \sin \theta$$
$$y^* = x \sin \theta + y \cos \theta \tag{8.10}$$

Since this rotation is around the z-axis, $z^* = z$. The transformation matrix for these rotations is

$$\mathbf{R}_\theta = \begin{bmatrix} \cos \theta & \sin \theta & 0 \\ -\sin \theta & \cos \theta & 0 \\ 0 & 0 & 1 \end{bmatrix} \tag{8.11}$$

so that $\mathbf{p}^* = \mathbf{p}\mathbf{R}_\theta$. This formulation of $\mathbf{R}$ assumes that both $\mathbf{p}$ and $\mathbf{p}^*$ are 1×3 matrices.

Generalize this procedure to obtain the rotation transformation matrices for rotations around the y- and x-axes $\mathbf{R}_\beta$ and $\mathbf{R}_\gamma$

$$\mathbf{R}_\beta = \begin{bmatrix} \cos \beta & 0 & -\sin \beta \\ 0 & 1 & 0 \\ \sin \beta & 0 & \cos \beta \end{bmatrix} \tag{8.12}$$

and

$$\mathbf{R}_\gamma = \begin{bmatrix} 1 & 0 & 0 \\ 0 & \cos \gamma & \sin \gamma \\ 0 & -\sin \gamma & \cos \gamma \end{bmatrix} \tag{8.13}$$

For $\theta = \gamma = 0$, $\mathbf{R}_\beta$ postmultiplying $\mathbf{p}$ transforms the components for a given rotation around the y-axis. Similarly, for $\theta = \beta = 0$, $\mathbf{R}_\gamma$ postmultiplying $\mathbf{p}$ transforms the components for a given rotation around the x-axis. What happens if we must rotate an object in such a way that $\theta, \beta, \gamma \neq 0$? We have already said that order is important—rotations must be taken in the order θ, β, γ. (Of course, we could have established another convention just as easily, but this one is consistent with a right-hand coordinate system.) This means that for general rotations with three components,

$$\mathbf{p}^* = \mathbf{p}\mathbf{R}_\theta\mathbf{R}_\beta\mathbf{R}_\gamma = \mathbf{p}\mathbf{R} \tag{8.14}$$

Find **R** by simply performing the indicated matrix multiplications

R =

$$
\begin{bmatrix}
\cos\theta\cos\beta & \sin\theta\cos\gamma + \cos\theta\sin\beta\sin\gamma & \sin\theta\sin\gamma - \cos\theta\sin\beta\cos\gamma \\
-\sin\theta\cos\beta & \cos\theta\cos\gamma - \sin\theta\sin\beta\sin\gamma & \cos\theta\sin\gamma + \sin\theta\sin\beta\cos\gamma \\
\sin\beta & -\cos\beta\sin\gamma & \cos\beta\cos\gamma
\end{bmatrix}
$$

$$(8.15)$$

This rotation matrix is applicable for any combination of component rotations $\mathbf{R}_\theta$, $\mathbf{R}_\beta$, and $\mathbf{R}_\gamma$ so long as they are taken in that order. Thus, if $\beta = 0$, merely substitute zero for β in the appropriate terms of the matrix elements, and the result applies to rotations around the x- and z-axes.

It is easy to show that the rotation matrix applies equally well to the algebraic and geometric coefficient matrices of curves and surfaces. Thus,

$$\mathbf{A}^* = \mathbf{AR} \quad \text{and} \quad \mathbf{B}^* = \mathbf{BR} \qquad (8.16)$$

With the techniques we have just developed, we can now derive transformation matrices for more general rigid-body movements. One of the most useful ways of rotating an object is around some given axis in space. This is illustrated in Fig. 8.4 for a curve. The problem is to find the transformation matrices and operations that will yield the points of a curve rotated from its

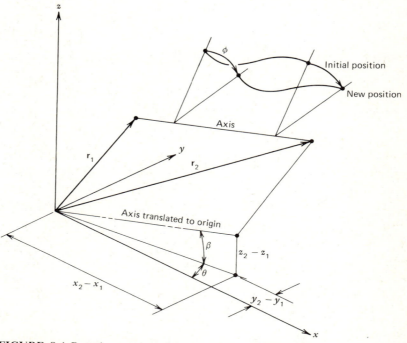

FIGURE 8.4 Rotating a curve about an arbitrary axis in space.

initial position to its final position through an angle ϕ around an axis whose end points are given by vectors $\mathbf{r}_1$ and $\mathbf{r}_2$.

An algorithm for doing this follows.

Step 1.   Translate the curve and axis so that one end of the axis, say, $\mathbf{r}_1$, is at the origin of the coordinate system. Thus, $\mathbf{t} = -\mathbf{r}_1$, and if $\mathbf{p}$ is any point on the curve, then

$$\mathbf{p}^* = \mathbf{p} - \mathbf{r}_1 \tag{8.17}$$

Step 2.   Rotate the curve and axis system so that the axis is colinear with the x-axis. Do this by rotating $-\theta$ degrees around the z-axis and then $-\beta$ degrees around the y-axis, where

$$\theta = \tan^{-1} \frac{y_2 - y_1}{x_2 - x_1} \tag{8.18}$$

and

$$\beta = \sin^{-1} \frac{z_2 - z_1}{|\mathbf{r}_2 - \mathbf{r}_1|} \tag{8.19}$$

Thus, $\mathbf{p}^*$ now becomes

$$\mathbf{p}^* = [\mathbf{p} - \mathbf{r}_1]\mathbf{R}_{\theta\beta} \tag{8.20}$$

Step 3.   Rotate the curve around the x-axis. After the preceding two steps, $\gamma = \phi$, to yield

$$\mathbf{p}^* = [\mathbf{p} - \mathbf{r}_1]\mathbf{R}_{\theta\beta}\mathbf{R}_\gamma \tag{8.21}$$

Step 4.   Reverse Step 2. Notice that there is a sign change in the rotation angles.

$$\mathbf{p}^* = [\mathbf{p} - \mathbf{r}_1]\mathbf{R}_{\theta\beta}\mathbf{R}_\gamma\mathbf{R}_{-\theta\beta} \tag{8.22}$$

Step 5.   Reverse Step 1 to obtain the final transformation

$$\mathbf{p}^* = [\mathbf{p} - \mathbf{r}_1]\mathbf{R}_{\theta\beta}\mathbf{R}_\gamma\mathbf{R}_{-\theta\beta} + \mathbf{r}_1 \tag{8.23}$$

The set of components resulting from these five operations defines the curve points in the new position. Again, it is usually more efficient to transform the **B** matrix of the object and then proceed to compute points or other properties in the transformed state. For a curve, these steps result in the following analogous transformation sequence:

$$\mathbf{B}^* = [\mathbf{B} + \mathbf{T}]\mathbf{R}_{\theta\beta}\mathbf{R}_\gamma\mathbf{R}_{-\theta\beta} - \mathbf{T} \tag{8.24}$$

where

$$\mathbf{T} = \begin{bmatrix} -\mathbf{r}_1 \\ -\mathbf{r}_1 \\ 0 \\ 0 \end{bmatrix} \tag{8.25}$$

Now, let $\mathbf{R}_{\theta\beta}\mathbf{R}_{\gamma}\mathbf{R}_{-\theta\beta} = \mathbf{R}_a$. Then

$$\mathbf{B}^* = [\mathbf{B} + \mathbf{T}]\mathbf{R}_a - \mathbf{T} \tag{8.26}$$

which expands to

$$\mathbf{B}^* = \mathbf{B}\mathbf{R}_a + \mathbf{T}\mathbf{R}_a - \mathbf{T} \tag{8.27}$$

Then let $\mathbf{T}\mathbf{R}_a - \mathbf{T} = \mathbf{T}_a$. Substitute into Eq. 8.27 to obtain

$$\mathbf{B}^* = \mathbf{B}\mathbf{R}_a + \mathbf{T}_a \tag{8.28}$$

Thus, the rigid-body transformations for a curve or any geometric object rotated around an arbitrary axis in space mathematically reduce to a rotation around the coordinate system axes and a subsequent translation.

As an interesting exercise, investigate other, perhaps more exotic, transformation situations, such as rotating a curve in space around a line joining its end points or around one of its tangent vectors. What is the effect on the transformed **B** matrix in either case?

There is one final transformation algorithm to explore. It, too, combines rotations and translations and is of considerable practical value to geometric modeling in CAD systems. It is called the three-point-to-three-point transformation, or the three-point transformation. The problem is as follows: Given a geometric model that contains points $\mathbf{p}_1, \mathbf{p}_2$, and $\mathbf{p}_3$ and given three other points $\mathbf{q}_1, \mathbf{q}_2$, and $\mathbf{q}_3$, find the total rigid-body transformation that (1) transforms $\mathbf{p}_1$ into $\mathbf{q}_1$; (2) transforms the vector $(\mathbf{p}_2 - \mathbf{p}_1)$ into the vector $(\mathbf{q}_2 - \mathbf{q}_1)$ (direction only); and (3) transforms the plane containing the three points $\mathbf{p}_1, \mathbf{p}_2$, and $\mathbf{p}_3$ into the plane containing $\mathbf{q}_1, \mathbf{q}_2$, and $\mathbf{q}_3$.

Note: $\mathbf{p}_1, \mathbf{p}_2$, and $\mathbf{p}_3$ can be reference points in the model system. It is their relationship to $\mathbf{q}_1, \mathbf{q}_2$, and $\mathbf{q}_3$ that determines the total rigid-body transformation applied to the geometric model. Figure 8.5 illustrates these conditions and relevant parameters.

The following algorithm produces the desired transformation.

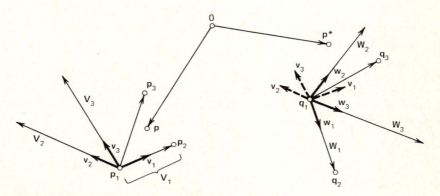

FIGURE 8.5 Three-point to Three-point transformation.

Step 1. Construct vectors $(\mathbf{p}_2 - \mathbf{p}_1)$, $(\mathbf{p}_3 - \mathbf{p}_1)$, $(\mathbf{q}_2 - \mathbf{q}_1)$, and $(\mathbf{q}_3 - \mathbf{q}_1)$.

Step 2. Let $\mathbf{V}_1 = \mathbf{p}_2 - \mathbf{p}_1$, $\mathbf{W}_1 = \mathbf{q}_2 - \mathbf{q}_1$. (*Note:* The convention established in Chapter 1 denoting vectors by lowercase letters only is temporarily waived here.)

Step 3. Construct $\mathbf{V}_3$ and $\mathbf{W}_3$ by

$$\mathbf{V}_3 = \mathbf{V}_1 \times (\mathbf{p}_3 - \mathbf{p}_1)$$
$$\mathbf{W}_3 = \mathbf{W}_1 \times (\mathbf{q}_3 - \mathbf{q}_1) \tag{8.29}$$

Step 4. Construct $\mathbf{V}_2$ and $\mathbf{W}_2$ by

$$\mathbf{V}_2 = \mathbf{V}_3 \times \mathbf{V}_1$$
$$\mathbf{W}_2 = \mathbf{W}_3 \times \mathbf{W}_1 \tag{8.30}$$

Clearly the vectors $\mathbf{V}_1$, $\mathbf{V}_2$, and $\mathbf{V}_3$ form a right-hand orthogonal system, as do $\mathbf{W}_1$, $\mathbf{W}_2$, and $\mathbf{W}_3$.

Step 5. Construct the unit vectors

$$\mathbf{v}_1 = \frac{\mathbf{V}_1}{|\mathbf{V}_1|}$$

$$\mathbf{v}_2 = \frac{\mathbf{V}_2}{|\mathbf{V}_2|} \tag{8.31}$$

$$\mathbf{v}_3 = \frac{\mathbf{V}_3}{|\mathbf{V}_3|}$$

and

$$\mathbf{w}_1 = \frac{\mathbf{W}_1}{|\mathbf{W}_1|}$$

$$\mathbf{w}_2 = \frac{\mathbf{W}_2}{|\mathbf{W}_2|} \tag{8.32}$$

$$\mathbf{w}_3 = \frac{\mathbf{W}_3}{|\mathbf{W}_3|}$$

Step 6. To transform any point $\mathbf{p}$ in the $\mathbf{v}$ system into the $\mathbf{w}$ system, use the transformation relationship

$$\mathbf{p}^* = \mathbf{p}\mathbf{R} + \mathbf{T} \tag{8.33}$$

Step 7. $[\mathbf{w}_1 \quad \mathbf{w}_2 \quad \mathbf{w}_3] = [\mathbf{v}_1 \quad \mathbf{v}_2 \quad \mathbf{v}_3]\mathbf{R}$, since $[\mathbf{w}]$ and $[\mathbf{v}]$ are the unit vector matrices. Then the required rotation matrix with respect to the $\mathbf{w}$ system is simply

$$\mathbf{R} = [\mathbf{v}]^{-1}[\mathbf{w}] \tag{8.34}$$

Step 8. Obtain the required translation matrix by substituting Eq. 8.34 for Eq. 8.33 and solving for $\mathbf{T}$ with $\mathbf{p}^* = \mathbf{q}_1$ and $\mathbf{p} = \mathbf{p}_1$

$$\mathbf{T} = \mathbf{q}_1 - \mathbf{p}_1[\mathbf{v}]^{-1}[\mathbf{w}] \tag{8.35}$$

Step 9.   Rewrite Eq. 8.33 as

$$\mathbf{p}^* = \mathbf{p}[\mathbf{v}]^{-1}[\mathbf{w}] - \mathbf{p}_1[\mathbf{v}]^{-1}[\mathbf{w}] + \mathbf{q}_1 \qquad (8.36)$$

This transformation is extremely useful for moving two solids into coincidence with each other and for simply repositioning elements in a geometric model.

EXERCISES

1. Verify and discuss the validity of Eq. 8.16.

2. Apply the five-step algorithm to the transformation of a bicubic patch given by $\mathbf{p}(u, w) = \mathbf{UMBM}^T\mathbf{W}^T$.

3. Assume the same conditions as Ex. 2, but apply them to a Bezier surface.

4. Show that the three-point transformation is equivalent to moving the $\mathbf{v}$ system to the origin, rotating it by $\mathbf{R} = [\mathbf{v}]^{-1}[\mathbf{w}]$, and then moving the result to $\mathbf{q}_1$.

5. Show that any rigid-body rotation matrix has a determinant value of 1.0.

6. Discuss the relative merits of rotation and translation transformation operations on a bicubic patch. Compare the algebraic, geometric, and 16-point forms. Do the same for a tricubic solid.

7. Find $\mathbf{R}$ if $\theta = 180°$, $\beta = 0$, $\gamma = 0$; see Eqs. 8.14 and 8.15.

8. Find $\mathbf{R}$ if $\theta = 90°$, $\beta = 90°$, $\gamma = 90°$.

9. Write a procedure to rotate the defining points of a curve or surface around the origin. Denote this as **ROTORG(NP, CI, TTA, BTA, GMA, CO)**, where

   NP is the input number of points to be rotated;

   CI(NP, 3) is the input array of coordinates of the points;

   TTA is the input angle of rotation around the z-axis;

   BTA is the input angle of rotation around the y-axis;

   GMA is the input angle of rotation around the x-axis; and

   CO(NP, 3) is the output array of coordinates of the rotated points.

10. Write a procedure to rotate the defining points of a curve or surface around an axis in space. Denote this as **ROTAXS(NP, CI, A1, A2, PHI, CO)**, where

   NP is the input number of points to be rotated;

   CI(NP, 3) is the input array of coordinates of the points;

   A1(3), A2(3) are the input arrays of coordinates of the end points of the axis of rotation;

   PHI is the input angle of rotation around the axis; and

   CO(NP, 3) is the output array of coordinates of the rotated points.

11. Write a procedure to translate and rotate the defining points of an object as specified by a three-point-to-three-point transformation. Denote this as **TPTP**(NP, CI, P, Q, CO), where

NP is the input number of points to be rotated;

CI(NP, 3) is the input array of coordinates of the points;

P(3, 3), Q(3, 3) are the input arrays of points defining the transformation; and

CO(NP, 3) is the output array of coordinates of the transformed points.

12. Show that by adding a third rotation component, say, $\mathbf{R}_y$, we introduce a redundancy that leads to order dependence.

3   SCALING

The absolute size of a geometric element, such as a curve or a surface, may be changed by multiplying its geometric coefficients by a scale factor. If we apply the same scale factor to each coordinate component, then the element will change in size, but not in shape. If we apply different scale factors to each of the components, then both size and shape will change. We will consider each of these cases.

The simplest kind of scaling occurs by applying a positive scalar multiplier to each of the geometric coefficients

$$\mathbf{B}^* = s\mathbf{B} \tag{8.37}$$

The scalar s is always positive. (If it is negative, we are then dealing with reflection, the subject of Section 4.) In Eq. 8.37, $\mathbf{B}$ is the matrix of geometric coefficients before scaling and $\mathbf{B}^*$ the scaled or transformed coefficients. According to matrix algebra, each element is multiplied by s. Thus, for a pc curve,

$$\mathbf{B}^* = \begin{bmatrix} s\mathbf{p}_0 \\ s\mathbf{p}_1 \\ s\mathbf{p}_0^u \\ s\mathbf{p}_1^u \end{bmatrix} \tag{8.38}$$

Figure 8.6 graphically shows the effect of this scaling transformation. The shapes of the curves $\mathbf{p}(u)$ and $\mathbf{p}^*(u)$ are identical, but curve $\mathbf{p}^*(u)$ is in this case larger by a factor of s. It is s times longer, and it occupies a different position in space, since the components of each point $\mathbf{p}^*$ are s times larger than the corresponding components of $\mathbf{p}$. The tangent vectors $\mathbf{p}_0^{*u}$ and $\mathbf{p}_1^{*u}$ are likewise proportionally greater than $\mathbf{p}_0^u$ and $\mathbf{p}_1^u$. The direction cosines at any point are unchanged, however. The pointwise transformation of a curve is clearly

$$\mathbf{p}^* = s\mathbf{p} \tag{8.39}$$

This transformation causes scaling, expansion or contraction, around the system's origin. We can adjust the equation, however, to cause scaling around

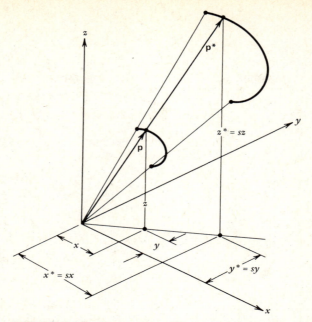

FIGURE 8.6 Scaling a curve with respect to the origin.

(with respect to) any point **q**. Figure 8.7 illustrates the effect of a curve on a point **p** when the curve is scaled around point **q**, where s is the scalar multiplier.

Vector arithmetic simplifies the derivation of the relationship between point **p** and its transformed counterpart **p**

$$\mathbf{p}^* = \mathbf{q} + s(\mathbf{p} - \mathbf{q}) \tag{8.40}$$

Rewrite Eq. 8.40 as

$$\mathbf{p}^* = s\mathbf{p} - (s - 1)\mathbf{q} \tag{8.41}$$

Figure 8.8 illustrates the effect of this type of scaling on the tangent vectors. From the properties of similar triangles,

$$\mathbf{p}^{*u} = s\mathbf{p}^u \tag{8.42}$$

By assembling these results, we obtain the transformation equation for scaling a pc curve around any point in space

$$\mathbf{B}^* = s\mathbf{B} + \mathbf{T}_s \tag{8.43}$$

where

$$\mathbf{T}_s = \begin{bmatrix} -\mathbf{q}(s - 1) \\ -\mathbf{q}(s - 1) \\ 0 \\ 0 \end{bmatrix} \tag{8.44}$$

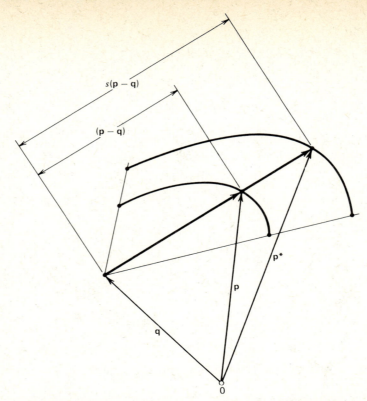

FIGURE 8.7 Scaling a curve with respect to an arbitrary point.

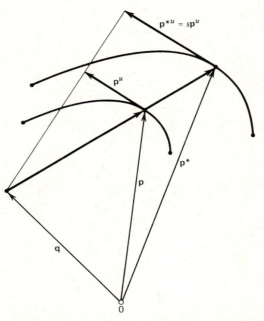

FIGURE 8.8 The effect of scaling on tangent vectors.

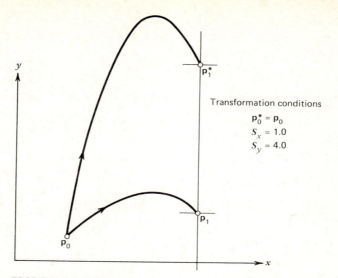

FIGURE 8.9 Differential scaling of a pc curve with respect to $\mathbf{p}_0$.

As in Eqs. 8.38 and 8.39, apply the same scale factor to each component p_x, p_y, and p_z. We are not constrained to do this in all cases. There are often situations in geometric modeling where it is necessary to scale (stretch or shrink) each component by a different factor. We denote these scale factors by s_x, s_y, and s_z. The operation of applying them is called **differential scaling**. Figure 8.9 shows an example of this for a curve in the x, y plane. Equation 8.45 shows how to use these scale factors to construct a general, differential scaling transformation matrix for a pc curve.

$$\mathbf{B}^* = \begin{bmatrix} s_x x_0 - q_x(s_x - 1) & s_y y_0 - q_y(s_y - 1) & s_z z_0 - q_z(s_z - 1) \\ s_x x_1 - q_x(s_x - 1) & s_y y_1 - q_y(s_y - 1) & s_z z_1 - q_z(s_z - 1) \\ s_x x_0^u & s_y y_0^u & s_z z_0^u \\ s_x x_1^u & s_y y_1^u & s_z z_1^u \end{bmatrix} \quad (8.45)$$

EXERCISES

1. Show that the scalar transformation in Eq. 8.38 does not change the values of the direction cosines of the tangent vector at a point.

2. What effect does differential scaling have on tangent vectors and direction cosines? Give an example.

3. Derive a scaling transformation for the bicubic patch similar to that for a pc curve given by Eq. 8.38.

4. Derive a differential scaling transformation for the bicubic patch similar to that for a pc curve given by Eq. 8.45.

5. Write a procedure to scale differentially the defining points of a curve or surface around any point. Denote this as **SCALE(NP, CI, P, SX, SY, SZ, CO)**, where

> NP is the input number of defining points;
>
> CI(NP, 3) is the input array of coordinates of the defining points;
>
> P(3) is the input array of coordinates of the point around which scaling is to be performed;
>
> SX, SY, SZ are the input scale factors; and
>
> CO(NP, 3) is the output array of coordinates of the scaled defining points.

6. Describe the effects of the shear transformation given by $\mathbf{p}^* = \mathbf{pA}$, where in two dimensions

$$\mathbf{A} = \begin{bmatrix} 1 & s \\ 0 & 1 \end{bmatrix} \quad \text{or} \quad \begin{bmatrix} 1 & 0 \\ s & 1 \end{bmatrix}$$

and s is a scalar constant.

4 SYMMETRY AND REFLECTION

A geometric element can be reflected through a plane, line, or point in space to create its reflected or symmetric counterpart. The necessary transformations are direct and easy to understand. Figure 8.10 illustrates some of the possible ways

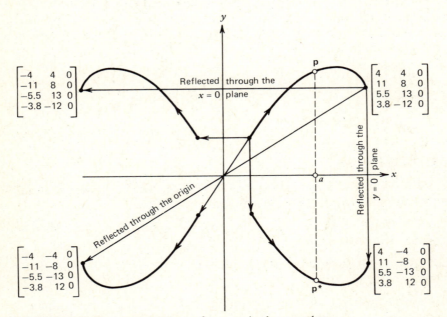

FIGURE 8.10 Various reflections of a curve in the x, y plane.

a curve can be reflected. Here, we have a pc curve that lies in the first quadrant of the x, y plane and three different reflected images—one in each of the other quadrants.

The procedure for constructing a reflected image of any object in space is simple. For example, to reflect the curve in the first quadrant through the $y = 0$ plane, transform each point **p** on it to its symmetric image **p*** located an equal distance from the plane but on the other side of it. The distance between point **a** and point **p** is equal to the distance between **a** and **p***. Point **a** lies on the straight line joining **p** and **p*** at its intersection with the $y = 0$ plane. Furthermore, the vector $(\mathbf{p}^* - \mathbf{p})$ is normal to this plane.

This transformation merely changes the algebraic signs of the y components of the curve coefficients. In fact, when the plane of symmetry is any one of the three principal planes, the reflection transformation is accomplished by changing the sign of the curve coefficients of the component corresponding to the plane of symmetry. Although the example shown in Fig. 8.10 is a plane curve, the same procedures apply to a curve that twists through space.

Reflection may also take place across any of the principal axes or through the origin. A procedure analogous to that for reflection through a plane of symmetry applies to reflection across a principal axis or through the origin. Figures 8.11 and 8.12 are more elaborate examples of the reflection of a curve. Careful study of these examples allows us to empirically deduce the derivation

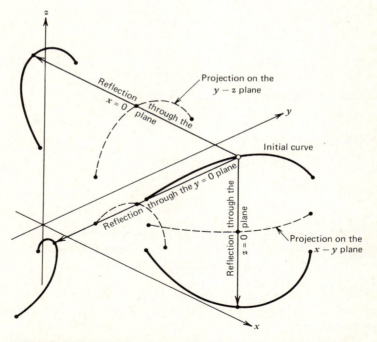

FIGURE 8.11 Reflecting a curve through the three principal planes.

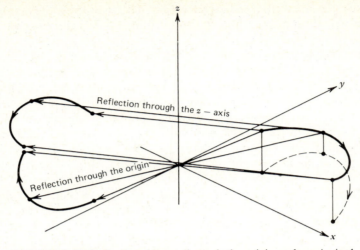

FIGURE 8.12 Reflecting a curve through the origin and a principal axis.

of the following relationship between the geometric coefficients of an initial pc curve and a reflected pc curve:

$$\mathbf{p}^* = \mathbf{pR}_f \tag{8.46}$$

where $\mathbf{R}_f$ is the transformation matrix for reflection, given by

$$\mathbf{R}_f = \begin{bmatrix} \pm 1 & 0 & 0 \\ 0 & \pm 1 & 0 \\ 0 & 0 & \pm 1 \end{bmatrix} \tag{8.47}$$

Rewrite Eq. 8.46 as follows:

$$\mathbf{p}^* = \mathbf{UMBR}_f \tag{8.48}$$

or

$$\mathbf{p}^* = \mathbf{UMB}^* \tag{8.49}$$

where

$$\mathbf{B}^* = \mathbf{BR}_f \tag{8.50}$$

Similar relationships are obtained for the algebraic form of a pc curve. For a bicubic surface, the mathematics is only slightly more complicated

$$\mathbf{p}^* = \mathbf{UMBM}^T\mathbf{W}^T\mathbf{R}_f \tag{8.51}$$

Remember that both $\mathbf{p}^*$ and the product $\mathbf{UMBM}^T\mathbf{W}^T$ are vectors, which we represent as 1×3 matrices. It is sometimes easy to lose sight of the dimensions of these matrices, since it is convenient to think of the product as

$$[1 \times 4][4 \times 4][4 \times 4][4 \times 4][4 \times 1]$$

which, superficially at least, results in a 1×1 matrix. However, this result is indeed a vector that has three components (in our usual three-dimensional Cartesian space). The mystery is solved when we recall that the **B** matrix is a 4×4 array of vectors. We could think of it as a $4 \times 4 \times 3$ array; ordinarily, we suppress this third dimension.

Instead of the pointwise transformation indicated by Eq. 8.51, compute **B*** as follows:

$$\mathbf{B^*} = [\mathbf{b}_{ij}\mathbf{R}_f] \tag{8.52}$$

We interpret this to mean that each vector element $\mathbf{b}_{ij}$ of the **B** matrix is transformed by $\mathbf{R}_f$ to construct and fill a new matrix **B***. A similar interpretation applies to Eq. 8.50. (See Ex. 1 in Section 2.)

Again, careful examination of Figs. 8.11 and 8.12 reveals that we can obtain any of the principal symmetric reflections by properly choosing the signs of the diagonal elements of the $\mathbf{R}_f$ transformation matrix. The following combinations are possible:

For reflection through the $x = 0$ plane,

$$\mathbf{R}_f = \begin{bmatrix} -1 & 0 & 0 \\ 0 & 1 & 0 \\ 0 & 0 & 1 \end{bmatrix} \tag{8.53}$$

For reflection through the $y = 0$ plane,

$$\mathbf{R}_f = \begin{bmatrix} 1 & 0 & 0 \\ 0 & -1 & 0 \\ 0 & 0 & 1 \end{bmatrix} \tag{8.54}$$

For reflection through the $z = 0$ plane,

$$\mathbf{R}_f = \begin{bmatrix} 1 & 0 & 0 \\ 0 & 1 & 0 \\ 0 & 0 & -1 \end{bmatrix} \tag{8.55}$$

For reflection through the x-axis,

$$\mathbf{R}_f = \begin{bmatrix} 1 & 0 & 0 \\ 0 & -1 & 0 \\ 0 & 0 & -1 \end{bmatrix} \tag{8.56}$$

For reflection through the y-axis,

$$\mathbf{R}_f = \begin{bmatrix} -1 & 0 & 0 \\ 0 & 1 & 0 \\ 0 & 0 & -1 \end{bmatrix} \tag{8.57}$$

For reflection through the z-axis,

$$\mathbf{R}_f = \begin{bmatrix} -1 & 0 & 0 \\ 0 & -1 & 0 \\ 0 & 0 & 1 \end{bmatrix} \tag{8.58}$$

For reflection through the origin,

$$\mathbf{R}_f = \begin{bmatrix} -1 & 0 & 0 \\ 0 & -1 & 0 \\ 0 & 0 & -1 \end{bmatrix} \qquad (8.59)$$

There are many other interesting reflectionlike transformations; we will consider two more. The first is the reflection of a curve through any point in space, illustrated in Fig. 8.13. The problem is: Given a point $\mathbf{p}$ on a curve and any point $\mathbf{q}$ through which it is to be reflected, find the transformation matrix $\mathbf{R}_q$ that will produce points $\mathbf{p}^*$ on the reflected curve.

Using the geometry shown in Fig. 8.13, we develop the following empirical solution. Through any point $\mathbf{p}$ on a curve, construct the vector $(\mathbf{q} - \mathbf{p})$. According to our understanding of reflection transformations, find the reflection of $\mathbf{p}$ through $\mathbf{q}$, or $\mathbf{p}^*$, by constructing $2(\mathbf{q} - \mathbf{p})$ at $\mathbf{p}$, with the following result:

$$\mathbf{p}^* = \mathbf{p} + 2(\mathbf{q} - \mathbf{p}) \qquad (8.60)$$

We find the transformation of the tangent vectors through the relationship of similar triangles, an easy graphic construction, which is omitted here. Interestingly enough, the magnitudes of the tangent vectors are unchanged; however, their directions are reversed, and we obtain

$$\mathbf{p}^{*u} = -\mathbf{p}^u \qquad (8.61)$$

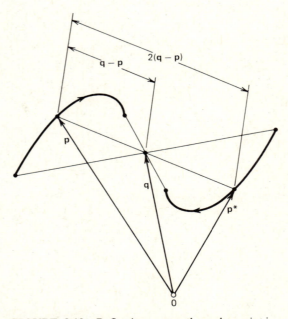

FIGURE 8.13   Reflecting a curve through a point in space.

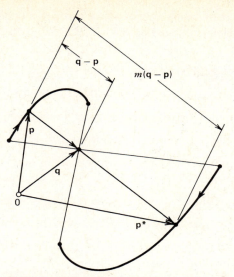

FIGURE 8.14 Reflecting and magnifying a curve through a point in space.

The **B*** matrix is

$$\mathbf{B^*} = \begin{bmatrix} 2\mathbf{q} - \mathbf{p}_0 \\ 2\mathbf{q} - \mathbf{p}_1 \\ -\mathbf{p}_0^u \\ -\mathbf{p}_1^u \end{bmatrix} \tag{8.62}$$

An analogous procedure can be applied to a combination of reflection with magnification (scaling) of a curve through any point in space; Fig. 8.14 shows the geometry for this. Using very elementary trigonometry, we can derive and verify the following relationships:

$$\mathbf{B^*} = \begin{bmatrix} (m + 1)\mathbf{q} - m\mathbf{p}_0 \\ (m + 1)\mathbf{q} - m\mathbf{p}_1 \\ -m\mathbf{p}_0^u \\ -m\mathbf{p}_1^u \end{bmatrix} \tag{8.63}$$

where **q** is the "focal" point and m is a scalar "magnification" factor.

Spend some time contemplating the meaning of Eqs. 8.62 and 8.63 and the associated figures. Try to conjure up images of lenses, focal points, and optical games and applications for them.

EXERCISES

1. Derive the transformation of **B** for the reflection of a bicubic surface through a point.

2. Follow the instructions in Ex. 1 for a Bezier surface.

3. Write a procedure to reflect a set of points through any of the seven principal coordinate system elements; see Eqs. 8.53–8.59. Denote this as **RFLCT**(NP, CI, IFLG, CO), where

NP is the input number of points to be reflected;

CI(NP, 3) is the input array of coordinates of points to be reflected;

IFLG is the input flag specifying the type of reflection; and

CO(NP, 3) is the output array of coordinates of the reflected points.

4. Write a procedure to reflect and magnify a set of points through any point in space. Denote this as **RFLMAG**(NP, CI, P, M, CO), where

NP is the input number of points to be reflected and magnified;

CI(NP, 3) is the input array of coordinates of the set of points;

P(3) is the input array of coordinates of the point through which reflection is to be performed;

M is the input magnification factor; and

CO(NP, 3) is the output array of coordinates of the reflected and magnified points.

5   HOMOGENEOUS TRANSFORMATIONS

Homogeneous coordinates are introduced to facilitate certain types of transformations. These coordinates are particularly useful for projective transformations, and we will later explore this application in both computer graphics and CAD/CAM. For the present, we will investigate the usefulness of homogeneous coordinates in translation, rotation, and scaling. Their great advantage is that they allow the concatenation of any number of individual transformations into a single transformation matrix, an operation that is not possible with the conventional transformations just discussed, since translations require matrix addition.

In homogeneous coordinates, we represent the position vector $\mathbf{p}$ in n-dimensional space by a vector of $(n + 1)$ components. In three dimensions, we represent the usual position vector $[x \quad y \quad z]$ by the four-component vector $[hx \quad hy \quad hz \quad h]$. The additional component in the vector acts as an additional coordinate, and the regular coordinates are related to homogeneous coordinates by

$$x = \frac{hx}{h}$$

$$y = \frac{hy}{h} \tag{8.64}$$

$$z = \frac{hz}{h}$$

There is no unique homogeneous coordinate representation of a point in three-dimensional space. For example, the homogeneous coordinates $[12, 8, 4, 4]$, $[6, 4, 2, 2]$, and $[3, 2, 1, 1]$ all represent the point $[3, 2, 1]$ in ordinary Cartesian space. For ease of computation, we usually choose $[x \quad y \quad z \quad 1]$ in homogeneous coordinates to represent the point $[x \quad y \quad z]$. The former we denote by $\mathbf{p}_h$ and the latter by $\mathbf{p}$.

We accomplish all transformations of homogeneous points by means of a 4×4 transformation matrix. It is here we see the advantage of introducing homogeneous coordinates, since in the 4×4 transformation matrix space is now available to include translation, rotation, and scaling. Let us investigate each of these transformations, first individually and then in combination.

We usually express these transformations as

$$\mathbf{p}_h^* = \mathbf{p}_h \mathbf{T}_h \tag{8.65}$$

where the subscript h denotes homogeneous coordinates for the points $\mathbf{p}_h^*$ and $\mathbf{p}_h$ and transformation matrix $\mathbf{T}_h$. In the following presentation, however, the vectors and transformation matrices are fully expanded.

Translate points with the following matrix product:

$$[x^* \quad y^* \quad z^* \quad 1] = [x \quad y \quad z \quad 1]\begin{bmatrix} 1 & 0 & 0 & 0 \\ 0 & 1 & 0 & 0 \\ 0 & 0 & 1 & 0 \\ t_x & t_y & t_z & 1 \end{bmatrix} \tag{8.66}$$

where t_x, t_y, and t_z are components of the vector $\mathbf{t}$ that describe the translation. Execute the indicated matrix product to obtain

$$[x^* \quad y^* \quad z^* \quad 1] = [x + t_x \quad y + t_y \quad z + t_z \quad 1] \tag{8.67}$$

from which we readily deduce the regular Cartesian transformation of the points; that is,

$$\begin{aligned} x^* &= x + t_x \\ y^* &= y + t_y \\ z^* &= z + t_z \end{aligned} \tag{8.68}$$

Rotate points around a principal axis with one of the three following matrix products:

For rotation around the z-axis,

$$[x^* \quad y^* \quad z^* \quad 1] = [x \quad y \quad z \quad 1]\begin{bmatrix} \cos\theta & \sin\theta & 0 & 0 \\ -\sin\theta & \cos\theta & 0 & 0 \\ 0 & 0 & 1 & 0 \\ 0 & 0 & 0 & 1 \end{bmatrix} \tag{8.69}$$

For rotation around the y-axis,

$$[x^* \quad y^* \quad z^* \quad 1] = [x \quad y \quad z \quad 1]\begin{bmatrix} \cos\beta & 0 & -\sin\beta & 0 \\ 0 & 1 & 0 & 0 \\ \sin\beta & 0 & \cos\beta & 0 \\ 0 & 0 & 0 & 1 \end{bmatrix} \quad (8.70)$$

For rotation around the x-axis.

$$[x^* \quad y^* \quad z^* \quad 1] = [x \quad y \quad z \quad 1]\begin{bmatrix} 1 & 0 & 0 & 0 \\ 0 & \cos\gamma & \sin\gamma & 0 \\ 0 & -\sin\gamma & \cos\gamma & 0 \\ 0 & 0 & 0 & 1 \end{bmatrix} \quad (8.71)$$

The pattern that emerges is quite clear; for example, the last equation can also be written as

$$[x^* \quad y^* \quad z^* \quad 1] = [x \quad y \quad z \quad 1]\begin{bmatrix} & & \vdots & 0 \\ & \mathbf{R}_y & \vdots & 0 \\ & & \vdots & 0 \\ \cdots & \cdots & \cdots & \cdots \\ 0 & 0 & 0 & \vdots & 1 \end{bmatrix} \quad (8.72)$$

where $\mathbf{R}_y$ is the rotation transformation matrix in Eq. 8.13. Furthermore, the general transformation of a homogeneous point for rotation around all three axes in θ, β, γ order is

$$[x^* \quad y^* \quad z^* \quad 1] = [x \quad y \quad z \quad 1]\begin{bmatrix} & & \vdots & 0 \\ & \mathbf{R}_{\theta\beta\gamma} & \vdots & 0 \\ & & \vdots & 0 \\ \cdots & \cdots & \cdots & \cdots \\ 0 & 0 & 0 & \vdots & 1 \end{bmatrix} \quad (8.73)$$

where $\mathbf{R}_{\theta\beta\gamma}$ is given by Eq. 8.15.

Differential scaling of homogeneous points is given by

$$[x^* \quad y^* \quad z^* \quad 1] = [x \quad y \quad z \quad 1]\begin{bmatrix} s_x & 0 & 0 & 0 \\ 0 & s_y & 0 & 0 \\ 0 & 0 & s_z & 0 \\ 0 & 0 & 0 & 1 \end{bmatrix} \quad (8.74)$$

where s_x, s_y, s_z are the scale factors.

To simply combine a set of individual homogeneous transformations into one equivalent transformation, take the product of the individual matrices in the appropriate order, thus combining translation, rotation, and scaling. The total matrix is

$$\mathbf{T}_h = \begin{bmatrix} a & b & c & \vdots & p \\ d & e & f & \vdots & q \\ g & i & j & \vdots & r \\ \cdots & \cdots & \cdots & \cdots & \cdots \\ l & m & n & \vdots & h \end{bmatrix} \quad (8.75)$$

The 3 × 3 submatrix generates the net rotations and scaling. The 1 × 3 row matrix generates the net translation. The 3 × 1 column matrix handles projection effects (we will investigate this later), and, of course, h is the homogeneous coordinate scale factor. If we must apply n transformations, then we simply concatenate the matrix operations as follows:

$$\mathbf{p}_h^* = \mathbf{p}_h \mathbf{T}_1 \mathbf{T}_2 \cdots \mathbf{T}_i \cdots \mathbf{T}_n \qquad (8.76)$$

How can we apply homogeneous coordinate transformations to the curves and surfaces we studied in earlier chapters? Perhaps one of the simplest and most direct ways is to represent these elements in their point format. This is no problem for the characteristic polygon and polyhedron vertex points representing Bezier curves and surfaces. For the Hermite polycubics, simply convert to the four-point, 16-point, or 64-point forms. Next, append the fourth or homogeneous coordinate to each point. Perform the required transformations, and then convert the resulting homogeneous points into three-dimensional Cartesian points.

EXERCISES

1. Using homogeneous coordinates, develop an algorithm for rotating an object around an arbitrary axis given by $\mathbf{a} = \mathbf{b} + u\mathbf{c}$ through an angle α.

2. Find $\mathbf{T}_h$ that translates $\mathbf{p}_0$ of a pc curve to the origin and rotates the curve so that $\mathbf{p}_1$ lies on the x-axis.

3. Repeat Ex. 2 to include a scale factor of 0.5 and reflection across the $x = 0$ plane.

4. Write a procedure to construct a homogeneous transformation matrix for translation, rotation, or scaling. Denote this as **TRANSH(IFLG, CI, TH)**, where

> IFLG is the input flag specifying the type of transformation: translation (IFLG = 1), rotation (IFLG = 2), scaling (IFLG = 3);

> CI(3) is the input array of components of the transformation. Thus, if IFLG = 1, then CI(3) = $[T_x \quad T_y \quad T_z]$, the translation components; or similarly for rotation and scaling; and

> TH(4, 4) is the output array of elements of the transformation matrix.

5. Write a procedure to compute the elements of a homogeneous transformation matrix that is the net result of a sequence of similar transformation matrices. Denote this as **THNET(NT, THI, THO)**, where

> NT is the input number of transformation matrices;

> THI(NT, 4, 4) is the input array of elements of the matrices; and

> THO(4, 4) is the output array of elements of the net transformation matrix.

6. Write a procedure to compute the homogeneous transformation of a set of points. Denote this as **THP(NP, P, TH, PSTR)**, where

NP is the input number of points in the set;

P(NP, 4) is the input array of homogeneous coordinates of the points;

TH(4, 4) is the input array of elements of the transformation matrix; and

PSTR(NP, 4) is the output array of coordinates of the transformed points.

Alternatively, write this procedure to accept points given by Cartesian coordinates and to output points in this same format.

6 COORDINATE SYSTEM AND NONLINEAR TRANSFORMATIONS

Many interesting and useful coordinate system transformations are associated with geometric modeling in computer graphics and CAD/CAM systems. We will investigate them as they arise in later chapters. Most of these transformations are easily developed, being merely the inverse of some rigid-body motion transformation we have already studied.

Such transformations as those from Cartesian to polar, spherical, or cylindrical systems are generally not useful in geometric modeling, since it is particularly awkward and difficult to support parametric representations in these systems. There are, however, useful applications for nonlinear coordinate system transformations; a two-dimensional example is given in Fig. 8.15. Here, we see the basic ingredients for the transformation of a curve in a rectangular coordinate system to a curved-axis coordinate system.

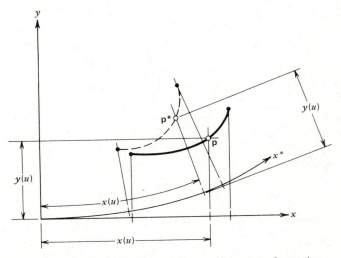

FIGURE 8.15 Curved axis coordinate system transformation.

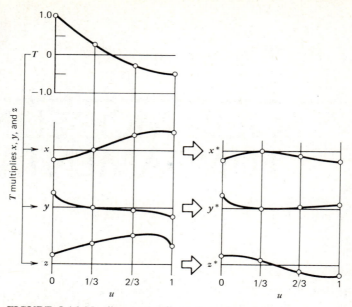

FIGURE 8.16 Nonlinear transformation of a curve.

To transform from the rectangular to the curved-axis system, for each point **p** on the original curve, find $x(u)$ and $y(u)$. In the curved-axis system, proceed along x^* a distance $x(u)$ and then a distance $y(u)$ perpendicular to x^* to locate **p***. *Note*: x^* may itself be given as a curve in the x, y coordinate system. By repeatedly applying this procedure to a sufficient number of points, we can express the transformed curve more completely. For example, if the initial curve is a parametric cubic curve, then transforming the four points $\mathbf{p}(0)$, $\mathbf{p}(1/3)$, $\mathbf{p}(2/3)$, and $\mathbf{p}(1)$ allows us to formulate the transformed curve in the four-point form.

This process is similar to transforming a curve (or any other parametric element) according to some general, perhaps nonlinear rule. The rule may be derived from some physical property inherent in the object that the curve represents. Here, the transformation of the curve corresponds to the response of the object to some condition applied to it. We can express such a rule parametrically and relate it to the curve through a one-to-one correspondence of the parametric variables. Figure 8.16 illustrates such a transformation. In this instance, we operate on the points corresponding to $u = 0$, 1/3, 2/3, and 1 to generate the transformed pc curve. Here, the operation is a simple multiplication; however, it could be additive or any analytically expressible rule. Furthermore, the rule could be different for each parametric variable. *Note*: It may take more than one curve to represent the transformation to a specified degree of accuracy.

9

SOLID MODELING FUNDAMENTALS

A solid geometric model is the unambiguous and informationally complete mathematical representation of the shape of a physical object in a form that a computer can process. The three principal types of solid models are wireframe, boundary surface, and solid. There are advantages and disadvantages to all three, which will emerge in the course of this and the next two chapters.

Solid modeling is an important aspect of geometric modeling that is used to create and communicate shape information. It involves creating and maintaining a solid model for future access and analysis. We can say that A is a model of B if A can be used to answer questions about B. In solid modeling, this means asking and answering questions about an object's volumetric properties, such as weight and moment of inertia, and about similarly appropriate topological properties, such as connectivity and containment relationships.

We will investigate various types of representation schemes. An important one, developed at the University of Rochester, is called **constructive solid geometry** (CSG), also known as unevaluated, implicit, direct, procedural, or algorithmic solid modeling. In CSG, an object is described in terms of elementary shapes (half-spaces) or primitives (bounded primitive solids). We represent the object by a binary tree of set or Boolean operations. The leaf nodes are solid primitive shapes sized and positioned in space, and the branch nodes are the set operators of union, difference, and intersection. In the case of half-spaces, we can represent a simple block as the intersection of six planar half-spaces.

Another important solid-representation method that we will encounter is the boundary model (sometimes called explicit, evaluated, or indirect solid modeling). Here, we represent a solid in terms of its spatial boundary, usually the enclosing surface with some convention to indicate on which side of the surface the solid lies. We can represent a solid as unions of faces, bound by edges, which are bound by vertices. Faces lie on surfaces, edges lie on curves, and vertices are at edge end points. A boundary model stores the mathematical data of the surface geometry on which the face lies, the curve geometry on which the edge lies, and the point geometry (coordinates) of the vertex.

Another scheme represents solids by means of a set of planar facets (sometimes called polyhedral or tessellated models); the facet edges are straight lines. This scheme is a special case of boundary representation where curved surfaces and edges are approximated by the planes and lines.

These schemes rely on a variety of construction operations. Set operations are common to many and use the previously mentioned Boolean operators: union, intersection, and difference. Sweep operators transform two-dimensional shapes into three-dimensional prismatic or axisymmetric shapes by extruding an object by means of a translational rule or a rotational sweep. There is a variety of hybrid schemes for making local changes to a solid model without changing the topology. Euler operators are often used to make faces, edges, and vertices and ensure topological consistency and validity.

Finally, a solid modeling system usually maintains two principal types of data describing a model—geometric data and topological data. Geometric data consist of the basic shape-defining parameters, for example, the geometric coefficients of a bicubic surface or the coordinates of the vertex points of the characteristic polyhedron of a Bezier surface. Topological data include the connectivity relationships among geometric components.

Before we begin constructing a solid model of some object, however, we must understand the principles that ensure a valid model. Although there are several major categories of techniques for model building, underlying all of them is a set of fundamental, almost universal, geometric principles to which all modeling schemes must conform. For example, all real, solid, three-dimensional objects are subject to certain topological restrictions; therefore, we will study the topology of closed paths, piecewise flat surfaces, and closed curved surfaces.

We will find that it is convenient to focus on the boundaries of objects, whether they are two-, three-, or n-dimensional. We should have, therefore, a general concept of a boundary. Set theory offers consistency and rigor in expressing both model topology and boundary relationships.

Model synthesis calls for a set of operators to accomplish the synthesis; the most popular and successful to date are the Boolean and Euler operators. We will look at these together with set-membership classification and formal modeling criteria.

1 TOPOLOGY OF CLOSED PATHS

In this section, we investigate the topology of closed paths and its relationship to problems in geometric modeling. We will develop the closed-path theorem, the effect of deformations on the topology of curves and planes, the Jordan curve theorem, and the concept of total curvature. We will find many useful applications for this material in model construction and evaluation.

D. Hilbert and S. Cohn-Vossen (1952), P. Alexandroff (1961), and, more recently, H. Abelson and A. Disessa (1981) have written thoroughly and imaginatively on this subject and the subjects discussed in Sections 2 and 3. These authors are the primary sources for Sections 1–3, and reference to their work is highly recommended. Abelson and Disessa, in particular, offer unique intuitive insights into fundamental topological principles highly relevant to geometric modeling.

1.1 Closed-Path Theorem

We begin our discussion with the **closed-path theorem**, which states: The total turning along any closed path is an integer multiple of 360°. The integer, denoted N_R, is called the **rotation number** of a path. It is an intrinsic property of a path and independent of where a path starts or how it is oriented; Fig. 9.1

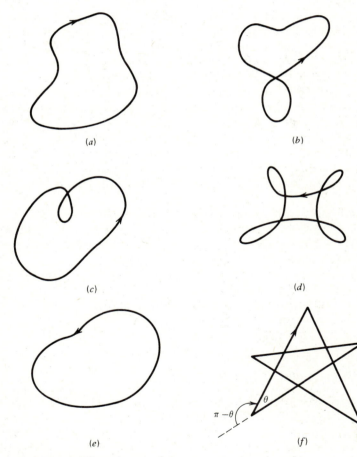

(a)

(b)

(c)

(d)

(e)

(f)

FIGURE 9.1 Rotation number.

provides several examples, including loops. Note that for the moment, we confine our discussion to closed paths on a plane surface.

Observe that simple polygons always appear to have total turning equal to $+360°$ or $-360°$, depending on the direction in which we traverse the path. Self-intersecting polygons, however, always have total turning different from $\pm 360°$. Observe that there are two different classes of paths: simple polygons, and self-intersecting polygons, often called star polygons.

The **simple-closed-path theorem** states that the total turning in a non-self-intersecting closed path is $\pm 360°$ (clockwise or counterclockwise). In other words, the rotation number N_R of any simple closed path is ± 1. Look at some examples of simple closed paths to convince yourself of the validity of this theorem, which, as simple as it appears, happens to be very difficult to prove rigorously. Note that the theorem implies that there is a relationship between the two properties of a closed path — the turning and the crossing points. This makes it considerably less obvious than the closed-path theorem but also much more powerful.

Curves defining closed paths that can be deformed into one another are curves that are topologically equivalent. Observe that total turning is a topological invariant for closed paths. Any two closed paths that are topologically equivalent must have the same total turning. Now, we must ask if the converse is true: If two paths have the same total turning, can they be deformed into one another? The answer is yes, and it was first proved in 1936 by H. Whitney and W. C. Graustein, who observed that two closed coplanar paths can be deformed into one another if, and only if, they have the same total turning.

The simple-closed-path theorem is more comprehensive than the closed-path theorem, and its proof is considerably more complicated. The simple-closed-path theorem is a link between local and global information. Since crossing points on a curve are nonlocal phenomena, there is no direct way for a traversal algorithm (that is, an algorithm proceeding stepwise around the path) to detect a crossing point. To do this requires "looking" at the entire curve at once. However, we can determine total turning by "local" computation, and this theorem relates total turning to the existence of crossing points, a powerful and useful capability.

Let us investigate an example of the link between local and global information contained in the theorem: Walk around a closed path, as defined by a set of written instructions, accumulating your total turning (take clockwise as positive). When you follow the path and return to your starting point, you will find that the total turning does not equal $\pm 360°$. Now, you can assert that somewhere the path must have at least one crossing point. Although you do not know where the crossing points are and you were unable to observe them while traversing the path, by applying the theorem, you can determine that one or more crossing points must exist. Thus, the simple-closed-path theorem is an example of a powerful principle: It is often possible to determine global properties accumulating local information. This theorem can be useful when we must determine the validity of a model or in the course of actual synthesis or model evaluation.

1.2 Deformation of Curves and Planes

Now, let us turn our attention to the consequences of deformations of curves and planes. Here, we develop proofs of the simple-closed-path theorem and some related theorems about the topology of simple closed curves.

We must show that, given a simple closed path (a path with no crossing points), then the total turning around the path is equal to $\pm 360°$. There are many paths for which this result is obvious; the simplest is the square. This suggests a strategy for proving the theorem by showing that any simple closed path can be deformed into a square. Since we already know that total turning is invariant under deformations, we can show that any simple closed path has the same total turning as a square—$\pm 360°$—depending on the direction around the square. This reduces the problem of proving the simple-closed-path theorem to showing that any simple closed curve can be deformed into a square.

In fact, we can go a bit farther: Given any simple closed path, not only can the path itself be deformed into a square, but we can imagine the deformation being done in a very special way. The entire plane can be deformed, pulled and stretched, so that the path becomes a square. For that matter, we even consider the plane itself deformed into a curved surface.

Until now, we have talked about deformations of paths. Now, we will also consider deformations of the plane. Imagine that the plane is an arbitrarily stretchable rubber sheet. Then any kind of stretching or shrinking that does

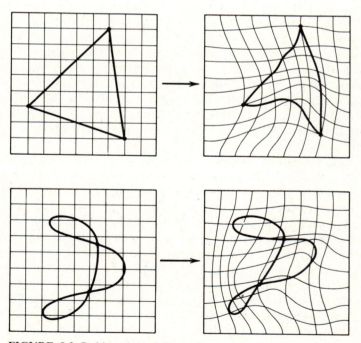

FIGURE 9.2 Rubber sheet deformation.

not result in cutting or tearing is a valid deformation. These two kinds of deformation are closely related. In particular, if we draw a closed path on the "rubber sheet" plane, then any deformation of the plane results in a deformation of the path (Fig. 9.2).

Imagine that the rubber sheet is stretched over a flat surface so that it remains flat. Clearly, straight-line segments may curve under such a deformation. Many of the changes that are legitimate for the deformation of curves, such as the overlap phenomenon, do not and cannot happen with plane deformation. We observe that crossing points are neither created nor destroyed during plane deformation.

Plane deformations are more conservative than path deformations. We see that every plane deformation is a path deformation, but path deformations that introduce crossovers are too violent to be plane deformations. The mathematical term for a path deformation is **regular homotopy**, while a rubber sheet deformation is an **ambient isotopy**.

The most important observation to make is that any simple closed path can be deformed into a square and, moreover, this can be done with a conservative deformation, that is a plane deformation.

We summarize our observations in this section by stating the deformation theorem for simple closed curves: For any simple closed curve in the plane, there is a rubber sheet deformation of the plane that reduces the curve to a square.

1.3   The Jordan Curve Theorem

Another result of the topology of simple closed paths that also follows from the deformation theorem is the **Jordan curve theorem**, which states: Any simple closed curve in the plane divides the plane into exactly two regions, an "inside" and an "outside." Such properties as "dividing the plane into two regions" and "having an inside and an outside" are invariant under rubber sheet deformations of the plane. If the deformed curve has these properties, then so must the original nondeformed curve.

The Jordan curve theorem probably seems obvious, but "simple" closed curves can be arbitrarily complex and convoluted. *Note*: The theorem is not true if we consider curves on surfaces other than the plane. For example, a simple closed curve can be drawn on a torus, and yet it does not divide the torus into two regions.

Not only does a closed curve have an inside and outside, but the inside itself is deformable, in the rubber sheet sense, into the interior of a square. A region that can be deformed into the interior of a square is called a **topological disk**. A topological disk has no holes or isolated points in it. *Note*: Deforming a rubber sheet region does not require keeping the region a part of the flat plane (that is, the region may be planar or nonplanar). Figure 9.3 shows a curve deformed into a rectilinear polygon, and we clearly see that the total turning is reduced to summing clockwise and counterclockwise 90° turns or angles.

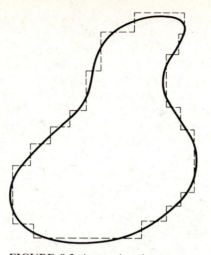

FIGURE 9.3 Approximating a curve by a rectilinear path on a grid.

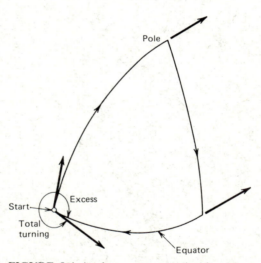

FIGURE 9.4 Angle excess associated with a closed path on a sphere.

1.4 Angle Excess

The concept mathematicians call **angle excess**, or simply **excess**, is by definition the total turning that some reference pointer undergoes when carried around a closed path. Excess is a rather general concept; it is an angle associated with a closed path on a curved surface. We can restate the closed-path theorem so that it holds for simple closed paths on arbitrary surfaces

$$T + E = 360° \tag{9.1}$$

where T = total turning along the path and E = excess along the path.

Figure 9.4 illustrates the angle excess associated with a closed path on a sphere. A trip along this path might start at the equator, proceed to the pole, then return to the equator along a different meridian, and continue along the equator to the starting point. Notice that the reference pointer is always transported parallel to itself relative to the surface in which the path lies. For a fixed path, the excess is the same no matter where the trip begins.

Angle excess is additive, and we find the following theorems apply:

1. If a triangle is subdivided into two subtriangles, then the excess of the triangle is the sum of the excesses of the pieces.

2. The excess of any polygon is the sum of the areas of the pieces in any polygonal subdivision.

3. For any topological disk on an arbitrary surface, the angle excess around the boundary is equal to the total curvature of the interior.

1.5   Total Curvature and Curvature Density

We can show that the total curvature for a sphere is 4π and the angle excess is proportional to the area $E = kA$. Thus, for any polygon on a sphere of radius r, $E = ka$, where $k = 1/r^2$ (if E is measured in radians). For a simple closed path on the surface of a sphere, total turning in radians is expressed as $T = 2\pi - A/r^2$, where A is the area enclosed by the path and r is the radius of the sphere.

In general, for an arbitrary surface, k is a measurement made at any point on the surface. The value of k at a point is the excess per unit area of a small patch of surface containing the point. We call k the **curvature density** of the surface at a point. We use the term *density* because k is "something per unit area" and the term *curvature density* rather than *excess density* because, although k is measured using excess, it tells us how "curved" the surface is at the point. To approximate a small patch of an arbitrary surface by a small patch of some surface we know very well, we observe that within a small enough area, almost any surface will appear planar.

We can make an even better approximation by using a small piece of a sphere. We will choose the approximating sphere to have the same excess per unit area as the patch of a surface. Therefore, if k does not change radically in the small patch, all the geometry there — angles, total turning, and so on — is then very close to the geometry on a sphere whose radius is determined by $k = 1/r^2$. The smaller the radius of the approximating sphere (that is, the greater the curvature density), the more curved we say the surface is. A football, for example, is not too curved in the middle; k is small there. But at the pointed ends, the football is curved as much as a sphere of small radius. From the middle to the pointed ends, k gradually increases.

Curvature density is a local quantity that is measured in the vicinity of a point on an arbitrary surface. There is a global version of curvature density called **total curvature**. Start with a region of an arbitrary surface and divide it into polygonal pieces. For each polygon, compute the excess; now, sum the excesses for all the pieces. The result is called the total curvature K of the region. Of course, in order for this definition to make sense, we must be sure that if we subdivide a region into polygons in two different ways, the sum of the excess of the pieces is the same in both cases.

If the initial region is itself a polygon, the additivity theorem implies that K is precisely equal to the excess around the boundary of the polygon. Therefore, think of K as a kind of excess over the region, with the condition that it is true for any region on the surface, not just for polygons.

1.6   Spheres, Tori, and Handles

Let us compute K for a region that has no boundary at all — the sphere. Divide a sphere into two pieces, for example, the northern and southern hemispheres. Each of them is bounded by the equator, which we know has excess 2π. Thus, the two hemispheres each have excess 2π, yielding 4π for the total curvature of the spheres. Observe that the curvature density k of a sphere depends on the radius, but the total curvature K is the same for all spheres.

It turns out that the total curvature K is a topological invariant for closed surfaces. For surfaces with a boundary, the total curvature is unchanged by deformations that do not affect the vicinity of the boundary of the surface.

Any torus has zero total curvature. We are not asserting that a torus is flat like a cylinder or a plane, which would imply that the curvature density is zero everywhere. We find, instead, that any torus has just as much negative curvature as positive curvature.

We now know the total curvature of two kinds of closed surfaces, spheres and tori. A sphere is not topologically the same as a torus, yet there is a relationship between them. The torus can be described as a sphere with a handle attached. The process of "handle attaching," though not a deformation, is a practical, general way of making new closed surfaces from old ones — practical because it is easy to keep track of how total curvature changes in the process. Let us explore this process a little more.

Start with a sphere, and flatten out two regions on it. From each region, cut out a flat disk; now, attach a "handle" in the holes left by the disks. The handle is topologically a cylinder, but the edges of the cylinder must be flared out to blend with the flat regions around the disks. The result is a torus; see Fig. 9.5. In other words,

$$\text{sphere} - 2 \text{ disks} + \text{handle} = \text{torus} \tag{9.2}$$

FIGURE 9.5 Sphere with handle.

Next, compute the total curvature on each side of the equation. The sphere has $K = 4\pi$; the disks were flat and thus have $K = 0$. The torus, as we just asserted, also has $K = 0$. Then, for the equation to be balanced, the handle must have total curvature $= -4\pi$.

With this process of flattening, cutting, and gluing, we can attach a handle to any surface. Since we know the curvature in the handle now, we see that it always decreases the total curvature by 4π. We find that the total curvature of a two-holed torus is the same as that of a torus with a handle attached or a sphere with two handles attached. This is expressed as

$$K_{\text{2-holed torus}} = 4\pi - 4\pi - 4\pi = -4\pi \qquad (9.3)$$

In general, for a surface that is topologically the same as a sphere with g handles attached, total curvature is given by

$$K_{\text{(sphere with } g \text{ handles)}} = 4\pi(1 - g) \qquad (9.4)$$

The importance of knowing about spheres with handles is: Any closed surface in three-dimensional space is topologically equivalent to a sphere with some number of handles attached. For example, the surface in Fig. 9.6a is equivalent to a sphere with six handles. What about the object in Fig. 9.6b?

We see from Eq. 9.4 that the total curvature of any closed surface in three-dimensional space is an integer multiple of 4π. This topological characteristic of closed surfaces is immediately applicable to the synthesis of solid models of complex objects, particularly boundary representations, and also as an analytic characteristic or determinant of topological model type.

There are closed surfaces that do not have a total curvature that is a multiple of 4π and, as a consequence, are too twisted to fit into three-dimensional space. As it turns out, any closed surface, including these twisted ones, must have total curvature equal to a multiple of 2π. This class of surfaces is not significant in geometric modeling.

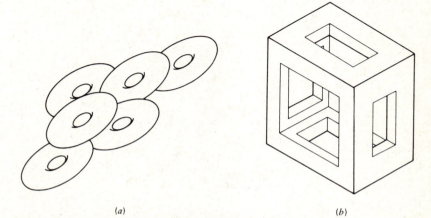

(a)   (b)

FIGURE 9.6 Topological equivalent of a sphere with six handles.

Remember, total turning for a closed curve path is always an integer multiple of 2π. Now, we find an analogous property for closed surfaces in three-dimensional space. All that arbitrary denting and bending and all those excess angles must somehow combine to give precisely an integer multiple of 4π. In changing the total curvature of a closed surface, we must change it by a multiple of 4π or not at all.

1.7 Knots

Before we end this exploration of the topological properties of closed paths, we will take a last look at the Jordan curve theorem and curves forming knots.

Consider a circle C in the plane. It is obvious that C divides the plane into two regions A and B, which are characterized by the property that any two points within any one of the regions (such as a, b and c, d) can be connected by a path that lies wholly within the region, whereas if we take a point from each of the regions (such as a and c), then every path between them will necessarily cut C; see Fig. 9.7. The bounded region A can be called the inside of C and B, the outside. The first proof of the now-famous Jordan curve theorem, that every simple closed curve in the plane divides the plane into two regions, appeared in 1893 in the important book, *Cours d'Analyse*, by C. Jordan (1838–1922), although Jordan did not solve the problem completely. Even though Jordan's original proof has by now been greatly simplified, it is still not easy to prove this apparently obvious fact. The well-defined notion of inside and outside for simple closed curves has direct bearing on many geometric-modeling problems. Incidentally, this fact is also useful for those who want to build or escape from mazes.

Jordan's theorem also has to do with imbedded circles in two-dimensional space. An imbedded circle in three-dimensional space is called a **knot**. The basic problem of knot theory is determining when two knots are equivalent, that is, when they can be transformed into each other by twisting and pulling. Figure 9.8 presents an example of two unequivalent knots. Try working out these ideas

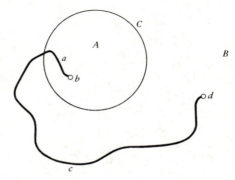

FIGURE 9.7 Division of a plane into two regions by a closed curve.

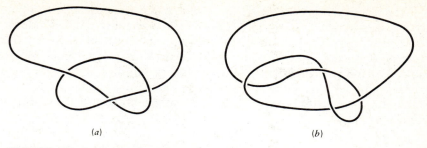

(a)   (b)

FIGURE 9.8 Jordan curve theorem and knots.

with pieces of string. To distinguish between different knot types in this context requires keeping the end points of the string fixed. By bringing the ends together to form a circle, we have another familiar problem, but one that is easier to handle mathematically.

Call a knot trivial or unknotted if it can be deformed into a conventional circle. A special case of the general problem is determining when a knot is unknotted. For example, if we are given a pile of string, how can we tell if there is a knot present? *Note*: We cannot distinguish between knots intrinsically because they are all the topological equivalent of circles. Our problem is to distinguish between various possible placements of the same circle in three-dimensional space, and thus this problem is different from other topological problems where we want to distinguish between objects.

Gauss made some brief remarks on knot theory at various times between 1823 and 1840. He attempted to classify singular curves in the plane, some of which arose from the projection of a knot, and in 1833, he found a formula connecting a certain double integral with the **linking number** of two knots, which is intuitively the number of times two knots are intertwined; see Fig. 9.9.

The first mathematician, however, to make a serious study of knots was Listing. His approach was to project the knot onto the plane and analyze it by considering the intersections, keeping track of over- and undercrossings. Such a method was also used later by P. G. Tait, T. P. Kirkman, and N. C. Little to classify the more elementary knots.

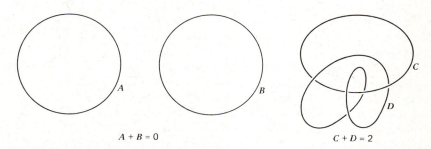

$A + B = 0$ $C + D = 2$

FIGURE 9.9 Linking numbers of knots.

The first invariant characteristics of knots that were effectively subjected to computation were found in the 1920s by J. W. Alexander and K. Reidemeist. These invariants were obtained by studying the complement of the knot in three-dimensional Euclidean space and its associated "knot group," which M. Dehn had defined in 1910. Knot theory, however, is very difficult, and many important unanswered questions remain. We have mentioned this topic only because it has another topological characteristic that might prove useful in geometric modeling.

EXERCISES

1. Discuss in terms of continuity any procedural rules necessary for traversing a simple closed path and determining its rotation number.

2. Discuss the implications of rubber sheet deformations on the topology of simple closed curves on a surface when the surface is subject to any transformations allowed by the homogeneous transformation matrix T_h. Can crossover points be induced? How is curve continuity affected? How are other local and global properties affected?

3. A more rigorous definition of a topological disk is: a surface region bounded by a closed curve that can be deformed and shrunk to a singular point without cutting the surface. Show that on a torus, there are two families of closed curves that are not boundaries of topological disks.

4. Demonstrate the validity of the additivity of angle excess using spherical triangles.

5. Show that angle excess is always zero around a closed path on a plane surface.

6. Show that the angle excess around the equator of a sphere is 2π. *Hint*: The total turning between any two points on a great circle of a sphere is zero.

2 PIECEWISE FLAT SURFACES

Many of the existing and potential future representation schemes for modeling solid objects use polyhedra and their supporting topology, so there are many opportunities for applying the material presented here. The classical polyhedron is an excellent example with which to begin a discussion of piecewise flat surfaces.

By **polyhedron** we mean an arrangement of polygons such that two and only two polygons meet at an edge, and it is possible to traverse the surface of the polyhedron by crossing its edges and moving from one polygonal face to another until all polygons have been traversed by this continuous path.

The simple polyhedra are the most important, since they are historically the source of topology's contribution to geometric modeling. The term **simple polyhedra** refers to all polyhedra that can be continuously deformed into a sphere.

Regular polyhedra are an example and subset of the simple polyhedra. Regular polyhedra have no reentrant edges; thus, they are convex. Furthermore, we apply the term **convex** to every polyhedron that lies entirely on one side of each of its polygonal faces. Although convexity is not a topological property, it does imply one: Every convex polyhedron is a simple polyhedron. A toroidal polyhedron is a nonsimple polyhedron.

There is a well-known relationship between the number of vertices, edges, and faces of a simple polyhedron, called Euler's formula for polyhedra. Let V, E, and F represent the number of vertices, edges, and faces of a polyhedron. Euler's formula for all simple polyhedra states that

$$V - E + F = 2 \tag{9.5}$$

This formula provides a direct and simple proof that there are only five regular polyhedra.

Let us take a slight digression and explore an application of Euler's theorem to geometric modeling of a more classical vintage. We will determine all possible regular polyhedra, that is, those polyhedra with every face having the same number of edges, say, h; every vertex having the same number of edges emanating from it, say, k; and every edge having the same length. Since every edge has two vertices and belongs to exactly two faces, it follows that $Fh = 2E = Vk$. Substitute this into Euler's formula

$$\frac{2E}{k} - E + \frac{2E}{h} = 2 \tag{9.6}$$

or

$$\frac{1}{E} = \frac{1}{h} + \frac{1}{k} - \frac{1}{2} \tag{9.7}$$

For a polyhedron, we safely assume that $h, k \geq 3$. On the other hand, if both h and k were larger than three, then the above equation would imply that

$$0 < \frac{1}{E} = \frac{1}{h} + \frac{1}{k} - \frac{1}{2} \leq \frac{1}{4} + \frac{1}{4} - \frac{1}{2} = 0 \tag{9.8}$$

which is obviously impossible. Therefore, either h or k equals 3. If $h = 3$, then

$$0 < \frac{1}{E} = \frac{1}{3} + \frac{1}{k} - \frac{1}{2} \tag{9.9}$$

implies that $3 \leq k \leq 5$. By symmetry, if $k = 3$, then $3 \leq h \leq 5$. Thus $(h, k, E) = (3, 3, 6)$, $(4, 3, 12)$, $(3, 4, 12)$, $(5, 3, 30)$, and $(3, 5, 30)$ are the only possibilities. They are, in fact, realized by the tetrahedron, the cube, the octahedron, the dodecahedron, and the icosahedron, respectively. Observe that we did not really use the fact that the edges of the polyhedron all have the same length, so that as long as the numbers h and k are constant, we still have only five possibilities (up to stretching or contracting).

Poincaré generalized Euler's formula to n-dimensional space. Instead of points, edges, and faces, he defined 0-, 1-, 2-,..., n-1-dimensional elements. He denoted the numbers of each of these elements present in a "hyper" polyhedron (or polytope) as N_0, N_1, N_2,..., N_{n-1}, respectively, and expressed Euler's formula as

$$N_0 - N_1 + N_2 - \cdots = 1 - (-1)^n \qquad (9.10)$$

For $n = 3$, this reduces to Euler's formula.

These regular polyhedra were known to Euclid, and he even claimed that there were no others. Euclid's definition of a "regular polyhedron" was, by today's standards, rather imprecise, however, and his claim is not quite correct in the generality with which he states it.

This is our cue to investigate nonsimple polyhedra. They are topological equivalents of any solid object with holes in it and are, therefore, of direct use to us in geometric modeling. In Fig. 9.10, there are some examples of nonsimple polyhedra. The simplest of these is a rectangular parallelepiped with a through hole, also in the shape of a parallelepiped whose sides are parallel to the outer faces. The two end faces are beveled, or "faceted," only to emphasize the properly interconnected grid of edges, vertices, and faces. The other faces are treated the same way but have more holes. None of them can be deformed into a sphere.

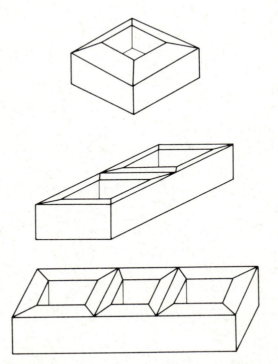

FIGURE 9.10 Examples of non-simple polyhedra.

We assign a connectivity number n to every polyhedron. If the surface of a polyhedron is divided into two separate regions by every closed path (loop) defined by edges of the polygons making up its faces, then we say the polyhedron has connectivity $n = 0$. All simple polyhedra have $n = 0$, because the surface of a sphere is divided into two parts by any closed curve lying on it. Conversely, any polyhedron with $n = 0$ can be deformed into a sphere.

We can define any orientable unbounded surface, that is, any closed non-self-intersecting surface, as a sphere with g handles. Thus, $g = 0$ for a sphere or any simply connected surface $g = 1$ for a torus and $g = 2$ for the surface of a solid figure eight. The number g is called the genus of the surface.

As we see in Fig. 9.10, there are closed loops of edges that do not divide the surface of a nonsimple polyhedron into two parts. We assign these polyhedra a connectivity number greater than one. We define the connectivity number n as the maximum number of distinct loops with or without common points (intersections) to be found on a polyhedron that do not divide its surface into two separate regions. We define the genus g of a polyhedron as the maximum number of nonintersecting loops to be found that do not divide its surface into two regions. Furthermore, we can extend Euler's formula to polyhedra of any connectivity n. Thus,

$$V - E + F = 2 - 2g \tag{9.11}$$

where $2g = n$, the connectivity number.

Conversely, by counting the vertices, edges, and faces, we can determine the connectivity and genus

$$n = -V + E - F + 2$$

$$g = \frac{(-V + E - F + 2)}{2} \tag{9.12}$$

Thus, we can deform simple polyhedra into a sphere and the rectangular parallelepiped with a through hole into a torus. We can also replace more complicated structures by topologically equivalent polyhedra.

We will now take a somewhat more general approach to developing additional topological properties. Surfaces formed by taking a collection of planar pieces and gluing them together along their edges are called **piecewise flat surfaces**. Any surface formed in this way will obviously be flat everywhere except possibly along the edges where the pieces are glued together. In fact, if all the planar pieces have straight edges, then the glued surface can have curvature at only the vertices.

The crucial step in designing a computer program to explore piecewise flat geometry is deciding how to represent these surfaces. The most straightforward way is simply to describe each face separately and keep track of which edges are adjoining. This kind of representation is called an **atlas**, Fig. 9.11 shows the atlas of a truncated pyramid. This topological atlas is similar to an ordinary road atlas, which is a collection of separate maps, each containing information directing the user to the next map.

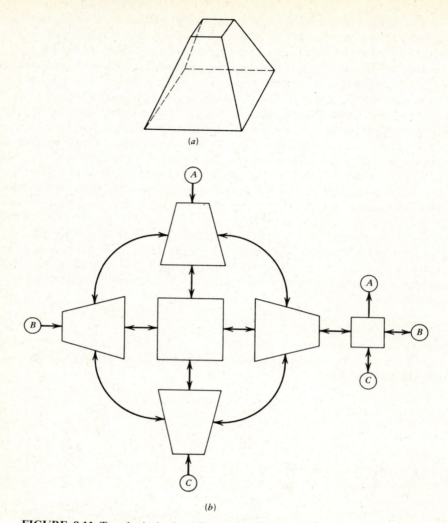

FIGURE 9.11 Topological atlas of a truncated pyramid.

Figure 9.12 illustrates two different ways of joining a pair of edges. The first, shown on the left, is called an **orientation-preserving** identification. The other possibility, shown on the right, is called an **orientation-reversing** identification. Each edge is numbered, and an arrow indicates the ascending direction of the numbers. Opposite edges are joined so that numbers match and the direction of the two arrows is the same. Either numbers or arrows can be used alone to identify the way the two edges are to be joined. Orientation-reversing identifications have both arrows pointing in the same rotational sense—either clockwise or counterclockwise—around their respective faces. However, if one arrow points clockwise and the other counterclockwise, the identification is

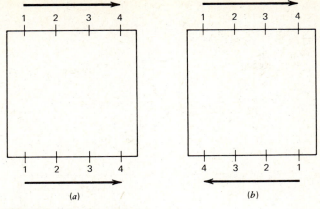

FIGURE 9.12 Orientation.

orientation preserving. Including orientation-reversing joints in an atlas enables us to construct new surfaces, including some mathematical curiosities called **nonorientable surfaces**.

Study the example in Fig. 9.13. Take the collection of matched-edge pairs as the atlas for a closed piecewise square surface. Each edge is labeled by a pair of numbers specifying the face and edge of the face of interest. Number the faces sequentially starting from 1, and number the four edges of each face clockwise from 0–3. For the cube shown, the atlas is

$$[(1, 0) \quad (2, 0)] \: [(1, 1) \quad (5, 0)] \: [(1, 2) \quad (4, 0)] \: [(1, 3) \quad (3, 0)]$$

$$[(2, 1) \quad (3, 3)] \: [(2, 2) \quad (6, 2)] \: [(2, 3) \quad (5, 1)] \: [(3, 1) \quad (4, 3)]$$

$$[(3, 2) \quad (6, 3)] \: [(4, 1) \quad (5, 3)] \: [(4, 2) \quad (6, 0)] \: [(5, 2) \quad (6, 1)]$$

Interpret this as edge 0 of face 1 matched with edge 0 of face 2, and so on.

Take a square and identify a pair of opposite sides in an orientation-preserving way to produce a cylinder, or, instead, identify the edges with the orientation reversed. Try this with a strip of paper. This identification amounts to making a half twist in the strip before gluing the edges together, resulting in a surface called a **Mobius strip**.

An inhabitant of a Mobius strip observes a curious phenomenon. If it starts out at some point and takes a trip all the way around the strip, when it gets back to its initial position, it finds that left and right are reversed. This happens because right and left are not intrinsically defined. Externally, whether a turn appears left or right depends on the side of the surface we are looking from. It is an extrinsic property of the surface and depends on setting up an external reference. Thus, we must specify that the observation is from the top side or the bottom side.

Right and left are characteristics of the two-dimensional Mobius strip inhabitant's motions, not characteristics of the surface. If we visualize ourselves as inhabitants of the Mobius strip, we can define right and left on the surface,

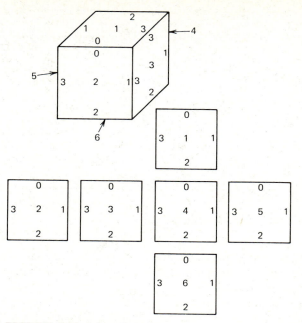

FIGURE 9.13 Atlas of a cube.

but the definition works only locally. We cannot look at some point on the surface, then decide a priori that one direction is either right or left. We must first move around on the surface. There is nothing to ensure that, as we move around the surface, the commands left and right will generate the same orientation each time we return to a given point. This potential reversal and confusion is exactly what happens on the Mobius strip.

A surface on which left and right are never reversed is called **orientable**, and we can establish a consistent left and right definition on the surface. If we find a path that confuses left and right, the surface is called **nonorientable**. The orientability of a surface is easily demonstrated in terms of another property, too. For example, an orientable surface is one on which we can define a notion of clockwise and counterclockwise in a consistent way. This means we choose a point p on the surface S and imagine standing at that point. Next, we decide which of the two possible rotations to call clockwise; this is called an orientation at p. Now, let q be another point on S. (**Note**: q may equal p.) As we walk to q along some path, we keep track of which rotation is defined as clockwise. This induces an orientation at q, that is, a sense of clockwise or counterclockwise for rotations at q.

There are obviously many paths from p to q. (Furthermore, there is not always a unique shortest path.) Different paths may induce different orientations. If we obtain the same orientation no matter which path we take, then the surface is orientable. However, if we walk around the meridian of a Mobius strip, we will end up with the opposite orientation from our original one. Therefore, we say that the Mobius strip is nonorientable, and our new definition agrees with the old. We conclude that orientability is an intrinsic property of surfaces.

As we learned in Section 1, any closed surface that fits into three-dimensional space must be topologically the equivalent of a sphere with g handles, where g can equal zero. Now, we add another condition, namely, the surface must be orientable. A sphere is orientable, and it is not hard to show that adding a handle cannot change that. Note that the Mobius strip does fit into three-dimesional space, but it is not closed. To transform the Mobius strip into a closed surface, we make an orientation-preserving identification of the top and bottom edges on the atlas exactly as we did when producing a torus from a cylinder (see Fig. 9.14). The resulting surface is called a **Klein bottle**, and like the Mobius strip, it is nonorientable.

Producing the Klein bottle involves sticking the bottle through itself—a rather drastic operation, but the best we can do in three dimensions. The Klein bottle will not fit into a three-dimensional space without self-intersections.

If we start with a Mobius strip and close it up by making a second orienta-tion-reversing identification, we produce another type of surface, called the

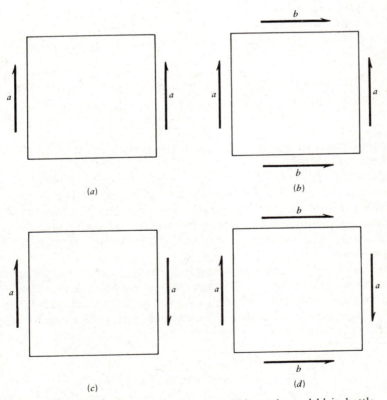

FIGURE 9.14 Atlas of a cylinder, torus, mobius strip, and klein bottle.

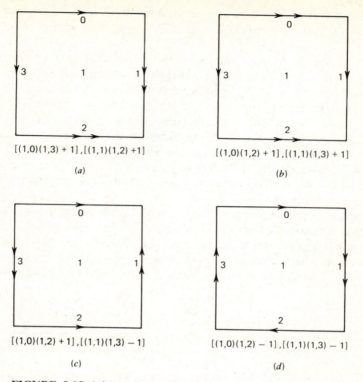

$[(1,0)(1,3) + 1],[(1,1)(1,2) +1]$

(a)

$[(1,0)(1,2) + 1],[(1,1)(1,3) + 1]$

(b)

$[(1,0)(1,2) + 1],[(1,1)(1,3) - 1]$

(c)

$[(1,0)(1,2) - 1],[(1,1)(1,3) - 1]$

(d)

FIGURE 9.15 Atlas and transition parity of a sphere, torus, klein bottle, and projective plane.

projective plane. The projective plane is nonorientable and topologically distinct from a Klein bottle.

The atlas must specify not only which edges are identified, but also whether the orientation is reversed. We denote this by including a $+1$ or -1, called the **transition parity**. Figure 9.15 presents several examples of a notation scheme including the transition parity number. It should now be obvious that simple "bookkeeping" changes produce dramatic changes in the topological properties of a surface.

So far, we have discussed Euler's formula, atlases, and orientability; now, let us turn our attention to the **curvature of piecewise flat surfaces**. All the curvature in a piecewise flat surface is concentrated at the vertices, which makes total curvature easy to compute. We need only sum up the angle excesses of small paths around each of the vertices; thus,

$$K = \sum_{i=1}^{v} E_i \qquad (9.13)$$

where E_i is the excess of a path around the vertex i.

Transform this expression into the simplest, most meaningful terms. Remember that excess is equal to 2π minus the total turning, and rewrite the equation as

$$K = \sum_{i=1}^{v} (2\pi - T_i) \tag{9.14}$$

where T_i is the total turning of a path around the vertex i. Factor out the 2π terms (one for each vertex) to obtain

$$K = 2\pi V - \sum_{i=1}^{v} T_i \tag{9.15}$$

where V is, of course, the number of vertices in the surface. We can further clarify this expression if we rewrite it using the fact that the total turning around a vertex is equal to the sum of all the interior angles meeting at that vertex. Summing over all the vertices gives us all the interior angles of all the faces in the surface. If we regroup these angles according to the faces they lie in, we compute total curvature as

$$K = 2\pi V - \sum_{i=1}^{F} f_i \tag{9.16}$$

where f_i is the sum of the interior angles of the face i.

This is a surprising result because we compute the second term without knowing how the edges are joined together. Therefore, if we have all the individual faces of a piecewise flat surface and we know V, we can compute total curvature without knowing anything about the atlas.

For piecewise square surfaces, the formula is even simpler, since the sum of interior angles of any face is 2π. Therefore, $K = 2\pi V - 2\pi F$ or $K = 2\pi (V - F)$. This equation for total curvature does not depend on angles at all! And we can do even better. Curiously enough, we can compute the total curvature of any closed piecewise flat surface without knowing any angles. We need only know the total number of faces and vertices and the number of edges. For a closed piecewise flat surface with V vertices, E edges, and F faces, the total curvature is

$$K = 2\pi (V - E + F) \tag{9.17}$$

Since we already know that $K = 2\pi V - \sum_{i=1}^{F} f_i$, to prove Eq. 9.17, we must show that the sum $\sum_{i=1}^{F} f_i$ can be expressed independently of the particular values of the angles. Now, we do not know much in general about the sum of the interior angles of a face, but we do know a closely related quantity—the sum of the exterior angles. Since the boundary of each face is a simple closed path, we know that the sum of the exterior angles is 2π. To relate the exterior angles to the interior angles, notice that each exterior angle pairs off with an interior angle to sum to $\pi = 180°$ and there are as many of these pairs as there are edges to a face. Therefore,

$$f_i = [\text{sum of } (\pi - \text{exterior angles})]$$
$$= \pi \times e_i - (\text{sum of exterior angles of the face}) \tag{9.18}$$
$$= \pi \times e_i - 2\pi$$

where e_i is the number of edges of the face i. Summing this quantity over all the faces gives

$$\sum_{i=1}^{F} f_i = \sum_{i=1}^{F} e_i - 2\pi F \qquad (9.19)$$

Now simplify $\sum_{i=1}^{F} e_i$. Note that in a closed surface each edge is shared by precisely two faces. Therefore, summing the number of edges in a face over all the separate faces counts each edge exactly twice; then,

$$\sum_{i=1}^{F} e_i = 2E \qquad (9.20)$$

Combining Eq. 9.20 with Eq. 9.19 yields

$$\sum_{i=1}^{F} f_i = 2\pi (E - F) \qquad (9.21)$$

And substituting Eq. 9.21 into Eq. 9.16 yields

$$K = 2\pi V - \sum_{i=1}^{F} f_i \qquad (9.22)$$

or

$$K = 2\pi (V - E + F) \qquad (9.23)$$

The quantity $(V - E + F)$ is called the **Euler characteristic** of a surface and is denoted by the Greek letter χ (chi). Using this notation, rewrite the curvature formula as

$$K = 2\pi\chi \qquad (9.24)$$

Think of Eq. 9.24 as a convenient way of calculating the total curvature of a piecewise flat surface. However, it also implies much more. As we have seen, K has special properties. First, K is defined for all surfaces, not only piecewise flat ones. Also, K is a topological invariant. Therefore, Eq. 9.24 leads us to suspect that the Euler characteristic also has these properties. This suggests the following questions: Can we show directly that $(V - E + F)$ is a topological invariant? Can we define $(V - E + F)$ for all surfaces, not just piecewise flat ones? If we can achieve these goals, we provide powerful tools for geometric modeling.

Look at topological invariance. What happens to $(V - E + F)$ when we deform a piecewise flat surface? Absolutely nothing! After all, $(V - E + F)$ is just a counting algorithm, and unless we cut a face in two, remove an edge, or perform some other nontopological transformation, none of the numbers can change. This then is our introduction to the next section.

EXERCISES

1. Is a star polyhedron a simple or nonsimple polyhedron?

2. For $n = 4$ in Eq. 9.10, write the complete equation, and discuss the meaning of N_0, N_1, N_2, and N_3.

3. How many regular polytopes are there in four-dimensional space?

4. Show that every face of a polyhedron must be a topological disk.

5. Construct the atlas for a tetrahedron.

6. Show that the total turning of a path around the vertex of a polyhedron is equal to the sum of all the interior angles meeting at that vertex.

7. Show that $K = 2\pi(V - F)$ for a cube.

8. Compute K for each of the five regular polyhedra.

9. For piecewise square surfaces, show why $K = 2\pi(V - E + F)$ reduces to $K = 2\pi(V - F)$.

10. Construct a toroidal polyhedron, and compute its Euler characteristic. Remember that each face of your construction must be a topological disk.

3 TOPOLOGY OF CLOSED CURVED SURFACES

Let us now try to define $(V - E + F)$ for an arbitrary surface. The image of a distorted piecewise flat surface suggests a new definition for the terms vertices, edges, and faces on a general surface. Therefore, we define faces on the surfaces not as flat polygons but as topological disks, and edges not as straight lines but as simple arcs with a vertex at either end. Thus, we define a **net** on a general surface as an arbitrary collection of simple arcs (terminated at each end by a vertex) that divide the surface everywhere into topological disks.

For convenience, we will continue to call the elements of a net edges, vertices, and faces. Given a net, we define the Euler characteristic χ by the same formula $\chi = V - E + F$.

The definition of the Euler characteristic raises a question: Since we draw an infinite number of different nets on a surface, how do we decide which net to use in computing the Euler characteristic? The answer is that it does not matter which one we choose, because **all nets on the same closed surface have the same Euler characteristic**.

To prove this, imagine starting with a particular net on a surface, then transforming this net into a different one by adding or deleting vertices or edges. We can do this in several ways, so we single out two **elementary net transformations**:

(1) Add (or delete) a face by drawing in (or erasing) an edge between existing vertices, or

(2) add (or delete) a vertex.

Now, let us see how these transformations affect the value of χ. If we add an edge, then E increases by 1. But F also increases by 1; hence $(F - E)$ is unchanged. And since V is unchanged, $(\chi = V - E + F)$ is unchanged by the first type of transformation. If we insert a new vertex into an edge, we produce not only a new vertex, but also a new edge. Thus, V and E each increase by 1, and F is unchanged; then $(V - E + F)$ is again unchanged. We conclude that χ is invariant under the two net transformations just defined. To complete the proof that any two nets on the same surface will give the same value of χ, we assert that, given any two nets, we can always get from one to the other by some sequence of these transformations.

Then, every surface has an Euler characteristic χ, and χ is a topological invariant. Now we have two topological invariants for surfaces—the Euler characteristic χ and the total curvature K. Furthermore, we know that they are related by $K = 2\pi\chi$ for piecewise flat surfaces. But does this hold for any surface? It does, and we may now assert the following theorem: For any closed surface, the total curvature and Euler characteristic are related by $K = 2\pi\chi$. This is called the **Gauss–Bonnet theorem**.

Let us discuss two different proofs of this theorem. The first is based on the fact that K and χ are both topological invariants. Start with an arbitrary surface and deform it into a piecewise flat surface. Since K and χ are both topological invariants, the deformation leaves them unchanged. Since we know that $K = 2\pi\chi$ for the piecewise flat surface, then this is true for the original surface as well. The only caveat in this proof is that we must show that any surface can be deformed into a piecewise flat surface. Intuitively, this seems possible. The approach is to flatten the surface piece by piece, pushing all the curvature into the edges between the pieces. Then flatten and straighten the edges piece by piece, pushing all the curvature into the vertices. The details of this procedure are difficult to express more rigorously, and they are not absolutely necessary here. A second proof that $K = 2\pi\chi$ is based on a direct computation of K for any surface and is similar to the way we proved the theorem for piecewise flat surfaces.

The Gauss–Bonnet theorem is important in the geometry of surfaces because it produces a relationship between quantities defined purely in terms of topology (such as the Euler characteristic) and quantities defined purely in terms of distances and angles (such as total curvature).

EXERCISES

1. Construct three distinct and different nets on a sphere, and compute the Euler characteristic for each one.

2. Repeat Ex. 1 for a tetrahedron and a cube.

3. Determine the effect of elementary net transformations by appropriately adding faces, edges, and vertices to the nets constructed for Exs. 1 and 2.

4. Construct a "two-hole" toroidal polyhedron, and compute its Euler characteristic and total curvature.

4   GENERALIZED CONCEPT OF A BOUNDARY

To begin the discussion of boundaries of physical objects, let us review certain ideas. We can denote Cartesian space of any dimension n by the symbol E^n. Thus, ordinary three-dimensional space is E^3, and two-dimensional space is

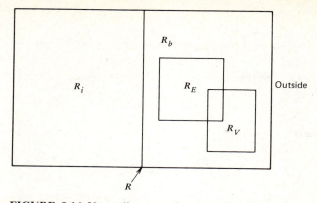

FIGURE 9.16 Venn diagram of points of a region.

E^2. Points in E^2 are defined by two real numbers, in E^3 by three real numbers, and in E^n by n real numbers. Coordinate space is unbounded, or unlimited, if you prefer, because the numbers defining coordinate points of the space can take any real values from $-\infty$ to $+\infty$.

A region R^n is a finite, bounded portion of space E^n. The points that comprise any region are individually characterized as either lying entirely within the region or on its boundary. Thus, the set of points denoted by R can be divided conveniently into two subsets R_i and R_b, where R_i is the set of points in the interior of a region and R_b is the set of points on its boundary. We write this as

$$R = [R_i, R_b] \tag{9.25}$$

All physical objects are three-dimensional regions R^3. The boundary of an R^3 region is a closed surface. (Later, we will explore in more detail how to define an object by defining its boundary surface.) The boundary surfaces of some objects contain slope discontinuities. These discontinuities may occur in such a way as to form an edge or a vertex. A set of points comprising edges is denoted R_E^3, and a set comprising vertices is R_V^3. The Venn diagram in Fig. 9.16 illustrates the possible relationships between subsets of points comprising a typical R^3.

Given any point in space, obviously it is either outside the region, in the boundary set R_b, or in the interior set R_i. If we ignore the refinement of the set R_b into subsets R_E and R_V, this applies to regions of any dimension.

A curve is a one-dimensional region R^1. It has two points in the set R_b^1 unless it is a closed curve; then there are no points in R_b^1. All other points of the curve are in R_i^1.

A surface is a two-dimensional region R^2. An ordinary open surface is always bounded by a closed curve. On the surface within this curve may occur one to n nonintersecting closed curves or loops. All the points on all the loops comprise the set R_b^2. All other points on the surface are in R_i^2; see Fig. 9.17.

If R^2 is a closed region, a variety of special conditions is possible. For example, consider the R^2 region defining the surface of a sphere. In it, all points are in R_i, and R_b is an empty set.

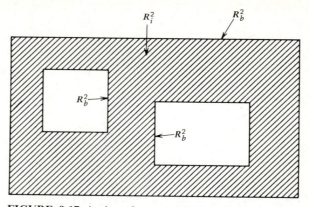

FIGURE 9.17 A plane figure and its boundaries.

A cone is an open R^2 region with R_b consisting of a vertex point and a base curve.

Now, we expand our horizons a little and also modify the notation scheme. Let $R^{m,n}$ be a region of E^n, where m = dimensionality of R and n = dimensionality of the space E in which R is located. Note that $m \leq n$. Then

$$R^{m,n} = [B^{m-1,n}, I^{m,n}] \tag{9.26}$$

where $B^{m-1,n}$ is the set of points on the boundary of $R^{m,n}$ and $I^{m,n}$ is the set of points in the interior of the region. The set $B^{m-1,n}$ is itself a proper region because it can be decomposed into a set of points on its boundary and a set on its interior. Table 9.1 defines allowable regions in E^3. Of course, we can construct a similar table for regions in spaces of any dimension. The general concept of a boundary is contained in the expression $B^{m-1,n}$.

TABLE 9.1 Allowable Regions in E^3

Order of $R^{m,n}$	Class Name	$B^{m-1,n}$	$I^{m,n}$
$R^{0,3}$	Point	The point itself	No interior points
$R^{1,3}$	Curve	The two end points	The set of points on the curve except for the two boundary points in $B^{0,3}$
$R^{2,3}$	Surface	One or more closed curves that define the boundaries of the surface	The set of all points on the surface except those on the bounding curves in $B^{1,3}$
$R^{3,3}$	Solid	One or more closed surfaces that define the boundaries of the solid	The set of all points within the solid except those on the bounding surfaces in $B^{2,3}$

Any point in space has one and only one of the following three properties with respect to any region $R^{m,n}$:

1. It is inside the region (that is, a member of the set $I^{m,n}$).
2. It is on the boundary of the region (that is, a member of the set $B^{m-1,n}$).
3. It is outside, not a member of, the set $R^{m,n}$.

For a homogeneous region, when $m = n$, then $I^{m,n}$ can be implied by an explicit formulation of $B^{m-1,n}$. Thus, for a homogeneous solid in E^3, the explicit definition of the $B^{2,3}$ of the solid is necessary and sufficient for the definition of $R^{3,3}$ of the solid. $B^{2,3}$ is called the outline or boundary surface of the solid, where points on the inside (that is, in $I^{3,3}$) are implied by $B^{2,3}$.

The concepts here appeal to our intuition. They lay the foundation for material in Sections 5 and 7, which approach this subject from slightly different directions and are more abstract and rigorous.

5 SET THEORY

In geometric modeling, when we combine simple shapes, often called **primitives**, in order to form more complex shapes, the application of **set theory** becomes useful. Solid modeling techniques, in particular, have drawn considerably from the axioms of set theory. For example, important work at the University of Rochester's Production Automation Project draws from point-set topology to establish more rigorous mathematical foundations for solid modeling.

Section 5 introduces concepts of set theory that are most important to solid modeling. We begin with some of the classical ideas of set theory and finish with a look at point-set topology, including set-membership classification and its relevance to solid modeling. (Incidentally, the first formal treatment of sets dates back to only the second half of the nineteenth century, when Cantor created the main body of the theory.)

The term **set** denotes any well-defined collection of objects. Objects belonging to a set are its elements or members. In geometric modeling, solid or otherwise, the basic element is the point.

The symbols $\{|\}$ are called **set-builder notation** and describe the set in terms of conditions on any arbitrary elenent of the set—conditions that every element of the set must meet. For example, $\{x | 2.5 < x < 3.5\}$ is a set consisting of all real numbers in the specified interval. On the left-hand side of the vertical line, we read, "the set of all $x \cdots$" On the right-hand side, we have the conditions for set membership.

Any set that contains all the elements of all the sets under consideration is the **universal set**, denoted by E. In other words, a universal set contains all the elements we are concerned with in a given situation. A deck of 52 playing cards may be the universal set if we are investigating the probabilities of certain combinations of cards in a poker hand. Conversely, a null set is a set which has no elements at all. It is denoted by ϕ and frequently referred to as the empty set or the void set.

In general, two sets A and B are equal, expressed as

$$A = B \tag{9.27}$$

whenever set A and set B contain exactly the same elements. Two sets A and B are placed in one-to-one correspondence if each element in A can be paired with exactly one element in B and each element in B can be paired with exactly one element in A. Two sets A and B are called **equivalent sets** if they can be placed in one-to-one correspondence.

Any set A is a subset of set B only if every element in A is also an element in B. The symbol $\subset$ indicates the subset relationship. Thus,

$$A \subset B \tag{9.28}$$

means that A is a subset of B. A is a **proper** subset of B if every element in A is contained in B and if B has at least one element not contained in A. Every set is a subset of itself, but not a proper subset.

New sets are formed by combining the elements in two or more sets in some fashion. Given sets A and B, let us construct a third set C whose elements are all the elements in A together with all the elements in B. We write this as

$$C = A \cup B \tag{9.29}$$

By the union of two sets, we mean the set of all elements in one set or in the other set or in both sets. For example, if $A = \{a, b, c\}$ and $B = \{c, d, e, f\}$, then $C = A \cup B = \{a, b, c, d, e, f\}$. Notice that there is no repetition of elements in C, even though element c is an element in both A and B. We may read $A \cup B$ as A or B. Here, the *or* is what logicians call the **inclusive or** because $A \cup B$ includes the elements that belong to A as well as B.

If we form a set D comprised of elements common to both A and B, set D is called the **intersection** of A and B, expressed as

$$D = A \cap B \tag{9.30}$$

By the intersection of two sets, we mean the set of all elements that are in both sets. Thus, if $A = \{a, b, c, d\}$, $B = \{c, d, e, f, g\}$, and $D = A \cap B$, then $D = \{c, d\}$.

The **complement** of a set A with respect to a universal set E is the set of all elements in E that are not elements in A, written cA.

Finally, if A and B are sets, then $A - B$ denotes the set of elements in A that are not also elements in B. Furthermore, if complements are formed with respect to a set E containing A and B as subsets, then $A - B = A \cap cB$. Clearly, $A - A = 0$. *Note*: This minus sign does not combine with $\cup$ in the way that a plus analogy might suggest. For example, $(A \cup A) - A = 0$, but $A \cup (A - A) = A$.

Sets and subsets are studied by means of Venn diagrams, which are particularly useful in representing such set relationships as equality and such set operations as union and intersection. Section 4 of this chapter described boundary properties with Venn diagrams. Figure 9.18 illustrates some set properties and operations by means of Venn diagrams. Set operations obey certain rules; that is, there are properties governing the ways we can combine sets; Table 9.2 lists these properties.

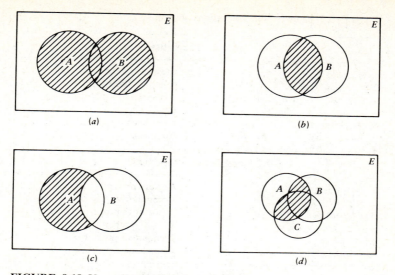

FIGURE 9.18 Venn diagrams and set theory.

There are useful geometric interpretations of the properties of, and operations on, sets, which are clearly suggested by the geometric nature of Venn diagrams, for example. For the purposes of geometric modeling, sets consist of points, and the universal set E is the set of points defining a Euclidean space with a dimension of our choosing. Set theory suggests methods for operating on these points and classifying them according to such properties as inside, outside, or on the boundary of a geometric solid.

Consider the real line defined by a continuous set of points in E^1, where E^1 is the universal set of points we are considering. Let a subset X of E^1 be defined by the line segment $a < X < b$, where a and b are **limiting** points of the set X. X is an **open set** because it does not contain its limit points. Conversely, if $a \leq X \leq b$, then X is a **closed set**; see Fig. 9.19, where the open circles at a and b symbolize that these points are not included in the set, and the solid circles in the lower figure indicate their inclusion. The **closure** of an open set is the union of the set with the set of all its limit points. These concepts extend to two- and three-dimensional spaces and point sets.

The **boundary** of a closed set is the set of all its limit points. Conversely, the **interior** of a closed set is the set of all points of a set not on its boundary. Thus,

$$X = bX \cup iX \qquad (9.31)$$

where bX denotes the set of boundary points and iX denotes the set of interior points.

We are now ready to see how these concepts allow us to use simple shapes to create more complex ones. To do this, we will use the set operators $\cup$, $\cap$, and difference $(-)$, called **Boolean operators**, and the rules for their application and combination, called Boolean algebra. Section 6 discusses Boolean operations on sets of points that might represent one-, two-, or three-dimensional objects.

TABLE 9.2 Properties of Operations on Sets

Union Properties

1. $A \cup B$ is a set	Closure property
2. $A \cup B = B \cup A$	Commutative property
3. $(A \cup B) \cup C = A \cup (B \cup C)$	Associative property
4. $A \cup \varnothing = A$	Identity property
5. $A \cup A = A$	Idempotent property
6. $A \cup cA = E$	Complement property

Intersection Properties

1. $A \cap B$ is a set	Closure property
2. $A \cap B = B \cap A$	Commutative property
3. $(A \cap B) \cap C = A \cap (B \cap C)$	Associative property
4. $A \cap E = A$	Identity property
5. $A \cap A = A$	Idempotent property
6. $A \cap cA = \varnothing$	Complement property

Distributive Properties

1. $A \cup (B \cap C) = (A \cup B) \cap (A \cup C)$	Union is distributive over intersection
2. $A \cap (B \cup C) = (A \cap B) \cup (A \cap C)$	Intersection is distributive over union

Complementation Properties

1. $cE = \varnothing$	The complement of the universal set is the empty set
2. $c\varnothing = E$	The complement of the empty set is the universal set
3. $c(cA) = A$	The complement of a complement of a set A is A
4. $c(A \cup B) = cA \cap cB$	DeMorgan's law
5. $c(A \cap B) = cA \cup cB$	De Morgan's law

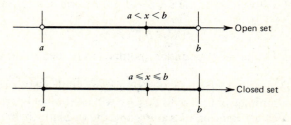

FIGURE 9.19 Open and closed sets.

EXERCISES

1. Show that for $A \subset B$, it is necessary and sufficient that $A \cup B = B$ or $A \cap B = A$.

2. Is the expression $A - (A - B) = A \cap B$ true?

3. Write a procedure to perform set operations on two finite lists of integer numbers. Assume two-digit decimal numbers. Denote this as **SETOP**(N1, L1, N2, L2, IOP, N3, L3), where

> N1, N2 are the input number of elements in each list;

> L1(N1), L2(N2) are the input lists of decimal integers;

> IOP is the input flag specifying the set operation to be performed. Use IOP = 1 for union, IOP = 2 for difference, and IOP = 3 for intersect. Since order is important, assume that L1 < IOP > L2.

> N3 is the output number of elements in L3; and

> L3(N3) is the output list of numbers resulting from the set operation on L1 and L2.

4. Repeat Ex. 3 for lists of eight-bit binary numbers.

6   BOOLEAN OPERATORS

A distinguishing feature of the geometric objects we will deal with here is that they are defined as closed sets of points having a boundary subset and an interior subset (refer to Section 9.4). Boolean operations similar to set intersection, union, and difference are used to combine simple objects to form more complex ones. The algorithms that perform these operations must produce as output objects that are also closed sets of points having boundary and interior subsets and preserve the dimensionality of the initial objects. The latter requirement means that in any Boolean operation, such as $A \cup B = C$, all the objects must be of the same spatial dimension.

Figure 9.20 demonstrates how the ordinary set-theoretic intersection of two well-defined two-dimensional objects produces a degenerate result. First, A and B are well defined because each possesses a boundary set bA and bB and an interior set iA and iB. Second, the resulting intersection is mathematically correct according to the tenets of set theory but geometrically incorrect, or improper, because C has no interior. Thus, C is not like A and B; it is not a two-dimensional object, and the intersection operation does not preserve dimensionality. We would prefer an operator that recognizes this condition and produces a null set. A. A. G. Requicha (1977) and others propose the use of **regularized** set operators, which preserve dimensionality and homogeneity (no dangling or disconnected parts of lower dimension). R. B. Tilove (1977) observes that "regularization amounts to taking what is inside a set and covering

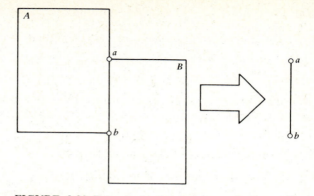

FIGURE 9.20 Degenerate intersection of two well-defined two-dimensional objects.

it with a tight skin." The importance of regularized operators will become apparent during our investigations.

Let us begin by combining two simple polygons A and B, shown in Fig. 9.21. Both A and B are two-dimensional objects with straight-line boundaries. The arrows indicate the direction of parametrization. Notice that both A and B are parametrized in a consistent direction. In this case, they are both counterclockwise. By maintaining consistency in orienting the polygons' edges, we can establish a convention defining which side contains points inside the polygon— on the left in our example.

We assume that transformations have already been applied to positions A and B. Now, let us look at an algorithm for finding $C = A \cup B$.

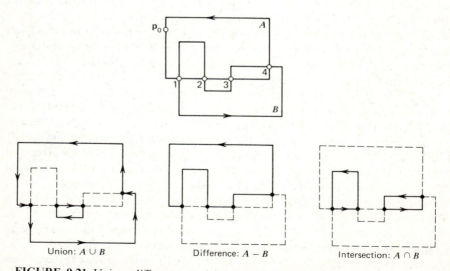

Union: $A \cup B$ Difference: $A - B$ Intersection: $A \cap B$

FIGURE 9.21 Union, difference, and intersection of two simple polygons.

1. Find all the intersection points of the edges of A and B; points 1, 2, 3, and 4.

2. Segment the edges of A and B. Thus, if the boundary polygon of A is parametrized from $u = 0$ to $u = 1$ and B from $v = 0$ to $v = 1$, then the boundary of A has four segments: $u \in [u_1; u_2]$, $u \in [u_2; u_3]$, $u \in [u_3; u_4]$, and $u \in [u_4; u_1]$. B also has four segments: $v \in [v_4; v_3]$, $v \in [v_3; v_2]$, $v \in [v_2; v_1]$, and $v \in [v_1; v_4]$.

3. Find a point on the boundary polygon of A that is outside B, say, $\mathbf{p}_0$. Then that segment is also outside B.

4. Starting at $\mathbf{p}_0$, trace the boundary of A to the next intersection point with B, point 1.

5. Find the intersecting segment of B, and trace along it in the direction of increasing v to its intersection with A, point 4. At this point, we discover we have traced back to the starting segment on A, but we have not yet exhausted the list of segments. We have found only one loop.

6. Find a point on one of the remaining segments of A that is outside B. Then that segment, too, is outside B.

7. Trace this boundary segment to the next intersection point with B, point 2.

8. Repeat step 5, tracing B to point 3. We have completed another loop. Since there are no more qualifying segments to trace, the union operation is complete. Notice that the second loop is parametrized in a clockwise direction, which means that it encloses a hole.

The boundary segments of A and B that comprise the loops are called the **active** segments; conversely, the others are **inactive**. If representations of the initial objects A and B are no longer necessary, discard them and delete them from the modeling data base. In any event, object C is represented by reparametrized concatenations of the appropriate active segments of A and B, forming two separate and distinct loops, as we have just seen.

The difference operation is the same as the union operation except that we trace the boundary segments of B clockwise (that is, in the direction of decreasing v). The intersection operation, too, is similar to union except that segment tracing must start from a point on the boundary of A that is inside B.

The algorithm just outlined requires several supporting algorithms. The principal ones are: a procedure for finding the intersection points; a procedure for determining if a point is inside or outside a closed polygon; and procedures tracing segments, forming loops, and reparametrizing the loops. The list will grow as we investigate more sophisticated problems involving Boolean operators.

The example we have just explored is very simple—too simple, in fact, to illuminate most situations we encounter when applying Boolean operators to modeling problems. However, the example demonstrates the basic approach: intersecting, testing, tracing, and sorting. Now, let us venture into deeper waters.

Consider the two-dimensional objects A and B in Fig. 9.22. Each is well defined, that is, closed and dimensionally homogeneous. Thus, A and B are expressed as

$$A = bA \cup iA \quad \text{and} \quad B = bB \cup iB \qquad (9.32)$$

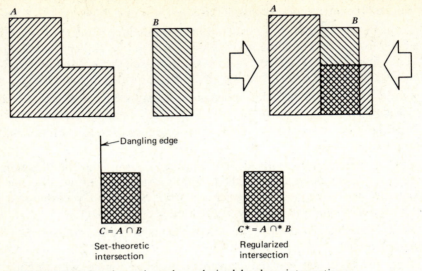

FIGURE 9.22 Set theoretic and regularized boolean intersections.

Next, translate A and B into position prior to combining them by the Boolean operation to form object C. First, perform the set-theoretic intersection, with the proper result shown in Fig. 9.22. Notice the dangling edge; this result is obviously not dimensionally homogeneous, yet it is the correct set-theoretic intersection. The result we seek is shown to the right. It is the regularized Boolean intersection denoted $A \cap^* B$, closed and dimensionally homogeneous.

Now, let us see how to achieve it. We begin by finding the regularized intersection, given the set-theoretic intersection,

$$C = A \cap B \qquad (9.33)$$

Rewrite this as

$$C = (bA \cup iA) \cap (bB \cup iB) \qquad (9.34)$$

which expands to

$$C = (bA \cap bB) \cup (iA \cap bB) \cup (bA \cap iB) \cup (iA \cap iB) \qquad (9.35)$$

The geometric interpretation of each of the four parenthetical terms is illustrated in Fig. 9.23. Notice, again, that the direction of parametrization is indicated by arrows in Fig. 9.23a. Since $C = bC \cup iC$, we must find the subsets of bC and iC that form a closed, dimensionally homogeneous object C^*. Candidates for C^* must be derived from the terms in Eq. 9.35. In Fig. 9.23d, we see the two-dimensional interior of C and correctly surmise that

$$iC = iC^* = iA \cap iB \qquad (9.36)$$

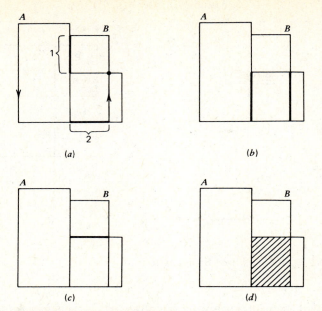

FIGURE 9.23 Candidate components of a regularized boolean intersection.

Next, we must determine bC^*, where $bC^* = \text{Valid}(bA \cup bB)$. Notice that the boundaries of any new object will always consist of boundary segments of the combining elements. We can generalize this observation as follows: Boundary points can become interior points, whereas interior points cannot become boundary points. Furthermore, for regularized intersections, we assert without proof that

(1) $iA \cap bB \subset bC^*$ and

(2) $bA \cap iB \subset bC^*$

So far, we have successfully accounted for Fig. 9.23b–Fig. 9.23d of the set-theoretic intersection in our quest for the regularized intersection. We must now analyze $(bA \cap bB)$ shown in Fig. 9.23a to determine which of its subsets are valid subsets of the boundary of C^*. The isolated point is a valid member of bC^*, since it must always be a member of both $(iA \cap bB)$ and $(bA \cap iB)$. We are left with two apparently identical overlapping intersections of A and B. These intersections are distinguished because they are interior to neither A nor B. Let us find some way, some test, to differentiate between them.

At some point $\mathbf{p}_1$ on segment 1, create a new point $\mathbf{p}_R$ offset perpendicularly from it a distance ε to the right (relative to the direction of parametrization), and a similar point $\mathbf{p}_L$ to the left (see Fig. 9.24). Do the same at a point $\mathbf{p}_2$ on segment 2. Next, construct a table for each segment to test whether each point $\mathbf{p}_R$ and $\mathbf{p}_L$ is inside A or B. For segment 1, neither test point is inside both A and B, whereas for segment 2, test point $\mathbf{p}_L$ is inside both A and B. This test

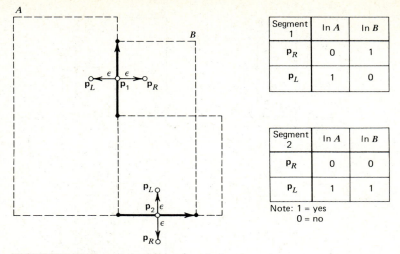

Segment 1	In A	In B
$\mathbf{p}_R$	0	1
$\mathbf{p}_L$	1	0

Segment 2	In A	In B
$\mathbf{p}_R$	0	0
$\mathbf{p}_L$	1	1

Note: 1 = yes
      0 = no

FIGURE 9.24 Regularized boundary test.

determines that segment 2 is a valid boundary of C^*. This procedure analyzes the neighborhood of points on the segment so that we can properly classify it.

Another, simpler classification test is available if we adopt a consistent direction-of-parametrization convention. At $\mathbf{p}_1$, compute the tangent vector $\mathbf{p}_1^u$ from the A boundary representation and $\mathbf{p}_1^v$ from the B boundary representation. In our example, they are in opposite directions. Do the same for segment 2, where the tangent vectors are in the same direction. From this we conclude: If the respective tangent vectors at a point of the overlapping boundaries of two intersection objects A and B are in the same direction, then the overlapping segment is a valid boundary of $C^* = A \cap^* B$; otherwise the segment is not a valid boundary.

Let us summarize the results. The valid regularized intersection of two objects A and B is given by

$$C^* = A \cap^* B \tag{9.37}$$

where

$$C^* = bC^* \cup iC^*$$

$$= \text{Valid}_b(bA \cap bB) \cup (iA \cap bB) \cup (bA \cap iB) \cup (iA \cap iB) \tag{9.38}$$

There is nothing in this expression that indicates dimensionality, so we conclude that it applies equally to one-, two-, three-, or n-dimensional objects. Later in this section, we will investigate the Boolean combination of three-dimensional solids. First, let us continue operations on the objects in Fig. 9.22.

The components of the set-theoretic union of A and B are shown in Fig. 9.25. Again, they are the full set of valid candidates for determining $C^* =$

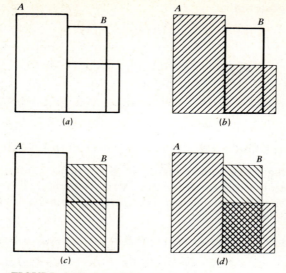

FIGURE 9.25 Candidate components of a regularized boolean union.

$A \cup^* B$. Begin, as in the preceding intersection, by expanding the set-theoretic expression

$$C = A \cup B \qquad (9.39)$$

$$C = (bA \cup iA) \cup (bB \cup iB) \qquad (9.40)$$

$$C = (bA \cup bB) \cup (iA \cup bB) \cup (bA \cup iB) \cup (iA \cup iB) \qquad (9.41)$$

Equation 9.41 is just as correct and subject to interpretation without the parenthetic separation of terms. Thus,

$$C = bA \cup bB \cup iA \cup b\underline{B} \cup b\underline{A} \cup iB \cup i\underline{A} \cup i\underline{B} \qquad (9.42)$$

The redundant elements are underlined. Delete them so that Eq. 9.42 becomes

$$C = A \cup B$$
$$= bA \cup bB \cup iA \cup iB \qquad (9.43)$$

which is identical to Eq. 9.40 with parentheses deleted. From these components, determine bC^* and iC^* and, hence, C^*. First, we observe that

$$iC^* = iA \cup iB \cup [\text{Valid}_i(bA \cap bB)] \qquad (9.44)$$

Note that some boundary points become interior points. If these points are not included, we will have what amounts to a "hole" in iC^*. Also, notice that it is redundant to add $\cup (bA \cap iB) \cup (bB \cap iA)$ to the right side of Eq. 9.44. Why is this not the case for $\cup [\text{Valid}_i(bA \cap bB)]$?

Next, we observe that

$$bC^* = \text{Valid}_b(bA \cup bB) \tag{9.45}$$

where

$$\text{Valid } bA = bA \text{ not in } iB \text{ and part on } bB \tag{9.46}$$

or

$$\text{Valid } bA = bA - [(bA \cap iB) \cup \text{Valid}_b(bA \cap bB)] \tag{9.47}$$

Similarly,

$$\text{Valid } bB = bB - [(bB \cap iA) \cup \text{Valid}_b(bA \cap bB)] \tag{9.48}$$

Again, notice that there is an ambiguity in $(bA \cap bB)$ that must be resolved by a test similar to that discussed for the intersect operator. If we discard all of $(bA \cap bB)$, then bC^* is incomplete. Therefore, the boundary of the regularized set C^* is

$$bC^* = bA \cup bB - [(bA \cap iB) \cup (bB \cap iA) \cup \text{Valid}_b(bA \cap bB)] \tag{9.49}$$

In Eqs. 9.44 and 9.49, we see that $(bA \cap bB)$ is subdivided and part assigned to iC^* and part to bC^*. Furthermore, nothing is lost, since we assert $\text{Valid}_i(bA \cap bB) = \text{Valid}_b(bA \cap bB)$.

Finally, consider Fig. 9.26. Here are the components of the difference operator $(A - B)$. Again, expand the set-theoretic expression

$$C = A - B \tag{9.50}$$

to yield

$$C = (bA - bB - iB) \cup (iA - bB - iB) \tag{9.51}$$

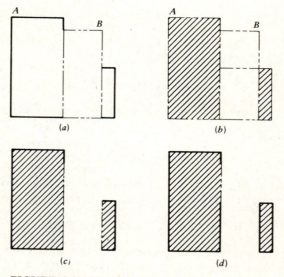

(a)   *(b)*   *(c)*   *(d)*

FIGURE 9.26 Candidate components of a regularized boolean difference.

Two things are immediately clear from Fig. 9.26. First, iC^* must equal $iA - bB - iB$; in the case of our example, two disjoint sets result. Second, $C^* \neq C$, since certain segments of bC^* are missing from C. If we add $iA \cap bB$ to C, as in Fig. 9.26d, the boundary is still incomplete. The missing segment is a subset of $bA \cap bB$. Here, again, a test must be performed to determine the valid subset. For the case of the difference operator, $\text{Valid}(bA \cap bB)$ are those segments adjacent to only iC^* or $(iA - iB)$. Thus,

$$bC^* = bC \cup (iA \cap bB) \cup \text{Valid}(bA \cap bB)$$

$$= (bA - bB - iB) \cup (iA \cap bB) \cup \text{Valid}(bA \cap bB) \tag{9.52}$$

Therefore, C^*, the regularized $(A - B)$, is

$$C^* = (bA - bB - iB) \cup (iA \cap bB) \cup \text{Valid}(bA \cap bB) \cup (iA - bB - iB) \tag{9.53}$$

Other relationships between A and B are possible; Fig. 9.27 is a particularly useful example. Here, object A completely encloses B. Use the equations just developed to verify the results. This is the way holes are modeled.

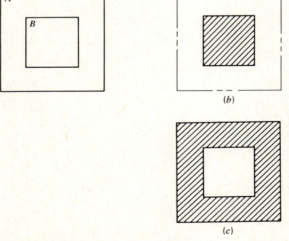
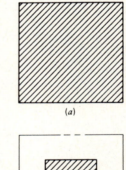

FIGURE 9.27 Examples of boolean operations.

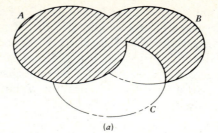

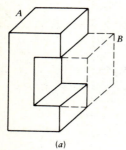

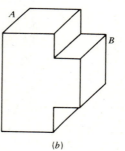

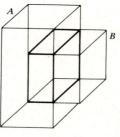

FIGURE 9.28 Order dependence of boolean operations.

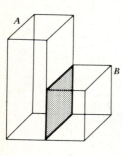

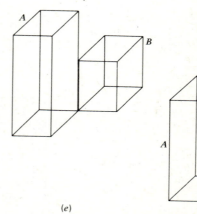

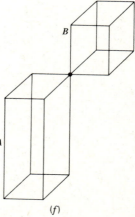

FIGURE 9.29 Boolean operations on a three-dimensional solid.

If we execute a sequence of two or more Boolean operations on a set of objects, then the result depends on the order of the sequence. In Fig. 9.28, three objects are combined, and the results are clearly dependent on the order.

Boolean operators apply to three-dimensional solids in exactly the same way that two-dimensional objects do. The regularized Boolean combining operations are the same, and closure and dimensional homogeneity are also necessary. Figure 9.29 illustrates the effect of various combining operations on two simple three-dimensional objects. In Fig. 9.29a–9.29c, nothing unusual results, but in Fig. 9.29d–9.29f, nonregularized operations produce intersections that are not three-dimensional. The regularized operators applied to these last three cases properly produce null results.

EXERCISES

1. Show that $A \cup B = A \cup^* B$.

2. Develop a simple algorithm for sorting and concatenating boundary segments of two-dimensional objects, produced by regularized Boolean operations, into closed loops. Suggest an appropriate reparametrization scheme.

7 SET-MEMBERSHIP CLASSIFICATION

In order to regularize sets resulting from combining operations on other sets, we have to determine whether a given point is inside, outside, or on the boundary of a given set. Three important subsets of any regularized set A are: the set of all its interior points, denoted iA; the set of all points on its boundary, denoted bA; and all points outside it, denoted cA. Assigning a specific point to one of these sets is called **set-membership classification**. Considerable work in formalizing and making rigorous set-membership classification for geometric modeling has been done by R. B. Tilove (1980) and others at the University of Rochester.

Tilove observes a fundamental similarity between certain types of geometric problems related to modeling:

1. *Point inclusion*: Given a solid and a point, is the point inside, outside, or on the boundary of the solid?

2. *Line/polygon clipping*: Given a polygon and a line segment, what part of the line is inside the polygon? What part is outside the polygon? What part is on the boundary of the polygon?

3. *Polygon intersection*: Given two polygons A and B, what is the intersection polygon $A \cap B$?

4. *Solid interference*: Given two solids, do they interfere (that is, intersect unintentionally)?

In each problem, two geometric elements are presented, with the requirement that we determine some inclusion relationship between them. Tilove proposes a set-membership classification function to unify the approach to these problems. He denotes this function as $M[\ \ ]$, which operates on two specified point sets—a reference set S and a candidate set X. Thus, $M[X, S]$ partitions X into subsets corresponding to their membership in bS, iS, or cS.

Let us look at several geometric classification relationships and consider some classification methods. Consider the possible relationships between sets taken from E^0, E^1, E^2, and E^3 universal sets, that is, points, lines or curves, surfaces, and solids.

Relationships involving points are shown in Fig. 9.30. The most basic, yet easily overlooked, classification relationship is between two points: A test point may be the same as a given point, or it may be different. A simple algorithm to compare coordinates with some degree of precision will produce the classification.

A point may be on or off a given curve; if on the curve, three subclasses are possible: on the initial point, on the end point, or on an intermediate point. For a straight line, an additional pair of classifications for a point exist—on the back extension of the line or on the forward extension of the line. Finally, a point can be to the right or left of a line or curve on a surface relative to its direction of parametrization. A simple test for right-side/left-side status is to first find the point on the given curve nearest the candidate point. Then take the vector product between the tangent vector to the curve at the nearest point and a vector constructed from the nearest point to the candidate point. Given a consistent convention, the sign of the resulting vector product indicates on which side of the curve we find the point.

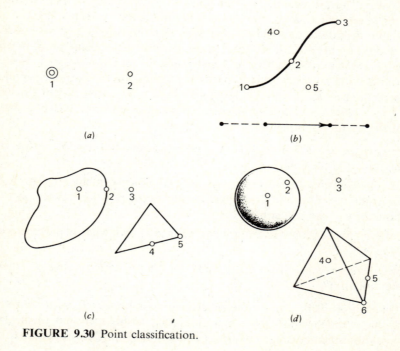

FIGURE 9.30 Point classification.

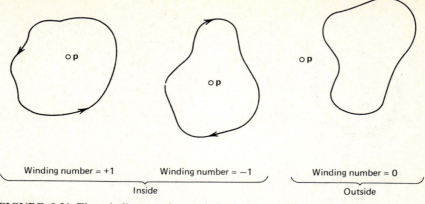

Winding number = +1 Winding number = −1 Winding number = 0

Inside Outside

FIGURE 9.31 The winding number and the inside-outside classification.

Given a bounded, simply-connected topological disc (for example, a closed, non-self-intersecting curve lying on a surface, of which a convex polygon on a plane is a special case), then a point may be inside the disc, on its boundary, or outside of it. In the special case of a straight-edged plane polygon, there are additional classes relative to the boundary—the point may be on a specific edge or a vertex.

One way of classifying a point as inside or outside a disc or polygon is by computing the winding number (see Apostol or any other calculus textbook). Notice in Fig. 9.31 that the sign of the winding number depends on the direction of parametrization. The numerical integration of the appropriate line integral and the direction of parametrization can be defined to yield either the signs shown in the figure or the reverse.

Given a solid, that is, a regularized three-dimensional set of points, a candidate point can again have the three basic relationships to it: inside, outside, or on its boundary. If the solid is a polyhedron, then the point on its boundary can be further described as being on an edge or a vertex. There are several ways of determining the inside/outside status; let us look at two different approaches.

Figure 9.32*a* shows an arbitrarily shaped solid whose boundary we will assume is a parametric surface. Given $\mathbf{p}$, compute the closest point $\mathbf{q}$ to it on the surface. Next, compute the surface normal $\mathbf{n}$ at $\mathbf{q}$, and compare its direction to the vector $(\mathbf{p} - \mathbf{q})$. Assuming the convention of an outward-pointing normal, then if $\mathbf{n}$ and $(\mathbf{p} - \mathbf{q})$ have the same sign, $\mathbf{p}$ is outside the solid; otherwise, it is inside (if not on the boundary, in which case, $\mathbf{p} - \mathbf{q} = 0$).

Figure 9.32*b* presumes a solid defined as the Boolean intersection of half-spaces in E^3. Here, we simply test $\mathbf{p}$ against the set of inequalities defining the half-spaces. As we proceed through an ordered list of these inequalities, we update a status flag on $\mathbf{p}$. As long as $\mathbf{p}$ satisfies each successive half-space definition, it is inside the solid and is so flagged. If $\mathbf{p}$ identically satisfies a half-space limit (for example, $\mathbf{p}_z = e$), then its status is changed to on the boundary.

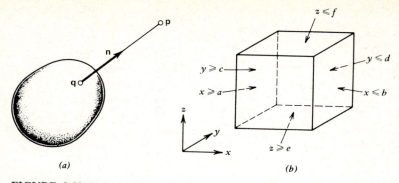

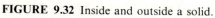

FIGURE 9.32 Inside and outside a solid.

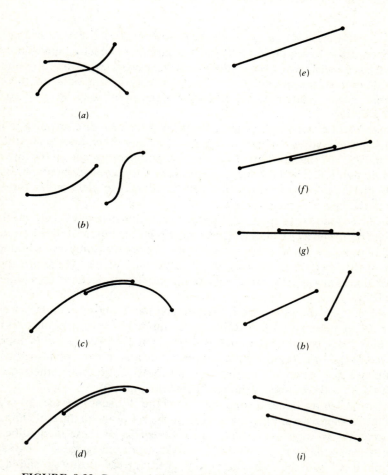

FIGURE 9.33 Curve and line segment classification.

If the point **p** fails any test, the algorithm terminates, and the point's status is flagged as outside the solid.

Set-membership classification also extends to curves and straight-line segments and, perhaps most importantly, to straight lines and polygons. Figure 9.33 shows the principal relationships between two curves and two straight lines. Detection and computation in Fig. 9.33*a* and 9.33*b* for curves, and Fig. 9.33*e*, 9.33*h*, and 9.33*i* for straight lines are fairly straightforward. The remaining cases are examples of overlap. For straight lines, detecting two points in common or a common point and common absolute direction of tangent vectors at the point indicates overlap. With curves, the situation is more complex. Here, we must accept an approximation of overlap because, unless forced to, most incidental matches will not mate identically.

Line and polygon classifications are shown in Fig. 9.34; the polygons represent regularized sets in E^2. A single line segment defining an edge of polygon B intersects polygon A in three different ways. The membership-classification function $M[\ \ ]$ subdivides and classifies this line: X_1 is a subset of the boundary of A, X_2 is a subset of the interior of A, or X_3 is a subset of the outside or complement of A. Several membership tests for these conditions were suggested in Section 6, so we will not discuss them again here.

In Fig. 9.35, we see the tangent vector convention, also mentioned in Section 9.6. This convention permits us to locate the direction of the inside of an object from any point on the boundary. We then extract information about its immediate neighborhood from the boundary representation. Tilove recognizes the ambiguity that arises when trying to classify point 2, say, with respect

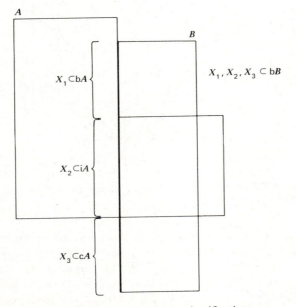

FIGURE 9.34 Line and polygon classifications.

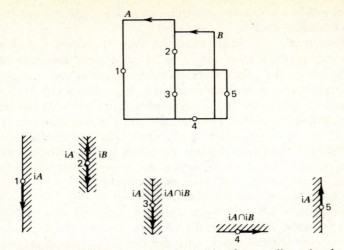

FIGURE 9.35 Tangent vector convention for two-dimensional objects.

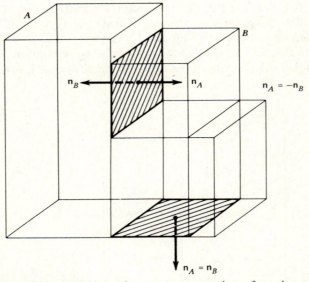

FIGURE 9.36 Normal vector convention for three-dimensional solids.

to $A \cap B$, given only the classification of the point with respect to A and B separately. He concludes that it is not possible to express $M[X, A|OP|B]$ solely in terms of $M[X, A]$ and $M[X, B]$. He suggests that we need more information to resolve ambiguities. (Here, the point is on bA and on bB, but not on bC, although $C = A \cap B$. See Eq. 9.38, and review the development in Section 6.) We see in Fig. 9.35 that we can use information embedded in the boundary itself to remove ambiguities. Of course, this does presuppose an analytical representation of the boundary.

Finally, consider the two intersecting solids in Fig. 9.36. If the surface of each one is represented analytically as a biparametric surface, then we can compute normal vectors at any points on them. We can use this information in a set-membership classification algorithm. Assume that the surfaces of the solids are parametrized in such a way that the normal at any point on them always points outward, away from the interior or material of the object. Then, on overlapping surfaces, we can easily classify a neighborhood. If $\mathbf{n}_A = \mathbf{n}_B$, the iA and iB are on the same side, and $iA \cap iB$ is not a null set. If $\mathbf{n}_A = -\mathbf{n}_B$, then iA and iB are on opposite sides and do not intersect in the neighborhood.

We have taken a rigorous approach in solving the problems of set-membership classification (see Tilove). Later sections will reveal the importance of other concepts that we touched on only briefly here. These concepts pervade computer graphics and CAD/CAM algorithms and other more abstract applications and offer ample opportunity for further study and research.

EXERCISES

1. Write a procedure to compute the winding number of a convex, *n*-sided regular polygon with respect to any given point in the plane of the polygon. Assume that the polygon lies in the *x, y* plane and is defined by an ordered list of the coordinates of its vertices.

2. Write a procedure to classify a point with respect to a solid defined by the intersection of half-spaces.

3. Write a procedure to classify line-segment intersections in a plane, given a list of pairs of points defining the line segments.

4. Classify the boundary segments of the two-dimensional shapes in Fig. 9.37*a* and *b* resulting from $A \cup B$, $A \cap B$, and $A - B$. Use the notation and labeling schemes in Figs. 9.34 and 9.35.

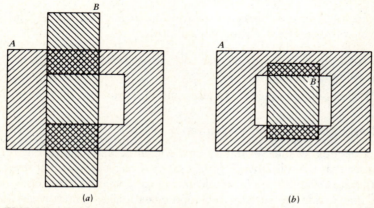

(a)   **(b)**

FIGURE 9.37 Problems for set membership classification.

5. Write a procedure to classify a point with respect to a closed planar Bezier curve. Denote this as **PCLASC(NP, BZ, P, IC)**, where

NP is the input number of control points;

BZ(NP, 3) is the input array of coordinates of the control points defining the Bezier curve;

P(3) is the input array of coordinates of the point to be classified; and

IC is the output flag specifying the point classification. IC = 1 if the point is inside the curve, IC = 2 if it is outside, IC = 3 if it is on the boundary.

6. Repeat Ex. 5 for pc and *B*-spline curves.

8   EULER OPERATORS

Euler objects always satisfy Euler's formula (Eq. 9.5). The processes that add or delete faces, edges, and vertices to create a new Euler object are called **Euler operators**. These operators provide a rational method for constructing solid, polyhedra-like objects and ensure that they are topologically valid (that is, closed and oriented).

The connectedness of the boundary surface of a solid is a property distinct from, and independent of, the enclosed interior points. Connectivity, orientation, and the characteristic of being non-self-intersecting are global properties of the surface and depend on all of its parts. Euler's formula asserts a quantitative relationship between these parts that allows us to assign certain distinctive global characteristics: the number of "handles" or through holes, total curvature, connectivity, and so on.

Since Euler's formula is not restricted to plane-faced polyhedra but also applies to any closed surface on which we can construct a proper net, the formula becomes a useful check on the topological validity of any solid whose surface can be expressed as a net of patches, curve segments, and vertices. To apply the formula $V - E + F = 2$, certain other conditions must be met:

1. All faces are simply connected, with no holes in them, and bounded by a single ring of edges. They are topological disks.

2. The solid object is simply connected (that is, its complement is connected) and has no holes through it.

3. Each edge adjoins exactly two faces and is terminated by a vertex at each end.

4. At least three edges meet at each vertex.

Let us look at some examples of the formula in action in Fig. 9.38. It is easy to see that the objects in Fig. 9.38 satisfy Euler's formula and the nets are proper, that is, a collection of simple arcs (edges), terminated at each end by a vertex, that divide the surface into topological disks. Can you also see that distorting these shapes by making the straight edges curves and the faces non-

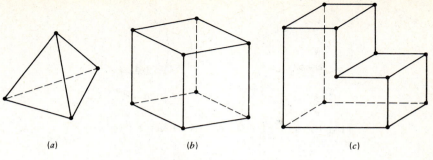

Euler's formula: $V - E + F = 2$

FIGURE 9.38 Euler's formula and simple polyhedra.

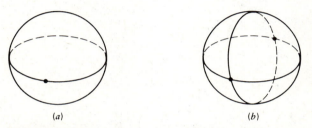

FIGURE 9.39 Euler's formula applied to a spherical net.

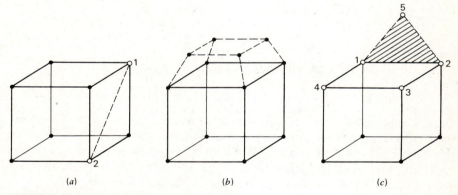

FIGURE 9.40 Euler operations on a cube.

planar surfaces does not change the applicability or validity of the formula? Figure 9.39 demonstrates this with two examples of spherical nets. What is special about Fig. 9.39*a*?

Vertices, edges, and faces added to a model must produce a result that satisfies both the Euler formula and the four conditions. In Fig. 9.40*a* and 9.40*b*, a cubical polyhedron is legitimately modified. In Fig. 9.40*c*, the formula is certainly satisfied, but notice that edges (1, 5) and (2, 5) do not adjoin two

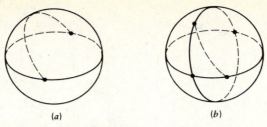

FIGURE 9.41 Modification of a Euler net on a sphere.

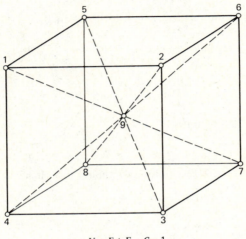

$$V - E + F - C = 1$$
$$9 - 20 + 18 - 6 = 1$$

FIGURE 9.42 Euler's formula and polyhedral cells.

faces. Furthermore, only two edges meet at vertex 5, and edge (1, 2) adjoins three faces. Therefore, as it stands, Fig. 9.40c is not a valid solid. We can remedy this situation by adding edges (3, 5) and (4, 5), resulting in a net gain of two edges and two faces. [Edges (1, 2), (2, 3), (3, 4), and (4, 1) no longer define a face.] Figure 9.41 shows valid modifications to the spherical nets.

An interesting modification of Euler's formula states that for a three-dimensional space divided into C polyhedral cells, the vertices, edges, faces, and cells are related by

$$V - E + F - C = 1 \qquad (9.54)$$

By adding a vertex, point 9, to the interior of a cube (shown in Fig. 9.42) and joining it to each of the other eight vertices with edges, we create a six-celled polyhedron. Verify the vertex, edge, face, and cell count.

Look at what happens to the polyhedral object with a hole through it in Fig. 9.43. Although the Euler formula is satisfied in Fig. 9.43a, two of the faces

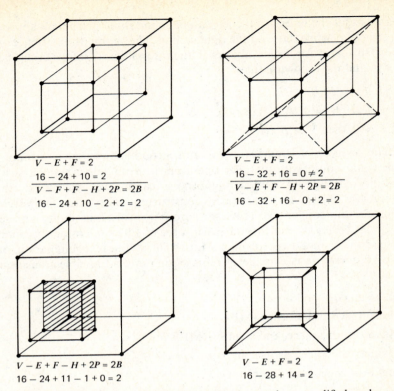

$V - E + F = 2$
$16 - 24 + 10 = 2$
$V - F + F - H + 2P = 2B$
$16 - 24 + 10 - 2 + 2 = 2$

$V - E + F = 2$
$16 - 32 + 16 = 0 \neq 2$
$V - E + F - H + 2P = 2B$
$16 - 32 + 16 - 0 + 2 = 2$

$V - E + F - H + 2P = 2B$
$16 - 24 + 11 - 1 + 0 = 2$

$V - E + F = 2$
$16 - 28 + 14 = 2$

FIGURE 9.43 Multiply-connected polyhedra and a modified euler formula.

are inadmissible according to our convention. The faces containing the entrance and exit to the hole are not topological disks, and therefore the surface net is not proper. This situation is corrected in Fig. 9.43*b* by adding the edges shown as dashed lines. But look at what happens to the formula. Modifying the formula and deleting conditions 1 and 2 accommodates such objects. Thus,

$$V - E + F - H + 2P = 2B \qquad (9.55)$$

where V, E, and F retain their usual meaning and denote the number of vertices, edges, and faces. H denotes the number of holes in faces; P denotes the number of passages (holes through the entire object); and B denotes the number of separate, disjoint bodies (objects). Another acceptable variation is shown in Fig. 9.43*c*; here, the through passage in Fig. 9.43*a* and *b* is now merely a concavity. If we add edges as in Fig. 9.43*d*, then we satisfy the original Euler formula. (The individual faces, as well as the interior, are each simply-connected entities.)

I. C. Braid, R. C. Hillyard, and I. A. Stroud (1978) arrange the terms in Eq. 9.55 in a slightly different way

$$V - E + F - H = 2(B - P) \qquad (9.56)$$

observing that the terms on the left can be found directly from an appropriate data structure, while those on the right cannot. They admit the existence of such a data structure for Euler objects if their topological information satisfies Eq. 9.56 and the following rules:

1. $V, E, F, H, P, B \geq 0$.
2. If $V, E, F, H = 0$, then $P, B = 0$.
3. If $B > 0$, then $V > B$ and $F > B$.

The research of Braid, Hillyard, and Stroud shows that to minimize the number and complexity of Euler operators in a modeling system, strict admissibility throughout a modeling process cannot be guaranteed (that is, inadmissible objects are temporarily constructed). They show, however, that by a proper choice of operators, a less rigorous form of admissibility is preserved.

Braid, Hillyard, and Stroud select a set of Euler operators by analyzing the six-dimensional hyperplane in Eq. 9.56. A linear combination of five primitive operators is capable of representing any admissible transition; the operators are

1. Make an edge and a vertex (MEV).
2. Make a face and an edge (MFE).
3. Make a body, a face, and a vertex (MBFV).
4. Make a cavity, or passage, and a body (MRB).
5. Make an edge and kill (delete) a hole (ME − KH).

Each of these five operators has a corresponding complementary operator: Kill an edge and vertex (KEV); kill a face and edge (KFE); kill a body, face, and vertex (KBFV); kill a cavity or passage and a body (KRB); and kill an edge and make a hole (KE − MH). The tabular array in Fig. 9.44 illustrates the relationship between these operators and the Euler formula. Thus, given a balanced set of coordinates V, E, F, H, B, and P, executing any operator does

	V 1	E −1	F 1	H −1	$2B$ 2	$2P$ −2
MEV	1	1	0	0	0	0
MFE	0	1	1	0	0	0
MBFV	1	0	1	0	1	0
MRB	0	0	0	0	1	1
ME − KH	0	1	0	−1	0	0
KEV	−1	−1	0	0	0	0
KFE	0	−1	−1	0	0	0
KBFV	−1	0	−1	0	−1	0
KRB	0	0	0	0	−1	−1
KE − MH	0	−1	0	1	0	0

FIGURE 9.44 Array of Euler operator sets.

not destroy this balance. For example, MEV adds one vertex $(+1)$ and one edge (-1), and the net change is zero. MBFV adds one vertex, one face, and one body to the data structure, the vertex and face adding $+2$ to the left side of the formula and the body adding a counterbalancing $+2$ to the right.

Consider the construction of a tetrahedron for which $V = 4$, $E = 6$, $F = 4$, $H = P = 0$, and $B = 1$. One of several possible sequences of Euler operations is shown in Fig. 9.45. Six operations are sufficient to define a valid data structure. The Euler formula is in balance at any intermediate stage,

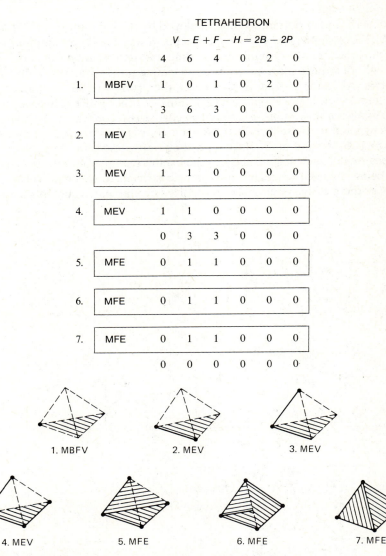

TETRAHEDRON

$$V - E + F - H = 2B - 2P$$

		4	6	4	0	2	0
1.	MBFV	1	0	1	0	2	0
		3	6	3	0	0	0
2.	MEV	1	1	0	0	0	0
3.	MEV	1	1	0	0	0	0
4.	MEV	1	1	0	0	0	0
		0	3	3	0	0	0
5.	MFE	0	1	1	0	0	0
6.	MFE	0	1	1	0	0	0
7.	MFE	0	1	1	0	0	0
		0	0	0	0	0	0

1. MBFV   2. MEV   3. MEV

4. MEV   5. MFE   6. MFE   7. MFE

FIGURE 9.45 Construction of a tetrahedron with Euler operators.

although a topologically valid solid is defined only after the last operation. Each operator is a complex algorithm, building a data structure by interpreting the input parameters. The first operator here is MBFV, which initializes the data structure, adds a vertex, face, and body. The coordinates of one vertex are specified through this operator as well as the surface description of one face (no edges, that is, surface boundaries, are specified yet), and a body counter is incremented by one. When this operation is complete, three vertices, six edges, and three faces remain to be constructed. The tetrahedron is completed by using MEV three times and MFE three times, each time specifying vertex, edge, and face data.

An early and important system developed by B. G. Baumgart (1974), called GEOMED, exhibited many of these concepts. It was a tool for computer vision research rather than a system for modeling and describing mechanical parts. The basic constructs, more primitive than simple polyhedra, were called **Euler primitives**. As in the preceding discussion, the primitive body contained only a face and a vertex. We should perhaps think of other Euler primitives in this system as algorithms, adding a face and an edge (MFE), and so on. Braid reports that implementing these primitives was not so useful as had been expected in evaluating the results of Boolean operations on polyhedra. Apparently, this was because the intermediate and transient states that an object passed through while being modeled could not always be restricted to Euler polyhedra. The lessons of these early systems have minimized or eliminated such problems.

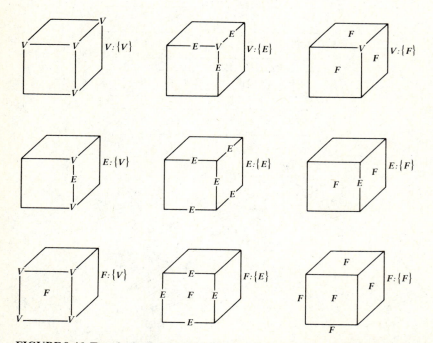

FIGURE 9.46 Topological relationships between pairs of polyhedron elements.

A polyhedron exhibits nine classes of topological relationships between pairs of the three types of elements: vertices, edges, and faces. A. Baer, C. Eastman, and M. Henrion (1979) illustrated these relationships in their 1979 survey, similar to that shown in Fig. 9.46. Thus, for example, a vertex is topologically characterized by adjacent vertices, edges, and faces. This survey found that for vector graphics (see Chapter 12), it is advantageous to use one of the relationships $V: \{V\}$, $E: \{V\}$, or $F: \{V\}$ to know how vertices are joined, whereas Boolean shape operations are facilitated by the ring of faces around the vertex $V: \{F\}$, and the Euler operators usually require adjacency among faces $F: \{F\}$.

9   FORMAL MODELING CRITERIA

We have explored the basic concepts underlying geometric modeling in Chapter 9. In the next chapter, we will examine the consequences of using these concepts in the major types of modeling systems. Until quite recently, the development and use of these concepts and their aggregation into systems was informal, ad hoc, and proceeded more or less empirically. There was no theory of geometric modeling, no universal criteria to judge the performance of existing modeling systems or to quide the design and development of more powerful new systems.

In the late 1970s, this situation began to change, largely through the efforts or Voelcker, Requicha, and others at the University of Rochester. Recognizing the need for formal modeling criteria in their own Production Automation Project and PADL system, they began to unify and make more rigorous the various modeling concepts, thus taking the first steps toward a theory of geometric modeling.

This process of unifying and formalizing is not yet complete, and there are ample opportunities for new approaches at this still early stage. A brief review of the principal ingredients of the emerging theory is appropriate now so that in subsequent chapters, we will be better equipped to judge the range and power of various modeling systems.

Let us first examine the motivation for a more formal theory of geometric modeling. Until very recently, most geometric-modeling systems relied on the user to verify the completeness, validity, and uniqueness of a model. Such verification becomes increasingly more difficult as the model becomes more complex, even with sophisticated computer graphic display techniques. On another front, a valid model may itself be used as input for applications that automatically generate modifications to the model, either transitory or permanent, as in some aspects of CAM. Finally, the day is not too far off when hybrid robotic-computer vision systems will not only require valid models of their "environment" on which to operate but must also construct these same models. The opportunity for human-user intervention to verify a geometric model is clearly decreasing inversely as the sophistication of the total modeling process increases. Thus, the need for self-verifying modeling is inevitable and must be preceded by a thorough understanding of just what constitutes a valid, complete, and unique model. Furthermore, a rigorous theory of modeling will allow us to use the full power of mathematics to investigate modeling processes and optimize modeling systems.

As a prelude to establishing formal modeling criteria, we abstract a set of geometric characteristics from the solid physical objects we see around us. Perhaps the most important of these characteristics are surface and shape. An object's shape is determined by the total spatial point set that defines it. The surface of an object is a subset of this point set. In geometric modeling, we are usually interested in only those shapes that are bounded and connected. A shape is bounded if it is definable within a finite space, and it is connected if every pair of points within its interior can be joined by a path without leaving the interior. A shape's surface must also satisfy certain conditions; it must be

1. closed,

2. orientable,

3. non-self-intersecting,

4. bounding, and

5. connected.

Every pair of points on a connected surface can be joined by a path lying on the surface. A bounding surface divides space into two disjoint domains, one of which is finite. A Klein bottle is an example of a self-intersecting and nonorientable closed surface. Being nonorientable means that a Klein bottle has no distinct inside and outside; the Mobius strip is another example of a one-sided nonorientable surface. In fact, Mobius suggested a way of determining whether a closed polyhedral surface is oriented or not. If we assign a consistent direction to the ring of edges surrounding each face of the polyhedron, say, counterclockwise when looking at the face from the outside, then each edge receives two directional arrows, one from each face adjoining it. Then a surface is orientable if, and only if, each edge contains one arrow in each direction; see Fig. 9.47.

FIGURE   9.47   The mobius rule for determining the orientability of a polyhedral surface.

A closed surface is ensured by certain topological relationships between the elements of a polyhedral net on the surface: each edge is adjoined by two and only two faces and vertices, there is an equal number of edges and vertices on the ring surrounding any face, and similarly there is an equal number of edges and faces surrounding any vertex (with the exception of the apex of a conical-type surface). These conditions rule out open, shell-like surfaces. The five conditions for a shape's surface determine its "well formedness," a set of criteria used in many modeling systems.

A. A. G. Requicha (1977a) and others assign to our abstract solid several additional properties that should be included in any solid modeling scheme:

6. rigidity,
7. homogeneous three-dimensionality,
8. closure under rigid motions and certain Boolean operations.
9. finite describability, and
10. boundary determinism.

Rigidity denotes an invariant shape, independent of a solid's location and orientation. Dimensional homogeneity means that the solid has a connected interior and no dangling or isolated boundary segments. Any combination of rotation and translation and any sequence of Boolean operations on a solid must produce other equally valid solids; this condition is known as *closure*. Finite describability requires the existence of some finite aspect of a solid to ensure that it can be represented by a finite statement or data in a computer (a finite number of enclosing faces, for example, is sufficient). Finally, boundary determinism requires that the bounding surface of a solid must distinguish analytically what is "inside" the solid, making the surface itself an adequate representation of the total solid.

Requicha identifies some fundamental questions that any theory of geometric modeling should address:

1. Does the boundary supply enough information for automatic computation of geometric properties, including properties required for display and plotting?

2. How can a geometric algorithm or system determine if it is operating on valid data?

3. Is there more than one way of representing a solid?

4. Is it possible to determine if different representations correspond to the same solid, and if so how?

5. Can some geometric properties be computed from one type of representation but not from another type?

6. Is there a "best" representation?

Next, he introduces some key formal properties of model representation schemes: domain, validity, completeness, and uniqueness.

Domain is the descriptive power of a representation scheme, the set of objects it is capable of modeling. A scheme's domain might include only objects that are plane faced, convex, and simply-connected polyhedra.

The set of valid forms of representation defines the range of a representation scheme. Criteria for determining **validity** are variable. In a more restricted or special-purpose system, multiply-connected objects may be invalid. In a CAM system designed to produce machinable parts, an object with an internal cavity is invalid. And, of course, there is a great range of what are called nonsense objects, which would be invalid in almost any modeling system.

Completeness is a measure of a model's (and, hence, representation scheme's) ability to respond to a broad range of analysis—its ability to answer geometric questions. For example, typical wireframe models cannot furnish answers about surface normals or volume, and they are therefore incomplete representations of a solid.

Uniqueness is a key factor in determining the equality of objects. Requicha points out that most representation schemes for geometric objects are nonunique; first, because substructures in a representation may be permuted (for example, $A \cup B \cup C = A \cup C \cup B$) and, second, because distinct representations may correspond to equivalent but differently positioned objects. Although these two examples of nonuniqueness are simple to state and understand, algorithms to detect the possible equivalence of two objects are computationally costly and sophisticated.

Conciseness, user friendliness, and efficacy are other less formal attributes of representation schemes. **Conciseness** refers to the amount of data required to define an object in a particular representation scheme. The more concise a model is, the more convenient it is to store or transmit, and the less redundant data it contains. The need for conciseness in a model must be balanced against the demands on downstream applications to elaborate and interpret the model data. **User friendliness** is a measure of the ease in creating valid models in either an interactive mode or batch mode. **Efficacy** measures how adequate and accommodating a model representation is to downstream applications as well as general geometric analysis. To date, little is known about how to quantify efficacy.

As we said earlier in this section, the process of unifying and formalizing a theory of geometric modeling is not complete. The foundations have been well established by the University of Rochester group and others. Further progress depends in part on greater experience with a broad range of contemporary modeling systems and the next generation of systems just now emerging.

SOLID MODEL CONSTRUCTION

There are six major categories of methods of constructing solid models: instances or parametrized shapes, cell decomposition (including spatial occupancy enumeration), sweep representations, constructive solid geometry, boundary representations, and wireframe representations. Underlying them are the concepts of graph-based models and Boolean models.

Of the six methods, three are important to contemporary modeling systems: sweep representations, constructive solid geometry, and boundary representations. Of these three, the last two will probably dominate in the future. All the methods offer advantages in certain applications, and even the most specialized methods may have supporting roles in future modeling systems.

We will begin with an overview of graph-based models and Boolean models, then explore each of the six major modeling methods. We will conclude with a brief review of the GMSolid and ROMULUS systems.

1   GRAPH-BASED MODELS

A geometric model emphasizing the topological structure, with data pointers linking together an object's faces, edges, and vertices, is a **graph-based model**. A solid object can be represented as a list of the object's faces and their respective surface equations. The edges of these faces are represented as curve equations, with pointers to their end-point vertices and adjoining faces. The vertices are represented as lists of coordinates, with pointers to the edges meeting at each vertex. Notice that there are two kinds of information—the pointers defining the topology or connectivity between vertices, edges, and faces and numerical data defining curve and surface equations and vertex coordinates. Subsequent modeling operations may alter just the pointers, just the numerical data, or both.

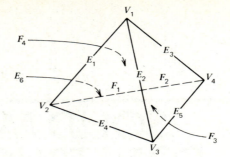

Vertices	Edges	Faces
V_1	E_1	F_1
V_2, V_3, V_4 E_1, E_2, E_3 F_1, F_2, F_4	V_1, V_2 E_2, E_3, E_4, E_6 F_1, F_4	V_1, V_2, V_3 E_1, E_4, E_2 F_2, F_3, F_4
V_2	E_2	F_2
V_1, V_3, V_4 E_1, E_4, E_6 F_1, F_3, F_4	V_1, V_3 E_1, E_3, E_4, E_5 F_1, F_2	V_1, V_3, V_4 E_2, E_5, E_3 F_1, F_3, F_4
V_3	E_3	F_3
V_1, V_2, V_4 $E_2, \ldots$	V_1, V_4 $E_1, E_2, \ldots$	V_4, V_3, V_2 $E_4, \ldots$

FIGURE 10.1 Graph-based model.

Figure 10.1 is an example of a graph-based model. The model contains redundant information, but in many situations, this is desirable. Some redundancy can speed up searching algorithms, thereby eliminating the need for global searches. Scaling and rigid-body transformations (translation and rotation) alter only numerical data, leaving the pointers unaffected. Also, wireframe computer graphic displays can be derived from a graph-based model in an amount of time proportional to the number of edges.

There are interesting differences among graph-based systems: In addition to differences in degree of redundancy, there are differences in accommodating the edges of a face. Some graph-based systems require a face to have a single boundary with no "hole." Others allow an outer boundary plus disjoint inner boundary loops representing holes. This characterizes the more highly structured systems and offers computational advantages for complex (multiply-connected) solids.

For solids represented as planar-faced polyhedra, many economies are available through graph-based systems. Since all edges are straight-line segments, they can be stored as vertex pairs, and face boundaries can be stored

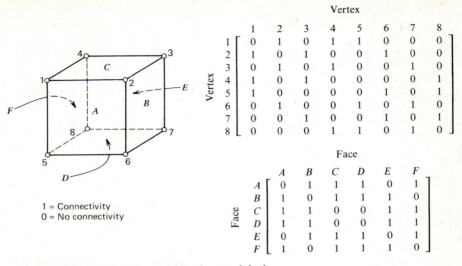

Vertex

	1	2	3	4	5	6	7	8
1	0	1	0	1	1	0	0	0
2	1	0	1	0	0	1	0	0
3	0	1	0	1	0	0	1	0
4	1	0	1	0	0	0	0	1
5	1	0	0	0	0	1	0	1
6	0	1	0	0	1	0	1	0
7	0	0	1	0	0	1	0	1
8	0	0	0	1	1	0	1	0

Face

	A	B	C	D	E	F
A	0	1	1	1	0	1
B	1	0	1	1	1	0
C	1	1	0	0	1	1
D	1	1	0	0	1	1
E	0	1	1	1	0	1
F	1	0	1	1	1	0

1 = Connectivity
0 = No connectivity

FIGURE 10.2 Connectivity matrices for a polyhedron.

as ordered lists (chains) of vertices. Or we can treat a polyhedron as a simple graph, listing the vertices with their coordinates in one array and their connectivity in another, which is called a **connectivity matrix** (see Fig. 10.2 for an example). The connectivity matrix is a binary matrix; zero-valued elements indicate no connectivity exists, and one-valued elements indicate connectivity exists between the pair of vertices. Such a matrix can be constructed for other elements, and one is shown in Fig. 10.2 for the connectivity between the faces of the polyhedron.

The connectivity matrix is also called an **adjacency matrix**. Such matrices are convenient for algorithms that determine whether an edge exists between two vertices. Let a_{ij} denote the element of the connectivity matrix describing the existence of an edge between vertex i and vertex j. Then an algorithm accesses a_{ij} and reads its value. The time needed to do this is independent of the number of vertices and the number of edges. The main disadvantage of a connectivity matrix is that it requires V^2 storage (V = number of vertices) even though most $a_{ij} = 0$. Note that if efficient packing is available the storage requirement can be reduced to V^2 bits.

A **graph** is a set of nodes (or points) connected by branches (lines). The number of branches at a node determines its degree. If any branch has a direction associated with it, the graph is called a **directed graph**; see Fig. 10.3. For directed graphs, we can assign an **in degree** and an **out degree** to each node by counting the branches pointing to it and away from it, respectively. A path from one node to another is the sequence of branches to be traversed. If the start and end nodes of a path are the same, the path is a **circuit**. If there is a path between any pair of nodes of a graph, the graph is **connected**. A **tree** is a connected graph without circuits. A graph containing some or all of the nodes and branches of a graph G and no other nodes or branches is a subgraph of G. A **spanning tree** of a

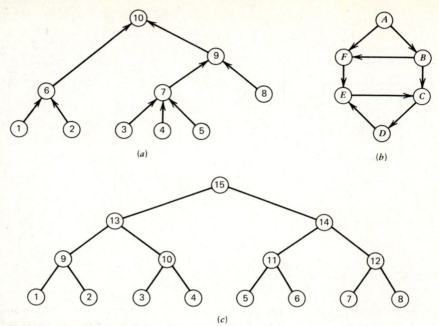

FIGURE 10.3 Examples of graphs.

connected graph G is a subgraph containing all its nodes and enough branches to maintain connectivity without creating any circuits.

There are many ways of representing a graph. One way involves using a list of vertices (nodes) and their connectivity matrix (branches), as we saw earlier. Another interesting way is to use bit vector matrix elements, often resulting in considerable efficiency for graph alogrithms. Lists may represent a graph by having a list of adjacent vertices associated with each vertex or node; Fig. 10.3*b* is a directed graph with six nodes. Its connectivity (or adjacency) matrix is

$$
\begin{array}{c c c c c c c}
 & A & B & C & D & E & F \\
A & 0 & 1 & 0 & 0 & 0 & 1 \\
B & 0 & 0 & 1 & 0 & 0 & 1 \\
C & 0 & 0 & 0 & 1 & 0 & 0 \\
D & 0 & 0 & 0 & 0 & 1 & 0 \\
E & 0 & 0 & 1 & 0 & 0 & 0 \\
F & 0 & 0 & 0 & 0 & 1 & 0
\end{array}
$$

Notice that for a directed graph, $a_{ij} = 1$ only if there is a directed edge *from* vertex i *to* vertex j. The six adjacency lists are

vertex A $[B, F]$ vertex D $[E]$

vertex B $[C, F]$ vertex E $[C]$

vertex C $[D]$ vertex F $[E]$

Edges of undirected graphs are represented twice in an adjacency matrix. If there is an undirected branch (or edge) connecting vertex i to vertex j, then $a_{ij} = a_{ji} = 1$. When we delete an edge from an undirected graph, we delete both a_{ij} and a_{ji} from the matrix. In the case of adjacency lists, we create a supplementary array to link the double entries of edges and speed the search process.

Again, a tree is a directed graph with no circuits. A tree has the following additional properties:

1. There is one and only one node, called the **root**, which no branches enter.

2. Every node except the root node has one and only one entering branch.

3. There is a unique path from the root node to each other node (shown in bold outline in Fig. 10.4).

Do not be alarmed by trees with roots at the top and leaves at the bottom; that is the convention. Given two adjacent nodes a and b, as in Fig. 10.4, if the branch is directed from a to b, then a is a proper ancestor of b (sometimes called the *father* of b), and b is a proper descendant of a (a son of a). A leaf node has no proper descendants.

There are several quantitative parameters associated with a tree, as in Fig. 10.4. Consider node b. The **depth** of node b in the tree is the length of the path (number of branches traversed) from the root R to b. The **height** of node b is the length of the longest path from b to a leaf. Notice that the height of the tree is the height of the root. The **level** of a node is the height of the tree minus the depth of the node.

If the descendants of each node are in order, say, from left to right, then the tree is **ordered**. When we study Boolean models in Section 2, we will use a special type of tree, called a **binary tree**, which is an ordered tree where each node has two descendants—a left descendant and a right one. The tree in Fig. 10.4 is a binary tree.

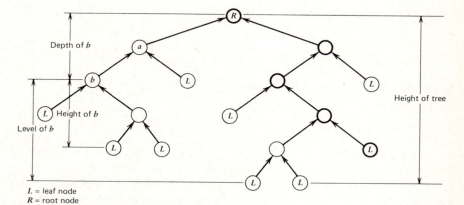

L = leaf node
R = root node

FIGURE 10.4 A tree graph.

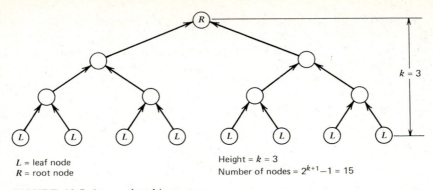

L = leaf node
R = root node

Height = k = 3
Number of nodes = $2^{k+1} - 1$ = 15

FIGURE 10.5 A complete binary tree.

Study the binary tree in Fig. 10.5; this tree is **complete**. A binary tree is complete if for some integer k every node of depth less than k has a left and right descendant and every node of depth k is a leaf. This is true for the nodes at depth $k = 3$ in Fig. 10.5. The total number of nodes in a complete binary tree of height k is $2^{k+1} - 1$.

Algorithms must often **traverse** a binary tree, that is, visit each node to get information about it or the path it lies on. We will investigate three systematic ways of doing this: preorder traversal, postorder traversal, and inorder traversal. Figure 10.6 illustrates these traversals on the binary tree in Fig. 10.5. First, define a node n and all its descendants as a subtree t_n of tree T. The node n is the root of t_n. Let r denote the root of the tree and d_L and d_R its immediate descendants. Then define the preorder traversal of T recursively as follows:

1. Visit the root r; and
2. visit in preorder the subtrees with roots d_L and d_R, in that order.

Define the postorder traversal of T recursively as follows:

1. Visit in postorder the subtrees with roots d_L and d_R, in that order; and
2. visit the root r.

Finally, define the inorder traversal recursively as follows:

1. Visit in inorder the left subtree of root r (if it exists);
2. visit r; and
3. visit in inorder the right subtree of r (if it exists).

If we assign a number in sequence to each node we visit, we notice some interesting properties. The inorder traversal numbers of the nodes of a binary tree occur in such a way that each node in the left subtree of a root node n_R has a number less than n_R, while each node in the right subtree has a number greater than n_R.

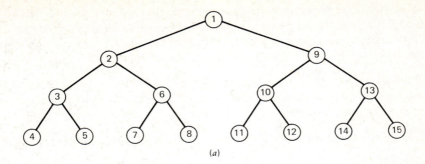

(a)

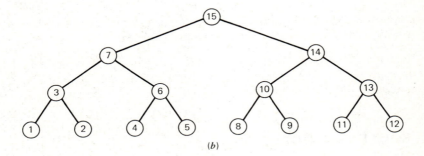

(b)

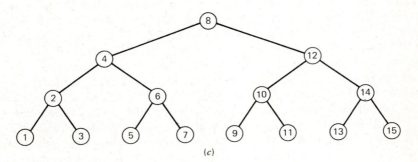

(c)

FIGURE 10.6 Traversals of a binary tree.

EXERCISES

1. What is the height of the tree in Figure 10.4? For the same figure, what are the height, depth, and level of node b?

2. Write out the connectivity matrix for the vertices and faces of a tetrahedron; for an octahedron.

3. For a complete binary tree of height $=4$; write a short subroutine to traverse it in preorder, postorder, and inorder.

4. Discover other properties of node numbering for the three types of traversals. (*Hint*: For a preorder traversal, count the number of descendants of each node. Will this determine if node n_i is a descendant of node n_j? Is this similar to postorder traversal?)

5. Verify the fact that a complete binary tree has $2^{k+1} - 1$ nodes. How many nodes are there in a complete **quadtree**? A complete **octree**? (How do you interpret the terms *quadtree* and *octree*?)

6. Develop a graph-based model for a cube. Follow the example in Fig. 10.1.

7. Develop two different graph-based models for a sphere. Again, follow the example in Fig. 10.1.

8. Develop a graph-based model that eliminates or minimizes the redundant information contained in the model developed in Ex. 6.

9. Derive a connectivity matrix for a cube, each face of which is triangulated by an edge, so that four edges meet at each of half the vertices and five edges meet at each of the other half of the vertices.

2   BOOLEAN MODELS

If a solid object is represented by the Boolean combination of two or more simpler objects, then the representation is a **Boolean model**. If A, B, and C denote solids and if $C = A\langle OP\rangle B$, where $\langle OP\rangle$ is any regularized Boolean operator, then $A\langle OP\rangle B$ is a Boolean model of C. Also, remember that A, B, and C must be of the same spatial dimension. From now on, we will assume that all Boolean operations are regularized, so that for conciseness, we can drop this term as a qualifier. The symbols $\cup$, $\cap$, and $-$ will denote the regularized operators. We will develop a more complicated specific model later, but first, let us look at an important feature of a Boolean model: It is a **procedural** model.

Study Fig. 10.7. Assume we know the size, position, and orientation of A, B, and C, perhaps as a list of vertex coordinates. The Boolean model of D is $D = (A \cup B) - C$. This Boolean statement defining D says nothing quantitative about the new solid it creates; it only specifies the combination of **primitive** solid constituents. It does not tell us the coordinates of the vertices of the new solid or anything about its edges or faces. We may know all there is to know geometrically and topologically about A, B, and C, but all we know about D is how to construct it. Therefore, we say that the Boolean model is a procedural representation; it is also called an **unevaluated** model. If we wish to know more, then we have to evaluate the Boolean model. We have to compute intersections to determine new edges and vertices. We have to analyze the connectivity of these new elements to determine the model's topological characteristics.

The boundary of D has 32 vertices, 48 edges, and 18 faces. The undirected connectivity matrix of the vertices is a 32×32 array, containing 1024 elements. Since each vertex in this model is connected to exactly three others, there are

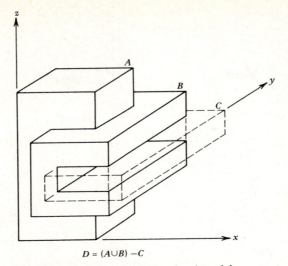

$$D = (A \cup B) - C$$

FIGURE 10.7 A simple procedural model.

only 96 nonzero elements in the matrix. A **boundary evaluator** routine uses this information to construct the boundary representation.

The binary tree for this model is shown in Fig. 10.8. The leaf nodes are the primitive solids, with Boolean operators at each internal node and the root. Each internal node combines the two objects immediately below it in the tree and, if necessary, transforms the result in readiness for the next operation. J. W. Boyse and J. E. Gilchrist (1982) note that the constructive representation of an object is a direct translation of the Boolean operators, as invoked by the users of their GMSolid, into a binary tree structure.

How are the primitives modeled? In many systems, they are stored as a graph-based model and become a unit template or parametrized shape to be scaled and positioned as a leaf node of the model's binary tree. Or a primitive may be a Boolean combination of directed surfaces or half-spaces. A directed surface is a surface whose normal at any point determines the inside and outside of the primitive solid. An unbounded surface divides Cartesian space into two unbounded regions; each region is called a **half-space**. The Boolean intersection of an appropriate set of half-spaces can form a closed three-dimensional solid.

The TIPS modeler developed at Hokaido University by N. Okino and H. Kubo (1973) uses Boolean combinations of directed surfaces defining half-spaces to construct the entire model. Each directed surface is given by an equation of the form $f(x, y, z) = 0$. The function is zero at the surface and positive inside the object. Thus, a complex object is defined by the union of the intersection of directed surfaces

$$F = \bigcup_{j=1}^{m} \left(\bigcap_{i=1}^{n} f_{ij} \right) \tag{10.1}$$

where the f_{ij} are directed surfaces or half-spaces.

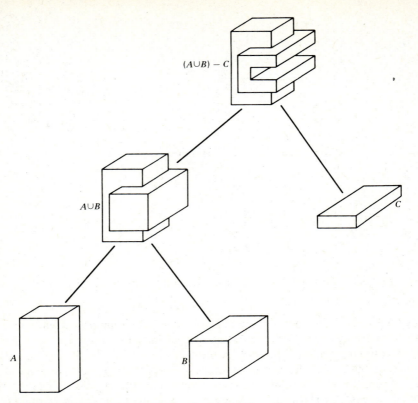

FIGURE 10.8 The binary tree for $D = (A \cup B) - C$.

Other systems, notably GMSolid, PADL (University of Rochester), and ROMULUS (Evans and Sutherland), operate with solid, bounded primitives. At any node of the binary tree, two valid solids combine to produce a third valid solid. The Boolean model of even relatively complex solids is generated quickly in these systems, and it has a very compact and concise data structure. In addition, GMSolid allows a user to retrace and revise the model at any point in its development; see Appendix A.

Figure 10.9 is another example of a Boolean model of a solid object. Let π denote a primitive object. Here, π_1 is a rectangular parallelepiped and π_2 is a right circular cylinder. T_i denotes a transformation that scales and positions the primitives.

First, the union operator combines $\pi_1 T_1$ and $\pi_2 T_2$ to produce S_1, a hublike shape with rectangular ears. Next, $\pi_1 T_3$ and $\pi_2 T_4$ combine to produce S_2, similar to S_1 but smaller in size. The Boolean difference between S_1 and S_2 creates a circular hole and pair of keyways in S_1 to produce S_3. Then the intersection of S_3 with $\pi_2 T_4$ reforms the outer surface of the ears, giving them a radius equal to that of $\pi_2 T_4$'s, producing S_4. S_4 and S_5 combine to produce S_6 and add a further refinement to the geometry of the ears. S_6 and S_7 combine

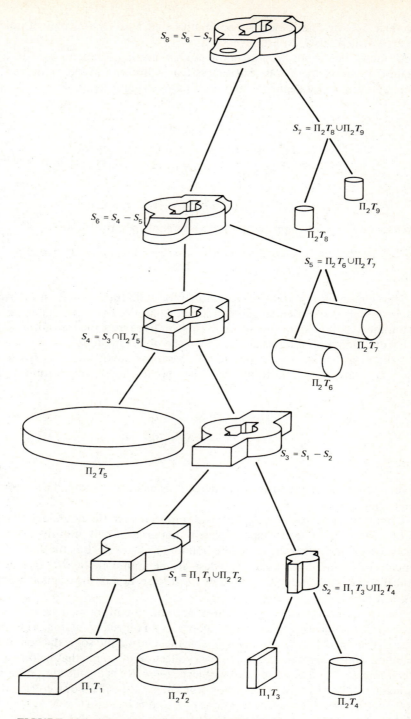

$S_8 = S_6 - S_7$

$S_7 = \Pi_2 T_8 \cup \Pi_2 T_9$

$\Pi_2 T_9$

$S_6 = S_4 - S_5$

$\Pi_2 T_8$

$S_5 = \Pi_2 T_6 \cup \Pi_2 T_7$

$S_4 = S_3 \cap \Pi_2 T_5$

$\Pi_2 T_7$

$\Pi_2 T_6$

$\Pi_2 T_5$

$S_3 = S_1 - S_2$

$S_1 = \Pi_1 T_1 \cup \Pi_2 T_2$

$S_2 = \Pi_1 T_3 \cup \Pi_2 T_4$

$\Pi_1 T_1$

$\Pi_2 T_2$

$\Pi_1 T_3$

$\Pi_2 T_4$

FIGURE 10.9 A boolean model of a mechanical part.

to produce S_8, the final object shape. Notice that S_7 defines a pair of holes; these are "drilled" in S_6 to produce S_8.

Although this model represents a modest accomplishment, we can appreciate the compactness and conciseness that a Boolean model achieves. To demonstrate this more explicitly, express S_8 as a string of symbols

$$
\begin{aligned}
S_8 &= S_6 - S_7 \\
&= (S_4 - S_5) - (\pi_2 T_8 \cup \pi_2 T_9) \\
&= (S_3 \cap \pi_2 T_5) - (\pi_2 T_6 \cup \pi_2 T_7) - (\pi_2 T_8 \cup \pi_2 T_9) \\
&= (S_1 - S_2) \cap \pi_2 T_5 - (\pi_2 T_6 \cup \pi_2 T_7) - (\pi_2 T_8 \cup \pi_2 T_9) \\
&= [(\pi_1 T_1 \cup \pi_2 T_2) - (\pi_1 T_3 \cup \pi_2 T_4)] \cap \pi_2 T_5 \\
&\quad - (\pi_2 T_6 \cup \pi_2 T_7) - (\pi_2 T_8 \cup \pi_2 T_9)
\end{aligned}
\tag{10.2}
$$

Then use the distributive property of the intersect operator to obtain

$$
\begin{aligned}
S_8 &= (\pi_1 T_1 \cap \pi_2 T_5 \cup \pi_2 T_2 \cap \pi_2 T_5) - (\pi_1 T_3 \cap \pi_2 T_5 \cup \pi_2 T_4 \cap \pi_2 T_5) \\
&\quad - (\pi_2 T_6 \cup \pi_2 T_7) - (\pi_2 T_8 \cup \pi_2 T_9)
\end{aligned}
\tag{10.3}
$$

This symbol string and the elements of the nine transformation matrices is the procedural, unevaluated geometric model. Notice that this symbol string is not unique. Although π_1 and π_2 are fixed primitive types, the transformations T_i vary depending on the size and shape of the object modeled.

Note also that for this example, all operations involving $\cap \pi_2 T_5$ except $\pi_1 T_1 \cap \pi_2 T_5$ are trivial, since all the other bodies lie entirely within $\pi_2 T_5$. Thus,

$$\pi_2 T_2 \cap \pi_2 T_5 = \pi_2 T_2$$

$$\pi_1 T_3 \cap \pi_2 T_5 = \pi_1 T_3$$

$$\pi_2 T_4 \cap \pi_2 T_5 = \pi_2 T_4$$

A sophisticated modeling system will detect this condition and will not process these trivial operations.

Computing intersections is of central importance in the boundary evaluation of a Boolean model. If the complete boundaries of two primitives or subobjects are defined, then the complete boundary of $A\langle OP \rangle B$ is the sum of the segments of A and B boundaries. These segments are the **active regions** on A and B and are themselves bounded by the intersection of the original boundaries of A and B.

Let us review the effect of the three Boolean operators on a set of two or three primitive objects. Remember, we consider an object, including primitives, to be a collection of points, some comprising the interior and some the boundary. The union of two objects, say, $A \cup B$, is a collection of points that belong to A, B, or A and B. The difference of two objects, $(A - B)$, is a collection of points that belongs to the surface and interior of A but is not inside B or a collection of points that belong to the surface regions of B and is inside A. Look at the two- and three-dimensional examples in Fig. 10.10.

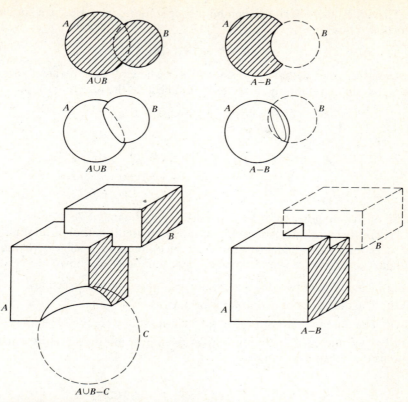

FIGURE 10.10 Examples of union and difference.

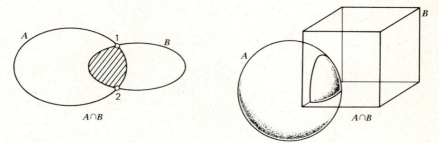

FIGURE 10.11 The intersection operation.

The operation $A \cap B$ is the set of points belonging to both A and B. These points are of three types: those interior to both A and B, those on boundary regions of A interior to B, and those on boundary regions of B interior to A; see Fig. 10.11 for examples. Only points 1 and 2 of $A \cap B$ in Fig. 10.11a are important, since they are the bounding points of the active regions of the bounding curves of A and B. There may be more than two points of intersection of the two-dimensional boundary curves, depending on their shapes.

The order of performing combining operations in a Boolean model is important. Thus,

$$A \cup B - C \neq A - B \cup C \qquad (10.4)$$

However, the order within a sequence subset where all the operations are the same type can be randomly mixed without affecting the resulting shape. Thus,

$$A \cup B - C = B \cup A - C \qquad (10.5)$$

or

$$\begin{aligned} A - B - C - D \cup E &= A - (B \cup C \cup D) \cup E \\ &= A - (B \cup D \cup C) \cup E \qquad (10.6) \\ &= A - (D \cup B \cup C) \cup E \end{aligned}$$

and so on. To understand order dependence better, study Fig. 9.28.

There are many interesting, difficult, and potentially pathological situations that a geometric-modeling system must accommodate when evaluating the boundary of a complex shape. Some of these situations are shown in Fig. 10.12.

10.12a. The union of two disjoint primitives (that is, nonintersecting).

10.12b. The difference of two disjoint primitives.

10.12c. The union of two primitives where one wholly contains the other.

10.12d. The difference of two primitives where the positive primitive wholly contains the negative primitive.

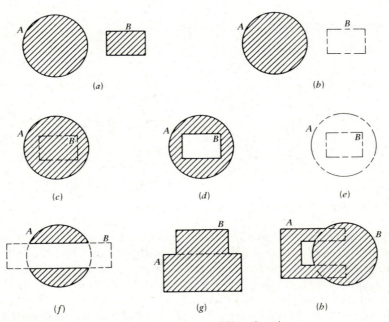

FIGURE 10.12 A variety of boolean modeling situations.

10.12*e*. The difference of two primitives where the negative primitive wholly contains the positive primitive.

10.12*f*. The difference of two primitives that creates two or more new objects.

10.12*g*. The union of two primitives that are tangent.

10.12*h*. The union of two primitives that creates inner loops or cavities (bubbles).

Not shown in Fig. 10.12 are the operations $A \cup A$ or $A - A$. Although the preceding situations are two-dimensional, each also has a direct three-dimensional analog.

Some generalizations are in order that Fig. 10.13 illustrates. Modeling systems use their algorithmic equivalents to check the validity of results or to speed computations.

10.13*a*. If two closed planar curves intersect, they will intersect at an even number of points. (A tangent condition is not counted as an intersection.)

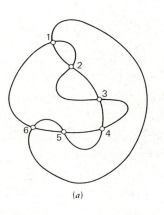

(a)

(b)

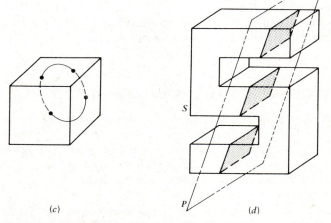

(c)　　　　(d)

FIGURE 10.13 Four general properties of boolean models.

10.13*b*. If two closed coplanar curves, say, *A* and *B*, do not intersect, then if a point on one, say, point 2 on *B* in the figure, is inside the other curve *A*, curve *B* is entirely inside curve *A*; the converse is true.

10.13*c*. If a closed curve intersects the bounding surface of a three-dimensional shape, it will intersect it at an even number of points. (Again, tangents are not counted.)

10.13*d*. If an unbounded plane *P* intersects the closed bounding surface of a three-dimensional shape *S*, then the intersection will consist of one or more closed, nonintersecting curves.

In Section 9, we will explore in more detail the problem of evaluating the boundary of a Boolean model.

EXERCISES

1. What distinguishes a primitive solid from other subobjects of a Boolean model?

2. Can you think of any useful geometric limitations on a primitive solid? Discuss them.

3. Discuss why creating a Boolean model is more than just "sticking bits and pieces together." Show how it is similar to many manufacturing processes.

4. Use the procedure implied by Eq. 10.1 to model a unit cube whose minimum and maximum vertices are $(0, 0, 0)$ and $(1, 1, 1)$, respectively.

5. Repeat Ex. 4 for a cylindrical solid of radius 0.5 whose axis coincides with the *z*-axis and is limited by the $z = 0$ and $z = 1$ planes.

6. Use Eq. 10.1 to express the solid resulting from the union of the cube and cylinder defined in Exs. 4 and 5.

7. Show how the Boolean difference operator can be incorporated into Eq. 10.1.

8. Can models created by the procedure impled by Eq. 10.1 be represented by a tree structure? If so, what rules would they follow?

3   INSTANCES AND PARAMETRIZED SHAPES

A direct way of defining a new shape is as a simple linear transformation of an existing one. Consider a unit cube. There are several ways of transforming it into a new shape by using scaling operators. Equal scaling of all three-dimensional components creates new cubes. Differential scaling creates a variety of rectangular solids. Each new cube or rectangular solid is a particular **instance** of the initial cube.

Figure 10.14 is an example of how an unlimited variety of specific instances of an original shape is created by simple scaling transformations. Notice that

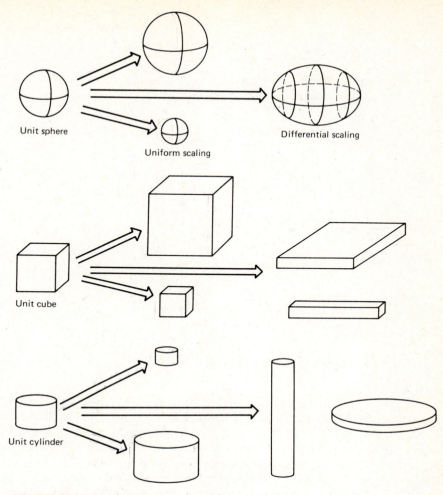

FIGURE 10.14 Instances.

such transformations affect the geometry but not the topology of a shape. The shapes in this figure are regular convex shapes.

Instancing an original shape is not limited to such simple originals—for example, notice the shape in Fig. 10.15. This Z section can be transformed by equal or differential scaling into an infinite number of instances. But there are certain restrictions: The widths of the top and bottom flanges will always be equal, and the thickness of the section will always be one-tenth the height. And look at what happens to the circular arcs defined by the bend radii: They become elliptic arcs. This is not a fatal flaw when instancing, but restricting shape change to three orthogonal coordinate scale factors limits the prospects of a modeling system based solely on these transformations. The use of fully parametrized shapes eliminates this limitation.

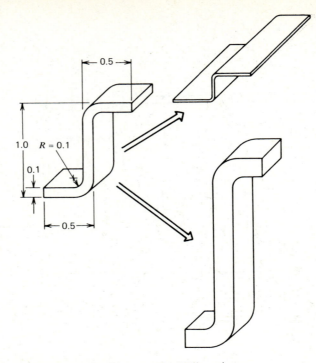

FIGURE 10.15 Instances of a "Z" section.

A small set of key dimensions is usually sufficient for defining the shape of a relatively simple class of objects. If each dimension is an independent variable, then we can produce a particular shape within a class by specifying a few key dimensions or parameters. Geometric-modeling algorithms then use these parameters to compute a more complete mathematical representation. In Fig. 10.16, we see the same basic shape that was presented in Fig. 10.15 but with much greater representational power.

It is not difficult to verify the validity of data specifying a model. For the Z section in Fig. 10.16, for example, we can apply the following restrictions and easily check them: $a, b, h, l, t > 0$, $b \leq a$, $a > 2t$, and $h > 4t$.

We can develop parametrized shapes with variable topologies, such as the n-celled structure in Fig. 10.17, but most parametrized shape procedures are restricted to a single topology.

A related technique, called **group technology**, was developed in concert with certain CAM techniques to encourage standardization in part design and production. Its central thesis is that many manufactured parts can be grouped into classes or families of similar shapes, where individual members of a family are distinguished by a few parameters (key dimensions). A single family of shapes is a **generic primitive**, and individual members are **primitive instances**. Figure 10.18 shows examples of generic primitives.

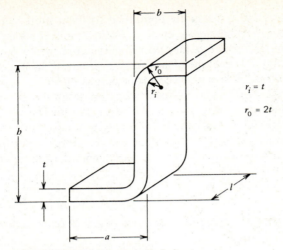

FIGURE 10.16 Parametrized shape.

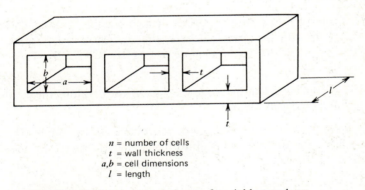

n = number of cells
t = wall thickness
a,b = cell dimensions
l = length

FIGURE 10.17 Parametrized shape of variable topology.

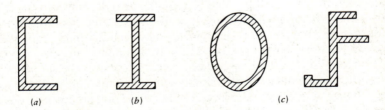

FIGURE 10.18 Typical generic primitives.

The instances are not usually combined to form more complex shapes, although no theoretical barriers prevent it. For example, it is easy to envision a modeling system that converts any primitive instance into a boundary representation (probably an envelope of surfaces), evaluates it to Boolean combinations of instances, and forms more complex shapes.

Parametrized-shape or group-technology models are easy to validate and use. They are unquestionably concise; however, the number of useful generic primitives, though large, is limited. Modeling systems built on this type of representation only are highly specialized; a very large repertoire of generic primitives is required for them to have a more general application.

EXERCISES

1. What is the effect of using one or more negative scale factors to create a shape instance?

2. The manufactured shapes discussed in the context of parametrized shapes, group technology, and generic primitives are all constant cross-section shapes. Can you think of other categories (*Hint*: fasteners, shoes)? What are the underlying geometric properties that distinguish the categories of shapes that are most amenable to parametrization?

3. Write a procedure to create instances of a unit cube. Assume the cube's edges are aligned with the principal axes, that it lies in the positive x, y, z octant, with a vertex at the origin. Denote this as **BLOCK**(L, W, H, BLOCK), where

> L, W, H are the input length (along the x-axis), width (along the y-axis), and height (along the z-axis); and

> BLOCK (8, 3) is the output array of coordinates of the block. The block should have the same relationship to the principal axes of the coordinate system as the cube.

4   CELL DECOMPOSITION AND SPATIAL-OCCUPANCY ENUMERATION

Take some common object, a coffee mug, for example, and imagine decomposing it into separate pieces so that each piece of the final decomposition is easier to describe than the original mug; use Fig. 10.19 as a guide. As the first step, detach the handle from the mug. This is a natural step and topologically reasonable, since there are then two parts, each with a simply-connected topology, instead of one multiply-connected object. Next, disconnect the bottom of the mug. Now there are two parts with a simply-connected topology and one with a multiply-connected topology. However, each of these parts is easier to describe than the mug. If necessary, continue to decompose any or all of these parts until some predetermined describability criterion is met.

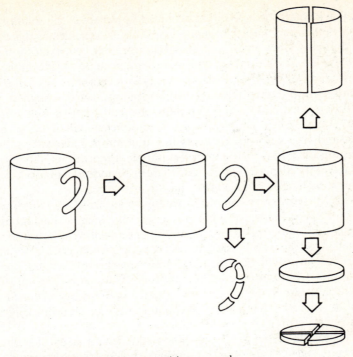

FIGURE 10.19 Cell decomposition example.

This process is **cell decomposition**. Any solid can be represented as the sum or union of a set of cells into which it is divided. The reason for using cell decomposition is that the total object may not be amenable to representation, but its cells are. There are many ways of decomposing a solid into constituent cells; none is unique, but all are unambiguous.

Cell decomposition, regularly used in structural analysis, is the basis for finite-element modeling. Advanced structural analysis of solids uses cells represented mathematically as parametric tricubic solids; see Chapter 4.

Spatial occupancy enumeration is a special case of cell decomposition where cells are cubical in shape and located in a fixed spatial grid. As the size of the cube decreases, this method approaches the representation of a solid body as a set of contiguous points in space. Defining a solid using spatial-occupancy enumeration requires a convenient way of representing this set of cubical cells. One way is simply by listing the coordinates of the centers of the cells; a solid object is thus a set of adjacent cells. Cell size determines the maximum resolution of the model.

A. Baer, C. Eastman, and M. Henrion (1979) note two advantages of representing a solid by spatial arrays: It is easy to access a given point, and spatial uniqueness is assured. There are also disadvantages: There is no explicit relationship between the parts of an object, and such schemes demand large amounts of data storage.

A cell in a spatial array is either occupied by a piece of the solid or it is not. A cell can be marked with the binary 1 or 0. Early schemes of this type were highly redundant, since all cells of an object were marked, even though it was highly likely that any particular cells had the same state as cells adjacent to it. Only close to an object's boundary do chances increase for a change of state.

Quadtrees and their three-dimensional analogs, **octrees**, suggest a way of using spatial-occupancy enumeration more efficiently. Let us look at quadtrees first. The quadtree representation of a two-dimensional object is based on recursive subdivision of a square array into quadrants. Each node represents a square region on a plane. In computer graphic applications, this may be the screen plane of the display. Notice that where each node of a binary tree has two descendants, each node of a quadtree has four.

In Fig. 10.20, we superimpose a square on an arbitrary two-dimensional object. If the object does not uniformly cover the square, then subdivide the square into four equal quadrants. If any of the resulting quadrants is full or empty, then no further subdivision of it is necessary, which is true for quadrant 3 in Fig. 10.20. If any of the resulting quadrants is partially full, subdivide it again into quadrants. Continue to subdivide partially full quadrants until the resulting regions are either full or empty or until some predetermined level of **resolution** is reached. The partially full quadrants can be arbitrarily declared full or empty according to a convention.

The root node of the quadtree represents the entire array; it is often called the **universe element**. Leaf nodes represent regions that require no further subdivision; such regions have standard sizes and positions related to powers of 2. The number of nested subdivisions from a given node to the universe element determines the node's level in the quadtree and also the size of the region represented by the node. If the height of the tree is n, then the maximum potential array size is $2^n \times 2^n$. For the example in Fig. 10.20, $n = 3$, and because of the efficiencies of quadtree representation, there are only 33 nodes instead of 64 (that is, $2^3 \times 2^3$). The resolution necessary to accurately represent an object depends on the difference in size of gross and fine features of the object and curvature of the boundary. The greater the magnitude of n, the greater the resolution. The process of reducing an object model into a quadtree representation is also called **quadtree encoding**.

Octree encoding is an extension into three dimensions of quadtree encoding. D. J. Meagher (1982) has developed a solid modeling method based on octree encoding for the high-speed manipulation, analysis, and computer graphic display of solid objects. His approach uses a spatially presorted 8-ary hierarchical tree structure, the octree, to represent solids. It also uses algorithms that grow only linearly with object complexity by taking advantage of the inherent data presorting in the tree structure. The octree encoding of a model proceeds similarly to quadtree encoding. Here, a cubical region is recursively subdivided into octants or eight cubical regions (see Fig. 10.21). Each node of an octree that is not a leaf node has eight descendants. If the height of the tree is n, then the maximum potential array size is $2^n \times 2^n \times 2^n$. For the example in Fig. 10.21, $n = 3$.

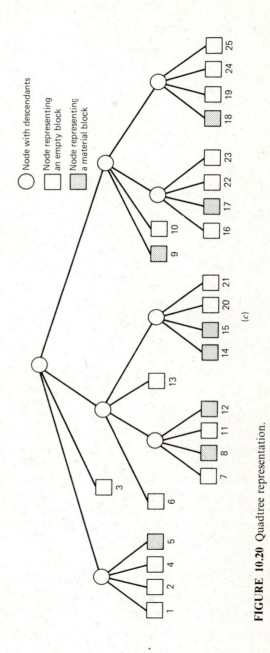

○ Node with descendants

□ Node representing an empty block

▨ Node representing a material block

FIGURE 10.20 Quadtree representation.

453

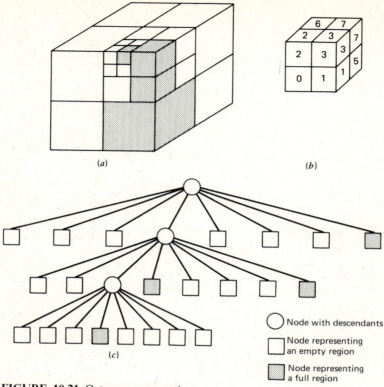

FIGURE 10.21 Octree representation.

Octree and quadtree encoding schemes suggest many interesting possibilities for representing, analyzing, and displaying complex solid objects. All computations on these models are based on integer arithmetic, which means that analysis algorithms are fast and amenable to parallel processing. Meagher has demonstrated algorithms that translate, rotate, and scale octree models, that combine them using Boolean operators, and that compute geometric properties and perform interference analysis. Variations of these schemes may furnish heuristic tools for rapid preprocessing or sorting models based on other techniques.

EXERCISES

1. Show that, for the octree representation of an object, the storage requirements are a function of surface area rather than volume.

2. Find the value of n required for the octree representation of an object whose gross dimensions are on the order of 50 cm and whose fine features are on the order of 0.05 cm. (*Hint*: First find an expression for the resolution r of an octree object. The resolution of an object is the edge size at the lowest level.)

3. Show that 90-degree rotations of an octree representation require only a simple reordering of the nodes.

4. Suggest a simple method for scaling octree models using powers of 2 (that is, binary manipulation).

5. Show how a consistent node-labeling scheme aids in hidden-surface removal for octree models.

6. Define the quadtree representation of a circular region that is accurate to one part in sixteen. Do this for only one quadrant of the universe element. Suggest an algorithm for continuing the construction of this representation to any level of accuracy.

7. Suggest procedures for finding the area and circumference of the model defined in Ex. 6. Show how the accuracy of these properties is related to the resolution.

8. Suggest a method for computing the tangent and normal at any point on the boundary of a quadtree model; of an octree model. How are these related to model resolution?

9. Suggest a method of measuring the complexity of an object; see D. J. Meagher (Aug. 1982) for one approach.

5   SWEEP REPRESENTATIONS

Sweep representations are based on the notion of moving a point, curve, or surface along some path. The locus of points generated by this process defines a one-, two-, or three-dimensional object. Sweep representations for modeling solids are simple to understand and execute yet offer a fertile field for developing new methods.

Sweep representations, used in many contemporary modeling systems, prove to be practical and efficient for modeling constant cross-section mechanical parts. They are also used to detect potential interference between parts of mechanisms. Here, a moving object O_1 collides with a fixed object O_2 if the volume swept by O_1 intersects O_2. A related use is in simulating and analyzing material-removal operations in manufacturing. Here, the volume swept by a tool moving along a predefined path intersects the raw stock for a part. The intersection volume represents material removed from the part.

For solid modeling, two ingredients are required—an object to be moved and a trajectory to move it along. The object can be a curve, surface, or solid, and the trajectory is an analytically definable path. Occasionally, the term **generator** is used to denote the object, and the term **director** is used to denote the trajectory.

Figure 10.22 shows examples of sweep representations. Two principal types of trajectories are depicted—translation and rotation. Notice that the director curve is not necessarily an element of the swept object. Also notice that

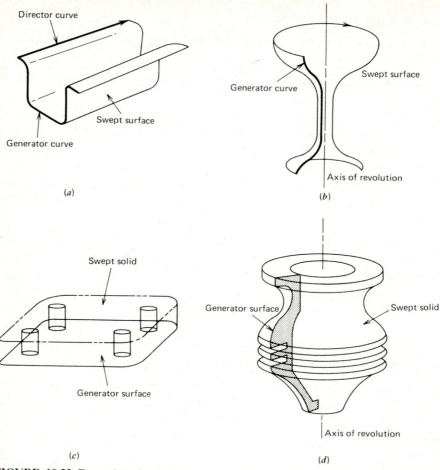

FIGURE 10.22 Examples of sweep representations.

the director of a rotational sweep amounts to an algorithm to move each point in the generator along a circular arc in a plane through the point perpendicular to the axis of rotation and with a radius defined as the perpendicular distance from the point to the axis. In all cases, the shape of the object being swept along does not change. Later, we will investigate more general, nonlinear sweep representations.

There are several obvious and some not-so-obvious ways of creating dimensionally nonhomogeneous objects; see Fig. 10.23. In Fig. 10.23a, the translational sweep of a curve creating a surface also creates two dangling edges. In Fig. 10.23b, two two-dimensional regions are connected by a one-dimensional structure. In Fig. 10.23c and Fig. 10.23d, creating solids using invalid or nonhomogeneous generators results in dimensionally nonhomogeneous solids and ambiguities. The rotational sweep of a generator passing

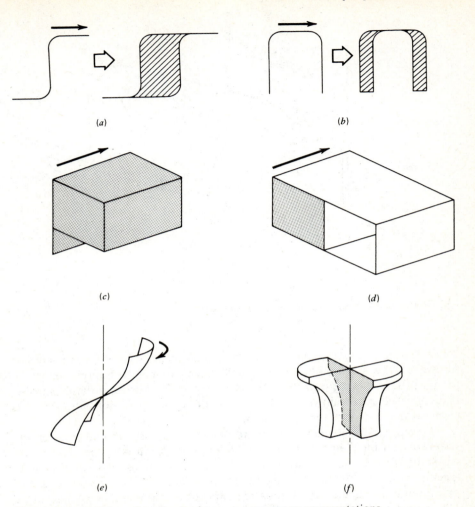

FIGURE 10.23 Dimensionally non-homogeneous sweep representations.

through the axis of rotation, as in Fig. 10.23*e* and Fig. 10.23*f*, produces a surface or solid with a singularity. These conditions produce unacceptable results for many applications, but sometimes these results may be expected and even desired. This is true for modeling systems requiring greater degrees of freedom— computer art, for example. By following criteria suggested in Fig. 10.23 and incorporating them in the design of sweep-representation-generating algorithms, we can create dimensionally homogeneous models.

D. L. Lossing and A. L. Eshleman (1974) developed a powerful technique for representing constant cross-section objects. Their approach emphasizes storing procedures that minimize data storage requirements. Lossing and Eshleman use a six-component trajectory and generator-orientation curve (called a **PD curve**, for *position* and *direction*). A closed planar curve defining

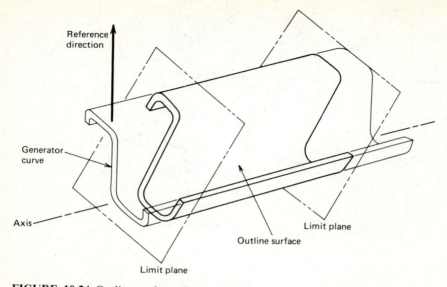

FIGURE 10.24 Outline surface of a constant-cross-section solid.

the cross section of a mechanical part sweeps along an axis to form the **outline surface** of a raw-stock model. This model is then trimmed by limiting planes to produce the bounding surfaces of the part; Fig. 10.24 illustrates these features. If the finished part followed a curved axis, a nonlinear transformation on the model would complete the simulation.

Lossing and Eshleman are able to define an almost unlimited variety of swept solids using their PD curve. They define a PD curve as a general form of a six-component curve, usually a parametric cubic curve that continuously specifies position and an associated direction. The first three components define a continuous parametric cubic equation of position in three-dimensional space. The second three components define a corresponding continuous parametric cubic equation for an associated direction vector. A common parametric variable associates the direction vector with a specific position on the curve. The PD curve in Fig. 10.25 defines a curved and twisting coordinate system.

We construct a local orthogonal system at $\mathbf{p}_i$ as follows: Compute $\mathbf{p}_i^u$, the tangent vector. From $\mathbf{p}_i^u$ and $\mathbf{d}_i$ (remember, $\mathbf{d}_i$ is given by another pc equation), find the orthogonal unit vectors $\mathbf{l}$, $\mathbf{m}$, and $\mathbf{n}$

$$\mathbf{l} = \frac{\mathbf{p}_i^u}{|\mathbf{p}_i^u|}$$

$$\mathbf{m} = \frac{\mathbf{d}_i \times \mathbf{p}_i^u}{|\mathbf{d}_i \times \mathbf{p}_i^u|} \qquad (10.7)$$

$$\mathbf{n} = \mathbf{l} \times \mathbf{m}$$

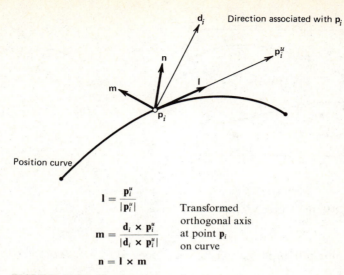

$$l = \frac{\mathbf{p}_i^u}{|\mathbf{p}_i^u|}$$

$$\mathbf{m} = \frac{\mathbf{d}_i \times \mathbf{p}_i^u}{|\mathbf{d}_i \times \mathbf{p}_i^u|}$$

Transformed
orthogonal axis
at point $\mathbf{p}_i$
on curve

$$\mathbf{n} = l \times \mathbf{m}$$

FIGURE 10.25 Characteristics of a pd curve.

The axes l and $\mathbf{n}$ define a direction plane in which $\mathbf{d}_i$ is located. This reference plane changes as the tangent vector and $\mathbf{d}_i$ change continuously along the PD curve. A PD curve defines continuous transformations for points on an outline curve or object to be swept.

Lossing and Eshleman describe how a PD curve can generate outline surfaces. Figure 10.26 illustrates a constant cross-section part that is not only curved but also twists. Figure 10.27 illustrates how to coordinate the two curves comprising a PC curve through a common value of the parametric variable and how to extract the elements. The authors also point out that two or more cross-section curves can be used with one PD curve (for example, a pipe or tube with an inner and outer cross-section curve). They also suggest multiple

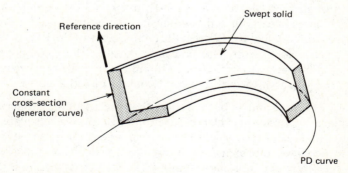

FIGURE 10.26 A constant-cross-section part that curves and twists.

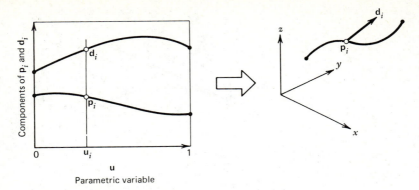

FIGURE 10.27 Components of a pd curve.

cross sections with multiple PD curves to generate a more complex or variable cross section.

Lossing and Eshleman use positive and negative generator curves to create complex cross sections with high variability in the axial direction. Multiple PD curves and associated generator curves create, in effect, half-spaces with directed normals and can be used to add or subtract "material" to a raw-stock model. These modeling procedures depend on the user for model validity.

A six-component curve trajectory for controlling the movement of a generator curve has obvious natural extensions. For example, three more components can be added to generate scale factors to apply to the generator curve, differentially expanding or contracting it and carefully coordinated with the motions imposed by the PD curve. This introduces nonrigid sweep representations, a field in its formative stages, which will not be discussed here.

EXERCISES

1. Define a PD curve where the position component is a parabolic curve segment through $\mathbf{p}_1 = [0\ 0\ 0]$, $\mathbf{p}_2 = [1\ 1\ 0]$, and $\mathbf{p}_3 = [2\ 0\ 0]$, and the direction component varies linearly in u along the curve from the positive z direction to the positive x direction at $\mathbf{p}_3$. Express the components of the PD curve in terms of the geometric coefficients. Use pc curves. Assume the y component of the direction curve is always zero. (*Hint*: Obtain a 4×3 array of geometric coefficients for each curve, the position curve, and the direction curve.)

2. Define a PD curve and an outline curve that sweep out the surfaces of a duct with a square cross section. The cross section must always be square in planes normal to the sweep axis or position curve, where the position curve is a composite pc curve approximating a circular arc in the x, y plane, with its center at the origin and a radius of 4 units. The duct cross section is defined by the four corner points $(0, 4, 0)$, $(0, 4, 1)$, $(0, 5, 1)$, and $(0, 5, 0)$. Use eight pc segments to form the composite curve and four to form the duct cross section.

The parametric variables of the position curve and direction curve, say, u, must be coordinated; parametrize each of these composite curves from 0 to 1. Tabulate the geometric coefficients as three sets of arrays—the position and direction curves each as a 32×3 array and the outline curve as a 16×3 array. Draw a sketch of the key geometric input parameters.

3. Repeat Ex. 2 using Bezier curves.

4. Repeat Ex. 3 using *B*-spline curves.

5. Compare and discuss the input, computations, and output model (conciseness, ease of construction, and so on) for Exs. 2–4.

6. Write a procedure to compute the geometric coefficients of a composite set of bicubic patches generated by a PD curve. Denote this as **PDSURF(NP, ND, NG, PCRV, DCRV, GENCRV, BCSURF)**, where

> NP, ND, NG are the input numbers of pc curve segments in the composite position, direction, and generator curves, respectively;
>
> PCRV(NP, 4, 3) is the input array of geometric coefficients of the position curve;
>
> DCRV(ND, 4, 3) is the input array of geometric coefficients of the direction curve;
>
> GENCRV(NG, 4, 3) is the input array of geometric coefficients of the generator curve; and
>
> BCSURF(NP, NG, 4, 4, 3) is the output array of geometric coefficients of the composite bicubic surface.

6 CONSTRUCTIVE SOLID GEOMETRY

Constructive solid geometry (CSG) is a term for modeling methods that define complex solids as compositions of simpler solids (primitives). Boolean operators are used to execute the composition.

Most of the concepts of CSG, often called **building-block geometry**, were first introduced by Voelcker, Requicha, and others on the Production Automation Project at the University of Rochester. These concepts include regularized Boolean operators, primitives, boundary-evaluation procedures, and point-membership classification. Although we have already discussed many of these concepts, we will review them here in the context of CSG.

A. A. G. Requicha (1980) views CSG as a generalization of cell decomposition. In cell decomposition models, individual cells are combined using a **gluing** operation, a limited form of the union operator where components are joined at only perfectly matched faces. Constructive solid geometry operators are more versatile, since boundaries of joined components (primitives) need not match and interiors need not be disjoint. Furthermore, CSG uses all the regularized Boolean operators: union, difference, and intersect, so that material can be added as well as removed.

Constructive solid geometry representations of objects are ordered binary trees whose leaf or terminal nodes are either primitives or transformation data for rigid-body motions. The nonterminal nodes are either regularized Boolean operators (union, difference, and intersect) or rigid-body motions (translation and/or rotation) that operate on their two subnodes (or subsolids). Each subtree of a node (not a transformation leaf) represents a solid resulting from the combination and transformation operations indicated below it. The root, of course, represents the final object. Let us look at a simple example.

In Fig. 10.28, the four leaf nodes represent the primitives π_1 and π_2 and the translation ΔX. The two internal nodes represent results of the operations $(\pi_1 - \pi_2)$ and $\pi_2[\Delta X]$. The root node represents the final object. Note that the primitive and intermediate objects are valid bounded solids. In addition, transformations are not limited to rigid motions. The full range of scaling and symmetry transformations is theoretically possible. Such transformations are limited by only the capabilities of subsequent Boolean algorithms.

If the primitive elements of a modeling system are valid bounded solids defined by the system and the combining operators are regularized, then the resulting solid models are valid and bounded. However, a modeling system is possible that allows user-defined primitives. Here, either the user or the system must verify the validity of a model.

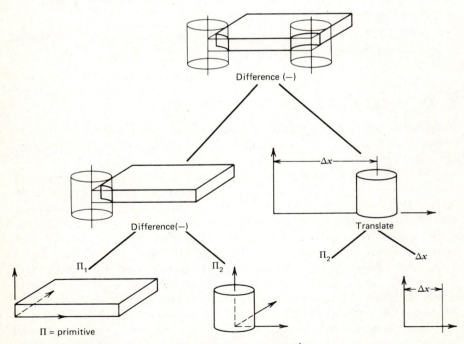

FIGURE 10.28 Constructive solid geometry representation.

The most common approach in contemporary modeling systems is to offer a finite set of concise, compact primitives whose size, shape, position, and orientation are determined by a small set of user-specified parameters. For example, most of these systems provide a block-type primitive, for which the user specifies the length, width, height, and initial position. The modeling system then checks the validity of the parameters (that is, the length, width, and height values must all be positive real numbers). Other common primitives are the cylinder, sphere, cone, and torus. These and other examples are shown in Fig. 10.29 with corresponding parameters (excluding position and orientation) and some of the modeling systems that use them.

The more sophisticated systems can also generate quasi primitives with sweeping- or extruding-type operators (such as GMSolid and ROMULUS).

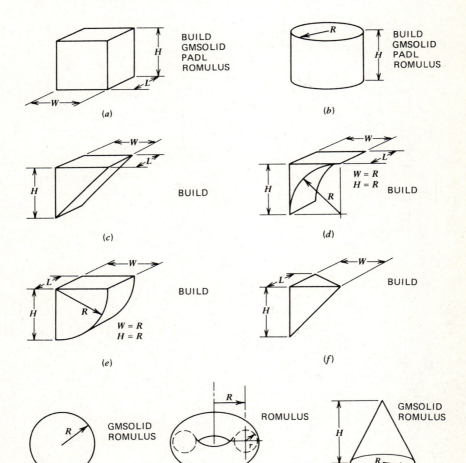

FIGURE 10.29 Primitive solids.

Even here, the system controls validity by checking input parameters. The number of primitives, however, is not a sign of the descriptive power of a modeling system. For example, the block and cylinder alone have the same descriptive power as the primitive set consisting of the block, cylinder, wedge, fillet, cylindrical segment, and tetrahedron if both sets have the same combining and transforming operators.

The primitives themselves are usually represented by the intersection of a set of curved or planar half-spaces. For example, the primitive block is represented by the regularized intersection of six planar half-spaces and the cylinder by the intersection of a cylindrical half-space and two planar half-spaces; see Fig. 10.30. Figure 10.30a shows only three of the six half-spaces defining the primitive block. Arrows indicate the direction of the material and not necessarily

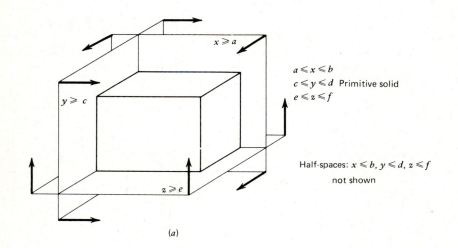

$$a \leqslant x \leqslant b$$
$$c \leqslant y \leqslant d \quad \text{Primitive solid}$$
$$e \leqslant z \leqslant f$$

Half-spaces: $x \leqslant b$, $y \leqslant d$, $z \leqslant f$
not shown

(a)

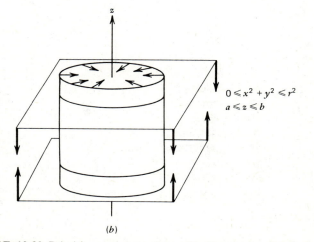

$$0 \leqslant x^2 + y^2 \leqslant r^2$$
$$a \leqslant z \leqslant b$$

(b)

FIGURE 10.30 Primitives as intersections of half-spaces.

the direction of the surface normals. (If normals are part of the representation, they can be assigned during the boundary-evaluation phase.) Purely graph-based primitives are also possible.

The Boolean operators used by CSG systems are the familiar threesome: union, difference, and intersect. However, some modelers include other operators, such as glue, which is a specialized version of the basic three. The complement operator is not used, for in an otherwise regularized scheme, it produces non-regular objects. Also, recall from set theory (Chapter 9, Section 5) and Boolean operators (Chapter 9, Section 6) that union and intersect are commutative (that is, $A \cup B = B \cup A$ and $A \cap B = B \cap A$), but the difference operator is not (that is, $A - B \neq B - A$).

More powerful modeling systems, such as Evans and Sutherland's ROMULUS and General Motors' GMSolid, generate two representations of a solid. The first is the constructive representation exemplified by a binary tree data structure linking primitives and successive subsolids by using combining operators and transformations, as we just discussed. The second is the boundary representation, which describes the faces, edges, and vertices of the boundary of the solid. This description itself has two forms—a topological representation of the connectivity of the boundary elements and numerical data describing the shape geometry and position of these elements. The data defining the object's topology usually point to the geometric shape data.

The boundary representation is computed from the constructive representation by a set of algorithms called the **boundary evaluator**. The boundary evaluator determines where component faces are truncated and new edges and vertices are created or deleted. Where boundary elements overlap or coincide, the evaluator merges them into a single element and thus maintains a consistent, nonredundant data structure representing a solid's boundary.

New edges are created where surfaces of two combined solids intersect. The boundary evaluator finds these intersections and then determines by set-membership classification which segments of the intersection are actual edges of the new solid. The new edges are terminated by new vertices, that is, where the edge intersects a surface. Again, the vertices must be classified. Let us look a little deeper into some of these concepts of boundary evaluation; we will use the example in Fig. 10.31.

First, intersect each surface of A with each surface of B; this operation assumes that both surfaces are unbounded. In our example, there are 36 intersection combinations, with only ten producing intersection edges. (Note that simple tests or heuristic techniques can dramatically reduce the number of candidates for intersection.) Thus, in Fig. 10.31, edge e_{ab} is produced by intersecting face a and face b. J. W. Boyse (1979) suggests listing the edges produced this way as tentative edges, or **t-edges**, for reasons that will soon be apparent. Consider the list of t-edges as a superset of the actual edges of the new solid C. Since the intersecting pair of surfaces defining face a and face b are unbounded, then so is t-edge e_{ab}.

Next, intersect all t-edges with all the faces to produce points (potential vertices) that segment the t-edges. Thus, e_{ab} is divided into three segments by two

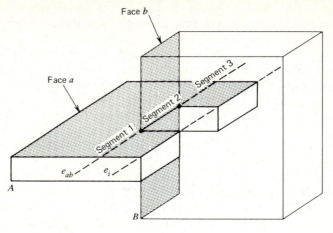

Face b

Face a

Segment 3

Segment 2

Segment 1

e_{ab} e_i

A

B

C = A − B

FIGURE 10.31 Boundary evaluation.

of the vertical plane faces of B. Now, classify these segments as outside, inside, or on the boundary of C. Only segments on the boundary are real edges.

Classify segments in steps: First classify each t-edge segment with respect to each primitive. Then use the classifications instead of the primitives, processing them through the model's binary tree and reclassifying them at each successive node. At the root node, the segments of t-edges on the boundary are real edges of C.

University of Rochester researchers found that we need to know which points near each segment of a t-edge are inside the solid and which are outside. Knowing only that a segment is on a primitive is inadequate for classifying it at higher levels in a model's binary tree. The researchers propose using a **neighborhood model**, consisting of points close to the segment, to indicate which points are inside and which ones are outside the solid.

Figure 10.32 shows a neighborhood model of segment 2 of e_{ab}. Here, a small disk symbolizes the neighborhood, and the shaded area indicates points inside the solid. Our perspective is from an end of the segment. The neighborhood model is defined by the faces that bound the model at this edge. We use neighborhood models of two primitives (or subsolids) and the corresponding operator to combine these neighborhoods to determine if the segment is outside, inside, or on the next highest level of the tree. More examples are presented in Fig. 10.33. In other words, create a neighborhood at any node in the CSG representation tree by applying the indicated operator to the neighborhoods of the two subnodes.

We can now make some generalizations. First, faces of a new solid are a subset of the faces of the combining solids. We can modify but not create faces (unless we admit sweep operators). However, we can create new edges and vertices, and we can delete any element type.

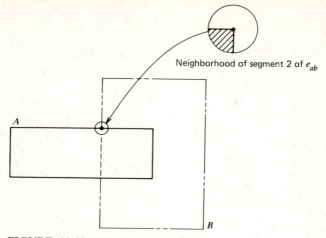

FIGURE 10.32 Neighborhood model.

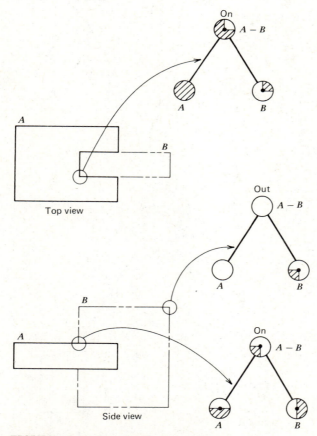

FIGURE 10.33 Combining neighborhood models.

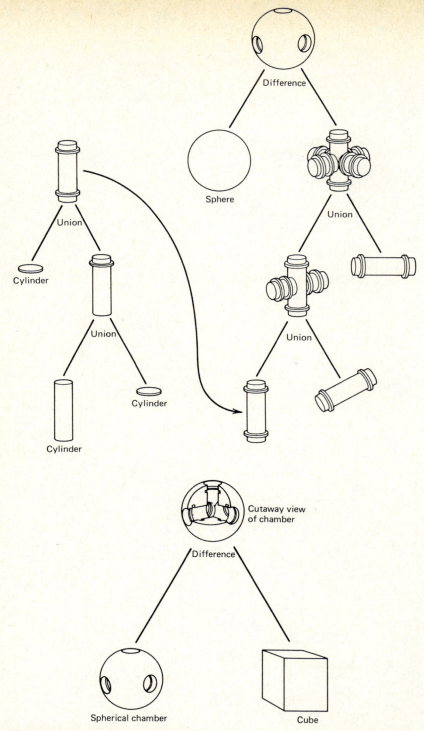

FIGURE 10.34 A ROMULUS-created solid.

In Section 7 (boundary representations) more concepts related to the boundary-evaluation problem are discussed. A review of the sections on set-membership classification and Boolean operators will be helpful.

We have seen that the constructive representation of a complex solid is compact and can be generated rapidly. This is true of the solid created by the ROMULUS system in Fig. 10.34. Extensive computations are necessary to generate the boundary representation from a constructive representation, but the former contains a wealth of useful information for computing global properties, graphic display models, and other applications. The constructive form is often called a **procedural** or **unevaluated** model, and the boundary form is often called an **evaluated** model. The most versatile and powerful modeling systems use both.

Section 9 reviews the two important modeling systems mentioned here, ROMULUS and GMSolid.

EXERCISES

1. Consider Fig. 10.31. Describe the neighborhood model of each segment of t-edge e_1.

2. Describe how to create a circular segment, such as the one shown in Fig. 10.29, using the block and cylinder and the three standard Boolean operators. How is a fillet created? A tetrahedron?

3. Is it possible to create concave primitive solids with a single set of summed intersections of half-spaces? (That is, can π_i be concave if $\pi_i = \bigcap_{j=1}^{n} H_{ij}$, where H_{ij} is a half-space?) Explain your answer.

4. Devise a boundary-evaluator algorithm for a glue operator. Remember, a glue operator is a special case of the union operator. In gluing two objects, only faces come in contact; the respective interiors are disjoint. Use the example of two blocks—a small one resting on a larger one.

5. Give examples of solids that cannot be created by using the primitives in Fig. 10.29. What is required to model these examples?

7 BOUNDARY REPRESENTATIONS

The boundary of a solid separates points inside from points outside the solid. The boundary is the primary interface between the solid and the surrounding environment. Reflection of light from a solid and, therefore, the solid's appearance are determined by the surface properties shape, color, and texture. Even the bounding surface of transparent objects influences light reflection. The bounding surface of a solid is where it contacts other solids. Manufacturing processes are concerned with the interaction of the surface envelope of a tool path with the surface of a part's raw-stock shape.

Boundary representations of solids must satisfy certain conditions. In Chapter 9, Section 9, we discussed conditions for a well-formed surface (closed, orientable, non-self-intersecting, bounding, and connected). We distinguished between the boundary of an object (including two-dimensional objects and objects of arbitrary dimensions) and its interior points in Chapter 9, Sections 4 and 5. Section 6 of Chapter 10 introduced a boundary evaluator for CSG representations. Thus, we have most of the ingredients for boundary representations.

Let us refine our notation of the bounding surface of a solid and then investigate a specific type of boundary representation. It is computationally convenient for us to segment the boundary surface of a solid into faces or patches, with each face or patch in turn bounded by a set of edges and vertices. A. A. G. Requicha (1980) observes that faces are representational artifacts that do not always correspond to our intuition. He suggests that faces satisfy the following minimal set of conditions:

1. A finite number of faces defines the boundary of a solid.
2. A face of a solid is a subset of the solid's boundary.
3. The union of all faces of an object defines its boundary.
4. A face is itself a subset or limited region of some more extensive surface.
5. A face must have a finite area and be dimensionally homogeneous.

These conditions are illustrated by the boundary surface of the simple cube and cylinder in Fig. 10.35. Both objects have a finite number of faces. The cube's boundary is conveniently segmented into six planar faces, defined by the discontinuities at the edges coinciding with our intuition. The cylinder's boundary is more arbitrarily segmented, with the end planar faces having the only natural segmentation and the cylindrical surface requiring a division into faces in any number of ways. (That is, assuming segmentation by vertical edges, we can have any number of faces of equal or unequal size.)

No single face of either the cube or the cylinder defines the entire object, and, therefore, every face is a subset of the complete boundary. Each face is also a subset of limited (bounded) region of a more extensive surface. Thus, each face of the cube is a bounded region of an infinite plane, and so are the circular disks or end faces of the cylinder. How about the cylindrical surface itself? Each face of the cube and cylinder has a finite area and is dimensionally homogeneous. (There are no dangling edges or isolated points.)

We can represent the planar faces, in turn, by their bounding edges. However, curved faces require more information. For example, the curved face may be a region of a Bezier surface, where this surface is defined by its geometric coefficients (that is, the points defining the characteristic polyhedron), and the curve on this surface that delimits the actual face is similarly defined (with its characteristic polygon). There are computational devices or conventions to indicate on which side of an edge or delimiting curve the face lies. We have investigated several, and Fig. 10.36 illustrates one such convention. Here, the face-bounding curve is parametrized in a consistent direction so that the vector $\mathbf{n} \times \mathbf{t}$ points to the face side of the curve.

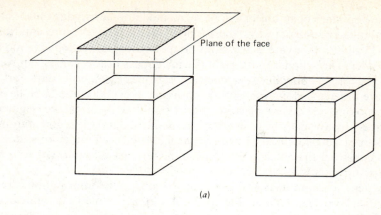

Plane of the face

(a)

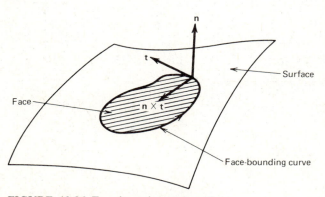

(b)

FIGURE 10.35 Faces defining the boundary of a solid.

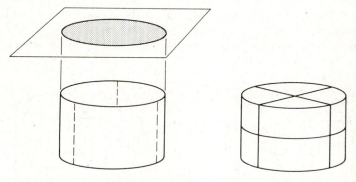

n

t

Surface

Face

n X t

Face-bounding curve

FIGURE 10.36 Face boundary convention.

The boundary of an object can be segmented into faces, edges, and vertices in an unlimited number of ways; there is no unique boundary representation. As Fig. 10.37 shows, there are many valid boundary representations for the cube and cylinder. The cylinder to the right of each cube is its topological equivalent, having the same boundary-segmentation scheme. Geometric data for each object are different, of course. Strictly speaking, a cylinder and cube have the same topology regardless of how their bounding surfaces are segmented. Sometimes, to avoid confusion when discussing boundary segmentation, we use the term **combinatorial structure** (particularly when Euler operators are used to construct the object). We can use the term **metric information** to mean geometric information.

In general, combinatorial structure and metric information are not independent. For example, in Fig. 10.38, if vertex 1 is made to coincide with vertex 2, then the object is no longer valid. We conclude that a valid combinatorial or topological structure does not in itself guarantee a valid object. Validity conditions must be established for both metric and combinatorial data. These conditions depend on general characteristics of the representation

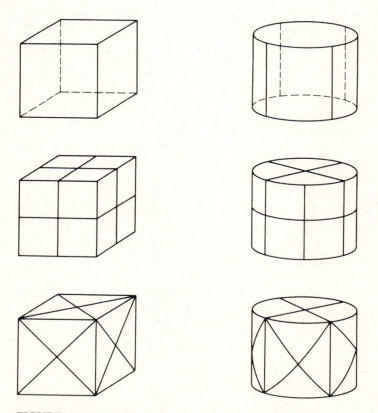

FIGURE 10.37 Boundary representations are not unique.

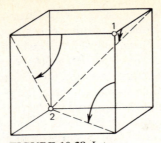

FIGURE 10.38 Interdependence of topology and geometry.

allowed by the construction processes (for example, many schemes are limited to faces defined by three edges).

There are powerful boundary-representation schemes where validity must be determined by means other than computing face, edge, and vertex relationships. Ordinarily, these schemes describe a solid as the union of very general faces embedded in extended surfaces, where the solid's edges are defined by the intersections of these surfaces. These schemes are not true directed half-spaces, since their net Boolean intersection is usually not related to the object model and surface behavior external to, and remote from, the face is not important. We can construct such models in two different ways: Assemble surfaces to create solid primitives that then, as in CSG schemes, combine to create more complex shapes, or omit the primitive phase altogether and construct complex shapes directly by assembling and intersecting appropriate surfaces. Let us look at some two- and three-dimensional examples.

The shape of a complex object is implied by

$$S = A \cup B \tag{10.7}$$

In boundary-representation schemes, this means that the bounding surface of S consists of a set of appropriate regions of the bounding surfaces of primitives A and B. These regions are called **active regions** and denoted by a_A and a_B. We determine the active regions with two conditions: First, the regions are bounded by the intersection of the boundary surfaces of A and B, $bA \cap bB$. In the two-dimensional example in Fig. 10.39, the intersection consists of points 1 and 2, which are obviously independent of the type of combining operation.

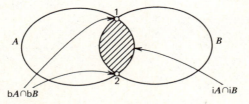

FIGURE 10.39 Boundary intersection.

Second, we select the active region of a specific primitive based on its being inside, outside, or on the boundary of other combining primitives.

Figure 10.40 gives an example of three primitives, and the combining sequence represented by the binary tree reduces to the linear sequence $A \cup B - C$. In Fig. 10.40a, each primitive is in position for performing the Boolean operations. Since this is a two-dimensional example, the primitives are closed parametric curves. The zero point of the parametric variable is shown along with an arrow indicating the direction of parametrization. Counterclockwise is the positive direction of parametrization for curve loops enclosing a "solid" region. Curve loops parametrized clockwise enclose or define holes.

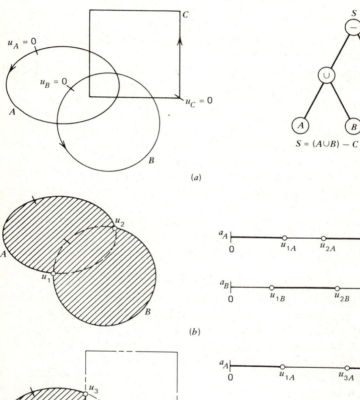

(a)

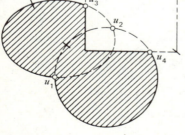

(b)

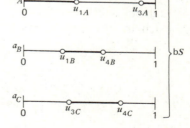

(c)

FIGURE 10.40 Two-dimensional boundary representation.

In Fig. 10.40*b*, find the union of A and B in two steps. First, find the intersection of the two bounding curves of A and B, points u_1 and u_2

$$bA \cap bB = \{u_1, u_2\} \qquad (10.8)$$

This step divides both curves into two segments. For the union operation, the active region of bA is that segment of bA lying outside B. The active region of bB is that segment of bB lying outside A. To the right in Fig. 10.40*b* are active regions on the unit parametric intervals of the respective boundary curves. The active region on bA, denoted a_A, consists of two intervals—one from 0 to u_{1A} and the other from u_{2A} to 1. There is one active interval a_B on bB—from u_{1B} to u_{2B}.

In Fig. 10.40*c*, complete the shape definition by combining the results of $A \cup B$ by taking the Boolean difference with C. Now find the active regions on bC as well as revised regions on bA and bB. Computing intersections is, again, the first step. In doing this, find only the intersection of bC with the active regions of bA and bB. For our example,

$$(bC \cap a_A) \cup (bC \cap a_B) = \{u_3, u_4\} \qquad (10.9)$$

Once u_3 and u_4 are found, select the proper interval a_C on bC and determine the effect on a_A and a_B. First, since this is a difference operation, a_C is that part of bC that is inside the result of $A \cup B$. Next, modify a_A and a_B by retaining only those portions lying outside C. The effect on a_A, a_B, and a_C is shown by the bold sections on the unit parametric intervals at the right in Fig. 10.40*c*. Notice that the union of the active intervals after any combining operation always defines one or more closed loops.

A similar situation occurs with three-dimensional objects. The active surface regions (faces) on all the primitives combined to define a solid will define a closed surface. Let us look at an example in Fig. 10.41. The shape consists of the union of a sphere with a skew-truncated cylinder. The cylinder is denoted by π_1 and the sphere by π_2.

Observe that the cylinder is defined by the concatenation of three surfaces whose mutual curves of intersection delimit face regions. The union of these face regions defines the boundary surface of the cylinder. Notice that points bound active regions on curves, and curves bound active regions on surfaces.

The active regions of the surfaces in this figure are shaded. The primitive π_1 (cylinder) is decomposed into its constituent surfaces, and each of these is mapped onto a unit square in parametric space. Their respective active regions are outlined. (Remember that loops defining holes are parameterized clockwise.) The primitive π_2 (sphere) is similarly shown.

The intersection of bπ_1 and bπ_2 is a closed space curve, which appears twice in the parametric space representations. First, it is associated with π_2 or surface D; second, it is associated with constituent surface A of π_1. In each case, the curve is a function of a third parametric variable t. The functions of t map the curve into the unit square in u, w space from the unit interval in t space.

Many unresolved issues remain in boundary-representation schemes. Chief among these is the question of validity: Is the representation closed,

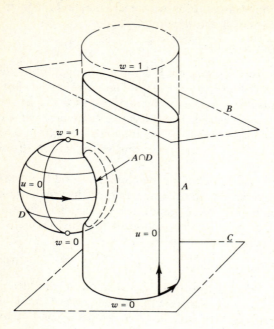

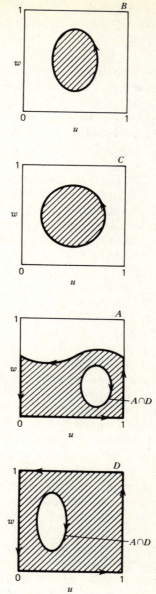

FIGURE 10.41 Three-dimensional boundary representation.

dimensionally homogeneous, oriented, and so on? Requicha suggests that in designing a boundary-representation scheme, we must consider the following questions:

1. What is a face?
2. How is a face represented?
3. What is an edge?
4. How are edges represented?

These issues are both resolved and made more complex by hybrid representation schemes. Both GMSolid and ROMULUS contain elements of constructive solid geometry and boundary representation, maintaining both types of models. Others, such as the McDonnell Douglas CADD system, have done boundary representations of solids with a wireframe modeler, which we explore in Section 8.

EXERCISES

1. Consider the interdependence of topology and geometry in Fig. 10.38. What test algorithm would you propose to detect and proscribe the situation that produces the invalid solid? Does your proposal work for the case of Fig. 10.42?

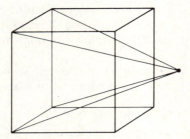

FIGURE 10.42 Invalid solid.

2. How might you computationally check the validity of the object in Fig. 10.41? How about such criteria as a finite volume or surface area or edges adjacent to two faces? Discuss these and others you may think of.

3. Develop the thesis that a boundary representation can be converted into an octree representation for more efficient preliminary computation of Boolean operations, hidden surfaces, and so on.

8 WIREFRAME SYSTEMS

Wireframe models represent the edges of an object and consist entirely of points, lines, and curves. Wireframe systems were developed in the early 1960s to automate design drafting. The first systems were only two-dimensional, with the user constructing a model point by point and line by line. Points, straight lines, circular arcs, and some conic curves were the basic elements. The more sophisticated systems, among them the first generation of McDonnell Douglas's CADD system, allowed the user to interrogate the model, and the modeling system responded with elementary geometric properties. These wireframe models did not represent solids analytically. The user had the burden of interpreting and ascribing solid properties to the model.

The first step toward a wireframe solid model was the introduction of three-dimensional constructions. Wireframe systems were still limited to the same point, line, and curve elements of the two-dimensional systems, but the model now had depth and was subject to three-dimensional translations and rotations, yielding a greater illusion of solidity. This relieved the user of some of the interpretive burden, but automatic volumetric analysis was still limited or nonexistent.

Some systems introduced planes and other simple surfaces, such as spheres and cylinders. The CADD system, which uses the bicubic patch, was among the first to add sculptured surface modeling capability. Within their limits, these systems became powerful modelers and left little to be desired when solving design and manufacturing descriptive-geometry problems. There were applications to other areas, too. A three-dimensional wireframe model was the ideal input format for finite-element analysis or numerical-control part programming in manufacturing. Printed circuit board design, and all that followed from this amazing technology, would have been severely handicapped without the wireframe model and computer graphic display. In fact, parallel and mutually reinforcing developments can be traced for wireframe modeling and computer graphics systems.

When we attempt to use pure wireframe representations to model three-dimensional solids, we are confronted with four problems:

1. The possibility of creating ambiguous models.

2. The possibility of creating nonsense objects.

3. The lack of graphic or visual coherence. (Profile lines or silhouettes are not usually derivable from the model.)

4. The possibility that the wireframe model approximating a solid is "verbose."

Many wireframe models are not ambiguous in representing solids, such as **two-and-a-half-dimension** models; see Fig. 10.43 for examples. It is difficult to misinterpret any of these examples; even Fig. 10.43d, with a through hole within a depression or cavity, can be interpreted only one way. Figure 1.1 is the classic example of an ambiguous wireframe model. What is not easy to determine is which class of potential objects is larger or which is more likely to occur. Similar observations apply to nonsense objects, as in Fig. 1.2.

The example in Fig. 1.2a is topologically appropriate for a solid, but, as we discussed in Section 7, its metric properties can invalidate a solid representation. Another type of nonsense object is possible: If one of the edges of the frame in Fig. 10.43a is deleted, the result is a nonsense object topologically.

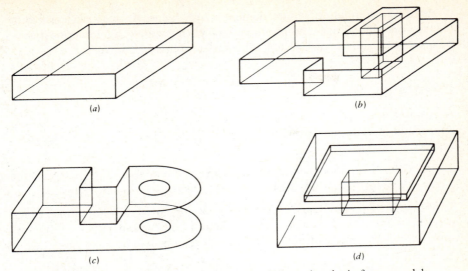

FIGURE 10.43 Non-ambiguous two-and-one-half-dimensional wireframe models.

A third shortcoming of wireframe representations is their lack of information for determining profile or silhouette lines. We saw this earlier in Fig. 1.3 and again in Fig. 10.43c. This deficiency affects not only graphic displays, but is especially critical in five-axes tool path computations or interference-detection studies.

The fourth problem concerns the amount of defining data we must specify. Wireframe models require us to supply "true" solid modelers; however, this requirement is not entirely clear. Take the example of a simple rectangular block. Using a solid modeler, we must specify length, width, and height as well as position and orientation. Using a wireframe modeler, we must specify at most the coordinates of the eight vertices and their edge connectivity. With a solid modeler, the representation may record an unevaluated model, and we must use a boundary evaluator. With the wireframe modeler, we create both the topological and the geometric data structures. For simple objects, the advantages and disadvantages of wireframe versus solid modeler are not so important as for complex object shapes. As a shape becomes more complex, the level of detail necessary increases rapidly.

Some modeling systems, such as CADD, are evolving boundary-representation schemes based on an underlying wireframe system. An approach by G. Markowsky and M. A. Wesley (1980) is an algorithm to discover all solid objects with a given wireframe. It is then up to us to select the appropriate solid objects from the candidates offered.

Wireframe representations will be useful in one capacity or another well into the foreseeable future, extending to solid modeling for limited classes of objects. Research in this area must address not only these four representational problems, but applications to mass-property calculation, cross sectioning, hidden-line elimination, finite-element mesh generation, surface-attribute association and storage, and multiaxis tool path generation and verification—all areas where current wireframe schemes are weak or inadequate.

EXERCISES

1. Apply Euler's law to each of the wireframe models in Fig. 10.43. Modify any of the models, as necessary, to satisfy Euler's law by adding edges and vertices.

2. Use CSG methods to represent the wireframe models in Fig. 10.43. Sketch a binary tree for each, showing primitives, transformations, and operators.

3. Show that sweeping each of the three orthogonal views of a conventional wireframe drawing of an object in a direction perpendicular to its plane and intersecting the three extruded solids is a close approximation of the intended solid.

9   REVIEW OF TWO SOLID MODELING SYSTEMS

Section 9 is a review of two contemporary solid modeling systems, ROMULUS and GMSolid. ROMULUS is a commercially available system developed by Shape Data Ltd., a subsidiary of Evans and Sutherland. GMSolid is a proprietary in-house system developed and used by General Motors Corporation. It was influenced by work of the University of Rochester's Production Automation Project, and ROMULUS by Braid's BUILD-1 and BUILD-2 system at the computer laboratory at Cambridge University.

Both of these systems use many of the concepts discussed in previous sections and Chapter 9 but in slightly different ways. We will take a broad view of how designers of these systems combine these concepts into effective geometric-modeling systems.

9.1   GMSolid

Users of GMSolid work directly with computer representations of solids rather than the points, lines, and curves of conventional wireframe CAD systems. GMSolid allows the user both to create and combine solids. A standard set of primitives is available, including block, cylinder, cone, and sphere. They are sized, positioned, and combined with other primitives to create more complex solids. Union, intersection, and difference are the combining operators. Since the primitives are valid solids and the combining operators are regularized, all solids the system produces are guaranteed to be valid.

When a GMSolid set operation combines two solids, it creates two representations. One is a constructive CSG representation linking the two original solids. This representation is a binary tree with primitive solids at the leaf nodes and set operations at all internal nodes and the root. A name is assigned to the solid produced at each node of the tree. The primitives at the leaf nodes are specified by both a type (block, cylinder, sphere, and so on) and a transformation matrix that sizes and positions the primitive. This CSG representation is a direct translation of the combining operations performed by the user into a binary tree structure.

GMSolid also creates a boundary representation, which records the metric and connectivity data of the surface elements, the faces, edges, and vertices as well as a directional parameter for each face defining inside and outside. A boundary-evaluator subsystem generates this data structure for combined solids.

The boundary evaluator determines where new edges and vertices must be created and existing ones deleted. Since edges on a solid occur at the intersection of two surfaces, the boundary evaluator finds tentative intersections that are then classified to determine which are actual edges of the solid. A similar intersection and classification process applies to edges and surfaces for finding new vertices. The boundary representation created by the boundary evaluator is then complete enough to compute an arbitrary point's location relative to the solid (that is, inside, outside, or on the surface).

J. W. Boyse and J. E. Gilchrist (1982) of General Motors report (Appendix B) that the constructive representation of a solid is very compact and quickly generated when two solids are combined. This is in contrast to the extensive computation required to generate a boundary representation. The boundary representation, however, contains useful data not immediately available in the CSG representation. By providing both forms of representation, GMSolid provides the best attributes of each.

9.2 ROMULUS

ROMULUS is a solid modeling system for building, maintaining, and accessing unambiguous geometric-shape information about physical objects. It stores a boundary representation of the solid model. ROMULUS represents the solid in terms of the solid's spatial boundary (enclosing surface) and indicates which side of the surface is solid. It thus represents a solid body as a union of faces, which are, in turn, represented as a loop of vertices with edges strung between them, and stores in the data base the mathematical data of the surface on which a face lies, the curve geometry on which an edge lies, and the point geometry of a vertex.

ROMULUS generates and stores separate topological and geometric models. The topological model contains the connectivity relationships between the geometric elements. These relationships include element bounds, for example, the edges surrounding a face or the vertices bounding an edge. The geometric model contains physical dimensions of the elements and their location in space. Except for vertices, the geometric definitions are unbounded.

Topological consistency is ensured by using Euler operators and applying Euler's rules. Examples of these operators are

MBFV: make body, face, and vertex

MEV: make edge and vertex

MEF: make edge and face

KEV: kill edge and vertex

and the appropriate Euler rules are

$$V - E + F = 2$$

and

$$V - E + F - H = 2(B - P)$$

where the familiar definitions apply.

The geometric elements available in ROMULUS are: the straight line, circle, ellipse, intersection curves, plane, cylinder, cone, sphere, and torus. There are several methods for building complex solids: three-dimensional edge-based procedures; linear sweeping; rotational sweeping; Boolean operations (union, intersection, and difference); and blending and chamfering. ROMULUS computes the following analytical properties: surface area, volume, center of gravity, the principal axes, and moments of inertia around any axis.

ROMULUS models complex solids in a general manner, independent of any application. The explicit data base is construction-path independent. Whether a cylinder is constructed by using the "cylinder" command, sweeping a circular face along an axis, or rotationally sweeping a straight line around an axis, the data base representation is identical. Thus, it is possible to solve the same modeling problem in many different ways with the same unambiguous result, providing great freedom in selecting a modeling approach.

ROMULUS checks model validity after nearly all modeling operations, particularly after those that alter the geometry or topology of the body. It checks Euler relationships after such commands as "combine," "create," "difference," "intersect," "merge," "unite," and "subtract." Some commands do not require model verification after their execution, such as "copy," "move," and "rotate."

Figures 10.44a and 10.44b are computer graphics displays of a wheel hub designed using ROMULUS. All hidden lines are displayed in 10.44a and removed in 10.44b. The high resolution and high vector display of the PS 330 make it an ideal tool for design and analysis; Fig. 10.45 is a photograph of the finished product. Figures 10.46a and 10.46b are computer graphics displays of casting molds, also designed using ROMULUS. These figures were photographed directly from the color raster display of a PS 340 high-performance computer graphics system. Appendix C presents a more detailed description of ROMULUS's design system and includes more examples of its capabilities.

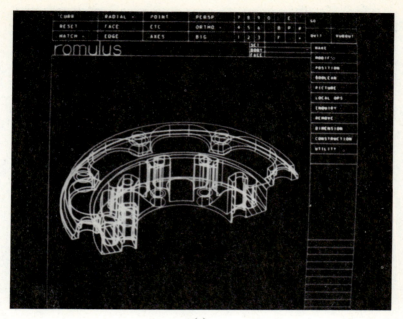

(a)

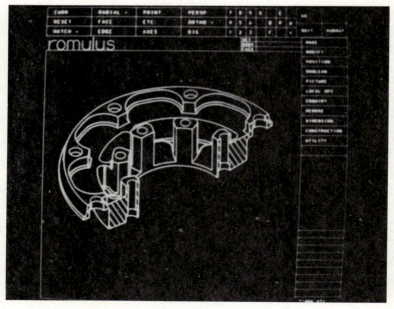

(b)

FIGURE 10.44 Display of wheel hub design using ROMULUS. (*Courtesy of Evans & Sutherland.*)

FIGURE 10.45 Wheel hub: the finished product. (*Courtesy of Evans & Sutherland.*)

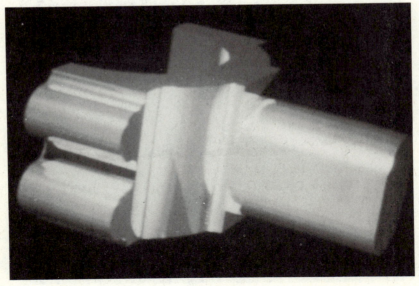

(a)

FIGURE 10.46 Color-shaded display of casting mold. (*Courtesy of Evans & Sutherland and Purdue University.*)

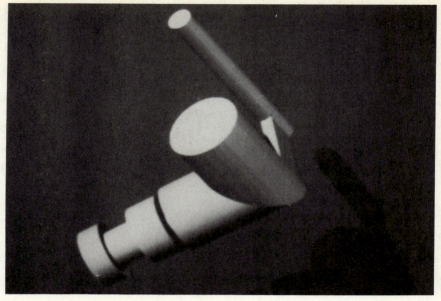

(b)

FIGURE 10.46 (*Continued*) Color-shaded display of casting mold. (*Courtesy of Evans & Sutherland and Purdue University.*)

11

GLOBAL PROPERTIES
OF SOLID MODELS

Global properties of solid models refer to volume, surface area, moment of inertia, center of gravity, and so on. In Chapter 5, Section 5, we investigated global properties of curves and surfaces, such as arc length and area and the volume of simple solids, by direct integration. This method is not effective, even if possible, for determining the global properties of complex solids, but other methods are available.

Computing global properties of a solid requires evaluating the triple integral

$$\phi = \int_{\text{Solid}} f(\mathbf{p}) \, dV \tag{11.1}$$

where ϕ is the property required, $f(\mathbf{p})$ is a vector function describing ϕ, and integration is over the entire volume of the solid. Certain theorems are useful for relating integrals of different types (line, surface, and volume). The two of interest to us are **Gauss's theorem** (also called the **divergence theorem**), which relates an integral over a closed surface to the integral over the corresponding enclosed volume, and **Green's theorem**, which relates a closed-path integral to the surface integral over the enclosed region. We will use these theorems particularly in the Timmer–Stern method (Section 2), but we will not explore their development, since it is treated in many calculus texts.

1 REPRESENTATION-DEPENDENT METHODS

Y. T. Lee and A. A. G. Requicha (1982) see a natural association between characteristic properties of a representation scheme and algorithms for computing the global properties of models produced by the scheme. Let us look at some examples they cite to support this important concept.

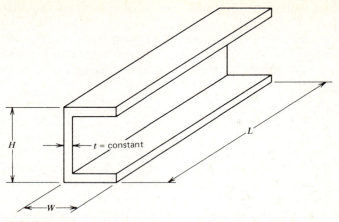

FIGURE 11.1 A parametric family of channel section shapes.

Instances and parametrized shapes (Chapter 10, Section 3) or primitive-instancing modeling schemes define families of objects using a finite number of parameters. Assigning a value to each of the parameters defines a member of the family. The global properties of a family are easily reduced to a special formula. Consider the channel section shape in Fig. 11.1. The volume of any instance of the family is

$$\text{Volume} = HWL - (H - 2t)(W - t)L \tag{11.2}$$

which simplifies to: $V = (2W + H - 2t)tL$. We can find similar formulas for other global properties of this shape.

A library of subroutines must contain the formula for each global property of each parametric family. The size of this library, of course, increases directly as the number and complexity of families increases. This is not necessarily a disadvantage, because these formulas are easily computed, whereas we will later discover that more general methods are often computationally costly. The only disadvantage here is the limited number of shapes capable of being represented.

We can analyze cell decomposition and spatial-occupancy enumeration (Chapter 10, Section 4) by the quasi-disjoint composition of such solids, expressed as

$$\text{Solids} = \bigcup_i \text{Cells}_i \tag{11.3}$$

so that any integral taken over the solid decomposes into a sum of integrals

$$\int_{\text{Solid}} f \, dV = \sum_i \int_{\text{Cell}_i} f \, dV \tag{11.4}$$

This summation is valid, since the cells have disjoint interiors. Each integral must be evaluated, which might require a special method—either simple or complex, depending on the nature of the cell. The simplest forms are the constant-sized cubical cells of spatial enumeration or the variable-sized cubical cells of octree compositions.

Consider evaluating the cellular decomposition of a composite tricubic solid. Here, the volume or triple integral of the deformed cells in object space or Cartesian coordinates are related to corresponding unit cells in parametric form as follows:

$$\iiint_{\text{Cell}} f \, ds \, dy \, dz = \iiint_{\text{Unit Cell}} f J \, du \, dv \, dw \tag{11.5}$$

where J denotes the Jacobian transformation.

Translational and rotational sweep representations are another example of a natural relationship between form and analysis. The path or trajectory geometry may be factored out, reducing the triple integral to a double integral. The integral may be further simplified, since the planar cross section that is translated or rotated is easily represented by its edges.

Boundary representations of solids naturally support evaluation of global properties by surface integrals, using either direct integration or Gauss's theorem. Certain simplifications are required to use direct integration. (See Apostol or another standard calculus text.) However, Gauss's theorem offers a more powerful approach. Let $\mathbf{p}$ be a continuous vector function, then

$$\int_{\text{Volume}} \nabla \cdot \mathbf{p} \, dV = \oint \mathbf{p} \cdot \mathbf{n} \, dA \tag{11.6}$$

where $\mathbf{n}$ is the unit surface normal. Take the integral on the right over the entire boundary surface of the volume V. Since a boundary representation usually consists of the union of several quasi-disjoint faces, F_i, write this integral as

$$\oint_{\text{Surface}} \mathbf{p} \cdot \mathbf{n} \, dA = \sum \int_{F_i} \mathbf{p} \cdot \mathbf{n}_i \, dF_i \tag{11.7}$$

H. G. Timmer and J. M. Stern (1980) use this theorem as well as Green's theorem to analyze solids whose bounding surfaces are subsets of parametrically defined patches. The Timmer–Stern method evaluates the contribution of each face by numerically integrating the line integral of the curve bounding the active portion of each patch, after transforming the dF_i variables. In Section 2, we explore this method in more detail.

Lee and Requicha suggest a "divide-and-conquer" strategy as the natural method for computing integral properties of CSG solids. An algorithm for this strategy consists of the following recursively applied formulas:

$$\int_{A \cup B} f \, dV = \int_A f \, dV + \int_B f \, dV - \int_{A \cap B} f \, dV$$

$$\int_{A - B} f \, dV = \int_A f \, dV - \int_{A \cap B} f \, dV \tag{11.8}$$

The Boolean operators are regularized set operators; Fig. 11.2 gives an example. The solid D results from the union of objects A and B, followed by the

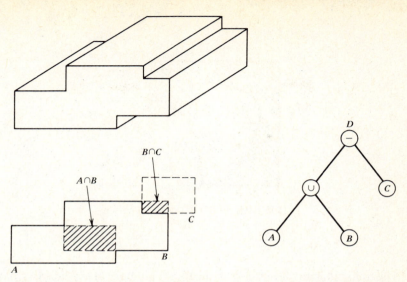

FIGURE 11.2 Divide-and-conquer strategy for computing global properties of CSG solids.

difference of C; I denotes the integral property. Applying the algorithm in Eq. 11.8, we obtain

$$I_D = I_A + I_B - I_{A \cap B} - I_{A \cap C} - I_{B \cap C} + I_{A \cap B \cap C} \qquad (11.9)$$

Notice, however, that $I_{A \cap C}$ and $I_{A \cap B \cap C}$ are null sets; therefore,

$$I_D = I_A + I_B - I_{A \cap B} - I_{B \cap C} \qquad (11.10)$$

Lee and Requicha point out that in the worst possible case, the number of integrals on the right side of Eq. 11.10 grows exponentially. This occurs when each primitive interacts with all or most other primitives in the construction. Most manufactured parts do not present this problem, so most interactions produce null results. However, from analyzing such a program at the University of Rochester, the authors found that the number of nonnull terms may still be too large to be "computationally attractive." Despite these and other drawbacks, it is too soon to abandon a method with such strong "natural" ties to the CSG representation scheme. Parallel processing techniques or new procedures may support its further implementation.

2   THE TIMMER–STERN METHOD

H. G. Timmer and J. M. Stern (1980) propose computing global properties of solid objects by operating directly on the boundary representation. Their method presupposes that each face of a solid is represented by a parametric equation of the form

$$\mathbf{p} = \mathbf{p}(u, w) \qquad u, w \in [0, 1] \qquad (11.11)$$

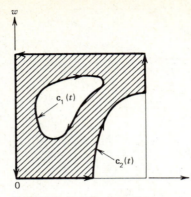

FIGURE 11.3 Parametric representation of a boundary face of a solid.

The union of many faces to form the total boundary of a solid (see Chapter 10, Section 7) and the combination of simple solids or primitives into a complex solid segments the unit square into active and inactive regions. The active regions of the surfaces define the faces of a solid. The closed curve, or loop, bounding each active region is defined parametrically on the unit square as

$$u = u_i(t) \qquad t \in [0, 1]$$
$$w = w_i(t) \tag{11.12}$$

where i identifies a particular loop. Figure 11.3 shows an example of a unit square segmented into active and inactive regions; the shaded area is active. The direction of parametrization of the loops indicates that the active portion is always to our left when we traverse the curve in the direction of increasing t.

As we have seen, global properties can be expressed as either surface integrals or volume integrals. The divergence theorem (Gauss's theorem) allows us to reduce volume integrals to surface integrals; see Eq. 11.7. If we use the notation of Timmer and Stern, this reduction is:

$$\iiint_R \nabla \cdot \Phi \, d\tau = \iint_S \Phi \cdot \mathbf{n} \, d\sigma \tag{11.13}$$

where ∇ denotes the operator $(\partial/\partial x, \partial/\partial y, \partial/\partial z)$ and $\mathbf{n}$ is the unit surface normal, taken as positive when directed outward from the interior of the object.

If the global property we want to compute is

$$\psi = \iiint F(x, y, z) \, d\tau \tag{11.14}$$

where $d\tau = dx \, dy \, dz$, then Φ must satisfy

$$\nabla \cdot \Phi = F \tag{11.15}$$

Φ is not unique. (Again, see any calculus text for a proof of this.) Timmer and Stern give the example of the moment of inertia around the x-axis, where

$$F(x, y, z) = y^2 + z^2 \tag{11.16}$$

For this case, $\Phi = [x(y^2 + z^2), 0, 0]$ or $\Phi = [0, y^3/3, z^3/3]$, for example, are equally valid. Thus, the computation is reduced to evaluating a surface integral

$$\psi = \iint_S G(x, y, z)\, d\sigma \tag{11.17}$$

Usually, the boundary of a solid is given by the union or sum of several, n, faces defined by the active portions of each surface. Rewrite Eq. 11.17 to reflect this

$$\psi = \sum_{i=1}^{n} \iint_{S_i} G(x, y, z)\, d\sigma = \sum_{i=1}^{n} \psi_i \tag{11.18}$$

Next, express the ψ_i in terms of the parametric variables (u, w)

$$\psi_i = \iint_{S_i} G[x(u, w), y(u, w), z(u, w)]\,|\mathbf{J}|\, dudw \tag{11.19}$$

or simply as

$$\psi_i = \iint_{S_i} H(u, w)\, dudw \tag{11.20}$$

where $|\mathbf{J}|$ in Eq. 11.19 is the Jacobian transformation

$$|\mathbf{J}| = \frac{\partial \mathbf{p}(u, w)}{\partial u} \times \frac{\partial \mathbf{p}(u, w)}{\partial w} \tag{11.21}$$

Using notation in Chapter 1, this becomes

$$|\mathbf{J}| = \|\mathbf{p}_{uw}^u \times \mathbf{p}_{uw}^w\| \tag{11.22}$$

Integrate over the active regions (that is, the shaded portion in Fig. 11.3). If the region corresponds identically to the unit square, then use double Gaussian quadrature, and approximate the integral by

$$\psi_i \cong \frac{1}{4} \sum_{j=1}^{m} \sum_{k=1}^{m} w_j w_k H(a_j, a_k) \tag{11.23}$$

where the w's and a's are the weights and zeros of a Legendre polynomial of order $m + 1$.

Usually, however, the active region is irregular, as in Fig. 11.3, making the computation more difficult. Thus, Timmer and Stern apply Green's theorem to transform the surface integral into a line integral to obtain

$$\iint_{S_i} \left[\left(\frac{\partial \alpha}{\partial u} \right) - \left(\frac{\partial \beta}{\partial w} \right) \right] du\, dw = \oint_{\Gamma_i} (\beta\, du + \alpha\, dw) \tag{11.24}$$

where $a = \alpha(u, w)$ and $\beta = \beta(u, w)$. In Fig. 11.3, the active region is bounded by the two closed curves $\mathbf{c}_1(t)$ and $\mathbf{c}_2(t)$. In a manner similar to that described in the divergence, α and β are

$$\left(\frac{\partial \alpha}{\partial u}\right) - \left(\frac{\partial \beta}{\partial w}\right) = H(u, w) \tag{11.25}$$

Timmer and Stern point out that $F(x, y, z)$ can be specified independently of an object's geometry, so that Φ can be determined by inspection. The function $H(u, w)$ is usually algebraically complex and not easy to evaluate. Therefore, we must approximate α and β.

If we assume that the surface S_i is reasonably smooth and H is also smooth and well behaved, then we can approximate H by the polynomial

$$H(u, w) = \sum_{j=1}^{M} \sum_{i=1}^{M} a_{ij} u^{i-1} w^{j-1} \tag{11.26}$$

Determining α and β so that they satisfy Eq. 11.25 is not unique, so we can assume without loss of generality that $\beta(u, w) = 0$, in which case

$$\alpha(u, w) = \sum_{i=1}^{M} \sum_{j=1}^{M} i^{-1} a_{ij} u^i w^{j-1} \tag{11.27}$$

Now rewrite Eq. 11.24 as follows:

$$\psi_i = \oint_{\Gamma_i} \alpha_i(u, w) \, dw \tag{11.28}$$

or as

$$\psi_i = \sum_{l=1}^{L} \int_0^1 \alpha_i[u_1(t), w_1(t)] \left(\frac{dw_1}{dt}\right) dt \tag{11.29}$$

where L is the number of closed curves or loops defining the active portion of S_i.

Timmer and Stern note that Eq. 11.29 can be evaluated exactly if the loops $\mathbf{c}_i(t)$ are polynomials, because the integrands are reducible to polynomials in t. However, the order of the polynomial may become intractably large. If the surfaces are bicubic polynomials and the curves are cubics, the resulting polynomial in t is of order 24.

Let us review some test cases executed by Timmer and Stern to verify their method. For these cases, they evaluated Eq. 11.29 using Gaussian quadrature. They constructed the boundary representation of the test solids as rectangular meshes of bicubic patches. They reported results for ten example objects, computing surface area, volume, centroids, and moments of inertia. Four of these results are shown in Figs. 11.4–11.7. Exact values for the properties are in parentheses for comparison.

It is interesting to review Timmer and Stern's analysis of the sources of computational inaccuracies; they divide them into two categories—ordinary computational problems and geometric approximations. In the first category are the usual errors inherent in Gaussian quadrature, truncation, and rounding off.

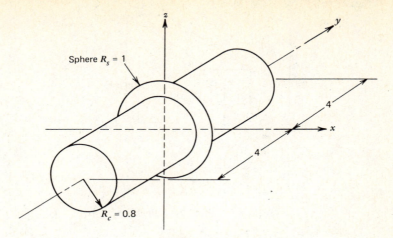

Cylinder $\cup$ sphere

$$\left. \begin{array}{l} 0 \le x^2 + z^2 \le 0.64 \\ -4 \le y \le 4 \end{array} \right\} \quad \text{Cylinder}$$

$$0 \le x^2 + y^2 + z_\bullet^2 \le 1 \quad \text{Sphere}$$

Surface area = 45.72(45.74)
Volume = 16.97(16.99)

FIGURE 11.4 Cylinder U sphere.

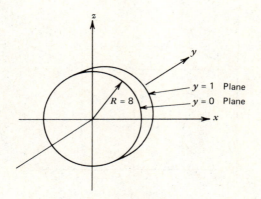

Cylinder

$$0 \le x^2 + z^2 \le 64$$
$$0 \le y \le 1$$

Surface area = 56.52(56.55)
Volume = 25.11(25.13)

FIGURE 11.5 Cylinder.

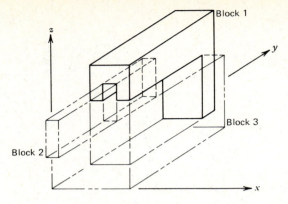

(Block 1 − Block 2) − Block 3

$1.5 \le x \le 4.5, 2 \le y \le 11, 1 \le z \le 9$: Block 1
$2.5 \le x \le 3.5, -4 \le y \le 5, 4.5 \le z \le 7.5$: Block 2
$0 \le x \le 6, 0 \le y \le 9, 0 \le z \le 6$: Block 3

Surface area = 185(185)
Volume = 106.5(106.5)

FIGURE 11.6 (Block 1 − Block 2)—Block 3.

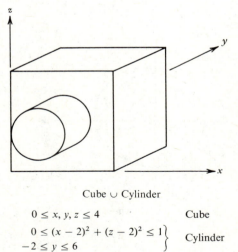

Cube ∪ Cylinder

$0 \le x, y, z \le 4$ — Cube

$\left.\begin{array}{l} 0 \le (x-2)^2 + (z-2)^2 \le 1 \\ -2 \le y \le 6 \end{array}\right\}$ Cylinder

Surface area = 121.137(121.133)
Volume = 76.56(76.57)

FIGURE 11.7 Cube U cylinder.

These computational problems are thoroughly discussed in any standard textbook on numerical analysis; for example, see F. S. Acton (1970).

Errors in geometric approximations arise when approximating an object's bounding surface by parametric bicubic polynomials. Other possible errors are related to the boundary loops, which are determined by the intersection of two bicubic surfaces. The algorithm that Timmer and Stern use to compute these intersections produces a string of discrete points guaranteed to be within a specified tolerance of the true intersection. Dual images are produced for each point (x, y, z), one on each surface (u_1, w_1) and (u_2, w_2). Separate curves are interpolated for each surface—$u_1(t)$, $w_1(t)$ and $u_2(t)$, $w_2(t)$. The authors point out that although this interpolation is done to a specified tolerance, it is carried out in parametric space and deviation from the true intersection is not controlled.

The Timmer–Stern method is computationally sophisticated, and even though the sources of error are subtle, they are easily identified and controlled. Since computations on each surface region are done independently, the algorithm is a good candidate for parallel processing techniques in future systems.

EXERCISES

1. Write a procedure based on the Timmer–Stern method to compute global properties: surface area, volume, centroid, and moments of inertia.

Use the procedure defined in Ex. 1 or another procedure to compute surface area, volume, centroid, and moments of inertia for the solids in Ex. 2–4. If the given half-space function $f(x, y, z) = 0$ is zero for a specified point, then the point is on the boundary; if positive, then inside the solid; if negative, then outside.

2. The unit cube,

$$(x = 0) \cap (1 - x = 0) \cap (y = 0) \cap (1 - y = 0) \cap (z = 0) \cap (1 - z = 0)$$

3. The unit sphere, $(1 - x^2 - y^2 - z^2 = 0)$

4. The difference between the cube and cylinder,

$$[(x = 0) \cap (4 - x = 0) \cap (y = 0) \cap (4 - y = 0) \cap (z = 0) \cap (4 - z = 0)$$
$$- [(1 - (x - 2)^2 - (z - 2)^2 = 0) \cap (y + 2 = 0) \cap (6 - y = 0)]$$

3   SPATIAL ENUMERATION BY POINT CLASSIFICATION

Computing global properties based on spatial enumeration by point classification first requires finding a rectangular volume enclosing the solid in question. We then partition this volume into cubical cells. We test each cell for inclusion in the solid by classifying a selected point in each cell as being in, on, or outside the

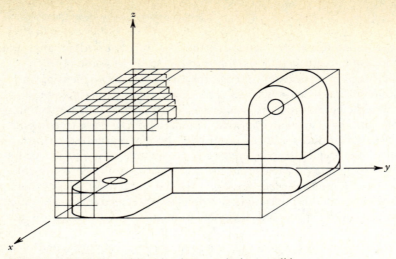

FIGURE 11.8 Cell-partitioned volume enclosing a solid.

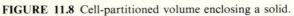

B

		In	On	Out
	In	In	On	Out
A	On	On	On/Out	Out
	Out	Out	Out	Out

Intersect

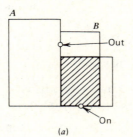

(a)

B

		In	On	Out
	In	In	In	In
A	On	In	In/On	On
	Out	In	On	Out

Union

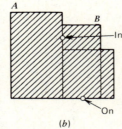

(b)

B

		In	On	Out
	In	Out	On	In
A	On	Out	On/Out	On
	Out	Out	Out	Out

Difference

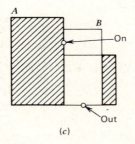

(c)

FIGURE 11.9 Point classification combinations.

solid. If these cells are small enough, the contribution of each to the global property is the product

$$I = \sum_i f_i \Delta V = \Delta V \sum_i f_i \tag{11.30}$$

where ΔV is the volume of a cell; see also Eq. 11.4. Figure 11.8 is an object enclosed in a volume arbitrarily partitioned into cells. Notice that in this case, the minimum and maximum coordinates of the object determine the extent of the volume.

The algorithm in Eq. 11.30 uses the CSG binary tree representation to process each cell through the tree to determine its classification. This algorithm is analogous to the boundary evaluator in Chapter 10, Section 6, including the use of neighborhood models to resolve ambiguities. Figure 11.9 presents the possible combinations and their classifications. Notice that it is the on-on combination that generates an ambiguous classification.

Y. T. Lee and A. A. G. Requicha (1982) report that tests comparing this algorithm with others show that it is slower, perhaps too slow for practical use on sequential machines.

4 BLOCK DECOMPOSITION BY CELL CLASSIFICATION

The spatial-enumeration algorithm described in Section 3 is significantly improved if we allow decomposition into variably sized blocks. The global properties are then summed over these blocks that comprise the solid. It is a very complex problem to decompose a CSG or boundary-representation model into such a form. The end result is analogous if not identical to octree construction; we will explore only the broad outlines of this problem and its solution here. Y. T. Lee and A. A. G. Requicha (1982), D. J. Meagher (1982), and others, present similar but more complete descriptions.

First, we find a cubical volume that just encloses a given solid and whose faces do not intersect the interior of this solid. We find the limiting planes defining the faces of this cube as follows: Find the minimum and maximum (extreme) coordinates of the solid. Then compute Δx, Δy, and Δz from $\Delta x = x_{max} - x_{min}$ and similarly for Δy and Δz. Use the maximum of these Δ's as the length of the enclosing cube's edge; see Fig. 11.10.

Next, divide the cube into eight equal subcubes. One way of deciding if a cubical cell is entirely inside or outside a solid is by testing for intersections between faces of a cell and faces of the solid. If there are no intersections within the appropriate domain, then clearly the cell is either entirely inside or outside the solid. Now, select any point known to be inside the cell (that is, its centroid) and classify it using the techniques we have discussed as inside or outside the solid. If the point is outside, then the cell is outside, since we know it does not intersect the solid; conversely if it is inside.

If the cell is outside the solid, discard it. If inside, store it and use it for computing the solid's global properties.

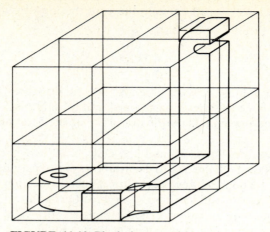

FIGURE 11.10 Block decomposition.

Again, subdivide those cells that intersect the solid into eight subcells, and repeat the preceding process. This algorithm continues recursively until we reach some predetermined cell size defining the resolution of the algorithm. This approach is usually too slow for sequential processing, and Lee and Requicha suggest a more sophisticated approach; see Y. T. Lee and A. A. G. Requicha (1982).

12

COMPUTER GRAPHICS

Although there are exceptions, advances in geometric modeling are motivated by advances in computer graphics. Interestingly enough, the converse is also true, since computer graphics systems have historically supported many geometric and quasi-geometric applications. A strong association, impossible to ignore, exists between these two technologies, and it declares, in effect, that graphics is geometry.

Computer graphics now encompasses many concepts—from the psychology and ergonomics of interactive environments to software and hardware design. All of these efforts are, in the end, directed toward creating a visual image. This image may appear as a sophisticated video display or merely the output of a simple plotter or somewhere in between. We will focus on the transformations required to convert a three-dimensional geometric model into a two-dimensional image—one that is convincingly three-dimensional in appearance.

First, we will briefly consider general demands of the display environment on the kinds of transformations and operations required to plot curves and surfaces. Next, we will investigate several types of projections, including orthographic and perspective. We will see that scene transformations require some special twists and interpretations of scaling, translating, and rotating transformations. We will discuss a surprisingly simple algorithm for computing outline curves of solids, with implications for solving the hidden-surface problem. Then, we will investigate the geometric principles for computing hidden surfaces and explore several algorithms to do this.

The realism of a computer graphic image of an object is enhanced by using certain geometric properties that affect display parameters: illumination and shadows, reflection, shading, texture, and so on. We will explore the relationship between these effects and the geometric properties of objects.

There is more "geometry" to computer graphics than will be presented here. Omitted are plotting techniques, clipping and windowing, display files, input and output devices, interactive techniques, and other such topics, so that we can focus more on concepts directly related to geometric modeling. These other topics are adequately covered in other texts, such as J. D. Foley and A. Van Dam (1982) and W. M. Newman and R. F. Sproull (1979).

1   DISPLAY ENVIRONMENT

The most obvious and most important characteristic of a computer graphic display environment is its two dimensionality, which determines the nature of the transformations from a three-dimensional model to a two-dimensional image. Before we discuss these transformations, let us consider two lesser, but nonetheless important, characteristics.

Most video-type—cathode ray tube (CRT)—display devices fall into one of two categories: **vector graphics** devices or **raster graphics** devices. Vector devices produce an image consisting solely of line segments "drawn" from point to point. Curves are displayed as a sequence of straight-line segments and surfaces by their bounding curves and perhaps a family of isoparametric curves lying in the surface. Raster devices produce an image by a scanning process similar to an ordinary television set. Small, discrete areas—*pixels*, short for picture elements—on the display screen are activated and illuminated to a controlled level of intensity. Strings of pixels approximate straight lines and curves; large groups of pixels approximate surfaces. Variation in brightness over a group can produce a strong perception of depth by simulating the effects of illumination on a curved surface. Figure 12.1 shows the distinguishing features of these two display types.

Vector displays were used extensively in the early CAD drafting systems, based first on two-dimensional and later on three-dimensional wireframe-modeling schemes. Current solid modelers, including ROMULUS, GMSolid, CADD, and others, still use vector devices for both constructing and displaying CSG and boundary-representation models. Vector displays offer a great deal of information that is amenable to interaction [see Foley and Van Dam (1982) or Newman and Sproull (1979)], and they also closely resemble the classical engineering line drawing.

Raster scan systems, on the other hand, are capable of producing very realistic images of three-dimensional objects. Although they are essentially identical in function to a conventional television set, adapting them to computer graphics by means of geometric modeling is more recent than adapting vector devices. To make full use of the capabilities of a raster scan system requires an adequate boundary representation of a three-dimensional object whose image is to be constructed. This is because illumination characteristics of the object, hence, visual appearance, depend on surface normal information.

Most of the transformations and operations we are about to explore are largely independent of either category of display device. These transformations and operations are presented in a context where the geometric content is

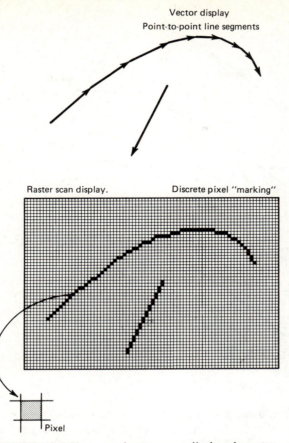

FIGURE 12.1 Vector and raster scan display characteristics.

highest and the concepts involved are not complicated by the idiosyncrasies of
display devices.

Clearly, to display any object, we must compute the images of its bounding
curves and surfaces. General-purpose curve and surface generators as integral
parts of computer graphic hardware are coming and will someday be a standard
part of all systems. These generators will process minimal curve- and surface-
defining parameters (for example, the vertex points of a Bezier curve polygon)
to produce an analog signal controlling and deflecting the electron beam of a
CRT through true curves. Until such hardware is available, we will be faced
with the problem of approximating surfaces with curves and curves, in turn,
with a sequence of straight-line segments, at least for vector devices.

A parametric curve is segmented into *n* equal parametric intervals. The
greater the number of intervals, the smoother the curve's displayed image;
however, computation demands also increase. Usually, model data contain

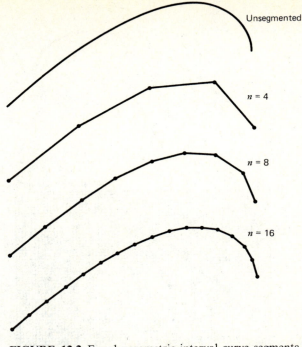

Unsegmented

$n = 4$

$n = 8$

$n = 16$

FIGURE 12.2 Equal parametric interval curve segmentation.

the beginning and end points of a curve, so the curve-defining polynomial must be evaluated $(n - 1)$ times to produce the n straight-line segments. Several examples are shown in Fig. 12.2. Notice the increase in apparent smoothness of the curve's image as n increases. This method is neat and direct, but not necessarily the most efficient, since it does not discriminate between a "flat" section of a curve, requiring fewer segments to approximate, and a highly curved one, requiring more segments.

An alternative segmentation method adjusts for changes in curvature. We do this by specifying a maximum change in the tangent vector direction between segments, denoted by $\Delta\theta$ in Fig. 12.3. The trick is to develop an algorithm guaranteed to be more efficient than the equal-parametric-interval method. A host of secondary factors may make such a guarantee impossible. However, one approach proceeds as follows: Compute the change in direction between the end-point tangent vectors. If this exceeds $\Delta\theta$, compute the tangent vector direction at $u = 0.5$ (assuming the curve is parametrized from $u = 0$ to $u = 1$), and compare this direction to that at each of the end points. Continue to subdivide and test segments until the change in direction between tangents at the ends of a segment is less than $\Delta\theta$. A variation of Timmer's step-size selection procedure (Chapter 7, Section 4.7) offers more direct computation of a sequence of points along a curve whose spacing is governed by curvature.

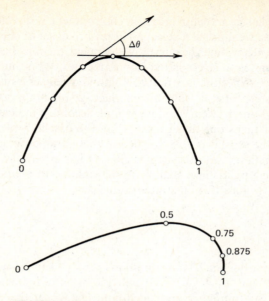

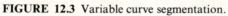

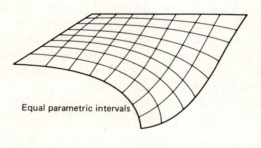

FIGURE 12.3 Variable curve segmentation.

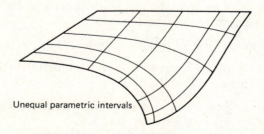

Equal parametric intervals

Unequal parametric intervals

FIGURE 12.4 Two surface display methods.

Surface display in a vector graphics environment poses slightly different problems, because we must first select an appropriate set of curves lying in the surface. Once we determine the curves, we segment them as before. By far the simplest set of curves to select is the isoparametrics. What interval scheme is most effective? Consider the fact that this set of curves serves two purposes in the displayed image. First, it "fills in" a surface, giving it a more "substantial" appearance than boundary curves alone would. Second, it indicates the surface's interior shape. It turns out that by satisfying the first in a pleasing although subjective way, the second is also satisfied. Equal parametric intervals in both the u and w directions yield the best results. Other criteria (yielding unequal spacing) produce a plaid effect that, from the standpoint of appearance alone, is less appealing. Figure 12.4 illustrates both cases.

EXERCISES

1. Given a pc curve defined by the four points $\mathbf{p}_1 = [0 \quad 0 \quad 0]$, $\mathbf{p}_2 = [1 \quad 1 \quad 0]$, $\mathbf{p}_3 = [2 \quad 1 \quad 0]$, and $\mathbf{p}_4 = [3 \quad 2 \quad 0]$, divide it into ten equal parametric segments. Find the minimum and maximum change in the direction of the tangent vectors between segments.

2. Given the pc curve defined in Ex. 1, segment it so that the maximum deviation of the chord of a segment from the true curve is 0.1.

3. Write a procedure to segment a pc curve based on some maximum allowable change in direction of the tangent from one end of the segment to the other. Assume the curve is planar, given, say, by the x, y components of its geometric coefficients. Use the method of successively halving the parametric interval of a segment until the tangent-deviation criterion is satisfied. Notice that this recursive binary-subdivision process can be viewed as generating a binary tree whose leaf nodes are curve segments represented by their parametric intervals that satisfy the deviation criterion. Clearly, not all leaf nodes will be on the same level in the binary tree. Be sure that the procedure you write properly segments the entire curve and does not terminate after finding the first acceptable segment. Denote this procedure as **BSEGPC(CI, MAXT, NP, CO)**, where

CI(4, 2) is the input array of geometric coefficients defining the pc curve;

MAXT is the input value of the maximum deviation of the tangent from one end of a segment to the other;

NP is the output number of points computed; and

CO(NP, 3) is the output array of x, y, u coordinates of the segmentation points on the curve.

4. Repeat Ex. 3 for a Bezier curve, denoting the procedure as **BSEGBZ**. Notice that the argument list will be somewhat different.

5. Repeat Ex. 3 for a B-spline curve, denoting the procedure as **BSEGBS**. Again, the argument list will be different.

2 PROJECTIONS

Projections are transformations that produce two-dimensional representations of three-dimensional objects. As a class, they are perhaps the most important kind of transformation, since without them, we could not construct the display image. Section 2 explores types of plane projections, that is, those projections formed on a flat, two-dimensional plane. The simplest is the **orthographic projection**.

2.1 Orthographic Projection

Consider the orthographic projection of any point **p** in space onto any arbitrary plane, as in Fig. 12.5. The projected image **p*** is found by constructing a line through **p** normal to the plane. This amounts to finding the minimum distance between **p** and the plane, and we may use Eq. 6.13. The intersection of this line with the plane is the point **p***. So **p*** must satisfy $(\mathbf{p^*} - \mathbf{p}) \times \mathbf{n} = 0$.

Find the orthogonal projection of **p** onto any of the three principal planes by simply setting the corresponding point coordinate equal to zero. Thus, the projection of the point (3, 6, 9) onto the $z = 0$ plane is (3, 6, 0). Orthographic projections are characterized by the perpendicular relationship of the line, or vector, between the initial point and the projected point and the plane.

Next, consider the orthogonal projection of parametric curves. Four projection problems of particular interest are

1. Projections onto the three principal planes.

2. Projections onto an arbitrary plane.

3. Projections onto an arbitrary plane by simultaneously rotating both the plane and the curve so that the plane becomes parallel to one of the principal planes.

4. Projections onto an arbitrary plane by projecting defining points.

The first three problems are solved specifically for the projection of parametric cubic curves, although the methods used apply to any of the parametric curves we have studied. The solution to the fourth problem is more general. We will assume that the geometric form defines the pc curve to be projected.

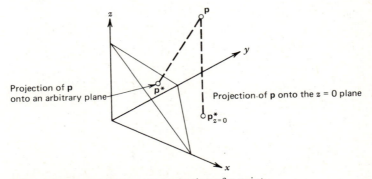

FIGURE 12.5 Orthographic projection of a point.

To project a pc curve onto one of the three principal planes, simply set equal to zero the four geometric coefficients of the coordinate corresponding to the axis normal to the principal plane of projection. Figure 12.6 shows this situation for a curve projected onto the $z = 0$ and $x = 0$ planes. Take the projection onto the $z = 0$ plane. Here, the z coordinates of all the points comprising the resulting projected curve are equal to zero. Therefore, boundary conditions or geometric coefficients related to z (in this example) must yield a zero contribution to the points in the set defining the new curve. We can generalize from this and say that geometric coefficients of the planar pc curve produced by projecting any general pc curve onto any plane coinciding with, or parallel to, any one of the three principal planes are given by one of the following three expressions:

$$\mathbf{B}^*_{x=a} = \mathbf{B} + \begin{bmatrix} a - x_0 & 0 & 0 \\ a - x_1 & 0 & 0 \\ -x_0^u & 0 & 0 \\ -x_1^u & 0 & 0 \end{bmatrix} \tag{12.1}$$

$$\mathbf{B}^*_{y=b} = \mathbf{B} + \begin{bmatrix} 0 & b - y_0 & 0 \\ 0 & b - y_1 & 0 \\ 0 & -y_0^u & 0 \\ 0 & -y_1^u & 0 \end{bmatrix} \tag{12.2}$$

$$\mathbf{B}^*_{z=c} = \mathbf{B} + \begin{bmatrix} 0 & 0 & c - z_0 \\ 0 & 0 & c - z_1 \\ 0 & 0 & -z_0^u \\ 0 & 0 & -z_1^u \end{bmatrix} \tag{12.3}$$

The asterisk denotes the transformed (projected) curve, and $\mathbf{a}$, $\mathbf{b}$, and $\mathbf{c}$ are vectors in the x, y, and z directions, respectively. These equations are simplified as follows:

$$\mathbf{B}^*_{x=a} = \begin{bmatrix} a & y_0 & z_0 \\ a & y_1 & z_1 \\ 0 & y_0^u & z_0^u \\ 0 & y_1^u & z_1^u \end{bmatrix} \tag{12.4}$$

$$\mathbf{B}^*_{y=b} = \begin{bmatrix} x_0 & b & z_0 \\ x_1 & b & z_1 \\ x_0^u & 0 & z_0^u \\ x_1^u & 0 & z_1^u \end{bmatrix} \tag{12.5}$$

$$\mathbf{B}^*_{z=c} = \begin{bmatrix} x_0 & y_0 & c \\ x_1 & y_1 & c \\ x_0^u & y_0^u & 0 \\ x_1^u & y_1^u & 0 \end{bmatrix} \tag{12.6}$$

To project a pc curve onto an arbitrary plane $Ax + By + Cz + D = 0$, apply the following transformations to the geometric coefficients of the initial curve (see Fig. 12.7):

$$\mathbf{B}^* = \mathbf{B} + \begin{bmatrix} d_1\mathbf{n} \\ d_2\mathbf{n} \\ (d_3 - d_1)\mathbf{n} \\ (d_4 - d_2)\mathbf{n} \end{bmatrix} \tag{12.7}$$

where $\mathbf{n}$ is the unit normal vector to the plane and d_1, d_2, d_3, and d_4 are the respective distances from the points $\mathbf{p}_0$, $\mathbf{p}_1$, $\mathbf{p}_0 + \mathbf{p}_0^u$, and $\mathbf{p}_1 + \mathbf{p}_1^u$ to the plane. Compute the unit normal and the required distances by using some simple relationships from analytic geometry. For example, the components of the unit normal are

$$\mathbf{n}_x = \frac{A}{\sqrt{A^2 + B^2 + C^2}}$$

$$\mathbf{n}_y = \frac{B}{\sqrt{A^2 + B^2 + C^2}} \tag{12.8}$$

$$\mathbf{n}_z = \frac{C}{\sqrt{A^2 + B^2 + C^2}}$$

and d_1 is

$$d_1 = \frac{Ax_0 + By_0 + Cz_0 + D}{\sqrt{A^2 + B^2 + C^2}} \tag{12.9}$$

Notice that d_1 is a directed distance. The algebraic sign associated with d_1 indicates that it is in the same direction $(+)$ as $\mathbf{n}$ or in the opposite direction $(-)$.

We can project a curve onto any plane in another way. First, rotate both the plane and the curve so that the plane is parallel to a selected principal plane. This just requires finding the rotation transformation that will rotate the normal to the plane into a direction parallel to the x, y, or z direction. Apply this transformation matrix to the initial curve. Second, find the projection of the curve onto the plane in this new orientation, using either Eq. 12.1, or 12.3. Finally, reverse the rotation transformation, applying it to the projected curve now lying in, or parallel to, one of the principal planes. This procedure gives the geometric coefficients of the curve in the required plane of projection; it is illustrated in Fig. 12.8.

A variation of this scheme will frequently simplify the computation of the projected image of a complex scene: Select a viewing direction (corresponding to the normal of the plane of projection), and apply a coordinate system transformation that brings this viewing direction into coincidence with, say, the z-axis. Then the x, y plane becomes the plane of projection or the screen plane of the display device.

The fourth and final method for orthographic curve projection that we will discuss is perhaps the simplest to perform. We will consider the four-point form

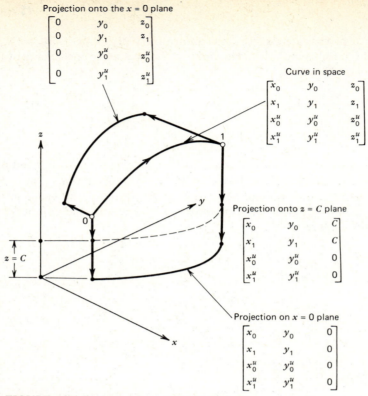

Projection onto the $x = 0$ plane

$$\begin{bmatrix} 0 & y_0 & z_0 \\ 0 & y_1 & z_1 \\ 0 & y_0^u & z_0^u \\ 0 & y_1^u & z_1^u \end{bmatrix}$$

Curve in space

$$\begin{bmatrix} x_0 & y_0 & z_0 \\ x_1 & y_1 & z_1 \\ x_0^u & y_0^u & z_0^u \\ x_1^u & y_1^u & z_1^u \end{bmatrix}$$

Projection onto $z = C$ plane

$$\begin{bmatrix} x_0 & y_0 & C \\ x_1 & y_1 & C \\ x_0^u & y_0^u & 0 \\ x_1^u & y_1^u & 0 \end{bmatrix}$$

Projection on $x = 0$ plane

$$\begin{bmatrix} x_0 & y_0 & 0 \\ x_1 & y_1 & 0 \\ x_0^u & y_0^u & 0 \\ x_1^u & y_1^u & 0 \end{bmatrix}$$

FIGURE 12.6 Projections onto principal planes.

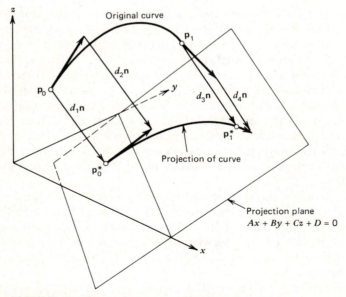

FIGURE 12.7 Projections onto an arbitrary plane.

508

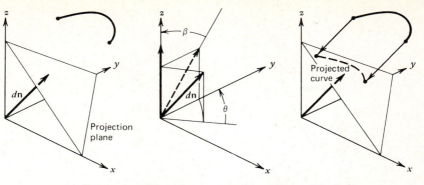

1. Rotary normal *dn* and curve around *z*-axis through angle θ.
2. Rotate results around *x*-axis through angle β.
3. Project transformed curve onto transformed plane $z = d$.
4. Rotate projected curve around *x*-axis through angle $-\beta$.
5. Rotate result from 4 around *z*-axis through angle $-\theta$.

FIGURE 12.8 Projections onto an arbitrary plane using simplifying rotation transformations.

of the pc curve, but the method applies directly to points defining the Bezier polygon or the *B*-spline and other curves. This method applies to pc curves because transformations performed on the four-point form produce the same result as if performed on the geometric or algebraic forms. Therefore, we can project a pc curve onto any plane by converting the geometric (or algebraic) coefficients into the four-point form, then simply project the four points onto the plane. Thus, for any point **p**, we obtain

$$\mathbf{p}^* = \mathbf{p} - d\mathbf{n} \tag{12.10}$$

where **n** is the normal to the plane and d is the distance from **p** to the plane. Figure 12.9 shows this situation for a five-point Bezier curve. Again, we can use the minimum-distance formula in Eq. 6.13 to calculate **p***.

We are now ready to take a more general approach to the projection problem. The basic ingredients are: the point **p** to be projected, the projection plane defined by $\mathbf{p}_0 + x^*\mathbf{u}_1 + y^*\mathbf{u}_2$, and a direction of projection $\mathbf{u}_3$; see Fig. 12.10. The vector $\mathbf{p}_0$ defines the origin of a coordinate system on the projection plane, with $\mathbf{u}_1$ and $\mathbf{u}_2$ as unit vectors defining the plane itself. If **p*** is the projection of **p** onto the plane, then

$$\mathbf{p}^* = \mathbf{p} - z^*\mathbf{u}_3 \tag{12.11}$$

where z^* is to be determined. This is clearly similar to Eq. 12.10. The projected point **p*** can be independently expressed in terms of vectors defining the coordinate system on the projection plane. Thus,

$$\mathbf{p}^* = \mathbf{p}_0 + x^*\mathbf{u}_1 + y^*\mathbf{u}_2 \tag{12.12}$$

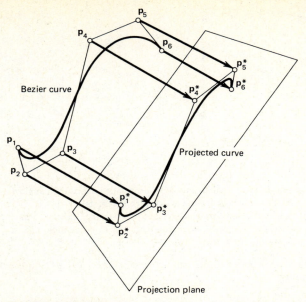

FIGURE 12.9 Point-wise projection of a Bezier curve onto an arbitrary plane.

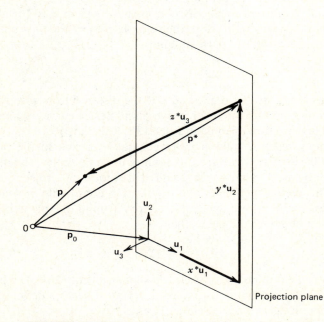

FIGURE 12.10 Vector geometry of the projection of a point.

Combining Eqs. 12.11 and 12.12 allows us to express the transformed co-ordinates as

$$\mathbf{p} - z^*\mathbf{u}_3 = \mathbf{p}_0 + x^*\mathbf{u}_1 + y^*\mathbf{u}_2 \qquad (12.13)$$

We solve for the coordinates x^*, y^*, and z^* by taking the appropriate scalar product of the equation (see Chapter 6). To solve for x^*, take the scalar product with $(\mathbf{u}_2 \times \mathbf{u}_3)$; for y^*, use $(\mathbf{u}_3 \times \mathbf{u}_1)$; and for z^*, use $(\mathbf{u}_1 \times \mathbf{u}_2)$. Doing this yields

$$x^* = (\mathbf{p} - \mathbf{p}_0) \cdot \frac{\mathbf{u}_2 \times \mathbf{u}_3}{\mathbf{u}_1 \cdot (\mathbf{u}_2 \times \mathbf{u}_3)}$$

$$y^* = (\mathbf{u} - \mathbf{u}_0) \cdot \frac{\mathbf{u}_3 \times \mathbf{u}_1}{\mathbf{u}_2 \cdot (\mathbf{u}_3 \times \mathbf{u}_1)} \qquad (12.14)$$

$$z^* = (\mathbf{p} - \mathbf{p}_0) \cdot \frac{\mathbf{u}_1 \times \mathbf{u}_2}{\mathbf{u}_3 \cdot (\mathbf{u}_1 \times \mathbf{u}_2)}$$

Notice that so far in this development, we have not specified a relationship between $\mathbf{u}_3$ and the plane defined by $\mathbf{u}_1$ and $\mathbf{u}_2$, nor have we specified any particular relationship between $\mathbf{u}_1$ and $\mathbf{u}_2$. Equations 12.14 are independent of the direction of these unit vectors. However, as we have seen, orthographic projection requires $\mathbf{u}_3$ to be perpendicular to the projection plane. Furthermore, it is convenient to choose $\mathbf{u}_1$ and $\mathbf{u}_2$ so that they, too, are mutually perpendicular. This means that we can simplify these equations dramatically, since, for example, $\mathbf{u}_1 \cdot (\mathbf{u}_2 \times \mathbf{u}_3) = 1$ and $\mathbf{u}_2 \times \mathbf{u}_3 = \mathbf{u}_1$. We can now write

$$x^* = (\mathbf{p} - \mathbf{p}_0) \cdot \mathbf{u}_1$$
$$y^* = (\mathbf{p} - \mathbf{p}_0) \cdot \mathbf{u}_2 \qquad (12.15)$$
$$z^* = (\mathbf{p} - \mathbf{p}_0) \cdot \mathbf{u}_3$$

Ordinarily, z^* helps provide information on visibility and obscurity conditions in $\mathbf{p}$ relative to other objects and the chosen projection plane. If we preserve z^*, we can use it to reconstruct the initial point $\mathbf{p}$ if necessary. But we do not use z^* directly in constructing the image on the plane, and we may not otherwise need to compute it. In Section 2.2, we will extend this approach to include perspective projection.

EXERCISES

1. Draw a sketch to interpret Eqs. 12.1–12.3.

2. Suppose a curve passes through the plane it is to be projected on. Does Eq. 12.7 still apply? If not, how should it be modified?

3. Describe two ways of constructing orthographic projections of surfaces.

4. Discuss the orthographic projection of octree encoded objects. Is it possible to transform the octree representation directly into the quadtree representation of the projection? Discuss ramifications of this idea.

5. Discuss the possibility of, and approach to, projecting an arbitrary parametric curve onto an arbitrary parametric surface. Show or develop the basic vector equations required to do this.

6. Discuss the projection of a curve onto another curve.

7. Write a procedure to project a set of points onto an arbitrary plane. Assume an orthogonal projection. Denote this as **PROJPT**(NP, PI, PL, PO), where

NP is the input number of points to be projected;

PI(NP, 3) is the input array of coordinates of the points;

PL(4) is the input array of plane coefficients A, B, C, D; and

PO(NP, 3) is the output array of coordinates of the projected points.

2.2 Perspective Projection

Perspective projections have been used for centuries by artists, architects, and illustrators to capture a more realistic two-dimensional image of three-dimensional objects. Such projections have been rarely used in engineering drawings, however, since they were time consuming and, hence, expensive to produce. Simple orthographic projections were adequate for communicating the necessary technical information between designer and shop. This situation is changing. The mathematics of geometric modeling and the computer's prodigious calculating abilities combine in computer graphics systems to make practical perspective images an effective alternative.

A curious phenomenon can occur. Artists have discovered that strict adherence to the rules of perspective projection often produces an image that does not look quite right. Therefore, they usually make subtle subjective

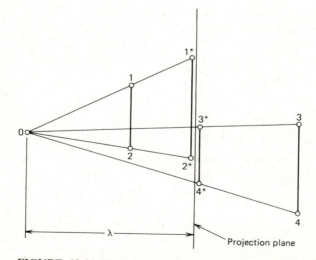

FIGURE 12.11 Elements of perspective projection.

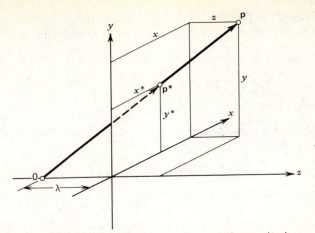

FIGURE 12.12 Geometry of the perspective projection of a point.

adjustments in rendering the image. This is relatively easy for the human artist, but a computer must follow the rigid dictates of a mathematically precise algorithm. Any alterations we choose to make must be expressed mathematically.

The perspective projection of an object approximates the way an observer forms a visual image of it. Objects in a scene are projected onto a plane from a central point, usually assumed to be the eye of the observer. The center of the projection plane is at the point of intersection between the plane and the normal to it from the observer at zero, which is the *center of projection*; see Fig. 12.11, which shows the projection of two line segments onto the plane of projection.

Now, let us consider the basic geometry. Let λ denote the normal distance from the viewer to the plane. Figure 12.12 presents the key geometric ingredients. Using the properties of similar triangles, we readily extract the following relationships between the coordinates of a point $\mathbf{p}$ and its projection $\mathbf{p}^*$

$$x^* = \frac{\lambda x}{z + \lambda}$$

$$y^* = \frac{\lambda y}{z + \lambda} \qquad (12.16)$$

$$z^* = 0$$

Homogeneous coordinates can expedite the transformation. Recall the total transformation matrix $\mathbf{T}_h$ for points given by homogeneous coordinates (Eq. 8.75)

$$\mathbf{T}_h = \begin{bmatrix} a & b & c & p \\ d & e & f & q \\ g & i & j & r \\ \hdashline l & m & n & h \end{bmatrix} \qquad (12.17)$$

When the elements in the last column of this matrix are not zero, a perspective transformation is formed. Consider the following transformation:

$$\mathbf{p}_h^* = \mathbf{p}_h \begin{bmatrix} 1 & 0 & 0 & 0 \\ 0 & 1 & 0 & 0 \\ 0 & 0 & 0 & r \\ 0 & 0 & 0 & 1 \end{bmatrix} \tag{12.18}$$

where the subscript h denotes homogeneous coordinates. If $\mathbf{p}_h = [x \quad y \quad z \quad 1]$, then for $\mathbf{p}_h^*$, we obtain

$$\mathbf{p}_h^* = [x \quad y \quad 0 \quad (rz + 1)] \tag{12.19}$$

The ordinary transformed coordinates are

$$\mathbf{p}^* = \left[\frac{x}{(rz + 1)} \quad \frac{y}{(rz + 1)} \quad \frac{0}{(rz + 1)} \right] \tag{12.20}$$

Now, how do we interpret this result? Refer again to Figure 12.12. Point $\mathbf{p}$ is transformed into point $\mathbf{p}^*$ by Eq. 12.20. Let $r = 1/\lambda$. Then, substituting into Eq. 12.20, we obtain

$$\mathbf{p}^* = \left[\frac{\lambda x}{(z + \lambda)} \quad \frac{\lambda y}{(z + \lambda)} \quad 0 \right] \tag{12.21}$$

These are, of course, the same values of x^*, y^*, and z^* that we derived in Eq. 12.16. From Fig. 12.12, we see that the center of projection is located at $[0 \quad 0 \quad -\lambda]$ and the plane of projection is the $z = 0$ plane. Observe that $\mathbf{p}$ and $\mathbf{p}^*$ coordinates share the same origin, since the transformation in Eq. 12.21 produces no translation. Notice that when λ approaches infinity, then r approaches zero, and the projection becomes axonometric (Section 2.4).

Points already in the $z = 0$ plane do not change by a projective transformation. From Eq. 12.20, we see that when $z = 0$, then $x^* = x$ and $y^* = y$. And this is true only when the homogeneous coordinate h is unity after we apply the projective transformation to $\mathbf{p}_h$ (that is, $[x \quad y \quad z \quad 1]$). A projective transformation may be preceded by any rotation and translation transformations, since they do not affect the value of the homogeneous coordinate. This means that we can obtain a perspective projection from an arbitrary viewpoint by first performing rigid-motion transformations. These will bring the z-axis into coincidence with the required line of sight and move the origin to the required distance λ from the center of projection.

In many situations, we must keep the z coordinate of points (for example, to compute display brightness intensities or position relationships). To preserve the z-coordinate information through the transformation, we modify Eq. 12.18 as follows:

$$\mathbf{p}_h^* = \mathbf{p}_h \begin{bmatrix} 1 & 0 & 0 & 0 \\ 0 & 1 & 0 & 0 \\ 0 & 0 & 1 & 1/\lambda \\ 0 & 0 & 0 & 1 \end{bmatrix} \tag{12.22}$$

changing the 0 in row 3, column 1 to a 1. Notice that the substitution $1/\lambda = r$ is included.

Let us see how all these rigid-motion and projection transformations apply to a specific example. Consider the unit cube in Fig. 12.13 lying in the positive octant with a vertex at the origin. To rotate the cube around the y-axis, translate the $[0 \quad 0 \quad 0]$ vertex to $[0 \quad m \quad n]$, and project the results onto the $z = 0$ plane from a center of projection at $[0 \quad 0 \quad -k]$, require the following product of transformation matrices:

$$
\begin{bmatrix}
\cos\theta & 0 & -\sin\theta & 0 \\
0 & 1 & 0 & 0 \\
\sin\theta & 0 & \cos\theta & 0 \\
0 & m & n & 1
\end{bmatrix}
\begin{bmatrix}
1 & 0 & 0 & 0 \\
0 & 1 & 0 & 0 \\
0 & 0 & 1 & 1/k \\
0 & 0 & 0 & 1
\end{bmatrix}
$$

(12.23)

$$
=
\begin{bmatrix}
\cos\theta & 0 & -\sin\theta & -\sin\theta/k \\
0 & 1 & 0 & 0 \\
\sin\theta & 0 & \cos\theta & \cos\theta/k \\
0 & m & n & (n/k) + 1
\end{bmatrix}
$$

Next, express the vertices of the cube as homogeneous coordinates, and transform them using the matrix determined by Eq. 12.23. Notice that since the projection is onto the $z = 0$ plane, the third column of this matrix can be omitted.

$$
\mathbf{p}_h^* = \mathbf{p}^*\mathbf{T}_h =
\begin{bmatrix}
0 & 0 & 0 & 1 \\
1 & 0 & 0 & 1 \\
1 & 1 & 0 & 1 \\
0 & 1 & 0 & 1 \\
0 & 0 & 1 & 1 \\
1 & 0 & 1 & 1 \\
1 & 1 & 1 & 1 \\
0 & 1 & 1 & 1
\end{bmatrix}
\begin{bmatrix}
\cos\theta & 0 & -\sin\theta/k \\
0 & 1 & 0 \\
\sin\theta & 0 & \cos\theta/k \\
0 & m & (n/k) + 1
\end{bmatrix}
$$

(12.24)

$$
\mathbf{p}_h^* =
\begin{bmatrix}
0 & m & (n/k) + 1 \\
\cos\theta & m & -\sin\theta/k + (n/k) + 1 \\
\cos\theta & 1 + m & -\sin\theta/k + (n/k) + 1 \\
0 & 1 + m & (n/k) + 1 \\
\sin\theta & m & \cos\theta/k + (n/k) + 1 \\
\cos\theta + \sin\theta & m & -\sin\theta/k + \cos\theta/k + (n/k) + 1 \\
\cos\theta + \sin\theta & 1 + m & -\sin\theta/k + \cos\theta/k + (n/k) + 1 \\
\sin\theta & 1 + m & \cos\theta/k + (n/k) + 1
\end{bmatrix}
$$

(12.25)

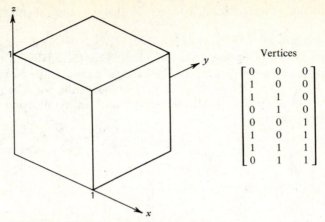

FIGURE 12.13 Transformations of a unit cube.

Using $\theta = 30°$, $m = -2$, $n = 2$, and $k = 2$, the transformation matrix $\mathbf{T}_h$ becomes

$$\mathbf{T}_h = \begin{bmatrix} 0.866 & 0 & -0.250 \\ 0 & 1.000 & 0 \\ 0.500 & 0 & 0.433 \\ 0 & -2.000 & 2.000 \end{bmatrix} \tag{12.26}$$

and the transformed homogeneous coordinates are

$$\mathbf{p}_h^* = \begin{bmatrix} 0 & -2 & 2 \\ 0.866 & -2 & 1.750 \\ 0.866 & -1 & 1.750 \\ 0 & -1 & 2 \\ 0.500 & -2 & 2.433 \\ 1.366 & -2 & 2.183 \\ 1.366 & -1 & 2.183 \\ 0.500 & -1 & 2.433 \end{bmatrix} \tag{12.27}$$

Remember, we have omitted the z coordinate, so the third column in the transformed matrix now consists of values of the homogeneous coordinate h. Converting these homogeneous points to ordinary points yields

$$\mathbf{p}^* = \begin{bmatrix} 0 & -1 \\ 0.495 & -1.143 \\ 0.495 & -0.571 \\ 0 & -0.500 \\ 0.206 & -0.822 \\ 0.626 & -0.916 \\ 0.626 & -0.458 \\ 0.206 & -0.411 \end{bmatrix} \tag{12.28}$$

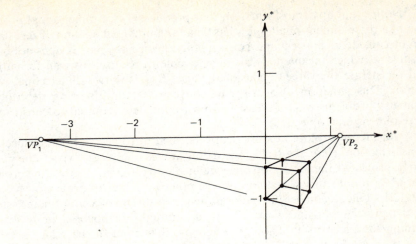

FIGURE 12.14 Perspective projection of a unit cube.

Figure 12.14 shows the projection of this cube. Edges of the cube originally parallel to the x- or z-axis now intersect at one of two vanishing points VP_1 or VP_2. Ordinarily, we think of parallel lines "meeting" at infinity.

Let us see how VP_1 and VP_2 have come to represent points at infinity. The homogeneous coordinates of the points at $x = \infty$, $y = \infty$, and $z = \infty$ are $\begin{bmatrix} 1 & 0 & 0 & 0 \end{bmatrix}$, $\begin{bmatrix} 0 & 1 & 0 & 0 \end{bmatrix}$, and $\begin{bmatrix} 0 & 0 & 1 & 0 \end{bmatrix}$. A value equal to infinity is obtained when the x, y, and z coordinates are divided by the zero-valued homogeneous coordinate. Let us apply the same transformation to these points that we used for the unit cube:

$$\mathbf{p}_h^* = \begin{bmatrix} 1 & 0 & 0 & 0 \\ 0 & 1 & 0 & 0 \\ 0 & 0 & 1 & 0 \end{bmatrix} \begin{bmatrix} 0.866 & 0 & -0.250 \\ 0 & 1 & 0 \\ 0.500 & 0 & 0.433 \\ 0 & -2 & 2 \end{bmatrix}$$

$$= \begin{bmatrix} 0.866 & 0 & -0.25 \\ 0 & 1 & 0 \\ 0.500 & 0 & 0.433 \end{bmatrix}$$

(12.29)

After dividing by the homogeneous coordinate, we find in ordinary coordinates

$$\mathbf{p}^* = \begin{bmatrix} -3.464 & 0 \\ 0 & \infty \\ 1.155 & 0 \end{bmatrix}$$

(12.30)

We interpret this as follows: The point at $x = \infty$ is transformed to $x^* = -3.464$, $y = 0$. This is VP_1 in Fig. 12.14. The point at $y = \infty$ is transformed to $y^* = \infty$, and the point at $z = \infty$ is transformed to $x^* = 1.155$, $y^* = 0$, which is VP_2 in Fig. 12.14.

Notice that any transformation that produces a two-dimensional projection of a three-dimensional object has what amounts to a column of zeros and can be considered a 4×3 matrix. As a consequence, the matrix does not have an inverse, which means that it is not possible to reverse this transformation and regenerate the three-dimensional object from the two-dimensional projection. This situation also arises in conventional drafting and descriptive geometry problems when at least two projections are required to construct a three-dimensional object.

Now, let us explore in more detail the vector geometry of the perspective transformation. Here, the basic ingredients are, as you might expect, the point $\mathbf{p}$ to be projected, the projection plane defined by $\mathbf{p}_0 + x^*\mathbf{u}_1 + y^*\mathbf{u}_2$, the projection $\mathbf{p}^*$ of $\mathbf{p}$ onto this plane, and an eye point or center of projection $\mathbf{p}_e$; see Fig. 12.15. Once again, let the vector $\mathbf{p}_0$ define the origin of a coordinate system on the projection plane, with $\mathbf{u}_1$ and $\mathbf{u}_2$ mutually orthogonal unit vectors defining the plane itself. The unit vector $\mathbf{u}_3$ is normal to the plane, so that $\mathbf{u}_1 \times \mathbf{u}_2 = \mathbf{u}_3$. From Fig. 12.15, we see that

$$\mathbf{p}^* = \mathbf{p}_0 + x^*\mathbf{u}_1 + y^*\mathbf{u}_2 \tag{12.31}$$

Since $\mathbf{p}^*$ is at the intersection of the projection plane and a line joining $\mathbf{p}_e$ and $\mathbf{p}$,

$$\mathbf{p}^* = z^*\mathbf{p} + (1 - z^*)\mathbf{p}_e \tag{12.32}$$

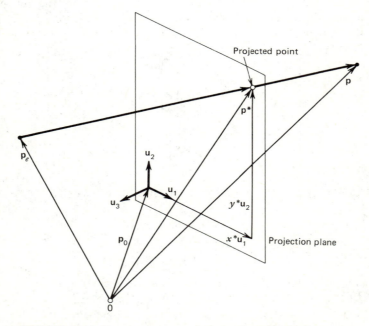

FIGURE 12.15 Vector geometry of perspective projection.

where z^* is some scalar. Combining these two equations, we obtain

$$\mathbf{p}_0 + x^*\mathbf{u}_1 + y^*\mathbf{u}_2 = z^*\mathbf{p} + (1 - z^*)\mathbf{p}_e \tag{12.33}$$

or

$$\mathbf{p}_0 + x^*\mathbf{u}_1 + y^*\mathbf{u}_2 = z^*(\mathbf{p} - \mathbf{p}_e) + \mathbf{p}_e \tag{12.34}$$

Following the example in Eq. 12.14, solve for the coordinates x^*, y^*, and z^* by taking the appropriate scalar product of the equation. To solve for x^*, take the scalar product with $\mathbf{u}_2 \times (\mathbf{p} - \mathbf{p}_e)$; for y^*, use $\mathbf{u}_1 \times (\mathbf{p} - \mathbf{p}_e)$; and for z^*, use $\mathbf{u}_1 \times \mathbf{u}_2$. Doing this for x^* yields

$$x^* = \frac{\mathbf{p}_e \cdot \mathbf{u}_2 \times (\mathbf{p} - \mathbf{p}_e) - \mathbf{p}_0 \cdot \mathbf{u}_2 \times (\mathbf{p} - \mathbf{p}_e)}{\mathbf{u}_1 \cdot \mathbf{u}_2 \times (\mathbf{p} - \mathbf{p}_e)} \tag{12.35}$$

Rewrite this expression using the property of triple vectors—$\mathbf{a} \cdot \mathbf{b} \times \mathbf{c} = \mathbf{a} \times \mathbf{b} \cdot \mathbf{c}$. Also, use the relationship $\mathbf{a} \times \mathbf{b} = -\mathbf{b} \times \mathbf{a}$ and, of course, the distributive properties of scalar and vector products. The reason for this will soon be apparent.

$$x^* = \frac{\mathbf{p}_e \times \mathbf{u}_2 \cdot (\mathbf{p} - \mathbf{p}_e) - \mathbf{p}_0 \times \mathbf{u}_2 \cdot (\mathbf{p} - \mathbf{p}_e)}{\mathbf{u}_1 \times \mathbf{u}_2 \cdot (\mathbf{p} - \mathbf{p}_e)} \tag{12.36}$$

Rearrange terms and simplify to obtain

$$x^* = \frac{(\mathbf{p} - \mathbf{p}_e) \cdot \mathbf{u}_2 \times (\mathbf{p}_0 - \mathbf{p}_e)}{(\mathbf{p} - \mathbf{p}_e) \cdot \mathbf{u}_1 \times \mathbf{u}_2} \tag{12.37}$$

This procedure also produces y^*

$$y^* = \frac{(\mathbf{p} - \mathbf{p}_e) \cdot \mathbf{u}_1 \times (\mathbf{p}_0 - \mathbf{p}_e)}{(\mathbf{p} - \mathbf{p}_e) \cdot \mathbf{u}_2 \times \mathbf{u}_1} \tag{12.38}$$

The computation of z^* is somewhat more direct and produces

$$z^* = \frac{(\mathbf{p}_0 - \mathbf{p}_e) \cdot \mathbf{u}_1 \times \mathbf{u}_2}{(\mathbf{p} - \mathbf{p}_e) \cdot \mathbf{u}_1 \times \mathbf{u}_2} \tag{12.39}$$

At this point we can further simplify by locating the origin of the coordinate system of the projection plane so that a line between it and the center of projection is normal to the plane and at a distance d from it. This geometry is illustrated in Fig. 12.16, where

$$\mathbf{p}_e = \mathbf{p}_0 + d\mathbf{u}_3 \tag{12.40}$$

Again, we construct the unit vectors so that

$$\mathbf{u}_1 = \mathbf{u}_2 \times \mathbf{u}_3$$
$$\mathbf{u}_2 = \mathbf{u}_3 \times \mathbf{u}_1 \tag{12.41}$$
$$\mathbf{u}_3 = \mathbf{u}_1 \times \mathbf{u}_2$$

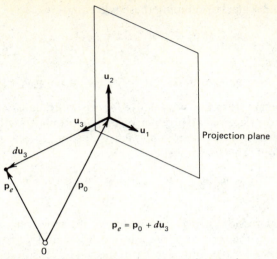

FIGURE 12.16 Locating the center of projection.

We use these relationships in Eqs. 12.37–12.39 to obtain

$$x^* = \frac{-d(\mathbf{p} - \mathbf{p}_0) \cdot \mathbf{u}_1}{(\mathbf{p} - \mathbf{p}_0) \cdot \mathbf{u}_3 - d}$$

$$y^* = \frac{-d(\mathbf{p} - \mathbf{p}_0) \cdot \mathbf{u}_2}{(\mathbf{p} - \mathbf{p}_0) \cdot \mathbf{u}_3 - d} \tag{12.42}$$

$$z^* = \frac{-d}{(\mathbf{p} - \mathbf{p}_0) \cdot \mathbf{u}_3 - d}$$

There is a clear relationship between Eqs. 12.42 and using homogeneous co-ordinates to compute a perspective transformation. Exercise 4 pursues this further.

EXERCISES

1. Write the transformation matrix for a perspective projection onto the $x = 0$ plane; onto the $y = 0$ plane.

2. What transformations will produce a perspective projection onto the $z = z_c$ plane, where z_c is a constant not equal to zero?

3. The transformation matrix

$$\mathbf{T}_h = \begin{bmatrix} 1 & 0 & 0 & p \\ 0 & 1 & 0 & q \\ 0 & 0 & 1 & r \\ 0 & 0 & 0 & 1 \end{bmatrix}$$

produces a **three-point perspective**; show that it maps $x = \infty$ into $x^* = 1/p$, $y = \infty$ into $y^* = 1/q$, and $z = \infty$ into $z^* = 1/r$. Perform this transformation on the unit cube rotated $\theta = 60°$ around the y-axis, with $l, m, n = 2$; $1/p = 4$,

$1/q = 8$, $1/r = 4$. Find the $z^* = 0$ projection, and accurately plot it on graph paper.

4. Show the relationship between Eqs. 12.42 and using homogeneous coordinates to compute a perspective transformation. (*Hint*: Notice that the denominator in all three equations is the same.) Express the relationship as the sequence of transformation matrices required to produce the proper orientation and projection.

5. Write a procedure to compute the perspective projection of a set of points onto an arbitrary plane and from an arbitrary view point. Denote this as **PRSPRJ**(NP, PT, VP, PL, PR), where

NP is the input number of points to be projected;

PT(NP, 3) is the input array of coordinates of the points to be projected;

VP(3) is the input array of coordinates of the view point;

PL(4) is the input array of projection-plane coefficients of A, B, C, D; and

PR(NP, 3) is the output array of coordinates of the projected points.

6. Show that the homogeneous perspective transformation can be applied to the defining geometric coefficients (matrix) of curves and surfaces (Hermite, Bezier, and *B*-spline) and the results define curves and surfaces that can then be projected orthographically to produce the desired perspective projection. (*Note*: In this case, we preserve the z coordinates.) This dramatically reduces the number of transformation computations required to plot an object.

2.3 The Photographic Projection

An interesting comparison can be made between perspective projections and optical projections. From elementary optics, we have the **thin-lens equation**

$$\frac{1}{z^*} + \frac{1}{z} = \frac{1}{f} \tag{12.43}$$

where f is the focal length of the lens, z^* is the location of the image point, and z is the location of the object point; see Fig. 12.17. We can rearrange Eq. 12.43 to obtain

$$z^* = \frac{fz}{z - f} \tag{12.44}$$

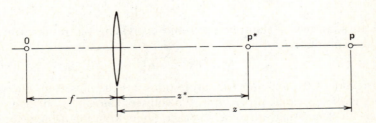

FIGURE 12.17 Thin lens geometry.

Clearly, this is similar to the results obtained for perspective projections. To produce a photographic projection, we modify the $\mathbf{T}_h$ matrix and rewrite Eq. 12.22:

$$\mathbf{p}_h^* = \mathbf{p}^* \begin{bmatrix} 1 & 0 & 0 & 0 \\ 0 & 1 & 0 & 0 \\ 0 & 0 & 1 & 1/f \\ 0 & 0 & 0 & -1 \end{bmatrix} \qquad (12.45)$$

A photographic projection preserves the third coordinate, usually lost in a perspective projection. Also, a photographic projection transforms from three dimensions into three dimensions, whereas a perspective projection transforms from three dimensions into two dimensions.

EXERCISES

1. How do photographic projections appear on "flat" film?

2. Show how to use the photographic transformation to provide a magnifying effect.

2.4   Axonometric Projection

Axonometric projections are classic methods of descriptive geometry used by engineers and draftsmen to simulate three-dimensionality in their drawings. The three most common types are trimetric, dimetric, and isometric projections. In an isometric projection, the coordinate system is oriented so that all three axes are equally foreshortened when projected, and, therefore, the same scale or measure ("isometric") can be used for all edges or lines parallel to these axes. In a dimetric projection, two of the three axes are equally foreshortened when projected. In a trimetric projection, all three axes are usually unequally foreshortened.

If the center of projection is placed at infinity (that is, $1/\lambda = 0$), then the perspective projection becomes an axonometric projection. The axonometric projection of a point $\mathbf{p}$ onto the $z = n$ plane is given by

$$\mathbf{p}_h^* = \mathbf{p}_h \begin{bmatrix} 1 & 0 & 0 & 0 \\ 0 & 1 & 0 & 0 \\ 0 & 0 & 0 & 0 \\ 0 & 0 & n & 1 \end{bmatrix} \qquad (12.46)$$

Performing matrix multiplication and expressing the result in ordinary coordinates produces

$$\mathbf{p}^* = [x \quad y \quad n] \qquad (12.47)$$

demonstrating that axonometric projections are a subset of the general category of orthographic projections. We will see that combined rotations are the basis for generating axonometric projections. Rotation around the y-axis followed by rotation around the *x*-axis is produced by the following matrix product:

$$\mathbf{p}_h^* = \mathbf{p}_h \begin{bmatrix} \cos\beta & 0 & -\sin\beta & 0 \\ 0 & 1 & 0 & 0 \\ \sin\beta & 0 & \cos\beta & 0 \\ 0 & 0 & 0 & 1 \end{bmatrix} \begin{bmatrix} 1 & 0 & 0 & 0 \\ 0 & \cos\theta & \sin\theta & 0 \\ 0 & -\sin\theta & \cos\theta & 0 \\ 0 & 0 & 0 & 1 \end{bmatrix} \quad (12.48)$$

Performing matrix multiplication yields

$$\mathbf{p}_h^* = \mathbf{p}_h \begin{bmatrix} \cos\beta & \sin\beta\sin\theta & -\sin\beta\cos\theta & 0 \\ 0 & \cos\theta & \sin\theta & 0 \\ \sin\beta & -\cos\beta\sin\theta & \cos\beta\cos\theta & 0 \\ 0 & 0 & 0 & 1 \end{bmatrix} \quad (12.49)$$

Apply this combined rotation matrix to a unit vector on the *x*-axis, $[1 \quad 0 \quad 0 \quad 1]$, to obtain

$$\mathbf{p}_h^* = [\cos\beta \quad \sin\beta\sin\theta \quad -\sin\beta\cos\theta \quad 1] \quad (12.50)$$

(Notice that points or vectors are dealt with in an identical manner.) Convert to ordinary coordinates and denote the result as $\mathbf{p}_1^*$

$$\mathbf{p}_1^* = [\cos\beta \quad \sin\beta\sin\theta \quad -\sin\beta\cos\theta] \quad (12.51)$$

Repeat this process for a unit vector on the *y*-axis, $[0 \quad 1 \quad 0 \quad 1]$, and denote the result as $\mathbf{p}_2^*$

$$\mathbf{p}_2^* = [0 \quad \cos\theta \quad \sin\theta] \quad (12.52)$$

Ignore the *z* coordinate of these two transformed vectors, since the projections are to be developed on the $z^* = 0$ plane. The magnitudes of these transformed vectors on the plane of projection are

$$|\mathbf{p}_1^*| = \sqrt{\cos^2\beta + \sin^2\beta\sin^2\theta} \quad (12.53)$$

and

$$|\mathbf{p}_2^*| = \sqrt{\cos^2\theta} = \cos\theta \quad (12.54)$$

$|\mathbf{p}_1^*|$ and $|\mathbf{p}_2^*|$ must be equal to produce a dimetric projection. So $|\mathbf{p}_1^*| = |\mathbf{p}_2^*|$, or

$$\cos^2\beta + \sin^2\beta\sin^2\theta = \cos^2\theta \quad (12.55)$$

or

$$\sin^2\beta = \frac{\sin^2\theta}{1 - \sin^2\theta} \quad (12.56)$$

We can now select a value for θ and using Eq. 12.56, easily compute β, which will produce the equal foreshortening required by the dimetric projection. An interesting way of doing this involves shortening the *z*-axis by some specified ratio. First, transform the unit vector on the *z*-axis, $[0 \quad 0 \quad 1 \quad 1]$, by the rotation matrix in Eq. 12.49

$$\mathbf{p}_h^* = [\sin\beta \quad -\cos\beta\sin\theta \quad \cos\beta\cos\theta \quad 1] \quad (12.57)$$

Convert Eq. 12.57 into ordinary coordinates, and denote the result as $\mathbf{p}_3^*$

$$\mathbf{p}_3^* = [\sin \beta \quad -\cos \beta \sin \theta \quad \cos \beta \cos \theta] \qquad (12.58)$$

Find $|\mathbf{p}_3^*|$, again ignoring the z coordinate

$$|\mathbf{p}_3^*| = \sqrt{\sin^2 \beta + \cos^2 \beta \sin^2 \theta} \qquad (12.59)$$

Now, for example, let the transformation foreshorten the z unit vector by 1/2

$$\sin^2 \beta + \cos^2 \beta \sin^2 \theta = (\tfrac{1}{2})^2 \qquad (12.60)$$

Again, applying a simple trigonometric identity, this becomes

$$\sin^2 \beta + (1 - \sin^2 \beta) \sin^2 \theta = \tfrac{1}{4} \qquad (12.61)$$

Substituting from Eq. 12.56, we obtain

$$8 \sin^4 \theta - 9 \sin^2 \theta + 1 = 0 \qquad (12.62)$$

which has roots of $\sin^2 \theta = \pm 1/8$ and ± 1. If $\sin \theta = \sqrt{1/8}$, then $\sin \beta = [(1/8)/(1 - 1/8)]^{1/2}$, and the dimetric projection transformation matrix is fully determined.

An isometric projection requires that all three coordinate axes be equally foreshortened; this occurs when

$$\cos^2 \beta + \sin^2 \beta \sin^2 \theta = \cos^2 \theta \qquad (12.63)$$

and

$$\sin^2 \beta + \cos^2 \beta \sin^2 \theta = \cos^2 \theta \qquad (12.64)$$

We have already simplified the first of these two equations, and we can rewrite the second as

$$\sin^2 \beta = \frac{1 - 2 \sin^2 \theta}{1 - \sin^2 \theta} \qquad (12.65)$$

If $\sin^2 \theta = 1/3$, then

$$\sin^2 \beta = \tfrac{1}{2} \qquad (12.66)$$

and

$$\cos^2 \beta = 1 - \sin^2 \beta = \tfrac{1}{2} \qquad (12.67)$$

and the angle α that the projected x-axis makes with the horizontal is

$$\alpha = \tan^{-1} \frac{\sin \beta \sin \theta}{\cos \beta} \qquad (12.68)$$

$$= \tan^{-1} \left(\frac{\sqrt{3}}{3} \right)$$

$$= 30°$$

which should be familiar to anyone with classical pencil-and-paper drafting experience, since two of the major orthogonal elements of an isometric projection make angles of 30° with the horizontal; see Fig. 12.18.

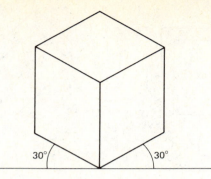

Isometric projection of a cube

FIGURE 12.18 Isometric projection geometry.

3 SCENE TRANSFORMATION

We can generate the view displayed of a geometric model by applying a sequence of transformation matrices to the defining point set of the model. The sequence may consist of a set of matrices establishing the view orientation, followed by the projection-transformation matrix. We can control view changes with algorithms that create or alter these matrices.

We must distinguish between **scene transformations** and **display transformations**. Scene transformations are characteristically three-dimensional and operate on the model data to alter the viewing orientation. Display transformations operate on the two-dimensional display data to change scale (zoom, for example) or rotate the display around the line of sight (in the plane of the display), for example. Display changes do not affect the projected view—only the viewer's relationship to the plane of the display. Figure 12.19 is an example of this difference.

We must also consider the relationship between the observer and the scene. Should the object or objects in the scene move, or should the eye move? Should the transformations be relative to a fixed global axis or relative to the current display axis? Once we transform and display an object, our next view of that object is produced by assuming that either the object itself moves or our eye moves. The difference is expressed by the algebraic sign of each of the elements of the transformation matrix. These issues deserve particular attention when designing the user interface of interactive geometric modeling systems. McDonnell Douglas's CADD and Townsend Microware's IMAGINATOR, the former a mainframe program and the latter a microcomputer three-dimensional graphics program, offer excellent examples of scene-transformation procedures.

We will explore several kinds of scene transformations: orbit, pan, and aim (as exemplified by IMAGINATOR), and scene fly-through. First, let us

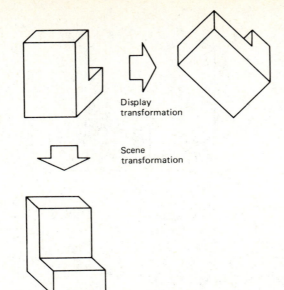

FIGURE 12.19 The distinction between scene and display transformations.

briefly review again the 4 × 4 transformation matrix $\mathbf{T}_h$ that applies to points given by homogeneous coordinates.

$$\mathbf{T}_h = \begin{bmatrix} a & b & c & p \\ d & e & f & q \\ g & i & j & r \\ l & m & n & h \end{bmatrix} \qquad (12.69)$$

Recall the effects produced by the elements in the four partitions of this matrix

Rotation

$$\begin{bmatrix} a & b & c & \vdots & 0 \\ d & e & f & \vdots & 0 \\ g & i & j & \vdots & 0 \\ \hdashline 0 & 0 & 0 & \vdots & 1 \end{bmatrix}$$

Scaling

$$\begin{bmatrix} s_x & 0 & 0 & \vdots & 0 \\ 0 & s_y & 0 & \vdots & 0 \\ 0 & 0 & s_z & \vdots & 0 \\ \hdashline 0 & 0 & 0 & \vdots & 1 \end{bmatrix}$$

Translation

$$\begin{bmatrix} 1 & 0 & 0 & \vdots & 0 \\ 0 & 1 & 0 & \vdots & 0 \\ 0 & 0 & 1 & \vdots & 0 \\ \hdashline t_x & t_y & t_z & \vdots & 1 \end{bmatrix}$$

Projection

$$\begin{bmatrix} 1 & 0 & 0 & \vdots & p \\ 0 & 1 & 0 & \vdots & q \\ 0 & 0 & 1 & \vdots & r \\ \hdashline 0 & 0 & 0 & \vdots & 1 \end{bmatrix}$$

We define or model objects in what we will call the **world coordinate** system We establish another coordinate system relative to the line of sight, with two axes corresponding to the x and y axes of a display screen lying in the projection plane and the third corresponding to the z-axis that we construct normal to the projection plane. This is called the **eye coordinate** system. Ordinarily, it is a left-handed system, so that on a display screen, the normal relationship between the x and y axes will be preserved: the positive x-axis pointing to the right and the positive y-axis pointing upward. This means that the positive z-axis points along the line of sight toward the object being viewed. The farther a point is from the viewer, the higher the value of its z coordinate.

The world coordinate is usually a right-handed one, so the transformation of object points to the eye coordinate system must also transform a right-handed to a left-handed system. We do this easily with a matrix that simply inverts the sign of the z coordinate

$$
\mathbf{T}_{RL} =
\begin{bmatrix}
1 & 0 & 0 & 0 \\
1 & 1 & 0 & 0 \\
0 & 0 & -1 & 0 \\
0 & 0 & 0 & 1
\end{bmatrix}
\tag{12.70}
$$

Suppose we must view an object from an arbitrary position in space. We interpret this to mean that the object remains "fixed" in its world coordinate system and our eye moves from point to point in space around the object. First, define the eye point, $\mathbf{p}_e$. Next, define a point toward which we are looking, $\mathbf{p}_v$; we will call this the view point. These two points define the line-of-sight vector $(\mathbf{p}_v - \mathbf{p}_e)$ and the colinear z direction, denoted by the unit vector $\mathbf{u}_3$, of the local eye coordinate system. (This system and the following vector geometry are illustrated in Fig. 12.20.) The projection plane is normal to the line of sight at

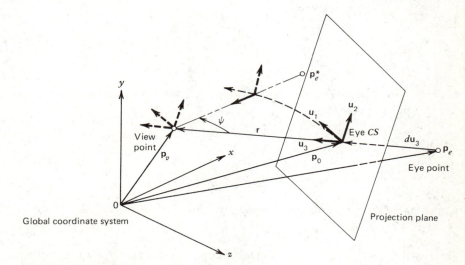

FIGURE 12.20 Orbit transformations.

$\mathbf{p}_0$, and the unit vectors $\mathbf{u}_1$ and $\mathbf{u}_2$ lying in it (corresponding to the x and y axes of a display screen) form an orthogonal triad with $\mathbf{u}_3$. We must specify our distance d from the projection plane. This means that we do not have to specify $\mathbf{p}_0$, since we can calculate it from the other inputs as follows:

$$\mathbf{p}_0 = \mathbf{p}_e + d\mathbf{u}_3 \tag{12.71}$$

Notice that the distance d may be a positive number, although the coordinates of the eye point in the eye-point system are interpreted as $(0, 0, -d)$. So we must first specify only $\mathbf{p}_e$, $\mathbf{p}_v$, and d.

Transformations to produce the initial view might proceed as follows: Construct a set of three orthogonal unit vectors $\mathbf{u}_1$, $\mathbf{u}_2$, $\mathbf{u}_3$ at the origin of the global coordinate system, aligned with the principal positive x, y, and z axes, respectively. Then "translate" them to $\mathbf{p}_v$ by means of $\mathbf{T}_v$, given by

$$\mathbf{T}_v = \begin{bmatrix} 1 & 0 & 0 & 0 \\ 0 & 1 & 0 & 0 \\ 0 & 0 & 1 & 0 \\ -x_v & -y_v & -z_v & 1 \end{bmatrix} \tag{12.72}$$

Notice the negative sign on x_v, y_v, and z_v. Since we are transforming or "moving" coordinate systems rather than moving the object, the negative signs are appropriate. Next, apply the transformation given in Eq. 12.70 to convert to a left-handed system. Then compute the angles of rotation that will align $\mathbf{u}_3$ with $(\mathbf{p}_v - \mathbf{p}_e)$, and apply the required rotation transformations to all three unit vectors to maintain their mutual orthogonal relationship. Usually, only two rotations are necessary, one around the x-axis and one around the y-axis, say $\mathbf{T}_\theta$ and $\mathbf{T}_\phi$. (*Caution*: Remember to multiply θ and ϕ by -1, since this is a co-ordinate-system transformation.) Then translate $[\mathbf{u}_1 \quad \mathbf{u}_2 \quad \mathbf{u}_3]$ to $\mathbf{p}_0$, which is simply given by

$$\mathbf{T}_0 = \begin{bmatrix} 1 & 0 & 0 & 0 \\ 0 & 1 & 0 & 0 \\ 0 & 0 & 1 & 0 \\ 0 & 0 & r & 1 \end{bmatrix} \tag{12.73}$$

where $r = |\mathbf{p}_v - \mathbf{p}_0|$. We are now ready to project the object, and we accomplish this by applying the matrix

$$\mathbf{T}_P = \begin{bmatrix} 1 & 0 & 0 & 0 \\ 0 & 1 & 0 & 0 \\ 0 & 0 & 1 & 1/d \\ 0 & 0 & 0 & 1 \end{bmatrix} \tag{12.74}$$

Concatenate these various transformations into a single expression

$$\mathbf{p}^* = \mathbf{p}\mathbf{T}_v\mathbf{T}_{RL}\mathbf{T}_\theta\mathbf{T}_\phi\mathbf{T}_0\mathbf{T}_P \tag{12.75}$$

Or treat $[\mathbf{u}_1 \quad \mathbf{u}_2 \quad \mathbf{u}_3]$ as an object, and apply these transformations to them with appropriate changes in the signs of the matrix elements and excluding

$\mathbf{T}_{RL}$, producing $[\mathbf{u}_1 \quad \mathbf{u}_2 \quad \mathbf{u}_3]^*$. Using this transformed set, $\mathbf{p}_0$, and d, apply Eq. 12.42 directly to the object points to be projected.

We can generate subsequent views by **orbiting** around point $\mathbf{p}_v$, creating **orbit-scene transformations**. This means that the line of sight is always toward $\mathbf{p}_v$, although we may vary both d and r. For example, if we orbit "to the right" relative to the eye coordinate system (in the direction of $\mathbf{u}_1$), then the eye point rotates about $\mathbf{p}_v$, say, through an angle ψ, in the plane defined by $\mathbf{u}_1$ and $\mathbf{u}_3$. This orbital motion can be described several ways, each leading to another set of transformation matrices appropriate for the new view.

Consider first rotating $\mathbf{p}_e$ around an axis in the direction of $\mathbf{u}_2$ and through $\mathbf{p}_v$ to produce $\mathbf{p}_e^*$, a new eye point. Repeat all the steps just outlined using $\mathbf{p}_e^*$ to create the new view. Or translate $[\mathbf{u}_1 \quad \mathbf{u}_2 \quad \mathbf{u}_3]$ back into $\mathbf{p}_v$. We need the following revised matrix $\mathbf{T}_v$ to perform this transformation in terms of the eye coordinate system:

$$\mathbf{T}_v = \begin{bmatrix} 1 & 0 & 0 & 0 \\ 0 & 1 & 0 & 0 \\ 0 & 0 & 1 & 0 \\ 0 & 0 & -r & 1 \end{bmatrix} \tag{12.76}$$

Again, this is the first of a set of matrices to be concatenated to produce a net coordinate system transformation, which is applied to the appropriate set of object points. Since it is a coordinate-system transformation, we use $-r$. Next, rotate the $[\mathbf{u}_1 \quad \mathbf{u}_2 \quad \mathbf{u}_3]$ system through angle ψ around $\mathbf{u}_2$. (The example in Fig. 12.20 shows this to be equal to rotating $\mathbf{u}_1$ into $\mathbf{u}_3$.) Finally, translate the resulting system a distance r along the new line of sight (same as Eq. 12.76 but with $+r$). Now, we can calculate a projection.

This orbiting type of eye point movement can, of course, include orbiting up or down by rotating around an axis through $\mathbf{p}_v$ and normal to the plane of $\mathbf{u}_2$ and $\mathbf{u}_3$ as well as variations in r and d. We can also rotate the view around $\mathbf{u}_3$, which does not change the projection. If we track the latest $\mathbf{p}_e^*$ relative to the global coordinate system and retain the initial eye point, then we can compute a "home" transformation that will restore the initial view. Notice that we can orbit to the far side of an object in order to obtain a rear view.

We produce **pan-scene transformations** from the initial display by equal vector translations of both the eye point $\mathbf{p}_e$ and the view point $\mathbf{p}_v$. We can pan right, left, up, or down. Figure 12.21 shows a pan right a distance c relative to $[\mathbf{u}_1 \quad \mathbf{u}_2 \quad \mathbf{u}_3]$. We can easily calculate the new eye point $\mathbf{p}_e^*$ and view point $\mathbf{p}_v^*$ from

$$\mathbf{p}_e^* = \mathbf{p}_e + c\mathbf{u}_1$$
$$\mathbf{p}_v^* = \mathbf{p}_v + c\mathbf{u}_1 \tag{12.77}$$

We compute the other scene-transformation components just as easily and thereby complete the input requirements for a subsequent projection transformation. Notice that $[\mathbf{u}_1 \quad \mathbf{u}_2 \quad \mathbf{u}_3]$ is translated parallel to its initial orientation, which also means that the projection plane is similarly translated to the new viewing position.

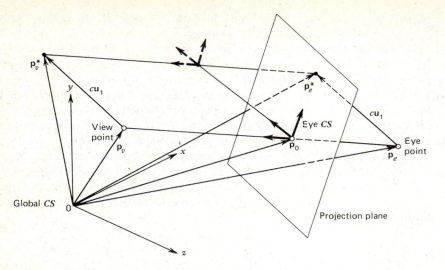

FIGURE 12.21 Pan transformations.

Figure 12.22 shows the effect of the pan transformations on the projection of a simple object. A cross marks the location of the view point relative to the object in each scene. The eye point is located some arbitrary distance away from the projection plane, the page, and on the normal through the view point. Note that the pan transformation can never generate a view of the rear of the object: A large enough pan motion will move the object out of the display in the opposite direction.

We produce **aim-scene transformations** by moving the view point while maintaining the eye point $\mathbf{p}_e$ in a fixed location. This is analogous to a person looking to the right or left, up or down, forward or backward, or any rotation in between, to view the surroundings. Relocate the view point by a sequence of rotations of the line of sight, $(\mathbf{p}_v - \mathbf{p}_e)$, around $\mathbf{p}_e$. Figure 12.23 shows the elements of this transformation. After constructing the initial display, define an aim movement by specifying an axis or direction of rotation and an angle. The example in Fig. 12.23 specifies a rotation ψ around the $\mathbf{u}_2$-axis. Thus, we determine new scene-transformation components by first translating $[\mathbf{u}_1 \quad \mathbf{u}_2 \quad \mathbf{u}_3]$ into $\mathbf{p}_e$ and then, in the case of the example, rotating these unit vectors and $(\mathbf{p}_v - \mathbf{p}_e)$ around $\mathbf{u}_2$ through an angle ψ. After this rotation, reverse the translation of $[\mathbf{u}_1 \quad \mathbf{u}_2 \quad \mathbf{u}_3]$, and all the vector elements are now in place for a new projection.

We usually need aim-scene transformations to develop views from an eye point position in the midst of a partially dispersed group of objects; this is also true for scene fly-through. In this case, a sequence of views is predetermined by defining a path or trajectory curve along which the eye point is to move as well as a varying line of sight coordinated with position on the path. Figure 12.24 shows the geometric elements of scene fly-through.

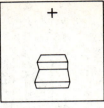

(a)

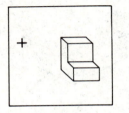

(b)

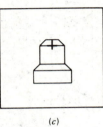

(c)

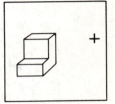

(d)

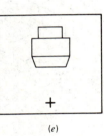

(e)

FIGURE 12.22 The effects of pan transformations.

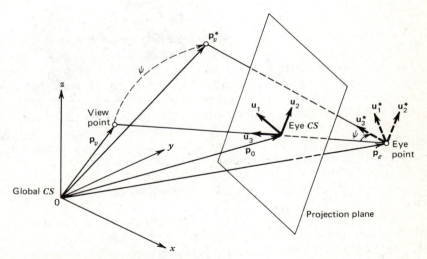

FIGURE 12.23 Aim transformations.

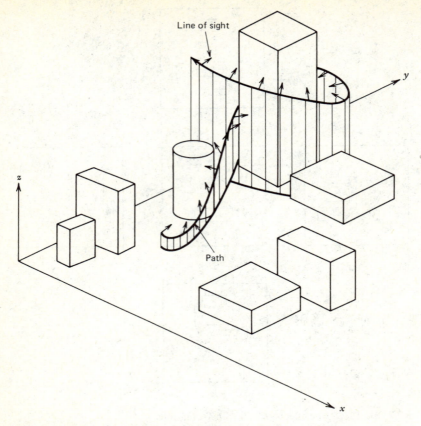

FIGURE 12.24 Scene fly-through transformation.

We can use a parametric curve to define the path we want through a group of objects. Then, we can use another parametric curve to define the set $[\mathbf{u}_1 \quad \mathbf{u}_2 \quad \mathbf{u}_3]$ that determines both the line of sight and the projection plane. The two curves are coordinated by equal values of their parametric variables. Lossing and Eshleman's six-component position-direction curve in Chapter 10, Section 5 offers an excellent way of performing this function. We can construct a new scene at specified equal increments of the parametric variable; we might also easily devise other "rules" for producing successive views. Animation requires some control of the time between increments to create the effect of fast, slow, or varying speed along the path. Another parametric component (a seventh component for a position-direction curve, for example), can be used as a timing device.

K. L. Shelley and D. P. Greenberg (1982) use a *B*-spline curve to define a path through a scene and another *B*-spline curve to define timing along the path. The line of sight is either automatically the tangent vector to the path or is specified interactively.

EXERCISES

1. Write a procedure to compute the net transformation matrix for a scene, given the eye and view points and distance from the plane of projection. Assume perspective projection. Denote this as **ORBIT**(PE, PV, D, T), where

> PE(3) is the input array of eye point coordinates;
>
> PV(3) is the input array of view point coordinates;·
>
> D is the input distance from the plane of projection; and
>
> T(4, 4) is the output array of elements of the transformation matrix.

2. Write a procedure to compute a sequence of orbit transformations from eye points lying on a *B*-spline curve. Assume that the view point and distance are fixed. Incorporate and invoke **ORBIT**, defined in Ex. 1, as necessary. Denote this as **PATH**(NP, P, K, U, PV, D, T), where

> NP is the input number of control points defining the B-spline curve;
>
> P(NP, 3) is the input array of control-point coordinates;
>
> K is the input degree of the B-spline curve;
>
> U is the input value of the parametric variable at the eye point;
>
> PV(3) is the input array of view point coordinates;
>
> D is the input distance from the plane of projection; and
>
> T(4, 4) is the output array of elements of the transformation matrix.

3. Write a procedure to compute a timed sequence of view points along a path. Assume that a second *B*-spline curve is used to define the relationship between time *T* and position on the path *U* (see **PATH** in Ex. 2).

4 SILHOUETTE CURVES

The problem of computing the silhouette curve of a solid object divides itself naturally into two cases, depending on the type of bounding surface. The simplest case is a solid bounded by plane polygonal faces. A more interesting case is a solid bounded by a curved surface.

A cube is a good example of the first case. Let us analyze the faces and edges of one and try to determine which geometric properties make some faces and edges visible and others not. Figure 12.25 illustrates a cube within the usual left-handed eye point coordinate system. A unit normal vector is associated with each of the cube's faces. In Fig. 12.25*b*, we see that each normal can be resolved into two components, one normal to the projection plane and one parallel to it. Those faces whose normal has a component in the $+z$ direction, which is pointing away from us, are not visible. Those faces whose normal has a component in the $-z$ direction, pointing toward us, or has a z component equal to zero, are visible and so are their bounding edges. This is also a solution

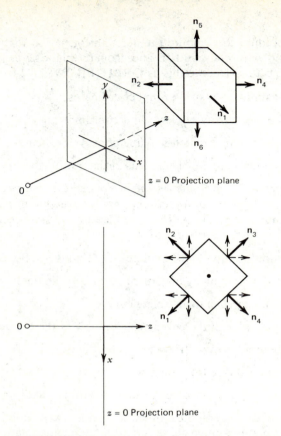

FIGURE 12.25 Face normals and the projection plane.

to the hidden-surface problem for a plane-faced convex polyhedral object not obscured by another object.

The next and final step is to find those visible edges that are part of the silhouette. Six of the nine visible edges of the cube shown in Fig. 12.26 define the silhouette; three of the edges do not contribute to its definition. Is there a mathematically testable difference between the two types of visible edges? Yes, each edge contributing to the silhouette has one visible adjacent face and one not visible. Notice that the three visible edges not contributing to the silhouette are each adjacent to two visible faces. Thus, we can design an algorithm that first determines the visibility of each face of a polyhedron for a particular view and then find the set of edges adjacent to one visible face and one invisible face.

Most objects, however, have some curved surfaces. Given an arbitrary well-behaved curved surface that is part of the total bounding surface of a solid, we determine if it is entirely visible, partially visible, or not visible. If a surface

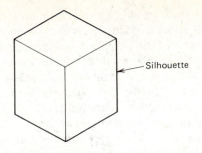

FIGURE 12.26 Silhouette edges of a polyhedron.

is only partially visible, then there is a curve on it that is part of the silhouette curve. The silhouette curve of a surface is that curve along which the z components of surface normals are zero. This, of course, assumes that the eye point is on the z-axis and that the surface normals point outward from the solid. Figure 12.27 illustrates the geometry of this condition.

To find the silhouette curve on a surface, we must find a more convenient way of expressing the normal vector at any point other than by computing $\mathbf{p}^u \times \mathbf{p}^w$ at each point of interest. For a bicubic surface, we have seen that the expression for the normal at any point is a quintic polynomial in u and w for each component (Chapter 3, Section 3). We learned that we can approximate this expression by a cubic polynomial in the absence of major surface undulations (in which case, we use subdivision). This approximation makes computations simpler and allows us to use many of the same algorithms for analyzing position. D. Schweitzer and E. S. Cobb (1982) present a derivation of this cubic approximation using patch corner point data similar to that developed in Chapter 3, Section 3. We will now look at a somewhat different derivation.

Given a bicubic surface, compute the unit normal vector at each of 16

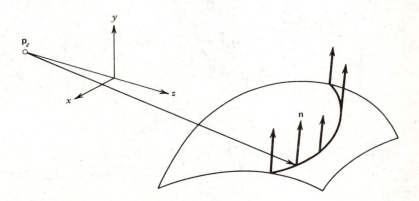

FIGURE 12.27 Silhouette curve on a surface.

points at u, $w \in [0: 1/3: 2/3: 1]$. Consider the unit normal vectors themselves to be points on some surface that we, in turn, represent as a bicubic surface by the equations for the 16-point form developed in Chapter 3, Section 8. This surface patch, call it the normals patch, lies on a unit sphere centered at the origin. (It should take only a little mental exercise to convince yourself that this is so.) Each u, w point on this normals patch yields the normal vector components at the same u, w point on the original surface. Assuming the line of sight is along the z-axis (the initial surface must be so transformed if necessary), then the intersection between this surface representing unit normals and the $z = 0$ plane is a curve on which the unit normals have a zero-valued z component. It is, in fact, the silhouette curve.

Since there is a one-to-one correspondence of u, w points between the surface representing the normals and the original surface comprising part of a solid's boundary, we readily plot this silhouette curve on the original surface. Just as easily, we plot this curve's x, y coordinates (see Chapter 3, Section 20). Figure 12.28 shows a schematic of this procedure.

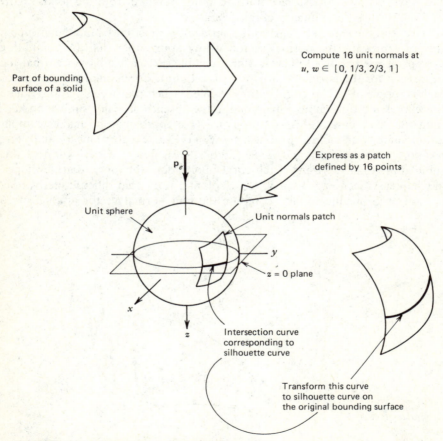

Part of bounding surface of a solid

Compute 16 unit normals at u, $w \in [0, 1/3, 2/3, 1]$

Express as a patch defined by 16 points

$\mathbf{p}_e$

Unit sphere

Unit normals patch

y

$z = 0$ plane

x

z

Intersection curve corresponding to silhouette curve

Transform this curve to silhouette curve on the original bounding surface

FIGURE 12.28 Computing a silhouette curve.

For the simple case of a convex solid with a nonplanar bounding surface, the complete set of normals patches will just cover the unit sphere. The intersection of these patches with the $z = 0$ plane will produce the total silhouette curve for the solid in a particular orientation to the line of sight. We can use techniques in Chapter 7 to produce the intersection-curve segments and string them together to form the complete and closed silhouette curve.

Notice that the normals patch for a plane face degenerates to a point. The normals patches of the six faces of a cube map onto the surface of the unit sphere as six points terminating six mutually orthogonal or colinear radii. If we proceed to intersect this set with the $z = 0$ plane, we are not likely to generate much of a curve. At best, we might find that the plane passes through two or four of the six points. However, we can say that the points on the unit sphere lying on the same side of the intersecting plane as the viewer represent the cube's visible faces. This, of course, is nothing more than another way of determining which face normals have z components pointing toward the viewer.

Certainly, many solids are bounded by surfaces comprised of both curved and planar faces. In this case, we treat the solid as a polyhedron with (potentially) curved faces and edges. We can compute the silhouette curves for those faces that are curved and subsequently considered to be new edges, dividing the faces in which they lie into two faces (one marked visible, the other marked hidden). The planar faces do not have to be subdivided; they are also marked visible or hidden. We now use the method described earlier of testing the faces adjacent to each edge. Edges adjacent to both a visible face and a hidden face are part of the solid's silhouette curve.

There are many approaches to this problem and many other complicating factors that we have not explored here: computing the silhouette of a solid with holes or with extreme folding or undulation of the shape (for example, tori or corrugated sheets). The hidden-surface problem obviously greatly overlaps the silhouette problem; the more complicated the shape of the solid, the greater the overlap. Research continues in this area, with many solution paths open to improvement and refinement. What we need are demonstrably complete solutions that are efficient to process.

EXERCISES

1. Given a bicubic patch (Hermite) whose 16-point form is

$$\mathbf{P} = \begin{bmatrix} (1,1,0) & (1,4,1) & (1,7,1) & (1,10,10) \\ (4,1,1) & (4,4,3) & (4,7,3) & (4,10,1) \\ (7,1,1) & (7,4,3) & (7,7,3) & (7,10,1) \\ (10,1,0) & (10,4,1) & (10,7,1) & (10,10,0) \end{bmatrix}$$

find its complete silhouette curve when viewed from $\mathbf{p}_e = \begin{bmatrix} -1 & -1 & 4 \end{bmatrix}$ toward $\mathbf{p}_v = \begin{bmatrix} 2 & 2 & 3 \end{bmatrix}$.

2. Repeat Ex. 1 for a cubic Bezier surface, using the same control points.

3. Repeat Ex. 1 for a cubic *B*-spline surface, again using the same control points.

4. Write a procedure to compute the silhouette curve of a bicubic patch when seen from a specified point in space. Denote this as **SILOET**(PE, PV, D, BC, NSEG, CRV), where

PE(3) is the input array of eye point coordinates;

PV(3) is the input array of view point coordinates;

D is the input distance of the eye point;

BC(4, 4, 3) is the input array of geometric coefficients defining the bicubic surface;

NSEG is the output number of segments of the silhouette curve; and

CRV(NSEG, 4, 3) is the output array of geometric coefficients of the silhouette curve. (***Note***: The geometric coefficients should be of the silhouette curve in the plane of projection.)

5   HIDDEN SURFACES

To produce a realistic display of a solid object, we must determine which curves and surfaces can be seen from a particular point of view and which cannot. This not only enhances the realism of a display but also removes any ambiguities, as we see in the example in Fig. 12.29. This is the **hidden-line** or **hidden-surface** problem. It has historically been and continues to be a very popular and a very important problem: popular perhaps because of the challenge of solving a difficult problem and important because, aside from the need for a realistic display, the solution algorithm must be computationally fast and efficient.

The hidden-surface problem obtained a solid theoretical foundation with the publication by I. E. Sutherland, R. F. Sproull, and R. A. Schumacker in 1974 of a comparative study of several hidden line and surface algorithms. The authors found that solution algorithms can be classified according to the frame of reference of the geometric computations: object space, image space, or some of both. This distinction will become clear as we study the subject. Furthermore, it happens that the hidden-line problem is associated with vector-type display devices, while the hidden-surface problem is associated with raster scan displays.

A rich and readily available literature exists on this aspect of computer graphics. Textbooks introducing this subject include those by J. D. Foley and A. Van Dam (1982) and W. M. Newman and R. F. Sproull (1979). The total problem abounds with many nongeometric and quasi-geometric aspects, including computational complexity and efficiency. Sorting is generally recognized as the major computational cost. Although it is carried out with geometric data, design and refinement of sorting algorithms are strictly nongeometric in nature. To present all of these subjects would require a separate volume,

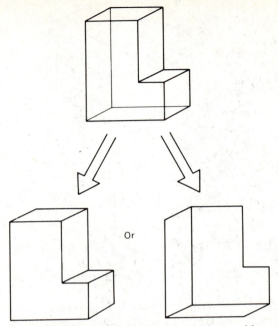

FIGURE 12.29 The hidden line and surface problem.

which is not the case here, therefore, we give only a broad introduction to the subject, problem formulation, and major geometric aspects.

Many geometric parameters contribute to the scope and complexity of the problem. How many distinct objects are there? What is their spatial distribution? Are the objects simple convex shapes or concave with multiply-connected bounding surfaces? Are the bounding surfaces planar or nonplanar? Is the display to be constructed from an eye point within a group of objects or from an external eye point? Is the display dynamic? Do all or only some of the objects move? Does the eye point change dynamically? Some algorithms solve the hidden-surface problem for only a very limited and simple subset of the possible conditions suggested by these questions. Other more ambitious algorithms attempt to solve the problem in its most general form and for all conditions. And, of course, many algorithms fall somewhere between these extremes. The degrees of success and efficiency are just as varied.

The problem size is reduced by half by finding and eliminating all backward-pointing surfaces. Many of the techniques for finding silhouette curves apply here; many other techniques are available for simplifying and reducing the problem size. For example, rather than treating each element of a scene as an independent, randomly located occurrence, hidden-surface algorithms are more effective when they account for element interrelationships. This interrelationship is called **coherence**, and coherence is present on two model-representation levels—object space and image space. We can identify many types of coherence.

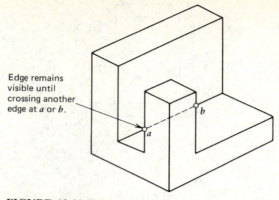

Edge remains
visible until
crossing another
edge at *a* or *b*.

FIGURE 12.30 Edge coherence.

Edge coherence: The visibility of an edge changes only when it crosses another edge; see Fig. 12.30 for an example.

Face coherence: If one part of a face is visible, then frequently the entire face is visible.

Object coherence: We can often determine the visibility of a complex object by computing the visibility of a simpler encompassing shape (for example, a box or circumscribing sphere). In Fig. 12.31, object 1 must be visible, since the projection of its circumscribing sphere at 1* does not overlap the other projected spheres. The projected circumscribing spheres of objects 2 and 3, however, do overlap on the projection plane. It is easy to determine whether or not two spheres intersect in object space. If spheres 2 and 3 do not intersect in object space, then, since the center of sphere 2 is closer to zero than the center of sphere 3, object 2 may obscure all or part of object 3. Conversely, object 3 cannot obscure any part of object 2.

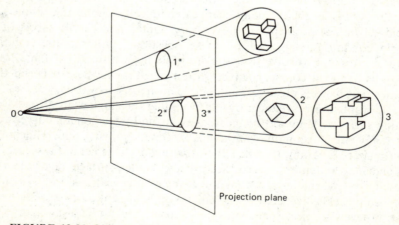

Projection plane

FIGURE 12.31 Object coherence.

Scanline coherence: Segments of an object visible on one scanline are apt to be visible on the next line.

Frame coherence: Either all or part of an image does not change from one display to the next (that is, from frame to frame). To be effective, a display program must keep track of, or differentiate between, those objects that change or move and those that are fixed or part of an unchanged background.

Geometric coherence: T. Pavlidis (1981) observes that geometric coherence has received relatively little attention in graphics, although widely discussed and applied in pattern recognition (scene analysis). Some of the concepts of geometric coherence produce very simple tests if, for example, we know in advance that a solid is convex. Specifically, if the maximum angle between adjacent edges at a vertex is greater than 180°, then the two extreme edges are visible, and the visibility of the other edges depends on their location relative to a plane defined by the first two. If no angle between adjacent edges at a vertex exceeds 180°, then all the edges have the same visibility. *Note*: These angles are measured in the projection plane. Figure 12.32 presents a very simple example of this coherence applied to the vertices of the projection of a cube.

If we construct a minimum-sized rectangle, in image space coordinates, around each polygon in a projection, then we can rapidly sort through these rectangles to find and distinguish between any overlapping polygons requiring additional visibility computations and spatially and visually isolated polygons requiring no further analysis (that is, they are entirely visible). Figure 12.33 depicts some of the possible conditions. Obviously, it is possible for rectangles to intersect while the polygons they contain do not. However, these "false" hits do not significantly alter the efficiency obtained by eliminating some polygons from any further computations.

This concept is closely related to another computation-minimizing technique. The extent of the visual field is also limited by the display **window**

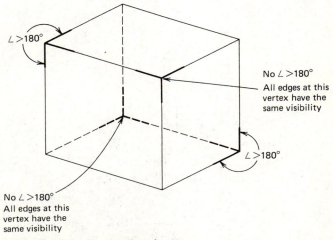

FIGURE 12.32 Geometric coherence.

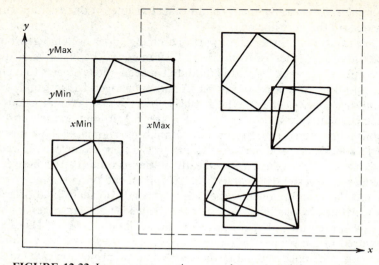

FIGURE 12.33 Image space sorting on polygon-containing rectangles.

defined. The window is that rectangular portion of the projection plane we select to fill all or part of the display screen. This is nothing more than another rectangle superimposed on the projection plane. The minimum vertex is usually at the origin, although not necessarily so. The dashed lines in Fig. 12.33 represent a display window. Both isolated and overlapping objects not inside this window are not visible (that is, not within the display or visual field) and, therefore, require no further computation. Notice that this window is actually the intersection of the projection plane with what we might call the solid cone or pyramid of vision (see Fig. 12.34). There are techniques, which we will not develop here,

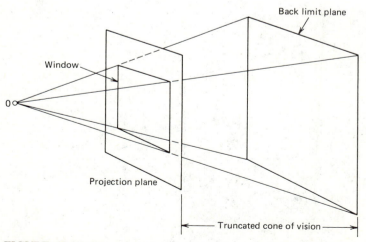

FIGURE 12.34 Cone of vision.

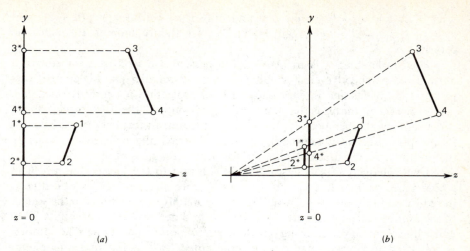

FIGURE 12.35 Depth comparison for parallel and perspective projection.

for culling and simplifying subsequent computations by not considering objects outside this cone of vision.

We must find and eliminate hidden surfaces before projection, since projection transformations may destroy the information needed for depth comparison. The problem of depth comparison is complicated somewhat because computations depend on the type of projection. This becomes clear by studying the different effects of projections on the same scene in Fig. 12.35. For the moment, we will ignore the x components and assume that the two lines are in the $x = 0$ plane. The parallel projections of these lines do not overlap, whereas the perspective projections do. For the parallel projection, we merely test y_1, $y_2 \in [y_4, y_3]$. If the test is positive, we proceed to check z coordinates to determine depth priority. For the perspective projection, we similarly test y_1^*, $y_2^* \in [y_4^*, y_3^*]$. However, we first have the additional burden of computing y_1^*, y_2^*, y_3^*, and y_4^*

$$y_1^* = \frac{dy_1}{d + z_1}$$

$$y_2^* = \frac{dy_2}{d + z_2}$$

$$y_3^* = \frac{dy_3}{d + z_3} \tag{12.78}$$

$$y_4^* = \frac{dy_4}{d + z_4}$$

In general, we observe that two points $\mathbf{p}_1$ and $\mathbf{p}_2$ overlap in their parallel projection if and only if $x_1 = x_2$ and $y_1 = y_2$ and the point with the lowest z value is visible. These same points overlap in their parallel projection if and only if $x_1^* = x_2^*$ and $y_1^* = y_2^*$; that is if and only if $x_1 = (d + z_1)x_2/(d + z_2)$ and $y_1 = (d + z_1)y_2/(d + z_2)$. Again, z_1 and z_2 are then compared to determine priority.

J. D. Foley and A. Van Dam (1982) present a matrix that transforms a normalized-perspective-view volume into a rectangular parallelepiped, distorting the objects and moving the center of projection to ∞ on the negative z-axis. The result of this is that the parallel projection of the transformed object is the same as the perspective projection of the untransformed object. The test for one point obscuring another is the same as for parallel projections, reducing the number of computations required for depth comparison.

D. R. Hedgely (1982) has a general solution to the hidden-line problem whose basic geometric concepts we will explore next. Hedgely's objective is to reduce computation time, which is approximately proportional to the square of the number of elements involved. His method determines the visibility of an entire line segment by choosing only a select few points on the line. The solution is restricted to objects that can be represented by straight lines and planar polygons, convex or concave, and with or without internal holes. He defines polygons by the coordinates of their vertices, and straight lines by their end points. The method involves two tasks — selecting points and determining their visibility.

We require the conventional equation of a plane

$$Ax + By + Cz + D = 0 \tag{12.79}$$

where, if $\mathbf{p}_1$, $\mathbf{p}_2$, and $\mathbf{p}_3$ are three noncolinear points, then

$$\begin{aligned}
A &= [(\mathbf{p}_2 - \mathbf{p}_1) \times (\mathbf{p}_3 - \mathbf{p}_2)]_x \\
B &= [(\mathbf{p}_2 - \mathbf{p}_1) \times (\mathbf{p}_3 - \mathbf{p}_2)]_y \\
C &= [(\mathbf{p}_2 - \mathbf{p}_1) \times (\mathbf{p}_3 - \mathbf{p}_2)]_z \\
D &= -(Ax_i + By_i + Cz_i)
\end{aligned} \tag{12.80}$$

where (x_i, y_i, z_i) are the coordinates of some point $\mathbf{p}_i$ on the plane (for example, $\mathbf{p}_1, \mathbf{p}_2$, or $\mathbf{p}_3$). Clearly, A, B, and C are components of a vector normal to the plane, and Eq. 12.79 is nothing more than the expression $(\mathbf{p} - \mathbf{p}_i) \cdot \mathbf{n} = 0$, where $\mathbf{n}$ is the unit normal vector (see also Section 2).

Consider the orthogonal projection of a closed planar polygon onto the $z = 0$ plane. Again, we will use a left-hand coordinate system and view the projection from the negative z side of the projection plane. If $\mathbf{p}^*$ is the projection of any point $\mathbf{p}$, then $x^* = x$, $y^* = y$, and $z^* = 0$. If $\mathbf{p}^*$ lies on the boundary of the projected polygon, then $\mathbf{p}^*$ is visible with respect to the polygon (that is, the polygon does not obscure it from view). If $\mathbf{p}^*$ lies outside the boundary of the polygon, then a straight line drawn from $\mathbf{p}^*$ in any direction to infinity will intersect the polygon an even number of times or not at all. Conversely, if $\mathbf{p}^*$ lies inside the boundary of the polygon, then a line drawn from $\mathbf{p}^*$ in any direction to infinity intersects the boundary of the polygon an odd number of times. See examples of each of these situations in Fig. 12.36. Notice that the odd–even criterion applies even if there are holes in the polygon and the polygon is con-

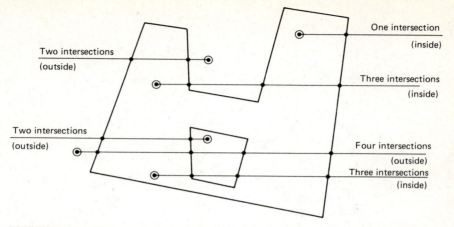

FIGURE 12.36 Odd-even intersection criteria for locating a point with respect to a planar polygon.

cave. Notice also that if the semi-infinite test line intersects a vertex, then another line must be chosen.

Using this odd–even intersection criterion, a point **p** is immediately visible with respect to a polygon if the projection of the point **p*** is outside or on the boundary of the projection of the polygon. If **p*** is inside the boundary, then another computation will determine on which side of the plane of the unprojected polygon the point **p** lies. Substitute the x, y coordinates for **p** into the equation of the plane of the polygon, and solve for z_p, where z_p is the z coordinate of the point on the plane on the intersection with the parallel projector through **p** (see Fig. 12.37).

$$z_p = \frac{-(Ax + By + D)}{C} \tag{12.81}$$

then

$$\text{if } z \leq z_p, \text{ then } \mathbf{p} \text{ is visible} \tag{12.82}$$

or

$$\text{if } z > z_p, \text{ then } \mathbf{p} \text{ is hidden by the polygon} \tag{12.83}$$

We must test the visibility of each selected point against every polygon in the display. If a point is visible with respect to every polygon, then it is visible. Otherwise, it is hidden.

Hedgely correctly observes that having a method for determining the visibility of points is not in itself sufficient for an efficient algorithm. A point-selection criterion is necessary to limit computation time. A brute-force approach that subdivides every line of a scene into many test points proves computationally inefficient. Instead, he suggests the following approach, which assumes that all intersections have been calculated between lines and planes.

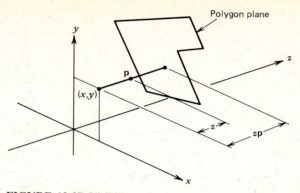

FIGURE 12.37 Visibility of a point with respect to a planar polygon.

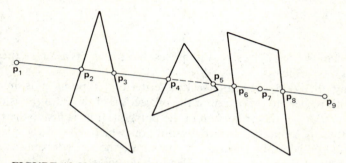

FIGURE 12.38 Visibility of a line.

Thus, all pertinent intersections have been computed for the line shown in Fig. 12.38, including its intersection with a plane at $\mathbf{p}_7$. These intersections divide the line into segments. It is obvious that if we know the visibility of a point on the interior of a segment, then we know that the segment has the same visibility. For convenience, we can test the midpoint of each segment. We do this for line and polygon edges in a scene, and when complete, all hidden lines will be found and deleted.

This method clearly solves the hidden-surface problem as well when the faces of solids are represented as planar polygons and the display is of the vector type. There is a convenient, if not elegant, extension of this and similar methods to solids with curved surfaces. We can approximate the curved surface by planar polygons, usually triangular or quadrilateral, obtained by simple geometric operations on a mesh of isoparametric curves on the surfaces; see Fig. 12.39.

It is unlikely that four mesh points defining a quadrilateral will be coplanar, so either one or more of the points must be adjusted or recourse must be made to creating triangular facets. An easy way to do this is to first establish a base quadrilateral and then connect the point at u_{min}, w_{min} to the point at u_{max}, w_{max}

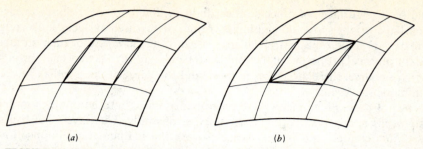

FIGURE 12.39 Approximation of a curved surface by planar polygons.

with a straight line to form two planar triangular polygons. We can control mesh spacing by slope or curvature criteria so that the spacing is greater where the surface is flatter. *Caution*: This is a brute-force approach and computationally costly in most situations. The literature contains many refinements of this approach that are not discussed here.

There are other approaches to the hidden-surface problem for solids with curved surfaces. One popular method requires creating a u, w mesh on each forward-facing surface in the display, where the mesh is sized so that on projection, it matches the size of the pixel that defines the resolution of the display. If more than one surface is projected onto a pixel, then a depth-checking algorithm is used to determine visibility. Here, again, various heuristic techniques are available for improving computational efficiency at both the image-space level and the object-space level.

There is no best hidden-surface algorithm. The possible applications are too varied to make general rankings possible or meaningful. Moreover, the field is a dynamic one: Computer graphic technology is evolving so rapidly that delegating computations between hardware and software is an increasingly difficult task.

EXERCISES

1. Given the radius and coordinates of the center point for two spheres, write a simple procedure for deciding whether or not they intersect.

2. Write a procedure for detecting intersecting rectangles given their maximum and minimum vertex coordinates (see Fig. 12.33). Extend your solution to three-dimensional boxes.

6   REALISM

How do we make the two-dimensional projected image of an object or group of objects look real? The next question is: What does "real" or "realism" mean in this context? Take a critical look at the objects around you now; the hidden-surface problem is neatly solved by the opaqueness of most materials. Notice that light and shadow contribute significantly to the perception of depth and solidity. The surfaces of objects reflect light in different ways: Shiny surfaces often have bright spots, or highlights; rough or textured surfaces look rough or textured. Some objects are transparent, and the curvature of their surfaces and the physical properties of their material bend light rays as they pass through. Then, of course, there are color, motion, and the effect of the parallax of your own binocular vision.

All of these phenomena have been incorporated into computer graphics systems if not practically, then at least experimentally. The physics, of course, is well understood and in most cases so are the mathematical implications. The greatest challenges are obtaining computationally fast and efficient algorithms and designing a proper division of labor between hardware and software.

Objects illuminated by a point source of light cast the geometrically simplest shadows. In fact, the mathematical description of this is exactly the same as for the hidden-surface problem. A hidden surface is a surface our eye cannot see, and we can think of a shadow as part of a surface a light source cannot see. Imagine being temporarily placed at a light source. The areas hidden from view are precisely the areas shaded by objects obscuring the light source. Objects with complex shapes will often shade various parts of themselves. In multi-object scenes, there may be many combinations of hidden-surface and shaded-surface intersections (see Fig. 12.40).

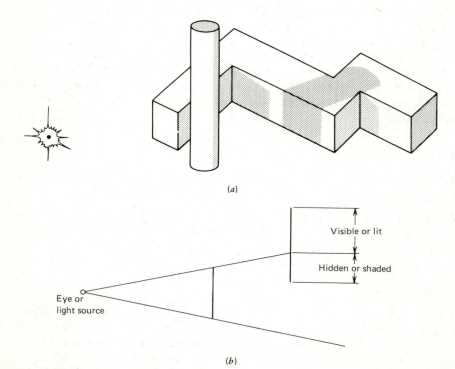

(a)

Visible or lit

Hidden or shaded

Eye or
light source

(b)

FIGURE 12.40 Shadows.

Notice that as long as the light source remains fixed relative to the objects in a scene, shadows do not change, even though the view itself may change. If there are many light sources, some of the shadows become reilluminated or canceled by another. While hidden surfaces are simply not displayed, shaded surfaces are displayed but with reduced intensity and muted color. Computational challenges become even greater for extended light sources, moving light sources, and even objects themselves that may "glow" or be an extended source of light.

The way an object reflects light depends on the characteristics of its surface material, the intensity of the light source, and the angle between the light source and the surface normal (see Fig. 12.41). Dull surfaces produce a diffuse reflection. They scatter light equally in all directions and thus appear equally bright from any viewing angle. A simple formula gives the intensity of diffuse light I_d reaching our eye

$$I_d = k_d I_p \cos \theta \tag{12.84}$$

where k_d is the reflection coefficient of the surface material, varying between 0 and 1; I_p is the intensity of a point source of light; and θ is the angle between the surface normal **n** and the direction **l** to the light source. In vector form, this expression becomes

$$I_d = k_d I_p (\mathbf{l} \cdot \mathbf{n}) \tag{12.85}$$

where **l** and **n** must be unit vectors.

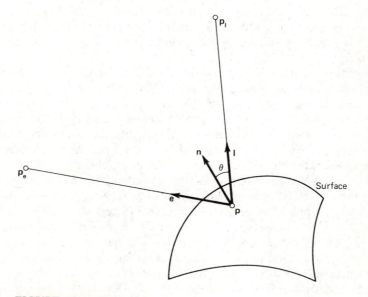

FIGURE 12.41 Reflection.

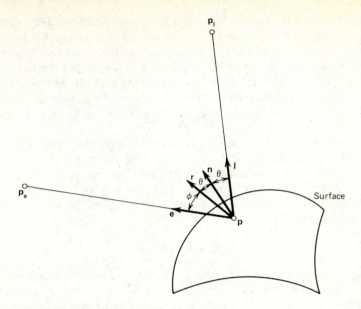

FIGURE 12.42 Specular reflection.

Shiny surfaces produce **specular reflection**, concentrated highlights or bright spots taking on the color of the light source. The surface is, in effect, producing a mirror image of the light source. The image is, of course, more or less distorted by the curvature and texture of the surface. As is true for any mirror-image effect, there must be a particular angular relationship between the viewer, the reflecting surface, and the image source. Figure 12.42 shows this relationship.

Specular reflection is visible only when ϕ approaches zero. Notice that **l**, **n**, and **r** are coplanar vectors. The relationship depends on the nature of the surface material and distortion due to its curvature. And, since real light sources are not geometric points but extended surfaces themselves, the specular reflection we observe around us every day is very difficult to describe mathematically in a simple way. J. D. Foley and A. Van Dam provide an excellent introduction to the subject of reflection for computer graphics.

The final subject we will consider here, and only briefly, is surface texture. To represent coarsely textured surfaces poses interesting problems. Obviously, representing each facet of an irregular surface (where the scale of the irregularities is very small compared to the overall object size) is computationally inefficient, if at all possible. Several schemes have been devised to simulate this roughness. One method randomly perturbs the surface normals to alter reflection characteristics of a surface. Another method uses the relatively new mathematics of fractals; see B. B. Mandelbrot (1983). Detailed descriptions of these and other methods are readily available in the literature.

EXERCISES

1. Write a procedure to compute the intensity of light reaching the eye of a viewer from a bicubic patch illuminated by a point source of light. Denote this as **LITE(PE, PL, KD, IP, BC, UVINCR, IPE)**, where

PE(3) is the input array of eye point coordinates;

PL(3) is the input array of point-light-source coordinates;

KD is the input value of the surface's reflection coefficient;

IP is the input value of the intensity of the light source;

BC is the input array of geometric coefficients defining the bicubic patch;

UVINCR is the input number of increments of the parametric variable u, v on the surface at which points a diffuse reflection intensity is computed; and

IPE(UVINCR + 1, UVINCR + 1, DI) is the output array of values of the diffusion intensity (DI) at points (u, v) on the surface.

If UVINCR $= 3$, then $u, w = 0, 1/3, 2/3, 1$, and another bicubic surface can be used to represent the diffuse reflection intensity at points on the object surface.

COMPUTER-AIDED DESIGN AND MANUFACTURING

Computer-aided design and computer-aided manufacturing (popularly known as CAD/CAM) are today broad-based systems of applications. As we shall see, they are closely interrelated. In the earliest stages of development of these applications strong inducement arose for the parallel developments of computer graphics and geometric modeling.

Our objective in this chapter is to explore some of the ways geometric modeling is used by CAD/CAM systems. We'll explore design and try to uncover what a designer does and how geometric modeling is important to the design process. This will lead to a look at mechanical parts, interference detection between objects, mechanisms, and tool path geometry—all aspects of CAD/CAM with strong ties to geometric modeling.

Before we investigate these more specific areas, let's take a wider view of the total CAD/CAM field. The types of CAD/CAM systems are distinguished by being graphic or nongraphic and conversational or nonconversational:

Graphic conversational: In this mode, we use a computer graphics terminal for input and output and operations directly involving the computer. We converse with the computer using a light pen, keyboard, digitizing tablet, and other devices. The emphasis is on graphic representation of mechanical-part geometry and direct interaction with it. The term *conversational* implies that, while we are sitting at a computer graphics terminal, the computer is processing our commands rapidly enough for us to feel it is worthwhile to wait for the response and rapidly enough not to break our train of thought.

Nongraphic conversational: This mode is the same as the one in conversational graphics in that the computer responds to commands rapidly enough to be comfortable and convenient for us and so that our thought processes are not interrupted by delays. Instead of being graphic terminals, the input and output devices are alphanumeric devices. They range from simple teletypewriters to typewriter and line printer combinations, including nongraphic video terminals. A nongraphic conversational program is invariably accessed and operated by some user-oriented system of commands. In terms of problem solving, these commands are difficult to define and develop primarily because the semantics of potential users vary so greatly depending on their particular disciplines.

Nongraphic nonconversational: The term *nonconversational* implies that a problem is run through the computer from beginning to end without breaking the computation into steps requiring user intervention or correction. This mode is commonly referred to as **batch processing**. We submit data for a particular problem in alphanumeric form to the computer (directly or remotely). We also get the response in alphanumeric form. The process takes so much time that our momentum is seriously interrupted. Usually, this is not due to the time it takes the computer to perform its function (which may be only a few seconds) but to the time it takes to get the job to the computer and then from the computer back to us. This amount of time is heavily dependent on other jobs waiting for processing and any intervening manual operations. The total elapsed time from job submittal to job return is called **turnaround** time. It is the difference in turnaround time from a few seconds to many hours that best distinguishes conversational from nonconversational computing.

Graphic nonconversational: This mode is the same as nongraphic, nonconversational except that we use graphic rather than alphanumeric input and output devices, which are auxiliary devices and not necessarily attached directly to the computer as in the case of a graphic conversational system. Hence, we do not converse with the computer but only introduce input data and receive data in a graphic form. In this case, "graphic" has more to do with the method of presenting data than with the design process itself. Auxiliary devices commonly used are digitizers for input and plotters for output. The technique is often referred to as **graphic batch**.

All of these permutations are possible, but it is the graphic conversational or rapid-response aspect that best characterizes contemporary CAD/CAM systems. Conversational or interactive graphics makes possible more rapid comprehension, formulation, and solution of design problems and enables us to explore a greater variety of solutions than was possible in the past. We can now interact more intimately with the algorithms of the computer, arranging, evaluating, directing, and redirecting them with our creative and heuristic abilities to achieve the best solutions.

Let us look at a specific example, representative of classical (pre-CAD/CAM) automotive and aerospace engineering practices. In the past, designers of structural components or systems of components constructed geometric **layouts**—large-scale, accurate drawings. They used them to study space allocation, structural arrangement, geometric envelopes, and cross-section properties defining strength and flexibility. Later in the design process, designers used these layouts to check compatibility with other systems, such as electrical and mechanical, to analyze tolerance build-up effects and to check assembly clearances and fit. The final design layout gave a complete description of a structural part or assembly of parts so that engineering production drawings could be made from it. Now, many aspects of layout development are incorporated into CAD systems, allowing a designer at a computer graphics terminal to create and analyze the geometry of various design solutions. Once a final design solution is selected, it is annotated and stored, perhaps as the model of a single part or an assembly or arrangement of many parts.

H. B. Voelcker (1974) and others have observed that these part models are the natural interface between CAD and CAM. A. Baer, C. Eastman, and M. Henrion (1979) view computers as a representational medium much like engineering drawings, and instead of merely computerizing verbatim the drafting process with its inherent descriptive limitations, we can now more directly represent parts and assemblies with geometric-modeling techniques. Advances in these modeling techniques, in combination with other information-transfer developments, may eventually completely eliminate drawings and prototypes, creating in their place paperless CAD/CAM systems.

In the first section of Chapter 13, we will elaborate a little more on the design process and discuss how we can create a geometric model.

1. DESIGN

Design is the name we give to the process of selecting, describing, and creating a form to fulfill a function. For example, the cantilevered ramps and walkways of a multistory shopping mall are supported by beams. An architect or engineer had to choose the shape and material of those beams. We say these beams were designed to fulfill the function of supporting the ramps and walkways. The function may be more complicated—convert crude oil into gasoline. In this case, an engineer must choose and combine many components and processes to create a form satisfying the function.

Design is also the name we give to the process of creating works for more aesthetic and less practical purposes. Again, the process produces a **form**, this time satisfying a less clear-cut function—producing aesthetic pleasure. In both kinds of design, form is the dominant feature of the final product. Form connotes geometry, and we quickly see the importance of geometric modeling in the design process (either kind). For the moment, let us look at the first type of design, commonly called engineering design to see how it operates.

Engineering design is an iterative process with brief, unpredictable moments of high creativity followed by longer periods of selecting, describing, modifying, and analyzing forms until the requirements of a product are met by the definition of some final form. A. L. Eshleman and H. D. Meriwether (1967) identify a successful design as the one more capable of fulfilling its intended functions for less cost than competing designs. They further observe that the design process is a race against time (market, customer, or resource imposed), where each iterative cycle of the process will generate a product superior to its predecessors. They add that ordinarily a design is not finalized until the cost of the next iterative cycle exceeds the expected benefits or the length of time necessary to execute the cycle is too long and would cause a default to the competition. During the

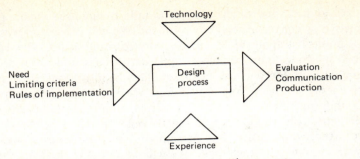

FIGURE 13.1 Design: from the abstract to the concrete.

design process, intuition and judgment are often used when quantitative or automatic techniques fail or are inappropriate.

The design process can also be thought of as a system to process information, graphic and analytic, for the purpose of describing some synthetic form. The initial information may be only a statement of need or the definition of a problem. Rules for how the need is to be satisfied are specified. To guide the process and minimize false starts and disastrous endings, recourse must be made to past experience and accumulated wisdom. Figure 13.1 illustrates the steps in the progress of a design from an abstract concept to concrete reality.

Another definition of design might be the synthesis of past experience, current technology, and future need—seasoned with creativity and intuition—that produces a new form to satisfy a practical function.

A major step toward a successful synthesis of engineering design and computer graphics to create a useful CAD system is simply being aware of the large amount of the total design time that involves graphics, particularly geometric modeling. The development of powerful geometric-modeling systems offers a way of increasing the number of design cycles possible in a given time span and also the productivity of each cycle.

Embedding a geometric-modeling system in a larger interactive CAD or CAD/CAM system is not particularly difficult—at least specifying the general operating characteristics or user interface is more or less straightforward. The early two-dimensional and three-dimensional design-drafting systems provided a good background.

The most successful design-drafting systems were less drafting tools than descriptive geometry tools. What we might call the third-generation CAD systems will be hybrid successors to the first-generation systems, which were drafting and wireframe-model oriented, and the second-generation systems, which are almost exclusively polyhedral solid modelers. To increase the usefulness of these somewhat limited solid modelers, they must be able to represent solids with sculptured surfaces and solve multidimensional descriptive geometry problems.

Let us specify some of the procedures we as users might require in a geometric-modeling system. As for the hardware, we assume that we will build a model at a graphics terminal equipped with a keyboard and light pen (or digitizer, joystick, and so on). We will want fast-response software to create an interactive or conversational environment. We will consider ways of constructing points, straight lines, curves, surfaces, and solids. To construct generalized solids, we must be able to construct and assemble faces defined by sculptured surfaces. To construct these surfaces in turn requires constructing arbitrary curves, straight lines, and points. We will consider only primitive elements and constructions; we can construct more complex solids (or any other geometric constructions) by using Boolean operators to combine these primitives. And we will learn only how to construct geometric elements, and not how to analyze or change them. We will assume that the computer takes care of verifying the solid model.

We can define and construct points in the following ways (see Fig. 13.2 for examples):

a. By keyboard entry of coordinates.

b. By the intersection of curvilinear elements.

c. Relative to a detected point by entry of Δx, Δy, and Δz offsets or ΔH (horizontal) and ΔV (vertical) offsets in the screen or previously defined construction-plane coordinate system.

(*Note*: Points or any other displayed elements are specified for input and processing with a light pen, cursor, digitizer, or equivalent device. In the absence of such a device, elements can be specified by identifying the name or number by keyboard entry.)

d. By the center of a detected arc or circle.

e. By being offset from a detected line with respect to a detected point and plane.

f. By being offset from a detected point in a direction relative to a detected line.

g. By the end point of a detected curve or line.

h. By the intersection of a detected curve and surface.

i. By the intersection of three detected surfaces.

j. By the end of a specified length along a detected curve relative to a detected initial point on the curve.

k. By the closest point on a detected surface to a detected point not on the surface.

l. By displayable spots.

m. By the midpoint of a detected line segment or curve (where the curve is thus bisected into two equal lengths as measured along the curve itself).

n. By the midpoint of a line segment defined by two detected end points.

o. By the ends of a concatenated set of chords of a detected curve, where the maximum deviation between each chord and the curve is within a specified tolerance.

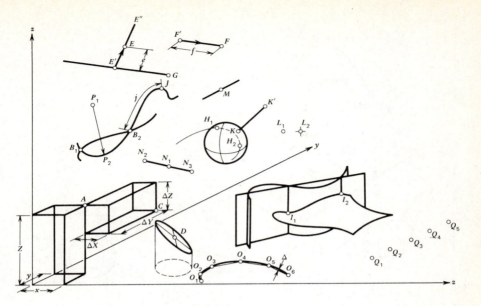

A. Point A generated by keying in X, Y, Z coordinates.

B. Points B_1 and B_2 generated at the intersections of two curves.

C. Point C constructed offset ΔX, ΔY, ΔZ from point A.

D. Point D generated as the center point of a circle in space.

E. Point E constructed in the plane $GE'E''$ as offset a distance e in the indicated direction from point E'.

F. Point F constructed at a given distance f from point F' in a direction parallel to line $E'G$.

G. Point G generated at the end of line segment $E'G$.

H. Points H_1 and H_2 generated at the intersections of the curved line and the spherical surface.

I. Points I_1 and I_2 generated at the intersection of three surfaces.

J. Point J generated at a given distance j along the curve from an initial point B_2.

K. Point K generated as the closest point on the spherical surface to point K'.

L. Points L_1 and L_2 displayed as a spot and a cross, respectively.

M. Point M is the midpoint of the line segment.

N. Point N_1 is the midpoint between points N_2 and N_3.

O. Points O_1 through O_6 satisfy the chordal tolerance Δ.

P. Point P_1 is the center of curvature at point P_2 of the curve.

Q. Points Q_1 through Q_5 are equally spaced.

FIGURE 13.2 Points.

p. By the center of curvature at any detected point on a curve.

q. At equal intervals between two points or along a line or curve.

We can define and construct straight lines in the following ways (see Fig. 13.3 for examples):

a. Between successively detected points.

b. Through a detected point and parallel, normal, or at a specified coplanar angle to a detected line.

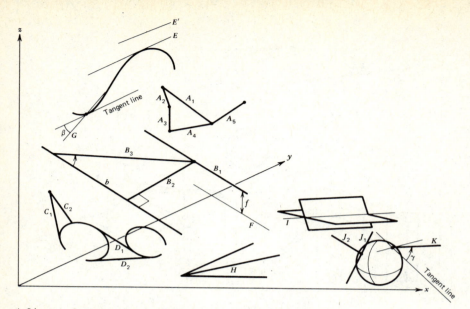

A. Lines A_1 through A_5 constructed through successively detected points.
B. Lines B_1, B_2 and B_3 constructed parallel, normal, and at an angle, respectively, to line b.
C. Lines C_1 and C_2 constructed through a detected point and, respectively, tangent and normal to a detected coplanar curve.
D. Lines D_1 and D_2 constructed tangent to two detected coplanar curves.
E. Line E constructed tangent to the curve and parallel to line E'.
F. Line F constructed parallel to line B_1 at offset f.
G. Line G constructed incident to the curve at an angle β.
H. Line H constructed as bisecting a detected angle.
I. Line I constructed as the intersection of two planes.
J. Lines J_1 and J_2 constructed tangent and normal, respectively, to the detected surface, through A detected point.
K. Line K constructed incident to a detected surface at an angle and through a detected point.

FIGURE 13.3 Straight lines.

c. Through a detected point and tangent or normal to a detected coplanar curve.

d. By a tangent to any two detected coplanar curves.

e. Parallel or normal to a detected line and tangent to a detected coplanar curve.

f. Parallel and offset relative to a detected coplanar line.

g. Through a detected point on a coplanar curve at an input angle to the tangent at the point.

h. Bisecting an angle between two detected lines.

i.   By the intersection of two detected planes.

j.   Tangent or normal to a detected surface through a detected point not necessarily on the surface. (Notice that in this and many subsequent constructions, we may be prompted to supply additional data to resolve ambiguous or incompletely defined construction operations.)

k.   Incident to a detected surface through a detected point not necessarily on the surface, at any specified set of angles or direction cosines in a specified construction plane.

l.   By an infinite or semiinfinite line initially defined as a line segment (not shown).

m.   Parallel coplanar midline between two parallel lines (not shown).

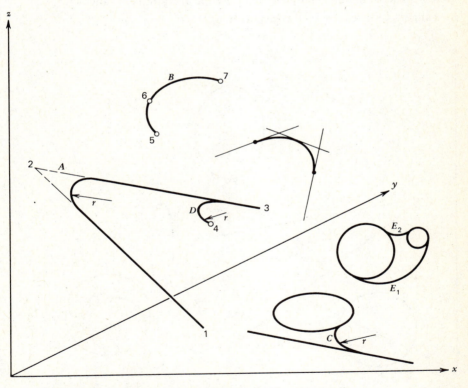

A. Arc *A* constructed tangent to lines 1–2 and 2–3 and with an input radius *r*.
B. Arc *B* constructed through points 5, 6, 7.
C. Arc *C* constructed tangent to the line and the circle and with an input radius *r*.
D. Arc *D* constructed through point 4, tangent to line 2–3, and with an input radius *r*.
E. Arcs E_1 and E_2 constructed tangent to the two circles, with input radius r_1 and r_2, respectively.
F. Tangent to three coplanar lines.

FIGURE 13.4 Circular arcs.

We can define and construct circular arcs in the following ways (see Fig. 13.4 for examples):

a. Tangent to two detected intersecting lines and having an input radius (where line segments between each point of tangency and the vertex are automatically truncated).

b. Through three detected points entered sequentially, with the first and last points at the ends of the arc.

c. Tangent to a detected line and coplanar curve and having an input radius, with the constructed arc automatically truncated at the points of tangency.

d. Through a detected point tangent to a detected line and having an input radius, with the constructed arc automatically truncated at the detected point and point of tangency with the line.

e. Tangent to two detected coplanar curves and having an input radius, with the constructed arc automatically truncated at the points of tangency.

f. Tangent to three detected coplanar lines.

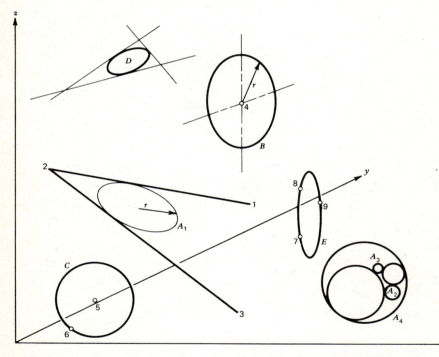

A. Circle A_1 constructed tangent to lines 1–2 and 2–3 with an input radius r. Circles A_2, A_3, and A_4 constructed tangent to two circles.
B. Circle B constructed with center at point 4, radius r, and a particular planar orientation.
C. Circle C constructed through point 6, with center at 5, and a particular planar orientation.
D. Circle D constructed tangent to three coplanar lines.
E. Circle E constructed through points 7, 8, 9.

FIGURE 13.5 Circles.

We can define and construct circles in the following ways (see Fig. 13.5 for examples):

a. Tangent to any two detected coplanar lines or curves having an input radius.

b. With a detected point as the center, an input radius, and a planar orientation.

c. By two detected points (one being the center and the other on the circumference) and a planar orientation.

d. Tangent to three detected coplanar lines.

e. Through three detected points.

We can define and construct conic curves in the following ways (see Fig. 13.6 for examples):

a. By five detected coplanar points, where the first and fifth points define the end slopes at the second and fourth input points. The curve passes through the second, third, and fourth points.

b. By input of end points, end slopes, and a rho value.

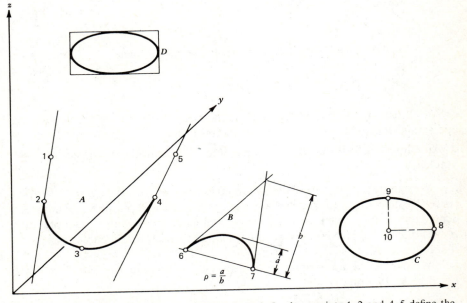

A. Conic *A* constructed by detecting points 1 through 5, where points 1–2 and 4–5 define the end slopes.
B. Conic *B* constructed by detecting points 6 and 7 and input of *Ae* value.
C. Ellipse *C* constructed by detecting points 8, 9, and 10, the end points, and intersection of the semimajor and minor axes.
D. Ellipse *D* constructed by detecting the circumscribing rectangle.

FIGURE 13.6 Conics.

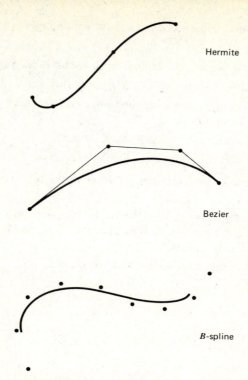

FIGURE 13.7 Hermite, Bezier, and *B*-spline cubic curves.

c. An ellipse, by three detected points: the two end points of the semimajor and semiminor axes and their point of intersection.

d. An ellipse, by four detected coplanar lines that form the circumscribed rectangle.

We can define and construct parametric cubic curves, Bezier, and *B*-spline curves by detecting n points. *B*-splines may require the input of K, defining the degree of the curve; see Fig. 13.7 for an example. (Note that any of the previous curves may be approximated by and encoded as parametric cubic, Bezier, or *B*-spline curves.)

We can define and construct arbitrary curves in the following ways (see Fig. 13.8 for examples):

a. By the curve resulting from a detected plane intersecting any detected surface.

b. By the intersection curve of two detected surfaces.

c. By a curve that is a segment of a previously defined curve.

d. By a composite curve that results from the concatenation of any number of already defined curves.

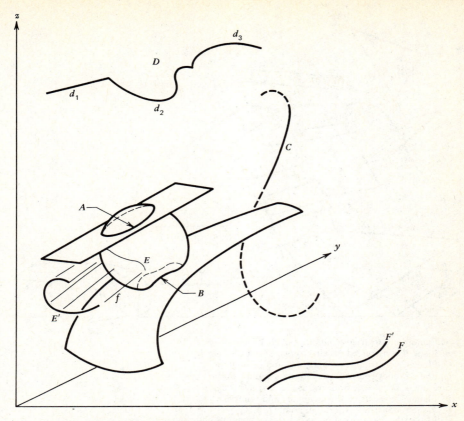

A. Curve *A* generated as the intersection of plane and sphere.
B. Curve *B* generated as the intersection of doubly curved surface and sphere.
C. Curve *C* generated as a portion of previously defined curve.
D. Curve *D* generated as the concatenation of curves d_1, d_2, d_3.
E. Curve *E* generated as the projection of curve *E′* on a spherical surface, where the direction cosines of the parallels or projection are given.
F. Curve *F* constructed offset from curve *F′*; thus, *F* is the locus of equal-length normals erected at successive points on *F′*.

FIGURE 13.8 Arbitrary curves.

e. By a curve resulting from the "parallel" projection of any curve onto any surface.

f. By a curve offset from a given curve.

We can define and construct planes in the following ways (see Fig. 13.9 for examples):

a. By input of coefficients of the equation $Ax + By + Cz + D = 0$.

b. By input of the distance and direction from a detected parallel plane.

c. Parallel to a detected plane through a detected point.

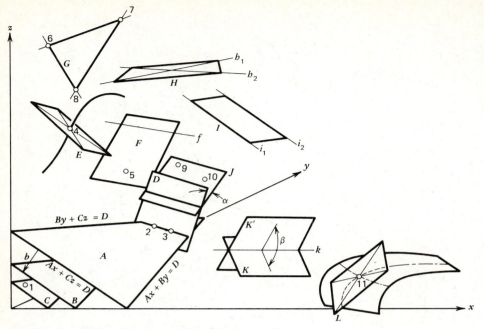

A. Plane A constructed from input of coefficients A, B, C, D.
B. Plane B constructed parallel to plane A at distance b.
C. Plane C constructed parallel to B and through point 1.
D. Plane D constructed through points 2 and 3 and perpendicular to plane A.
E. Plane E constructed normal to curve and through point 4 on curve.
F. Plane F constructed through line f and point 5.
G. Plane G constructed through points 6, 7, 8.
H. Plane H constructed through two intersecting lines h_1 and h_2.
I. Plane I constructed through two parallel lines i_1 and i_2.
J. Plane J constructed through points 9 and 10 and at angle α to plane D.
K. Plane K constructed through line k in plane K' and at angle β with plane K'.
L. Plane L constructed tangent to surface at point 11.

FIGURE 13.9 Planes.

d. Perpendicular to a detected plane and through two detected points on it.

e. Perpendicular to a detected curve through a detected point.

f. Through a detected point and a noncoincident detected line.

g. Through three detected noncolinear points.

h. By two detected intersecting lines.

i. By two detected parallel lines.

j. Through two detected points and an input angle to a detected plane.

k. At an input planar angle to a detected plane and through a line in the plane.

l. Tangent to a detected surface at a detected point on the surface.

We can define and construct general cylindrical surfaces through a planar curve and normal to its plane (see Fig. 13.10).

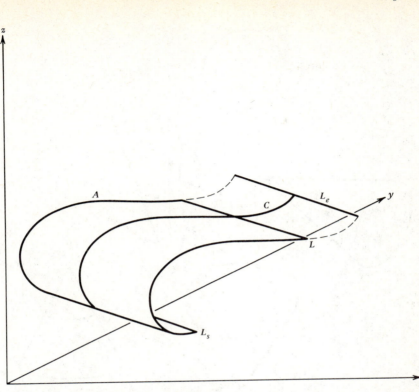

A. Cylindrical surface *A* generated by specifying curve *C* and the starting (L_s) and ending (L_e) positions of line L.

FIGURE 13.10 Cylindrical surfaces.

We can define and construct conical surfaces in the following ways (see Fig. 13.11 for examples):

a. By a detected axis-line segment and an input-base radius (right circular cone).

b. By a detected circle (base) and point (vertex).

c. By the revolution of a detected isosceles triangle around the perpendicular bisector of its base.

d. By two detected offset, parallel, concentric circles with unequal radii.

e. By any general conical surface by specifying a base curve (closed or open) and a vertex point.

We can define and construct spheres (surfaces or solids) in the following ways (see Fig. 13.12 for examples):

a. By a detected center point and an input radius.

b. Tangent to three detected intersecting planar surfaces and with an input radius.

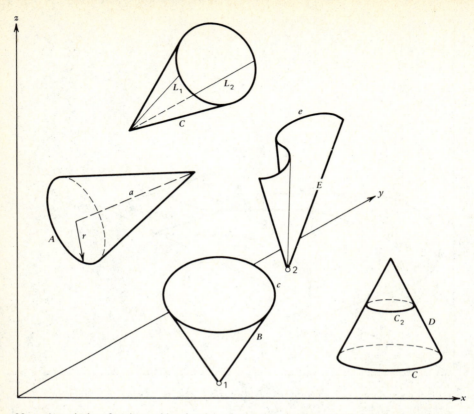

Note: A conical surface is considered to be only the sloped sides, excluding the base (surface).
    A. Cone *A* generated by specifying the axis line *a* and the base radius *r*.
    B. Cone *B* generated by specifying the vertex point 1 and the base circle *c*.
    C. Cone *C* generated by specifying an equilateral angle, $L_1 - L_2$.
    D. Cone *D* generated by specifying two offset, parallel, and concentric circles C_1 and C_2.
    E. Conical surface is generated by specifying curve *e* and vertex point 2.

FIGURE 13.11 Conical surface.

c. By a detected center point and a plane of tangency.

d. By a detected center point and a surface point.

e. By a minimum-diameter sphere enveloping a set of detected points.

We can define and construct ruled surfaces between two detected curves or a single curve and a line segment through it (see Fig. 13.13*a* and 13.13*b*).

We can define and construct translation surfaces by specifying a generator curve and a director curve (see Fig. 13.14).

We can define and construct surfaces of revolution by specifying any planar curve, an axis of revolution, a direction of rotation, and an angle of rotation (see Fig. 13.15).

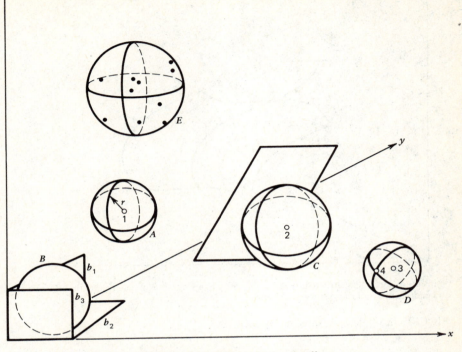

A. Sphere A constructed by specifying the center at point 1 and radius r.
B. Sphere B constructed by specifying three tangent planes b_1, b_2, b_3, and a radius.
C. Sphere C constructed by specifying the center at point 2 and a tangent plane.
D. Sphere D constructed by specifying the center at point 3 and a surface point 4.
E. Minimum sphere E constructed by specifying the set of points to be enveloped.

FIGURE 13.12 Spheres.

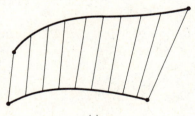

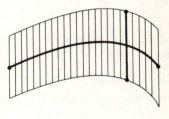

(a)   (b)

FIGURE 13.13 Ruled surface.

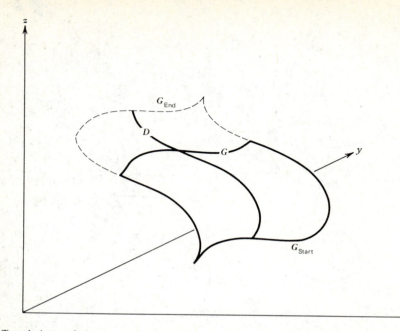

Translation surface constructed by specifying a directrix curve D, a generatrix curve G, and start and end positions of the generatrix curve.

FIGURE 13.14 Position-direction surface.

We can construct bicubic surfaces in the following ways (see Fig. 13.16 for examples). Note that any of the preceding surfaces may be approximated by and encoded as bicubic surfaces.

a. By detecting 16 points. The order of detection establishes predefined (default) u, w values for the points.

b. By detecting four boundary curves. Each curve is assumed to be an appropriate constant u or w curve determined by order of detection.

c. By detecting four curves interpreted to be u or $w = 0$, 1/3, 2/3, 1, or other specified values.

d. By detecting a rectangular mesh of curves.

e. By detecting a rectangular mesh of points.

We can define and construct Bezier or B-spline surfaces by detecting a rectangular mesh of vertex points (see Fig. 13.17). (Note that B-spline surfaces may have default or specified K values.)

We can define and construct primitive solids in the following ways (see Fig. 13.18 for examples):

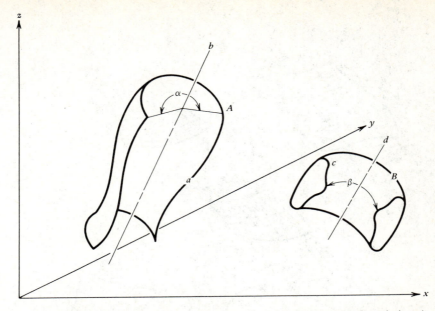

A. Surface of revolution *A* generated by specifying curve *a*, axis *b*, and angle of revolution *α'*.

B. Toroidal surface of revolution *B* generated by specifying closed planar curve *c*, axis *d*, and angle of revolution *β*.

FIGURE 13.15 Surface of revolution.

a. A block by detecting three orthogonal line segments representing length, width, and height.

b. A cylinder by detecting two orthogonal line segments representing the radius and simultaneously the height and normal to the plane of the circular faces.

c. A wedge by detecting three orthogonal line segments representing length, width, and height.

d. A fillet or circular segment by detecting three orthogonal line segments representing length, width, and height. (Note that the segments representing width and height must be equal.) The width or height is also the required radius.

e. A sphere by any of the previously mentioned ways of defining a spherical surface.

f. A cone by detecting a base circle and a vertex point.

g. A torus by detecting two orthogonal line segments representing the major and minor radii, with the initial plane of the rotated circle defined by the two line segments.

h. A polyhedron by sequentially detecting vertex points defining each polygonal face.

i. By detecting a set of intersecting surfaces that form the bounding surface of a solid.

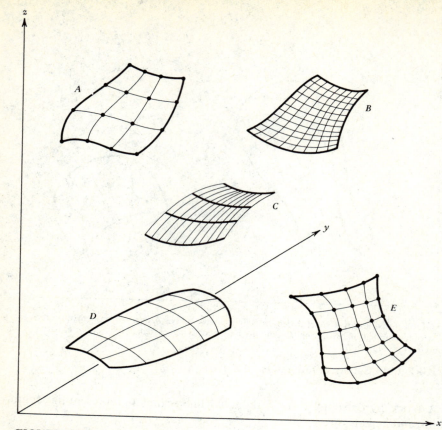

FIGURE 13.16 Bicubic surface.

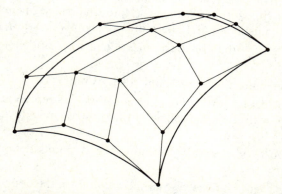

FIGURE 13.17 Bezier surface.

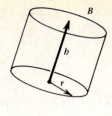

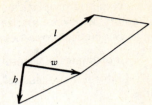

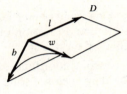

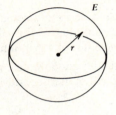

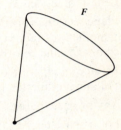

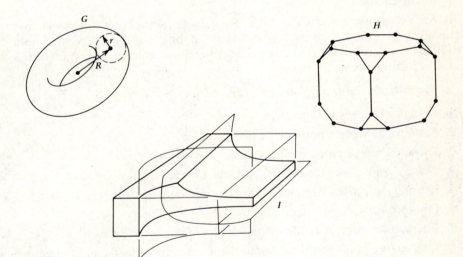

FIGURE 13.18 Primitive solids.

The value of a geometric model is severely limited unless we can extract the geometric properties useful to us. Many of them were discussed in earlier chapters; let us summarize and review them here.

Geometric properties that we can use are:

1. Coordinates of a point.

2. End-point coordinates, slope, true length, and projected length of a straight line.

3. End-point coordinates, center point, arc length, and arc angle of a circular arc.

4. Coordinates and tangent at any point on a curve.

5. Length of any curve segment.

6. Plane section properties, such as peripheral length, area, moment of inertia, polar moment of inertia, section modulus, radius of gyration, and location of principal axes and centroid coordinates of any bounded planar surface.

7. Surface area, wetted area, enclosed volume, mass-centroid coordinates, mass moment of inertia, roll moment of inertia, and weight of an enclosed solid.

8. Any projected area.

9. For a space curve, all points of zero and infinite radius of curvature or the location of all maximum and minimum radii of curvature, their values, direction, and radius center point as well as maximum and minimum values of x, y, and z for any orientation.

10. Surface properties, such as area, normals, tangent plane at a point; and curvature (or radius of curvature) at detected points; maximum and minimum radii of curvature anywhere on the surface; and maximum and minimum values of x, y, z for any orientation.

Geometric relationships that we can use are:

1. Distance between two points.

2. Distance between a point and a line.

3. Distance between a point and a plane.

4. Normal and closest distance(s) between a point and any curve.

5. Normal and closest distance(s) between a point and a surface.

6. Distance between two parallel lines.

7. Intersection-point coordinates and angles of two intersecting lines.

8. Intersecting-point coordinates and angle between a plane and intersecting line.

9. Intersection-point coordinates and angle between local tangent line to any curve and an intersecting line.

10. Intersection-point coordinates and angle between local tangent plane to any surface and an intersecting line.

11. Normal distance between two parallel planes.

12. Normal distance between two parallel surfaces.

13. Angle between two intersecting planes.

14. Normal and closest distance(s) between a curve and a surface.

15. Normal and closest distance(s) between two surfaces.

EXERCISES

1. Given two coplanar nonintersecting circles, how many unique common tangent lines are there? Identify them in a sketch. Can you think of any way the interactive process of creating a single common tangent between these two circles can be designed to distinguish between the alternative solutions?

2. Given two points p_1 and p_2, find the coordinates of the intermediate points for n equal intervals. Outline a procedure to do this, with p_1, p_2, and n as input arguments and p_1–p_{n+1} as output arguments.

3. Describe a procedure to determine if a set of points is coplanar.

4. Write a procedure to compute the center point of a circular arc represented by a pc curve. Denote this as **ARCNTR(CRV, CP)**, where

> CRV(4, 3) is the input array of geometric coefficients defining the pc curve and
>
> CP(3) is the output array of center-point coordinates.

5. Write a procedure to compute the coordinates of a point on a pc curve a specified distance away as measured along the curve from another point. Denote this as **PTPCRV(CRV, U, L, P)**, where

> CRV(4, 3) is the input array of geometric coefficients defining the pc curve;
>
> U is the input value of the parametric variable at the "from" point;
>
> L is the input distance. Use a negative sign with L to indicate that the point to be created is in the direction of decreasing value of the parametric variable from the given point; and
>
> P(4) is the output array of x, y, z, u coordinates of the new point.

Note that since it is generally not possible to achieve an exact solution, some ε should be specified to limit computation.

6. Assume the same conditions as in Ex. 5 but for a *B*-spline curve.

7. Write a procedure to compute the midpoint of a pc curve segment (where the curve is thus bisected into two equal lengths as measured along the curve itself). Denote this as **MIDCRV(CRV, MP)**, where

> CRV(4, 3) is the input array of geometric coefficients defining the pc curve and
>
> MP(4) is the output array of x, y, z, u coordinates of the midpoint.

Control solution accuracy with some specified tolerance ε.

8. Assume the same conditions as in Ex. 7 but for a Bezier curve.

9. Write a procedure to compute the center of curvature of a pc curve at any point on the curve. Denote this as **CNTCRV(CRV, U, P)**, where

CRV(4, 3) is the input array of geometric coefficients defining the pc curve;

U is the input parametric coordinate of a point on the curve; and

P(4) is the output array of x, y, z coordinates and r of the center of curvature.

10. Given a point and a closed Bezier curve lying in the same plane, write a procedure to compute the nearest normal to the curve from the point and all tangents to the curve from the point. Denote this as **PTNCRV(NP, BZ, P, PN, N, NT, TN, PT)**, where

NP is the input number of control points defining the Bezier curve;

BZ(NP, 3) is the input array of control-point coordinates;

P(3) is the input array of point coordinates;

PN(4) is the output array of x, y, z, u coordinates of the point of intersection of the normal line and curve;

N(3) is the output array of direction cosines of the normal line from the point to the curve;

NT is the output number of tangent lines;

TN(NT, 3) is the output array of direction cosines of the tangent lines; and

PT(NT, 4) is the output array of x, y, z, u coordinates of the tangent points on the curve.

Control solution accuracy desired with either an input ε value or an incorporated default value.

11. Write a procedure to compute the geometric coefficients of a pc curve representing a circular arc tangent to two intersecting straight lines. Assume that the straight lines are finite and terminate at their point of intersection. Denote this as **ARCLNS(P1, P2, PI, R, ARC)**, where

P1(3), P2(3) are the input arrays of coordinates of the line's end points;

PI(3) is the input array of coordinates of the point of intersection of the two lines;

R is the input radius of the arc; and

ARC(4, 3) is the output array of geometric coefficients defining the pc curve representing a circular arc.

Consider the modifications necessary to allow multiple pc curve segments, depending on the angle subtended by the arc and the accuracy required.

12. Write a procedure to compute the geometric coefficients of a composite string of pc curves representing a circle. Assume the circle is defined by three points on its circumference. Use eight equal pc curve segments (that is, subtending arcs of 45°). Denote this as **PCCRCL**(P1, P2, P3, CRC), where

P1(3), P2(3), P3(3) are the input arrays of coordinates of the three points and

CRC(8, 4, 3) is the output array of geometric coefficients of the eight pc curves.

13. Devise a way of using a Bezier curve (or curves) to represent a circle. How do you control the representational accuracy?

14. Devise a way of using *B*-spline curves to represent a circle. How do you control the representational accuracy?

15. Write a procedure to compute the coefficients of a plane through a given point and parallel to a given plane. Denote this as **PARPLN**(P, PLN, COEF), where

P(3) is the input array of coordinates of the point;

PLN(4) is the input array of plane coefficients A, B, C, D; and

COEF(4) is the output array of plane coefficients.

16. Write a procedure to compute the coefficients of a plane through two points and perpendicular to another plane. Denote this as **PERPLN**(P1, P2, PLN, COEF), where

P1(3), P2(3) are the input arrays of coordinates of the two points;

PLN(4) is the input array of plane coefficients A, B, C, D; and

COEF(4) is the output array of plane coefficients.

2. MECHANICAL PARTS

Computer-aided design and CAM systems are concerned with describing mechanical parts. Mechanical parts are solid, three-dimensional objects that we can see and touch. They are also artifacts, that is, manufactured objects as opposed to trees, flowers, cats, planets, or other naturally occurring objects.

In addition to these rather broad characteristics, there are aspects of mechanical parts that influence the way they are represented by geometric models. We will explore some of these aspects in Section 2, including typical geometric characteristics of mechanical parts, design details, dimensions, tolerances, and process planning.

Tools and the ways they must be used determine the geometric characteristics of mechanical parts. A critical look at the manufactured objects around you will convince you that this is not a serious limitation, since we are surrounded by artifacts of a seemingly unlimited variety of sizes and shapes. A

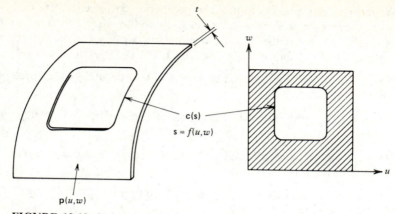

FIGURE 13.19 A sheet metal part.

closer look, however, reveals many common characteristics. Recognizing and accounting for this commonality will help make a geometric-modeling system more effective and efficient.

In the mid-seventies, an industrial-part survey was conducted by N.M. Samuel (1976) and others at the University of Rochester, who analyzed 128 mechanical parts of a contemporary office machine. The survey revealed that fully 30% of these parts could be represented by various Boolean combinations of orthogonally positioned blocks and cylinders (that is, the block edges and cylinder axis were aligned parallel to the coordinate axes). If a wedge-shaped primitive were included, then 85% of the parts could be represented. This study further revealed that 35% of the parts were of sheet metal construction, 35% cast or molded, and 14% turned.

We can easily represent the geometric model of a sheet-metal part by a surface—planar or perhaps sculptured—a set of closed curves representing the outer boundary and boundaries of interior cutouts, and by a thickness t; see Fig. 13.19 for an example. These design parameters might form the input for a solid-modeler algorithm that would compute and organize the details of the entire set of bounding surfaces.

We can also treat an extruded part just as easily and directly. Such parts have a constant cross section (see Fig. 13.20). Thus, a closed planar curve suffices to define the cross section, and a specified length completes the definition. Again, we can devise a solid-modeler algorithm to create the complete boundary representation.

Turned parts are axisymmetric. They may be represented simply as a surface of revolution (Fig. 13.21). Here, we would define an outline curve and an axis of revolution, and the solid modeler would complete the boundary-surface definition.

The simplicity and commonality of these forms raise important questions: Why are the shapes of most parts composed of simple rectangular blocks and circular cylinders? Is this due to the limitations of our design and manufacturing

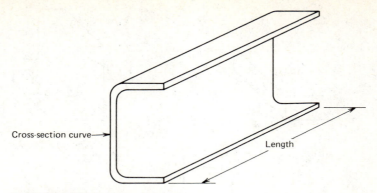

Cross-section curve→

Length

FIGURE 13.20 An extruded part.

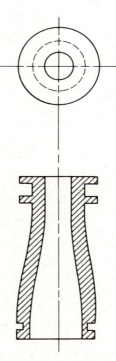

FIGURE 13.21 A turned part.

tools? Or do these shapes arise so frequently because they are the best design solutions, independent of the tools that produce them? If the answer to the second question is yes, then we must develop geometric-modeling systems beyond those merely duplicating current capabilities. These and other questions will be pursued further in the discussion of the frontiers of geometric modeling in Chapter 14.

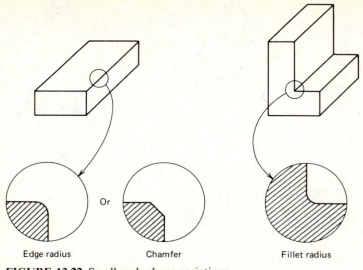

FIGURE 13.22 Small-scale shape variations.

Manufacturing, assembly, and handling requirements introduce small-scale variations in the shape of a part. For example, it is difficult and impractical to cast or forge metal parts with sharp edges. Adjacent faces must blend smoothly into one another through some small radius of curvature. Even machined parts usually require convex edges and corner radii or at least chamfered edges (see Fig. 13.22), and inside corners and edges are produced with fillet radii. Many modeling systems provide alternative means of representing these details, usually as miscellaneous data appended to the main body of geometric data in much the same way that this information appears as notes on conventional engineering drawings.

Here, again, we must decide how powerful we require a geometric modeling system to be. The ability to model these details is clearly not an issue, since most current modelers do this quite easily. It is more a matter of processing speed and efficiency as well as demands on data storage and retrieval facilities. On a still more detailed geometric level, we encounter threads, knurls, serration patterns, dimpling, and other types of surface finishes. These nearly microscopic geometric details are usually not modeled directly.

Manufacturing processes impose certain limits on the geometric model. For example, it is not possible to make something exactly 2 in. long (or "exactly" any size or shape). A variety of physical limitations on manufacturing and assembly processes, as well as material properties, conspire to limit the precision with which we can make things. And the more precise the size and shape of a part must be, the more expensive it is to produce. Therefore, we assign a **tolerance** or range of acceptable values to each dimension of a part. If the size and shape of a part are not within the maximum and minimum envelope defined by these tolerances, the part is unacceptable, and we must rework it or scrap it.

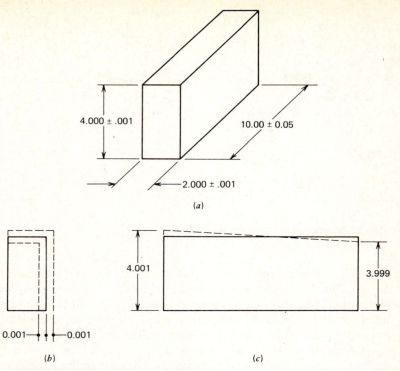

FIGURE 13.23 Dimensions and tolerances.

Let us look at a few examples of how to dimension parts and the effect of tolerances on their size and shape. The bar in Fig. 13.23 is 10 in. long, 2 in. wide, and 4 in. high; they are the **nominal** dimensions. The plus and minus numbers following the nominal dimensions are the tolerances. Notice that the tolerances are not all the same: Tolerances for the height and width are much smaller than the tolerance for the length. Perhaps the designer was concerned about the bar achieving a good fit in a channel. In any case, Fig. 13.23 shows two of many ways the shape of a part may vary.

The part in Fig. 13.24 illustrates yet another level of dimensioning complexity in the geometric model of a mechanical part. Only two of the three dimensions are necessary. Including all three will result in **double dimensioning** or **over dimensioning** and can lead to ambiguous or conflicting interpretation during manufacturing and inspection. We can control the location of the step in this part's profile by specifying any two of the three dimensions. Each possible combination of two dimensions allows us to control different characteristics of the part's geometry. Selecting dimensioning alternative (*b*) or (*c*) allows more control of the total length and step location. Alternative (*a*) allows more variation in the total length while controlling the step lengths.

There are many established conventions for dimensioning and specifying tolerances for parts. We can control "flatness" of a plane surface with a special

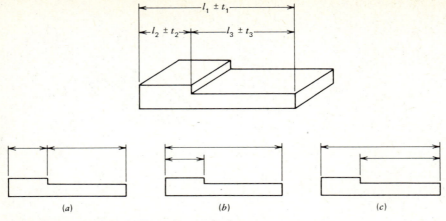

FIGURE 13.24 Multiple dimensions and tolerances.

tolerance call-out. We can control parallelism between two plane surfaces. We can specify a tolerance for the allowable "ovality" of nominally circular shells or tubes. The list of tolerance types is quite long, and they must all be accounted for somehow in the geometric model, as either an integral part of the geometric data itself or supplementary data. The technical issues in accomplishing either alternative are still not completely resolved. How do we incorporate and use tolerance information within the nominal geometric model of a mechanical part?

Consider the stepped part with a circular hole shown in Fig. 13.25. If the geometric model of this part consists solely of its boundary representation, then it is difficult to identify the hole and determine its size. Ordinarily, a hole is dimensioned by its radius, axis orientation, and location, none of which are

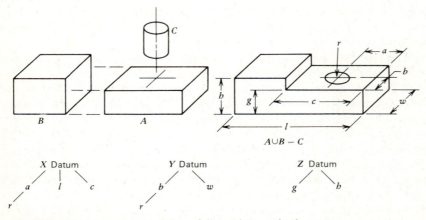

FIGURE 13.25 Tree representation of dimensions and tolerances.

directly accessible from a boundary representation. A procedural form of representation, such as that exhibited by a constructive solid geometry system, quite naturally includes this type of information, and it is easily extracted after design by subsequent planning and manufacturing functions. As we saw in the example in Fig. 13.24, there are subtle interrelationships or dependencies between dimensions. These dependencies may be represented by a tree or graph form of data structure (see Fig. 13.25). A loop or cycle in such a graph is created by overdimensioning. It is quite easy to associate tolerances with their respective dimensions and include them in the graph. R. C. Hillyard and I. C. Braid (1978), as well as A. A. G. Requicha (1977) and others, have studied this problem. A review of their work is an excellent introduction to this very complex aspect of modeling mechanical parts.

Finally, the geometric model is becoming more important in **process planning**, a description of the manufacturing process required for a part. Thus, a CAD/CAM modeling system must be able to represent the effects of tools and processes. Ideally, these effects are represented as mathematical transformations acting on the geometry of the raw stock or initial material destined to become a finished mechanical part. Such transformations include selecting the size and shape of the raw stock; selecting appropriate jigs or fixtures to hold the raw stock at appropriate locations during machining, cutting, or other shape-changing processes; and representing important intermediate shapes.

H. B. Voelcker and others (1974) give a good, simple definition of mechanical part manufacturing: "the transformation of stock (solid objects) into parts (solid objects) in accordance with descriptions (of solid objects)."

They emphasize that manufacturing is the "transformation" in the definition, and this transformation process is the target for automation. The principal input to this transformation is the geometric model. Understanding and mathematically formalizing the transformation process and appropriately engaging it according to definitions of specific input models is at the farthest frontier of geometric modeling and offers unlimited exercises for the imagination and creative talents of all those in this field.

3.   INTERFERENCE DETECTION

When we bring together two or more separate parts to form an assembly, we must check for proper fit. If there is insufficient clearance, the parts would have to interpenetrate, which is physically impossible for real parts. A set of parts must not only fit in their final assembled positions, but each part must be able to be moved to its assembled position along a collision-free path. A geometric-modeling system as part of a CAD/CAM system should be able to help us answer the following questions.

1. Do all parts fit in an assembly without interference?

2. Can parts be moved to their positions within an assembly along collision-free paths?

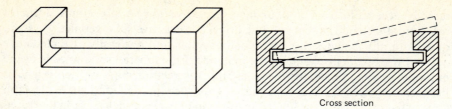

Cross section

FIGURE 13.26 Static and dynamic interference.

3. What are the maximum and minimum clearances between parts in an assembly? Is there room for a wrench, rivet gun, or welding equipment?

4. How do tolerances affect the clearances?

The first question is concerned with static interference, the second with dynamic interference. As illustrated in Fig. 13.26, two or more parts may not exhibit interference, but it may be impossible to move them to this position.

Computationally, interference detection amounts to looking for the presence or absence of intersections between parts. Static interference is the easiest to detect. For two objects A and B, simply compute $A \cap B$. One of the following conditions must result (see also Fig. 13.27):

$$A \cap B = \varnothing$$
$$A \cap B = A$$
$$A \cap B = B \qquad (13.1)$$
$$A \cap B = C$$

The null result indicates the absence of interference. If $A \cap B = A$ or $A \cap B = B$, then one part is contained within the other. This is physically impossible, since A and B are solid objects, so interference is present. In some

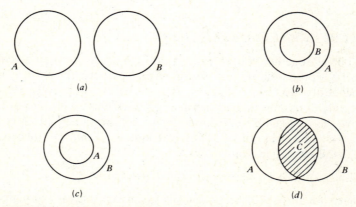

FIGURE 13.27 Possible intersections of two objects.

cases, however, interference is not present. (For example, this is true when A or B does not represent a solid but is itself a clearance envelope.) Finally, if $A \cap B = C$, then interference is clearly indicated. We can use Monte Carlo methods to modify the nominal sizes and shapes of objects and thus represent the effects of tolerances. We then apply the same interference tests to these modified objects.

Dynamic interference detection is somewhat more difficult. Let us look at just a few of the many possible situations where this type of interference occurs. The path or trajectory of the moving object must be described mathematically as well as the object's orientation. The path may be merely a translation or rotation, or it may be an arbitrary general curve through space. The position direction curve we studied earlier is an excellent way of representing the trajectory and orientation of an object.

A solid moving through space defines a volume. Rather than searching for intersections of two objects for many sequential positions of the moving object, we compute instead intersections of the stationary object(s) and the volume defined by the moving object. First, we must compute and represent the defined volume. To do this, we observe that the cross section of the volume at any particular point on the path is given by the silhouette of the object as projected onto a plane normal to the tangent to the path at the point (see Fig. 13.28). If the moving object does not change its orientation relative to the direction of the path, then the defined volume is bounded by a relatively simple

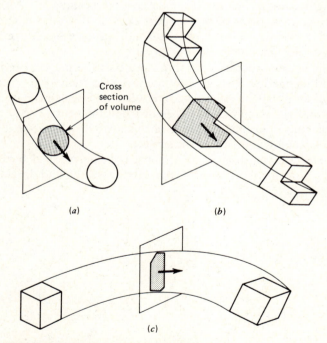

FIGURE 13.28 Object sweeping a volume.

and constant cylindrical surface. However, if the orientation of the object changes relative to its trajectory, the defined volume may be bounded by a much more complex surface.

The surface of the volume defined by the sphere, shown in Fig. 13.28*a*, is obviously independent of the sphere's orientation along the path. The step-shaped object in Fig. 13.28*b* maintains a constant orientation relative to its path and therefore generates a cylindrical, although curved, bounding surface. The cube in Fig. 13.28*c*, on the other hand, changes its orientation, which results in a more subtle bounding surface.

M. A. Wesley (1980) and others have suggested that in a CAD/CAM environment, it would be useful to represent the assembly process symbolically as a set of procedures to be preserved and stored in a CAD/CAM data base as part of the complete assembly description. Any time we redesign the shape of one of the parts of the assembly, we would execute the assembly procedures to verify the presence or absence of static or dynamic interference. The PD curve allows us to control not only the gross path of an object but also the orientation so that we can twist and turn a part as it moves into its final position.

4. MECHANISMS

Many CAD/CAM systems help us design and analyze mechanisms. A **mechanism** is a collection of objects joined in a way that permits only highly constrained relative motions; see Fig. 13.29 for an example. Usually, mechanisms are designed as highly idealized geometric models of objects and joints. More realistic mechanism design appears to be a logical extension of solid modeling. R. B. Tilove (1983) has described how GMSolid was successfully extended to mechanism design and kinematic simulation.

Such simulation permits tests to ensure that the mechanism can be assembled and its moving parts do not interfere with each other. Furthermore, geometric clearances can be checked between the mechanism parts and its fixed structural surroundings. The detailed development of this application of solid modeling is beyond the scope of this textbook. However, Tilove provides an excellent introduction to the subject.

An interesting variation on this theme is being pursued by Badler and others at the University of Pennsylvania. Here, research and development are

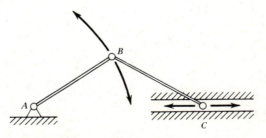

FIGURE 13.29 Idealized mechanism.

underway on human movement simulation. N. I. Badler and others (1983) describe a system called TEMPUS, which will simulate graphically the motions of humans or robots in work environments.

5. TOOL PATH GEOMETRY

Machines programmed to automatically cut shaped parts from raw stock are now commonplace. We will frequently encounter the acronym **NC** in the literature, which is shorthand for **numerical control** and synonymous with **programmed**, as in programmed machine tools. We might refer to NC machines or NC processes. A programming language has evolved for these machines called **APT**, automatic programmed tool.

A manufacturing engineer or technician, called a part programmer, writes a program based on the geometry of a part defined by an engineering drawing. This program includes specifying the size and shape of the raw stock of material from which the part is to be cut, type(s) and size(s) of cutting tools, tool speed (rotational) and feed rate (translation), and tool path or trajectory. The engineer verifies the program by cutting a test part or by more abstract means (for example, using a plotting device to "draw" the tool path), and then loads the verified program into the appropriate NC machine tool and correctly positions the raw stock. The part is then automatically machined, with perhaps only a few scheduled interruptions to reposition the part. Once a program is verified, it can produce an unlimited number of identical parts.

Conventional engineering drawings of mechanical parts are rapidly being replaced by equivalent mathematical representations or geometric models constructed by a CAD/CAM system and stored in a CAD/CAM data base for subsequent retrieval. An important use of the part's geometric model is as input to programs designed to automatically generate instructions for manufacturing the part.

Two important categories of manufacturing processes are additive and subtractive processes and deformational processes. Examples of the first category are machining, cutting, and welding. In Section 5, we will focus our attention on these additive and subtractive processes. Examples of deformational processes are bending and forging; this latter category is much more difficult to model. However, parametric tricubic solids may offer an interesting approach to solutions in the near future.

H. B. Voelcker and others (1981) at the University of Rochester have pioneered in studying the geometric effects of various manufacturing processes. Two important uses of geometric modeling technology have emerged—modeling or simulating machining processes (for example, automatic generation of NC programs) and automatic verification of NC programs.

The machining process begins with a simple stock shape (for example, a block, bar, or plate) and then proceeds to remove (subtract) material with variously shaped cutting tools until some final designed shape is achieved. Clearly, there is a parallel between this "subtractive" machining process and the regularized difference operator in constructive solid geometry. And, in

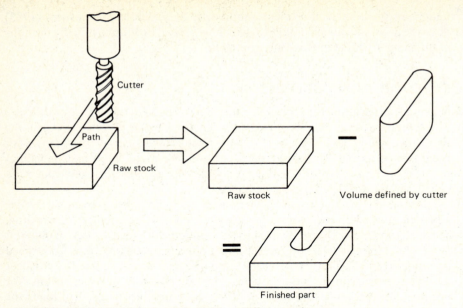

FIGURE 13.30 Machining and the Boolean difference operator.

fact, we can simulate the successive "passes" of the cutter of a machine tool by a sequence of difference operations on the initial model of the part's raw stock shape.

A simple example is presented in Fig. 13.30, where the finished part will be a slotted plate. A single pass with a cylindrically shaped cutting tool will do the job. Geometrically, we can represent this as the difference between the raw stock model and the volume defined by the motion of the cutter.

The problem becomes more complex as the number of passes with a cutter increases and the complexity of the paths themselves increase. For example, in Fig. 13.31, we want to machine the sculptured surface S into block B by using cutter C. Several questions immediately arise: Where should the cutting start? How deep should the first pass be? (This implies creating many intermediate shapes before we produce the final shape.) What path should the tool take as it cuts the shape? Two important computations and transformations are aligning the tool axis appropriately with the surface normal and finding the distance between the tool and the surface.

Finally, when the model of the process is complete, we must verify that the result is identical to our goal, the original geometric model of the part. This is computationally the most difficult part of the problem; Fig. 13.32 presents a simple example. Let $C = A \cup B$ in Fig. 13.32a represent the geometric model of a mechanical part and let $G = D - E - F$ in Fig. 13.32b represent the interpretation of this part by a machining-process model. Now, the question is: Are C and G equivalent geometric models? Determining equivalence is an important research topic today, and so far the only reliable way of verifying

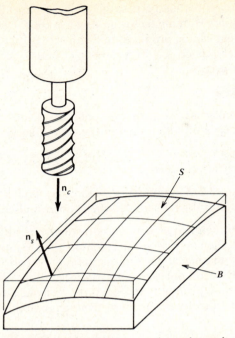

FIGURE 13.31 Cutter and sculptured surface.

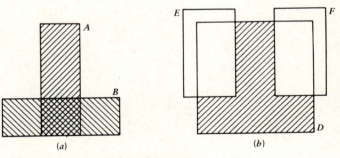

FIGURE 13.32 Equivalence of models.

equivalence is to test for a null result when taking the so-called *symmetric difference* of C and G, which is expressed as

$$(C - G) \cup (G - C) = \varnothing \tag{13.2}$$

The problems of tool path geometry and developing automatic modeling and verifying algorithms for manufacturing processes become even more sophisticated (and interesting) when more of the physics involved is taken into consideration. This includes tool vibration, wear, cutting speed, depth of cut, and deformation under load. These factors have direct impact on the economics and profitability of manufacturing because they determine the number of parts per hour that can be produced and still satisfy quality control requirements. Unfortunately, space prohibits pursuing these topics here.

EXERCISES

1. List and briefly discuss the pros and cons of various types of geometric models based on their ability to represent the machining process.

2. Show that in Eq. 13.2, either $C - G = 0$ or $G - C = 0$ alone is not sufficient to guarantee geometric equivalence.

3. Show that for Eq. 13.2 to apply, the two candidate models must be identically aligned within their coordinate systems when superimposed.

4. Show that if two identical objects do not exhibit the alignment required in Eq. 13.2 (see Ex. 3), then achieving this alignment is a nontrivial problem. For example, given the mathematical representations of two blocks of identical size and shape but with different spatial orientation, show that they are identical. How does the form of representation affect the difficulty of the problem? How does the difficulty increase for identical sculptured shapes?

14

FRONTIERS OF GEOMETRIC MODELING

The synthesis of computer and computational geometry has produced geometric-modeling systems whose powers to represent both concrete and abstract phenomena far surpass those of a designer or draftsman at a drawing board. Further extensions of these powers seem to be limited by only the scope of our imagination. Although we can certainly expect many surprises along the way, recent history and current trends do offer insight into at least some future advances in geometric modeling.

There will, of course, be extensions of current applications, most notably CAD/CAM. Here, we can expect solid modelers to move well beyond the realm of simple blocks, cylinders, wedges, and the like to offer the capability of representing assemblages of very general shapes whose faces are defined by sculptured surfaces. New, more sophisticated shape transformations will be available to design engineers, allowing a mechanical part representation to be literally sculpted, stretched, scaled, locally flattened or rounded, or smoothed to produce the desired shape. Many mechanical parts are made of composite materials whose properties vary with direction and depth of the part's volume. The step from a simple volumetric model of a solid to a full three-dimensional model that includes physical properties is at least as great as the step from the designer's drawing board to present day modelers, but it is a step sure to be taken.

The ability of a robot to interpret, move through, and manipulate its environment is directly related to its ability to develop and analyze a geometric model of that environment. Such a model must be continually updated, not only because the robot's relation to its environment changes as it moves through and manipulates that environment but also because other forces may be at work altering the scene. A model must be rich enough in detail to permit a wide range of robot functions, yet not so rich that it overwhelms the on-board model analysis facilities. Computer vision systems, scene analysis, and pattern recognition are closely related to robotics, particularly through the mediating geometric model. Overarching all this is the field of artificial intelligence, where we can expect geometric modeling to play a key role and make important contributions.

A possible application drawing on all these fields—one with enough significant problems in geometric modeling for several doctoral dissertations—involves a hypothetical air defense weapon system that must identify potential targets. The system must store a complete geometric model of each possible friendly or enemy aircraft or missile. Assume that radar or laser-ranging systems spot a potential target and, at a minimum, are capable of describing the instantaneous target silhouette. A mathematical representation of the silhouette is input to a special algorithm that must find a match between it and one of the stored models.

We must obviously ask: Can an object be identified merely by its silhouette? Note that any object has an infinite number of unique silhouettes, each arising from a different view point. If possible target shapes consisted of only cubes and spheres, the problem would be trivial. But what about discriminating between cubes and rectangular solids with square cross sections? These two different objects present identical silhouettes when the cube is viewed normal to any of its faces and the rectangular solid is viewed normal to its cross-section plane, which yields a square.

Any silhouette may belong to a constant-cross-section object viewed along the appropriate axis. At least one other view or some additional information is required to eliminate this possibility. Now, we can expand our original question to ask: Given a silhouette known to belong to a specific object whose model representation is contained in the data base, can we find the orientation of the object that produces the given silhouette? And can this be done in a rapid systematic way other than by trial and error? This problem is not especially unique to a hypothetical weapon system. Sophisticated collision-avoidance and scene-analysis problems must also address it. It is a typical problem at the frontier of geometric modeling, and it does not end here.

How can we develop meaningful geometric models of natural objects—plants and animals, for example? We must seek not only graphic rendering, but articulated and analytically complex representations, representations of which we can ask useful questions. And let us not exlude other physical phenomena, such as clouds, flow, turbulence, swarms of bees. P. S. Stevens (1974) offers a sampling of the challenge that lies ahead, and W. T. Reeves (1983), with his particle systems, offers a thought-provoking approach and solution to at least a part of that challenge. Appendix D presents Reeves's description of his work in this area.

Geometric modeling promises to become an important artistic tool. We have already witnessed more than a decade of growth in computer graphic art. Modeling techniques have created animated human figures to study and choreograph dance movements; Twyla Tharp's production of *The Catherine Wheel* is an interesting example. Several feature-length films have been produced, and more are on the way that are partially, if not completely, computer animated. One company, Digital Productions, founded by John Whitney, Jr., and Gary Demos, has taken a giant step beyond mere animation with its introduction and use of a Cray supercomputer to produce previously unobtainable levels of realism in computer simulation. This company's most recent accomplishments include *The Last Starfighter* for Lorimar Productions and *Dune* for Dino DeLaurentiis Productions.

Perhaps an unexpected but pleasant surprise that may result from the power and scope of future geometric modeling facilities in CAD/CAM systems will be the ability to break away from merely utilitarian solutions to even simple mechanical design problems. The promise of highly automated CAD/CAM systems does not mean that the human element could or should be eliminated from the design process. It does mean that we can take advantage of these systems to reach higher levels of innovation and also to endow even simple mechanical parts and mechanisms with aesthetic appeal. The human race has never suffered from too much beauty.

Finally, we also see computer hardware evolving in directions supportive of advances in geometric modeling. For example, parallel processors will be immensely useful in intersection and hidden-surface algorithms. It is entirely possible that we will soon see "geometry engines," special-purpose computer hardware—mainframe, microcomputer, or peripheral—for processing and analyzing geometric models, making possible model-driven robots, real-time highly realistic simulation and animation generators. [See J. H. Clark (1982).]

We cannot be sure when or how mankind first constructed a geometric model. Perhaps it was one of our ancestors sketching in the sand along the shore of some prehistoric lake, trying to represent the surrounding countryside so that she could point out new hunting and gathering grounds. Perhaps it was an Egyptian surveyor laying out a new town or monument or resolving a boundary dispute. Geometry is truly the oldest science, yet it has never slowed or stagnated in its development over thousands of years. Thales and Pythagoras gave us the beginnings of demonstrative geometry, the culmination of millennia of intuitive and empirical geometry. Euclid's *Elements* some 300 years later codified and extended the earlier work with such rigor and thoroughness that his name is synonymous with geometry.

Today, geometric modeling, although fundamentally an applied science, is advancing our understanding of many abstract concepts. We have certainly not exhausted the potential contributions of projective geometry, topology, and set theory. Karl Menger in 1949 observed that the most striking development of geometry during the last 2000 years is the continual expansion of the concept of a "geometric object." He noted that this concept began by comprising only the few curves and surfaces of Greek synthetic geometry. During the Renaissance, the concept expanded to encompass the entire domain of objects defined by analytic geometry, and geometry now extends over the "boundless universe treated by point-set theory."

The frontier moves forward, and possibilities seem limitless. Mastering the fundamentals presented in this textbook marks the first steps of what can be a long and fascinating journey. Perhaps Galileo said it best:

Philosophy is written in this great book (by which I mean the universe) which stands open to our view, but it cannot be understood unless one first learns how to comprehend the language and interpret the symbols in which it is written, and its symbols are triangles, circles, and other geometric figures, without which it is not humanly possible to comprehend even one word of it; without these one wanders in a dark labyrinth.

Galileo Galilei (1623)

CADD

This appendix consists of a reprint of an article by A. E. Doelling, entitled, "CAD/CAM Computer-Aided Technology."

ABSTRACT

The use of Computer Aided Technology in the design and manufacturing processes of the aerospace industry is expanding at an accelerating rate. This session will discuss an integrated CAD/CAM approach for cost-effective utilization of computer technology. A "total systems approach" encourages optimization of the company's overall operations, rather than undue concentration on isolated aspects which may account for only small savings potential. Basic to the integrated approach is the design, control, and use of a common data base that reduces redundant efforts, assures validity of data and shortens calendar time. A comprehensive review will be given of CAD/CAM Computer Aided Technology at McDonnell Aircraft including activities in aircraft configuration development, advanced design, production design, and numerical control manufacturing and quality assurance.

INTRODUCTION

During the last decade or so most aerospace companies have applied computer technology to various applications in the design and manufacturing process. McDonnell Aircraft (MCAIR) in the mid-1950's used one of the first CPC (Card Programmed Computer—Vacuum Tube) to perform design computations in the engineering structures design area. In 1959 MCAIR Loft developed the capability of storing the math definition of the exterior aircraft shape for subsequent retrieval and processing of data for design and manufacturing use. Since then, the developments of new applications have appeared at an increasing rate. (Figure 1)

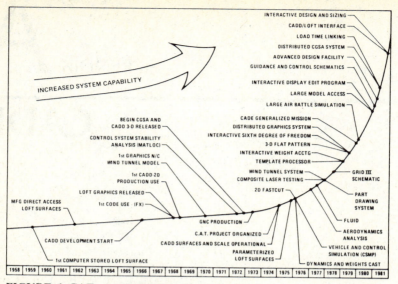

FIGURE 1 CAT systems development.

In 1974 MCAIR organized the Computer Aided Technology (CAT) project pulling together all of the engineering disciplines under one Project management and providing for complete interface coordination with Computer Aided Manufacturing (CAM) activities. (Figure 2)

Heavy emphasis is given to the manufacturing as well as engineering portion of company activities since the manufacturing portion of total system costs is greater across a broad spectrum of program types. (Figure 3)

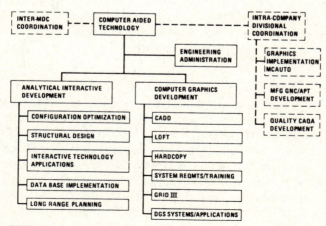

FIGURE 2 CAT organization chart.

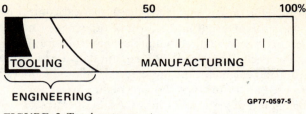

FIGURE 3 Total system cost.

The objectives of the CAT activities are to—

☐ coordinate and promote efficient use of computer resources
☐ coordinate software development insuring proper interfaces between disciplines
☐ provide for shared use of subroutines
☐ develop plans for efficient use of CAT systems on project; i.e., desired capabilities, procedures required, and personnel training.

The present interactive graphics capability ranges from advanced design sizing and performance analysis to quality assurance inspection data. (Figure 4)

The primary graphics module in this system is CADD (Computer Aided Design Drafting) after which all the others are patterned. All are resident on the central computer complex and hence accessible to central data base files. These files are created by each discipline and then accessed as required by the other disciplines. For example, the loft information is directly accessible by the

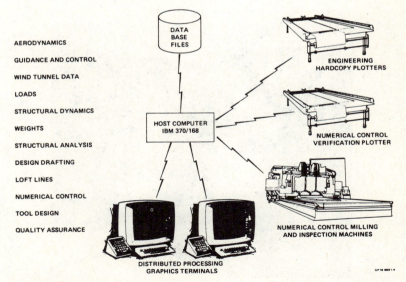

FIGURE 4 McDonnell aircraft computer aided technology.

structures analysts, the design engineers, the numerical control programmers, and the part inspectors. Output in the form of drawings or numerical control milling instructions is sent by direct numerical control wire (DNC) to plotters and milling and inspection machines. Redundant data and efforts are kept to a minimum; errors and calendar time are reduced. All of the interactive graphic display devices used are of the type shown. (Figure 5)

The following is a description of how CADD operates.

CADD—Computer Aided Design Drafting. The cost element in design development is geometry definition. Nearly all the studies, analyses, and definition aspects of a system directly or indirectly relate to the geometry of that system. The CADD module is a very generalized, versatile system for creating, interfacing and storing geometric data in mathematical form in the computer. Hence, the CADD system is the focal point for a wide range of development and fabrication capability at MCAIR. To understand the many possible relationships, a description of the CADD system is helpful.

The engineer interfaces with the computer using the graphics console in three ways. He can use the fiber optics light pen to address the face of the display tube with the computer programmed to respond to the function located at the tube face addressed by the light pen. This is the prime interface used in CADD. The alpha/numerical keyboard below the display surface is used to key-in specific data or text information. The third interface is the function keyboard to the left of the screen. The keyboard (Figure 5) consists of 32 keys which initiate computer commands for frequently used functions. Three types of functions are on this keyboard: create type functions (e.g., point, line, arc, conic, etc.); manipulate functions (e.g., translate/rotate, flip, depth, intersection); and administrative type functions such as reject, delete, display type, etc.

FIGURE 5

By these interfaces the engineer can access and create data in a very generalized, versatile manner permitting real time solutions of geometry problems which are fast, accurate and cost effective.

ADVANCED DESIGN: CONFIGURATION SYNTHESIS

The CADD system is effective as a design tool in Advanced Design. The aircraft configurationist is able to rapidly and accurately construct the vehicle, using time-saving CADD capabilities such as the automated wing planform routine, the kinematics routine for defining the landing gear and controls spatial mechanism geometry, the vision plot routine for ascertaining the limits of the pilot's vision from the cockpit, the vision of a missile's seeker head when mounted as an external store, and the retrieval of engine and weapons geometry, previously created and stored in the computer files. This eliminates the need for recreating these common geometry data each time they are needed for a different vehicle configuration. The designer is also able to define the surfaced outline of the vehicle, the fuel tanks and other internal volumes. This provides him the capability to take section cuts of the vehicle to determine area distribution curves, fuel volumes, and other properties of the configuration that affect the vehicle's performance and mission adequacy.

A typical sequence of design events, iluustrated in Figure 6, is discussed in the following paragraphs to show in more detail how CADD is currently being used in this process.

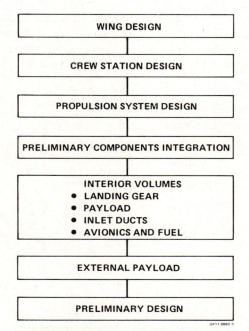

FIGURE 6 Preliminary configuration layout flow chart.

Wing Design

The optimization of a vehicle configuration requires that many different wings be considered and evaluated for performance adequacy. The study, analysis, and test of various combinations of geometric considerations such as wing area, aspect ratio, leading edge, angle, taper ratio, and variations of wing twist, camber, and sweep are necessary to assure the best geometric and aerodynamic wing design is selected. For example, Figure 7 shows but a few of the wings planforms considered for a modern air superiority fighter.

Today, with the use of CADD, this design and selection process provides a compression of the study's time frame at less cost. To establish a wing planform, the designer needs only to light-pen detect TAPERED WING from the menu displayed on the CRT, and type in values for wing area, aspect ratio, taper ratio, sweep angle, and apex origin. The wing planform, mean aerodynamic chord, and quarter chord are immediately created and displayed as shown in Figure 8. Different types of airfoils, previously created and stored in the computer, are available to the designer for merging with the wing planform to give it a three-dimensional definition in a wireframe form. After selecting the desired airfoil, he scales it to the proper chord lengths for the root and tip chords and translates them to their respective positions on the wing (Figure 9). By detecting the two airfoils with the light pen and depressing the RULED SURFACE function key, the wing becomes fully surface-defined with a parametric rules surface (PRS). Should he desire to define the wing with parametric cubic (PC) patches, he converts the airfoils to PC entities, light-pen detects them, and depresses the SURFACE function key. Either of these routines provides the designer the capability to determine the volume of the wing, accomplished by light-pen detecting the surface of the wing and depressing the GEOMETRIC PROPERTY function key. Internal volumes,

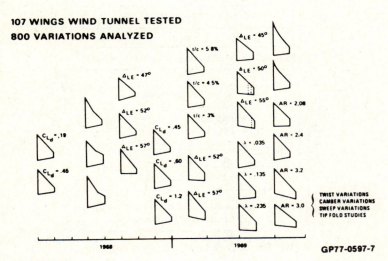

107 WINGS WIND TUNNEL TESTED
800 VARIATIONS ANALYZED

GP77-0597-7

FIGURE 7

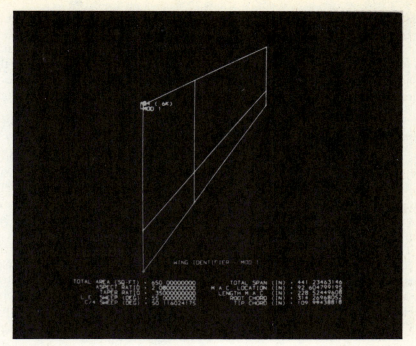

FIGURE 8

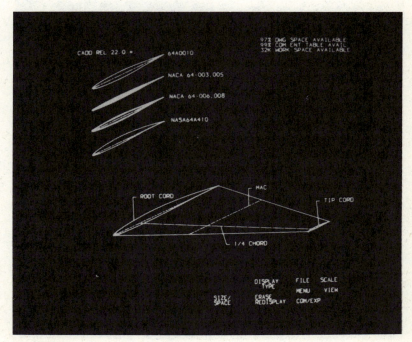

FIGURE 9

such as fuel tanks, can be determined in the same manner, after defining the tanks with surfaces.

This automated and interactive routine allows the designer to consider many different wings in this phase of the design activities and assures him that he has adequately defined a wing that meets all design and performance requirements.

Crew Station Design

The cockpit layout is generally based on specifications, number of crew, method of crew egress, etc. The specification requirements include pilot vision limits, percentile man, instruments required, and cockpit clearance. The methods of egress could be with ejection seat or escape module, with the selection normally resulting from tradeoffs between survivability requirements and minimum weight of the system.

By positioning the pilot's eye at the desired location, the designer is able to construct the canopy and establish the fore, aft, and over-the-side vision constraints as shown in Figure 10. With this established, he can then construct the sill width, console width, and sidewall thickness. The forward fuselage can then be completed by constructing the fuselage half breadth line, the upper

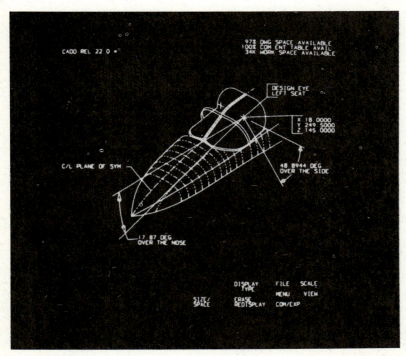

FIGURE 10

and lower moldline, several fuselage station cuts, then transforming the wire-frame geometry into a fully surfaced definition in a manner similar to that previously discussed for the wing design.

Systems and Structural Component Integration

The remainder of the fuselage structure and other structural and system components comprising the airplane are then constructed and integrated into a complete definition of the airframe. These components include the tail surface, inlet and propulsion system, landing gear system, fuel tanks, avionics bays, and weapons systems. The tail surfaces are constructed in the same manner used for constructing the wing. Figure 11 shows the wing and tail surfaces. For the propulsion system design, the designer must first size the engine based on aerodynamic inputs (T/W vs TOGW) using propulsion scaling curves. The base engine configuration, defined and stored in the computer, is retrieved from the computer files, displayed on the CRT (Figure 12), and translated to a position established by Weights department personnel in their preliminary balance of the vehicle. The inlet system is then constructed and merged with the engine as shown in Figure 13.

The kinematics routine in the CADD system is ideally suited for solving landing gear mechanism geometry problems. The location of a skewed axis landing gear trunnion can be determined readily by establishing the extended and retracted positions of the landing gear wheel. In addition, the path of

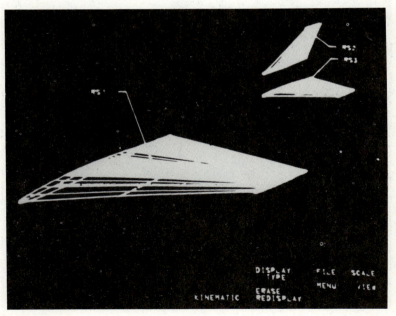

FIGURE 11

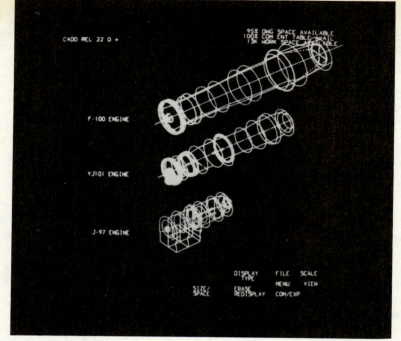

FIGURE 12 Engine.

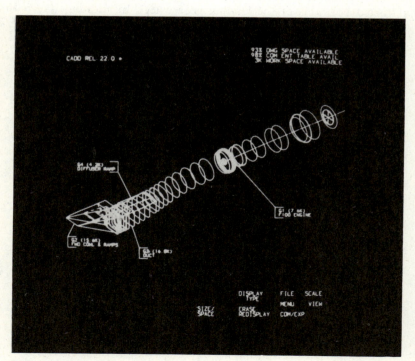

FIGURE 13 Propulsion system.

travel of the gear during retraction is displayed in as many increments as desired, allowing the designer to determine landing gear door opening requirements and minimum clearances with aircraft structure and external stores in the landing gear retraction/extension cycle. It is also beneficial for solving spatial four bar linkage problems (Figure 14) that are time consuming and fraught with tolerance errors when solved manually. This routine provides a graphic display of the positions of the driver, coupler, and driven bellcrank of a spatial mechanism with its intermediate incremental positions displayed, thus providing the designer mechanical advantages and load/stroke data for the mechanism throughout its motion, merely by interrogating the computer.

Several types of external and internal weapons are always considered when defining combat mission scenarios for military aircraft. In the past, it was necessary to trace or redraw the weapons at various locations on the wing and fuselage for each of the various tradeoff configurations, which was a time consuming and noncreative task. Currently, the designer can retrieve from the computer the weapons file (Figure 15) that contains various military inventory weapon geometric descriptions, and merge the selected weapons with the vehicle description displayed on the CRT at any location he desires. Figure 16 shows a missile that has been placed at the airplane's wing tip. This capability saves countless hours of prosaic drafting that would be required if computers were not used.

When the "preliminary design" (Figure 17) configuration is completed, it is ready for aerodynamic analysis to determine if its mission and performance requirements can be met. If analysis shows that these requirements have not been met, ICADE may be used to size the vehicle and determine the sensitivity of the aircraft size to incremental changes in the various vehicle components.

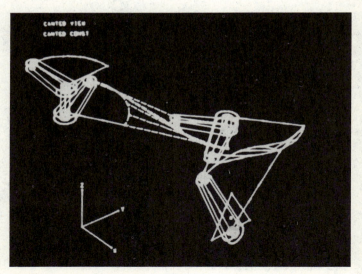

FIGURE 14 Mechanisms.

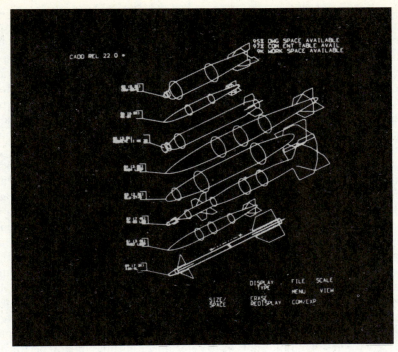

FIGURE 15

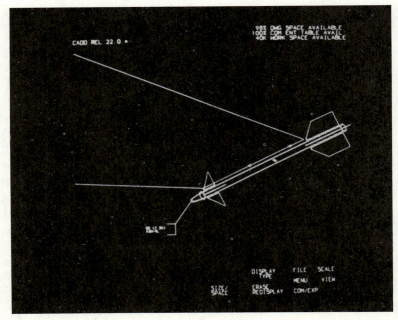

FIGURE 16

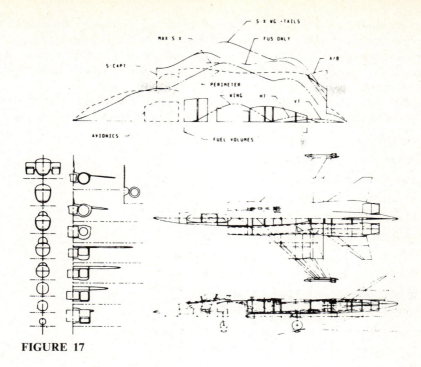

FIGURE 17

Interactive Computer Aided Evaluation (ICADE)

When an aircraft configuration has been synthesized on the CRT and stored in the computer, members of the design team responsible for performance analysis can access geometric data at the same console used to create the configuration, and apply an interactive parametric sizing analysis program called ICADE (pronounced eye′-kaydee), to determine if the flight performance requirements and vehicle size are compatible. ICADE has the capability of sizing the aircraft configuration to satisfy a specified set of flight performance requirements, i.e., size wing, engine, and/or fuel volumes to meet or exceed performance requirements and determine concomitant changes in the aircraft geometry, mass properties, aerodynamic characteristics, installed engine performance, and life-cycle cost characteristics. The system is quite effective when used to conduct studies for determining the effect of parametric variations in aircraft design parameters (e.g., wing geometry, wing loading, thrust-to-weight ratio, structural load factors, etc.) on the aircraft size while holding the performance constant, or determining the effect of parametric changes in performance parameters (e.g., mission radius, specific excess power, etc.) on aircraft size. In addition, ICADE can be used to evaluate specific design or performance tradeoffs; determine the sensitivity of aircraft size to incremental changes in engine size, fuel quantity, fixed weight, aerodynamic drag, specific fuel consumption, etc., while holding the performance constant; and generate data required for the Parametric Design Analysis Procedure, a separate batch analysis program, which is used to determining "optimum" design parameter values which maximize or minimize a measure of aircraft effectiveness (e.g., take-off gross weight, life-cycle cost, etc). Figure 18 depicts the data flow of the program.

All input data required by ICADE have been divided into thirteen categories to facilitate the editing process. These data categories are:

- ☐ Design parameters
- ☐ Program control indicators
- ☐ Sensitivity parameters
- ☐ Mission performance parameters
- ☐ Maneuver performance parameters
- ☐ Miscellaneous performance parameters
- ☐ Geometric characteristics—baseline aircraft
- ☐ Weight characteristics—baseline aircraft
- ☐ C.G. characteristics—baseline aircraft
- ☐ Aerodynamic characteristics—baseline aircraft
- ☐ Propulsion System characteristics—reference engine
- ☐ Cost characteristics—baseline aircraft
- ☐ Payload characteristics

The analyst initiates the editing process by detecting with the light pen the data category to be edited from a menu display. The data from the selected category are then displayed in card image form identical to that used for the batch version of the program. A menu of commands is also displayed on the CRT.

During execution of the ICADE sizing iteration, the iteration number and current values of fundamental aircraft sizing parameters are displayed. These data provide the analyst with information on the convergence behavior of the sizing iteration, but do not allow him to interact with this phase of program execution.

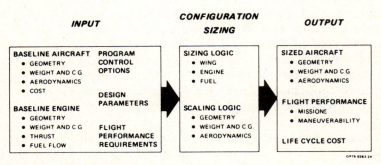

FIGURE 18 ICADE program operation.

After the sizing iteration has converged to a solution and the cost and mission performance characteristics have been calculated, the following summary of the sized aircraft characteristics is displayed:

- □ Geometric characteristics (comparison with baseline aircraft)
- □ Weight characteristics (comparison with baseline aircraft)
- □ C.G. characteristics (optional)
- □ Mission performance
- □ Maneuver performance
- □ Cost characteristics (optional)
- □ Aerodynamic characteristics

With the rapid output of these data from the computer to the advanced design team, a more thorough evaluation of the study vehicle can be made, attributed largely to the increased number of design iterations possible in a given time frame, and due to increased visibility and recognition of convergence to the desired performance results.

COMPUTER GRAPHICS AERODYNAMICS ANALYSIS (CGAA)

This module has been developed to provide a data bank of vehicle geometry information for input to all of the aerodynamic analysis methods used to predict vehicle performance. The module is designed to be applied in the early phase of advanced design.

In the earliest phases of a vehicle design study, configurations are evaluated by manual layout which form the basis of the studies to establish wing, fuselage, and control surfaces locations as well as vehicle volume. CGAA uses these layouts as an input through the digitizing capability of the 4014 Tektronix system. CGAA operates using a Tektronix 4014 terminal, two digitizing tables (12″ × 12″ and 48″ × 36″) and a 4631 hardcopy device and uses the IBM/TSO system.

The geometry input requirements for aerodynamic analysis programs are assembled in CGAA in a fashion to simplify the data management. These data files for fuselage, wing, control surfaces (horizontal and vertical tails, canards, etc.), engine pods, and fuel tanks are maintained in the file structure so that the user can modify, merge, delete and create a geometric data base through CGAA. Figure 19 shows the menu table installed on the digitizer and Figure 20 shows typical geometry data as displayed by the program.

The aero analysis programs such as Vortex Lattice (Reference 1), Wave Drag (Reference 2) and other procedures which use vehicle geometry as input are coupled to CGAA with pre-processors and post-processors to prepare and check input and display output results of each calculation method. Figures 21 and 22 show typical input to the Vortex Lattice method and displays of results. The program provides a significant reduction in manpower and span

1 FILE MANIPU-LATION	2 WING	3 H TAIL	4 V TAIL	5 CANARD	6 FUSELAGE	7 POD	8 NACEL	9	10 END
11 RETURN	12 CREATE	13 MODIFY	14 DISPLAY	15 DELETE	16 INTERR	17 SECTION	18 PLAN-FORM	19 SEGMENT	20 REJECT
21 ACCEPT	22 KEYINN	23 DIGITIZE	24 TRANS-LATE	25 SCALE	26	27	28	29	30
31 A	32 B	33 C	34 D	35	36	37	38	39	40

GP76 0710 5

FIGURE 19

time required to perform estimates of aerodynamic performance and allows a rapid refinement of the vehicle geometry.

Computer Graphics Structural Analysis (CGSA)

CGSA is a module that has been developed to aid structural analysis and design. CGSA is used as an aid in providing internal loads, stresses, and strains for structural sizing, and documentation of structural integrity. This system functions in conjunction with the CADD system. A special overlay is used on the function keyboard when the CGSA module is in operation.

CGSA is used to prepare three-dimensional finite element model input data for either the CASD or NASTRAN structural analysis programs (Figure 23). (CASD, an acronym for Computer Aided Structural Design, is a structural analysis program developed by Douglas Aircraft Company prior to the development of NASTRAN by NASA.) Basic geometry is developed using CADD from loft data, conventional engineering drawings, or the three-dimensional geometry of the vehicle constructed and filed in the computer by

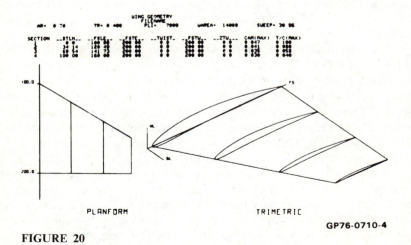

PLANFORM   TRIMETRIC

GP76-0710-4

FIGURE 20

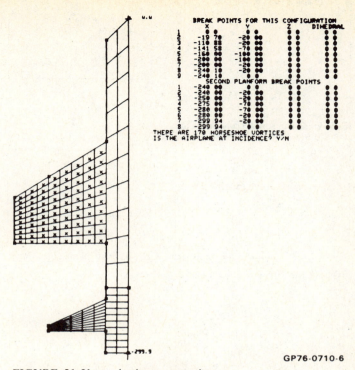

FIGURE 21 Vortex lattice program input.

GP76-0710-6

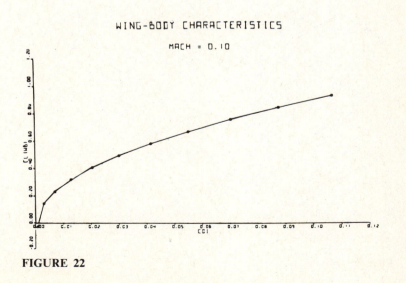

FIGURE 22

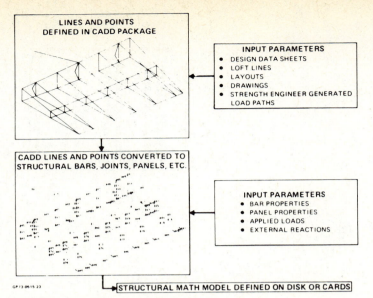

FIGURE 23 Structural idealization.

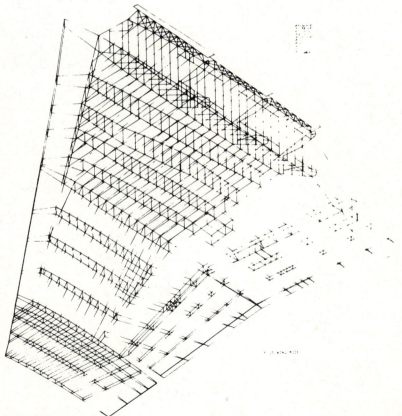

FIGURE 24

the designer (Figure 24). This point-line model is converted to a finite element model with CGSA by defining bar/panel/membrane element properties, applied loads, and external boundary conditions. The model may be viewed, checked, and edited at the console for verification of the input data. Computer-drawn hardcopies can be produced displaying input properties and element labels (Figure 25) and the model is then stored in the CADD drawing files for future retrieval. The CASD or NASTRAN input card images are automatically created on a disc data set for submittal to the structural analysis program. The utilization of CGSA for model generation eliminates the need for manually creating the pictures, input sheets, and punched cards.

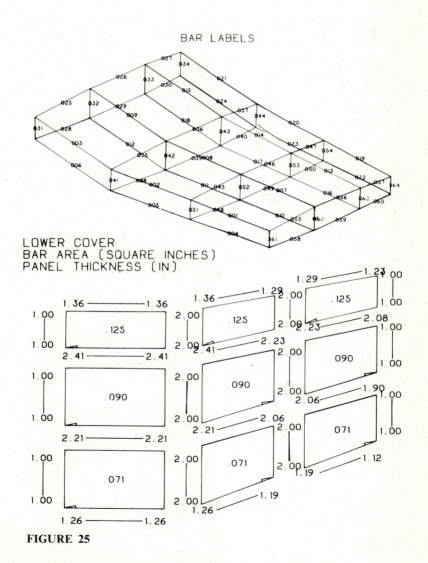

FIGURE 25

Upon completion of the CASD or NASTRAN structural analysis program (which performs a minimum strain energy solution), the results are stored with the model in the CADD drawing file and are also output in the more usual manner on an alphanumeric printer. The answers may be displayed at the console and hardcopied by the computer for all structural elements. The display options include magnitudes of bar load, bar stress, panel shear flow, panel shear stress, membrane stress, membrane strain, and membrane running loads. Also the plotting of bar loads or bending moments, and deflected shapes are available (Figure 26).

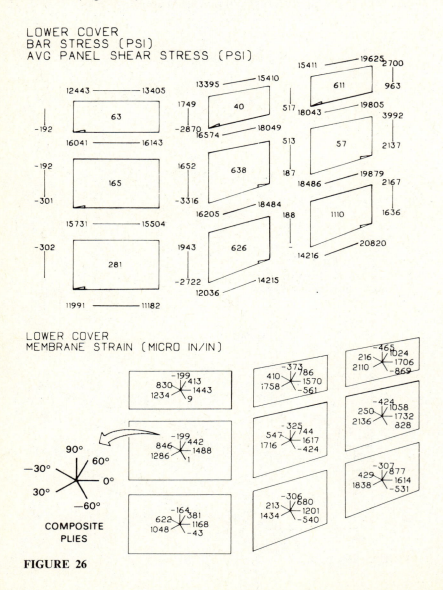

FIGURE 26

The displays and hardcopies in the form of load sheets are used by engineering supervision to gain an overall feel for the structural analysis, and also by the detail stress analyst as inputs for his computations. This type of graphical presentation eliminates the need to manually prepare load sheets. This was previously done by reading the printed output data and associating it with the pictures of the structural idealization.

CGSA is effective when used in both Advanced Design and Project environments. Advanced Design requires smaller models, quick turnaround, and good visibility, and CGSA adds to the technical credibility of technical proposals. In a Project environment, it is effective in handling large quantities of data (the F-15 wing model has 10,000 card images) and in providing Design and Strength department personnel with essential data when responding to design changes and updates.

INTEGRATION OF STRUCTURAL ANALYSIS PROCEDURES

The efficiency of the structural design process has been significantly improved by integrating structural analysis methods on the computer. This allows direct data transfers between computer programs used by individuals in strength, loads, weights, and structural dynamic disciplines.

In the past, it was not uncommon for computer generated data to be printed, transferred to input sheets, key-punched, and loaded back into the computer. Figure 27 shows an example of computer controlled data in which the Loads department is providing the Strength department external loads for finite element analysis. The Loads department receives geometry, structural stiffness and degree of freedom data from the Strength model. The external loads are computed and stored with the Strength model. This avoids manual intervention and assures compatibility of load application points with structural analysis models.

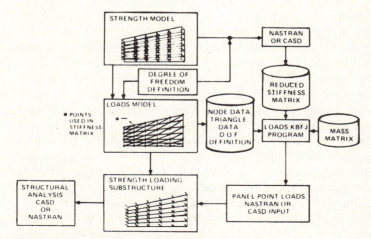

FIGURE 27 Strength/Loads.

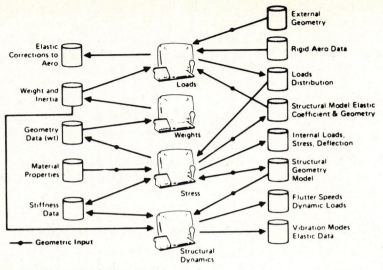

FIGURE 28 Structural technology.

Figure 28 schematically illustrates the coupling which has been accomplished to integrate structure technology programs.

Field of Vision Plots

Investigation of the pilot's field of vision is a customer requirement. In the past, the results were illustrated on two-dimensional diagrams culminating from geometry data generated by conventional drawing board techniques. With the construct capabilities and the three-dimensional aspects of the CADD package, however, the vision limitation angles can be determined directly (Figure 29). Geometric diagrams made on the CRT can then be merged with a standard vision plot format retrieved from the files in the computer (Figure 30). Hardcopies of the completed vision plot can then be made for the final report.

The configuration at the end of the optimization cycle becomes a point-design configuration (Figure 31), and copies of the configuration may be retrieved by the designers from the computer files for auditing or for making more detailed design studies, such as internal structural arrangements, landing gear mechanism geometry synthesis, field of vision variations, etc. The point designs are representative of established mission areas in the study matrices. There are as many baseline designs generated and optimized as are practical to obtain a comprehensive cross section for point design selection. The baseline designs are investigations of tradeoffs of such things as variations of wing sweep and aspect ratio (Figure 32) and fixed vs variable sweep wing geometry (Figure 33). Some points in a matrix may be completely different design approaches to preclude the omission of any configuration possibility. In the end, the study program manager must select the vehicle configuration that

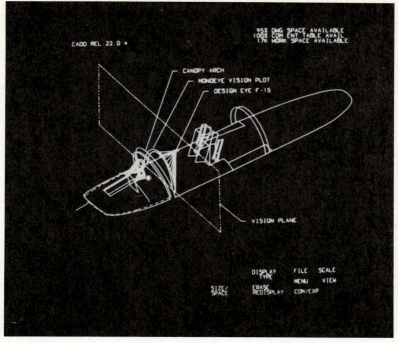

FIGURE 29

best meets the mission requirements stipulated by the customer and establish it as the "point design."

Design Aids

Design aid models of plywood or similar material can be quickly and accurately fabricated by using a Fastcut module that accesses the CADD drawing and automatically computes N/C milling instructions that are sent via DNC to a special two-axis N/C profiler (Figure 34). The individual pieces are then assembled, erector set fashion, into structure mockups (Figure 35).

Loft Lines

When the configuration is established to the point that many disciplines need access to the current external shape (loft lines), a mathematical definition of external surface patches of the aircraft is generated and computer stored (Figure 36). The surface patches are given identification numbers which serve as a label for retrieval by using disciplines (Figure 37). The user need only define the labels of the surfaces of interest and the kind of information desired, i.e., section cuts, bend angles. numerical control machining path, etc.

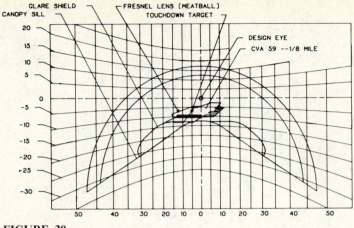

FIGURE 30

Most of the surface types used are of the "generated" variety. A surface is generated by a curve type moving along directrix control lines (Figure 38). The mathematical curve types used can vary from 1st degree (straight line) equations to 6th degree polynomials (Figure 39). These surfaces are processed using parametric mathematics, utilizing standard evaluators and general processors. This allows loft to generate and process parametric bicubic surfaces when required, which is usually for advanced design projects.

The CALL module has been developed to utilize interactive computer graphics in the definition and check-out of these surfaces. Several surface analysis options are available to assist the loftsman in verifying the lofted contour. These include a mathematical surface-surface intersection algorithm (Figure 40), a true parallel surface option, and planar cutting logic that is not dependent on boundary intersections (Figure 41).

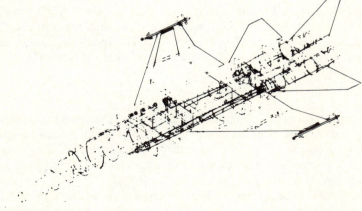

FIGURE 31

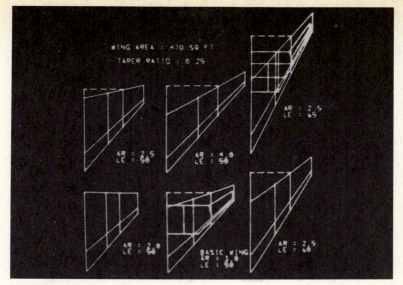

FIGURE 32

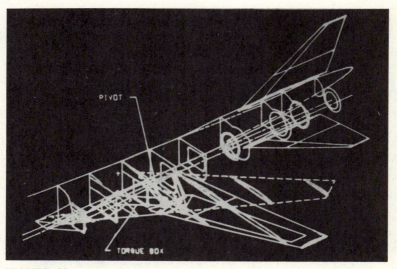

FIGURE 33

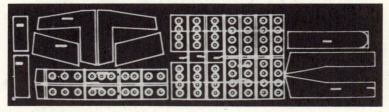

FIGURE 34

FIGURE 35

To facilitate communications between the designer and loftsman, compatibility with the CADD drawing files is maintained; this is important when configurations are being iterated and data must be transferred between two application modules. Communications are also maintained with the batch and time-sharing systems; this allows the loftsman options to utilize the most efficient mode of computing for preliminary preprocessing.

Several surface processing options are also available to furnish feedback to the designer; the most used of these are wetted area, volume, and field-of-vision plots (Figure 42). These vision plots have been adapted to obtain clearance plots for laser pods (Figure 43), cameras, etc., and also to make shadow

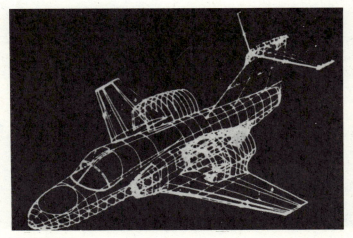

FIGURE 36

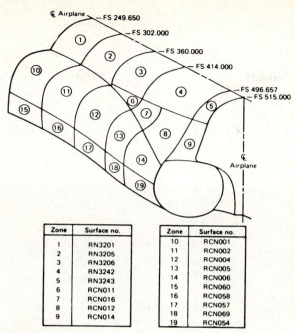

Zone	Surface no.
1	RN3201
2	RN3205
3	RN3206
4	RN3242
5	RN3243
6	RCN011
7	RCN016
8	RCN012
9	RCN014

Zone	Surface no.
10	RCN001
11	RCN002
12	RCN004
13	RCN005
14	RCN006
15	RCN060
16	RCN058
17	RCN057
18	RCN069
19	RCN054

FIGURE 37

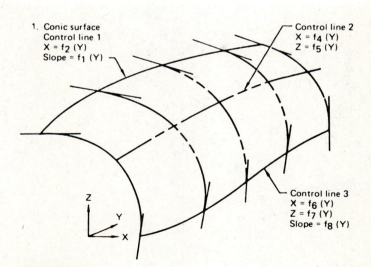

1. Conic surface
Control line 1
$X = f_2 (Y)$
Slope $= f_1 (Y)$

Control line 2
$X = f_4 (Y)$
$Z = f_5 (Y)$

Control line 3
$X = f_6 (Y)$
$Z = f_7 (Y)$
Slope $= f_8 (Y)$

FIGURE 38

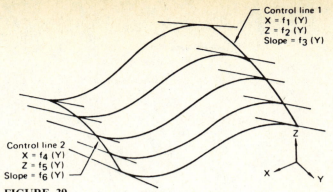

Control line 1
$X = f_1 (Y)$
$Z = f_2 (Y)$
Slope $= f_3 (Y)$

Control line 2
$X = f_4 (Y)$
$Z = f_5 (Y)$
Slope $= f_6 (Y)$

FIGURE 39

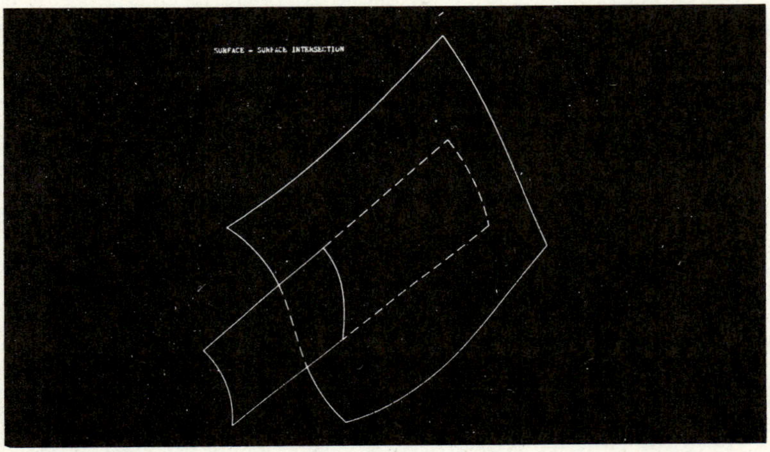

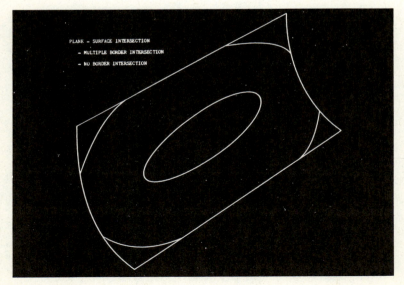

SURFACE - SURFACE INTERSECTION

FIGURE 40

PLANE - SURFACE INTERSECTION

- MULTIPLE BORDER INTERSECTION

- NO BORDER INTERSECTION

FIGURE 41

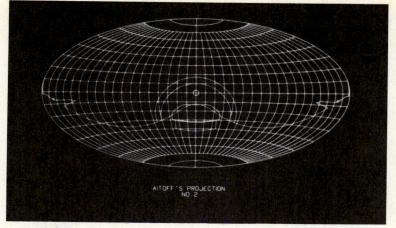

FIGURE 42

plots. The data for these plots can be displayed in either perspective projection, Aitoff's equal-area spherical projection or in degrees of azimuth versus elevation. Plots of this nature are not only used for design criteria but also to satisfy customer requirements.

Loft has the responsibility for maintaining the plotting facilities (Figure 44) and the hardcopy module that is used by MCAIR. The hardcopy module interfaces all graphic application modules that have hardcopy requirements. It computes and files the plotting data in a queue for subsequent transmission

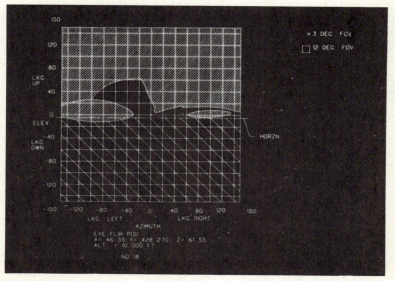

FIGURE 43

FIGURE 44

to a plotting facility. Each plotter has sufficient mini-computing capacity to generate the pen motions required so that certain identities can be set as symbols or code. This minimizes data transmission and reduces execution time on the 370/168. These mini-computers also allow the hard-copy module to be machine-independent by acting as an interface.

Wind Tunnel Model Fabrication

In the process of converging on a configuration, many variations must be extensively tested in wind tunnels. A disassembled wind tunnel model (shown in Figure 45) gives an indication of the variations of wings that are tested on each model. In a large development program such as the F-15, over 100 different wing configurations consisting of variations of twist, camber, and sweep were tested on models of different scale resulting in the requirement for fabricating 550 (Figure 46) unique model wings of solid aluminum or stainless steel.

FIGURE 45

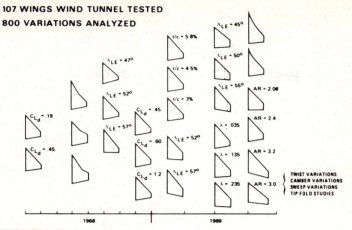

FIGURE 46 Wing development.

MCAIR applied N/C Manufacturing techniques in 1958 to the fabrication of model parts as a cost and time savings measure. The demanding requirement of quickly producing a large number of unique, but similar, parts by N/C led to the development of a specialized computer program that will produce the N/C machining instructions from a minimum of program input.

The model parts are sculptured by N/C from metal plates of standard thickness (Figure 47). The diagram of the program operation (Figure 48) indicates on the left side the "standard" input items such as planform geometry, model scale, cutter radius, etc. Then depending on the shape desired, the equations of various control sections are also input.

The program then constructs a mathematical model of the surface and generates the cutter path for the desired size of cutter. The output is fully

FIGURE 47

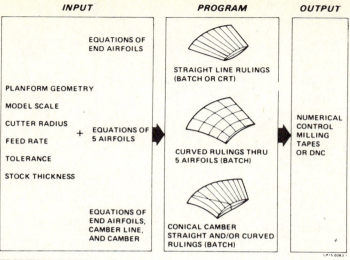

FIGURE 48 Wing mill.

postprocessed N/C milling machine instructions in the form of punched tape or Direct Numerical Control (DNC) instructions.

Other more complex shapes such as the F-111 escape capsule (Figure 49) are programmed conventionally and sculptured by N/C.

These methods allow much more accurate parts to be generated quickly and less expensively. To a project trying to find optimum shape configurations, this means more iterations in a limited time and hence a better design.

FIGURE 49

Product Design Applications

The designer can use loft data as discrete intersections, or he can use a series of intersections to generate a surface that approximates the lofted surface. The most widely used applications of CADD are geometry studies that involve determining the relationship of one set of geometry to another set. Figure 50 shows the outline of a hydraulic actuator nested inside a modeled approximation of loft surface. It is a very easy task to interrogate the geometry to determine the distance between the actuator and the approximated loft surface. Figure 51 shows the F-18 Inflight Refueling Probe nested into the refueling drogue. A short line modeled at the aft inboard corner of the drogue envelope shows the minimum distance between the drogue and the fuselage during refueling. A logical next step to this design would be to determine the linkages mechanism to drive the probe from retract to extend and back again. CADD kinematics routines will automatically develop equivalent positions along the rotation of a series of cans, bell cranks, etc. Kinematics can be used to determine the arc path required for a given bell crank, the arc length for a bell crank, the length of control linkages, and the on-center or locked position for many control systems without restriction as to size, complexity, or location in real space.

After a design is proven to be feasible, it becomes necessary to communicate that design to other people for coordination. Figure 52 shows the Design Data Sheet to define the mounting provisions for a radar set. The point coordinate data are automatically generated.

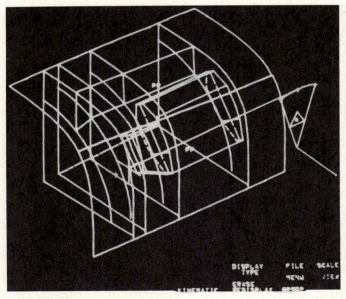

FIGURE 50

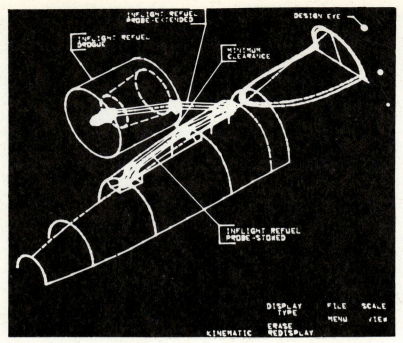

FIGURE 51

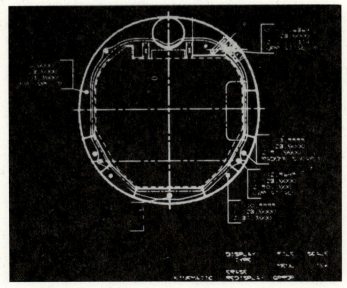

FIGURE 52

Three-dimensional surface definitions in CADD permit full definition of machined parts as shown in Figure 53. The surface definition on this part can be used by design to insure proper fit with mating part, and by manufacturing and inspection as direct input to numerically controlled machine part programs. For two-dimensional blueprint documentation of the part definition, automated routines will generate and arrange section cuts through the part as shown. The labels are the identification of the surface from which the section cut was taken. The dimensions were added interactively with the value automatically calculated.

CADD has also been an invaluable tool in the design of laminated composite structure such as the F-15 speedbrake. The structure is made up of a series of 0.0054 inch thick graphite composite plies that are preimpregnated with epoxy. There are seventy-six plies at one section of the speedbrake, each of which is moldline related. In the past, it was necessary for the designer to describe each ply in the installed configuration, but he was unable to define the flat pattern dimensions required by manufacturing to cut out the plies from raw stock. Personnel from the Master Layout section of the Loft department would have to manually lay out each ply which was costly and time consuming. With the Triangulated Flat Pattern routine in the CADD system, our designers now define the installed, curved configuration of each ply, then instruct the computer to unroll them into a flat pattern. In an instant, the ply is displayed on the CRT screen. Each of the flat patterned plies is then converted to numerical control instructions and set via DNC to an automated laser cutting machine for cutting out the material. Thus, manual layout errors are eliminated and the design/fabricate cycle time is reduced. Figure 54 shows a typical example of a flat pattern that has been rolled out from its curved installation position.

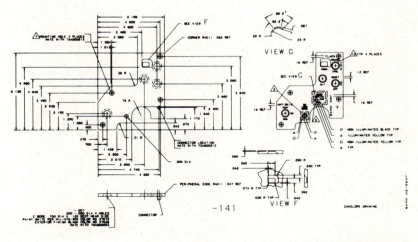

FIGURE 53 Automated dimensioning.

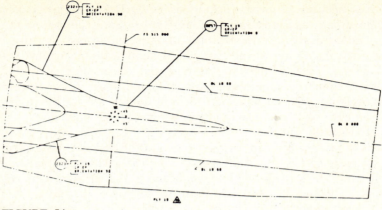

FIGURE 54

Conventional sheet metal parts are easily defined with CADD. By light-pen detection, a CADD operator can input the contour and flange limits of the part. In turn the computer asks tutorial questions that may be answered by selecting menu items displayed on the CRT screen or by the alphanumeric keyboard. The variables requiring answers are flange width, material thickness, flange bend radius, and rivet size and spacing. From these inputs, the program develops a completed flat pattern of the part showing joggles, developed flanges, rivet locations, bend angles, form block lines, and bend tangent lines. This

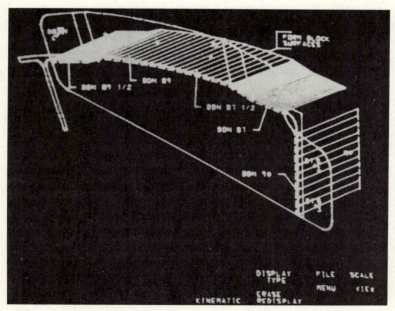

FIGURE 55 CADD sheet metal design.

routine compresses the design time required for sheet metal parts. Figure 55 shows a typical flat pattern sheet metal part designed with this routine.

Many other types of design tasks are accomplished daily using the CADD system, including complex spatial kinematics synthesis problems, the routing of hydraulic and fuel lines, general equipment installations, determination of clearance problems of moving elements with respect to structure, and generally, any type of layout work conceivable. All of these tasks are accomplished faster and with more fidelity than was ever possible prior to the advent of interactive computer graphics.

Manufacturing Interface

The mathematical models of structure that are created by Design using the CADD system and stored in the Engineering Data Base serve as the vital link in the Design-Manufacturing interface. The manufacturing programmer using CADD retrieves a copy of the CADD part model and adds tooling lugs, clamps, and other manufacturing requirements.

The N/C parts programmer, using Graphics Numerical Control (GNC), then accesses the revised CADD drawing, and using the graphics display interactively, he creates a N/C source program which describes the manner for cutter to sculpture the part using the CADD labels for identification. Also included are the N/C milling machine to be used, cutter size, feeds, speeds, etc. After the programmer is satisfied with the cutter motions graphically depicted (Figure 56), he processes the source program through the APT program which retrieves the engineering geometry mathematics and generates the cutter path. The parts programmer need only concern himself with the geometry labels and need not be involved with the complex mathematics which define the part. If changes in cutter sizes or part geometry are made, the source

FIGURE 56

FIGURE 57

program can be easily changed from an alphanumeric terminal (or batch) and the source program can be reprocessed through APT for revised machine cutting instructions. The part is then machined on one of the 90 N/C milling machines at MCAIR (Figure 57).

Composites Fabrication

The trend toward more extensive use of composite materials is clearly being established by the F-18, Harrier and other pending programs. These advanced materials require controls and fabrication techniques quite foreign to traditional manufacturing processes.

The typical composite assembly is made up of a honeycomb core, a torque box with spars and ribs, and a thin machined frame on which graphite or boron plys are layed-up (Figures 58 and 59). These components are then bonded together to form an integral assembly. Viewing the final article is likely to give a false impression of its complexity since it is apparently very simple, yet each component must be fabricated with great care and precision to assure that the assembly goes together and performs as required.

One of the most critical operations is cutting the many plys which form the outside, visible layer. The DNC laser cutting machine (Figure 60) is evidence of the new technology being employed in the fabrication of these advanced components. These machines operate at high speeds while burning through the thin broadgood material. Each ply is a little thicker than the paper on which this is written, so it is easy to see that many plys are required to make a structurally sound assembly. The fabric is constructed so that it has strength in only one direction—along its grain. To compensate for this, each ply is oriented with its grain in some specific direction. If we could look through the layers

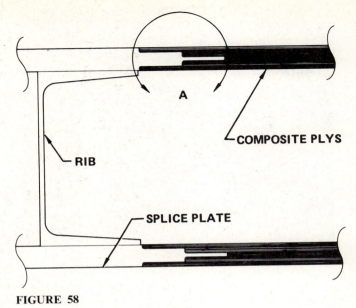

FIGURE 58

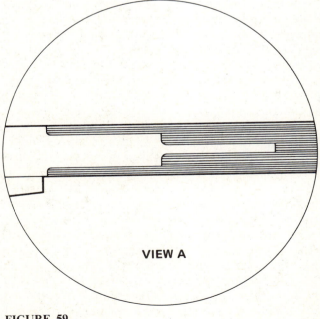

VIEW A

FIGURE 59

FIGURE 60

we would see a crosshatch pattern which gives the required strength in each direction. This means that even though two adjoining plys have the same peripheral shape, the two cannot be fabricated the same. In many cases the orientation of the grain will result in the condition where the shape will not fit within the width of the material. When this happens, the detail can be split along the grain to be joined together later when the plys are layed-up.

Because of the complexity and not unimportantly the high cost of these materials, it becomes imperative that advanced techniques be developed to support the cutting and fabrication processes. With this in mind, MCAIR initiated a program to define, develop and implement computer-assisted systems for fabricating these plys.

The key to properly controlling the flow and utilization of this high cost material is a technique called nesting. Nesting is quite simple in concept. Most people have put together a picture puzzle. For all practical purposes a puzzle is a nest where the pieces are contrived in such a way that when the puzzle is done all of the pieces fit within the borders leaving no holes. This results in 100% utilization of material. Industrial nesting could be compared to taking one piece from a number of different puzzles and then placing them together the best possible way. A lot of holes will result, but, if enough time is expended, a fairly high yield will result (Figure 61). Except for some very unusual situations, it is not possible to arrive at any single solution for a given set of shapes. In general, the more time spent in producing a nest, the better will be the yield. The advantage in using a computer is that in a given period of time, more tries can be made to improve the nest. A yield on the order of 60–90% can be expected in most cases (Figure 62). The actual yield is influenced by a number of factors such as complexity of the shapes, number of details, raw material dimensions and the kind of constraints imposed upon the nest. The kind of constraints most frequently encountered are related to matching

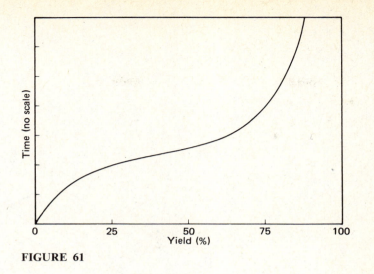

FIGURE 61

grain direction, patterns, and keeping the right side up. Graphics is used to add visibility during the nesting process and to provide a degree of immediate verification not possible in so-called black box automated systems.

The Graphics Nesting Program (GNP) is the result of this effort. GNP utilizes existing IBM 2250 graphics display hardware to facilitate rapid implementation. The CADD file management and display routines were used "as is" to assure compatibility with design users. See Figure 63 for GNP system flow.

CAD/CAM Tubing System

This system, which employs the use of a common data base of part-number related data, is jointly accessed by Engineering, Manufacturing and Q/C personnel to aid the design and manufacturing of tubing within MDC.

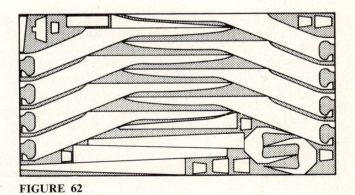

FIGURE 62

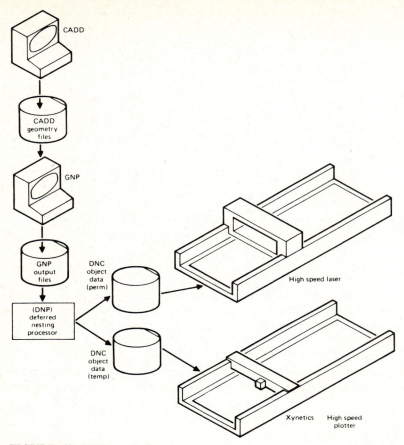

FIGURE 63 GNP system flow.

Within an aircraft such as the F-18 there are approximately 1200 tubes. In an aircraft such as the DC-10 there are about 3000 tubes requiring three miles of tubing.

If we take a closer look at a DC-10 main landing gear wheel well (Figure 64), we'll see a maze of tubing—tubes that must be bent within many constraints in order to fit. A missile or a fighter aircraft has fewer tubes but in more restricted space. These conditions plus the fact that a considerable amount of more rigid and more expensive titanium tubing is being used, have made accurate tube production a highly desirable goal from a cost standpoint.

One of the major factors that paved the way for a computer-aided system was the new generation of measuring and bending equipment which became commercially available in 1974. This bending equipment is designed to produce tubes that match predefined coordinates. The process control minicomputer, which is the controlling part of the measuring and bending equipment, also has the software to compensate for springback.

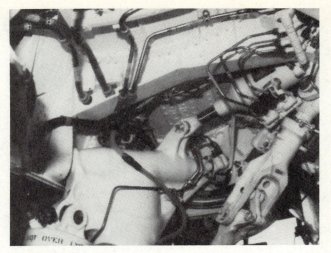

FIGURE 64

With this equipment bent tubes can be repetitively and accurately fabricated and will conform to the mathematical definition which is stored in the form of three-dimensional coordinates resident in the process minicomputer.

The measuring equipment accurately measures tubes either for the purpose of establishing the coordinates so that duplication by the bending equipment is possible, or for the purpose of comparing the bent tube to predefined coordinates to ascertain acceptability of the tube.

The following is a description of how this newly developed computer-aided system works for tubes that are completely defined by Engineering.

Beginning with the CADD interactive graphics system, the design engineer (Figure 65) develops the layout of a system of tubing on a 2250 terminal using the centerline of the tubing and testing for distance between centers where it looks like spacing may be a problem. When he is satisfied with the layout, he selects a portion of the layout which he determines will constitute one tube, and enters the description of this tube into a CADD program function called FLUID, or Formed Lines Using Interactive Data. This newly developed function is the interface between the CADD geometric file model and the "Tube File."

The FLUID software prompts the engineer by displaying appropriate menus on the graphics CRT (Figure 66), allowing him to lightpen pick many of the elements of data. Through built-in software controls, one and only one version of any tube data is stored in the Tube File at any given time. The Tube File is the data base of all the pertinent information about a tube—its shape, end fittings, material, pressure test requirements, and other items. From the data in the Tube File, the application programs generate reports, forms and data required by the users of the system. A data identifier, a two-digit suffix to the tube part number is used to control configuration during development

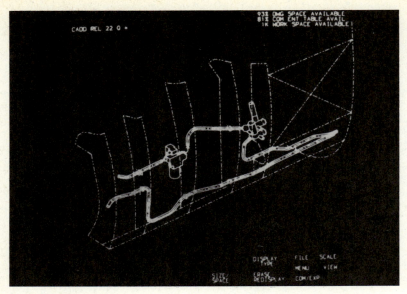

FIGURE 65

FIGURE 66

of the tube. This identifier is automatically incremented whenever any data defining the tube is changed. When Engineering, Quality, and Manufacturing have signed off on a new tube configuration, the data for that tube are then identified by Engineering Release as "firm-release" and from then on are protected from any further change—a safeguard built into the system.

By using a Tektronix terminal in the Manufacturing Planning Department, a Production Planner will be able to review any tube in the file by having a Tube Fabrication Instruction Sheet (TFIS) (Figure 67) displayed on his CRT. A press of a button and a hardcopy of the TFIS pictorial planning will be produced for filing and attaching to work orders as they are released to the shop. This new form is computer generated from the data in the Tube File and is designed to eliminate the need for both the master tube and the manually maintained bending data called a "chart card." When we eliminate the master tube we eliminate this storage area also.

Another application, using data in the Tube File, is the generation of contractually specified customer drawings (Figure 68). In the case of military aircraft, the customer drawing will be produced by a batch run of an application program on an as-required basis. The generation of the customer drawing is truly a by-product of the system.

In the tube manufacturing area, data for the measuring and bending equipment are requested via a communications minicomputer from the host computer, where the Tube File resides. Data required for bending the tube are copied from the file and transmitted back to the minicomputer for intermediate storage, where data for approximately 50 tubes can be stored. This is how tube data are moved from the Tube File to the communcation minicomputer.

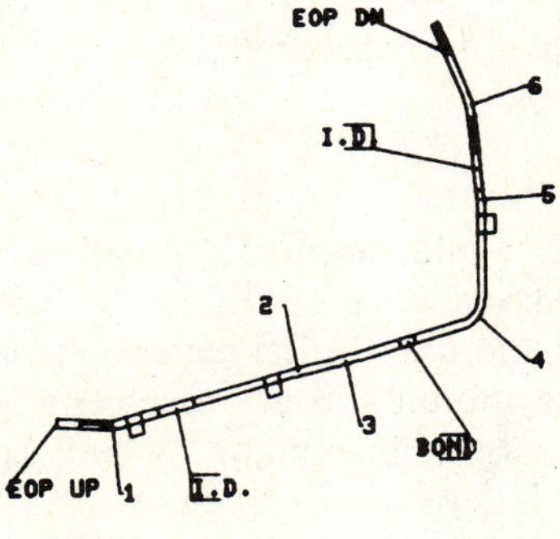

TFIS

BEND	DBB	POB	DOB	BEND	DBB	POB	DOB
1-UP	1.83	.0	22.6	9			
2	6.01	-140.9	49.1	10			
3	1.64	180.0	70.9	11			
4	3.98	-99.1	79.2	12			
5	2.77	85.5	14.4	13			
6	2.30	-139.2	61.1	14			
7-DN	2.36						
8							

END-TO-END LENGTH 17.41

TI3AL-2.5V.250 X.028X25.620
END FITTING XMAT
UP 7M523T4 NONE
DN PERMSWG 1.0
BENDER TF55 BR-.750 WD-NR
PD-NONE CB-.750 PT-12000 PSI
EC-NR ST9M389HA151 STF-1301

TOL: LGTH-.05 OFST-.07 | TFIS-88A693725-1001 -01 | WMP REV A 08/10/81

FIGURE 67

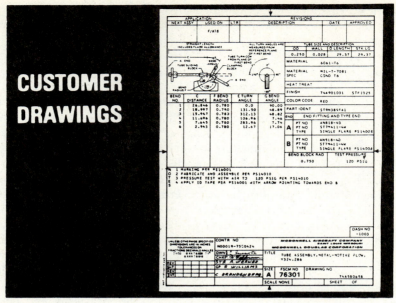

CUSTOMER DRAWINGS

FIGURE 68

Today, lengthly Engineering calculations and manual summarizations are required to obtain tubing weights and center-of-gravity (C.G.) data. Under the CAD/CAM Tubing System this requirement will be satisfied by inputting liquid density and calculated C.G. coordinates into the Tube File via the FLUID module and running an application program to obtain weights and C.G.s of tubes or groups of tubes. The total tubing system includes operations from Design through Quality (Figure 69).

S.M.D.C.

The most recent FLUID application is for creating S.M.D.C. (Shielded Mild Detonating Cord) tube drawings. S.M.D.C. is an explosive stimulus contained in stainless steel tubing and used in the aircraft emergency escape system for transfer of the explosive stimulus from one escape system device to another. It is independent of electrical, pneumatic or hydraulic power sources which in the event of a system failure would preclude the safe ejection of the aircraft crew. The fluid routine is used for data storage of SMDC tube data and geometry in the tube file in the same manner as a regular tube, but with the added feature of producing a computer generated drawing (Figure 70) which is included in a SMDC Procurement Specification Control Drawing. This drawing provides all the centerline coordinate data of the tube geometry along with other pertinent data which the vendor uses to manufacture the tube on the same type automated measuring and bending equipment used for a conventional fluid tube.

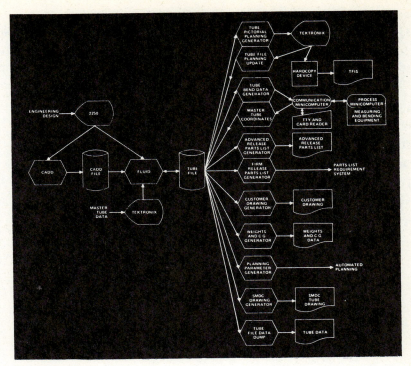

FIGURE 69

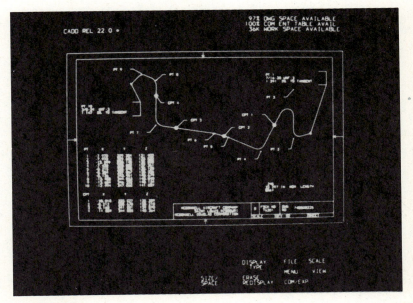

FIGURE 70

Standard Parts/Items File

The CADD Graphics User has the option to access a standard part/item file. Within this file are categorized sub files of many of the most commonly used standard parts and items. A brief list of some of the contents in these sub files follows:

☐ Standard Parts—clamps, fittings, bolts, nuts, etc.

☐ Weapons Library—missiles, bombs, guns, etc.

☐ Equipment Library—antennae, equipment envelopes, etc.

☐ Drafting Aids—drawing forms, aircraft profiles, etc.

☐ Tooling—angles, tooling tabs, slots, etc.

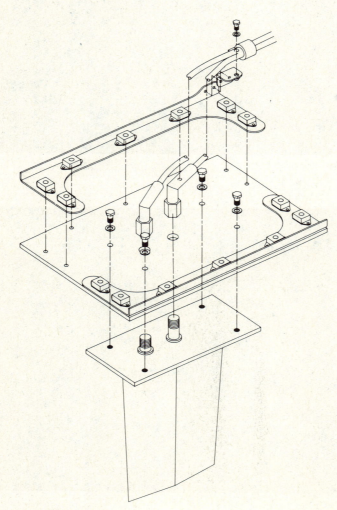

FIGURE 71

These Standard Parts/Items are accurate, CADD defined, three dimensional models stored in a protected file to insure their integrity.

The capability to access these files allows the user to place any standard part/item into a graphics model at any desired orientation. This is accomplished by one of three input methods. The sequence of detected inputs made by the user determines the orientation of the incoming standard part/item.

Figure 71 (exploded for clarity) is a typical graphics 3D structural assembly which contains 25 standard parts/items. These parts can be interrogated to assure proper form, fit and function.

The weapons library enables the user to place a variety of weapons, external fuel tanks, etc., into a model which contains wing and/or fuselage geometry.

Once again, the model can be interrogated for proper clearances and orientation, adjustments can be made if required, with a minimum of effort.

The advantages of having access to the standard part/item files are:

☐ Current, uniform and accurate data is on line and available for all users

☐ The ability to place items at any desired orientation

☐ A reduction in time to create a drawing containing a multitude of standard parts/items

Quality Assurance

An integral requirement to automatic fabrication is to also consider the necessary Inspection process.

The Quality Assurance programmer using CAQA accesses the engineering model and lightpen detects the surfaces of the part for which inspection data are to be generated (Figure 72). The CAQA module analyzes the surfaces and

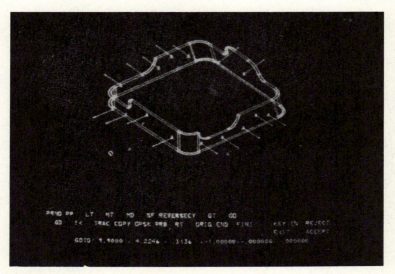

FIGURE 72

FIGURE 73

FIGURE 74

determines the location and quantity of inspection points necessary for an optimum inspection. These data are stored in the computer for subsequent retrieval by the Inspector, and via DNC the data are transmitted to a coordinate measuring machine for physical probing of the part (Figures 73 and 74).

After verification that the GNC machining program can produce an acceptable part, the machine instructions are placed in a read-only computer file to prevent unauthorized changes that could affect Quality Assurance responsibility for the part.

An additional capability of the CAQA module allows the programmer to retrieve the selected inspection points in X, Y, Z coordinate data. These can be used as input for manual coordinate measuring machines or other less-sophisticated measuring equipment.

Graphic portions of the CRT display can also be hardcopied on stable material for subsequent use as optical comparator charts or film overlays to accurately check the dimensional characteristics of machined parts.

REFERENCES

1. R. J. Margason and J. E. Lamar, Vortex Lattice Fortran Program for Estimating Subsonic Aerodynamic Characteristics of Complex Planforms, NASA TN D-6142, Feb 1971.
2. R. V. Harris, Jr., An Analysis and Correlation of Aircraft Wave Drag, NASA TMX-947, Mar 1964.

GMSOLID

This appendix consists of a reprint of an article by J. W. Boyse and J. E. Gilchrist, entitled, "GMSolid: Interactive Modeling for Design and Analysis of Solids." © 1982 IEEE. Reprinted by permission from *IEEE Computer Graphics and Applications* **2(2)**:27–42, Mar. 1982.

For some years now, General Motors engineers have used the company's several hundred graphics consoles to improve and speed up the design of sheet metal parts. GMSolid is an interactive graphics system that improves GM's computer-assisted design and analysis capabilities for functional parts, such as engine blocks, crankshafts, and dies. It allows an engineer to design with solids—cubes, cylinders, cones, and spheres.

GMSolid represents a significant technical advance over conventional wireframe graphics systems because it provides for a complete solid description[1] of the object to be stored in the computer. These complete descriptions make possible such important engineering analyses as direct computation of the area, volume, and mass properties of parts modeled with GMSolid. In contrast, more conventional line-oriented computer graphics systems are generally direct replacements of drafting methods. As such, they cannot support automatic computation of solid object properties because drawings are typically ambiguous and require human interpretation.

Because GMSolid embodies significant new concepts in computer-aided design and analysis, it required substantial research and development in a number of areas.[2–7] This article focuses on its structure, human interface, and analysis capabilities.

SYSTEM STRUCTURE

GMSolid was developed to improve General Motors' ability to use computer aids in the design, analysis, and manufacture of automotive components and tooling. It had to fit smoothly with General Motors' Corporate Graphics System and build upon that system's existing graphics capabilities. CGS is an open-ended system that hosts several other graphics applications, including body surfacing and finite element modeling. Because it is part of the same system, GMSolid can communicate with these applications and their data bases.

GMSolid differs from conventional wireframe CAD systems in that its users design components by working directly with computer representations of solids rather than with points, lines, and surfaces. Furthermore, the solids remain valid as the user performs operations on them. This guarantee of validity can be important for analysis.

Two basic capabilities, the creation and combination of solids, underlie GMSolid. The user can create any basic primitive solid—block, cylinder, cone, or sphere—directly by specifying its size, location, and other parameters. The user can combine any two solids with the set operators *union*, *intersection*, and *difference*.

Set operations. Several geometric modeling systems have been based on set operations that build representations of complex solids from simple solids.[8-10] Most have been experimental systems developed in university research projects. GMSolid builds on some of the work done by Voelcker, Requicha, and Tilove at the University of Rochester. Their research dealt with the use of regularized set operations in Euclidean three-space[11,12] where regularity guarantees that valid solids always result from set operations.

Solid primitives and set operations comprise a natural and powerful design mode because set operations have natural analogs in manufacturing. When a designer wants a hole drilled in a part, what could be more natural than to place a cylinder in the desired hole position and then subtract it from the part?

Figure 1 shows a simple example of the set operations applied to a block and a cylinder. The internal model kept by the computer is a true solid representation, not just the wireframe shown in the figure. Figure 2 is a more complicated example of a solid model built with GMSolid. Because the computer stores a true solid model, the crankshaft can be displayed with hidden edges automatically removed or as a shaded picture on a raster graphics device.

Constructive and boundary representations. When a set operation combines two solids, two representations of that solid are generated. The first is a constructive representation, which links the two original solids through a set operation. This constructive representation is a binary tree with primitive solids at the leaf nodes and set operations at all internal nodes and at the root, as shown in Figure 3. Each node in the tree represents a solid and is assigned a name. Leaf nodes are represented by a type (block, cylinder, etc.), as well as by a matrix that transfers the primitive solid from a template space (e.g., a unit-radius and unit-length cylinder on the positive z axis) to its proper position and scale in the solid. Each internal node shows the set operation to be applied

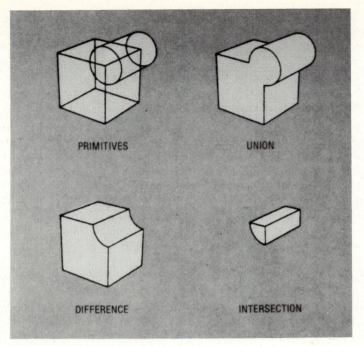

FIGURE 1 Primitives and set operations.

to the two subsolids below it in the tree. The constructive representation is a direct translation of the set operations performed by the user into a tree structure.

The second solid representation maintained by GMSolid is the boundary representation. It explicitly contains the faces, edges, and vertices that bound the solid, as well as the connectivity among these elements. It also keeps track of which side of each face lies inside the solid. The boundary representation is generated either directly by special procedures for the primitives or by a key GMSolid subsystem called the *boundary evaluator*. The latter is employed for solids that have been combined with set operations.

The process of evaluating the boundary of a solid includes determining where new edges and new vertices must be created and where old edges and vertices must be deleted. When surfaces, edges, or vertices coincide, the boundary evaluator merges them into a single entity. Thus, it maintains a consistent, nonredundant data structure for the solid's boundary. Edges on a solid occur where pairs of surfaces intersect. Therefore, the boundary evaluator generates new potential or tentative edges by finding the curves of intersection for pairs of surfaces of the two solids being combined by a set operation. A curve classification process is then used to determine which curves or parts of curves are actual edges of the solid. Similarly, vertices occur at points where three or more surfaces intersect. These intersection points are computed, and a classi-

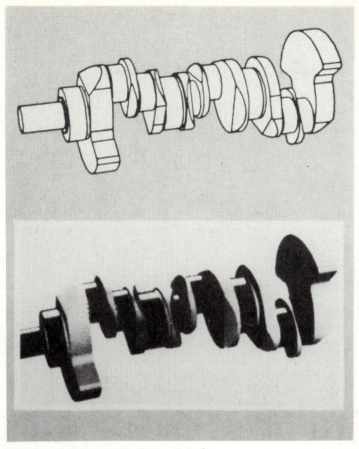

FIGURE 2 Solid model of a crankshaft.

fication process determines which are true vertices to be added to the data structure and which can be discarded. Figure 4 is a simple example of the creation of new edges and vertices, deletion of some edges and vertices of subsolids, and merging of coincident boundary elements.

Figure 5 is a more complex example. A cylinder is first subtracted from a cone, and then a sphere is subtracted from the result. Figure 5 deals exclusively with nonplanar curves and surfaces. Therefore, we must characterize the curves of intersection that occur between these surfaces and find points where the three nonplanar surfaces intersect.

Surfaces in GMSolid are quadric. They can be represented by an equation of the form $F(x, y, z) = 0$, where F is a second-order polynomial in x, y, and z. Algebraic methods[7] are used to determine whether intersection curves between quadrics are planar or nonplanar and to determine just what type of planar— line, ellipse, hyperbola, or parabola—or nonplanar curve exists. For example,

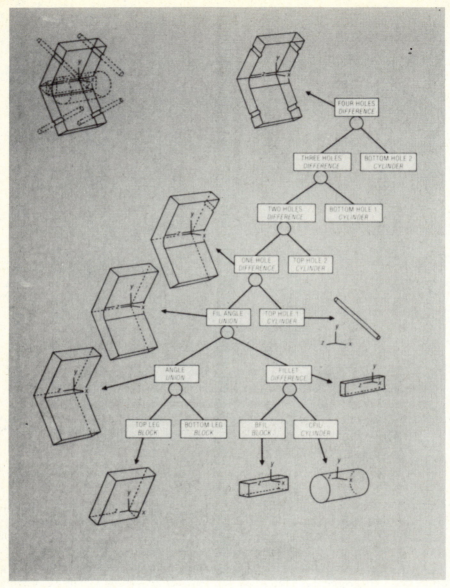

FIGURE 3 Constructive representation of solids.

to determine whether the curve of intersection between quadrics $F_1(x, y, z)$ and $F_2(x, y, z)$ is planar, form the linear combination $G(x, y, z) = F_1(x, y, z) + kF_2(x, y, z)$. The intersection curve is planar if and only if there exists a value for k such that $G(x, y, z)$ can be factored as the product of two planes. Algorithms in GMSolid find such a k or show that it does not exist. Vertices can occur where three quadrics intersect. Finding these points requires finding

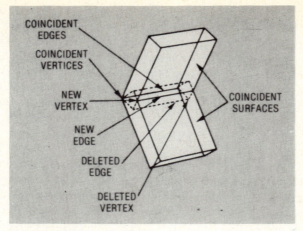

FIGURE 4 Determination of boundary elements on a solid created by union of blocks.

all real solutions to three second-order polynomials in three unknowns. Relatively new mathematical theory, known as a *homotopy continuation*, is used to compute these points.[6] This technique is guaranteed to find all solutions. Without this guarantee, the modeler would not be robust.

The boundary evaluator generates a data structure for the boundary of a solid. This boundary representation contains enough information to allow

FIGURE 5 Complex intersections required for boundary evaluation.

determination of where an arbitrary point lies: inside, outside, or on the solid's surface. Figure 6 shows a simplified diagram of one piece of the network data structure generated by the boundary evaluator. The figure shows four of the several entity types. Each entity contains a number of attributes that together completely describe the boundary of the solid. The edge entity contains flags that determine which side of each of the two surfaces bounded by the edge lies inside the solid. The vertex entity contains spatial coordinates for the vertex. Each surface entity contains data on the type and location of the surface. Although not shown in the data structure diagram, each vertex and surface link with several other edges.

The boundary representation is a complex network data structure. It is built, accessed, and maintained with Associative Programming Language/ Virtual Associative Access Method, or APL/VAAM. APL, a powerful associative data-base manager within CGS, consists of statements and functions added to PL/1. These statements are used to create, connect, and access the entities that represent elements of the CSG tree and boundary structure in GMSolid. VAAM and its precursor, VDAM,[13] support the virtual memory files that

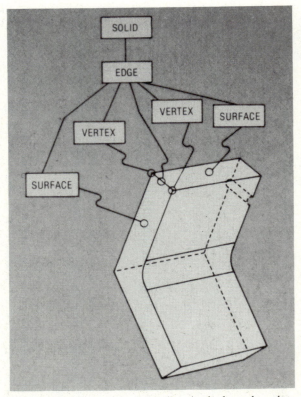

FIGURE 6 Elements and relations in the boundary data structure.

contain the APL data structures. VAAM uses the paging mechanism of the operating system to access these files directly in virtual memory. Therefore, programs that access the data base do no I/O, in the conventional sense of explicit reads and writes. In a dynamic sense, the data structures are very flexible. For example, an attribute of an edge, such as a fillet radius, can be inserted by adding a new attribute to the edge entity or by adding a new fillet entity and attaching it to appropriate edge entities. In either event, existing programs that build and access the data base need not be affected.

The constructive representation of a solid is very compact and can be generated quickly when two solids are combined by set operations. On the other hand, it takes significant computation to generate a boundary representation from a constructive representation. The boundary representation, however, contains useful data not directly available in a constructive representation. For example, vertices and edges in the boundary data structure can be used during design for placing one solid in relation to another. Attributes such as surface finish can be attached to faces in a boundary representation. Which representation is best? This can only be answered in the context of a specific application. GMSolid provides access to both.

USER INTERFACE

The human interface is extremely important to the acceptance and usefulness of a solid modeling system. Important human factors include ease of use, friendliness, visualization, reliability, performance, and flexibility. The graphics interface is often more important in gaining user acceptance than the technical sophistication of the modeling data base, especially in the early stages of use.

The GM Corporate Graphics System, or CGS, currently supports many graphic consoles driven by IBM 3033 computers under the MVS operating system. A typical work station consists of a vector-refresh display tube (an IBM 2250/3250, a DEC GT48/GT62, or an Adage 4250), a light pen for interaction with the display, a keyboard, and 32 function buttons. The function buttons are under application control.

Screen layout. The screen layout for GMSolid, shown in Figure 7, is divided into four basic areas.

(1) The *data area* is for display of solids, surfaces, lines, points, and vectors in current use. Specific data needed to fill in menus can be selected with the light pen.

(2) The *menu area* (lower right) is for input of data needed by the computer to perform desired tasks. For example, the CUT SOLIDS menu in Figure 7 is used to section solids. The user selects NEW to create a new sectioned solid. Otherwise, the system modifies the existing solid by sectioning it. When enough data has been specified to allow the section to be cut, PROCEED brightens; if the user is satisfied with the specified input data, he selects PROCEED to cause the sectioning to occur.

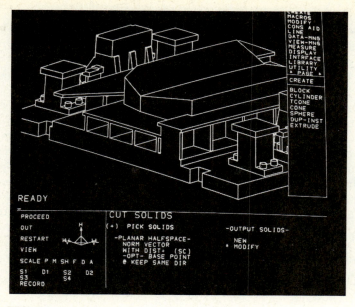

FIGURE 7 CGS graphic screen layout for GMSolid.

(3) The *oplist area* (optionally on right or left for right- or left-handed people) is for lists that allow the user to direct the system to the next desired menu. The user's selection of a set of menus from the upper portion of the list causes a list of specific menus to appear below. A menu selected from this lower list then appears in the menu area.

(4) The *control area* (lower left) is for a number of functions that are useful across many menus—zoom-in, for example.

Common capabilities. A number of CGS capabilities are available to all applications. These include:

☐ *Storage of current job status between sessions.* Few jobs are completed in one day; most take weeks or even months. Each session at the terminal starts with the data base and graphics displays in the same state they were in at the conclusion of the user's previous session.

☐ *Display controls.* A complete set of windowing controls allows the user to zoom in, zoom out, or fill the screen with all the data at any time. Other controls allow the user to determine the number, type, and placement of views of his data base during any phase of the design process.

☐ *Screen recordings.* At any time, the user can request a hard copy of everything on the screen by selecting the RECORD control.

☐ *Labeling, dimensioning, and drawing.* The user can label and dimension any data, then send it to a drawing machine to produce a high-quality drawing.

☐ *Function buttons.* GMSolid uses the function buttons for direct access to the most frequently used operations. They are shortcuts for the more experienced users.

☐ *Menu control.* GMSolid leads the user through the process of satisfying a menu. The system checks input for proper and sufficient content before signaling that execution constraints have been satisfied.

Display options. The graphic display helps the user to visualize a solid. Visualization consists of building a mental 3-D solid model from a 2-D image. As shown in Figure 8, view-dependent profile (silhouette) curves are always generated and shown, to add realism to the display of solids with curved surfaces. Like drawings, GMSolid presents multiple views to eliminate ambiguities. To help the user visualize large objects, GMSolid allows division of the data area into multiple windows (Figure 9). He can then zoom in on various details of the object in each window. Each window can also be used to display data in different views and to display different collections of data. Menu questions can be answered by using the light pen to select data items displayed in any view in any window.

A solid representation in the computer makes it possible to delete hidden edges from the image automatically. This has become the normal mode of operation for most users, due to its high value and low cost. It is also possible to display all edges of a solid, including the profile curves, with no hidden-edge removal. This produces something like an enhanced wireframe picture.

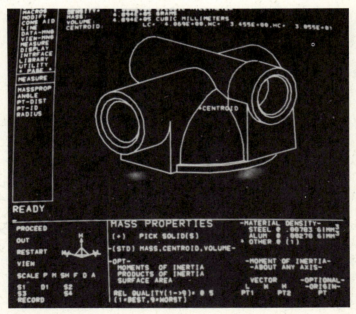

FIGURE 8 Silhouette curves automatically generated and displayed.

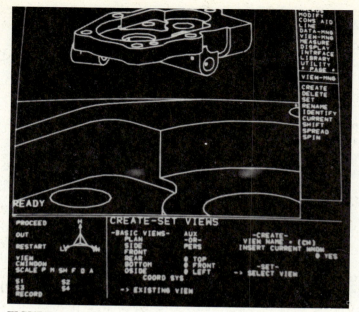

FIGURE 9 Multiple windows.

Another display option allows the user to see current feature points, such as vertices or surface centers. In the case of a cylindrical hole, the displayed point is where the axis of the cylinder would strike the skin of the solid if the cylinder were not there (Figure 10). In all cases in which solid feature points are being displayed, their images are suppressed if the solid hides them from view. When the solid is rotated, a different set of points is automatically presented.

Response time. Feedback time is an important human factor, especially for very simple requests. The data area display is updated whenever anything happens that should change it, such as a change in the data or the way it is presented. For example, users frequently scale (blow up) an area of the screen or fill the screen with all of the data. Additional information stored with the solid makes this type of operation very fast.

The light pen. Most user interaction with the system is through the light pen. Light pen interaction reduces the need to remember and key in point coordinates and names. The light pen is used for selecting solids and the feature points on them, such as vertices and hole centers. When the current menu question is expecting a point in Euclidean three-space, the user selects a visible position feature on the solid with the light pen. A point is extracted from the data base and displayed to the user.

"Thinking solid." GMSolid encourages the user to "think solid," that is, to fully understand the three-dimensional, solid-volume nature of objects.

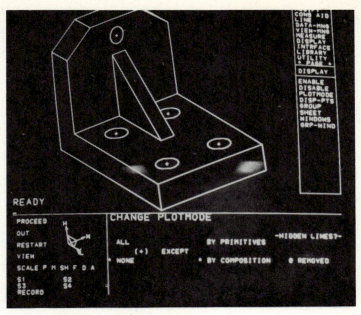

FIGURE 10 Selectable feature points

Though they find this approach a little foreign at first, designers soon prefer it to wireframes or drafting representations.

Because modifications are made at the solid level rather than at the boundary (face, edge, and vertex) level, any model that can be constructed is guaranteed to represent a physically realizable solid. The GMSolid user starts with primitive solids, which are legal by definition. He is allowed to modify them only in ways that preserve the integrity of the solid model. This guarantee of integrity is particularly important for engineering analyses that depend on the geometry in the GMSolid data base.

HOW SOLIDS ARE BUILT

A number of methods are available to the user for building solid models. The most basic of these are the menus that create the four primitive solids (Figure 11). The user provides the dimensions, location, and orientation, tells the system to proceed, and the primitive solid appears.

One important technique for communicating location and orientation information is a vector or a coordinate system, which consists of three mutually perpendicular vectors (labeled *A*, *B*, and *C*, from left to right) and a base point. A number of tools are provided to easily move vectors or coordinate systems by shifting, rotating, and aligning them with other data. Should the user create a solid that appears in the wrong position, he can correct the situation by means of a complete set of transformation capabilities that includes mirroring and scaling.

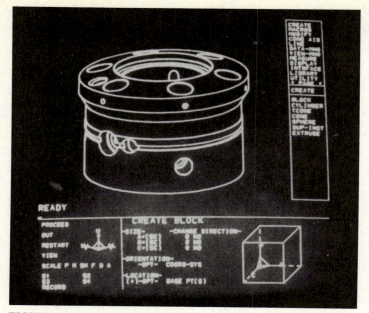

FIGURE 11 Primitive solid creation.

The process of supplying information for a menu that creates solids is aided by operations that support the extraction of attributes, locations, and distances from the current design scene. The results of these operations are converted to displayable information on the terminal. This information can be selected by the user to satisfy other menu questions. For example, the user can measure the distance between two vertices on a solid (Figure 12). He can save this on the screen and later select the number to specify the length of another object he wishes to create. This process minimizes the need to jot down information, and it supports the needed accuracy of double precision for the mathematical algorithms.

After the user creates and positions several solids, the next step is usually to combine them with a union, intersection, or difference operation (Figure 13). The user selects the solids to be combined, the operation, and the proceed control. The selected (input) solids are replaced on the screen by the new combined solid. Of course, the combined solid is not always what the user has in mind, so tools are provided to make it possible to back up and break combined solids into their constituent parts. As in all operations that modify solids, the user can protect his original object by duplicating the input solids and carrying out the requested modification on the duplicate.

We stated above that the constructive representation on GMSolid is always a binary tree in which two solids and one set operation define a new solid. This restriction, however, is not placed on the user by the graphics interface. The user can select any number of objects at one time and describe an operation

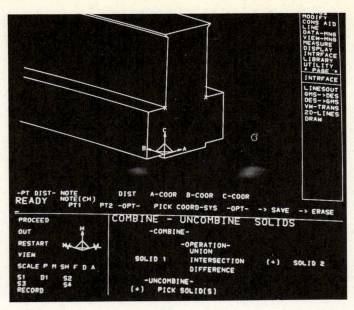

FIGURE 12 Generation, from light pen selections, of the distance between vertices.

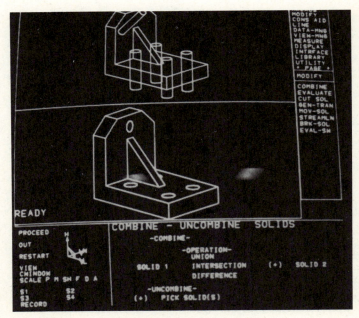

FIGURE 13 Combining solids with set operations.

to act on this family of objects. For example, the user can apply the union operation to several solids in one step or subtract several solids from a given solid. This approach gives the user community more flexibility and reduces computer processing time by suppressing display of and user interaction with intermediate results.

Beyond the menu and light pen. We have been discussing GMSolid as one would interact with it graphically via menus and light pen. The user can also work at an alphanumeric terminal to access GMSolid through a command language, which includes display commands to produce images on a storage tube. The language can be used interactively or to define background requests. The graphic and command language interfaces to GMSolid are compatible— the user can change methods at will for any given job. He can define and launch complex background jobs through the graphics interface for later execution, terminate the session, and then review the results before continuing design work during his next session at the graphics terminal. This allows for the best possible employment of the user's time during procedures that require extensive computations.

The boundary structure of a solid object is a desirable representation scheme for many, but not all, design and analysis functions. Because the generation of a boundary sometimes requires considerable computer resources, GMSolid allows it to be postponed when objects are combined. This keeps the graphics session more interactive. GMSolid informs the user of the necessity

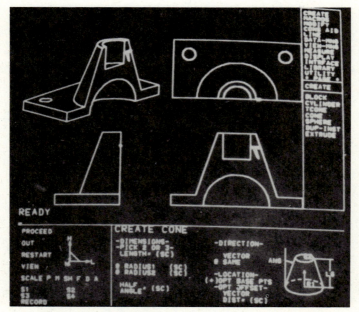

FIGURE 14 Truncated cone.

to boundary-evaluate the solid, should he try to perform an operation that requires the boundary representation. Through a menu, the user can generate the boundary in either the foreground or background. If it is run in the background, the user can continue his interactive tasks while the computer is carrying out the boundary evaluation.

This design approach is a powerful one; almost any complex part can be represented by solids designed from primitives. It is not, however, the only way to construct a solid. GMSolid offers several other techniques with which to define and build complex solids, including extruded solids, macro solids, and referencing of existing objects from a corporate data base. These solids are restricted to planar, cylindrical, conical, and spherical surfaces.

Although the underlying system supports only cuboids, cylinders, cones, spheres, and extruded solids as primitives, menus exist for wedges and truncated cones (Figure 14) because they are frequently requested. The system automatically generates the desired wedge or truncated cone by means of set operations on the underlying system primitives.

USE OF NONSOLID DATA

Because of our large base of existing applications and users, it is desirable for GMSolid to appear as an evolutionary enhancement to the CGS. This requires solid modeling to coexist with other forms of data, such as space curves (lines), sculptured surfaces, and the application programs that operate on these data. We have dealt with this requirement in several ways. They include providing capabilities for the user to change applications, the ability to exchange data with other application data bases, and operations within GMSolid that deal with many forms of data.

One set of menus in GMSolid allows the user to deal with the world outside the GMSolid application. Some of these menus allow the user to input data from other applications or send data to them. There is no restriction on the type or format of this information. One can, for example, use GMSolid to design a wiring harness that included lines as well as solids in the engine compartment. The user can then bring all this information together, to resolve a packaging problem, for example. One menu in this set allows the user to convert a solid into a wireframe representation. The user can request that the wireframe model be sorted into visible edges, hidden edges, and profile curves for any specified views.

Another set of menus deals with lines or line-solid combinations. The operations that deal only with lines are a small but popular subset of the functions available within the line application. The more powerful operations available in this package deal with functions that work on line/solid combinations:

☐ The *classify line* menu identifies the segments of a space curve as in, on, or out of a given solid. Each part of the space curve is then displayed differently to provide immediate, graphic feedback to the user.

☐ The *visible lines* menu subdivides lines into their visible and hidden parts, given a family of solids and views of interest. The results are displayed to the user in the appropriate line font (dashed or solid).

☐ The *solid/plane intersection* menu generates lines that are the intersections of a family of planes and the boundaries of a family of solids.

☐ The *distance* menu locates the point on a selected solid and the point on an input space curve at which the two items are nearest. The output of this menu consists of the two points and the distance between them.

A production-quality solid modeling system must coexist with current modeling methods and data. This is essential for broad user acceptance of new technology.

ORGANIZING GMSOLID DATA

GMSolid and similar applications have many data types and the probability of large numbers of each type. Therefore, they must give the user aids with which to deal effectively with all this information. Available graphics terminals pose further problems, because they are small compared to lofting tables. This limitation requires methods to help the user limit the graphics display to data that contributes to the current design steps.

Data management skills are very different from those needed in the conventional design environment. In an industrial design system, the function and packaging of data management schemes are among the most important issues governing user acceptance of any new computer procedure. One of the most powerful features of the solid model representation is that it manages all the information about the skin and interior of an object. This frees a user to approach problems at a much higher level, since he can make references to a single input, the solid object. In addition to the solid itself, GMSolid supports several user methods of dealing with data, including assemblies, groups, sheets, and libraries. All of these capabilities are simply methods of sorting data into collections; they allow user menus to easily identify input data items.

Assemblies. Assemblies are collections of solids or of other assemblies. They can be used as input for most operations that accept solids. For example, the user can request hidden-line displays for assemblies, a section can be cut through an assembly, or its volume and surface area can be calculated.

Groups. A group is a collection of any combination of data types. Once defined, the groups give the GMSolid user control over the display environment; he can choose to display or not to display groups in individual windows in the data area of the graphics terminal.

Sheets. A sheet is a collection of data (groups) and graphical presentation methods (windows and views) for the data. In conventional practice, a designer usually expresses a design intent with multiple drawings, such as assembly drawings, detail drawings, layout studies, and prototype designs. In GMSolid, a sheet is a logical piece of paper in the computer that can be used for any of the above functions. Menus allow the user to specify a displayed sheet and to transfer data from one sheet to another. For example, one might work out a design on an assembly sheet and then create other sheets on which to detail the individual components by assigning them labels and dimensions.

ANALYSIS CAPABILITIES

A solid representation is a powerful basis for applications besides graphics interaction during the design process. Many engineering analyses are difficult or impossible with wireframe models, but become straightforward when a solid model exists. With GMSolid, one can section a solid, compute its area, and compute mass properties. All these applications use a basic capability inherent in solid modeling: the ability to determine which parts of a given line fall inside, outside, and on the surface of a solid.

Sectioning. A section view of any solid model can be created simply by first positioning a large block so that one of its faces forms the sectioning plane and then subtracting the block from the solid. Crosshatching is carried out by generating hatch lines in the sectioning plane and then drawing the segments of the lines that are determined to lie on the sectioned solid (Figure 15).

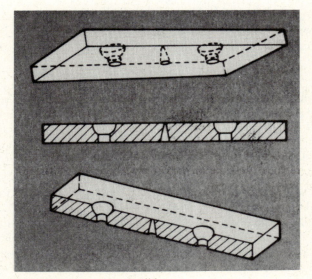

FIGURE 15 Sectioned solid.

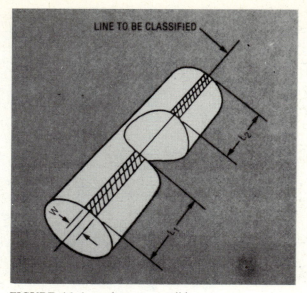

FIGURE 16 Area element on solid.

Area computation. Surface area is computed through an integration method. An element of area that contributes $dA = (L1 + L2) * w$ to the total area of the solid is shown in Figure 16. To determine the area of this element, we generate a line that is on the cylindrical surface and inside the element, as shown. The part of this line that lies on the solid ($L1 + L2$ in the figure) determines the area of the element. All surfaces that bound the solid are easily retrieved from the data structure for the solid. Therefore, it is a straightforward task to generate a set of elements that completely covers the surface of the solid.

Computation of mass properties. Volume and mass properties are also found through numerical integration. Conceptually, we cut up the solid by pushing it through a sieve, something like cutting a potato into french fries by pushing it through a slicer. Each piece of the solid is then a long, thin bar—approximately a parallelepiped with a square cross section. We compute the mass properties for each bar and then sum to obtain the mass properties for the solid. Solids are assumed to be homogeneous, although an assembly of solids with differing mass properties is possible.

A complete representation of a solid and the ability to easily create and modify that representation opens the door to many special applications. One example is combustion-chamber analysis. Investigation of alternative geometric configurations and their effect on combustion is an important part of combustion-chamber design. One very useful thermodynamic model for combustion analysis assumes that the flame front is spherical and its center is at the spark plug.[14] The thermodynamic model requires the following geometric quantities as input:

(1) *Burned volume*: volume inside the spherical flame front.
(2) *Area of flame front*: area of the spherical flame front in the combustion chamber.
(3) *Wetted area of head*: area of the combustion chamber head inside spherical flame front.
(4) *Wetted area of liner*: area of the combustion chamber cylinder wall inside the spherical flame front.
(5) *Wetted area of piston crown*: area of the piston crown inside the spherical flame front.

These quantities must be determined for a given set of values of two independent variables: flame front radius and piston position.

Because combustion chambers have irregular shapes, these geometric quantities cannot be computed analytically. If a model of the combustion chamber is developed with GMSolid, however, the numerical integration methods outlined above can be used to compute the needed volume and surface areas. Furthermore, GMSolid is a tool that makes it easy to vary the chamber and thus to study the effects of geometric changes on the combustion process. Figure 17 shows a combustion-chamber model. Figure 18 shows the results for this chamber, which were computed with GMSolid.

This example is important because it shows that specific applications come easily and naturally once basic solid modeling capabilities exist.

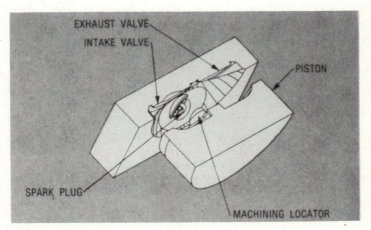

FIGURE 17 Model of a combustion chamber.

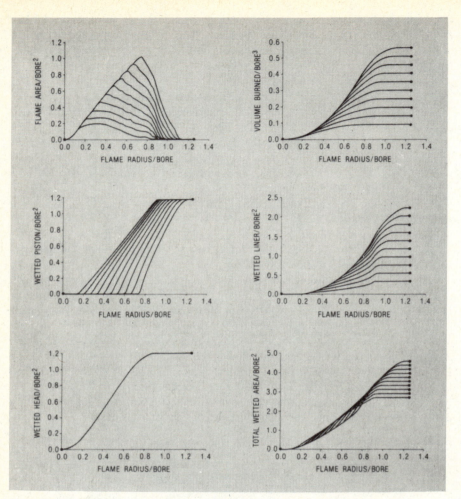

FIGURE 18 Volume/area results for combustion chamber in Figure 17.

REFERENCES

1. H. B. Voelcker and A. A. G. Requicha, "Geometric Modeling of Mechanical Parts and Processes," *Computer*, Vol. 10, No. 12, Dec. 1977, pp. 48–57.
2. J. W. Boyse, "Preliminary Design for a Geometric Modeller," Report GMR-2768, Computer Science Dept., General Motors Research Laboratories, Warren, Mich., July 1978.
3. J. W. Boyse, "Interference Detection Among Solids and Surfaces," *Comm. ACM*, Vol. 22, No. 1, Jan. 1979, pp. 3–9.
4. J. W. Boyse, "Data Structure for a Solid Modeller," *Proc. Workshop Representation of Three-Dimensional Objects*, May 1979.
5. S. D. Roth, "Ray Casting as a Method for Solid Modeling," Report GMR-3466, Computer Science Dept., General Motors Research Laboratories, Warren, Mich., Oct. 1980.

6. A. P. Morgan, "A Method for Computing All Solutions to Systems of Polynomial Equations," Report GMR-3651, Mathematics Dept., General Motors Research Laboratories, Warren, Mich., July 1981.

7. R. F. Sarraga, "Algebraic Methods for Intersections of Quadric Surfaces in GMSolid," Report GMR-3944, Computer Science Dept., General Motors Research Laboratories, Warren, Mich., Jan. 1982.

8. A. Baer, C. Eastman, and M. Henrion, "Geometric Modeling: A Survey," *Computer-Aided Design*, Vol. 11, No. 5, Sept. 1979, pp. 253–272.

9. G. Spur, F. L. Krause, R. Mayr, G. Müller, and W. Schliep, "A Survey about Geometric Modeling Systems," *Annals CIRP*, Vol. 28, No. 2, 1979.

10. A. A. G. Requicha, "Representations for Rigid Solids: Theory, Methods, and Systems," *Computing Surveys*, Vol. 12, No. 4, Dec. 1980, pp. 437–464.

11. A. A. G. Requicha and H. B. Voelcker, "Constructive Solid Geometry," Technical Memo 25, Production Automation Project, University of Rochester, Rochester, N.Y., Nov. 1977.

12. A. A. G. Requicha and R. B. Tilove, "Mathematical Foundations of Constructive Solid Geometry: General Topology of Closed Regular Sets," Technical Memo 27, Production Automation Project, University of Rochester, Rochester, N.Y., Mar. 1978.

13. D. R. Warn, "VDAM—A Virtual Data Access Manager for Computer-Aided Design," *Proc. Workshop Data Bases for Interactive Design*, 1975, pp. 104–111.

14. J. H. Lienesch, "Engine Simulation Identifies Optimal Combustion Chamber Design," Report GMR-3206, Engine Research Dept., General Motors Research Laboratories, Warren, Mich., Feb. 1980.

APPENDIX **C**

ROMULUS

This article by Dr. A. W. Bishop describes the ROMULUS design system, a successful contemporary example of Evans & Sutherland's and Shape Data Ltd.'s approach to geometric modeling. Figures C.1–C.16 following this article show a sequence of computer graphics displays of the parts of a table-top vise created using the ROMULUS system.

COMPUTER SIMULATION OF THE REAL WORLD FOR THE MECHANICAL AND INDUSTRIAL DESIGNER

DR. A. W. BISHOP

MARKETING MANAGER

SHAPE DATA LTD.

(A wholly owned subsidiary of Evans & Sutherland)

Designers must think in three dimensions in order to produce objects for the real world. Computer aids to the designer must therefore simulate 3D solid objects, and enable the designer to interact with the computer representation in a natural manner. The Romulus2 design system provides such facilities and its main features are described.

Introduction

The designer faces the demanding job of creating a 3D component or assembly which must perform a specified task. The product will have to meet geometric and functional constraints, and may need to be styled attractively. So the designer works essentially in 3D. Although he is forced to record his design in 2D (on a drawing board) he will often be seen producing models in paper and wood, and sometimes has to resort to making expensive mock-ups. Naturally, there is a tendency to produce items which are essentially 2D, but increasing competitive pressure is forcing more complex components to be designed. For example, the electronics industry is moving quickly from sheet metal cabinets (with many components and expensive assembly) to plastic mouldings (Fig. 1). But these mouldings are very complex geometrically, and have caused severe design problems.

What the Designer Wants

The designer has two requirements of any design system, which he will regard as crucial:

☐ Firstly, he needs to be given an effective means of interacting with the model which is stored in the computer. One user has expressed this as "electronic metal". In other words, he can almost feel the object which is in front of him on the screen. He must be able to "walk around" the object. He wants it to have a feeling of solidity, and if he wants to change its shape, he expects to be able to point at the feature and move it directly. Changing a design (Fig. 2) must be possible without having to reconstruct the object, as this implies a knowledge of how it was initially designed. The system must support a style of graphical interaction which is natural and meaningful to the designer.

☐ Secondly, a complete accurate model of the component must be produced. The user wants a total "product description" including geometry, surface finish, dimensions and other attributes of the product. This implies that every face and edge of the model should be accurately represented and that other attributes can be attached to such entities in the data structure.

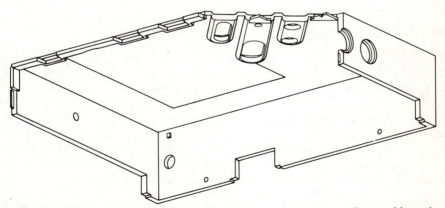

FIGURE 1 A moulded plastic component—a complex part creating problems in design and manufacture.

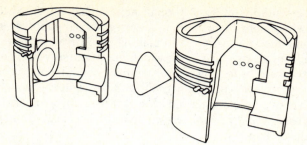

FIGURE 2 Variational design with Romulus2.

The designer will also expect to be able to obtain the mass properties of the product, check for interference between components, animate mechanisms, and produce high-quality pictures for sales and technical literature.

What the Organisation Wants

The designer cannot work in isolation. The product description which he generates has to be used by many other specialists. It has been said that the greatest asset of a company is its data. Organising a smooth flow of data from department to department is crucial to efficiency.

Most companies will need to integrate several packages, some brought in and some developed in-house.

In addition, the organisation may impose further constraints, such as the type of computer and display hardware which may be used, and the networking system which will enable all departments to access the data. Thus Romulus2 is normally used as part of a wider CAE system, of which Fig. 3 is a typical example.

How Romulus Does It

The choice of data structure is important in any application. In solid modelling it has proved to be crucial to the success of the product. The data must represent every item of interest to the designer such as assemblies, components, faces, edges and vertices. It must enable the designer to select these items using a graphical input device, and must allow the system to generate suitable pictures quickly. It must also allow the user a very wide variety of modelling operations. Not only are the traditional drafting techniques required for producing profiles, but also methods for generating simple solid shapes (such as cones and cubes) together with means of combining such objects (the so-called "Boolean Operations").

There are two major approaches used in current solid modellers. The first is the constructive solid geometry (CSG) data structure which models the simple

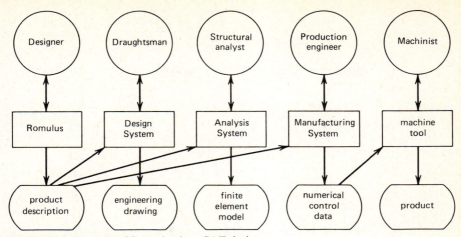

FIGURE 3 The role of Romulus in a CAE design.

elements needed to build the component, together with the operations to combine them. Such a data structure is very compact but does not represent the final edges and faces which are produced after combining the simple elements. The complete model has to be re-evaluated after every design change (which could take many minutes) and interaction with such a system becomes very difficult. The associated problem of re-evaluating the object for every picture is sometimes overcome by adding a boundary file (see below). Many existing modellers are of this type, as they are relatively simple to implement, but their use is limited to producing pictures of objects which have been designed some other way.

Romulus2 takes an alternative approach, using the "boundary file representation" (or b-rep) data structure. Using this approach, every face, edge and vertex of the final object is represented. The data structure must define how these details are connected together (the "topology") with the geometry of each detail. This provides fast picture generation, direct user interaction, and a very wide variety of modelling operations (Fig. 4).

However, the b-rep approach has proved to be exceptionally difficult to implement, which explains why there are so few modellers of this type on the market. The major practical problems in implementing a useful solid modelling system have turned out to be reliability and efficiency.

☐ Reliability—As users of earlier types of CAD system will know, it can be quite difficult to get a system to do one surface/surface intersection reliably, and it quite often requires interactive help. Romulus2 does many such intersections with each command, and has to do it automatically. This has to be done in the face of numerical errors generated by the computer, and only a deep understanding of the problems, and a long period of hard engineering produces a reliable system.

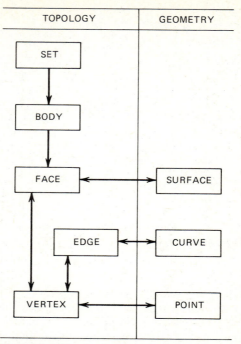

FIGURE 4 The basic Romulus2 data structure.

☐ Efficiency—Romulus2 does an enormous amount of work automatically, unlike a drafting system which does only one simple thing in response to each command. A section is produced automatically with a single command, and once the designer gets used to this, he would like to have the result instantly! A great deal of effort has been applied to improving search algorithms, and this has led to some dramatic improvements in speed over earlier versions. Solid modellers used to have a reputation for requiring a mainframe computer, but Romulus2 is in use today on single-chip processors such as the Apollo Domain.

Visualisation

"Seeing" the object is just part of the story. The designer must also be able to interact with the visual representation by pointing at features which he wants to change. Also these processes are intimately linked with the way in which commands are issued, changes highlighted and warnings made.

Once a solid model is available in the computer, any standard of picture can be generated. The trade-off is principally one of time versus the completeness of the representation. As the algorithms are quite simple, specialised hardware is widely used to speed up processing.

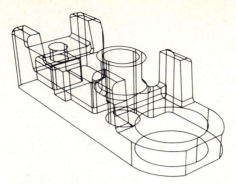

FIGURE 5 Wire frame representation of an object.

Wire Frame

The simplest and quickest way to generate an image is to draw its edges (a "wire-frame" picture, Fig. 5). This is well suited to vector refresh displays with real time 3D rotation and translation (such as the Evans and Sutherland PS300) which enable mechanisms, robots and tool paths to be simulated with animated sequences. Such displays also have "depth modulation", which displays nearer edges more brightly. This overcomes the problem of depth inversion, and largely replaces the need for hidden line removal.

Another major practical problem is the over-abundance of information which appears on the screen when complete assemblies are displayed. A structured display file will enable components to be temporarily deleted. Z clipping planes will remove foreground or background detail. Dynamic zoom will allow specific areas to be enlarged. All these facilities are necessary for practical design. Colour can also be used effectively to distinguish between bodies, construction lines and other details.

Hidden Line

The removal of hidden edges, and the addition of silhouette lines provides a more realistic representation (Fig. 6), but takes more time. It is useful for periodic checks, and for hard copy output which will be distributed to others. Specialized hardware can be built for hidden-line processing, but this is comparable in complexity to that required for shaded pictures (see below) which are much more useful. There are partial (approximate) hidden-line algorithms which are widely used because they are substantially faster.

Colour Shaded Images

If each face of the object can be given a colour, then the intensity at each point on the surface can be calculated for given light sources. Other properties can be added to define surface texture and optical qualities of the material. These

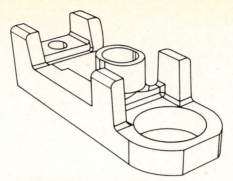

FIGURE 6 Hidden line representation of an object.

enable a photographic level of representation to be approached. Such pictures are essential for the assessment of aesthetic qualities in the design of consumer goods and buildings. If available in real time, such pictures could be invaluable for improving interaction and for animating mechanisms. Such displays do exist (e.g., the Evans and Sutherland CT5 used for pilot and engineering simulation) but they are generally too costly. Simpler displays are available for CAD, but they are severely limited by speed and by the use of a facetted approximation of the object.

Conclusion

Romulus2 uses the above techniques to give the designer the feeling that he is "in touch" with the computer model which he is building. Shape Data has broken away from the idea that a "drawing board emulator" is the natural way to design 3D objects. Designers who have used Romulus2 remark on their improved ability to think in 3D. This allows them to design the more complex and better packaged products which are now required by the forces of competition.

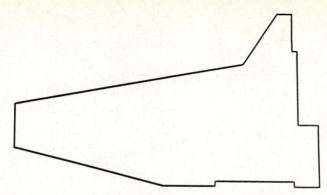

FIGURE C.1 A two-dimensional profile of a part for a tabletop vise design, created using 2-D geometry techniques of ROMULUS on the PS 300 high-performance computer graphics system. The model represents one of various ways solid parts can be created. ROMULUS is a 3-D solid geometric modeling tool for Shape Data, Ltd., Cambridge, England, a wholly-owned subsidiary of Evans & Sutherland. (*Courtesy of Evans & Sutherland.*)

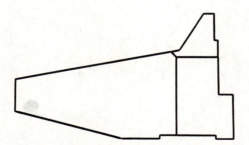

FIGURE C.2 A three-dimensional part for a tabletop vise design sectioned with infinitely large planes, creating new faces with angular offsets from the original. This represents sectioning, another method by which parts may be constructed using ROMULUS on the PS 300 high-performance computer graphics system. (*Courtesy of Evans & Sutherland.*)

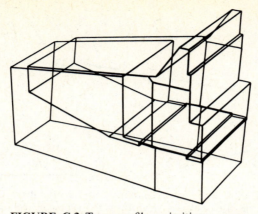

FIGURE C.3 Two profile primitives, representing parts for a tabletop vise design, are positioned prior to connecting them through the ROMULUS intersection operation. The design was created using ROMULUS, a 3-D solid geometric modeling tool, on the PS 300 high-performance computer graphics system. (*Courtesy of Evans & Sutherland.*)

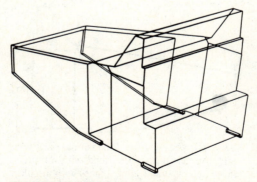

FIGURE C.4 Using the intersection command of ROMULUS, two profile primitives are connected, thus forming a finished wireframe object of a part for a tabletop vise design. (*Courtesy of Evans & Sutherland.*)

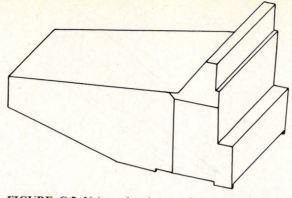

FIGURE C.5 Using the intersection command of ROMULUS, two profile primitives are connected, thus forming a finished wireframe of a part for a tabletop vise design. It is shown here in a full hidden-line version. (*Courtesy of Evans & Sutherland.*)

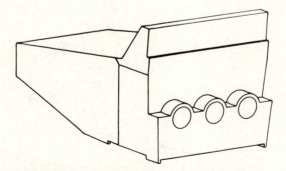

FIGURE C.6 Surface features have been created and added to a part for a tabletop vise design using the edged base construction feature of ROMULUS. Edged base construction was used on a particular face, then the faces were swept along a linear axis. (*Courtesy of Evans & Sutherland.*)

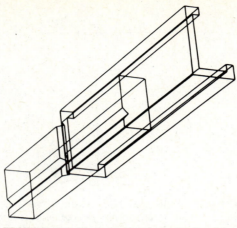

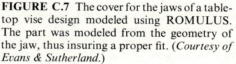

FIGURE C.7 The cover for the jaws of a table-top vise design modeled using ROMULUS. The part was modeled from the geometry of the jaw, thus insuring a proper fit. (*Courtesy of Evans & Sutherland.*)

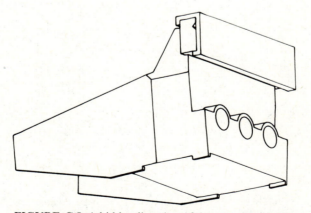

FIGURE C.8 A hidden-line view of the completed body for a tabletop vise design. The view shows the completed body displaying the proper placement of the components in relation to each other. (*Courtesy of Evans & Suther-land.*)

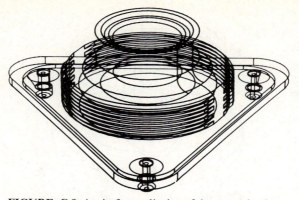

FIGURE C.9 A wireframe display of the mounting base for a tabletop vise design. The part was modeled using a combination of 2-D geometry and sweeping operations of ROMULUS. (*Courtesy of Evans & Sutherland.*)

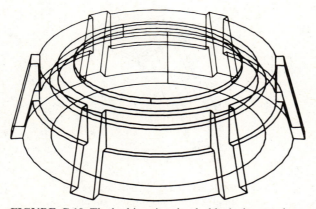

FIGURE C.10 The locking ring that holds the base and top part of a tabletop vise together. The part was constructed using boolean operations of cylindrical and conical primitives, in combination with rotation of a 2-D profile about the axes of symmetry. (*Courtesy of Evans & Sutherland.*)

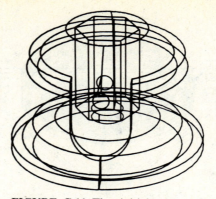

FIGURE C.11 The initial steps of creating a part within a part, verifying that geometric constraints are maintained and incorporated into the design. The piece is part of a tabletop vise model. (*Courtesy of Evans & Sutherland.*)

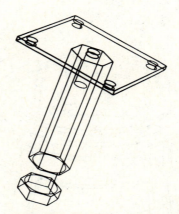

FIGURE C.12 This hexagonal prism was united to a plate onto which the body of a tabletop vise design will be secured to a universal ball joint. (*Courtesy of Evans & Sutherland.*)

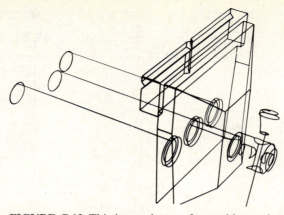

FIGURE C.13 This jaw and cover for a tabletop vise design are merely copied and mirrored from original parts modeled using ROMULUS. All components are individual pieces, allowing for later kinematic and dynamic analysis in real time. (*Courtesy of Evans & Sutherland.*)

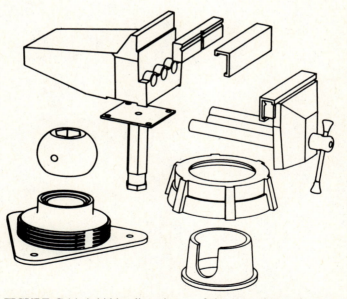

FIGURE C.14 A hidden-line picture of the parts that make up a tabletop vise. This exploded view details the various key components that make up the vise. (*Courtesy of Evans & Sutherland.*)

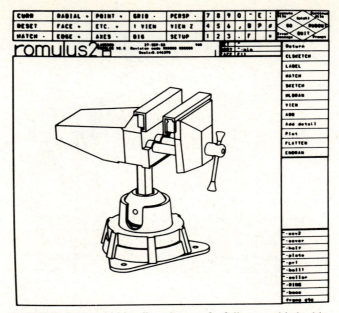

FIGURE C.15 A hidden-line picture of a fully assembled table-top vise. The vise was modeled using ROMULUS. The photo shows the tree-structured menu of ROMULUS. (*Courtesy of Evans & Sutherland.*)

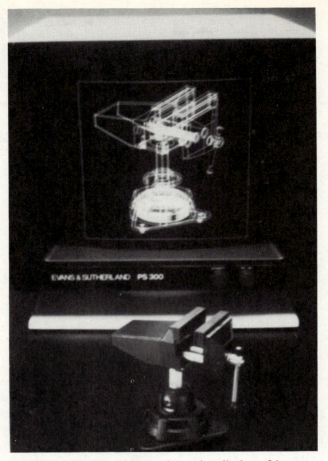

FIGURE C.16 The tabletop vise and a display of its geometric model constructed using ROMULUS. (*Courtesy of Evans & Sutherland.*)

PARTICLE SYSTEMS

1 INTRODUCTION

Modeling phenomena such as clouds, smoke, water, and fire has proved difficult with the existing techniques of computer image synthesis. These "fuzzy" objects do not have smooth, well-defined, and shiny surfaces; instead their surfaces are irregular, complex, and ill defined. We are interested in their dynamic and fluid changes in shape and appearance. They are not rigid objects nor can their motions be described by the simple affine transformations that are common in computer graphics.

This paper presents a method for the modeling of fuzzy objects that we call particle systems. The representation of particle systems differs in three basic ways from representations normally used in image synthesis. First, an object is represented not by a set of primitive surface elements, such as polygons or patches, that define its boundary, but as clouds of primitive particles that define its volume. Second, a particle system is not a static entity. Its particles change form and move with the passage of time. New particles are "born" and old particles "die." Third, an object represented by a particle system is not deterministic, since its shape and form are not completely specified. Instead, stochastic processes are used to create and change an object's shape and appearance.

In modeling fuzzy objects, the particle system approach has several important advantages over classical surface-oriented techniques. First, a particle (for now, think of a particle as a point in three-dimensional space) is a much simpler primitive than a polygon, the simplest of the surface representations.

Therefore, in the same amount of computation time one can process more of the basic primitives and produce a more complex image. Because a particle is simple, it is also easy to motion-blur. Motion-blurring of fast-moving objects for the removal of temporal aliasing effects has been largely ignored in computer image synthesis to date. A second advantage is that the model definition is procedural and is controlled by random numbers. Therefore, obtaining a highly detailed model does not necessarily require a great deal of human design time as is often the case with existing surface-based systems. Because it is procedural, a particle system can adjust its level of detail to suit a specific set of viewing parameters. As with fractal surfaces [5], zooming in on a particle system can reveal more and more detail. Third, particle systems model objects that are "alive," that is, they change form over a period of time. It is difficult to represent complex dynamics of this form with surface-based modeling techniques.

Modeling objects as collections of particles is not a new idea. Fifteen years ago, the earliest computer video games depicted exploding spaceships with many little glowing dots that filled the screen. Point sources have been used as a graphics data type in many three-dimensional modeling systems (e.g., the early Evans and Sutherland flight simulators), although there are few real references to them in the literature. Roger Wilson at Ohio State [4] used particles to model smoke emerging from a smokestack. There were neither stochastic controls nor dynamics in his model. Alvy Ray Smith and Jim Blinn used particles to model star creation and death in galaxies for the Cosmos series [11]. Alan Norton [9] used particles to generate and display three-dimensional fractal shapes. Jim Blinn [3] discussed light reflection functions for simulating light passing through and being reflected by layers of particles. His technique was used to produce images of the rings of Saturn. Blinn did not address the fuzzy object modeling problem which is the topic of this paper. Volumetric representations have also been proposed as viable alternatives to surface representations. Solid modeling [13] is a form of volumetric representation, as is the work of Norm Badler and Joe O'Rourke on "bubbleman" [2]. The use of stochastic modeling relates our work to the recent advances in fractal modeling [5].

Section 2 describes the basic framework of particle systems in more detail. Section 3 examines how particle systems were used to produce the fire element in the Genesis Demo sequence from the movie *Star Trek II: The Wrath of Khan* [10]. Section 4 presents several other applications of particle systems, and Section 5 discusses ongoing and future research in his area.

2   BASIC MODEL OF PARTICLE SYSTEMS

A particle system is a collection of many minute particles that together represent a fuzzy object. Over a period of time, particles are generated into a system, move and change from within the system, and die from the system.

To compute each frame in a motion sequence, the following sequence of steps is performed: (1) new particles are generated into the system, (2) each new particle is assigned its individual attributes, (3) any particles that have existed within the system past their prescribed lifetime are extinguished, (4) the remaining particles are moved and transformed according to their dynamic attributes, and finally (5) an image of the living particles is rendered in a frame buffer. The particle system can be programmed to execute any set of instructions at each step. Because it is procedural, this approach can incorporate any computational model that describes the appearance or dynamics of the object. For example, the motions and transformations of particles could be tied to the solution of a system of partial differential equations, or particle attributes could be assigned on the basis of statistical mechanics. We can, therefore, take advantage of models which have been developed in other scientific or engineering disciplines.

In the research presented here, we use simple stochastic processes as the procedural elements of each step in the generation of a frame. To control the shape, appearance, and dynamics of the particles within a particle system, the model designer has access to a set of parameters. Stochastic processes that randomly select each particle's appearance and movement are constrained by these parameters. In general, each parameter specifies a range in which a particle's value must lie. Normally, a range is specified by providing its mean value and its maximum variance.

The following subsections describe in more detail the basic model for particle systems, and how they are controlled and specified within the software we have written.

2.1   Particle   Generation

Particles are generated into a particle system by means of controlled stochastic processes. One process determines the number of particles entering the system during each interval of time, that is, at a given frame. The number of particles generated is important because it strongly influences the density of the fuzzy object.

The model designer can choose to control the number of new particles in one of two ways. In the first method, the designer controls the mean number of particles generated at a frame and its variance. The actual number of particles generated at frame f is

$$NParts_f = MeanParts_f + Rand(\ ) \times VarParts_f,$$

where *Rand* is a procedure returning a uniformly distributed random number between -1.0 and $+1.0$, $MeanParts_f$ the mean number of particles, and $VarParts_f$ its variance.

In the second method, the number of new particles depends on the screen size of the object. The model designer controls the mean number of particles generated per unit of screen area and its variance. The procedural particle system can determine the view parameters at a particular frame, calculate the approximate screen area that it covers, and set the number of new particles accordingly. The corresponding equation is

$$NParts_f = (MeanParts_{sa_f} + Rand(\ ) \times VarParts_{sa_f}) \times ScreenArea,$$

where $MeanParts_{sa_f}$ is the mean per screen area, $VarParts_{sa_f}$ its variance, and *ScreenArea* the particle system's screen area. This method controls the level of detail of the particle system and, therefore, the time required to render its image. For example, there is no need to generate 100,000 particles in an object that covers 4 pixels on the screen.

To enable a particle system to grow or shrink in intensity, the designer is able to vary over time the mean number of particles generated per frame (i.e., $MeanParts_f$ and $MeanParts_{sa_f}$ are, as used above, functions of frame number). Currently, we use a simple linear function

$$MeanParts_f = InitialMeanParts + DeltaMeanParts \times (f - f_0)$$

or

$$MeanParts_{sa_f} = InitialMeanParts_{sa} + DeltaMeanParts_{sa} \times (f - f_0),$$

where f is the current frame, f_0 the first frame during which the particle system is alive, *InitialMeanParts* the mean number of particles at this first frame, and *DeltaMeanParts* its rate of change. The variance controls, $VarParts_f$ and $VarParts_{sa_f}$, are currently constant over all frames. More sophisticated quadratic, cubic, or perhaps even stochastic variations in both the mean and variance parameters would be easy to add.

To control the particle generation of a particle system, therefore, the designer specifies f_0 and either the parameters *InitialMeanParts*, *DeltaMeanParts*, and *VarParts*, or the parameters $InitialMeanParts_{sa}$, $DeltaMeanParts_{sa}$, and $VarParts_{sa}$.

2.2 Particle Attributes

For each new particle generated, the particle system must determine values for the following attributes:

(1) initial position,

(2) initial velocity (both speed and direction),

(3) initial size,

(4) initial color,

(5) initial transparency,

(6) shape,

(7) lifetime.

Several parameters of a particle system control the initial position of its particles. A particle system has a position in three-dimensional space that defines its origin. Two angles of rotation about a coordinate system through this origin give it an orientation. A particle system also has a *generation shape* which defines a region about its origin into which newly born particles are randomly placed. Among the generation shapes we have implemented are: a sphere of radius r, a circle of radius r in the x–y plane of its coordinate system, and a rectangle of length l and width w in the x–y plane of its coordinate system.

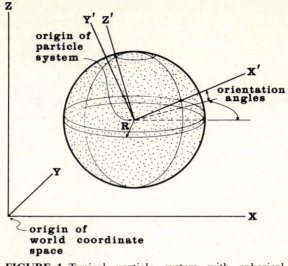

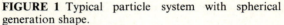

FIGURE 1 Typical particle system with spherical generation shape.

Figure 1 shows a typical particle system with a spherical generation shape. More complicated generation shapes based on the laws of nature or on chaotic attractors [1] have been envisioned but not yet implemented.

The generation shape of a particle system also describes the initial direction in which new particles move. In a spherical generation shape, particles move outward away from the origin of the particle system. In a circular or rectangular shape, particles move upward from the x–y plane, but are allowed to vary from the vertical according to an "ejection" angle, which is another parameter (see Figure 3). The initial speed of a particle is determined by

$$InitialSpeed = MeanSpeed + Rand(\ ) \times VarSpeed,$$

where *MeanSpeed* and *VarSpeed* are two other parameters of the particle system, the mean speed and its variance.

To determine a particle's initial color, a particle system is given an average color,[1] and the maximum deviation from that color. Particle transparency and particle size are also determined by mean values and maximum variations. The equations are similar to the one given above for initial speed.

A particle system has a parameter that specifies the shape of each of the particles it generates. The particle shapes implemented so far are spherical, rectangular, and streaked spherical. The latter is used to motion-blur particles— a very important feature when modeling fast-moving objects. We discuss streaking particles in more detail in Sections 2.5 and 3.

[1] In more detail, average red, green, and blue values are specified.

The number of possible attribute control parameters and their variants is endless. We are presenting those that we have found to be most useful and interesting.

2.3   Particle Dynamics

Individual particles within a particle system move in three-dimensional space and also change over time in color, transparency, and size.

To move a particle from one frame to the next is a simple matter of adding velocity vector to its position vector. To add more complexity, a particle system also uses an acceleration factor to modify the velocity of its particles from frame to frame. With this parameter the model designer can simulate gravity and cause particles to move in parabolic arcs rather than in straight lines.

A particle's color changes over time as prescribed by the rate-of-color change parameter. The transparency and size of particles are controlled in exactly the same way. In our implementation, these rates of change are global for all particles in a particle system, but one can easily imagine making this parameter stochastic too.

2.4   Particle Extinction

When it is generated, a particle is given a lifetime measured in frames. As each frame is computed, this lifetime is decremented. A particle is killed when its lifetime reaches zero.

Other mechanisms, if enabled, arrange for particles to be killed because they can contribute nothing to the image. If the intensity of a particle, calculated from its color and transparency, drops below a specified threshold, the particle is killed. A particle that moves more than a given distance in a given direction from the origin of its parent particle system may also be killed. This mechanism can be used to clip away particles that stray outside a region of interest.[2]

2.5   Particle Rendering

Once the position and appearance parameters of all particles have been calculated for a frame, the rendering algorithm makes a picture. The general particle-rendering problem is as complicated as the rendering of objects composed of the more common graphical primitives, such as polygons and curved surfaces. Particles can obscure other particles that are behind them in screen depth. They can be transparent and can cast shadows on other particles. Furthermore, particles can coexist in a scene with objects modeled by surface-based primitives, and these objects can intersect with the particles.

In our existing system, two assumptions allow us to simplify the rendering algorithm. First, we assume that particle systems do not intersect with other

---

[2] Note that this clipping is performed in modeling space—to a given plane for example. Clipping to the viewing frustum occurs later in the rendering stage and is discussed below.

surface-based modeling primitives, and hence our rendering algorithm need only handle particles. Objects modeled using other techniques are composited together with particle system objects in a postrendering compositing stage. In order for a particle system to intersect or be behind other objects, the rendering system will split the image of a particle system into subimages based on clipping planes defined in the model coordinate space. These subimages are then combined with other images in the compositing stage.

The other simplifying assumption made in our current rendering system is that each particle can be displayed as a point light source.[3] With this assumption, determining hidden surfaces is no longer a problem. Each particle adds a bit of light to the pixels that it covers. A particle behind another particle is not obscured but rather adds more light to the pixels covered. The amount of light added, and its color depend on the particle's transparency and color. Currently, the amount of light added does not depend on the distance between the particle and the viewing position, but that would be an easy extension. The viewing transformation, the particle's size, and its shape determine which pixels are covered. All particle shapes are drawn antialiased in order to prevent temporal aliasing and strobing. Light may be added to a pixel by many particles, so the rendering algorithm clamps the individual red, green, and blue intensities at the maximum intensity value of the frame buffer instead of letting them wrap around.

With this algorithm and assumptions, no sorting of the particles is needed. They are rendered into the frame buffer in whatever order they are generated. Shadows are no longer a problem, since particles do not reflect but emit light.

2.6   Particle Hierarchy

Our system has a mechanism that supports the formation and control of particle system hierarchies. The model designer creates a particle system in which the particles are themselves particle systems. When the parent particle system is transformed, so are all of its descendant particle systems and their particles. The parent particle system's mean color and its variance are used to select the mean color and variance of the offspring particle systems using the same equations presented earlier. The number of new particle systems generated at a frame is based on the parent's particle generation rate. The other parameters of the parent similarly affect those of its children. The data structure used to represent the hierarchy is a tree.

A hierarchy can be used to exert global control on a complicated fuzzy object that is composed of many particle systems. For example, a cloud might be composed of many particle systems, each representing a billowing region of water particles. A parent particle system could group these all together and control the cloud's global movement and appearance as influenced by the wind and terrain.

---

[3] Explosions and fire, the two fuzzy objects we have worked with the most, are modeled well with this assumption. Other fuzzy objects, such as clouds and water, are not. Section 5 discusses rendering algorithms for these objects.

3 USING PARTICLE SYSTEMS TO MODEL A WALL OF FIRE AND EXPLOSIONS

The Genesis Demo sequence [14] from the movie *Star Trek II*: *The Wrath of Khan* [10] was generated by the Computer Graphics project of Lucasfilm Ltd. The sequence depicts the transformation of a dead, moonlike planet into a warm, earthlike planet by an experimental device called the Genesis bomb. In a computer-simulated demonstration, the bomb hits the planet's surface and an expanding wall of fire spreads out from the point of impact to eventually "cleanse" the entire planet. The planet's surface begins to buckle, mountains grow, and oceans, vegetation, and an atmosphere form to produce an earth-like environment.

The wall-of-fire element in the Genesis Demo was generated using a two-level hierarchy of particle systems. The top-level system was centered at the impact point of the genesis bomb. It generated particles which were themselves particle systems. Figure 2 illustrates the positions of these second-level particle systems and how they formed expanding concentric rings on the surface of the planet. The number of new particle systems generated in each ring was based on the circumference of the ring and a controlling density parameter. New particle systems were spaced randomly around the ring. Particle systems over-lapping others in the same or adjacent rings gave the ring a solid, continuous look.

The second-level particle systems began generating particles at varying times on the basis of their distance from the point of impact. By varying the

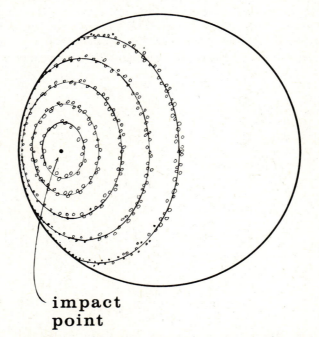

impact point

FIGURE 2 Distribution of particle systems on the planet's surface.

starting times of the particle systems, the effect of an expanding wall of fire was produced.

The second-level particle systems were modeled to look like explosions. Figure 3 shows an example. The generation shape was a circle on the surface of the planet. Each particle system was oriented so that particles, generated at random positions within the circle, flew upward away from the planet's surface. The initial direction of the particles' movement was constrained by the system's ejection angle to fall within the region bounded by the inverted cone shown in Figure 3. As particles flew upward, the gravity parameter pulled them back down to the planet's surface, giving them a parabolic motion path. The number of particles generated per frame was based on the amount of screen area covered by the particle system.

The individual particle systems were not identical. Their average color and the rates at which the colors changed were inherited from the parent particle system, but varied stochastically. The initial mean velocity, generation circle radius, ejection angle, mean particle size, mean lifetime, mean particle generation rate, and mean particle transparency parameters were also based on their parent's parameters, but varied stochastically. Varying the mean velocity parameter caused the explosions to be of different heights.

All particles generated by the second-level particle systems were predominately red in color with a touch of green. Recall from Section 2.5 that particles are treated as point light sources and that colors are added, not matted, into a pixel. When many particles covered a pixel, as was the case near the center and base of each explosiion, the red component was quickly clamped at full intensity and the green component increased to a point where the resulting color was orange and even yellow. Thus, the heart of the explosion had a hot yellow-orange glow which faded off to shades of red elsewhere. Actually, a small blue component caused pixels covered by very many particles to appear white. The

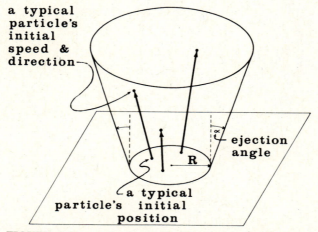

FIGURE 3 Form of an explosion-like particle system.

rate at which a particle's color changed simulated the cooling of a glowing piece of some hypothetical material. The green and blue components dropped off quickly, and the red followed at a slower rate. Particles were killed when their lifetime expired, when their intensity fell below the minimum intensity parameter, or if they happened to fall below the surface of the planet.

A quickly moving object leaves a blurred image on the retina of the human eye. When a motion picture camera is used to film live action at 24 frames per second, the camera shutter typically remains open for 1/50 of a second. The image captured on a frame is actually an integration of approximately half the motion that occurred between successive frames. An object moving quickly appears blurred in the individual still frames. Computer animation has traditionally imaged scenes as indidvidual instants in time and has ignored motion blur. The resulting motion often exhibits temporal aliasing and strobing effects that are disconcerting to the human eye. Motion blur is a complex topic that is beginning to appear in the literature [7, 12].

The particles in our wall-of-fire element are motion-blurred. Three-dimensional positions are calculated for a particle at the beginning of a frame and about halfway through the frame, and an antialiased straight line is drawn between the corresponding screen coordinate positions in the frame buffer.[1] Antialiased lines are used to prevent staircasing (moving jaggies) and strobing (popping on and off) effects. To be perfectly correct, screen motion due to movement of the camera should be considered when calculating where to blur a particle. One can also argue that simulating the imperfect temporal sampling of a movie camera is not ideal and that motion blur should really simulate what happens in the human eye. This is a good area for future research.

In the finished sequence, the wall of fire spread over the surface of the planet both in front of and behind the planet's limb (outer edge). The rendering algorithm generated two images per frame—one for all particles between the camera's position and the silhouette plane of the planet, and one for all particles on the other side of this clipping plane. These two elements were composited with the barren moonlike planet element and the stars element in back to front order—stars, background fires, planet, and foreground fires.

Because the wall of fire was modeled using many small light-emitting particles, light from the fire should have reflected off the planet's surface. Our current implementation of particle systems does not handle light reflection on surface-based objects. To achieve this effect, Lucasfilm team member Tom Duff added an additional strong local light source above the center of the rings of fire when he rendered the planet's surface. This produced the glow that circles the ring of fire on the planet's surface. (This glow is visible in Figure 5.)

Figure 4 is a frame showing the initial impact of the Genesis bomb. It was generated from one very large particle system and about 20 smaller ones about its base. About 25,000 particles exist in this image. Figure 5 occurs partway through the first half of the sequence. It contains about 200 particle systems and

---

[1] A particle's trajectory is actually parabolic, but the straight-line approximation has so far proved sufficient.

FIGURE 4 Initial explosion.

75,000 particles. Figure 6 shows the ring of fire extending over and beyond the limb of the planet. It is formed from about 200 explosions and 85,000 particles. Figure 7 shows the wall of fire just before it engulfs the camera; in Figure 8 the camera is completely engulfed. Both employ about 400 particle systems and contain over 750,000 particles. The textures in Figure 8 are completely synthetic and yet have a "natural" and highly detailed appearance that is uncommon in most computer graphics images. These images are interesting statically, but they

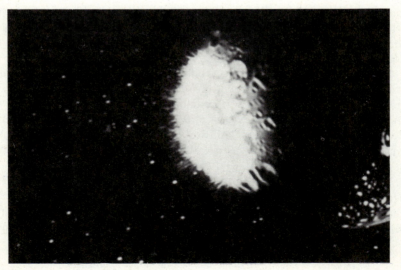

FIGURE 5 Expanding wall of fire.

FIGURE 6 Wall of fire over limb of planet.

only really come alive on the movie screen. It is interesting to note that this is also the case for many of the best traditional (i.e., non-computer generated) special effects shots where motion blur is an important factor.

A few points concerning random numbers are of interest from a production point of view. The random number routine we use is based on [6], and generates numbers uniformly in the range [0.0, 1.0]. It is an incremental algorithm based on updating a table of seed values. To checkpoint a production, all that need be

FIGURE 7. Wall of fire about to engulf camera.

FIGURE 8 Wall of fire completely engulfing camera.

saved is this random number table—we do not save all the parameters of 750,000 particles. To restart a computation at frame n, the closest preceding frame p is found that cannot contribute particles to frame n (this is determined from the lifetime parameters of all the active particle systems). Frame $p + 1$'s random number table is then read, and particle generation can begin from there. No particles are drawn until the simulation reaches frame n, so this backing up and restarting usually takes only a few minutes.

Particles moving off screen or being extinguished for any reason do not affect the randomness of other particles. This is because all stochastic decisions concerning a particle are performed when it is generated. After that, its motion is deterministic. If stochastic elements were to be used to perturbate the dynamics of a particle (e.g., to simulate turbulence), more care would have to be taken when checkpointing a frame and killing particles. In that case, it would probably be better to use a more deterministic and reproducible random number generator.

4   OTHER APPLICATIONS OF PARTICLE SYSTEMS

4.1   Fireworks

We are currently using particle systems to model fireworks. The fireworks differ from the Genesis Demo in that the control parameters of the particle systems vary more widely, and streaking is more predominant. Figure 9 shows two red explosions superimposed. One explosion is tall, thin, and near the end of its lifetime, and the other is short, fat, and building up to full steam. Figure 10 shows

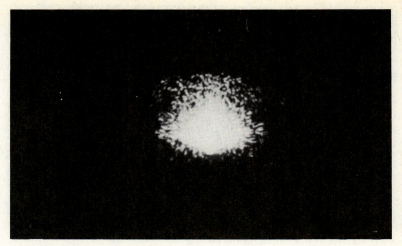

FIGURE 9 Two red fireworks.

several green explosions dying off and blue spherical explosion starting up. Figure 11 contains overlapping, multicolored explosions formed with different generation shapes and ejection angles. Again, these images only really come alive when projected at 24 frames per second.

4.2 Line Drawing Explosions

Particle systems are being used to model exploding objects in a computer-simulated tactical display for a scene from the movie *Return of the Jedi* [8]. In this case, the particle systems are implemented on a line-drawing display. In

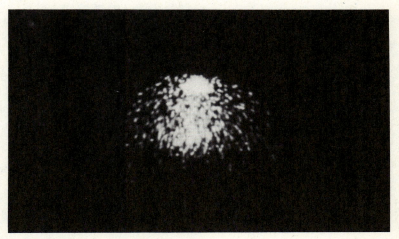

FIGURE 10 Green and blue fireworks.

FIGURE 11 Multicolored fireworks.

order to simulate motion blur, the particles are drawn as very small straight lines instead of as points. The texturing effects that are evident in the previous examples are lost on a line-drawing display, and yet the motion still looks real and the sequence gives the viewer the impression that something is exploding. This is because the model is dynamic—it moves well.

4.3 Grass

To model grass, we use an explosive type of particle system, similar to that used in the Genesis Effect. Instead of drawing particles as little streaks, the parabolic

FIGURE 12 *white sand.*

trajectory of each particle over its entire lifetime is drawn. Thus, the time-domain motion of the particle is used to make a static shape. Grasslike green and dark green colors are assigned to the particles which are shaded on the basis of the scene's light sources. Each particle becomes a simple representation of a blade of grass and the particle system as a whole becomes a clump of grass. Particle systems randomly placed on a surface and overlapping one another are used to model a bed or patch of grass.

Figure 12 is a picture entitled *white sand* by Alvy Smith of Lucasfilm. The grass elements of this image were generated as described above. The plant elements were generated using a partially stochastic technique similar to particle systems.

5 ONGOING RESEARCH IN PARTICLE SYSTEMS

A logical extension of this research will be to use particle systems to model fuzzy objects in which the individual particles can not be rendered as point light sources, but must be rendered as individual light-reflecting objects.

To this end, we have begun to investigate the modeling of clouds. Clouds are difficult for several reasons. First, the shape and form of clouds are complex depending on many factors such as wind direction, temperature, terrain, and humidity. The atmospheric literature abounds with cloud models that are simple in concept but computationally difficult, since most are based on partial differential equations. Second, clouds are difficult because they can throw shadows on themselves. This property is very important in making a cloud look like a cloud. Third, the number of particles needed to model a cloud will be very large. This will require an efficient rendering algorithm.

6 CONCLUSIONS

We have presented particle systems, a method for the modeling of a class of fuzzy objects, and have shown how they were used in making the fire element of the Genesis Demo sequence for the movie *Star Trek II: The Wrath of Khan*. Particle systems have been used as a modeling tool for other effects and appear promising for the modeling of phenomena like clouds and smoke.

Particles, especially when modeled as point light sources or as streaks of light, have proved efficient to render—they are merely antialiased lines. Because they are so simple, they lend themselves to a hardware or firmware implementation. With a hardware antialiased line-drawing routine, the computation of our wall-of-fire element would have been two to three times faster.

Particle systems are procedural stochastic representations controlled by several global parameters. Stochastic representations are capable of producing minute detail without requiring substantial user design time. The textures in the fire sequence could not have been modeled with other existing methods. Fire images, scanned in from a photograph or painted, could have been texture mapped, but they would still have been static. Another advantage of a procedural representation is its ability to adapt to several different viewing environments. For example, procedural representations can generate only as much detail as is needed in a frame, potentially saving significant amounts of computation time.

Having finally come to grips with spatial aliasing, it is now time for computer image synthesis to begin to investigate and solve temporal aliasing problems. The Genesis Demo is the first "big screen" computer-synthesized sequence to include three-dimensional dynamic motion blur. The particle system can easily be motion-blurred because they are so simple. A great deal of work remains to be done in this area—blurring particles is much easier than blurring curved surface patches.

Particle systems can model objects that explode, flow, splatter, puff up, and billow. These kinds of dynamics have not been produced with surface-based representations. The most important aspect of particle systems is that they move: good dynamics are quite often the key to making objects look real.

7   ACKNOWLEDGEMENTS

The author gratefully acknowledges the suggestions and encouragement of all members of the graphics project at Lucasfilm Ltd, especially those who worked on the Genesis Demo sequence: Loren Carpenter, Ed Catmull, Pat Cole, Rob Cook, David DiFrancesco, Tom Duff, Rob Poor, Tom Porter, and Alvy Ray Smith. The crusade for motion blur and antialiasing in computer image synthesis is a goal of the entire graphics project and Lucasfilm as a whole. One of the referees deserves credit for pointing out several extensions and improvements to the motion blurring discussion. Finally, thanks to Ricki Blau for editorial and photographic assistance.

REFERENCES

1. Abraham, R., and Shaw, C. *DYNAMICS—The geometry of Behavior*. City on the Hill Press, Santa Cruz, Calif., 1981.
2. Badler, N. I., O'Rourke, J., and Toltzis, H. A spherical human body model for visualizing movement. *Proc. IEEE* **67**, 10 (Oct. 1979).
3. Blinn, J. F. Light reflection functions for simulation of clouds and dusty surfaces. Proc. SIGGRAPH '82, In *Comput. Gr.* **16**, 3 (July 1982), 21–29.
4. Csuri, C., Hackathorn, R., Parent, R., Carlson, W., and Howard, M. Towards an interactive high visual complexity animation system. Proc. SIGGRAPH 79. In *Comput. Gr* **13**, 2 (Aug. 1979), 289–299.
5. Fournier, A., Fussel, D., and Carpenter, L. Computer rendering of stochastic models. *Commun. ACM* **25**, 6 (June 1982), 371–384.
6. Knuth, D. E. *The Art of Computer Programming*, vol. 2. Addison-Wesley, Reading, Mass., (1969), p. 464.
7. Korein, J., and Badler, N. I. Temporal anti-aliasing in computer generated animation. To appear in Proc. SIGGRAPH '83 (July 1983).
8. Lucasfilm. *Return of the Jedi* (film), May 1983.

9. Norton, A. Generation and display of geometric fractals in 3-D. Proc. SIGGRAPH '82. In *Comput. Gr.* **16**, 3 (July 1982), 61–67.

10. Paramount. *Star Trek II: The Wrath of Khan* (film), June 1982.

11. PBS. *Carl Sagan's Cosmos Series.* (television series), Public Broadcasting System, 1980.

12. Potmesil, M., and Chakravarty, I. Modeling motion blur in computer-generated images. To appear in Proc. SIGGRAPH '83 (July 1983).

13. Requicha, A. A. G., and Voelcker, H. B. Solid modeling: A historical summary and contemporary assessment. *IEEE Comput. Gr. Appl.* (March 1982).

14. Smith, A. R., Carpenter, L., Catmull, E., Cole, P., Cook, R., Poor, T., Porter, T. and Reeves, W. *Genesis Demo Documentary* (film), June 1982, Lucasfilm Ltd.

SUMMARY OF IMPORTANT EQUATIONS

VECTORS

$$\mathbf{p} = [p_x\mathbf{i} + p_y\mathbf{j} + p_z\mathbf{k}]$$

$$\mathbf{p} = [p_x \quad p_y \quad p_z][\mathbf{i} \quad \mathbf{j} \quad \mathbf{k}]^T$$

or more simply

$$\mathbf{p} = [p_x \quad p_y \quad p_z]$$

$$|\mathbf{p}| = \sqrt{p_x^2 + p_y^2 + p_z^2}$$

$$\mathbf{n}_p = \frac{\mathbf{p}}{|\mathbf{p}|}$$

$$\mathbf{p} \cdot \mathbf{r} = p_x r_x + p_y r_y + p_z r_z$$

$$\mathbf{p} \times \mathbf{r} = \begin{vmatrix} \mathbf{i} & \mathbf{j} & \mathbf{k} \\ p_x & p_y & p_z \\ r_x & r_y & r_z \end{vmatrix}$$

PLANE EQUATIONS

$$Ax + By + Cz + D = 0$$

unit normal

$$n_x = \frac{A}{\sqrt{A^2 + B^2 + C^2}}$$

$$n_y = \frac{B}{\sqrt{A^2 + B^2 + C^2}}$$

$$n_z = \frac{C}{\sqrt{A^2 + B^2 + C^2}}$$

Normal distance to plane from origin

$$d = \frac{D}{\sqrt{A^2 + B^2 + C^2}}$$

Normal vector from origin to plane

$$\mathbf{d} = [n_x d \quad n_y d \quad n_z d]$$

Given three points on a plane

$$\mathbf{p}_1 = [a \quad 0 \quad 0]$$
$$\mathbf{p}_2 = [0 \quad b \quad 0]$$
$$\mathbf{p}_3 = [0 \quad 0 \quad c]$$

then

$$\frac{x}{a} + \frac{y}{b} + \frac{z}{c} = 1$$

pc CURVES

$$\mathbf{p}(u) = \mathbf{a}_3 u^3 + \mathbf{a}_2 u^2 + \mathbf{a}_1 u + \mathbf{a}_0 \qquad u \in [0, 1]$$

$$\mathbf{p}(u) = \mathbf{UA}$$

$$\mathbf{U} = [u^3 \quad u^2 \quad u \quad 1]$$

$$\mathbf{A} = [\mathbf{a}_3 \quad \mathbf{a}_2 \quad \mathbf{a}_1 \quad \mathbf{a}_0]^T$$

$$\mathbf{p}(u) = \mathbf{UMB}$$

$$\mathbf{M} = \begin{bmatrix} 2 & -2 & 1 & 1 \\ -3 & 3 & -2 & -1 \\ 0 & 0 & 1 & 0 \\ 1 & 0 & 0 & 0 \end{bmatrix}$$

$$\mathbf{B} = \begin{bmatrix} \mathbf{p}_0 \\ \mathbf{p}_1 \\ \mathbf{p}_0^u \\ \mathbf{p}_1^u \end{bmatrix}$$

$$\mathbf{A} = \mathbf{MB}$$

$$\mathbf{B} = \mathbf{M}^{-1}\mathbf{A}$$

$$\mathbf{M}^{-1} = \begin{bmatrix} 0 & 0 & 0 & 1 \\ 1 & 1 & 1 & 1 \\ 0 & 0 & 1 & 0 \\ 3 & 2 & 1 & 0 \end{bmatrix}$$

$$\mathbf{F} = \mathbf{UM}$$

$$\mathbf{p}(u) = \mathbf{FB}$$

$$F_1 = 2u^3 - 3u^2 + 1$$

$$F_2 = -2u^3 + 3u^2$$

$$F_3 = u^3 - 2u^2 + u$$

$$F_4 = u^3 - u^2$$

$$\mathbf{p}^u = \mathbf{F}^u \mathbf{B}$$

$$F_1^u = 6u^2 - 6u$$

$$F_2^u = -6u^2 + 6u$$

$$F_3^u = 3u^2 - 4u + 1$$

$$F_4^u = 3u^2 - 2u$$

$$\mathbf{p}^{uu} = \mathbf{F}^{uu} \mathbf{B}$$

$$F_1^{uu} = 12u - 6$$

$$F_2^{uu} = -12u + 6$$

$$F_3^{uu} = 6u - 4$$

$$F_4^{uu} = 6u - 2$$

$$\mathbf{B} = \mathbf{KP}$$

$$\mathbf{K}^{-1}\mathbf{B} = \mathbf{P}$$

$$\mathbf{P} = [\mathbf{p}(0) \quad \mathbf{p}(1/3) \quad \mathbf{p}(2/3) \quad \mathbf{p}(1)]^T$$

$$\mathbf{p}(u) = \mathbf{UMKP}$$

$$\mathbf{K} = \begin{bmatrix} 1 & 0 & 0 & 0 \\ 0 & 0 & 0 & 1 \\ -11/2 & 9 & -9/2 & 1 \\ -1 & 9/2 & -9 & 11/2 \end{bmatrix}$$

$$\mathbf{K}^{-1} = \begin{bmatrix} 1 & 0 & 0 & 0 \\ 20/27 & 7/27 & 4/27 & -2/27 \\ 7/27 & 20/27 & 2/27 & -4/27 \\ 0 & 1 & 0 & 0 \end{bmatrix}$$

$$\mathbf{N} = \mathbf{MK}$$

$$\mathbf{p}(u) = \mathbf{UNP}$$

$$\mathbf{N} = \begin{bmatrix} -9/2 & 27/2 & -27/2 & 9/2 \\ 9 & -45/2 & 18 & -9/2 \\ -11/2 & 9 & -9/2 & 1 \\ 1 & 0 & 0 & 0 \end{bmatrix}$$

pc CONIC CURVES

$$\mathbf{p}(u) = F[\mathbf{p}_0 \quad \mathbf{p}_1 \quad 4\rho(\mathbf{p}_2 - \mathbf{p}_0) \quad 4\rho(\mathbf{p}_1 - \mathbf{p}_2)]^T$$

Ellipse $\qquad\qquad\qquad \rho \in (0, 0.5)$

Parabola $\qquad\qquad\qquad \rho = 0.5$

Hyperbola $\qquad\qquad\qquad \rho \in (0.5, 1)$

Circular arc $\qquad\qquad\qquad \rho = \dfrac{\cos \theta}{1 + \cos \theta}$

BEZIER CURVES

$$\mathbf{p}(u) = \sum_{i=0}^{n} \mathbf{p}_i B_{i,n}(u) \qquad u \in [0, 1]$$

$$B_{i,n}(u) = C(n, i)u^i(1 - u)^{n-1}$$

$$C(n, i) = \frac{n!}{i!(n - i)!}$$

Three points, $n = 2$

$$\mathbf{p}(u) = (1 - u)^2 \mathbf{p}_0 + 2u(1 - u)\mathbf{p}_1 + u^2 \mathbf{p}_2$$

Four points, $n = 3$

$$\mathbf{p}(u) = (1 - u)^3 \mathbf{p}_0 + 3u(1 - u)^2 \mathbf{p}_1 + 3u^2(1 - u)\mathbf{p}_2 + u^3 \mathbf{p}_3$$

$$\mathbf{p}(u) = [u^3 \quad u^2 \quad u \quad 1] \begin{bmatrix} -1 & 3 & -3 & 1 \\ 3 & -6 & 3 & 0 \\ -3 & 3 & 0 & 0 \\ 1 & 0 & 0 & 0 \end{bmatrix} \begin{bmatrix} \mathbf{p}_0 \\ \mathbf{p}_1 \\ \mathbf{p}_2 \\ \mathbf{p}_3 \end{bmatrix}$$

Five points, $n = 4$

$$\mathbf{p}(u) = (1 - u)^4\mathbf{p}_0 + 4u(1 - u)^3\mathbf{p}_1 + 6u^2(1 - u)^2\mathbf{p}_2$$
$$+ 4u^3(1 - u)\mathbf{p}_3 + u^4\mathbf{p}_4$$

B-SPLINE CURVES

$$\mathbf{p}(u) = \sum_{i=0}^{n} \mathbf{p}_i N_{i,k}(u)$$

For $n + 1$ control points,

$k - 1$ degree of polynomial in u, $u \in [0, n - k + 2]$

$$N_{i,1}(u) = 1 \qquad \text{if } t_i \leq u < t_{i+1}$$
$$= 0 \qquad \text{otherwise}$$

$$N_{i,k}(u) = \frac{(u - t_i)N_{i,k-1}(u)}{t_{i+k-1} - t_i} + \frac{(t_{i+k} - u)N_{i+1,k-1}(u)}{t_{i+k} - t_{i+1}}$$

For open curves,

$$t_i = 0 \qquad \text{if } i < k$$
$$t_i = i - k + 1 \qquad \text{if } k \leq i \leq n$$
$$t_i = n - k + 2 \qquad \text{if } i > n$$

OPEN PERIODIC B-SPLINE CURVES

For $k = 3$,

$$\mathbf{p}_i(u) = \tfrac{1}{2}[u^2 \quad u \quad 1]\begin{bmatrix} 1 & -2 & 1 \\ -2 & 2 & 0 \\ 1 & 1 & 0 \end{bmatrix}\begin{bmatrix} \mathbf{p}_{i-1} \\ \mathbf{p}_i \\ \mathbf{p}_{i+1} \end{bmatrix} \qquad i \in [1:n-1]$$

For $k = 4$,

$$\mathbf{p}_i(u) = \tfrac{1}{6}[u^3 \quad u^2 \quad u \quad 1]\begin{bmatrix} -1 & 3 & -3 & 1 \\ 3 & -6 & 3 & 0 \\ -3 & 0 & 3 & 0 \\ 1 & 4 & 1 & 0 \end{bmatrix}\begin{bmatrix} \mathbf{p}_{i-1} \\ \mathbf{p}_i \\ \mathbf{p}_{i+1} \\ \mathbf{p}_{i+2} \end{bmatrix} \qquad i \in [1:n-2]$$

CLOSED PERIODIC B-SPLINE CURVES

For $k = 3$,

$$\mathbf{p}_i(u) = \mathbf{U}_3\mathbf{M}_3\begin{bmatrix} \mathbf{p}_{(i-1)\bmod(n+1)} \\ \mathbf{p}_{i\bmod(n+1)} \\ \mathbf{p}_{(i+1)\bmod(n+1)} \end{bmatrix} \qquad i \in [1:n+1]$$

For $k = 4$,

$$\mathbf{p}_i(u) = \mathbf{U}_4\mathbf{M}_4 \begin{bmatrix} \mathbf{P}_{(i-1)\bmod(n+1)} \\ \mathbf{P}_{i\bmod(n+1)} \\ \mathbf{P}_{(i+1)\bmod(n+1)} \\ \mathbf{P}_{(i+2)\bmod(n+1)} \end{bmatrix} \qquad i \in [1{:}n+1]$$

BICUBIC SURFACE (HERMITE)

$$\mathbf{p}(u, w) = \mathbf{U}\mathbf{A}\mathbf{W}^T \qquad u, w \in [0, 1]$$

$$\mathbf{U} = [u^3 \quad u^2 \quad u \quad 1]$$

$$\mathbf{W} = [w^3 \quad w^2 \quad w \quad 1]$$

$$\mathbf{A} = \begin{bmatrix} \mathbf{a}_{33} & \mathbf{a}_{32} & \mathbf{a}_{31} & \mathbf{a}_{30} \\ \mathbf{a}_{23} & \mathbf{a}_{22} & \mathbf{a}_{21} & \mathbf{a}_{20} \\ \mathbf{a}_{13} & \mathbf{a}_{12} & \mathbf{a}_{11} & \mathbf{a}_{10} \\ \mathbf{a}_{03} & \mathbf{a}_{02} & \mathbf{a}_{01} & \mathbf{a}_{00} \end{bmatrix}$$

$$\mathbf{p}(u, w) = \mathbf{U}\mathbf{M}\mathbf{B}\mathbf{M}^T\mathbf{W}^T$$

$$\mathbf{B} = \begin{bmatrix} \mathbf{p}_{00} & \mathbf{p}_{01} & \mathbf{p}_{00}^w & \mathbf{p}_{01}^w \\ \mathbf{p}_{10} & \mathbf{p}_{11} & \mathbf{p}_{10}^w & \mathbf{p}_{11}^w \\ \mathbf{p}_{00}^u & \mathbf{p}_{01}^u & \mathbf{p}_{00}^{uw} & \mathbf{p}_{01}^{uw} \\ \mathbf{p}_{10}^u & \mathbf{p}_{11}^u & \mathbf{p}_{10}^{uw} & \mathbf{p}_{11}^{uw} \end{bmatrix}$$

$$\mathbf{n} = \frac{\mathbf{p}^u \times \mathbf{p}^w}{|\mathbf{p}^u \times \mathbf{p}^w|}$$

$$\mathbf{p}(u, w) = \mathbf{U}\mathbf{N}\mathbf{P}\mathbf{N}^T\mathbf{W}^T$$

$$\mathbf{p} = \begin{bmatrix} \mathbf{p}(0, 0) & \mathbf{p}(0, 1/3) & \mathbf{p}(0, 2/3) & \mathbf{p}(0, 1) \\ \mathbf{p}(1/3, 0) & \mathbf{p}(1/3, 1/3) & \mathbf{p}(1/3, 2/3) & \mathbf{p}(1/3, 1) \\ \mathbf{p}(2/3, 0) & \mathbf{p}(2/3, 1/3) & \mathbf{p}(2/3, 2/3) & \mathbf{p}(2/3, 1) \\ \mathbf{p}(1, 0) & \mathbf{p}(1, 1/3) & \mathbf{p}(1, 2/3) & \mathbf{p}(1, 1) \end{bmatrix}$$

$$\mathbf{B} = \mathbf{L}\mathbf{P}\mathbf{L}^T$$

$$\mathbf{L} = \mathbf{M}^{-1}\mathbf{N} = \begin{bmatrix} 1 & 0 & 0 & 0 \\ 0 & 0 & 0 & 1 \\ -11/2 & 9 & -9/2 & 1 \\ 1 & 9/2 & -9 & 11/2 \end{bmatrix}$$

BEZIER SURFACE

$$\mathbf{p}(u, w) = \sum_{i=0}^{m} \sum_{j=0}^{n} \mathbf{p}_{ij} B_{i,m}(u) B_{i,n}(w) \qquad u \in [0, 1]$$

For $m = n = 3$,

$$\mathbf{p}(u, w) = [(1 - u)^3 \quad 3u(1 - u)^2 \quad 3u^2(1 - u) \quad u^3] \, \mathbf{P} \begin{bmatrix} (1 - w)^3 \\ 3w(1 - w)^2 \\ 3w^2(1 - w) \\ w^2 \end{bmatrix}$$

$$\mathbf{P} = \begin{bmatrix} \mathbf{p}_{11} & \mathbf{p}_{12} & \mathbf{p}_{13} & \mathbf{p}_{14} \\ \mathbf{p}_{21} & \mathbf{p}_{22} & \mathbf{p}_{23} & \mathbf{p}_{24} \\ \mathbf{p}_{31} & \mathbf{p}_{32} & \mathbf{p}_{33} & \mathbf{p}_{34} \\ \mathbf{p}_{41} & \mathbf{p}_{42} & \mathbf{p}_{43} & \mathbf{p}_{44} \end{bmatrix}$$

B-SPLINE SURFACE

$$\mathbf{p}(u, w) = \sum_{i=0}^{m} \sum_{j=0}^{n} \mathbf{p}_{ij} N_{i,k}(u) N_{j,l}(w)$$

$$\mathbf{p}_{s,t}(u, w) = \mathbf{U}_k \mathbf{M}_k \mathbf{P}_{k,l} \mathbf{M}_l^T \mathbf{W}_l^T \qquad s \in [1 : m + 2 - k]$$

$$t \in [1 : n + 2 - l]$$

$$u, w \in [0, 1]$$

PARAMETRIC SOLID

$$\mathbf{p}(u, v, w) = \sum_{i=1}^{4} \sum_{j=1}^{4} \sum_{k=1}^{4} \mathbf{a}_{ijk} u^{4-i} v^{4-j} w^{4-k}$$

or

$$\mathbf{p}_{uvw} = \mathbf{a}_{ijk} u^{4-i} v^{4-j} w^{4-k}$$

$$\mathbf{p}_{uvw} = F_i(u) F_j(v) F_k(w) \mathbf{b}_{ijk}$$

$$\mathbf{a}_{ijk} = \mathbf{M}_{il} \mathbf{M}_{jm} \mathbf{M}_{kn} \mathbf{b}_{lmn}$$

A LIST OF GEOMETRIC-MODELING PROCEDURES

The following procedures are described in exercises throughout the text. Refer to the index for their exact location.

ARCL	Computes the length of a pc curve.
ARCLNS	Computes the geometric coefficients of a pc curve representing a circular arc tangent to two intersecting straight lines.
ARCNTR	Computes the center point of a circular arc represented by a pc curve.
ARCPC	Computes the geometric coefficients of a pc curve representing a circular arc.
AREA	Computes the surface area of a bicubic patch.
ATG	Converts pc curve coefficients in algebraic form into geometric form.
ATGS	Converts bicubic patch coefficients in algebraic form into geometric form.
ATGSOL	Converts tricubic solid coefficients in algebraic form into geometric form.
BF	Computes the blending functions of a pc curve.
BFU	Computes first-derivative blending functions of a pc curve.
BFUU	Computes second-derivative blending functions of a pc curve.
BLNDC	Computes the geometric coefficients of a pc curve that joins two disconnected pc curves.
BLOCK	Creates instances of a unit cube.
BN	Computes the geometric coefficients of the unit normals patch for a specified bicubic patch.
BSEGBS	Segments a B-spline curve for subsequent plotting.
BSEGBZ	Segments a Bezier curve for subsequent plotting.
BSEGPC	Segments a pc curve for subsequent plotting.
BSPLNC	Computes the x, y, z coordinates of a point on a B-spline curve.

BSPLNS	Computes the x, y, z coordinates of a point on a *B*-spline surface.
BZCRV	Computes the x, y, z coordinates of a point on a Bezier curve.
BZSRF	Computes the x, y, z coordinates of a point on a Bezier surface.
CDU	Creates a pc curve representing the first derivative of any point on an initial pc curve.
CNTCRV	Computes the center of curvature at a point on a pc curve.
CONSRF	Computes the geometric coefficients of a bicubic patch that approximates a conic surface.
CRTCRV	Computes the curvature, radius of curvature, and torsion at a point on a pc curve.
CRVINV	Computes the parametric variable at a point on a pc curve (inverse point solution).
CRVSRF	Computes the intersection of a pc curve and a bicubic patch.
CRV2BS	Computes the intersection of two *B*-spline curves.
CRV2BZ	Computes the intersection of two Bezier curves.
CRV2PC	Computes the intersection of two pc curves.
CUBIC	Computes the roots of a cubic equation.
CYLSRF	Computes the geometric coefficients of a cylindrical bicubic patch.
C4SRF	Computes the geometric coefficients of a bicubic patch given four pc curves.
DETVAL	Evaluates a 3×3 determinant.
EFGLMN	Computes the coefficients of the first and second fundamental forms at a point on a cubic surface.
FPTOG	Computes the geometric coefficients of a pc curve given four points.
GTAS	Converts bicubic patch coefficients in geometric form into algebraic form.
GTASOL	Converts tricubic solid coefficients in geometric form into algebraic form.
GTOFP	Converts the geometric coefficients of a pc curve into the four-point form.
G16P	Converts the geometric form of a bicubic patch into the 16-point form.
INVCRV	Reverses the direction of parametrization of a pc curve.
INVSRF	Reverses the direction of parametrization of a bicubic patch.
LITE	Computes the intensity of light reaching a viewer from a bicubic patch illuminated by a point of light.
LNCRV	Computes the intersection of a straight line and a pc curve.
LNLN	Computes the intersection of two straight lines.
LNSRF	Computes the intersection of a straight line and a bicubic patch.
MATINV	Inverts a 4×4 matrix.
MATRNS	Transposes a 4×4 matrix.
MIDCRV	Computes the midpoint of a pc curve segment.
MINDCC	Computes the minimum distance between two pc curves.
MINDCS	Computes the minimum distance between a pc curve and a bicubic patch.

MINDPC	Computes the minimum distance between a point and a pc curve.
MINDPS	Computes the minimum distance between a point and a bicubic path.
NORPLN	Computes the coefficients of the normal, osculating, and rectifying planes at a point on a pc curve.
ORBIT	Computes the net transformation matrix for a scene.
PARPLN	Computes the coefficients of a plane through a given point and parallel to a given plane.
PATH	Computes a sequence of orbit transformations.
PCCRCL	Computes the geometric coefficients of a composite string of pc curves representing a circle.
PCLASC	Classifies a point with respect to a closed planar curve.
PCLASS	Classifies a point on a bicubic surface.
PCLIN	Computes the geometric coefficients of a pc straight line.
PCNC	Computes the geometric coefficients of a pc curve representing a conic.
PCRSOL	Computes the x, y, z, u, v, w coordinates of a point on a curve in a tricubic solid.
PCRV	Computes the x, y, z coordinates of a point u on a pc curve, using the geometric coe cients.
PCRVA	Computes the x, y, z coordinates of a point u on a pc curve, using the algebraic coefficients.
PDSURF	Computes the geometric coefficients of a composite set of bicubic patches generated by a PD curve.
PERPLN	Computes the coefficients of a plane through two points and perpendicular to another plane.
PLN	Computes the geometric coefficients of a planar bicubic patch.
PLNBCS	Computes the intersection of a plane and a bicubic surface.
PLNBSC	Computes the intersection of a plane and a nonperiodic B-spline curve.
PLNBSS	Computes the intersection of a plane and a periodic B-spline surface.
PLNBZC	Computes the intersection of a plane and a Bezier curve.
PLNBZS	Computes the intersection of a plane and a Bezier surface.
PLNLN	Computes the intersection of a plane and a straight line.
PLNPC	Computes the intersection of a plane and a pc curve.
PLNPLN	Computes the line of intersection of two planes.
PNC	Computes the principal normal curvatures at a point on a bicubic surface.
POCOS	Computes the x, y, z, u, w coordinates of a point on a pc curve on a bicubic surface.
POICRV	Computes the points of inflection on a pc curve.
PPCRV	Computes the x, y, z coordinates of points on a pc curve at equal intervals on the parametric variable.
PPSRF	Computes the x, y, z coordinates of a rectangular array of points on a bicubic patch.

PROJPT	Computes the orthogonal projection of a set of points.
PRSPRJ	Computes the perspective projection of a set of points.
PSOLID	Computes the x, y, z coordinates of a point on or in a tricubic solid.
PSRF	Computes the x, y, z coordinates of a point on a bicubic patch given u and w, using the geometric form.
PSRFA	Computes the x, y, z coordinates of a point on a bicubic patch given u and w, using the algebraic form.
PSRF16	Computes the x, y, z coordinates of a point on a bicubic patch using the 16-point form.
PSRSOL	Computes the x, y, z, u, v, w coordinates of a point on a surface in a tricubic solid.
PTGSOL	Converts tricubic solid coefficients in point form to geometric form.
PTNCRV	Computes the nearest normal and tangents from a point to a closed planar Bezier curve.
PTPCRV	Computes the coordinates of a point on a pc curve at a specified distance from another point as measured along the curve.
PUM	Computes the magnitudes of the tangent vectors of a planar pc curve whose end points, slopes, and two intermediate points are specified.
P16G	Converts the 16-point form of a bicubic patch to the geometric form.
RFLCT	Reflects a set of points through any of the seven principal coordinate-system elements.
RFLMAG	Reflects and magnifies a set of points through any point in space.
ROTAXS	Rotates points about an axis in space.
ROTORG	Rotates points about the origin.
RPRCRV	Reparametrizes a pc curve.
RPSRF	Reparametrizes a bicubic patch.
RULSRF	Computes the geometric coefficients of a ruled bicubic patch.
SCALE	Scales a set of defining points for a curve or surface.
SETOP	Performs set operations on two finite lists of integer numbers.
SILOET	Computes the silhouette curve of a bicubic patch.
SPCRV	Computes the geometric coefficients of a pc space curve.
SPHSRF	Computes the geometric coefficients of a spherical bicubic patch.
SPLNPC	Fits a string of pc curves to a set of points to form a composite spline curve.
SPROD	Computes the scalar product of two vectors.
SRFINV	Computes the parametric variable at a point on a bicubic surface (inverse point solution).
SRFREV	Computes the geometric coefficients of a bicubic surface of revolution.
SUBSRF	Subdivides and reparametrizes a bicubic patch.
TANPLN	Computes the coefficients of the tangent plane to a point on a bicubic surface.

THNET	Computes a net homogeneous transformation matrix.
THP	Computes the homogeneous transformation of a set of points.
TNBCRV	Computes the unit tangent, normal, and binormal vectors at a point on a pc curve.
TPTP	Transforms points according to the three-point-to-three-point transformation.
TRANSH	Constructs a homogeneous transformation matrix.
TRNBZC	Translates a Bezier curve.
TRNCRV	Truncates and reparametrizes a pc curve.
TRNPC	Translates a pc curve.
TVCRV	Computes the parametric slopes and direction cosines at a point on a pc curve.
TVSRF	Computes the tangent vectors at a point on a bicubic patch.
TWSTV	Computes the twist vector at a point on a bicubic patch.
UNV	Computes the unit normal vector at a point on a bicubic patch.
VANG	Computes the angle between two vectors.
VMAG	Computes the magnitude of a vector.
VPROD	Computes the vector product of two vectors.
VSUM	Computes the sum of two vectors.
VUNIT	Computes the unit vector of a vector.

BIBLIOGRAPHY

Abelson, H., and A. Disessa. *Turtle Geometry: The Computer as a Medium for Exploring Mathematics.* MIT Press, Cambridge, MA, 1981.

Acton, F. S. *Numerical Methods that Work.* Harper and Row, New York, 1970.

Adams, J. A. Geometric concepts for computer graphics. Engineering and Weapons Report No. EW-72-4. U.S. Naval Academy, Annapolis, MD, Sept. 1972.

Agin, G. J., and T. O. Binford. Computer descriptions of curved objects. *IEEE Transactions on Computers* **C-25**:439, 1976.

Agoston, M. K. *Algebraic Topology.* Marcel Dekker, New York, 1976.

Aho, A. V., J. E. Hopcroft, and J. D. Ullman. *The Design and Analysis of Computer Algorithms.* Addison-Wesley, Reading, MA, 1976.

Ahuja, D. V. An algorithm for generating splinelike curves. *IBM Systems Journal,* **17(3)**:206–17, 1968.

Ahuja, D. V., and S. A. Coons. Interactive graphics in data processing: geometry for construction and display. *IBM System Journal,* **3, 4**:188, 1968.

Albus, J. S., and J. M. Evans, Jr. Robot systems. *Scientific American,* 76, August 1976.

Alexandroff, P. *Elementary Concepts of Topology.* Dover, New York, 1961.

ANSI Standard Y14.26.1 (proposed). Digital representation of physical object shapes. *ASME* New York, 1978.

Apostol, T. M. *Calculus, Vol. 2: Calculus of Several Variables with Applications to Probability and Vector Analysis.* Blaisdell, New York, 1978.

Badler, N. I., and R. Bajcsy. Three-dimensional representations for computer graphics and computer vision. *Computer Graphics* **12(3)**:153–60, Aug. 1978.

Badler, N. I., and J. O'Rourke. Representation of articulable, quasi-rigid, three-dimensional objects. Technical Report MS-CIS-79-24. Computer and Information Science Department, Moore School of Electrical Engineering, University of Pennsylvania, Philadelphia, Apr. 1979.

Badler, N. I., B. L. Webber, J. U. Korein, and J. Korein. TEMPUS: a system for the design and simulation of mobile agents in a workstation and task environment. *Proceedings of the IEEE Trends and Applications Conference*, 1983.

Baer, A., C. Eastman, and M. Henrion. Geometric modelling: a survey. *Computer-Aided Design*, **11(5)**:253–72, Sept. 1979.

Ball, A. A. A simple specification of the parametric cubic segment, *Computer-Aided Design*, **10(3)**:181–82, 1978.

Barnhill, R. E. Blending-function interpolation: a survey and some new results. Numerical Analysis Report No. 9. Department of Mathematics, University of Dundee, Dundee, Scotland, July 1975.

———. A survey of the representation and design of surfaces. *IEEE Computer Graphics and Applications* **3(7)**:9–16, Oct. 1983.

Barnhill, R. E., G. Birkhoff, and W. J. Gordon. Smooth interpolation in triangles. *Journal of Approximation Theory* **8**:114–28, 1973.

Barnhill, R. E., and R. F. Riesenfeld, eds. *Computer-Aided Geometric Design*. Academic, New York, 1974.

Barsky, B. A. The beta-spline: a local representation based on shape parameters and fundamental geometric measures. Ph.D. dissertation, Department of Computer Science, University of Utah, Salt Lake City, Utah, Dec. 1981.

———. A study of the parametric uniform B-spline curve and surface representation. Technical Report CSD 83/118, Computer Science Division, University of California, Berkeley, California, May 1983.

———. A description and evaluation of various 3-D models. *IEEE Computer Graphics and Applications* **4(1)**:38–52, Jan. 1984.

Barsky, B. A., R. H. Bartels, and J. C. Beatty. An introduction to the use of splines in computer graphics. CS-83-9, Department of Computer Science, University of Waterloo, Waterloo, Ontario, Canada, 1983.

Barsky, B. A. and J. C. Beatty. Varying the betas in beta-splines. Technical Report CSD 82/112, Computer Science Division, University of California, Berkeley, California, Dec. 1982.

———. Local control of bias and tension in beta-splines. ACM Transactions in Graphics, **8(2)**: 109–34, April 1983.

Barton, E. E., and I. Buchanan. The polygon package. *Computer-Aided Design.* **12(1)**:3–11, Jan. 1980.

Baumgart, B. G. Winged-edge polyhedron representation. STAN-CS-320. Computer Science Department, Stanford University, Palo Alto, CA, 1972.

———. GEOMED—a geometric edition. Stanford Artificial Intelligence Laboratory Report No. AIM-232. Computer Science Department, Stanford University, Palo Alto, CA, May 1974.

————. Geometric modelling for computer vision. *Stanford Artificial Intelligence Research Memo No. 249.* Computer Science Department, Stanford University, Palo Alto, Oct. 1974.

Beyond special effects: an interview with Gary Demos. *Computer Graphics World,* 63–66, Jan. 1983.

Bezier, P. *Numerical Control: Mathematics and Applications.* Wiley, London, 1972.

————. Mathematical and practical possibilities of UNISURF. In *Computer-Aided Geometric Design.* R. E. Barnhill and R. F. Riesenfeld, eds., Academic, New York, 1974.

————. Degenerate surfaces and particularly about three-sided patches. Renault, Boulogne-Billancourt, France, Apr. 1975.

Binford, T. O., and G. J. Agin. Computer description of curved objects. *Proceedings of the Third International Conference on Artificial Intelligence,* 629–40, 1973.

Binford, T. O., R. A. Brooks, and D. G. Lowe. Image understanding via geometric models. *Proceedings of the Fifth International Conference on Pattern Recognition.* 364–69, 1980.

Blinn, J. F. Models of light reflection for computer-synthesized pictures. *Computer Graphics* **11(2)**:192, summer 1977.

————. Computer display of curved surfaces. Technical Report No. 1060. Jet Propulsion Laboratory, Pasadena, CA, Oct. 1978.

————. Simulation of wrinkled surfaces. *Computer Graphics* **12(3)**:286, Aug. 1978.

————. A generalization of algebraic surface drawing. *ACM Transactions on Graphics* **1(3)**:235–56, July 1982.

————. Light reflection functions for simulation of clouds and dusty surfaces. *Computer Graphics* **16(3)**:21–29, July 1982.

Blinn, J. F., and M. E. Newell. Texture and reflection in computer-generated images. *Communications of the ACM* **19(10)**:542, Oct. 1976.

Boden, M. *Artificial Intelligence and Natural Man.* Basic Books, New York, 1976.

Bohn, W. Inserting new knots into *B*-spline curves. *Computer-Aided Design* **12(4)**:199–201, 1980.

Boyse, J. W. Preliminary design for a geometric modeller. Research Publication GMR-2768. Computer Science Department, General Motors Research Laboratories, Warren, MI, 1978.

————. Interference detection among solids and surfaces. *Communications of the ACM* **21**:3–9, 1979.

Boyse, J. W. Data structure for a solid modeller. Research Publication GMR-2933. Computer Science Department, General Motors Research Laboratories, Warren, MI, 1979.

Boyse, J. W., and J. E. Gilchrist, GMSolid: interactive modeling for design and analysis of solids. *IEEE Computer Graphics and Applications* **2(2)**:27–40, Mar. 1982.

Braid, I. C. Designing with volumes. CAD Group. Cambridge University, Cambridge, England, 1973.

————. The synthesis of solids bounded by many faces. *Communications of the ACM* **18(4)**:209–16, Apr. 1975.

Braid, I. C. Six systems for shape design and representation: a review. CAD Group Document No. 87. Cambridge University, Cambridge, England, May 1975.

————. A new shape design system. CAD Group Document No. 89. Cambridge University, Cambridge, England, Mar. 1976.

————. On storing and changing shape information. *Computer Graphics* **12(3)**: 234–38, 1978.

————. New directions in geometric modeling. *Proceedings of the Workshop on Geometric Modeling*, CAM-I, Inc., Mar. 1978.

Braid, I. C., R. C. Hillyard, and I. A. Stroud. Stepwise construction of polyhedra in geometric modelling. CAD Group Document No. 100. Cambridge University, Cambridge, England, 1978.

Braid, I. C., and C. Lang. The design of mechanical components with volume building blocks. In *Computer Languages for Numerical Control*, J. Hatvany, ed. North-Holland, Amsterdam, Holland, 1973.

Brodlie, K., ed. *Mathematical Methods in Computer Graphics and Design*. Academic, New York, 1980.

Brown, C. M. Some mathematical and representational aspects of solid modelling, *IEEE Transactions on Pattern Analysis and Machine Intelligence* **PAMI-3(4)**: 444–53, July 1981.

————. Structured interactive design: a goal for CAD systems. *Proceedings of the 1981 National Computer Graphics Association Conference*, Baltimore, June 1981.

————. PADL-2: a technical summary. *IEEE Computer Graphics and Applications* **2(2)**: 69–84, Mar. 1982.

Brown, C. M., A. A. G. Requicha, and H. B. Voelker. Geometric-modelling systems for mechanical design and manufacturing. *Proceedings of the 1978 Annual Conference of ACM*, 770, Washington, D.C., 1978.

Brown, C. M., and H. B. Voelker. The PADL-2 project. *Proceedings of the Seventh NSF Conference on Production Research Technology*, F1, Ithaca, New York, 1979.

Brown, K. Q. Comments on algorithms for reporting and counting geometric intersections. *IEEE Transactions on Computing* **C-30(2)**: 147–48, Feb. 1981.

Burkley, R. M., and J. P. Mayfield. A numerical geometry system. *Proceedings of the Society of Manufacturing Engineers*, 1975.

Burton, F. W. and M. M. Huntbach, Lazy evaluation of geometric objects. *IEEE Computer Graphics and Applications*, **4(1)**: 28–33, Jan. 1984.

Burton, W. Representation of many-sided polygons and polygonal lines for rapid processing. *Communications of the ACM* **20(3)**: 166, Mar. 1977.

Butterfield, K. R. The computation of all the derivatives of a *B*-spline basis. British Leyland, Oxford, England, May 1974.

————. CAM-I sculptured surface documentation. Publication No. PS-80-SS-01, vols. 1–3. Computer-Aided Manufacturing International, Inc., Arlington, TX, 1980.

Catmull, E. E. Computer display of curved surfaces. *Proceedings of the IEEE Conference on Computer Graphics, Pattern Recognition, and Data Structure*, 11, May 1975.

————. A hidden-surface algorithm with anti-aliasing. *Computer Graphics* **12(3)**:6, Aug. 1978.

————. A subdivision algorithm for computer display of curved surfaces. Department of Computer Science, University of Utah, Salt Lake City, 1981.

Catmull, E. E., and J. Clark, Recursively generated *B*-spline surfaces on arbitrary topological meshes. *Computer-Aided Design*. **10(6)**, Nov. 1978.

Catmull, E. E., and R. Rom, A class of local interpolating spline. In *Computer-Aided Geometric Design*, R. E. Barnhill and R. F. Riesenfeld, eds. Academic, New York, 1974.

Chaikin, G. M. Geometric description and generation of surfaces. Report No. CRL-348, AD A032692. Division of Applied Science, New York University, New York, Jan. 1975.

————. An algorithm for high-speed curve generation. *Computer Graphics and Image Processing*, June 1975.

Chasen, S. H. *Geometric Principles and Procedures for Computer Graphic Applications*. Prentice-Hall, Englewood Cliffs, NJ, 1978.

Chazelle, B., and D. P. Dobkin. Decomposing a polygon into its convex parts. *Proceedings of the* 11*th Annual ACM Symposium on Theory of Computation*, 1979.

————. Detection is easier than computation. *Proceedings of the 12th Annual ACM Symposium on Theory of Computing*, 146–53 Los Angeles, Apr. 1980.

Clark, J. H. Some properties of *B*-splines. *Second USA–Japan Computer Conference Proceedings*, 542–45, 1975.

————. Designing surfaces in 3-D. *Communications of the ACM* **19(8)**:454–60, Aug. 1976.

————. Hierarchial geometric models for visible surface algorithms. *Communications of the ACM* **19(10)**:547, Oct. 1976.

————. Parametric curves, surfaces, and volumes in computer graphics and computer-aided geometric design. NASA Ames Research Center, 1978.

————. The geometry engine: a VLSI geometry system for graphics. *Computer Graphics* **16(3)**:127–33, July 1982.

Claytor, R. N. Specifications for the geometric-modeling project (addendum 3, geometric-modeling proposal). CAM-I Geometric-Modeling Interest Group Meeting, Arlington, TX, Mar. 1976.

Cohen, E. Some mathematical tools for a modeler's workbench. *IEEE Computer Graphics and Applications* **3(7)**:63–66, Oct. 1983.

Cohen, E., T. Lyche, and R. Riesenfeld. Discrete *B*-splines and subdivision techniques in computer-aided geometric design and computer graphics. *Computer Graphics and Image Processing* **14(2)**:87–111, Oct. 1980.

Cohen, J., and T. Hickey. Two algorithms for determining volumes of convex polyhedra. *Journal of the ACM* **26**:401, 1979.

Comba, P. G. A procedure for detecting intersections of three-dimensional objects. *Journal of the ACM* **15(3)**:354–66, July 1968.

Cook, R. L., and K. E. Torrance, A reflectance model for computer graphics. *SIGGRAPH* '81, 307–16, Dallas, TX, Aug. 1981.

Coons, S. A. An outline of the requirements for a computer-aided design system. MIT Memo ESL-TM-169. Massachusetts Institute of Technology, Cambridge, MA, Mar. 1963.

———. Surfaces for computer-aided design of space figures. MIT Memo, Massachusetts Institute of Technology, Cambridge, MA, 1965.

———. Surfaces for computer-aided design of space forms. MIT Project MAC, MAC-TR-41. Massachusetts Institute of Technology, Cambridge, MA, June 1967.

———. Surface patches and *B*-spline curves. In *Computer-Aided Geometric Design*, R. E. Barnhill and R. F. Riesenfeld, eds. Academic, New York, 1974.

Cox, M. G. The numerical evaluation of *B*-splines. National Physical Laboratory DNAC 4, Aug. 1971.

Coxeter, H. M. S. *Regular Polytopes*. Dover, New York, 1963.

Crow, F. C. Shaded computer graphics in the entertainment industry. *Computer* **11(2)**:242, summer 1977.

Davis, P. J. *Interpolation and Approximation*. Ginn-Blaisdell, New York, 1963 (also Dover, 1975).

deBoor, C. On calculating with *B*-splines. *Journal of Approximation Theory* **6**:50–62, 1972.

———. Good approximation by splines with variable knots. *ISNM* **21**:57–72, 1973.

———. Splines as linear combinations of *B*-splines, a survey. In *Approximation Theory II*, G. G. Lorentz, et al., eds. Academic, New York, 1976.

———. Package for calculating with *B*-splines. *J. of Numerical Analysis* **14(3)**:441–72, 1977.

———. *A Practical Guide to Splines*. Springer-Verlag, New York, 1978.

Dewhirst, D. L., and R. C. Hillyard. Application of volumetric modelling to mechanical design and analysis. *Proceedings of the Design Automation Conference 18th*, Nashville, TN, June 1981.

———. Digital representation of physical object shapes. American Society of Mechanical Engineers, United Engineering Center, New York.

Dimsdale, B. Bicubic patch bounds. *Computing and Mathematics with Applications* **3(2)**:95–104. 1977.

Dimsdale, B., and K. Johnson. Multiconic surfaces. *IBM J. of Research and Development* 523–29, Nov. 1975.

Dimsdale, B., and R. M. Burkley. Bicubic patch surfaces for high-speed numerical control processing. *IBM J. of Research and Development* 358–67, July 1976.

Doctor, L. J., and J. G. Torborg. Display techniques for octree-encoded objects. *IEEE Computer Graphics and Applications* **1(3)**:29–38, July 1981.

Dube, R. P. Univariate blending functions and alternatives. *Computer Graphics and Image Processing* **6**:394–408, 1977.

Dyer, C. R., A. Rosenfeld, and H. Samet. Region representation: boundary codes from quadtrees. *Communications of the ACM* **23**:171, 1980.

Eastman, C. M. Representations for space planning. *Communications of the ACM* **13(4)**:242–50, Apr. 1970.

————. General-purpose building description systems. *Computer Aided Design* **8(1)**:17–26, Jan. 1976.

Eastman, C. M., et al. An outline of the building description system. Institute of Physical Planning Research Report No. 50. Carnegie-Mellon University, Pittsburgh, Sept. 1974.

Eastman, C. M., and J. Lividini. Spatial search. Institute of Physical Planning Research Report No. 55. Carnegie-Mellon University, Pittsburgh, May 1975.

Eastman, C. M., and K. Weiler. Geometric modelling using the Euler operators. Institute of Physical Planning Research Report No. 78. Carnegie-Mellon University, Pittsburgh, 1979.

Eastman, C. M., and C. I. Yessios. An efficient algorithm for finding the union, intersection, and difference of spatial domains. Department of Computer Science, Carnegie-Mellon University, Pittsburgh, 1971.

Encarnacao, J., ed. *Computer-Aided Design Modeling, Systems Engineering, CAD Systems.* Springer-Verlag, New York, 1981.

Eshleman, A. L. Informal notes on parametric cubic curves, undated.

Eshleman, A. L., and H. D. Meriwether. Animated display of dynamic characteristics of complex structures. Presented to the 1966 UAIDE annual meeting, San Diego, CA, Oct.–Nov. 1966.

————. Graphic applications to aerospace structural design problems. Presented to SHARE, Fourth Annual Design Automation Workshop, Los Angeles, June 1967.

Farin, G. Bezier polynomials over triangles and the construction of piecewise C^r polynomials. Report TR/91. Department of Mathematics, Brunel University, Uxbridge, Middlesex, England, Jan. 1980.

Faux, I. D., and M. J. Pratt. *Computational Geometry for Design and Manufacture.* Halsted, Chichester, England, 1980.

Feng, D. Y., and R. F. Riesenfeld. Some new surface forms of computer-aided geometric design. *Computing Journal* **23(4)**:324–31, 1980.

Ferguson, J. C. Multivariable curve interpolation. *Journal of the ACM* **11**:221–28, 1964.

Firth, I. M. *Holography and Computer-Generated Holograms.* Mills and Boon, 1972.

Fishback, W. T. *Projective and Euclidean Geometry.* Wiley, New York, 1969.

Fitzgerald, W., F. Gracer, and R. Wolfe. GRIN: Interactive Graphics for Modelling Solids. *IBM Journal of Research and Development* **25(4)**:281–94, July 1981.

Foley, J. D., and A. Van Dam. *Fundamentals of Interactive Computer Graphics*. Addison-Wesley, Reading, MA, 1982.

Forrest, A. R. Curves and surfaces for computer-aided design. Ph.D. diss., Cambridge University, Cambridge, England 1968.

———. Coons surfaces and multivariate functional interpolation. Cambridge University, Cambridge, England, Feb. 1979.

———. Computational geometry. *Proceedings of the Royal Society of London* **A,321**:187–95, 1971.

———. A new curve form for computer-aided design. CAD Group Document No. 66. Cambridge University, Cambridge, England, June 1972.

———. On Coons and other methods for representation of curved surfaces. *Computer Graphics and Image Processing* **1(4)**:341–89, Dec. 1972.

———. Interactive interpolation and approximation by Bezier polynomials. *Computing Journal* **15**:71–79, 1972.

———. Computer-aided design of three-dimensional objects: a survey. *Proceedings of the ACM/AICA International Computing Symposium*, Venice, Italy, 1972.

———. Computational geometry—achievements and problems. In *Computer Aided Geometric Design*. R. E. Barnhill and R. F. Riesenfeld, eds. Academic, New York, 1974.

———. On the rendering of surfaces. *Computer Graphics* **13(3)**:253, Aug. 1979.

———. Recent work on geometric algorithms. In *Mathematical Methods in Computer Graphics and Design*, K. Brodlie, ed. Academic, New York, 1980.

———. The twisted cubic curve: a computer-aided geometric design approach. *Computer Aided Design* **12(4)**:165–72, July 1980.

Forsythe, E. A., et al., *Computer Science: A First Course*. Wiley, New York, 1975.

Freeman, H. Computer processing of line-drawing images. *Computing Surveys* **6(1)**: 57–97, Mar. 1974.

Gellert, G. O. Geometric computing, electronic geometry for semi-automated design. *Machine Design* part 1, 152–59, Mar. 1965; part 2, 94–100, Apr. 1965.

Giloi, W. K. *Interactive Computer Graphics*. Prentice-Hall, Englewood Cliffs, NJ, 1978.

Goldman, R. N. An urnful of blending functions. *IEEE Computer Graphics and Applications* **3(7)**:49–54, Oct. 1983.

Goldstein, R. Defining the bounding edges of a synthavision solid model. *Proceedings of the Design Automation Conference 18th*, Nashville, TN, June–July 1981.

Goldstein, R., and L. Malin. 3-D modelling with the synthavision system. *Proceedings of the First Annual Conference on Computer Graphics in CAD/CAM Systems*, 244, Cambridge, MA, 1979.

Gordon, W. J. Distributive lattices and the approximation of multivariate functions. *Proceedings of the Symposium on Approximation with Special Emphasis on Splines.* I. J. Schoenberg, ed. University of Wisconsin Press, Madison, pp. 223–77, 1969.

Gordon, W. J. An Operator calculus for surface and volume modeling. *IEEE Computer Graphics and Applications.* **3(7)**:18–22, Oct. 1983.

Gordon, W. J., and R. F. Riesenfeld. Bernstein-Bezier methods for the computer-aided design of freeform curves and surfaces. *Journal of the ACM* **21(2)**:293–310, Apr. 1974.

————. *B*-spline curves and surfaces. *Computer Aided Geometric Design*, R. E. Barnhill and R. F. Riesenfeld, eds. Academic, New York, 1974.

Graham, N. *Artificial Intelligence.* TAB Books, 1979.

Grayer, A. R. Alternative approaches in geometric modelling. *Computer Aided Design* **12(4)**:189–92, July 1980.

Griffiths, J. G. A data structure for the elimination of hidden surfaces by patch subdivision. *Computer Aided Design* **7(3)**:171–78, July 1975.

————. A surface display algorithm. *Computer Aided Design* **10(1)**:65–73, Jan. 1978.

————. Bibliography of hidden-line and hidden-surface algorithms. *Computer Aided Design* **10(3)**:203–06, July 1978.

Grossman, D. D. Procedural representation of three-dimensional objects. *IBM Journal of Research and Development* **20**:582, 1976.

————. Monte Carlo simulation of tolerancing in discrete part manufacturing and assembly. Stanford Artificial Intelligence Laboratory Memo AIM-280. Stanford University, Palo Alto, CA, May 1976.

Guzman, A. Decomposition of a visual field into three-dimensional bodies. In *Automatic Interpretation and Classification of Images*, A. Grasselli, ed. Academic, New York, pp. 243–76, 1969.

————. Analysis of curved line drawings using context and global information. *Machine Intelligence* **6**:325–76, 1971.

Ham, I. Selected references on group technology in english. Society of Mechanical Engineers Technical Paper MR71-284, 1971.

Hall, R. A., and D. P. Greenberg. A testbed for realistic image synthesis. *IEEE Computer Graphics and Applications* **3(8)**:10–20, Nov. 1983.

Hanna, S. L., J. F. Abel, and D. P. Greenberg. Intersection of parametric surfaces by means of look-up tables. *IEEE Computer Graphics and Applications* **3(7)**:39–48, Oct. 1983.

Hedelman, H. A data flow approach to procedural modeling, *IEEE Computer Graphics and Applications* **4(1)**:16–26, Jan. 1984.

Hedgely, D. R. A general solution to the hidden-line problem. *NASA Reference Publication 1085*, Mar. 1982.

Hilbert, D., and S. Cohn-Vossen. *Geometry and the Imagination.* Chelsea, New York, 1952.

Hillyard, R. C. Dimensions and tolerances in shape design. *Computer Laboratory Report No. 8.* Cambridge University, Cambridge, England, 1978.

———. The BUILD group of solid modelers. *IEEE Computer Graphics and Applications* **2(2)**:43–52, Mar. 1982.

Hillyard, R. C., and I. C. Braid. The analysis of dimensions and tolerances in computer-aided shape design. *Computer Aided Design* **19**:161–66, June 1978.

———. Characterizing nonideal shapes in terms of dimensions and tolerances. *Computer Graphics* **12(3)**:234–38, 1978.

Hinds, J. K. Methods of representing conic segments by means of rational parametric cubic equations. Northern Illinois University, De Kalb, IL. Informal notes, undated.

Horn, B. Obtaining shape from shading information. In *The Psychology of Computer Vision*, P. H. Winston, ed. McGraw-Hill, New York, pp. 115–56, 1975.

Hyde, M. O. *Computers Who Think: The Search for Artificial Intelligence.* Enslow, 1982.

Inselberg, A. Cubic splines with infinite derivatives at some knots. *IBM Journal of Research and Development* 430–36, Sept. 1976.

———. IPAD geometry standard research. Prepared under Contract No. NAS1-14700. By Boeing Commercial Airplane Co., Seattle, for Langley Research Center, NASA, Dec. 1981.

Jackins, C. L., and S. L. Tanimoto. Oct-trees and their uses in representing three-dimensional objects. *Computer Graphics and Image Processing* **14(3)**:249–70, Nov. 1980.

Jones, L. J. Solid modelling: the future for graphics. *Proceedings of CAM-I International Spring Seminar*, P-80-MM-01, 21–31, Cam-I, Inc., Denver, Apr. 1980.

Julesz, B. *Foundations of Cyclopean Perception.* University of Chicago Press, Chicago, 1970.

———. Experiments in the visual perception of texture. *Scientific American*, **232**:34–43, 1975.

Kinnucan, P. Computer-aided manufacturing aims for integration. *High Technology* 49–56, May–June 1982.

Knuth, D. E. *The Art of Computer Programming, Vol. 3: Sorting and Searching.* Addison-Wesley, Reading, MA, 1973.

Lafue, G. Recognition of three-dimensional objects from orthographic views. *Computer Graphics* **10(2)**:103–08, summer 1976.

Lane, J. M., and R. F. Riesenfeld. A theoretical development for computer generation and display of piecewise polynomial surfaces. *IEEE Transactions on Pattern Analysis and Machine Intelligence* **PAMI-2**, 35–46, 1980.

Lane, J. M., L. C. Carpenter, T. Whitted, and J. F. Blinn. Scan line methods for displaying Parametrically defined surfaces. *Communications of the ACM* **23**:23–34, 1980.

Lane, E. P. *Metric Differential Geometry of Curves and Surfaces.* The University of Chicago Press, Chicago, 1940.

Lee, Y. T., and A. A. G. Requicha. Algorithms for computing the volume and other integral properties of solid objects. Technical Memo No. 35. Production Automation Project, University of Rochester, Rochester, NY 1980.

————. Algorithms for computing the volume and other integral properties of solids. Part 1, known methods and open issues. *Communications of the ACM* **25(9)**: 635–41, Sept. 1982.

————. Algorithms for computing the volume and other integral properties of solids. Part 2, a family of algorithms based on representation conversion and cellular approximation. *Communications of the ACM* **25(9)**: 642–50, Sept. 1982.

Levin, J. A parametric algorithm for drawing pictures of solid objects composed of quadric surfaces. *Communications of the ACM* **19(10)**: 555–63. Oct. 1976.

Levoy, M. A color animation system based on the multiplane technique. *Computer Graphics* **11(2)**: 65–71, summer 1977.

Lewell, John. Turning dreams into reality with super computers and super visions. *Computer Pictures* 40–47, Jan.–Feb. 1983.

Lieberman, H. How to color in a coloring book. *SIGGRAPH '78* 111–16, Aug. 1978.

Lieberman, L. I. Model-driven vision for industrial automation. Presented at the IBM International Symposium on Advances in Digital Image Processing, Bad Neuenahr, Germany, Sept. 1978.

Lieberman, L. I., and M. A. Wesley. AUTOPASS: an automatic programming system for computer-controlled mechanical assembly. *IBM Journal of Research and Development*, **21(4)**: 321–33, July 1977.

Lieberman, L. I., M. A. Wesley, and M. A. Lavin. A geometric-modeling system for automated mechanical assembly. IBM Watson Research Center Report RC7089. Yorktown Heights, NY, 1978.

Liming, R. A. *Mathematics for Computer Graphics*. Aero, Fallbrook, CA, 1979.

Lorentz, G. G., et al. *Approximation Theory II*. Academic, New York, 1976.

Lores, M. E., S. H. Chasen, and J. M. Garner. Evaluation of 3-D graphics software: a case study. *IEEE Computer Graphics and Applications* **3(8)**: 73–77, Nov. 1983.

Lossing, D. L., and A. L. Eshleman. Planning a common data base for engineering and manufacturing. *SHARE XLIII*, Chicago, Aug. 1974.

Loutrel, P. A solution to the hidden-line problem for computer-drawn polyhedra. *IEEE Transactions on Computers* **C-19(3)**: 205–13, Mar. 1970.

Lozano-Perez, T., and M. A. Wesley. An algorithm for planning collision-free paths among polyhedral objects. *Communications of the ACM* **22(10)**: 560–70, Oct. 1979.

Lyche, T., and L. L. Schumaker. Algorithm 408: procedure for computing, smoothing, and interpolating natural splines (EI). *Communications of the ACM* **17(8)**, Aug. 1974.

Machover, C. A guide to sources of information about computer graphics. *IEEE Computer Graphics and Applications* **1**: 73–85, Jan. 1981.

Mandelbrot, B. B. *Fractal Geometry of Nature.* W. H. Freeman, San Francisco, 1983.

Markowsky, G., and M. A. Wesley. Fleshing out wireframes. IBM Watson Research Center, Research Report No. RC8124. Yorktown Heights, NY, 1980.

Matsushita, Y. Hidden-line elimination for a rotating object. *Communications of the ACM* **15(4)**:245, Apr. 1972.

McLeod, R. J., and E. L. Wachpress, eds. Frontiers of applied geometry. In *Proceedings of Symposium, Las Cruces, New Mexico.* Pergamon, Elmsford, NY, 1980.

Meagher, D. J. Octree encoding: a new technique for the representation, manipulation, and display of arbitrary three-dimensional objects by computer. Technical Report IPL-TR-80,111. Image Processing Laboratory, Rensselaer Polytechnic Institute, Troy, NY, Oct. 1980.

————. Octree generation, analysis, and manipulation. Technical Report IPL-TR-027. Image Processing Laboratory, Rensselaer Polytechnic Institute, Troy, NY, Apr. 1982.

————. Efficient synthetic image generation of arbitrary 3-D objects. *Proceedings of the IEEE Computer Society Conference on Pattern Recognition and Image Processing*, June 1982.

————. Geometric modeling using octree encoding. *Computer Graphics and Image Processing* **19**:129–47, June 1982.

————. Computer software for robotic vision. *SPIE*, Aug. 1982.

————. The octree encoding method for efficient solid modeling. Technical Report IPL-TR-032. Image Processing Laboratory, Rensselaer Polytechnic Institute, Troy, NY, Aug. 1982.

Mendelson, B. *Introduction to Topology*, 3d. Allyn and Bacon, Boston, 1975.

Meriwether, H. D. Informal notes, undated.

Metelli, F. The perception of transparency. *Scientific American* **230(4)**:90–98, Apr. 1974.

Miller, R. E., and J. W. Thatcher, eds. *Complexity of Computer Computations.* Plenum, New York, 1972.

Minsky, M. Form and content in computer science. *J. of the ACM* **17(2)**:197–215, 1972.

Mitchell, W. J. *Computer-Aided Architectural Design.* Petrocelli-Charter, New York, 1977.

Munchmeyer, F. C., C. Shubert, and H. Nowacki. Interactive design of fair hull surfaces using *B*-splines. *Computers in Industry* **1(2)**:77–86, 1979.

Myers, R. E. *Microcomputer Graphics.* Addison-Wesley, Reading, MA, 1982.

Myers, W. An industrial perspective on solid modeling. *IEEE Computer Graphics and Applications* **2(2)**:86–97, Mar. 1982.

Newell, M. E., R. G. Newell, and T. L. Sancha. A new approach to the shaded-picture problem. *Proceedings of the ACM National Conference*, 1973.

Newman, W. M., and R. F. Sproull. *Principles of Interactive Computer Graphics.* McGraw-Hill, New York, 1979.

Okino, N., and H. Kubo. Technical information system for computer-aided design, drawing, and manufacturing. *Proceedings of the Second PROLAMAT 73*, 1973.

Okino, N., et al. Geometry data base for multiparts in CAD/CAM system—TOPS/GDB. Technical Paper Series MS No. 79-149. Society of Mechanical Engineers, 1979.

Okino, N., and H. Kubo. Geometric modelling in CAD/CAM. *Information Processing Society of Japan* **21(7)**:725–33, 1980.

O'Neill, B. *Elementary Differential Geometry*. Academic, New York, 1966.

———. Computer graphics as an operational tool. Douglas Paper 5672. Presented to Computer Graphics 70 International Symposium, Brunel University, Uxbridge, England, Apr. 1970.

Palacol, E. L., and E. L. Stanton. Anisotropic parametric plate discrete elements. *International J. of Numerical Methods in Engineering* **4**:413–25, 1973.

Parent, R. E. A system for sculpting 3-D data. *Computer Graphics* **11(2)**:138–47, summer 1977.

Pavlidis, T. *Algorithms for Graphics and Image Processing: Computer Graphics and Pictorial Information Processing*. Computer Science, 1981.

Peters, G. J. Interactive computer graphics application of the parametric bicubic surface. In *Computer-Aided Geometric Design*, R. E. Barnhill and R. F. Riesenfeld, eds. Academic, New York, 1974.

Phong, B. T. Illumination for computer-generated images. *Communications of the ACM* **6**:311–17, June 1975.

Posdamer, J. L. A vector development of the fundamentals of computational geometry. *Computer Graphics and Image Processing* **6**:382–93, 1977.

Pressman, R. S., and J. E. Williams. *Numerical Control and Computer Aided Manufacturing*. Wiley, New York, 1977.

Reddy, D. R., and S. Rubin. Representation of three-dimensional objects. Report CMU-CS-78-113. Computer Science Department, Carnegie-Mellon University, Pittsburg, 1978.

Reeves, W. T. Particle systems—a technique for modeling a class of fuzzy objects. *Computer Graphics* **17(3)**:359–76, July 1983.

Requicha, A. A. G. Mathematical models of rigid solid objects. Technical Memo No. 28. Production Automation Project, University of Rochester, Rochester, NY, 1977.

———. Part of assembly description languages. I. dimensioning and tolerancing. Technical Memo No. 19. Production Automation Project, University of Rochester, Rochester, NY, 1977.

———. Representations for rigid solids: theory, methods, and systems. *Computing Surveys* **12(4)**:437–64, Dec. 1980.

Requicha, A. A. G., N. Samuel, and H. Voelcker. Part and assembly description languages—part 2. Technical Memo 20b. Production Automation Project, University of Rochester, Rochester, NY, 1974.

Requicha, A. A. G., and R. B. Tilove. Mathematical foundations of constructive solid geometry: general topology of regular closed sets. Technical Memo No. 27. Production Automation Project, University of Rochester, Rochester, NY, 1978.

Requicha, A. A. G., and H. B. Voelcker. Constructive solid geometry. Technical Memo No. 25. Production Automation Project, University of Rochester, Rochester, NY, 1977.

————. Geometric modelling of mechanical parts and machining processes. *COMPCONTROL*, Sopron, Hungary, Nov. 1979.

————. An introduction to geometric modelling and its applications. In *Advances in Information Systems Sciences*. J. T. Tou, ed. **8**, Plenum, New York, pp. 293–328, 1981.

————. Solid modelling: a historical summary and contemporary assessment. *IEEE Computer Graphics and Applications* **2(2)**:9–24, Mar. 1982.

————. A graduate program in programmable automation. Presented at the First Annual Workshop on Interactive Computing, CAD/CAM, and Electrical Engineering Education. University of Virginia, Charlottesville, Oct. 1982.

————. Solid modeling: current status and research directions. *IEEE Computer Graphics and Applications* **3(7)**:25–37, Oct. 1983.

Reynolds, R. A. *Computer Methods for Architects*. Butterworth, Woburn MA, 1980.

Riesenfeld, R. F. Applications of *B*-spline approximation to geometric problems of computer-aided design. Ph.D. diss., Syracuse University, Syracuse, NY, 1972. Also published as University of Utah report UTEC-CSc-73-126.

————. Aspects of modelling in computer-aided geometric design. *Proceedings of the National Computer Conference*, AFIPS Press, 1975.

————. Homogeneous coordinates and projective planes in computer graphics. *IEEE Computer Graphics and Applications* **1**:50–55, Jan. 1981.

Roberts, L. G. Homogeneous matrix representation and manipulation of *N*-dimensional constructs. MS-1405, M.I.T. Lincoln Lab, Cambridge, MA, May 1965.

Ross, D. T. The AED approach to generalized computer-aided design. *Proceedings of the ACM National Conference*, 1967.

Ross, D. T., and J. E. Rodriguez. Theoretical foundations for the computer-aided design system. *SJCC* 1963.

Rowin, M. S. Conic, cubic, and *T*-conic segments. Document No. D2-23252. The Boeing Company, Seattle, 1964.

Sabella, P., and M. J. Wozny. Toward fast color-shaded images of CAD/CAM geometry. *IEEE Computer Graphics and Applications* **3(8)**:60–71, Nov. 1983.

Sammet, H., and A. Rosenfeld. Quadtree representations of binary images. *Proceedings of the Fifth International Conference on Pattern Recognition, Miami Beach*, 815–18, Dec. 1980.

Samuel, N. M., A. A. Requicha, and S. Elkind. Methodology and results of an industrial parts survey. Technical Memo No. 21. Production Automation Project, University of Rochester, Rochester, NY, July 1976.

Schweitzer, D., and E. S. Cobb. Scanline rendering of parametric surfaces. *Computer Graphics* **16(3)**:265–71, July 1982.

Semple, J. G., and G. T. Kneebone. *Algebraic Projective Geometry*. Oxford University Press, Oxford, England, 1952.

Shamos, M. I. Geometric complexity. *Proceedings of the Seventh Annual ACM Symposium on Theory of Computing*, 224–33, 1975.

———. *Computational Geometry*. Springer-Verlag, New York, 1979.

Shamos, M. I., and D. Hoey, Geometric intersection problems. *Proceedings of the 17th Annual IEEE Symposium on Foundations of Computer Science*, 208–15, Oct. 1976.

Shelley, K. L., and D. P. Greenberg. Path specification and path coherence. *Computer Graphics* **16(3)**:157–66, July 1982.

Shepard, R. N., and B. Metzler. Mental rotation of three-dimensional objects. *Science* **171**:701–03, 1971.

Shoenberg, I. J. Cardinal spline interpolation. *SIAM*, Philadelphia, 1973.

Shu, H. Synthesis of 3-D objects having complex surface boundaries. IITRI. Management and Computer Sciences Division, Chicago, 1976.

Shu, H., and O. Oyake. Recent research in geometric modelling with primitive solids. *Proceedings of the CAM-I International Seminar*, Atlanta, Apr. 1976.

Silva, C. E. Alternative definitions of faces in boundary representations of solid objects. Technical Memo No. 36. Production Automation Project, University of Rochester, Rochester, NY, 1981.

Sorensen, P. Movies, computers, and the future. *American Cinematographer*, Jan. 1983.

Spur, G., et al. Survey about geometric modelling systems. *Annual CIRP* **28(2)**: 519–38, 1979.

Srihari, S. N. Representation of three-dimensional digital images. *Computing Surveys* **13(4)**:399–424, Dec. 1981.

Stanton, E. L. Untitled, informal notes on tricubic interpolation and modeling, 1976.

Stanton, E. L., and L. M. Crain. PATCHES-III User's Manual. McDonnell Douglas Astronautics Company Report No. MDC G5538, Nov. 1974.

Stanton, E. L., L. M. Crain, and T. F. Neu. A parametric cubic modeling system for general solids of composite material. *International Journal of Numerical Methods in Engineering* **11**:653–70, 1977.

Stevens, P. S. *Patterns in Nature*. Little, Brown, Boston, 1974.

Sutherland, I. E. Sketchpad: a man-machine graphical communication system. *Proceedings of the SJCC* **23**:329–49, 1963.

———. Ten unsolved problems in computer graphics. *Datamation* **12(5)**:22, May 1966.

Sutherland, I. E., R. F. Sproull, and R. A. Schumacker. A characterization of ten hidden-surface algorithms. *Computing Surveys* **6**:1–56, Mar. 1974.

Thornton, R. Interactive modelling in three dimensions through two dimensional windows, *Proceedings of CAD 78*. IPC Business Press, England, Mar. 1978.

Tiller, W. Rational *B*-splines for curves and surface representation. *IEEE Computer Graphics and Applications* **3(6)**:61–69, Sept. 1983.

Tilove, R. B. A study of geometric set-membership classification. Technical Memo No. 30. Production Automation Project, University of Rochester, Rochester, NY, Nov. 1977.

———. Set-membership classification: a unified approach to geometric intersection problems. *IEEE Transactions on Computers* **C-29**, 1980.

———. Line/polygon classification: a study of the complexity of geometric computation. *IEEE Computer Graphics and Applications* **1(2)**:75–78:80–82:84, Apr. 1981.

———. Exploiting spatial and structural locality in geometric modelling. Technical Memo No. 38. Production Automation Project, University of Rochester, Rochester, NY, 1981.

———. Extending solid modeling systems for mechanism design and kinematic simulation. *IEEE Computer Graphics and Applications* **3(3)**:9–19, May–June 1983.

Tilove, R. B., and A. A. G. Requicha. Closure of Boolean operations on geometric entities. *Computer Aided Design*, **12(5)**: 219–20, Sept. 1980.

Timmer, H. G. Spline-blended interpolation of a network of curves. Douglas Aircraft Company Technical Memorandum C1-250-CAT-76-044, Aug. 1976.

———. Analytical background for computation of surface intersections. Douglas Aircraft Company Technical Memorandum C1-250-CAT-77-036, Apr. 1977.

———. Informal notes, undated.

Timmer, H. G., C. L. Arne, T. R. Stokes, Jr., and H. H. Tang. Ablation aerodynamics for slender reentry bodies. AFFDL-TR-70-27, vol. 1, Mar. 1970.

Timmer, H. G., and J. M. Stern, Computation of global geometric properties of solid objects. *Computer Aided Design* **12(6)**, Nov. 1980.

Traub, J. F., ed. *Complexity of Sequential and Parallel Numerical Algorithms*. Academic, New York, 1973.

Udupa, S. Collision detection and avoidance in computer-controlled manipulators. *Proceedings of the Fifth International Joint Conference on Artificial Intelligence, Cambridge, Massachusetts*, 737–48, 1977.

Vernon, M., G. Ris, and J. P. Musse. Continuity of biparametric surface patches. *CAD* **8(4)**: 167–73, 1976.

Voelcker, H. B., A. E. Middleditch, and P. R. Zuckerman. Discrete part manufacturing: theory and practice. Production Automation Project, University of Rochester, Rochester, NY, 1974.

Voelcker, H. B., and W. A. Hunt. The role of solid modelling in machining-process modelling and NC verification. *Proceedings of the 1981 International Congress of the Society of Automotive Engineers, Detroit, MI*, Feb. 1981.

Voelcker, H. B., and A. A. G. Requicha. Geometric modelling of mechanical parts and processes. *Computer* **10(48)**, 1977.

———. Geometric-modelling research at the University of Rochester. Presented at the First Annual Conference on Computer Graphics in CAD/CAM Systems, MIT, Cambridge, MA, Apr. 1979.

Voelcker, H. B., et al. The PADL-1.0/2 system for defining and displaying solid objects. *Computer Graphics* **12**:257, 1978.

Voelcker, H. B., and A. A. G. Requicha. Boundary-evaluation procedures for objects defined via constructive solid geometry. Technical Memo No. 26. Production Automation Project, University of Rochester, Rochester, NY, 1980.

Wesley, M. A. High-level languages and modelling. Research Report RC8195. IBM Thomas J. Watson Research Center, Yorktown Heights, NY, Mar. 1980.

———. Construction and use of geometric modelers. In *Computer-Aided Design*, J. Encarnacao, ed. Springer-Verlag, New York, pp. 79–136, 1980.

———. Robotics and geometric modelling. *Proceedings of COMPSAC 81*, 211–12, 1981.

Wesley, M. A., et al. A geometric-modelling system for automated mechanical assembly. *IBM Journal of Research and Development* **24**:64, 1980.

Wesley, M. A., and G. Markowsky. Fleshing out projections. Computer Science Department Research Report 8884. IBM Thomas J. Watson Research Center, Apr. 1981.

Whitted, T. An improved illumination model for shaded display. *Communications of the ACM* **23**:343–49, 1980.

Willet, K. G. *Hidden-Surface Elimination in Computer Graphics*. Publisher unknown, 1981.

Williams, L. Casting curved shadows on curved surfaces. *Computer Graphics* **12(3)**:270, Aug. 1978.

Williams, R. A. A survey of data structures for computer graphics systems. *Computing Surveys* **3(1)**:1–21, Mar. 1971.

Wordenweber, B. Surface triangulation for picture production. *IEEE Computer Graphics and Applications* **3(8)**:45–51, Nov. 1983.

Yamaguchi, K., T. L. Kunii, K. Fujimura and H. Toriya. Octree-related data structures and algorithms. *IEEE Computer Graphics and Applications* **4(1)**:53–59, Jan. 1984.

Yamaguchi, K., T. L. Kunii, D. F. Rogers, S. G. Satterfield, and F. A. Rodriguez. Computer-integrated manufacturing of surfaces using octree encoding, *IEEE Computer Graphics and Applications* **4(1)**:60–65, Jan. 1984.

Yessios, C. I. A notation and system for 3-D constructions. *Proceedings of the 15th Design Automation Conference, Las Vegas*, 125, 1978.

Youngblood, G. Next . . . total scene simulation. *Video Systems* **9(2)**:18–27, Feb. 1983.

ANSWERS TO SELECTED EXERCISES

CHAPTER 1

Section 3

7. $d^2r/dt^2 = 6at + 2b$

9. $\partial^2 r/\partial t^2 = 2cs + 6\,dt + 2f$

11. $\mathbf{AB} = [(a_{11}b_{11} + a_{12}b_{21} + a_{13}b_{31})(a_{11}b_{12} + a_{12}b_{22} + a_{13}b_{32})]$

13. $\mathbf{C} = \begin{bmatrix} 8 & 8 & 5 \\ -2 & -7 & 9 \end{bmatrix}, \qquad \mathbf{D} = \begin{bmatrix} 6 & -2 & -7 \\ 6 & -3 & 3 \end{bmatrix}$

15. $\mathbf{C} = \begin{bmatrix} 76 & 38 \\ 48 & 8 \\ 12 & 6 \end{bmatrix}$

17. $\mathbf{A}^T = \begin{bmatrix} 4 & 5 \\ 0 & 1 \\ 7 & 2 \end{bmatrix}$

27. *Hint*: $(d\mathbf{n} - \mathbf{p}_0) \cdot d\mathbf{n} = 0$; and for any point $\mathbf{p}$, $(\mathbf{p} - \mathbf{p}_0) \cdot \mathbf{n} = 0$.

Section 7

1. $y = 0.0992x^3 - 1.2699x^2 + 3.1707x + 6$

2. $y = 0.07407x^3 - 0.72217x^2 + 0.22210x + 8.4260$

7. Same curve results, but the direction of parametrization is reversed.

CHAPTER 2

1. $a = b = c = 0, l = m = 4, n = 3$; at $u = 0.5, x = 2, y = 2, z = 1.5$

2. $a = b = 1, c = 2, l = m = 0, n = -6$; at $u = 0.5\ x = 1, y = 1, z = -1$

Section 1

1. From Eq. 2.18, $\mathbf{a}_1 = \mathbf{a}_2 = \mathbf{a}_3 = 0$, $\mathbf{a}_0 = [2 \quad 3.5 \quad -1.2]$.

5. The geometric coefficients are $\mathbf{p}_0 = [10 \quad 2 \quad 2]$, $\mathbf{p}_1 = [10 \quad 10 \quad 2]$, $\mathbf{p}_0^u = [0 \quad 32 \quad 64]$, $\mathbf{p}_1^u = [0 \quad 32 \quad -64]$. Compute the algebraic coefficients by appropriate substitution of these geometric coefficients into Eqs. 2.21 to obtain

$$\mathbf{a}_0 = [10 \quad 2 \quad 2], \mathbf{a}_1 = [0 \quad 32 \quad 64], \mathbf{a}_2 = [0 \quad -72 \quad -64], \mathbf{a}_3 = [0 \quad 0 \quad 0].$$

7. The geometric coefficients are

$$\mathbf{p}_0 = [10 \quad 2 \quad 2], \mathbf{p}_1 = [10 \quad 10 \quad 2], \mathbf{p}_0^u = [0 \quad 64 \quad 128], \mathbf{p}_1^u = [0 \quad 8 \quad -16];$$

the corresponding algebraic coefficients are

$$\mathbf{a}_0 = [10 \quad 2 \quad 2], \mathbf{a}_1 = [0 \quad 64 \quad 128], \mathbf{a}_2 - [0 \quad -112 \quad -240], \mathbf{a}_3 = [0 \quad 56 \quad 112].$$

9. There are 18 degrees of freedom available in the parametric quintic polynomials. The algebraic expression of these polynomials is

$$x(u) = a_{5x}u^5 + a_{4x}u^4 + a_{3x}u^3 + a_{2x}u^2 + a_{1x}u + a_{0x}$$
$$y(u) = a_{5y}u^5 + a_{4y}u^4 + a_{3y}u^3 + a_{2y}u^2 + a_{1y}u + a_{0y}$$
$$z(u) = a_{5z}u^5 + a_{4z}u^4 + a_{3z}u^3 + a_{2z}u^2 + a_{1z}u + a_{0z}$$

Section 2

1. If these three vectors are linearly dependent, then

$$k_1(\mathbf{p}_1 - \mathbf{p}_0) + k_2\mathbf{p}_0^u + k_3\mathbf{p}_1^u = 0$$

where k_1, k_2, and k_3 are arbitrary scalar values not all equal to zero. It is easy to show (a sketch will do) that if these vectors are linearly dependent, then these vectors must be coplanar and produce a planar curve. Conversely, if these vectors are linearly independent, a nonplanar curve is produced.

3. Begin with the expression

$$\lambda(u)(\mathbf{p}_1 - \mathbf{p}_0) + \mu(u)\mathbf{t}_0 + v(u)\mathbf{t}_1 = \mathbf{p}(u) - \mathbf{p}_0$$

To isolate $\lambda(u)$, take the scalar product of both sides of the equation with $(\mathbf{t}_0 \times \mathbf{t}_1)$ to obtain

$$\lambda(u)(\mathbf{p}_1 - \mathbf{p}_0) \cdot (\mathbf{t}_0 \times \mathbf{t}_1) + \mu(u)\mathbf{t}_0 \cdot (\mathbf{t}_0 \times \mathbf{t}_1)$$
$$+ v(u)\mathbf{t}_1 \cdot (\mathbf{t}_0 \times \mathbf{t}_1) = [\mathbf{p}(u) - \mathbf{p}_0] \cdot (\mathbf{t}_0 \times \mathbf{t}_1)$$

Since $\mathbf{t}_0$ and $\mathbf{t}_1$ are perpendicular to $\mathbf{t}_0 \times \mathbf{t}_1$, this reduces to

$$\lambda(u) = \frac{[\mathbf{p}(u) - \mathbf{p}_0] \cdot (\mathbf{t}_0 \times \mathbf{t}_1)}{(\mathbf{p}_1 - \mathbf{p}_0) \cdot (\mathbf{t}_0 \times \mathbf{t}_1)}$$

To isolate $\mu(u)$, take the scalar product of both sides of the equation with $\mathbf{t}_1 \times (\mathbf{p}_1 - \mathbf{p}_0)$ to obtain

$$\mu(u) = \frac{[\mathbf{p}(u) - \mathbf{p}_0] \cdot \mathbf{t}_1 \times (\mathbf{p}_1 - \mathbf{p}_0)}{(\mathbf{p}_1 - \mathbf{p}_0) \cdot (\mathbf{t}_0 \times \mathbf{t}_1)}$$

Finally, to isolate $v(u)$, take the scalar product of both sides of the equation with $(\mathbf{p}_1 - \mathbf{p}_0) \times \mathbf{t}_0$ to obtain

$$v(u) = \frac{[\mathbf{p}(u) - \mathbf{p}_0] \cdot (\mathbf{p}_1 - \mathbf{p}_0) \times \mathbf{t}_0}{(\mathbf{p}_1 - \mathbf{p}_0) \cdot (\mathbf{t}_0 \times \mathbf{t}_1)}$$

5. Eliminate F_1 in Eq. 2.25 with the substitution $(F_1 = 1 - F_2)$ to obtain, after rearranging terms,

$$\mathbf{p} = \mathbf{p}_0 + F_2(\mathbf{p}_1 - \mathbf{p}_0) + F_3 \mathbf{p}_0^u + F_4 \mathbf{p}_1^u$$

Since $\mathbf{p}_0^u = a\mathbf{t}_0^u$ and $\mathbf{p}_1^u = b\mathbf{t}_1^u$, with further substitution, we find that

$$\mathbf{p} = \mathbf{p}_0 + F_2(\mathbf{p}_1 - \mathbf{p}_0) + aF_3 \mathbf{t}_0 + bF_4 \mathbf{t}_1$$

7. By definition, we are only interested in points on a pc curve segment in the interval $u = 0$ to $u = 1$. Therefore, only roots of the cubic in this interval are acceptable. The unit tangent vectors $\mathbf{t}_0$ and $\mathbf{t}_1$ define not only the slope of the curve at each end point but also the direction of parametrization along the curve. If a or b is negative, then the direction of $\mathbf{t}_0$ or $\mathbf{t}_1$ is reversed, producing a different curve from the one intended.

Section 3

1. $\mathbf{p}_0 = [1 \quad 2 \quad 3]$, $\mathbf{p}_1 = [1 \quad 2 \quad 3]$, $\mathbf{p}_0^u = \mathbf{p}_1^u = 0$
3. $\mathbf{p}_0 = [1 \quad 3 \quad 1]$, $\mathbf{p}_1 = [2 \quad 2 \quad 4]$, $\mathbf{p}_0^u = \mathbf{p}_1^u = [1 \quad -1 \quad 3]$
5. $\mathbf{p}_0 = [4 \quad 2 \quad 4]$, $\mathbf{p}_1 = [1 \quad 4 \quad 1]$, $\mathbf{p}_0^u = [-3 \quad 2 \quad 0]$, $\mathbf{p}_1^u = [-3 \quad 2 \quad 0]$

Section 4

1. Since $F_1 = 2u^3 - 3u^2 + 1$, $dF_1/du = F_1^u = 6u^2 - 6u$
$$F_2 = -2u^3 + 3u^2, \quad F_2^u = -6u^2 + 6u$$
$$F_3 = u^3 - 2u^2 + u, \quad F_3^u = 3u^2 - 4u + 1$$
$$F_4 = u^3 - u^2, \quad F_4^u = 3u^2 - 2u$$

and $\mathbf{F}^u = [F_1^u \quad F_2^u \quad F_3^u \quad F_4^u]$

Similarly, $F_1^{uu} = 12u - 6$
$$F_2^{uu} = -12u + 6$$
$$F_3^{uu} = 6u - 4$$
$$F_4^{uu} = 6u - 2$$

and $\mathbf{F}^{uu} = [F_1^{uu} \quad F_2^{uu} \quad F_3^{uu} \quad F_4^{uu}]$

3. $x = -6.5u^3 + 9u^2 + 0.5u + 1$

$$F_1 x_0 = 2u^3 - 2u^2 + 1$$
$$F_2 x_1 = 4(-2u^3 + 3u^2)$$
$$F_3 x_0^u = 0.5(u^3 - 2u^2 + u)$$
$$F_4 x_1^u = -(u^3 - u^2)$$

$$F_1^u x_0 = 6u^2 - 6u$$
$$F_2^u x_1 = 4(-6u^2 + 6u)$$
$$F_3^u x_0^u = 0.5(3u^2 - 4u + 1)$$
$$F_4^u x_1^u = -(3u^2 - 2u)$$

$$F_1^{uu} x_0 = 12u - 6$$
$$F_2^{uu} x_1 = 4(-12u + 6)$$
$$F_3^{uu} x_0^u = 0.5(6u - 4)$$
$$F_4^{uu} x_1^u = -(6u - 2)$$

Similar expressions are found for y and z.

Section 5

1. If $\mathbf{p}(u) = \mathbf{a}_3 u^3 + \mathbf{a}_2 u^2 + \mathbf{a}_1 u + \mathbf{a}_0$ and if $u = bv + c$, then substituting, we obtain

$$\mathbf{p}(v) = \mathbf{a}_3(bv + c)^3 + \mathbf{a}_2(bv + c)^2 + \mathbf{a}_1(bv + c) + \mathbf{a}_0$$

which is still a cubic polynomial. However, if $u = b_n v^n + b_{n-1} v^{n-1} + \cdots + c$, then

$$\mathbf{p}(v) = \mathbf{a}_3(b_n v^n + b_{n-1} v^{n-1} + \cdots + c)^3$$
$$+ \mathbf{a}_2(b_n v^n + b_{n-1} v^{n-1} + \cdots + c)^2$$
$$+ \mathbf{a}_1(b_n v^n + b_{n-1} v^{n-1} + \cdots + c) + \mathbf{a}_0$$

which is clearly not cubic when $n > 1$.

3. If $u = av + b$, then $du/dv = a$. The tangent vectors of the two parametrizations are related by a constant scalar magnitude. If the scalar is negative, then the direction of the tangent vectors is reversed, but the tangent line is unchanged.

Section 6

1. $\mathbf{p}^v = (u_j - u_i)(v_j - v_i)^{-1} \mathbf{p}^u$ or $\mathbf{p}^v = 0.5 \mathbf{p}^u$, thus

$$\mathbf{B}_v = [\mathbf{p}_{0.2} \quad \mathbf{p}_{0.7} \quad 0.5\mathbf{p}_{0.2}^u \quad 0.5\mathbf{p}_{0.7}^u]^T$$

Section 7

1.
$$\begin{bmatrix} x_i - x_0 \\ y_i - y_0 \end{bmatrix} = \begin{bmatrix} x_1 - x_0 \\ y_1 - y_0 \end{bmatrix} F_2(u_i) + \begin{bmatrix} t_{x0} + t_{x1} \\ t_{y0} + t_{y1} \end{bmatrix} \begin{bmatrix} aF_3(u_i) \\ bF_4(u_i) \end{bmatrix}$$

$$\begin{bmatrix} t_{x0} + t_{x1} \\ t_{y0} + t_{y1} \end{bmatrix} \begin{bmatrix} aF_3(u_i) \\ bF_4(u_i) \end{bmatrix} = \begin{bmatrix} x_i - x_0 \\ y_i - y_0 \end{bmatrix} - \begin{bmatrix} x_1 - x_0 \\ y_1 - y_0 \end{bmatrix} F_2(u_i)$$

$$\begin{bmatrix} aF_3(u_i) \\ bF_4(u_i) \end{bmatrix} = \begin{bmatrix} t_{x0} + t_{x1} \\ t_{y0} + t_{y1} \end{bmatrix}^{-1} \left\{ \begin{bmatrix} x_i - x_0 \\ y_i - y_0 \end{bmatrix} - \begin{bmatrix} x_1 - x_0 \\ y_1 - y_0 \end{bmatrix} F_2(u_i) \right\}$$

3. If $\mathbf{p}_1 - \mathbf{p}_0$, $\mathbf{p}_0^u$, and $\mathbf{p}_1^u$ are linearly independent, the pc curve will not self-intersect.

Section 8

2.
$$\mathbf{K} = \begin{bmatrix} 0 & 32/3 & -32/3 & 16/3 \\ 0 & -56/3 & 40/3 & -16/3 \\ 0 & 8 & -8/3 & 1 \\ 0 & 0 & 0 & 0 \end{bmatrix}$$

5.
$$\mathbf{K} = \begin{bmatrix} -6 & 6 & -3 & -3 & -0.5 & 0.5 \\ 15 & -15 & 8 & 7 & 1.5 & -1 \\ -10 & 10 & 6 & 4 & -1.5 & 0.5 \\ 0 & 0 & 0 & 0 & 0.5 & 0 \\ 0 & 0 & 1 & 0 & 0 & 0 \\ 1 & 0 & 0 & 0 & 0 & 0 \end{bmatrix}^{-1}$$

$$\times \begin{bmatrix} 0 & 0 & 0 & 0 & 0 & 1 \\ (0.2)^5 & (0.2)^4 & (0.2)^3 & (0.2)^2 & 0.2 & 1 \\ (0.4)^5 & (0.4)^4 & (0.4)^3 & (0.4)^2 & 0.4 & 1 \\ (0.6)^5 & (0.6)^4 & (0.6)^3 & (0.6)^2 & 0.6 & 1 \\ (0.8)^5 & (0.8)^4 & (0.8)^3 & (0.8)^2 & 0.8 & 1 \\ 1 & 1 & 1 & 1 & 1 & 1 \end{bmatrix}^{-1}$$

Section 9

1. The diagonals of a parallelogram intersect at their midpoints. Note that

$$\mathbf{p}_0 + (\mathbf{p}_1 - \mathbf{p}_0)/2 = (\mathbf{p}_0 + \mathbf{p}_1)/2$$

3. The coordinates of the point at 45° are [0.707 0.707], and at 22.5° they are [0.924 0.383]. The coordinates of $\mathbf{p}_{0.5}$ on the pc curve are [0.707 0.707] and $\mathbf{p}_{0.25} =$ [0.921] 0.389].

Section 10

1. Differentiate Eq. 2.123 to obtain

$$\mathbf{p}'' = [3(a + b - 2)u^2 - 2(2a + b - 3)u + a](\mathbf{p}_1 - \mathbf{p}_0)$$

At $u = 0$ and $u = 1$,

$$\mathbf{p}_0'' = a(\mathbf{p}_1 - \mathbf{p}_0) \text{ and } \mathbf{p}_1'' = b(\mathbf{p}_1 - \mathbf{p}_0)$$

For condition 1, $\mathbf{p}_0'' = \mathbf{p}_1'' = \mathbf{p}_1 - \mathbf{p}_0$
For condition 2, $\mathbf{p}_0'' = \mathbf{p}_1'' = 0$
For condition 3, $\mathbf{p}_0'' = \mathbf{p}_1'' = \mathbf{p}_0 - \mathbf{p}_1$

3. From $x = 9u^3 - 14u^2 + 4u + 2$
$$y = -27.9u^3 + 43.4u^2 - 12.4u + 1.2$$
$$z = -9u^3 + 14u^2 - 4u + 5.4$$

Compute $x^u = 27u^2 - 28u + 4$

$$y^u = -83.7u^2 + 86.8u - 12.4$$
$$z^u = -27u^2 + 28u - 4$$

Set $x^u = y^u = z^u = 0$ and solve for the roots of u to find $u = 0.171, 0.866$. They are the u values at the turnaround points.

Section 11

5. $\mathbf{p}^u_{0.5} = (1.5 - \rho)(\mathbf{p}_1 - \mathbf{p}_0)$, therefore $\mathbf{p}^u_{0.5}$ is parallel to $(\mathbf{p}_1 - \mathbf{p}_0)$.

Section 12

1. $\mathbf{p}_i(1) = \mathbf{p}_{i+1}(0)$

$\mathbf{p}^u_i(1) = k_1 \mathbf{p}^u_{i+1}(0)$

$\mathbf{p}^{uu}_i(1) = k_2 \mathbf{p}^{uu}_{i+1}(0)$

$\mathbf{p}^{uuu}_i(1) = k_3 \mathbf{p}^{uuu}_{i+1}(0)$

and the osculating planes must coincide.

5. $\mathbf{B}_i = [\mathbf{p}_{i-1} \quad \mathbf{p}_i \quad k_{i-1,i}\mathbf{t}_{i-1} \quad k_{i,i}\mathbf{t}_i]^T$

6. There are n points.

$$\mathbf{B}_1 = [\mathbf{p}_0 \quad \mathbf{p}_1 \quad k_{0,1}\mathbf{t}_0 \quad k_{1,1}\mathbf{t}_1]^T$$
$$\mathbf{B}_n = [\mathbf{p}_{n-1} \quad \mathbf{p}_0 \quad k_{n-1,n}\mathbf{t}_{n-1} \quad k_{0,n}\mathbf{t}_0]^T$$

Section 14

2. Substitute appropriately into Eq. 2.235 to obtain

$$\mathbf{p}(u) = (1 - 2u + u^2)\mathbf{p}_0 + 2(u - u^2)\mathbf{p}^* + u^2\mathbf{p}_3$$

One way to verify that this is a segment of a parabola is to test the equation with selected values of the control-point vectors. Try $\mathbf{p}_0 = [-1 \quad 1 \quad 0]$, $\mathbf{p}^* = [0 \quad 0 \quad 0]$, and $\mathbf{p}_3 = [1 \quad 1 \quad 0]$.

3. $\mathbf{p}^u_0 = 5(\mathbf{p}_1 - \mathbf{p}_0)$

$\mathbf{p}^{uu}_0 = 20(\mathbf{p}_0 - 2\mathbf{p}_1 + \mathbf{p}_2)$

$\mathbf{p}^{uuu}_0 = 60(-\mathbf{p}_0 + 3\mathbf{p}_1 - 3\mathbf{p}_2 + \mathbf{p}_3)$

$\mathbf{p}^u_1 = 5(\mathbf{p}_5 - \mathbf{p}_4)$

$\mathbf{p}^{uu}_1 = 20(\mathbf{p}_3 - \mathbf{p}_4 + \mathbf{p}_5)$

$\mathbf{p}^{uuu}_1 = 60(-\mathbf{p}_2 + 3\mathbf{p}_3 - 3\mathbf{p}_4 + \mathbf{p}_5)$

Section 15

1. $n \geq k - 1$, otherwise Eq. 2.259 breaks down.

4. Substitute $u = u + 1$ into $N_{i+1,3}(u)$ and $u = u + 2$ into $N_{i+2,3}(u)$. These equations are now identical to $N_{i,3}(u)$, and we conclude that the curves are identical but shifted away from each other by $u = 1$.

8. Both cases are degenerate, since they result in $n - k + 2 \le 0$.

10. $\mathbf{p}_i(u) = \mathbf{U}_k \mathbf{M}_k \mathbf{P}_k$ $i \in [1 : n + 1]$

where $\mathbf{U}_k$ and $\mathbf{M}_k$ are as defined for open curves and

$$\mathbf{P}_k = [\mathbf{p}_{j \bmod(n+1)}]   j \in [i - 1 : i + k - 2]$$

CHAPTER 3

1. $\mathbf{p} = \mathbf{p}(u, w)$

3. $x = -1 + 2 \cos u \cos w$

   $y = 1 + 2 \cos u \sin w$ $u \in [-\pi/2, \ \pi/2]$,

   $z = 4 + 2 \sin u$ $w \in [0, \ 2\pi]$

The equation of the equator is found at $u = 0$; thus,

$$x = -1 + 2 \cos w$$
$$y = 1 + 2 \sin w$$
$$z = 4$$

Section 1

1. $\mathbf{p}(u, w) = \sum\limits_{i=0}^{5} \sum\limits_{j=0}^{5} \mathbf{a}_{ij} u^i w^j$

There are $6 \times 6 \times 3$ algebraic coefficients.

3. $\mathbf{p}_{1w}^u = \mathbf{F}[\mathbf{p}_{10}^u \quad \mathbf{p}_{11}^u \quad \mathbf{p}_{10}^{uw} \quad \mathbf{p}_{11}^{uw}]^T$

   $\mathbf{p}_{0u}^w = \mathbf{F}[\mathbf{p}_{00}^w \quad \mathbf{p}_{01}^w \quad \mathbf{p}_{00}^{uw} \quad \mathbf{p}_{01}^{uw}]^T$

   $\mathbf{p}_{1u}^w = \mathbf{F}[\mathbf{p}_{10}^w \quad \mathbf{p}_{11}^w \quad \mathbf{p}_{10}^{uw} \quad \mathbf{p}_{11}^{uw}]^T$

Why is $\mathbf{F}$, and not $\mathbf{F}^u$, a term?

Section 6

1.

$$\mathbf{B^*} = \begin{bmatrix} \mathbf{p}_{00} & \mathbf{p}_{01} & 3\mathbf{p}_{00}^w & 3\mathbf{p}_{01}^w \\ \mathbf{p}_{10} & \mathbf{p}_{11} & 3\mathbf{p}_{10}^w & 3\mathbf{p}_{11}^w \\ 3\mathbf{p}_{00}^u & 3\mathbf{p}_{01}^u & 9\mathbf{p}_{00}^{uw} & 9\mathbf{p}_{01}^{uw} \\ 3\mathbf{p}_{10}^u & 3\mathbf{p}_{11}^u & 9\mathbf{p}_{10}^{uw} & 9\mathbf{p}_{11}^{uw} \end{bmatrix}$$ $u^*, w^* \in [0, 1/3]$

Section 7

1.   The geometric coefficients of the central patch in the 3×3 array of subpatches are

$$\mathbf{B} = \begin{bmatrix} \mathbf{p}(1/3, 1/3) & \mathbf{p}(1/3, 2/3) & (1/3)\mathbf{p}^w(1/3, 1/3) & (1/3)\mathbf{p}^w(1/3, 2/3) \\ \mathbf{p}(2/3, 1/3) & \mathbf{p}(2/3, 2/3) & (1/3)\mathbf{p}^w(2/3, 1/3) & (1/3)\mathbf{p}^w(2/3, 2/3) \\ (1/3)\mathbf{p}^u(1/3, 1/3) & (1/3)\mathbf{p}^u(1/3, 2/3) & (1/9)\mathbf{p}^{uw}(1/3, 1/3) & (1/9)\mathbf{p}^{uw}(1/3, 2/3) \\ (1/3)\mathbf{p}^u(2/3, 1/3) & (1/3)\mathbf{p}^u(2/3, 2/3) & (1/9)\mathbf{p}^{uw}(2/3, 1/3) & (1/9)\mathbf{p}^{uw}(2/3, 2/3) \end{bmatrix}$$

3.
$$
B = \begin{bmatrix}
\mathbf{p}(0.1, 0.6) & \mathbf{p}(0.1, 0.8) & 0.2\mathbf{p}^w(0.1, 0.6) & 0.2\mathbf{p}^w(0.1, 0.8) \\
\mathbf{p}(0.4, 0.6) & \mathbf{p}(0.4, 0.8) & 0.2\mathbf{p}^w(0.4, 0.6) & 0.2\mathbf{p}^w(0.4, 0.8) \\
0.3\mathbf{p}^u(0.1, 0.6) & 0.3\mathbf{p}^u(0.1, 0.8) & 0.06\mathbf{p}^{uw}(0.1, 0.6) & 0.06\mathbf{p}^{uw}(0.1, 0.8) \\
0.3\mathbf{p}^u(0.4, 0.6) & 0.3\mathbf{p}^u(0.4, 0.8) & 0.06\mathbf{p}^{uw}(0.4, 0.6) & 0.06\mathbf{p}^{uw}(0.4, 0.8)
\end{bmatrix}
$$

Section 8

1. The patch approximates a region of a spherical surface.

Section 9

1. From Section 1, the geometric coefficients of the boundary curves are (in terms of the patch coefficients)

$$
\begin{aligned}
\text{For } u = 0, \quad \mathbf{B} &= [\mathbf{p}_{00} \quad \mathbf{p}_{01} \quad \mathbf{p}_{00}^w \quad \mathbf{p}_{01}^w] \\
\text{For } u = 1, \quad \mathbf{B} &= [\mathbf{p}_{10} \quad \mathbf{p}_{11} \quad \mathbf{p}_{10}^w \quad \mathbf{p}_{11}^w] \\
\text{For } w = 0, \quad \mathbf{B} &= [\mathbf{p}_{00} \quad \mathbf{p}_{10} \quad \mathbf{p}_{00}^u \quad \mathbf{p}_{10}^u] \\
\text{For } w = 1, \quad \mathbf{B} &= [\mathbf{p}_{10} \quad \mathbf{p}_{11} \quad \mathbf{p}_{10}^w \quad \mathbf{p}_{11}^w]
\end{aligned}
$$

Section 10

1. Use Eq. 3.89 to obtain

$$
\begin{aligned}
\mathbf{p}_{00} &= [1 \quad 1 \quad 0] & \mathbf{p}_{01} &= [1 \quad 2 \quad 4] \\
\mathbf{p}_{10} &= [5 \quad 1 \quad 0] & \mathbf{p}_{11} &= [5 \quad 2 \quad 4] \\
\mathbf{p}_{00}^u &= [4 \quad 0 \quad 0] & \mathbf{p}_{01}^u &= [4 \quad 0 \quad 0] \\
\mathbf{p}_{10}^u &= [4 \quad 0 \quad 0] & \mathbf{p}_{11}^u &= [4 \quad 0 \quad 0] \\
\\
\mathbf{p}_{00}^w &= [0 \quad 1 \quad 4] & \mathbf{p}_{01}^w &= [0 \quad 1 \quad 4] \\
\mathbf{p}_{10}^w &= [0 \quad 1 \quad 4] & \mathbf{p}_{11}^w &= [0 \quad 1 \quad 4] \\
\\
\mathbf{p}_{00}^{uw} &= 0 & \mathbf{p}_{01}^{uw} &= 0 \\
\mathbf{p}_{10}^{uw} &= 0 & \mathbf{p}_{11}^{uw} &= 0
\end{aligned}
$$

2. Use Eq. 3.90 to obtain

$$
\begin{aligned}
\mathbf{p}_{00} &= [0 \quad 0 \quad 3] & \mathbf{p}_{01} &= [0 \quad 3 \quad 0] \\
\mathbf{p}_{10} &= [3 \quad 0 \quad 0] & \mathbf{p}_{11} &= [3 \quad 3 \quad -3] \\
\\
\mathbf{p}_{00}^u &= [3 \quad 0 \quad -3] & \mathbf{p}_{01}^u &= [3 \quad 0 \quad -3] \\
\mathbf{p}_{10}^u &= [3 \quad 0 \quad -3] & \mathbf{p}_{11}^u &= [3 \quad 0 \quad -3] \\
\\
\mathbf{p}_{00}^w &= [0 \quad 3 \quad -3] & \mathbf{p}_{01}^w &= [0 \quad 3 \quad -3] \\
\mathbf{p}_{10}^w &= [0 \quad 3 \quad -3] & \mathbf{p}_{11}^w &= [0 \quad 3 \quad -3] \\
\\
\mathbf{p}_{00}^{uw} &= 0 & \mathbf{p}_{01}^{uw} &= 0 \\
\mathbf{p}_{10}^{uw} &= 0 & \mathbf{p}_{11}^{uw} &= 0
\end{aligned}
$$

Section 12

1. *Hint*: Reparametrize one or both curves so that their respective parametric variables have the same range.

3. It is a ruled surface, and the boundary curves $u = 0$, $u = 1$, and $w = 0$ are straight lines.

Section 13

1. *Hint*: What effect do the auxiliary curves have on the interior of a surface of revolution with $\mathbf{p}^{uw} = 0$ or with $\mathbf{p}^{uw} \neq 0$?

3. *Hint*: Rotate the characteristic polygon (control points). Can you find another method, say, using Eq. 3.97?

Section 15

1.
$$\mathbf{B} = \begin{bmatrix} [0 \ 0 \ 0] & [0 \ 4 \ 0] & [0 \ 4 \ -4] & [0 \ 4 \ -4] \\ [4 \ 0 \ 0] & [4 \ 4 \ 0] & [0 \ 4 \ 4] & [0 \ 4 \ -4] \\ [4 \ 0 \ 0] & [4 \ 0 \ 0] & [0 \ 0 \ 0] & [0 \ 0 \ 0] \\ [4 \ 0 \ 0] & [4 \ 0 \ 0] & [0 \ 0 \ 0] & [0 \ 0 \ 0] \end{bmatrix}$$

Section 16

3. 14 degrees of freedom for a bicubic array of patches with C^0 and C^1 continuity.

Section 17

5. Appropriately differentiate $\mathbf{p}(u, w)$ as given by Eq. 3.136b to obtain

$$\mathbf{p}^{uw}(u, w) = [(-3 + 6u - 3u^2) \ (3 - 12u + 9u^2) \ (6u - 9u^2) \ 3u^2]\mathbf{P} \begin{bmatrix} (-3 + 6w - 3w^2) \\ (3 - 12w + 9w^2) \\ (6w - 9w^2) \\ 3w^2 \end{bmatrix}$$

From this equation, compute

$$\mathbf{p}^{uw}_{00} = [-3 \ \ 3 \ \ 0 \ \ 0]\mathbf{P} \begin{bmatrix} -3 \\ 3 \\ 0 \\ 0 \end{bmatrix} = 9(\mathbf{p}_{11} - \mathbf{p}_{21} - \mathbf{p}_{12} + \mathbf{p}_{22})$$

$$\mathbf{p}^{uw}_{10} = [0 \ \ 0 \ \ -3 \ \ 3]\mathbf{P} \begin{bmatrix} -3 \\ 3 \\ 0 \\ 0 \end{bmatrix} = 9(\mathbf{p}_{31} - \mathbf{p}_{41} - \mathbf{p}_{32} + \mathbf{p}_{42})$$

$$\mathbf{p}_{01}^{uw} = [-3 \quad 3 \quad 0 \quad 0]\mathbf{P}\begin{bmatrix} 0 \\ 0 \\ -3 \\ 3 \end{bmatrix} = 9(\mathbf{p}_{13} - \mathbf{p}_{23} - \mathbf{p}_{14} + \mathbf{p}_{24})$$

$$\mathbf{p}_{11}^{uw} = [0 \quad 0 \quad -3 \quad 3]\mathbf{P}\begin{bmatrix} 0 \\ 0 \\ -3 \\ 3 \end{bmatrix} = 9(\mathbf{p}_{33} - \mathbf{p}_{43} - \mathbf{p}_{34} + \mathbf{p}_{44})$$

The results agree with Eq. 3.150.

Section 18

2. There are 49 control points required for a 2×2 array patch, 64 for a 3×3 array, and 81 for a 4×4 array. We find that $n_p = (n_a + 5)^2$.

3. $n_p = (n_a + k - 1)^2$, where $k = 1$.

4. $n_p = n_a(n_a + k - 1)$.

Section 19

2. If you increase the twist vectors sufficiently, a self-intersecting patch will be produced. The boundary curves are unchanged. If the boundary curves form a square, as stated in the exercise, then one set of conditions for self-intersection to occur is

$$\mathbf{p}_{00}^{uw} \neq \mathbf{p}_{10}^{w} - \mathbf{p}_{00}^{w}, \quad \neq \mathbf{p}_{01}^{u} - \mathbf{p}_{00}^{u}$$

$$\mathbf{p}_{10}^{uw} \neq \mathbf{p}_{10}^{w} - \mathbf{p}_{00}^{w}, \quad \neq \mathbf{p}_{11}^{u} - \mathbf{p}_{10}^{u}$$

$$\mathbf{p}_{01}^{uw} \neq \mathbf{p}_{11}^{w} - \mathbf{p}_{01}^{w}, \quad \neq \mathbf{p}_{01}^{u} - \mathbf{p}_{00}^{u}$$

$$\mathbf{p}_{11}^{uw} \neq \mathbf{p}_{11}^{w} - \mathbf{p}_{01}^{w}, \quad \neq \mathbf{p}_{11}^{u} - \mathbf{p}_{10}^{u}$$

4.
$$\mathbf{B} = \begin{bmatrix} \mathbf{p}_{00} & \mathbf{p}_{00} & 0 & 0 \\ \mathbf{p}_{10} & \mathbf{p}_{10} & 0 & 0 \\ \mathbf{p}_{00}^{u} & \mathbf{p}_{01}^{u} & \mathbf{p}_{00}^{uw} & \mathbf{p}_{01}^{uw} \\ \mathbf{p}_{10}^{u} & \mathbf{p}_{11}^{u} & \mathbf{p}_{10}^{uw} & \mathbf{p}_{11}^{uw} \end{bmatrix}$$

Notice that although $\mathbf{p}_{01} = \mathbf{p}_{00}$ and $\mathbf{p}_{11} = \mathbf{p}_{10}$, in general, $\mathbf{p}_{01}^{u} \neq \mathbf{p}_{00}^{u}$ and $\mathbf{p}_{11}^{u} \neq \mathbf{p}_{10}^{u}$. Furthermore, the twist vectors are not necessarily equal to zero.

CHAPTER 4

Section 1

1. Triquadric, $\mathbf{p} = \mathbf{a}_{ijk} u^{3-i} v^{3-j} w^{3-k}$ $i, j, k \in [1:3]$
    Triquartic, $\mathbf{p} = \mathbf{a}_{ijk} u^{5-i} v^{5-j} w^{5-k}$ $i, j, k \in [1:5]$
    Triquintic, $\mathbf{p} = \mathbf{a}_{ijk} u^{6-i} v^{6-j} w^{6-k}$ $i, j, k \in [1:6]$

2. Triquadric, 27 points
   Triquartic, 125 points
   Triquintic, 216 points

Section 7

1. $p_i(u_1, u_2, \ldots, u_n) = \sum\limits_{j_1=1}^{6} \sum\limits_{j_2=1}^{6} \cdots \sum\limits_{j_n=1}^{6} a_{ij_1 j_2 \cdots j_n} u_1^{6-j_1} u_2^{6-j_2} \cdots u_n^{6-j_n}$

3. $p_i(u_1, u_2, \ldots, n) = \sum\limits_{j_1=0}^{m_1} \sum\limits_{j_2=0}^{m_2} \cdots \sum\limits_{j_n=0}^{m_n} P_{ij_1 j_2 \cdots j_n} B_{j_1 3}(u_1) B_{j_2 3}(u_2) \cdots B_{j_n 3}(u_n)$

CHAPTER 5

Section 1

1. The osculating plane at any point on a plane curve coincides with the plane of the curve. The osculating plane at any point on a straight line is not defined; since any three neighboring points on a straight line are colinear, they do not unambiguously define a plane.

5. The curvature is zero, since $\mathbf{p}_{0.5}^{uu} = 0$.

6. There is an inflection point at $u = 0.5$, since $\mathbf{p}_{0.5}^{uu} = 0$.

8. Compute $|\mathbf{p}_{0.5}^u|^3 = 3929$; then use Eq. 5.24 to compute the curvature $(1/\rho)_{0.5}$

$$\left(\frac{1}{\rho}\right)_{0.5} = \frac{|[14 \quad 7 \quad -2] \times [-12 \quad 24 \quad 18]|}{3929}$$
$$= 0.129$$

The radius of curvature is $(0.129)^{-1} = 7.752$. The center of curvature is the vector $\mathbf{p}_{0.5} = \rho_{0.5} \mathbf{n}_{0.5} = [3.347 \quad 4.016 \quad 4.889]$. Compute the torsion by using Eq. 5.30 to obtain

$$\tau_{0.5} = -0.108$$

11. Use Eq. 5.24 to find the curvature

$$\left(\frac{1}{\rho}\right)_{0.5} = 0.276$$

The radius of curvature is

$$\mathbf{p}_{0.5} = 3.623$$

The curvature vector is

$$k_{0.5} = \rho_{0.5} \mathbf{n}_{0.5} = [-0.355 \quad -3.344 \quad 1.348]$$

The center of curvature is

$$\mathbf{p}_i + \mathbf{k}_i = [6.603 \quad 5.489 \quad 3.369]$$

Use Eq. 5.30 to compute the torsion

$$\tau_{0.5} = \frac{\begin{vmatrix} 2.25 & -1 & 2 \\ -1.50 & -4 & 4 \\ -3.125 & 2.5 & 7 \end{vmatrix}}{380.562} = -0.305$$

Section 2

4. For $u = 0.25$ and $w = 0.5$, compute the following vector properties:

$$\mathbf{p} = [3 \quad 6 \quad 3.938] \qquad \mathbf{p}^{uw} = [0 \quad 0 \quad 0]$$

$$\mathbf{p}^u = [12 \quad 0 \quad 4.5] \qquad \mathbf{p}^{uu} = [0 \quad 0 \quad -18]$$

$$\mathbf{p}^w = [0 \quad 12 \quad 0] \qquad \mathbf{p}^{ww} = [0 \quad 0 \quad -18]$$

Next, compute the unit normal using Eq. 5.58.

$$\mathbf{n} = [-0.351 \quad 0 \quad 0.936]$$

Use Eq. 5.59 to compute the tangent plane

$$\begin{vmatrix} x - 3 & 12 & 0 \\ y - 6 & 0 & 12 \\ z - 3.938 & 4.5 & 0 \end{vmatrix} = 0$$

or $x - 2.67z + 7.50 = 0$.

To compute the principal curvatures and directions, first compute the coefficients of the first and second fundamental forms

$$E = \mathbf{p}^u \cdot \mathbf{p}^u = 164.25 \qquad L = \mathbf{p}^{uu} \cdot \mathbf{n} = -16.848$$
$$F = \mathbf{p}^u \cdot \mathbf{p}^w = 0 \qquad M = \mathbf{p}^{uw} \cdot \mathbf{n} = 0$$
$$G = \mathbf{p}^w \cdot \mathbf{p}^w = 144 \qquad N = \mathbf{p}^{ww} \cdot \mathbf{n} = -16.848$$

Use Eq. 5.64 to compute the principal curvatures

$$23{,}652 \, \kappa^2 + 5193.4\kappa + 283.86 = 0$$

$$\kappa = -0.103, \ -0.117$$

Compute the principal directions from Eq. 5.65

$$h = 0$$

Therefore the principal directions are aligned with $\mathbf{p}^u$ and $\mathbf{p}^w$, with the $w = 0.5$ curve having the maximum curvature.

For $u = 0$ and $w = 0.5$, compute the following vector properties:

$$\mathbf{p} = [0 \quad 6 \quad 2.25] \qquad \mathbf{p}^{uw} = [0 \quad 0 \quad 0]$$

$$\mathbf{p}^u = [12 \quad 0 \quad 9] \qquad \mathbf{p}^{uu} = [0 \quad 0 \quad -18]$$

$$\mathbf{p}^w = [0 \quad 12 \quad 0] \qquad \mathbf{p}^{ww} = [0 \quad 0 \quad -18]$$

Next, compute the unit normal using Eq. 5.58.

$$\mathbf{n} = [-0.6 \quad 0 \quad 0.8]$$

Use Eq. 5.59 to compute the tangent plane

$$\begin{vmatrix} x - 0 & 12 & 0 \\ y - 6 & 0 & 12 \\ z - 2.25 & 9 & 0 \end{vmatrix} = 0$$

or $x - 1.333z + 3 = 0$

To compute the principal curvatures and directions, first compute the coefficients of the first and second fundamental forms

$$E = \mathbf{p}^u \cdot \mathbf{p}^u = 225 \qquad L = \mathbf{p}^{uu} \cdot \mathbf{n} = -14.4$$
$$F = \mathbf{p}^u \cdot \mathbf{p}^w = 0 \qquad M = \mathbf{p}^{uw} \cdot \mathbf{n} = 0$$
$$G = \mathbf{p}^w \cdot \mathbf{p}^w = 144 \qquad N = \mathbf{p}^{ww} \cdot \mathbf{n} = -14.4$$

Use Eq. 5.64 to compute the principal curvatures

$$32{,}400\kappa^2 + 5313.6\kappa + 207.36 = 0$$

$$\kappa = -0.064, \quad -0.100$$

Compute the principal directions from Eq. 5.65

$$h = 0$$

Therefore, the principal directions are aligned with $\mathbf{p}^u$ and $\mathbf{p}^w$, with the $w = 0.5$ curve having the maximum curvature.

5. Classify these points by using the expression $LN - M^2$.

For $u = 0.5$, $w = 0.5$, $LN - M^2 > 0$. Therefore, it is an elliptic point. However, since $\kappa =$ constant, it is also an umbilical point.

For $u = 0.25$, $w = 0.5$, $LN - M^2 > 0$. Therefore, it is an elliptic point.

For $u = 0$, $w = 0.5$, $LN - M^2 > 0$. Therefore, it is an elliptic point.

Section 3

1. Use Eq. 5.74b to find that

$$\begin{vmatrix} x^u & x^{uu} & x^{uuu} \\ y^u & y^{uu} & y^{uuu} \\ z^u & z^{uu} & z^{uuu} \end{vmatrix} \neq 0$$

Therefore, the curve is not a plane curve. Since a straight line is a plane curve, the curve is not a straight line.

4. We can test whether $\kappa_n \doteq 0$ at all points on the surface, or we can perform a set of simpler tests. First, determine if the four corner points are coplanar. To do this, compare unit normal vectors

$$\mathbf{n}_1 = \frac{(\mathbf{p}_{10} - \mathbf{p}_{00}) \times (\mathbf{p}_{11} - \mathbf{p}_{00})}{|(\mathbf{p}_{10} - \mathbf{p}_{00}) \times (\mathbf{p}_{11} - \mathbf{p}_{00})|}$$

to

$$\mathbf{n}_2 = \frac{(\mathbf{p}_{11} - \mathbf{p}_{00}) \times (\mathbf{p}_{01} - \mathbf{p}_{00})}{|(\mathbf{p}_{11} - \mathbf{p}_{00}) \times (\mathbf{p}_{01} - \mathbf{p}_{00})|}$$

If they are equal, then the four points are coplanar. In this exercise,

$$\mathbf{n}_1 = \mathbf{n}_2 = [2/3 \quad 2/3 \quad 1/3].$$

This is a necessary, but not sufficient, condition for a plane surface. Next, we check the tangent vectors, taking the vector product of those adjacent at each corner and comparing the unit vector of the result to $\mathbf{n}_1$ or $\mathbf{n}_2$. For example, let

$$\mathbf{n}_{00} = \frac{\mathbf{p}_{00}^u \times \mathbf{p}_{00}^w}{|\mathbf{p}_{00}^u \times \mathbf{p}_{00}^w|} \qquad \mathbf{n}_{10} = \frac{\mathbf{p}_{10}^u \times \mathbf{p}_{10}^w}{|\mathbf{p}_{10}^u \times \mathbf{p}_{10}^w|}$$

$$\mathbf{n}_{01} = \frac{\mathbf{p}_{01}^u \times \mathbf{p}_{01}^w}{|\mathbf{p}_{01}^u \times \mathbf{p}_{01}^w|} \qquad \mathbf{n}_{11} = \frac{\mathbf{p}_{11}^u \times \mathbf{p}_{11}^w}{|\mathbf{p}_{11}^u \times \mathbf{p}_{11}^w|}$$

And in this exercise, we find that

$$\mathbf{n}_{00} = \mathbf{n}_{01} = [0.4729 \quad 0.7811 \quad 0.3941] \neq \mathbf{n}_1, \mathbf{n}_2$$

$$\mathbf{n}_{10} = \mathbf{n}_{11} = [2/3 \quad 2/3 \quad 1/3] = \mathbf{n}_1, \mathbf{n}_2$$

We conclude that the surface is not a plane.

Finally, use Eq. 5.77 to determine if the surface is developable. More simply, in this exercise, boundary curves $\mathbf{p}_{u0} = \mathbf{p}_{u1}$; and not only does $\mathbf{p}_{0w} = \mathbf{p}_{1w}$, but these two boundary curves are straight lines. We conclude that the surface is developable.

Section 4

1. $(\cdots(((a_n u + a_{n-1})u + a_{n-2})u + a_{n-3})u + \cdots)u + a_0$

Section 5

1. $L = 21.668$

3. $L = 23.876$

8. $L = 60.$ Since $K = 2$, the curve is merely a series of straight lines connecting the control points.

10. $L = 16.224$

CHAPTER 6

Section 1

2. $7.3809 < L_g < 7.6262$

Section 2

1. From Chapter 2, Section 10, use the following expression for a straight line: $\mathbf{p} = \mathbf{p}_0 + u(\mathbf{p}_1 - \mathbf{p}_0)$. This means that $\mathbf{p}^u = \mathbf{p}_1 - \mathbf{p}_0$. Substitute these equations into Eq. 6.3 to obtain

$$[\mathbf{p}_0 + u(\mathbf{p}_1 - \mathbf{p}_0) - \mathbf{q}] \cdot (\mathbf{p}_1 - \mathbf{p}_0) = 0$$

This is a linear equation, so we can solve for u directly after carrying out the required scalar product. Thus,

$$[x_0 + u(x_1 - x_0) - q_x](x_1 - x_0)$$
$$+ [y_0 + u(x_1 - x_0) - q_y](y_1 - y_0)$$
$$+ [z_0 + u(z_1 - z_0) - q_z](z_1 - z_0) = 0.$$

Section 3

1. The plane is defined by the three points $\mathbf{p}_1 = \begin{bmatrix} 7 & 0 & 0 \end{bmatrix}$, $\mathbf{p}_2 = \begin{bmatrix} 0 & 10 & 0 \end{bmatrix}$, and $\mathbf{p}_3 = \begin{bmatrix} 0 & 0 & 5 \end{bmatrix}$. The unit normal to the plane is proportional to $(\mathbf{p}_2 - \mathbf{p}_1) \times (\mathbf{p}_3 - \mathbf{p}_1)$, since we know that $(\mathbf{p}_2 - \mathbf{p}_1)$ and $\mathbf{p}_3 - \mathbf{p}_1)$ lie in the plane.

    The unit normal to the plane is

    $$\mathbf{n} = \frac{(\mathbf{p}_2 - \mathbf{p}_1) \times (\mathbf{p}_3 - \mathbf{p}_1)}{|(\mathbf{p}_2 - \mathbf{p}_1) \times (\mathbf{p}_3 - \mathbf{p}_1)|}$$

    and the normal distance $|\mathbf{d}|$ is

    $$|\mathbf{d}| = \mathbf{p}_1 \cdot \mathbf{n} = \mathbf{p}_2 \cdot \mathbf{n} = \mathbf{p}_3 \cdot \mathbf{n}$$

    The plane coefficients are

    $$A = n_x$$
    $$B = n_y$$
    $$C = n_z$$
    $$D = -|\mathbf{d}|$$

    For the plane specified, $|\mathbf{d}| = 3.769$ and $\mathbf{d} = \begin{bmatrix} 2.029 & 1.420 & 2.841 \end{bmatrix}$.
    The unit normal is $\mathbf{n} = \begin{bmatrix} 0.538 & 0.377 & 0.754 \end{bmatrix}$.

Section 5

1. The line segment can be parallel to the plane, lying entirely in the plane, or outside it. In either case, all points on the line are equidistant from the plane. Alternatively, the line can be askew to the plane, intersecting the plane at a point, or lying entirely outside it. If the line intersects the plane, then obviously the point of intersection is closest. Otherwise, one of the end points of the line segment is closest to the plane.

Section 6

1. Equations 6.22 and 6.23 define a system of six nonlinear equations in four unknowns: u, w, s, and t. Equation 6.22 produces three nonlinear polynomials of degree eight in u and w, and degree three in s and t. Equation 6.23 produces three nonlinear polynomials of degree eight in s and t and degree three in u and w.

Section 7

1. $(2n - 4)$ vertices, $(3n - 6$ edges), including the edges and vertices required to close the assembly. Notice that if two vertices coincide, then a vertex and an edge are lost in the count.

1. *Hint*: Use perpendicular bisecting planes to the lines joining points in the array. This creates polyhedral cells around each point.

CHAPTER 7

Section 1

1a. No intersection.
 b. Overlapping intersection: $u_i \in [0.5, 1]$, $w_i \in [0, 0.5]$
 c. No intersection.
 d. Intersect at $u = 0.8$, $w = 0.25$, $\mathbf{p} = [6 \quad 0 \quad 0]$

2. Use Eq. 7.5 with $\mathbf{a} = [7 \quad 0 \quad 0]$, $\mathbf{b} = [-7 \quad 4 \quad 0]$, and $\mathbf{c} = [-7 \quad 0 \quad 7]$. Since the plane is unbounded, we are interested in only t from Eq. 7.7.

a. Intersection at $t = -0.435$, outside $t \in [0, 1]$;
 intersection at $t = 0.130$, $\mathbf{p} = [-0.739 \quad 3.130 \quad 2.261]$.

b. Intersection at $t = -0.435$, outside $t \in [0, 1]$;
 intersection at $t = -0.935$, outside $t \in [0, 1]$.

c. Intersection at $t = -0.435$, outside $t \in [0, 1]$;
 intersection at $t = -0.067$, outside $t \in [0, 1]$.

d. Intersection at $t = 0.873$, $\mathbf{p} = [3.382 \quad 7.109 \quad 1.745]$;
 intersection at $t = 0.125$, $\mathbf{p} = [5.5 \quad 0 \quad 1.5]$.

Section 2

1. $\mathbf{p}_1(u, w)$ intersects $\mathbf{p}_2(u, w)$ at u_1, w_1, $u_2 = 0$, $w_2 = 1$, $\mathbf{p} = [0 \quad 0 \quad 0]$ and $u_1 = 0.5$, $w_1 = 0.6$, $u_2 = 0.75$, $w_2 = 0$, $\mathbf{p} = [5 \quad 4 \quad 6]$.
 $\mathbf{p}_1(u, w)$ intersects $\mathbf{p}_3(u, w)$ at $u_1 = 0$, $w_1 = 0.4$, $u_3 = 0.2$, $w_3 = 0.8$, $\mathbf{p} = [0 \quad 0 \quad 4]$ and $u_1 = 0.8$, $w_1 = 0.4$, $u_2 = 0.84$, $w_2 = 0$, $\mathbf{p} = [8 \quad 6.4 \quad 4]$.
 $\mathbf{p}_2(u, w)$ intersects $\mathbf{p}_3(u, w)$ at $u_2 = 0.889$, $w_2 = 0.333$, $u_3 = 0.467$, $w_3 = 0$, $\mathbf{p} = [8 \quad 2.667 \quad 4]$ and $u_2 = 0.056$, $w_2 = 0.333$, $u_3 = 0.467$, $w_3 = 1$, $\mathbf{p} = [2 \quad 2.667 \quad 4]$.

CHAPTER 8

Section 1

1. $\mathbf{a}_0^* = \mathbf{a}_0 + \mathbf{t}$, $\mathbf{a}_1^* = \mathbf{a}_1$, $\mathbf{a}_2^* = \mathbf{a}_2$, $\mathbf{a}_3^* = \mathbf{a}_3$

3. Bezier curve: $\mathbf{p}^*(u) = \displaystyle\sum_{i=0}^{n} (\mathbf{p}_i + \mathbf{t})B_{i,n}(u)$

 B-spline curve: $\mathbf{p}^*(u) = \displaystyle\sum_{i=0}^{n} (\mathbf{p}_i + \mathbf{t})N_{i,k}(u)$

5. $\mathbf{B}^* = [(0, 0, 0) \quad (4, 7, 5) \quad (4, 3, 0) \quad (8, 10, 0)]^T$

Section 2

1. *Hint*: First verify that $\mathbf{B}^* = \mathbf{BR}$, then use $\mathbf{B}^* = \mathbf{M}^{-1}\mathbf{A}^*$.

3. $\mathbf{p}^*(u, w) = \displaystyle\sum_{i=0}^{m} \sum_{j=0}^{n} \mathbf{p}_{ij}^* B_{i,m}(u)B_{j,n}(w)$

 where $\mathbf{p}_{ij}^* = [\mathbf{p}_{ij} + \mathbf{T}]\mathbf{R}_{\theta\beta}\mathbf{R}_{\gamma}\mathbf{R}_{-\theta\beta} - \mathbf{T}$

and

$$\mathbf{T} = \begin{bmatrix} -\mathbf{r}_1 & -\mathbf{r}_1 & 0 & 0 \\ -\mathbf{r}_1 & -\mathbf{r}_1 & 0 & 0 \\ 0 & 0 & 0 & 0 \\ 0 & 0 & 0 & 0 \end{bmatrix}$$

5.   Use Eq. 8.15.

$$|\mathbf{R}| = R_{11} \cos \theta \cos \beta - R_{21} \sin \theta \cos\beta + R_{31} \sin \beta$$
$$= \cos^2 \theta \cos^2 \beta + \sin^2 \theta \cos^2 \beta + \sin^2 \beta$$
$$= 1$$

7. $\mathbf{R} = \begin{bmatrix} -1 & 0 & 0 \\ 0 & -1 & 0 \\ 0 & 0 & 1 \end{bmatrix}$

Section 3

1.   Show that $\mathbf{p}^{*u} = s\mathbf{p}^u = [sx^u \quad sy^u \quad sz^u]$. Then use Eq. 1.3 to show that the scale factor s cancels out.

3. $\mathbf{B}^* = s\mathbf{B}$

Section 4

1. $\mathbf{B}^* = -\mathbf{B}$

2.   Change the sign of the coordinates of the control points $\mathbf{p}_{ij}^* = -\mathbf{p}_{ij}$.

Section 5

2.   Compute

$$\theta = \tan^{-1}\left(\frac{y_1 - y_0}{x_1 - x_0}\right)$$

$$\beta = \tan^{-1} \frac{z_1 - z_0}{\sqrt{(x_1 - x_0)^2 + (y_1 - y_0)^2}}$$

Then

$$T_1 = \begin{bmatrix} 1 & 0 & 0 & 0 \\ 0 & 1 & 0 & 0 \\ 0 & 0 & 1 & 0 \\ -x_0 & -y_0 & -z_0 & 1 \end{bmatrix}$$

$$T_2 = \begin{bmatrix} \cos \theta & \sin \theta & 0 & 0 \\ -\sin \theta & \cos \theta & 0 & 0 \\ 0 & 0 & 1 & 0 \\ 0 & 0 & 0 & 1 \end{bmatrix}$$

$$T_3 = \begin{bmatrix} \cos \beta & 0 & -\sin \beta & 0 \\ 0 & 1 & 0 & 0 \\ \sin \beta & 0 & \cos \beta & 0 \\ 0 & 0 & 0 & 1 \end{bmatrix}$$

and $\mathbf{T}_h = \mathbf{T}_1 \mathbf{T}_2 \mathbf{T}_3$

or

$$
\mathbf{T}_h =
\begin{bmatrix}
\cos\theta\cos\beta & \sin\theta & -\cos\theta\sin\beta & 0 \\
-\sin\theta\cos\beta & \cos\theta & \sin\theta\sin\beta & 0 \\
\sin\beta & 0 & \cos\beta & 0 \\
-x_0\cos\theta\cos\beta + y_0\sin\theta\cos\beta - z_0\sin\beta & -x_0\sin\theta - y_0\cos\theta & x_0\cos\theta\sin\beta - y_0\sin\theta\sin\beta - z_0\cos\beta & 1
\end{bmatrix}
$$

CHAPTER 9

Section 2

1. Yes, because it can be deformed into a sphere.
2. $N_0 - N_1 + N_2 - N_3 = 0$

   N_0 = vertices

   N_1 = edges

   N_2 = faces

   N_3 = three-dimensional regions bounding a four-dimensional polytope. See H. M. S. Coxeter (1963).
3. There are 16 regular polytopes in four-dimensional space. The advanced student should refer to H. M. S. Coxeter (1963).
5. $[(1,0) \quad (3,2)]$ $[(1,1) \quad (4,1)]$ $[(1,2) \quad (2,0)]$ $[(2,1) \quad (4,0)]$ $[(2,2) \quad (3,0)]$
   $[(3,1) \quad (4,2)]$
7. For a cube, $V - E + F = V - F$.
9. Each face has four edges, and each edge is shared by two faces. Therefore, $E = 2F$, so that $V - E + F$ reduces to $V - F$.

Section 5

1. *Hint*: Draw a Venn diagram of $A \subset B$, which also demonstrates

$$
A \cup B = B \quad \text{and} \quad A \cap B = A.
$$

CHAPTER 10

Section 1

1. Height of the tree $= 5$;
   height of $b = 2$;
   depth of $b = 2$;
   level of $b = 3$.

5. Each node of a quadtree except a leaf node has four descendants. Each node of an octree except a leaf node has eight descendants. A complete quadtree of height k has $\sum_{i=0}^{k} 4^i$ nodes; a complete octree has $\sum_{i=0}^{k} 8^i$ nodes.

Section 2

5. $F = (0.25 - x^2 - y^2 = 0) \cap (z = 0) \cap (1 - z = 0)$
6. $F = (0.25 - x^2 - y^2 = 0) \cup (x = 0) \cap (1 - x = 0) \cap (y = 0) \cap (1 - y = 0)$
   $\cap (z = 0) \cap (1 - z = 0)$

Section 4

2. $r = (1/2)^n$; for $r = 0.001, n = 10$
7. Each full node at the highest level contributes $(1/2)^2$ of the area of the total quadrant. The next lower level contributes $(1/2)^4$, the next $(1/2)^6$, and so on, to the lowest level. The area of the figure represented by the quadtree is

$$\left(\frac{1}{2}\right)^2 + 7\left(\frac{1}{2}\right)^4 + 8\left(\frac{1}{2}\right)^6 + 6\left(\frac{1}{2}\right)^8$$

Section 5

1.
$$\mathbf{B}_P = \begin{bmatrix} 0 & 0 & 0 \\ 1 & 0 & 0 \\ 2 & 2 & 0 \\ 2 & -2 & 0 \end{bmatrix} \qquad \mathbf{B}_D = \begin{bmatrix} 0 & 0 & 1 \\ 1 & 0 & 0 \\ 1 & 0 & -1 \\ 1 & 0 & -1 \end{bmatrix}$$

3. Duct cross-section curves

$$\mathbf{B}_{z1} = \begin{bmatrix} 0 & 4 & 0 \\ 0 & 4 & 1 \end{bmatrix} \qquad \mathbf{B}_{z2} = \begin{bmatrix} 0 & 4 & 1 \\ 0 & 5 & 1 \end{bmatrix}$$

$$\mathbf{B}_{z3} = \begin{bmatrix} 0 & 5 & 1 \\ 0 & 5 & 0 \end{bmatrix} \qquad \mathbf{B}_{z4} = \begin{bmatrix} 0 & 5 & 0 \\ 0 & 4 & 0 \end{bmatrix}$$

The direction curve

$$\mathbf{B}_{zD} = \begin{bmatrix} 0 & 0 & 0 \\ 0 & 0 & 1 \end{bmatrix}$$

The position curves (use the results from Ex. 2 and Eqs. 2.246 and 2.247 to compute Bezier control points $\mathbf{p}_1$ and $\mathbf{p}_2$)

$$\mathbf{B}_{zP1} = \begin{bmatrix} 0 & 4 & 0 \\ 1.061 & 4 & 0 \\ 2.078 & 3.578 & 0 \\ 2.828 & 2.828 & 0 \end{bmatrix} \qquad \mathbf{B}_{zP2} = \begin{bmatrix} 2.828 & 2.828 & 0 \\ 3.578 & 2.078 & 0 \\ 4 & 1.061 & 0 \\ 4 & 0 & 0 \end{bmatrix}$$

$$\mathbf{B}_{zP3} = \begin{bmatrix} 4 & 0 & 0 \\ 4 & -1.061 & 0 \\ 3.578 & -2.078 & 0 \\ 2.828 & -2.828 & 0 \end{bmatrix} \qquad \mathbf{B}_{zP4} = \begin{bmatrix} 2.828 & -2.828 & 0 \\ 2.078 & -3.578 & 0 \\ 1.061 & -4 & 0 \\ 0 & -4 & 0 \end{bmatrix}$$

$$\mathbf{B}_{zP5} = \begin{bmatrix} 0 & -4 & 0 \\ -1.061 & -4 & 0 \\ -2.078 & -3.578 & 0 \\ -2.828 & -2.828 & 0 \end{bmatrix} \qquad \mathbf{B}_{zP6} = \begin{bmatrix} -2.828 & -2.828 & 0 \\ -3.578 & -2.078 & 0 \\ -4 & -1.061 & 0 \\ -4 & 0 & 0 \end{bmatrix}$$

$$\mathbf{B}_{zP7} = \begin{bmatrix} -4 & 0 & 0 \\ -4 & 1.061 & 0 \\ -3.578 & 2.078 & 0 \\ -2.828 & 2.828 & 0 \end{bmatrix} \qquad \mathbf{B}_{zP8} = \begin{bmatrix} -2.828 & 2.828 & 0 \\ -2.078 & 3.578 & 0 \\ -1.061 & 4 & 0 \\ 0 & 4 & 0 \end{bmatrix}$$

Section 6

3. Yes. Try $(4 - x^2 - y^2 = 0) \cap (x^2 + y^2 - 1 = 0) \cap (x = 0) \cap (z = 0) \cap (1 - z = 0)$

Section 8

1a. $V - E + F = 2; 8 - 12 + 6 = 2$

b. $V - E + F - H + 2P = 2B; 28 - 42 + 17 - 3 + 2 = 2$

c. $V - E + F - H + 2P = 2B; 14 - 21 + 9 - 2 + 2 = 2.$

Note that a vertex must be added to each of the two circles defining the through hole, and they are then joined by an edge.

d. $V - E + F - H + 2P = 2B; 24 - 36 + 15 - 3 + 2 = 2$

CHAPTER 11

Section 2

4. Surface area = 114.85
   Volume = 51.43
   Centroid = [2   2   2]
   Moments of inertia: $I_{xx} = 150.77$, $I_{yy} = 164.38$, $I_{zz} = 150.77$

CHAPTER 12

Section 1

2. Equation 12.7 applies to all plane and curve relationships. The distances d_1, d_2, d_3, and d_4 are directed distances; that is, they carry an algebraic sign determined by the point coordinates and plane coefficients in Eq. 12.9 and similar equations for the other distances. No modification is required.

Section 2.2

1. For projection onto the $x = 0$ plane,

$$\mathbf{p}_h^* = \mathbf{p}_h \begin{bmatrix} 0 & 0 & 0 & 1/\lambda \\ 0 & 1 & 0 & 0 \\ 0 & 0 & 1 & 0 \\ 0 & 0 & 0 & 1 \end{bmatrix}$$

For projection onto the $y = 0$ plane,

$$\mathbf{p}_h^* = \mathbf{p}_h \begin{bmatrix} 1 & 0 & 0 & 0 \\ 0 & 0 & 0 & 1/\lambda \\ 0 & 0 & 1 & 0 \\ 0 & 0 & 0 & 1 \end{bmatrix}$$

2. For projection onto the $z = z_c$ plane,

$$\mathbf{p}_h^* = \mathbf{p}_h \begin{bmatrix} 1 & 0 & 0 & 0 \\ 0 & 1 & 0 & 0 \\ 0 & 0 & 0 & 1/\lambda \\ 0 & 0 & z_c & 1 \end{bmatrix}$$

CHAPTER 13

Section 1

2. $\mathbf{p}_i = (i - 1)\left(\dfrac{\mathbf{p}_2 - \mathbf{p}_1}{n}\right), \quad i \in [1\!:\!n + 1]$

INDEX